Some Useful Constants and Physical Measurements*

astronomical unit	1 A.U. = 1.496×10^8 km (1.5×10^8 km)
light year	1 ly = 9.46×10^{12} km (10^{13} km; 6 trillion miles)
parsec	1 pc = 3.09×10^{13} km = 3.3 ly
speed of light	c = 299,792.458 km/s (3×10^5 km/s)
Stefan-Boltzmann constant	σ [Greek sigma] = 5.67×10^{-8} W/m$^2 \cdot$ K^4
Planck's constant	h = 6.63×10^{-34} Js
gravitational constant	G = 6.67×10^{-11} Nm^2/kg^2
mass of Earth	$M_\oplus$ = 5.97×10^{24} kg (6×10^{24} kg; about 6000 billion billion tons)
radius of Earth	$R_\oplus$ = 6378 km (6500 km)
mass of the Sun	$M_\odot$ = 1.99×10^{30} kg (2×10^{30} kg)
radius of the Sun	$R_\odot$ = 6.96×10^5 km (7×10^5 km)
luminosity of the Sun	$L_\odot$ = 3.90×10^{26} W
effective temperature of the Sun	$T_\odot$ = 5778 K (5800 K)
Hubble constant	$H_0 \approx$ 70 km/s/Mpc
mass of the electron	m_e = 9.11×10^{-31} kg
mass of the proton	m_p = 1.67×10^{-27} kg

The rounded-off values used in the text are shown in parentheses.

Conversions between English and Metric Units

1 inch	=	2.54 centimeters (cm)		1 mile	=	1.609 kilometers (km)
1 foot (ft)	=	0.3048 meters (m)		1 pound (lb)	=	453.6 grams (g) or 0.4536 kilograms (kg) (on Earth)

Main-Sequence Stellar Properties by Spectral Class

Spectral Class	Typical Surface Temperature	Color	Mass*	Luminosity*	Lifetime*	Familiar Examples
	(K)		($M_\odot$)	($L_\odot$)	(10^6 yr)	
O	>30,000	Electric Blue	>20	>100,000	<2	Mintaka (O9)
B	20,000	Blue	7	500	140	Spica (B2)
A	10,000	White	3	60	500	Vega (A0) Sirius (A1)
F	7,000	Yellow - white	1.5	7	2000	Procyon (F5)
G	6,000	Yellow	1.0	1.0	10,000	Sun (G2) Alpha Centauri (G2)
K	4,000	Orange	0.8	0.3	30,000	Epsilon Eridani (K2)
M	3,000	Red	0.1	0.00006	16,000,000	Proxima Centauri (M5) Barnard's Star (M5)

*Approximate values for stars of solar composition

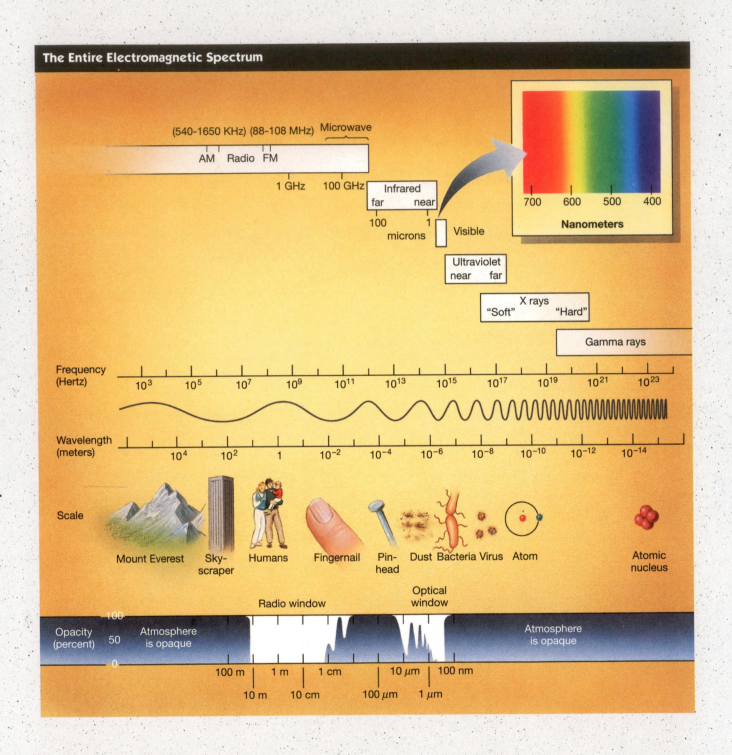

The Entire Electromagnetic Spectrum

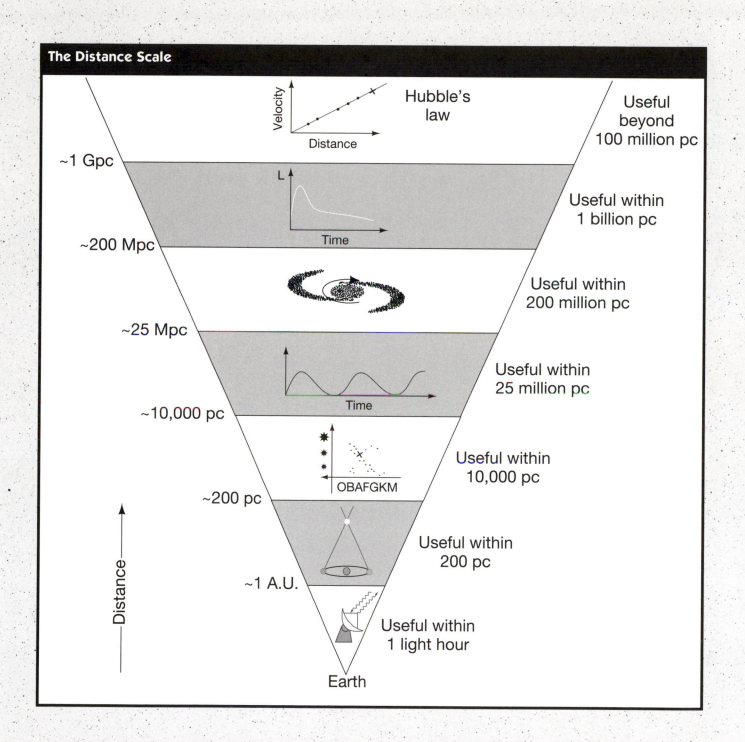

YOUR ACCESS TO SUCCESS

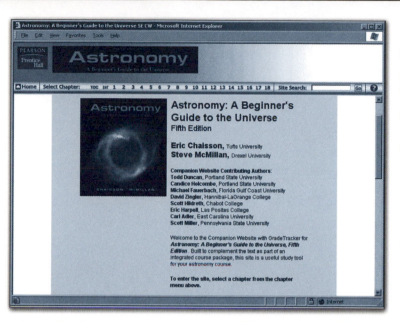

Students with new copies of Chaisson/McMillan, *Astronomy: A Beginner's Guide to the Universe, Fifth Edition,* have full access to the book's Companion Website with GradeTracker— a 24/7 study tool with Tutorials, Animations and Videos, quizzes, an eBook, and other features designed to help you make the most of your limited study time.

Just follow the easy website registration steps listed below...

Registration Instructions for the *Astronomy: A Beginner's Guide to the Universe, Fifth Edition* Companion Website with GradeTracker

1. Go to *www.prenhall.com/chaisson*.
2. Click the cover for *Astronomy: A Beginner's Guide to the Universe, Fifth Edition.*
3. Click on the link to the Companion Website with GradeTracker.
4. Click "Register".
5. Using a coin, scratch off the metallic coating below to reveal your Access Code.
6. Complete the online registration form, choosing your own personal Login Name and Password.
7. Enter your pre-assigned Access Code exactly as it appears below.
8. Complete the online registration form by entering your School Location information.
9. After your personal Login Name and Password are confirmed by e-mail, repeat steps 1-3 above, and enter your new Login Name and Password to log in.

Your Access Code is:

PSWAGB-LEVEE-AISLE-ARLON-CENTO-VEXES

If there is no metallic coating covering the access code above, the code may no longer be valid and you will need to purchase online access using a major credit card to use the website. To do so, go to *www.prenhall.com/chaisson*, click the cover for *Astronomy: A Beginner's Guide to the Universe, Fifth Edition*, and follow the instructions to purchase access to the Companion Website with GradeTracker.

Minimum system requirements

PC Operating Systems:

Windows 2000/XP

Pentium II 233 MHz processor. 64 MB RAM In addition to the minimum memory required by your OS.

Internet Explorer™ 5.5 or 6, Netscape™ 7, Firefox 1.0.

Macintosh Operating Systems:

Macintosh Power PC with OS X (10.2 and 10.3)

In addition to the RAM required by your OS, this application requires 64 MB RAM, with 40MB Free RAM, with Virtual Memory enabled.

Netscape™ 7, Safari 1.3, Firefox 1.0.

Macromedia Shockwave™

8.50 release 326 plugin

Macromedia Flash Player 6.0.79 & 7.0

Acrobat Reader 6.0.1

4x CD-ROM drive

800 x 600 pixel screen resolution

Important: Please read the Subscription and End-User License Agreement located on the "Log In" screen before using the *Astronomy: A Beginner's Guide to the Universe, Fifth Edition* Companion Website with GradeTracker. By using the website, you indicate that you have read, understood, and accepted the terms of the agreement.

Technical Support Call 1-800-677-6337. Phone support is available Monday–Friday, 8am to 8pm and Sunday 5pm to 12am, Eastern time. Visit our support site at *http://247.pearsoned.com.* E-mail support is available 24/7.

ASTRONOMY

ASTRONOMY

A Beginner's Guide to the Universe

FIFTH EDITION

Eric Chaisson

Tufts University

Steve McMillan

Drexel University

PEARSON

Prentice
Hall

Upper Saddle River, New Jersey 07458

Library of Congress Cataloging-in-Publication Data

Chaisson, Eric.
  Astronomy : a beginner's guide to the universe / Eric Chaisson, Steve
  McMillan.—5th ed.
    p. cm.
  ISBN 0-13-187165-X (pbk.)
  1. Astronomy—Textbooks.   I. McMillan, S. (Stephen),   II. Title.
  QB43.3.C43 2006
  520—dc22

2006000630

Senior Editor: Erik Fahlgren
Editor in Chief, Science: Dan Kaveney
Production Editor: Shari Toron
Electronic Production Specialist/
  Electronic Pagemakeup Specialist: Joanne Del Ben
Executive Managing Editor: Kathleen Schiaparelli
Assistant Managing Editor: Beth Sweeten
Formatting Manager: Allyson Graesser
Editorial Assistant: Jessica Berta
Director of Marketing: Patrick Lynch
Editor in Chief, Development: Carol Trueheart
Development Editor: Anne Scanlan-Rohrer
Director of Creative Services: Paul Belfanti
Creative Director: Juan López
Senior Managing Editor, Visual Assets: Patricia Burns
Manager, Production Technologies: Matthew Haas
Managing Editor, Visual Assets: Abigail Bass

Visual Assets Production Editor: Connie Long
Assistant Manager and Illustration Art Director: Sean Hogan
Assistant Production Manager, Illustrator: Ronda Whitson
Lead Illustrator: Mark Landis
Art Director: John Christiana
Interior and Cover Designer: Solid State Graphics
Cover Photo: **A New View of the Helix Nebula** *NASA, ESA, C.R.*
  O'Dell (Vanderbilt University), M. Meixner and P. McCullough (STScI)
Manufacturing Buyer: Alan Fischer
Manufacturing Manager: Alexis Heydt-Long
Managing Editor, Science and Math Media: Nicole M. Jackson
Media Production Editor: Tyler Suydam
Media Editor: Michael J. Richards
Associate Editor: Christian Botting
Image Resource Center Director: Beth Brenzel
Image Coordinator: Debbie Latronica
Copy Editor: Jane Loftus

Pearson Education LTD., *London*
Pearson Education Australia PTY, Limited, *Sydney*
Pearson Education *Singapore*, Pte. Ltd
Pearson Education North Asia Ltd, *Hong Kong*
Pearson Education Canada, Ltd., *Toronto*
Pearson Educación de *Mexico*, S.A. de C.V.
Pearson Education—*Japan*, Tokyo
Pearson Education *Malaysia*, Pte. Ltd

Brief Contents

Contents

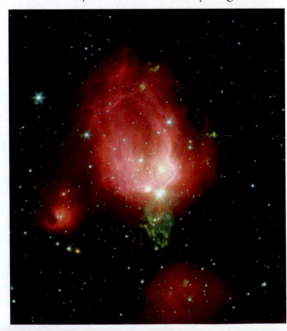

PART 2 OUR PLANETARY SYSTEM

4 THE SOLAR SYSTEM
Interplanetary Matter and the Birth of the Planets 98

5 EARTH AND ITS MOON
Our Cosmic Backyard 132

13 NEUTRON STARS AND BLACK HOLES

Strange States of Matter 344

PART 4 GALAXIES AND THE UNIVERSE

14 THE MILKY WAY GALAXY

A Grand Design 374

15 NORMAL AND ACTIVE GALAXIES

Building Blocks of the Universe 400

17 COSMOLOGY

The Big Bang and the Fate of the Universe 454

18 LIFE IN THE UNIVERSE

Are We Alone? 482

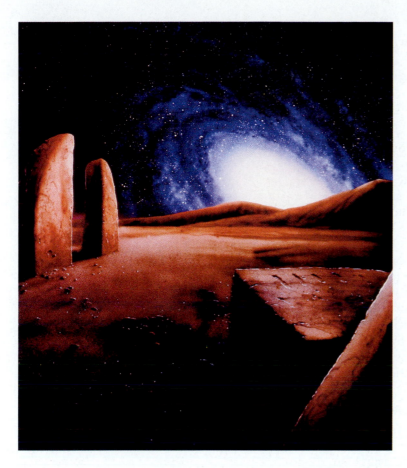

16 GALAXIES AND DARK MATTER

The Large-Scale Structure of the Cosmos 430

About the Authors

ERIC CHAISSON

Eric holds a doctorate in astrophysics from Harvard University, where he spent ten years on the faculty of Arts and Sciences. For five years, Eric was a senior scientist and director of educational programs at the Space Telescope Science Institute and adjunct professor of physics at Johns Hopkins University. For the past decade he has been at Tufts University, where he is professor of physics, professor of education, and director of the Wright Center for Innovative Science Education. He has written nine books on astronomy, which have received such literary awards as the Phi Beta Kappa Prize, two American Institute of Physics Awards, and Harvard's Smith-Weld Prize for Literary Merit. He has published more than 100 scientific papers in professional journals and has also received Harvard's Bok Prize for original contributions to astrophysics.

STEVE MCMILLAN

Steve holds a bachelor's and master's degree in mathematics from Cambridge University and a doctorate in astronomy from Harvard University. He held post-doctoral positions at the University of Illinois and Northwestern University, where he continued his research in theoretical astrophysics, star clusters, and numerical modeling. Steve is currently Distinguished Professor of Physics at Drexel University and a frequent visiting researcher at Princeton's Institute for Advanced Study and the University of Tokyo. He has published more than 50 scientific papers in professional journals.

Preface

We are pleased to have the opportunity to present in this book a representative sample of the known facts, evolving ideas, and frontier discoveries in astronomy today.

Astronomy: A Beginner's Guide to the Universe has been written and designed for students who have taken no previous college science courses and who will likely not major in physics or astronomy. We present a broad view of astronomy, straightforwardly descriptive and without complex mathematics. The absence of sophisticated mathematics, however, in no way prevents discussion of important concepts. Rather, we rely on qualitative reasoning as well as analogies with objects and phenomena familiar to the student to explain the complexities of the subject without oversimplification. We have tried to communicate the excitement that we feel about astronomy and to awaken students to the marvelous universe around us.

We are very gratified that the first four editions of this text have been so well received by many in the astronomy education community. In using those earlier texts, many of you—teachers and students alike—have given us helpful feedback and constructive criticisms. From these, we have learned to communicate better both the fundamentals and the excitement of astronomy. Many improvements inspired by your comments have been incorporated into this edition.

Organization and Approach

As in previous editions, our organization follows the popular and effective "Earth out" progression. We have found that most students, especially those with little scientific background, are much more comfortable studying the relatively familiar solar system before tackling stars and galaxies. With Earth and Moon as our initial planetary models, we move through the solar system. Integral to our coverage of the solar system is a discussion of its formation. This line of investigation leads directly into a study of the Sun.

With the Sun as our model star, we broaden the scope of our discussion to include stars in general—their properties, their evolutionary histories, and their varied fates. This journey naturally leads us to coverage of the Milky Way Galaxy, which in turn serves as an introduction to our treatment of other galaxies. Finally, we reach cosmology and the large-scale structure and dynamics of the universe as a whole. Throughout, we strive to emphasize the dynamic nature of the cosmos—virtually every major topic, from planets to quasars, includes a discussion of how those objects formed and how they evolve.

We place much of the needed physics in the early chapters—an approach derived from years of experience teaching thousands of students. Additional physical principles are developed as needed later, both in the text narrative and in the *Discovery* and *More Precisely* boxes (described on p. xvi). We have made the treatment of physics, as well as the more quantitative discussions, as modular as possible, so that these topics can be deferred if desired. Instructors presenting this material in a one-quarter course, who wish to (or have time to) cover only the essentials of the solar system before proceeding on to the study of stars and the rest of the universe, may want to teach only Chapter 4 and then move directly to Chapter 9 (the Sun).

Improvements

Astronomy is a rapidly evolving field, and almost every chapter in the fifth edition has been updated with new and late-breaking information. Several chapters have also seen significant internal reorganization in order to streamline the overall presentation, strengthen our focus on the process of science, and reflect new understanding and emphases in contemporary astronomy. Among the many improvements are the following:

- The single-column design allows for a more readable presentation of the material and was the winner of the 18th annual New York Book Show sponsored by the Bookbinder's Guild of New York.

- New chapter-opening images reflecting the latest discoveries.

- Streamlining of the art program to provide more direct and accurate representations of astronomical objects, including numerous "visual analogies" connecting the concepts presented in the text to everyday events and circumstances familiar to the reader.

- Restructuring of the text to further emphasize the process of science and explain how astronomers "know what they know."

- Expansion of the *More Precisely* boxes wherever possible to contain worked examples.

- Reworking of the end-of-chapter summaries to tie them directly to the learning goals and to link them to key figures in the chapter.

- Reorganization of the self-test section at the end of each chapter to include seven to eight multiple choice questions, two of them tied to the comprehension of figures in the text.

- Expanded material in Chapter 3 on interferometry and the latest developments in telescope design, plus new coverage of the *Spitzer* mission.

- Updates in Chapter 4 on the *Stardust* and *Deep Impact* missions, along with expanded material on observations of extrasolar planets and their implications for competing theories of planet formation.

- New coverage in Chapters 4 and 8 of Sedna, 2003UB3113, and Pluto's planetary status.

- Greatly expanded discussion in Chapter 6 of recent missions to Mars and their most important findings, particularly in relation to the evolution of the Martian climate.

- A final update in Chapter 7 on the *Galileo/GEM* mission.

- Coverage in Chapters 7 and 8 of the *Cassini* mission to Saturn, including the findings of the *Huygens* probe to Titan.

- Reorganization of presentation in Chapter 9 on the Sun and an update on neutrino oscillations.

- Consistent and up-to-date stellar properties in text and examples throughout Part 3.

- New material in Chapter 11 on competitive accretion and collisions in star formation.

- Added discussions in Chapter 12 of the evolution of low-mass stars, mass loss in high-mass stars.

- Substantial update in Chapter 13 of the two most likely explanations for gamma-ray bursts.

- Expansion of special and general theories of relativity now covered in Chapter 13.

- The latest results in Chapter 14 on Sgr A* and the Galaxy's central black hole.

- Updates and reorganization in Chapter 15 further emphasizing the connection between normal and active galaxies.

- Substantial reorganization of the discussions in Chapter 16 of black holes and galaxy evolution and the large-scale structure of the universe.

- Introduction of the Sloan Digital Sky Survey as a primary research tool for astronomers.

- Revision of the cosmological discussion in Chapter 17 to emphasize the importance of geometry and curvature in the universe and of the Big Bang and inflation as testable theories.

- Updated coverage of Europa, Mars, and extremophilic life in Chapter 18.

- An expanded glossary, which now includes over 300 additional terms used in the text, but not identified explicitly as keywords.

- New detailed seasonal star charts, courtesy of *Astronomy* magazine.

The Illustration Program

Visualization plays an important role in both the teaching and the practice of astronomy, and we continue to place strong emphasis on this aspect of our book. We have tried to combine aesthetic beauty with scientific accuracy in the artist's conceptions that enrich the text, and we have sought to present the best and latest imagery of a wide range of cosmic objects. Each illustration has been carefully crafted to enhance student learning; each is pedagogically sound and tightly tied to nearby discussion of important scientific facts and ideas.

Full-Spectrum Coverage and Spectrum Icons

Increasingly, astronomers are exploiting the full range of the electromagnetic spectrum to gather information about the cosmos. Throughout this book, images taken at radio, infrared, ultraviolet, X-ray, or gamma-ray wavelengths are used to supplement visible-light images. As it is sometimes difficult (even for a professional) to tell at a glance which images are visible-light photographs and which are false-color images created with other wavelengths, each photo in the text is accompanied by an icon that identifies the wavelength of electromagnetic radiation used to capture the image.

Other Pedagogical Features

As with many other parts of our text, instructors have helped guide us toward what is most helpful for effective student learning.

Learning Goals Studies indicate that beginning students often have trouble prioritizing textual material. For this reason, a few (typically five or six) well-defined learning goals are provided at the start of each chapter. These help students structure their reading of the chapter and then test their mastery of key concepts. The goals are numbered and keyed to the items in the Chapter Summary, which in turn refer back to passages in the text. This highlighting of the most important aspects of the chapter helps students prioritize information and also aids in their review. The goals are organized and phrased in such a way as to make them objectively testable, affording students a means of gauging their own progress.

∞ **Concept Links** The connection between the astronomical material and the physical principles set forth early in the text is crucial. It is important that students, when they encounter, say, Hubble's law in Chapter 15, recall what they learned about spectral lines and the Doppler shift in Chapter 2. Similarly, the discussions of the masses of binary star components (Chapter 10) and of galactic rotation (Chapter 14) both depend on the discussion of Kepler's and Newton's laws in Chapter 1. Throughout, discussions of new astronomical objects and concepts rely heavily on comparison with topics introduced earlier in the text.

It is important to remind students of these links so that they can recall the principles on which later discussions rest and, if necessary, review them. To this end, we have inserted "Concept Links" throughout the text—symbols that mark key intellectual bridges between material in different chapters. The links, denoted by the symbol ∞, signal students that the topic under discussion is related in some significant way to ideas developed earlier and direct them to material that they might wish to review before proceeding.

Key Terms Like all subjects, astronomy has its own specialized vocabulary. To aid student learning, the most important astronomical terms are boldfaced at their first appearance in the text. Boldfaced key terms in the Chapter Summary are linked with the page number where the term was defined. In addition, a full alphabetical glossary, defining each key term and locating its first use in the text, appears at the end of the book.

Concept Checks We incorporate into each chapter a number of "Concept Checks"—key questions that require the reader to reconsider some of the material just presented or attempt to place it into a broader context.

> **CONCEPT CHECK ▶▶▶**
>
> Why do astronomers regard the disk and halo as distinctly different components of our Galaxy?

H–R Diagrams and Acetate Overlays All of the book's H–R diagrams are drawn in a uniform format, using real data. In addition, a unique set of transparent acetate overlays dramatically demonstrates to students how the H–R diagram helps us organize our information about the stars and track their evolutionary histories.

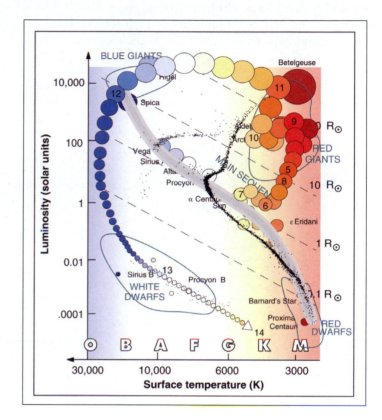

---

tidal bulge, however, always points directly toward Jupiter, so it moves back and forth across Io's surface as the moon wobbles. The conflicting forces acting on Io result in enormous tidal stresses that continually flex and squeeze its interior.

Just as repeated back-and-forth bending of a wire can produce heat through friction, Io is constantly energized by this ever-changing distortion. Researchers estimate that the total amount of heat generated within Io as a result of tidal flexing is about 100 million megawatts—more than five times the total power consumption of all the nations on Earth. The large amount of heat generated within Io causes huge jets of gas and molten rock to squirt out of the surface. It is likely that much of Io's interior is soft or molten, with only a relatively thin solid crust overlying it.

Europa: Liquid Water Locked in Ice

Europa (Figure 8.5) is a world very different from Io. Europa has relatively few craters on its surface, again suggesting that recent activity has erased the scars of ancient meteoric impacts. However, the surface displays a vast network of lines crisscrossing bright, clear fields of water ice. Many of these linear bands extend halfway around the satellite.

Based on the *Voyager* images, planetary scientists theorized that Europa is covered by an ocean of liquid water whose surface layers are frozen. More recent *Galileo* images seem to support this idea. Figure 8.5(b) is a high-resolution image of this weird moon, showing ice-filled surface cracks similar to those seen in Earth's polar ice floes. Figure 8.5(c) shows what appear to be icebergs—flat chunks of ice that have been broken apart and reassembled, perhaps by the action of water below. Mission scientists speculate that Europa's ice may be several kilometers thick and that the ocean below it may be 100 km deep. As on Io, the ultimate source of the energy that keeps the ocean liquid and also drives the internal motion leading to the cracks in the icy surface is Jupiter's gravity. The heating mechanism is essentially similar to that on Io, but the effects are less extreme on Europa because of its greater distance from the planet.

Other detailed images of the surface support this hypothesis. Figure 8.5(d) shows a region where Europa's icy crust appears to have been pulled apart and new material has filled in the gaps between the separating ice sheets. Elsewhere on the surface, *Galileo* found what appeared to be the icy equivalent of lava flows on Earth—regions where water apparently erupted through the surface and flowed for many kilometers before solidifying. The scarcity of impact craters on Europa implies that the processes responsible for these features did not stop long ago. Rather, they must be ongoing.

Further evidence comes from studies of Europa's magnetic field. Magnetic measurements made by *Galileo* on repeated flybys of the moon revealed that Europa has a weak magnetic field that constantly changes strength and direction. This finding is consistent with the idea that the field is generated by the action of Jupiter's magnetism on a shell of electrically conducting fluid about 100 km below Europa's surface—in other words, the salty liquid water layer suggested by the surface observations. These results convinced quite a few skeptical scientists of the reality of Europa's deep ocean.

▶ **Figure 8.5 Europa** (a) A *Voyager 1* mosaic of Europa. Resolution is about 5 km. (b) Europa's icy surface is only lightly cratered, indicating that some ongoing process must be obliterating impact craters soon after they form. The origin of the cracks crisscrossing the surface is uncertain. (c) At 50-m resolution, this image from the *Galileo* spacecraft shows a smooth yet tangled surface resembling the huge ice floes that cover Earth's polar regions. This region is called Conamara Chaos. (d) This detailed *Galileo* image shows "pulled apart" terrain that suggests liquid water upwelling from the interior and freezing, filling in the gaps between separating surface ice sheets. (NASA)

Compound Art It is rare that a single image, be it a photograph or an artist's conception, can capture all aspects of a complex subject. Wherever possible, multiple-part figures are used in an attempt to convey the greatest amount of information in the most vivid way:

- Visible images are often presented along with their counterparts captured at other wavelengths.

- Interpretive line drawings are often superimposed on or juxtaposed with real astronomical photographs, helping students to really "see" what the photographs reveal.

- Breakouts—often multiple ones—are used to zoom in from wide-field shots to close-ups, so that detailed images can be understood in their larger context.

More Precisely Boxes These boxes provide more quantitative treatments of subjects discussed qualitatively in the text. Removing these more challenging topics from the main flow of the narrative and placing them within a separate modular element of the chapter design allow instructors greater flexibility in setting the level of their coverage.

17-1 MORE PRECISELY

Curved Space

Euclidean geometry is the geometry of flat space—the geometry taught in high schools everywhere. Set forth by Euclid, one of the most famous of the ancient Greek mathematicians, it is the geometry of everyday experience. Houses are built with flat floors. Writing tablets and blackboards are also flat. We work easily with flat, straight objects because a straight line is the shortest distance between any two points.

In constructing houses or any other straight-walled buildings on the surface of Earth, some other basic axioms of Euclid's geometry also apply: parallel lines never meet even when extended to infinity; the angles of any triangle always sum to 180°; the circumference of a circle equals π times its diameter. If these rules were not obeyed, walls and roof would never meet to form a house!

In reality, though, the geometry of Earth's surface is not flat. It is curved. We live on a sphere, and on its surface Euclidean geometry breaks down. Instead, the rules are those of *Riemannian geometry*, named after the nineteenth-century German mathematician Georg Friedrich Riemann. There are no parallel lines (or curves) on a sphere's surface—any "straight" lines drawn on the surface and around the full circumference will eventually intersect. The sum of a triangle's angles always *exceeds* 180°. (For example, in the 90°-90°-90° triangle shown in the figure, the sum is 270°.) And the circumference of a circle is *less* than π times its diameter.

Riemannian geometry on the curved surface of a sphere differs greatly from the flat-space geometry of Euclid. The two are approximately the same only if we confine ourselves to a small patch on the surface. If the patch is much smaller than the sphere's radius, the surface surrounding any point looks flat, and Euclidean geometry is approximately valid. This is why we can draw a usable map of our home, our city, even our state, on a flat sheet of paper, but an accurate map of the entire Earth must be drawn on a globe.

When we work with larger portions of Earth, we must abandon Euclidean geometry. World navigators are well aware of this. Aircraft do not fly along what you might regard as a straight-line path from one point to another. Instead they follow a *great circle*, which, on the curved surface of a sphere, is the shortest distance between two points. For example, a flight from Los Angeles to London does not proceed directly across the United States and the Atlantic Ocean as you might expect from looking at a flat map. Instead, it goes far to the north, over Canada and Greenland, above the Arctic Circle, finally coming in over Scotland for a landing at London. This is the shortest path between the two cities. Look at a globe of Earth (or see the accompanying figure) to verify that this is indeed the case.

The *positively curved* space of Riemann is not the only possible departure from flat space. Another is the *negatively curved* space first studied by Nikolai Ivanovich Lobachevsky, a nineteenth-century Russian mathematician. In this geometry, there are an *infinite* number of lines through any given point parallel to another line, the sum of a triangle's angles is *less* than 180° (see the figure), and the circumference of a circle is *greater* than π times its diameter. Instead of the flat surface of a plane or the curved surface of a sphere, this type of space is described by the surface of a curved saddle. It is a difficult geometry to visualize!

Even though the universe on the largest scales may be curved as we have just described, most of the local realm of the *three*-dimensional universe (including the solar system, the neighboring stars, and even our Milky Way Galaxy) is correctly described by Euclidean geometry.

11-1 DISCOVERY

Ultraviolet Astronomy and the "Local Bubble"

The ultraviolet is that part of the spectrum where we expect to witness events and see objects involving temperatures in the hundreds of thousands, even millions, of kelvins—hot regions like the seething atmospheres or eruptive flares of stars; violent events, such as massive stars caught in the act of exploding; and active galaxies whose spinning hearts may harbor black holes. Yet ultraviolet astronomy has also contributed greatly to the study of the interstellar medium, by allowing a unique mapping of our local cosmic neighborhood.

Only 30 years ago astronomers had presumed that the gases filling the spaces among the stars would absorb virtually all short-wavelength UV radiation before it had a chance to reach Earth. But in 1975, during a historic linkup of the *Apollo* (U.S.) and *Soyuz* (USSR) space capsules, the onboard astronauts and cosmonauts performed a key experiment—they used a small telescope to detect extreme ultraviolet radiation from a few nearby, very hot stars. Shortly thereafter, theoretical ideas changed as astronomers began to realize that interstellar gas is distributed very unevenly in space, in cool dense clumps interspersed by irregularly shaped regions of hot, low-density gas.

Since then, space-based ultraviolet observations have shown that some regions of interstellar space are much thinner (5000 atoms/m^3) and hotter (500,000 K) than previously expected. Part of the space among the dust clouds and the emission nebulae seems to contain extremely dilute yet seething plasma, probably the result of the concussion and expanding debris from stars that exploded long ago. These superheated interstellar "bubbles," or *intercloud medium*, may extend far into interstellar space beyond our local neighborhood and conceivably into the even vaster spaces among the galaxies.

The Sun seems to reside in one such low-density region—a huge cavity called the *Local Bubble*, sketched in the accompanying figure. Only because we live within this vast, low-density bubble can we detect so many stars in the far UV; the hot, tenuous interstellar gas is virtually transparent to this radiation. The Local Bubble contains about 200,000 stars and extends for hundreds of trillions of kilometers, or nearly 100 pc. It was probably carved out by a supernova explosion that occurred in these parts roughly 300,000 years ago. ∞ *(Sec. 12.4)* Our distant ancestors must have seen it—a stellar catastrophe as bright as the full Moon.

Discovery Boxes Exploring a wide variety of interesting topics, *Discovery* boxes provide the reader with insight into how scientific knowledge evolves and emphasizes the process of science.

Chapter Summaries The chapter summaries, a primary review tool for the student, have been expanded and improved for the fifth edition. All key terms introduced in each chapter are listed again, in context and in boldface, along with key figures and page references to the text discussion.

CHAPTER REVIEW

Summary

1 A core-collapse supernova may leave behind an ultracompressed ball of material called a **neutron star** (p. 346). This is the remnant of the inner core that rebounded and blew the rest of the star apart. Neutron stars are extremely dense and, at formation, are predicted to be extremely hot, strongly magnetized, and rapidly rotating. They cool down, lose much of their magnetism, and slow down as they age.

2 According to the **lighthouse model** (p. 347), neutron stars, because they are magnetized and rotating, send regular bursts of electromagnetic energy into space. The beams are produced by charged particles confined by the strong magnetic fields. When we can see the beams from Earth, we call the source neutron star a **pulsar** (p. 346). The pulse period is the rotation period of the neutron star.

3 A neutron star that is a member of a binary system can draw matter from its companion, forming an accretion disk, which is usually a strong source of X-rays. As gas builds up on the star's surface, it eventually becomes hot enough to fuse hydrogen. When hydrogen burning starts on a neutron star, it does so explosively, and an **X-ray burster** (p. 350) results. The rapid rotation of the inner part of the accretion disk causes the neutron star to spin faster as new gas arrives on its surface. The eventual result is a very rapidly rotating neutron star—a **millisecond pulsar** (p. 350). Many millisecond pulsars are found in the hearts of old globular clusters. They cannot have formed recently, and must have been spun up by interactions with other stars. Careful analysis of the radiation received has shown that some millisecond pulsars are orbited by planet-sized objects.

4 **Gamma-ray bursts** (p. 352) are very energetic flashes of gamma rays that occur about once a day and are distributed uniformly over the entire sky. In some cases, their distances have been measured, placing them at very large distances and implying that they are extremely luminous. The leading theoretical models for these explosions involve the violent merger of neutron stars in a distant binary system, or the recollapse and subsequent violent explosion following a "failed" supernova in a very massive star.

5 Einstein's special theory of relativity deals with the behavior of particles moving at speeds comparable to the speed of light. It agrees with Newton's theory at low velocities, but makes many very different predictions for high-speed motion. All of its predictions have been repeatedly verified by experiment. The modern replacement for Newtonian gravity is Einstein's **general theory of relativity** (p. 357), which describes gravity in terms of the warping, or bending, of space by the presence of mass. The more mass, the greater the warping. All particles—including photons—respond to that warping by moving along curved paths.

6 The upper limit on the mass of a neutron star is about three solar masses. Beyond that mass, the star can no longer support itself against its own gravity, and it collapses to form a **black hole** (p. 357), a region of space from which nothing can escape. The most massive stars, after exploding in a supernova, form black holes rather than neutron stars. Conditions in and near black holes can only be described by general relativity. The radius at

End-of-Chapter Questions and Problems Many elements of the end-of-chapter material have seen substantial reorganization:

- Each chapter has 15 Review and Discussion Questions, which may be used for in-class review or for assignment. As with the Self-test Questions, the material needed to answer review questions may be found within the chapter. The Discussion Questions explore particular topics more deeply, often asking for opinions, not just facts. As with all discussions, these questions usually have no single "correct" answer.

- Each chapter now incorporates 15 Self-test Questions, roughly divided between true/false and multiple choice formats, designed to allow students to assess their understanding of the chapter material. Two of the multiple choice questions in each chapter are tied directly to a specific figure or diagram in the text, to test students' comprehension of the material presented there. Answers to all these questions appear at the end of the book.

- The end-of-chapter material also includes 10 Problems based on the chapter contents and requiring some numerical calculation. In many cases the Problems are tied directly to quantitative statements made (but not worked out in detail) in the text. The solutions to the Problems are not contained verbatim within the chapter, but the information necessary to solve them has been presented in the text. Answers appear at the end of the book.

Instructor Resources

Instructor's Resource Manual

Revised by William Alexander (James Madison University), this manual provides: sample syllabi and course schedules, an overview of each chapter, pedagogical tips, useful analogies, suggestions for classroom demonstrations, writing questions, answers to the end-of-chapter Review and Discussion Questions, Conceptual Self-test Questions, and Problems, selected readings, and additional references and resources. ISBN 0-13-187173-0

Transparency Acetates Over 270 figures and tables from the text are available as a package of color acetates. ISBN 0-13-187176-5

Test Item File An extensive file of approximately 2500 test questions, newly compiled and revised for the fifth edition by Gregory Dolise (Harrisburg Area Community College), including many new multiple choice and essay questions for added conceptual emphasis. The questions are organized and referenced by chapter section and by question type. This manual is also available electronically both within TestGenerator and as chapter-by-chapter Microsoft Word files on the *Instructor Resource Center on CD-ROM*. ISBN 0-13-187175-7

Instructor Resource Center on CD-ROM This multiple CD-ROM set provides virtually every electronic asset you'll need in and out of the classroom. Though you can navigate freely through the CDs to find the resources you want, the software allows you to browse and search through the catalog of assets. The CD-ROMs are organized by chapter and contain all text illustrations, all tables, and many photos in jpeg and PowerPoint formats, as well as the animations, videos, and Interactive Tutorials, from the **Companion Website with GradeTracker** (see description below). The IRC/CDs also contain TestGenerator, an easy to use, fully networkable program for creating tests ranging from short quizzes to long exams. Questions from the *Test Item File* are supplied, and professors can use the question editor to modify existing questions or create new questions. This CD-ROM set also contains chapter-by-chapter lecture outlines in PowerPoint, conceptual "clicker" questions in PowerPoint, and electronic versions of the *Instructor's Resource Manual*, the *Test Item File,* and instructor notes to the SkyChart III Projects. ISBN 0-13-187167-6

Comets Only Prentice Hall provides you with an annual media update. This unique service is available at the beginning of each academic year and available to qualified adopters of Chaisson/McMillan. *Comets* is a library of new videos and slides for you to use in class. The slide kit contains new slides from sources such as NASA, JPL, STScI, GSFC, *HST* Comet LINEAR Investigation Team, APL, JPL, ESA, *Hubble Heritage* Team, IPAC, European Southern Observatory, SDSS/Astrophysical Research Consortium, and the U.S. Department of Defense. Custom animations prepared by the Wright Center for Science Visualization and many other videos of new discoveries and animations from various sources, including NASA, STScI, APL/NRL, ESA, Stanford Lockheed Institute for Space Research, and JPL, are provided in both CD and DVD formats. A newsletter provides a cross reference between all the materials in the *Comets* kit and corresponding chapters of the fifth edition, as well as annotations describing the subject and source of each slide and animation in the kit.

Course and Homework Management Tools

OneKey Blackboard

Blackboard™ is a comprehensive and flexible e-learning software platform that delivers a course management system, customizable institution-wide portals, online communities, and an advanced architecture that allows for Web-based integration of multiple administrative systems.

Features:

- Course management system, including progress tracking, class and student management, gradebook, communication, assignments, and reporting tools.

- Testing programs that allow instructors to create online quizzes and tests and automatically grade and track results. All tests can be entered into the gradebook for easy course management.

- Communications tools such as the Virtual Classroom™ (chat rooms, WhiteBoard, slides), document sharing, and bulletin boards.

OneKey CourseCompass
Powered by **Blackboard**

With the highest level of service, support, and training available today, CourseCompass combines tested, quality, online course resources with easy-to-use online course management tools. CourseCompass is designed to address the individual needs of instructors, who will be able to create an online course without any special technical skills or training.

Features:

- Flexibility—Instructors can adapt *Astronomy: A Beginner's Guide to the Universe, Fifth Edition* content to match their own teaching goals.

- Assessment, customization, class administration, and communication tools.

- Point-and-click access—Prentice Hall's vast educational resources are available to instructors at a click of the mouse.

- A nationally hosted and fully supported system that relieves individuals and institutions of the burdens of troubleshooting and maintenance.

OneKey WebCT WebCT offers a powerful set of tools that enables you to create practical Web-based educational programs—the ideal resources to enhance a campus course or to construct one entirely online. The WebCT shell and tools, integrated with the *Astronomy: A Beginner's Guide to the Universe, Fifth Edition* content, results in a versatile, course-enhancing teaching and learning system.

Features:

- Page tracking, progress tracking, class and student management, gradebook, communication, calendar, reporting tools, and more.

- Communication tools including chat rooms, bulletin boards, private e-mail, and WhiteBoard.

- Testing tools that help create and administer timed online quizzes and tests and automatically grade and track all results.

OneKey content. Along with all the material from the Companion Website (http://www.prenhall.com/chaisson), our course management cartridges include additional materials available only through Blackboard, CourseCompass, or WebCT, and include

- test bank questions, converted from our TestGen *Test Item File*

- resources from the *Instructor Resource Center* on CD-ROM.

WebAssign (http://www.webassign.com)

WebAssign's homework delivery service harnesses the power of the Internet and puts it to work for you. Collecting and grading homework becomes the province of the WebAssign service, which gives you the freedom to get back to teaching. Create assignments from a database of end-of-chapter problems and *Test*

Item File questions from *Astronomy: A Beginner's Guide to the Universe, Fifth Edition* or write and customize your own. You have complete control over the homework your students receive, including due date, content, feedback, and question formats.

Learner-Centered Astronomy Teaching: Strategies for ASTRO 101

by Timothy F. Slater, University of Arizona, and
Jeffrey P. Adams, Montana State University

Strategies for ASTRO 101 is a guide for instructors of the introductory astronomy course for non-science majors. Written by two leaders in astronomy education research, this book details various techniques instructors can use to increase students' understanding and retention of astronomy topics, with an emphasis on making the lecture a forum for active student participation. Drawing from the large body of recent research to discover how students learn, this guide describes the application of multiple classroom-tested techniques to the task of teaching astronomy to predominantly non-science students.
ISBN 0-13-046630-1

Peer Instruction for Astronomy

by Paul Green, Harvard Smithsonian Center for Astrophysics

Peer instruction is a simple yet effective method for teaching science. Techniques of peer instruction for introductory physics were developed primarily at Harvard and have aroused interest and excitement in the physics education community. This approach involves students in the teaching process, making science more accessible to them. Peer instruction is a new trend in astronomy that is finding strong interest and is ideally suited to introductory astronomy classes. This book is an important vehicle for providing a large number of thought-provoking, conceptual short-answer questions aimed at a variety of class levels. While significant numbers of such questions have been published for use in physics, *Peer Instruction for Astronomy* provides the first such compilation for astronomy.
ISBN 0-13-026310-9

Student Resources

Companion Website with GradeTracker
(http://www.prenhall.com/chaisson)

The text-specific Companion Website with GradeTracker for *Astronomy: A Beginner's Guide to the Universe, Fifth Edition* organizes material from a variety of sources on a chapter-by-chapter basis, providing students with a variety of inquiry, visual, and interactive tools of each chapter's topics, and provides gradebook functionality for instructors who wish to use it. The Companion Website for the fifth edition includes:

- The newly assembled **Animations and Videos** gallery. This collection of the most up-to-date, illustrative, and dynamic pieces on key concepts and recent events helps to create a more scientifically literate astronomy student. See the associated icons within the text at point of use.

- The revamped and amplified **Tutorials** gallery. These animated, interactive Flash™ files bring concepts from the text to life, allowing students to explore ideas in depth. Students are engaged in the thought process as they answer questions and change parameters in these exploratory activities. Also included are gradable follow-up multiple choice **Tutorials quiz questions**, which will check the student's understanding of the tutorial. See the associated icons within the text at point of use.

- Now online for the fifth edition, the **eBook** is a full electronic version of the text, with key term hyperlinks and imbedded media elements.

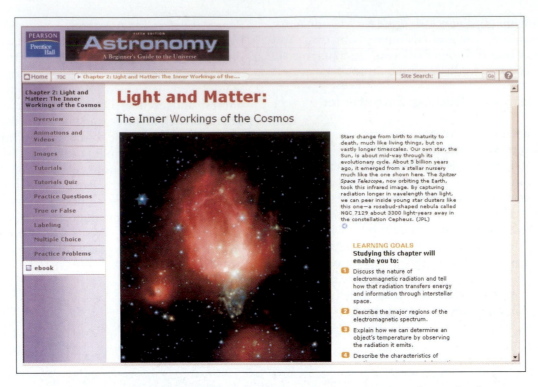

- A collection of the most current and astonishing **images** from the best sources on the World Wide Web.

- **Practice questions, multiple choice,** and **true or false** quizzes with hints and instant feedback, including links to relevant sections from the eBook.

- Algorithmically generated versions of the end-of-chapter **practice problems** from the text.

- **Labeling** exercises, where students identify various components of a fundamental figure from the textbook.

- The **SkyChart III** Planetarium software.

- Instructor tools include online grading capabilities and a syllabus manager.

Collaborators for this edition's Website include Todd Duncan and Candice Holcombe (Portland State University), Michael Fauerbach (Florida Gulf Coast University), David Ziegler (Hannibal-LaGrange College), Scott Hildreth (Chabot College), Eric Harpell (Las Positas College), Carl Adler (East Carolina University), and Scott Miller (Pennsylvania State University).

Student Lecture Notebook This lecture notebook contains all art, tables, and many photos from *A Beginner's Guide to the Universe, Fifth Edition,* including note-taking sections, so that students spend time taking notes during lecture and not redrawing figures. ISBN 0-13-198980-4

Observation, Research, Starry Night™ Pro, and SkyChart III Projects This supplement contains observation and research projects authored by Steve McMillan, as well as projects for the **Starry Night™ Pro** planetarium software by Erin O'Connor (Santa Barbara City College) and projects for the SkyChart III Student Version planetarium software by Erik Bodegom and Sean P. Goe (Portland State University). ISBN 0-13-187171-4

Starry Night™ Pro Student CD-ROM This best-selling planetarium software lets you escape the Milky Way and travel within 700 million light years of space. View more than 16 million stars in stunningly realistic star fields. Zoom in on thousands of galaxies, nebulae, and star clusters. Move through 200,000 years of time to see key celestial events in a dynamic and ever-changing universe. Blast off from Earth and see the motions of the planets from a new perspective. Hailed for its breathtaking realism, powerful suite of features, and intuitive ease-of-use, **Starry Night™ Pro** lives up to its reputation as astronomy software's brightest star … night after night.

Edmund Scientific Star and Planet Locator

The famous rotating roadmap of the heavens shows the location of the stars, constellations, and planets relative to the horizon for the exact hour and date you determine. This 8-inch square star chart was plotted by the late astronomer and cartographer George Lovi. The reverse side of the locator is packed with additional data on the planets, meteor showers, and bright stars. Included with each star chart is a 16-page, fully-illustrated, pocket-size instruction booklet.

Norton's Star Atlas and Reference Handbook, 20th edition
by Ian Ridpath

Now in a superbly redesigned, two-color landmark 20th edition, this combination star atlas and reference work has no match in the field. First published in 1910, *Norton's* owes much of its legendary success to its unique maps, arranged in slices known as gores, each covering approximately one-fifth of the sky. Every star visible to the naked eye under the clearest skies—down to magnitude 6.5—is charted, along with star clusters, nebulae, and galaxies. Extensive tables of data on interesting objects for observation accompany each of the precision-drawn maps. Preceding the maps is the unique and authoritative reference handbook covering timekeeping and positional measurements on the celestial sphere; the Sun, Moon, and other bodies of the solar system; telescopes and other equipment for observing and imaging the sky; and stars, nebulae, and galaxies. Throughout, succinct fundamental principles and practical tips guide the reader into the night sky. The appendices Units and Notation, Astronomical Constants, Symbols and Abbreviations, and Useful Addresses complete what has long been the only essential reference for the stargazer. ISBN 0-13-145164-2

Lecture Tutorials for Introductory Astronomy
by Jeffrey P. Adams, Montana State University,
Edward E. Prather, University of Arizona,
Timothy F. Slater, University of Arizona, and
the CAPER (Conceptual Astronomy and Physics Education Research) Team

Funded by the National Science Foundation, Lecture-Tutorials for Introductory Astronomy are designed to help make large lecture-format courses more interactive. Each of the 29 lecture-tutorials is presented in a classroom-ready format, asks students to work in groups of two or three, takes between 15 and 20 minutes, challenges students with a series of carefully designed questions that spark classroom discussion, engages students in critical reasoning, and requires no equipment. ISBN 0-13-147997-0

Acknowledgments

Throughout the many drafts that have led to this book, we have relied on the critical analysis of many colleagues. Their suggestions ranged from the macroscopic issue of the book's overall organization to the minutiae of the technical accuracy of each and every sentence. We have also benefited from much good advice and feedback from users of the first four editions of the text and our more comprehensive text, *Astronomy Today*. To these many helpful colleagues, we offer our sincerest thanks.

Reviewers of the Fifth Edition
Stephen G. Alexander, *Miami University*
George E. McCluskey, Jr., *Leigh University*
L. Kent Morrison, *University of New Mexico*
Jon Pedicino, *College of the Redwoods*
Philip J. Siemens, *Oregon State University*
Peter Stine, *Bloomsburg University of Pennsylvania*
Robert Potter, *Bowling Green State University*
Doug Franklin, *Western Illinois University*
Cynthia W. Peterson, *University of Connecticut*
F. Bary Malik, *Southern Illinois University-Carbondale*
Scott Miller, *Pennsylvania State University*
Jonathan Tan, *University of Florida*
Gregory R. Taylor, *California State University, Chico*
George Tremberger, Jr., *Queensborough Community College*
Donald Witt, *Nassau Community College*

Reviewers of Previous Editions
Nadine G. Barlow, *University of Central Florida*
Michael L. Broyles, *Collin County Community College*
Juan Cabanela, *Haverford College*
Michael L. Cobb, *Southeast Missouri State University*
Anne Cowley, *Arizona State University*
Manfred Cuntz, *University of Texas at Arlington*
James R. Dire, *U.S. Coast Guard Academy*
John Dykla, *Loyola University–Chicago*
Tina Fanetti, *Western Iowa Tech Community College*
Peter Garnavich, *University of Notre Dame*
Richard Gelderman, *Western Kentucky University*

Martin Goodson, *Delta College*
David J. Griffiths, *Oregon State University*
Doug Ingram, *Texas Christian University*
Susan Hartley, *University of Minnesota–Duluth*
Marvin Kemple, *Indiana University–Purdue University, Indianapolis*
Linda Khandro, *Shoreline Community College*
Mario KlariÊ, *Midlands Technical College*
Andrew R. Lazarewicz, *Boston College*
M.A.K. Lodhi, *Texas Tech University*
Fred Marschak, *Santa Barbara College*
John Mattox, *Francis Marion University*
Richard Nolthenius, *Cabrillo College*
Edward Oberhofer, *University of North Carolina–Charlotte*
Robert S. Patterson, *Southwest Missouri State University*
Heather L. Preston, *U.S. Air Force Academy*
Andreas Quirrenbach, *University of California–San Diego*
James Regas, *California State University, Chico*
Tim Rich, *Rust College*
Frederick A. Ringwald, *California State University, Fresno*
Gerald Royce, *Mary Washington College*
Rihab Sawah, *Moberly Area Community College*
John C. Schneider, *Catonsville Community College*
C. Ian Short, *Florida Atlantic University*
Earl F. Skelton, *George Washington University*
Don Sparks, *Los Angeles Pierce College*
Angela Speck, *University of Missouri–Columbia*
Phillip E. Stallworth, *Hunter College of CUNY*

Irina Struganova, *Barry University*
Jack W. Sulentic, *University of Alabama*
Craig Tyler, *Fort Lewis College*
J. Wayne Wooten, *University of West Florida*
Garett Yoder, *Eastern Kentucky University*
David C. Ziegler, *Hannibal-LaGrange College*

We would also like to express our gratitude to Ray Villard (Space Telescope Science Institute) for compiling the Comets supplement and to Companion Website contributors Todd Duncan and Candice Holcombe (Portland State University), Michael Fauerbach (Florida Gulf Coast University), David Ziegler (Hannibal-LaGrange College), Scott Hildreth (Chabot College), Eric Harpell (Las Positas College), Carl Adler (East Carolina University), and Scott Miller (Pennsylvania State University).

The publishing team at Prentice Hall has assisted us at every step along the way in creating this text. Much of the credit goes to our senior editor, Erik Fahlgren, who has successfully navigated us through the twists, turns, and "absolute final deadlines" of the publishing world, all the while managing the many variables that go into a multifaceted publication such as this. Anne Scanlan-Rohrer, our development editor, has skillful-ly helped us update and improve this edition. Production editor Shari Toron has done an excellent job of tying together the threads of this very complex project, made all the more complex by the necessity of combining text, art, and electronic media into a coherent whole. Special thanks are also in order to the art director, John Christiana, for making the fifth edition look spectacular; the associate editor, Christian Botting, for managing the ancillary and Course Management System programs; the media editor, Michael J. Richards, for managing the creation of the tutorials and extensive materials available online; and the editorial assistant, Jessica Berta, for her attention to detail and deadlines.

Finally, we would like to express our gratitude to renowned space artist Dana Berry for allowing us to use many of his beautiful renditions of astronomical scenes and to Lola Judith Chaisson for assembling and drawing all the H–R diagrams (including the acetate overlays) for this edition. We are interested in your feedback on this text. Please email us at physics_service@prenhall.com if you find any errors or have comments for the authors.

Eric Chaisson
Steve McMillan

ASTRONOMY

PART 1 Foundations

Mountains $\sim 10^4$ m

Cells $\sim 10^{-5}$ m

Astronomy is the study of the universe—the totality of all space, time, matter, and energy. It is a subject like no other, for it requires us to profoundly change our perspective and to consider sizes, scales, and times quite unfamiliar to us from everyday experience.

In this book, we shall study objects housing billions upon billions of atoms, even billions of whole stars. We likewise encounter very long durations of time, some spanning millions and billions of not merely seconds or hours but billions of years. The universe is *astronomical*, both in space and in time. To appreciate the subject, we must broaden our views and expand our minds. We must think big!

Recognize that a million is much larger than a thousand, and a billion, much, much larger still: One thousand (10^3 in scientific notation) is easy enough to understand—at the rate of one number per second, you could count to a thousand in about 15 minutes. But if you wanted to count to a million (10^6), it would take you two weeks of counting at the rate of one number per

Earth is neither central nor special; we inhabit no unique place in the universe.

second (assuming you counted for 16 hours each day and allowed for 8 hours of sleep). And at the same rate, about 50 years would be needed just to count to a billion (10^9)—try it! As for counting to a trillion (10^{12}), you couldn't do it.

Part 1 of this book presents the basic methods used by astronomers to chart the space around us. We describe the slow progress of scientific knowledge, from stories of chariots and gods to today's well-tested ideas of planetary motion and quantum physics. We also delve into the microscopic realm of atoms and molecules, whose properties hold keys to understanding the universe on macroscopic scales.

The figures on this page illustrate the range of scales encountered in Part 1, from atoms and humans to Earth itself. In later parts of this book, we shall study much larger objects, lying at incredible distances from us and existing far back in time—lighting up the far away and the long ago. Our ability to see, examine, and decipher extraterrestrial phenomena is one of the great achievements of modern science.

Earth ~10^7 m

Humans ~2 m

Atoms ~10^{-10} m

Exploring the Heavens

An Introduction to Astronomy

Stars are perhaps the most fundamental component of the universe that we inhabit. Roughly as many stars reside in the observable universe as grains of sand in all the beaches of the world. Here, we see displayed one of the most easily recognizable star fields in the winter sky—the familiar constellation Orion. Near center are three blue-white diagonally aligned stars hotter than our Sun, below which is a red-white nebulosity where stars are now forming. At top left is an old giant red star probably about to explode. The entire field of view spans about 100 light-years, or 10^{15} kilometers. (J. Sanford/Astrostock-Sanford)

Nature offers no greater splendor than the starry sky on a clear, dark night. Silent and jeweled with the constellations of ancient myth and legend, the night sky has inspired wonder throughout the ages—a wonder that leads our imaginations far from the confines of Earth and out into the distant reaches of space and time. Astronomy, born in response to that wonder, is built on two basic traits of human nature: the need to explore and the need to understand. People have sought answers to questions about the universe since the earliest times. Astronomy is the oldest of all the sciences, yet never has it been more exciting than it is today.

LEARNING GOALS

Studying this introductory chapter will enable you to:

1 Describe the concept of the celestial sphere and how we use angular measurement to locate objects in the sky.

2 Account for the apparent motions of the Sun and the stars in terms of the actual motion of Earth, and explain why our planet experiences seasonal changes.

3 Discuss the changing appearance of the Moon and how the relative motions of Earth, the Sun, and the Moon lead to eclipses.

4 Explain the simple geometric reasoning that allows astronomers to measure the distances and sizes of otherwise inaccessible objects.

5 Describe how scientists combine observation, theory, and testing in their study of the universe.

Visit www.prenhall.com/chaisson for additional annotated images, animations, and links to related sites for this chapter.

E.1 The "Obvious" View

Our Place in Space

Of all the scientific insights achieved to date, one stands out boldly: Earth is neither central nor special. We inhabit no unique place in the universe. We live on an ordinary rocky **planet** called Earth, one of nine known planets orbiting an average **star** called the Sun, a middle-aged star near the edge of a huge collection of stars called the Milky Way **galaxy**, one galaxy among countless billions of others spread throughout the observable **universe**. Figure E.1 places these very different objects in perspective, illustrating the enormous range of scales involved (see also the Part 1 opening discussion on page 1).

Today our scientific understanding of the cosmos extends literally as far as we can see, spanning the vast reaches of intergalactic space. But the modern view of the universe depicted here is in many ways the "punch line" of the story presented in this text. It is the culmination of countless scientific discoveries, large and small, representing the work of generations of astronomers who have contributed to our common understanding of the heavens. How have we come to know the universe around us? How have astronomers achieved the perspective sketched in Figure E.1? Our study of the universe, the science of **astronomy**, begins by examining the sky.

Constellations in the Sky

Between sunset and sunrise on a clear night, we can see some 3000 points of light. Include the view from the opposite side of Earth, and nearly 6000 stars are visible to the unaided eye. A natural human tendency is to see patterns and relationships between objects even when no true connection exists, and people long ago connected the brightest stars into configurations called **constellations**. In the Northern Hemisphere, most constellations were named

Universe

x10⁴

Galaxy

x10⁸

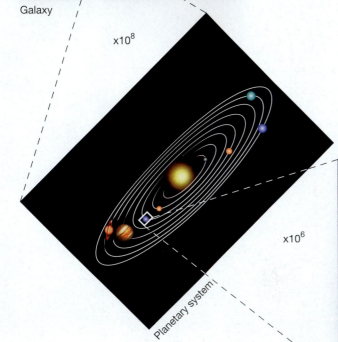

Planetary system

▼ **Figure E.1 Size and Scale in the Universe**   This artist's conception presents the modern view of scale in the universe. Bottom right of this figure shows humans on Earth, a view that widens progressively in each of the next four scenes illustrated from bottom to top—Earth, the planetary system, a galaxy, and truly deep space. The numbers within the dashed zooms indicate approximately the increase in scale between successive images: Earth is ten million times larger than humans, our solar system in turn is some ten million times larger than Earth, and so on. (NASA; J. Lodriguss)

x10⁶

Earth

x10⁷

Humans

▶ **Figure E.2 Constellation Orion** (a) A photograph of the group of bright stars that make up the constellation Orion. (See the Preface, p. xiv, for an explanation of the icon at the bottom, which simply indicates that this image was made in visible light.) (b) The stars connected to show the pattern visualized by the Greeks: the outline of a hunter. The Greek letters serve to identify some of the brighter stars in the constellation (see Figure E.3 and p. 6). You can easily find Orion in the northern winter sky by identifying the line of three bright stars in the hunter's "belt." (J. Sanford/Astrostock)

(a)

R I V U X G

(b)

after mythological heroes and animals. Figure E.2 shows a constellation especially prominent in the northern night sky from October through March: the hunter Orion, named for a mythical Greek hero famed, among other things, for his amorous pursuit of the Pleiades, the seven daughters of the giant Atlas. According to Greek mythology, in order to protect the Pleiades from Orion, the gods placed them among the stars, where Orion now nightly stalks them across the sky. Many other constellations have similarly fabulous connections with ancient cultures.

The stars making up a particular constellation are generally not close together in space. They merely are bright enough to observe with the naked eye and happen to lie in the same direction in the sky as seen from Earth. Figure E.3 illustrates this point for Orion, showing the true relationships between that constellation's brightest stars. Yet despite the fact that constellation patterns have no real significance, the terminology is still used today. Constellations provide a convenient means for astronomers to specify large areas of the sky, much as geologists use continents or politicians use voting precincts to identify certain localities on Earth. In all, there are 88 constellations, most of them visible from North America at some time during the year.

The Celestial Sphere

Over the course of a night, the constellations appear to move across the sky from east to west. However, ancient sky-watchers noted that the *relative* positions of stars (to each other) remained unchanged as this nightly march took place. It was natural for those first astronomers to conclude that the stars were attached to a **celestial sphere** surrounding Earth—a canopy of stars like an

What we see projected on sky

|—————————1,000 light-years—————————|

◀ **Figure E.3 Orion in 3D** The true three-dimensional relationships among the most prominent stars in Orion. The distances in light-years were determined by the *Hipparcos* satellite in the early 1990s. (See Chapter 10.)

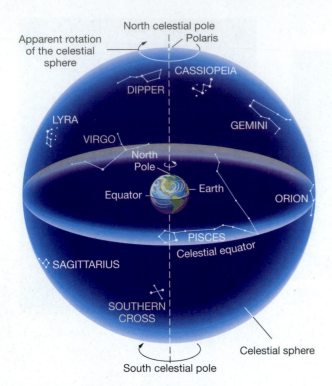

▲ Figure E.4 The Celestial Sphere   Planet Earth sits fixed at the hub of the celestial sphere, which contains all the stars. This is one of the simplest possible models of the universe, but it doesn't agree with the facts that astronomers now know about the cosmos.

▲ Figure E.5 The Northern Sky   A time-lapse photograph of the northern sky. Each trail is the path of a single star across the night sky. The duration of the exposure is about five hours. (You can see this by noting that the trails form arcs of circles spanning roughly 75 degrees, and recalling that Earth rotates through 360 degrees in 24 hours, or 15 degrees per hour). The center of the concentric circles is near the North Star, Polaris, whose short, bright arc is prominently visible. (AURA)

astronomical painting on a vast heavenly ceiling. Figure E.4 shows how early astronomers pictured the stars as moving with this celestial sphere as it turned around a fixed, unmoving Earth. Figure E.5 shows how stars appear to move in circles around a point in the sky very close to the star Polaris (better known as the Pole Star or the North Star). To the ancients, this point represented the axis around which the celestial sphere turned.

From our modern standpoint, the apparent motion of the stars is the result of the spin, or **rotation**, not of the celestial sphere but of Earth. Polaris indicates the direction—due north—in which Earth's rotation axis (the line through the center around which the planet rotates) points. Even though we now know that a revolving celestial sphere is an incorrect description of the heavens, astronomers still use the idea as a convenient fiction that helps us visualize the positions of stars in the sky. The point where Earth's axis intersects the celestial sphere in the Northern Hemisphere is known as the **north celestial pole**; it is directly above Earth's North Pole. In the Southern Hemisphere, the extension of Earth's axis in the opposite direction defines the **south celestial pole**. Midway between the north and south celestial poles lies the **celestial equator**, representing the intersection of Earth's equatorial plane (the plane through Earth's center, perpendicular to the rotation axis) with the celestial sphere.

Celestial Coordinates

The simplest method of locating stars in the sky is to specify their constellation and then rank the stars in it in order of brightness. The brightest star is denoted by the Greek letter α (alpha), the second brightest by β (beta), and so on. The two brightest stars in the constellation Orion—Betelgeuse and Rigel—are therefore also known as α Orionis and β Orionis, respectively (see Figures E.2 and E.3). (Precise observations show that Rigel is actually brighter than Betelgeuse, but the names are now permanent.) Because there are many more stars in any given constellation than there are letters in the Greek alphabet, this method is of limited use. However, for naked-eye astronomy, where only bright stars are involved, it is quite satisfactory.

(a)

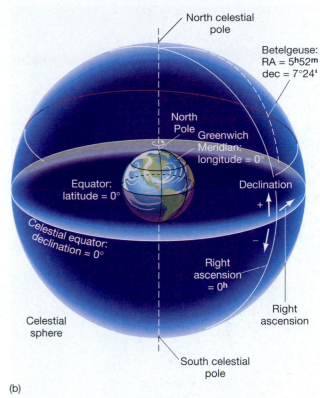

(b)

For more precise measurements, astronomers find it helpful to lay down a system of **celestial coordinates** on the sky. If we think of the stars as being attached to the celestial sphere centered on Earth, then the familiar system of angular measurement on Earth's surface—latitude and longitude (Figure E.6a)—extends quite naturally to the sky. The celestial analogs of latitude and longitude are called **declination** and **right ascension**, respectively (Figure E.6b). Just as latitude and longitude are tied to Earth, right ascension and declination are fixed on the celestial sphere. Although the stars appear to move across the sky because of Earth's rotation, their celestial coordinates remain constant over the course of a night.

Declination (dec) is measured in *degrees* (°) north or south of the celestial equator, just as latitude is measured in degrees north or south of Earth's equator (see *More Precisely E-1*). The celestial equator is at a declination of 0°, the north celestial pole is at +90°, and the south celestial pole is at −90° (the plus sign here just means "north of the celestial equator"; minus means "south"). Right ascension (RA) is measured in angular units called *hours, minutes,* and *seconds,* and it increases in the eastward direction. Like the choice of the Greenwich Meridian as the zero-point of longitude on Earth, the choice of zero right ascension is quite arbitrary—it is conventionally taken to be the position of the Sun in the sky at the instant of the vernal equinox (to be discussed in the next section).

▲ **Figure E.6 Right Ascension and Declination**
(a) Longitude and latitude allow us to locate a point on the surface of Earth by specifying its distance (as an angle) east or west of the Greenwich Meridian, and north or south of the equator. For example, to find Washington, D.C., on Earth, look 77° west of Greenwich and 39° north of the equator.
(b) Similarly, right ascension and declination specify locations on the sky. To locate the star Betelgeuse on the celestial sphere, look 5^{h}52^m east of the vernal equinox (the line on the sky with a right ascension of zero) and 7°24′ north of the celestial equator.

◀◀◀ **CONCEPT CHECK**

Earth isn't really enclosed in a sphere with stars attached. Why then do astronomers find it convenient to retain the fiction of the celestial sphere? What vital piece of information about stars is lost when we talk about their positions "on" the sky?

E.2 Earth's Orbital Motion

Day-to-Day Changes

We measure time by the Sun. The rhythm of day and night is central to our lives, so it is not surprising that the period of time from one sunrise (or noon, or sunset) to the next, the 24-hour **solar day**, is our basic social time unit. As we have just seen, this apparent daily progress of the Sun and other stars across the sky (known as *diurnal motion*) is a consequence of Earth's rotation. But

◄ **Figure E.7** **Solar and Sidereal Days**
A sidereal day is Earth's true rotation period—the time taken for our planet to return to the same orientation in space relative to the distant stars. A solar day is the time from one noon to the next. The difference in duration between the two is easily explained once we understand that Earth revolves around the Sun at the same time as it rotates on its axis. Frames (a) and (b) are one sidereal day apart. During that time, Earth rotates exactly once on its axis and also moves a little in its solar orbit—approximately 1°. Consequently, between noon at point A on one day and noon at the same point the next day, Earth actually rotates through about 361° (frame c), and the solar day exceeds the sidereal day by about four minutes. Note that the diagrams are not drawn to scale; the true 1° angle is in reality much smaller than shown here.

the stars' positions in the sky do not repeat themselves exactly from one night to the next. Each night, the whole celestial sphere appears shifted a little compared with the night before—you can confirm this for yourself by noting over the course of a week or two which stars are visible near the horizon just after sunset or just before dawn. Because of this shift, a day measured by the stars—called a **sidereal day** after the Latin word *sidus*, meaning "star"—differs in length from a solar day.

The reason for the difference in length between a solar day and a sidereal day is sketched in Figure E.7. It is a result of the fact that Earth moves in two ways simultaneously: it rotates on its central axis while at the same time **revolving** around the Sun. Each time Earth rotates once on its axis, it also moves a small distance along its orbit. Each day, therefore, Earth has to rotate through slightly more than 360° in order for the Sun to return to the same apparent location in the sky. As a result, the interval of time between noon one day and noon the next (a solar day) is slightly greater than the true rotation period (one sidereal day). Our planet takes 365 days to orbit the Sun, so the additional angle is 360°/365 = 0.986°. Because Earth takes about 3.9 minutes to rotate through this angle, the solar day is 3.9 minutes longer than the sidereal day.

Seasonal Changes

Because Earth revolves around the Sun, our planet's darkened hemisphere faces in a slightly different direction each night. The change is only about 1° per night (Figure E.7)—too small to be easily discerned with the naked eye from one evening to the next. However, the change is clearly noticeable over the course of weeks and months, as illustrated in Figure E.8. In six months Earth moves to the opposite side of its orbit, and we face an entirely different group of stars and constellations at night. Because of this motion, the Sun appears, to an observer on Earth, to move relative to the background stars over the course of a year. This apparent motion of the Sun on the sky

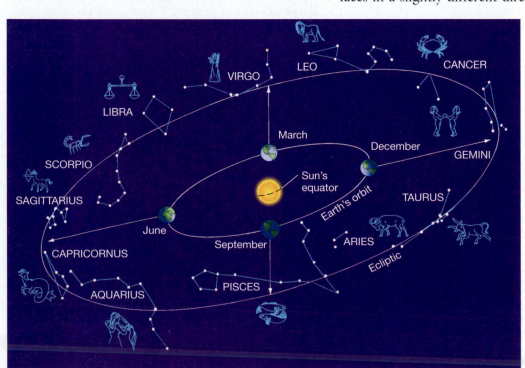

◄ **Figure E.8** **The Zodiac** The view of the night sky changes as Earth moves in its orbit about the Sun. As drawn here, the night side of Earth faces a different set of constellations at different times of the year. The twelve constellations named here make up the astrological zodiac. The arrows indicate the most prominent zodiacal constellations in the night sky at various times of year. For example, in June, when the Sun is "in" Gemini, Sagittarius and Capricornus are visible at night.

▶ **Figure E.9 Ecliptic**   The apparent path of the Sun on the celestial sphere over the course of a year is called the ecliptic. In this picture of the heavens, the seasons result from the changing height of the Sun above the celestial equator. At the summer solstice, the Sun is at its northernmost point on its path around the ecliptic. It is therefore highest in the sky, as seen from Earth's Northern Hemisphere, and the days are longest. The reverse is true at the winter solstice. At the vernal and autumnal equinoxes, when the Sun crosses the celestial equator, day and night are of equal length.

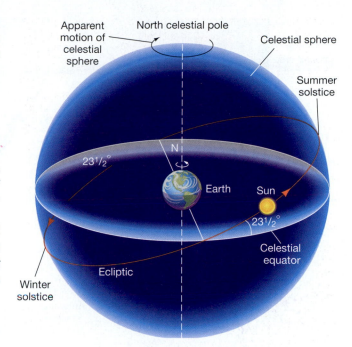

traces out a path on the celestial sphere known as the **ecliptic**. The 12 constellations through which the Sun passes as it moves along the ecliptic—that is, the constellations we would see looking in the direction of the Sun if they weren't overwhelmed by the Sun's light—had special significance for astrologers of old. They are collectively known as the **zodiac**.

As illustrated in Figure E.9, the ecliptic forms a great circle on the celestial sphere, inclined at an angle of about 23.5° to the celestial equator. In reality, as shown in Figure E.10, the plane defined by the ecliptic is *the plane of Earth's orbit around the Sun*. Its tilt is a consequence of the *inclination* of our planet's rotation axis to its orbital plane.

E-1 MORE PRECISELY

Angular Measure

The size and scale of astronomical objects are often specified by measuring lengths and angles. The concept of length measurement is fairly intuitive. The concept of angular measurement may be less familiar, but it too can become second nature if you remember a few simple facts:

- A full circle contains 360 arc degrees (or 360°). Therefore, the half-circle that stretches from horizon to horizon, passing directly overhead and spanning the portion of the sky visible to one person at any one time, contains 180°.
- Each 1° increment can be further subdivided into fractions of an arc degree, called arc minutes; there are 60 arc minutes (60′) in one arc degree. Both the Sun and the Moon project an angular size of 30 arc minutes on the sky. Your little finger, held at arm's length, does about the same, covering about a 40-arc-minute slice of the 180° horizon-to-horizon arc.
- An arc minute can be divided into 60 arc seconds (60″). Put another way, an arc minute is 1/60 of an arc degree, and an arc second is 1/60 × 1/60 = 1/3600 of an arc degree. An arc second is an extremely small unit of angular measure—it is the angular size of a centimeter-size object (a dime, say) at a distance of about 2 kilometers (a little over a mile).

The accompanying figure illustrates this subdivision of the circle into progressively smaller units.

One final note: Arc degrees have nothing to do with temperature, and arc minutes and arc seconds have nothing to do with time. However, the angular units used to measure right

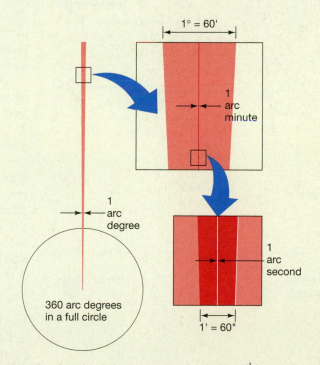

ascension (and only right ascension)—hours (^{h}), minutes (^{m}), and seconds (^{s})—*are* constructed to parallel the units of time, the connection being provided by Earth's rotation. In 24 hours Earth rotates once on its axis, or through 360°. In a time interval of one hour, Earth rotates through 360°/24 = 15°, or 1^h. In one minute of time Earth rotates through 15°/60 = 0.25° = 15′ = 1^m; in one second Earth rotates through 15′/60 = 0.25′ = 15″ = 1^s. Just bear in mind that angular hours, minutes, and seconds refer *only* to right ascension and you should avoid undue confusion.

Smaller ground area covered — Light from Sun

Light from Sun — Larger ground area covered

Vernal equinox (Mar. 21)

Summer solstice (June 21)

Sun

Autumnal equinox (Sept. 21)

Winter solstice (Dec. 21)

Animation/Video
Summer Solstice

Animation/Video
Winter Solstice

Vernal equinox

Summer solstice

Winter solstice

Sun

Autumnal equinox

▲ **Figure E.11 Earth's Orbit**   Seen face on, Earth's orbit around the Sun is almost indistinguishable from a perfect circle. The distance from Earth to the Sun varies only slightly over the course of a year and is not the cause of the seasonal temperature variations we experience.

◄ **Figure E.10 Seasons**   In reality, the Sun's apparent motion along the ecliptic is a consequence of Earth's orbital motion around the Sun. The seasons result from the inclination of our planet's rotation axis with respect to the orbit plane. The summer solstice corresponds to the point on Earth's orbit where our planet's North Pole points most nearly toward the Sun. The opposite is true of the winter solstice. The vernal and autumnal equinoxes correspond to the points in Earth's orbit where our planet's axis is perpendicular to the line joining Earth and the Sun. The insets show how rays of sunlight striking the ground at an angle (for example, during northern winter) are spread over a larger area than rays coming nearly straight down (e.g., during northern summer). As a result, the amount of solar heat delivered to a given area of Earth's surface is greatest when the Sun is high in the sky.

The point on the ecliptic where the Sun is at its northernmost point above the celestial equator (see Figure E.9) is known as the **summer solstice** (from the Latin words *sol*, meaning "sun," and *stare*, "to stand"). As indicated in Figure E.10, it represents the point on Earth's orbit where our planet's North Pole is oriented closest to the Sun. This occurs on or near June 21—the exact date varies slightly from year to year because the actual length of a year is not a whole number of days. As Earth rotates on that date, points north of the equator spend the greatest fraction of their time in sunlight, so the summer solstice corresponds to the longest day of the year in Earth's Northern Hemisphere and the shortest day in Earth's Southern Hemisphere.

Six months later the Sun is at its southernmost point below the celestial equator (Figure E.9)—or, equivalently, Earth's North Pole is oriented farthest from the Sun (Figure E.10). We have reached the **winter solstice** (December 21), the shortest day in Earth's Northern Hemisphere and the longest in the Southern Hemisphere.

The tilt of Earth's rotation axis relative to the ecliptic is responsible for the **seasons** we experience—the marked difference in temperature between the hot summer and cold winter months. As illustrated in Figure E.10, two factors combine to cause this variation. First, there are more hours of daylight during the summer than in winter. To see why this is, look at the yellow lines on the surfaces of the Earths in the figure. (For definiteness, they correspond to a latitude of 45 degrees—roughly that of the Great Lakes or the south of France.) A much larger fraction of the line is sunlit in the summertime, and more daylight means more solar heating. Second, as illustrated in the insets in Figure E.10, when the Sun is high in the sky in summer, rays of sunlight striking Earth's surface are more concentrated—spread out over a smaller area—than in winter. As a result, the Sun feels hotter. Therefore summer, when the Sun is highest above the horizon and the days are longest, is generally much warmer than winter, when the Sun is low and the days are short.

A popular misconception is that the seasons have something to do with Earth's distance from the Sun. Figure E.11 illustrates why this is *not* the case. It shows Earth's orbit "face on," instead of almost edge-on, as in Figure E.10. Notice that the orbit is almost perfectly circular, so the distance from Earth to the Sun varies very little (in fact, by only about 3 percent) over the course of a year—not nearly enough to explain the seasonal changes in temperature. What's more, Earth is actually *closest* to the Sun in early January, the dead of winter in the Northern Hemisphere, so distance from the Sun cannot be the main factor controlling our climate.

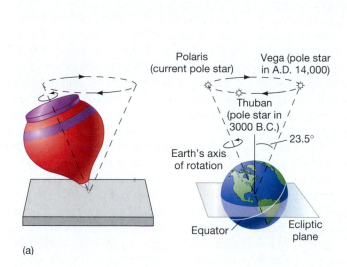

The two points where the ecliptic intersects the celestial equator (Figure E.9)—that is, where Earth's rotation axis is perpendicular to the line joining Earth to the Sun (Figure E.10)—are known as **equinoxes**. On those dates, day and night are of equal duration. (The word *equinox* derives from the Latin for "equal night.") In the fall (in Earth's northern hemisphere), as the Sun crosses from the northern into the southern celestial hemisphere, we have the **autumnal equinox** (on September 21). The **vernal equinox** occurs in spring, on or near March 21, as the Sun crosses the celestial equator moving north. The vernal equinox plays an important role in human timekeeping. The interval of time from one vernal equinox to the next—365.242 solar days—is known as one **tropical year**.

Long-Term Changes

The time required for Earth to complete exactly one orbit around the Sun, relative to the stars, is called a **sidereal year**. One sidereal year is 365.256 solar days long, about 20 minutes longer than a tropical year. The reason for this slight difference is a phenomenon known as **precession**. Like a spinning top that rotates rapidly on its own axis while that axis slowly revolves about the vertical, Earth's axis changes its direction over the course of time, although the angle between the axis and a line perpendicular to the plane of the ecliptic always remains close to 23.5°. Figure E.12 illustrates Earth's precession, which is caused by the combined gravitational pulls of the Moon and the Sun (see Chapter 1). During a complete cycle of precession, taking about 26,000 years, Earth's axis traces out a cone. Because of this slow shift in the orientation of Earth's rotation axis, the vernal equinox, which defines the tropical year, drifts slowly around the ecliptic over the course of the precession cycle.

The tropical year is the year our calendars measure. If our timekeeping were tied to the sidereal year, the seasons would slowly march around the calendar as Earth precessed—13,000 years from now, summer in the Northern Hemisphere would be at its height in mid-February! By using the tropical year instead, we ensure that July and August will always be (northern) summer months. However, in 13,000 years' time, Orion, now a prominent feature of the northern winter sky, will be a summer constellation.

▲ **Figure E.12 Precession** (a) Earth's axis currently points nearly toward the star Polaris. Some 12,000 years from now—nearly halfway through one cycle of precession—Earth's axis will point toward a star called Vega, which will then be the "North Star." Five thousand years ago the North Star was a star named Thuban in the constellation Draco. (b) The circle shows the precessional path of the north celestial pole among some prominent Northern stars. Tick marks indicate intervals of 1000 years.

Animation/Video
The Equinoxes

Animation/Video
The Earth's Seasons

◄◄◄ **CONCEPT CHECK**

Earth is actually farthest from the Sun during northern summer. Why, then, is it hottest in North America during this season?

E.3 The Motion of the Moon

Early astronomers had very practical reasons for studying the sky. Some stars (such as Polaris) served as navigational guides, while others served as primitive calendars to predict planting and harvesting seasons. By observing repeating patterns in the sky and associating them with events on Earth, astronomers began to establish concrete connections between celestial events and everyday life and took the first steps toward true scientific understanding of the heavens. In a real sense, then, human survival depended on astronomical knowledge. The ability to predict and even explain astronomical events was undoubtedly a highly prized, and perhaps jealously guarded, skill.

The Moon also played an important role in ancient astronomy. Calendars and religious observances were often tied to its phases and cycles, and even today the calendars of most of the world's major religions are still based wholly or partly on the lunar orbit. The Moon's regularly changing appearance (as well as its less regular, but much more spectacular, eclipses) was an integral part of the framework within which ancient astronomers sought to understand the universe. We will study the Moon's physical properties in more detail in Chapter 5. Here we continue our inventory of the sky with a brief description of the motion of our nearest neighbor in space.

Lunar Phases

Apart from the Sun, the Moon is by far the brightest object in the sky. Like the Sun, the Moon appears to move relative to the background stars. Unlike the Sun, however, the explanation for this motion is the obvious one—the Moon really does revolve around Earth.

The Moon's appearance undergoes a regular cycle of changes, or **phases**, taking a little more than 29 days to complete. (The word *month* is derived from the word Moon.) Figure E.13(a) illustrates the appearance of the Moon at different times in this monthly cycle. Starting from the **new Moon**, which is all but invisible in the sky, the Moon appears to *wax* (grow) a little each night and is visible as a growing *crescent* (panel 1 of Figure E.13a). One week after new Moon, half of the lunar disk (the circular face we would see if the Moon were completely illuminated) can be seen (panel 2). This phase is known as a **quarter Moon**. During the next week, the Moon continues to wax, passing through the *gibbous* phase (more than half of the lunar disk visible, panel 3) until, two weeks after new Moon, the **full Moon** (panel 4) is visible. During the next two weeks, the Moon *wanes* (shrinks), passing in turn through the gibbous, quarter, and crescent phases (panels 5–7), eventually becoming new again.

The location of the Moon in the sky, as seen from Earth, depends on its phase. For example, the full Moon rises in the east as the Sun sets in the west, while the first quarter Moon actually rises at noon, but often only becomes visible late in the day as the Sun's light fades. By this time the Moon is already high in the sky. These connections between lunar phase and rising/setting times are indicated on Figure E.13(a).

Unlike the Sun and the other stars, the Moon emits no light of its own. Instead, it shines by reflected sunlight, giving rise to the phases we see. As indicated in Figure E.13(a), half of the Moon's surface is illuminated by the Sun at any moment, but not all of the Moon's sunlit face can be seen because of the Moon's position with respect to Earth and the Sun. When the Moon is full, we see the entire "daylit" face because the Sun and the Moon are in opposite directions from Earth in the sky. The Sun's light is not blocked by Earth at the full phase because, as shown in Figure E.13(b), the Moon's orbit is inclined at

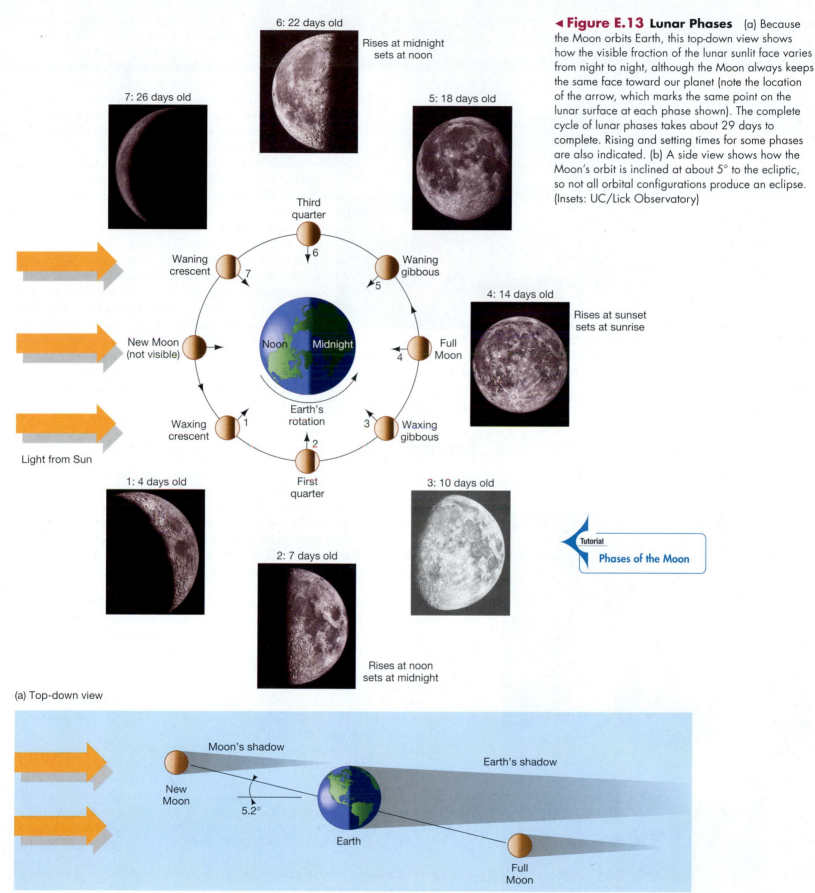

6: 22 days old

Rises at midnight
sets at noon

7: 26 days old

5: 18 days old

◀ **Figure E.13** **Lunar Phases**   (a) Because the Moon orbits Earth, this top-down view shows how the visible fraction of the lunar sunlit face varies from night to night, although the Moon always keeps the same face toward our planet (note the location of the arrow, which marks the same point on the lunar surface at each phase shown). The complete cycle of lunar phases takes about 29 days to complete. Rising and setting times for some phases are also indicated. (b) A side view shows how the Moon's orbit is inclined at about 5° to the ecliptic, so not all orbital configurations produce an eclipse. (Insets: UC/Lick Observatory)

Third
quarter

6

Waning
crescent

7

Waning
gibbous

5

4: 14 days old

Rises at sunset
sets at sunrise

Light from Sun

New Moon
(not visible)

Noon   Midnight

4

Full
Moon

Waxing
crescent

1

Earth's
rotation

2

3

Waxing
gibbous

1: 4 days old

First
quarter

3: 10 days old

Tutorial

Phases of the Moon

2: 7 days old

Rises at noon
sets at midnight

(a) Top-down view

Moon's shadow

Earth's shadow

New
Moon

5.2°

Earth

Full
Moon

(b) Side view

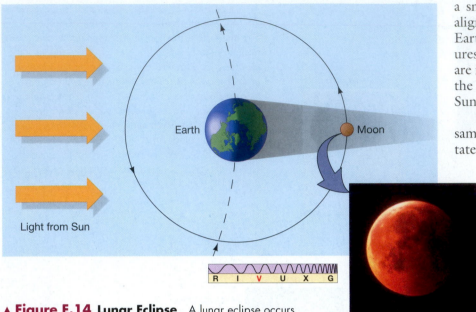

R I V U X G

▲ **Figure E.14** **Lunar Eclipse** A lunar eclipse occurs when the Moon passes through Earth's shadow. At these times we see a darkened, copper-colored Moon, as shown by the partial eclipse in the inset photograph. The coloration is caused by sunlight deflected by Earth's atmosphere onto the Moon's surface. (Note that this figure is not drawn to scale.) (Inset: G. Schneider)

a small angle (5.2°) to the plane of the ecliptic, so the alignment of the three bodies is not perfect. (The sizes of Earth and the Moon are greatly exaggerated in these figures.) In the case of a new Moon, the Moon and the Sun are in nearly the same part of the sky, and the sunlit side of the Moon is oriented away from us—at new Moon the Sun is almost behind the Moon, from our perspective.

Notice, by the way, that the Moon always keeps the same face toward Earth—as indicated on the figure, it rotates on its axis in exactly the same time it takes to orbit Earth. This is called *synchronous rotation*. We will discuss the reason for it in Chapter 5.

As it revolves around Earth, the Moon's position in the sky changes with respect to the stars. In one **sidereal month** (27.3 days), the Moon completes one revolution and returns to its starting point on the celestial sphere, having traced out a great circle in the sky. The time required for the Moon to complete a full cycle of phases, one **synodic month**, is a little longer—about 29.5 days. The synodic month is a little longer than the sidereal month for the same basic reason that a solar day is slightly longer than a sidereal day (Figure E.7): because of Earth's motion around the Sun, the Moon must complete slightly more than one full revolution to return to the same phase in its orbit.

Eclipses

From time to time—but only at new or full Moon—the Sun, Earth, and the Moon line up precisely and we observe the spectacular phenomenon known as an **eclipse**. When the Sun and the Moon are in exactly *opposite* directions as seen from Earth, Earth's shadow sweeps across the Moon, temporarily blocking the Sun's light and darkening the Moon in a **lunar eclipse**, as illustrated in Figure E.14. From Earth we see the curved edge of Earth's shadow begin to cut across the face of the full Moon and slowly "eat" its way into the circular lunar disk.

Usually the alignment of the Sun, Earth, and the Moon is imperfect, so the shadow never completely covers the Moon. Such an occurrence is known as a **partial eclipse**. Occasionally, however, the entire lunar surface is obscured in a **total eclipse** (such as that shown in the inset in Figure E.14). Total lunar eclipses last only as long as is needed for the Moon to pass through Earth's shadow—no more than about 100 minutes. During that time, the Moon often acquires an eerie, deep red coloration, the result of a small amount of sunlight being refracted (bent) by Earth's atmosphere onto the lunar surface, preventing the shadow from being completely black.

When the Moon and the Sun are in exactly the *same* direction as seen from Earth, an even more awe-inspiring event occurs. The Moon passes directly in front of the Sun, briefly turning day into night in a **solar eclipse** (Figure E.15). In a *total solar eclipse*, when the alignment is perfect, planets and some stars become visible in the daytime as the Sun's light is blocked. By pure chance the Sun and Moon have almost exactly the same angular size as seen

R I V U X G

◄ **Figure E.15** **Solar Eclipse** During a total solar eclipse, the Sun's corona becomes visible as an irregularly shaped halo surrounding the blotted-out disk of the Sun. This was the August 1999 eclipse, as seen from the banks of the Danube River near Sofia, Bulgaria. (Bencho Angelov)

from Earth—the Sun is much bigger than the Moon, but it also lies much farther away (see *More Precisely E-1*). Consequently, during a total solar eclipse, we can often see the Sun's ghostly outer atmosphere, or *corona*, its faint glow becoming temporarily visible when the rest of the Sun's glare is obscured. In a *partial solar eclipse*, the Moon's path is slightly off center, and only a portion of the Sun's face is covered.

Unlike a lunar eclipse, which is simultaneously visible from all locations on Earth's night side, a total solar eclipse can be seen from only a small portion of the daytime side. The Moon's shadow on Earth's surface is about 7000 kilometers wide—roughly twice the diameter of the Moon. Outside that shadow, no eclipse is seen. However, only within the central region of the shadow, in the **umbra**, is the eclipse total. The umbra is the part of a shadow (in this case, the Moon's) where all light from the source is blocked. Within the shadow but outside the umbra, in the **penumbra**, some but not all of the Sun's light is blocked and the eclipse is partial, with less and less of the Sun being obscured the farther one travels from the shadow's center.

The connections between the umbra, the penumbra, and the relative locations of Earth, Sun, and Moon are illustrated in Figure E.16. The umbra is always very small—even under the most favorable circumstances, its diameter never exceeds 270 kilometers. Because the Moon's shadow sweeps across Earth's surface at a speed of more than 1700 kilometers per hour, the duration of a total solar eclipse at any given point can never exceed 7.5 minutes (270 km divided by 1700 km/hour, times 60 minutes per hour).

The Moon's orbit around Earth is not exactly circular. As a result, the Moon may be far enough from Earth at the moment of an eclipse that its disk fails to cover the disk of the Sun completely, even though their centers coincide. In that case, there is no *region of totality*—the umbra never reaches Earth

Animation/Video
Solar Eclipse in Indiana

◄◄◄ **CONCEPT CHECK**

If Earth's distance from the Sun were to double, would you still expect to see total solar eclipses? What if the distance became half its present value?

◄ **Figure E.16 Solar Eclipse Types** (a) The Moon's shadow consists of two parts: the umbra, where no sunlight is seen, and the penumbra, where a portion of the Sun is visible. (b) If we are in the umbra, we see a total eclipse; in the penumbra, we see a partial eclipse. (c) If the Moon is too far from Earth at the moment of the eclipse, the umbra does not reach Earth and there is no region of totality; instead, an annular eclipse is seen. (Note that these figures are not drawn to scale.) (Insets: G. Schneider)

▲ Figure E.17 Eclipse Geometry An eclipse occurs when Earth, Moon, and Sun are precisely aligned. If the Moon's orbital plane lay in exactly the plane of the ecliptic, this alignment would occur once a month. However, the Moon's orbit is inclined at about 5° to the ecliptic, so not all configurations are actually favorable for producing an eclipse. For an eclipse to occur, the line of intersection of the two planes must lie along the Earth–Sun line. For clarity, only the umbra of each shadow is shown (see Figure E.16).

at all, and a thin ring of sunlight can still be seen surrounding the Moon. Such an occurrence, called an **annular eclipse** (from the word *annulus*, meaning "ring"), is depicted at the bottom of Figure E.16. Roughly half of all solar eclipses are annular.

Because the Moon's orbit is slightly inclined to the plane of the ecliptic (Figure E.13b), we don't see a lunar eclipse at every full Moon. The Moon is usually above or below the ecliptic when a full Moon occurs and hence is untouched by Earth's shadow. Similarly, most new Moons do not result in solar eclipses. The chance that a new (or full) Moon will occur just as the Moon happens to cross the ecliptic plane (so Earth, Moon, and Sun are perfectly aligned, as illustrated in Figure E.17) is quite low. As a result, eclipses are relatively rare events. Nevertheless, because we know the orbits of Earth and the Moon to great accuracy, we can predict eclipses far into the future. Figure E.18 shows the location and duration of all total and annular eclipses of the Sun between 2000 and 2020.

▲ Figure E.18 Eclipse Tracks Regions of Earth that saw or will see total (red) or annular (blue) solar eclipses between the years 2000 and 2020. Each track represents the path of the Moon's umbra across Earth's surface during an eclipse. High-latitude tracks are broader because sunlight strikes Earth's surface at an oblique angle near the poles and because of the projection of the map.

▶ **Figure E.19 Triangulation** Surveyors often use simple geometry to estimate the distance to a faraway object by triangulation.

E.4 The Measurement of Distance

So far we have considered only the *directions* to the Sun, Moon, and stars, as seen from Earth. But knowing the direction in which an object lies is only part of the information needed to locate it in space. Before we can make a systematic study of the heavens, we must find a way of measuring *distances*, too. One distance-measurement method, called **triangulation**, is based on the principles of Euclidean geometry and finds widespread application today in both terrestrial and astronomical settings. Modern engineers, especially surveyors, use this age-old geometrical idea to measure indirectly the distance to faraway objects. In astronomy it forms the foundation of the family of distance-measurement techniques that together make up the **cosmic distance scale**.

Imagine trying to measure the distance to a tree on the other side of a river. The most direct method would be to lay a tape across the river, but that's not always practical. A smart surveyor would make the measurement by visualizing an imaginary triangle (hence the term *triangulation*), sighting the tree (measuring its direction) on the far side of the river from two positions on the near side, as illustrated in Figure E.19. The simplest triangle is a right triangle, in which one of the angles is exactly 90°, so it is often convenient to set up one observation position directly opposite the object, as at point A, although this isn't necessary for the method to work. The surveyor then moves to another observation position at point B, noting the distance covered between A and B. This distance is called the **baseline** of the imaginary triangle. Finally, the surveyor, standing at point B, sights toward the tree and notes the angle formed at point B by the intersection of this sightline and the baseline. No further observations are required. Knowing the length of one side (AB) and two angles (the right angle at A and the angle at B) of the triangle, and having some knowledge of elementary trigonometry, the surveyor can construct the remaining sides and angles and so establish the distance from A to the tree.

Obviously, for a fixed baseline, the triangle becomes longer and narrower as the tree's distance from A increases. Narrow triangles cause problems because it becomes hard to measure the angles at A and B with sufficient accuracy. The measurements can be made easier by "fattening" the triangle—that is, by lengthening the baseline—but in astronomy there are limits on how long a baseline we can choose. For example, consider an imaginary triangle extending from Earth to a nearby object in space, perhaps a neighboring planet. The triangle is now extremely long and narrow, even for a relatively nearby object (by cosmic standards). Figure E.20(a) illustrates a case in which the longest baseline possible on Earth—Earth's diameter, measured from point A to point B—is used. In principle, two observers could sight the planet from opposite sides of Earth, measuring the triangle's angles at A and B. However, in practice it is easier to measure the third angle of the imaginary triangle. Here's how.

The observers each sight toward the planet, taking note of its position *relative to some distant stars* seen on the plane of the sky. The observer at A sees the planet at apparent location A′ (pronounced "A prime") relative to those stars, as indicated in Figure E.20(a). The observer at B sees the planet at location B′. If each observer takes a photograph of the same region of the sky, the

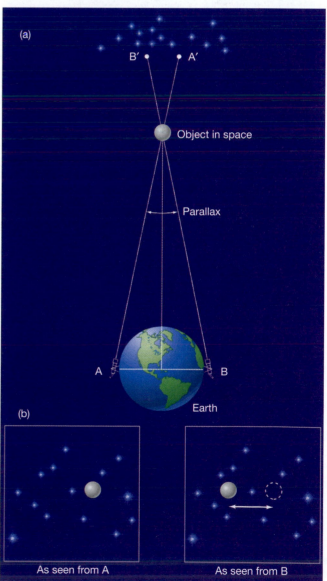

▶ **Figure E.20 Parallax** (a) This imaginary triangle extends from Earth to a nearby object in space (such as a planet). The group of stars at the top represents a background field of very distant stars. (b) Hypothetical photographs of the same star field showing the nearby object's apparent displacement, or shift, relative to the distant undisplaced stars.

CONCEPT CHECK ▶▶▶

Why is elementary geometry essential for measuring distances in astronomy?

Tutorial

Stellar Parallax

planet will appear at slightly different places in the two images, as shown in Figure E.20(b). (The positions of the background stars appear unchanged because of their much greater distance from the observer.) This apparent displacement (shift) of a foreground object relative to the background as the observer's location changes is known as **parallax**. The size of the shift in Figure E.20(b), measured as an angle on the celestial sphere, is equal to the third angle of the imaginary triangle in Figure E.20(a).

The closer an object is to the observer, the larger the parallax. To see this for yourself, hold a pencil vertically just in front of your nose (see Figure E.21). Look at some far-off object—a distant wall, say. Close one eye, then

E-2 MORE PRECISELY

Measuring Distances with Geometry

We can convert baselines and parallaxes into distances using a simple argument made by the Greek geometer Euclid. The first accompanying figure represents Figure E.20(a), but we have changed the scale and added the circle centered on the target planet and passing through our baseline on Earth.

Circumference = $2\pi \times$ distance

360°

Baseline (known)

A

Parallax

B

Planet

Distance (unknown)

To compute the planet's distance, we note that the ratio of the baseline AB to the circumference of the large circle shown in the figure (2π times the distance to the planet) must be equal to the ratio of the parallax angle to one full revolution, 360°:

$$\frac{\text{baseline}}{2\pi \times \text{distance}} = \frac{\text{parallax}}{360°},$$

from which we find

$$\text{distance} = \text{baseline} \times \frac{(360°/2\pi)}{\text{parallax}}.$$

For example, two observers 1000 km apart looking at the Moon might measure (using the photographic technique described in the text) a parallax of 9.0 arc minutes—that is, 0.15°. It then follows that the distance to the Moon is 1000 km × (57.3°/0.15°) ≈ 380,000 km. (More accurate measurements, based on laser ranging using equipment left on the lunar surface by *Apollo* astronauts, yield a distance of 384,000 km.)

Knowing the distance to an object, we can then determine many other properties. For example, by measuring the object's *angular diameter*, we can compute its size. The second figure illustrates the geometry involved. Notice that this is basically the same diagram as the previous one, except that now the angle (the angular diameter) and the distance are known, instead of the angle (the parallax) and the baseline. The same reasoning as before allows us to calculate the diameter:

$$\frac{\text{diameter}}{2\pi \times \text{distance}} = \frac{\text{angular diameter}}{360°},$$

so

$$\text{diameter} = \text{distance} \times \frac{\text{angular diameter}}{57.3°}.$$

Continuing our previous example, the Moon's angular diameter is measured to be about 31 arc minutes—a little over half a degree. From the preceding discussion, it follows that the Moon's actual diameter is 380,000 km × (0.52°/57.3°) ≈ 3450 km. A more precise measurement gives 3476 km.

Distant object

Observer

Angular diameter

360°

Distance (known)

Diameter (unknown)

You should study the above reasoning carefully. Although the observations are straightforward and the geometrical reasoning elementary, simple measurements such as these form the basis for almost every statement made in this book about size and scale in the universe.

▶ **Figure E.21 Parallax Geometry**   Parallax is inversely proportional to an object's distance. An object near your nose has a much larger parallax than an object held at arm's length.

Parallax for nearby pencil

Parallax for distant pencil

Baseline

open it while closing the other. You should see a large shift in the apparent position of the pencil relative to the wall—a large parallax. In this example, one eye corresponds to point A in Figure E.20, the other eye to point B. The distance between your eyeballs is the baseline, the pencil represents the planet, and the distant wall the remote field of stars. Now hold the pencil at arm's length, corresponding to a more distant object (but still not as far away as the distant stars). The apparent shift of the pencil will be smaller. By moving the pencil farther away, you are narrowing the triangle and decreasing the parallax. If you were to paste the pencil to the wall, corresponding to the case where the object of interest is as far away as the background star field, blinking would produce no apparent shift of the pencil at all.

The amount of parallax is inversely proportional to an object's distance. Small parallax implies large distance. Conversely, large parallax implies small distance. Knowing the amount of parallax (as an angle) and the length of the baseline, we can easily derive the distance through triangulation. *More Precisely E-2* explores the connection between angular measure and distance in a little more detail.

Surveyors of the land routinely use these simple geometric techniques to map out planet Earth (*Discovery E-1* presents an early example). As surveyors of the sky, astronomers use the same basic principles to chart the universe.

E.5 Scientific Theory and the Scientific Method

The scientific facts just presented did not come easily or quickly. Progress in science is often slow and intermittent, and may require a great deal of patience before significant progress is made. The earliest known descriptions of the universe were based largely on imagination and mythology and made little attempt to explain the workings of the heavens in terms of testable earthly experience. However, history shows that some early scientists did come to realize the importance of careful observation and testing to the formulation of their ideas. The success of their approach changed, slowly but surely, the way science was done and opened the door to a fuller understanding of nature. Experimentation and observation became central parts of the process of inquiry.

To be effective, a **theory**—the framework of ideas and assumptions used to explain some set of observations and make predictions about the real world—must be continually tested. If experiments and observations favor it, a theory can be further developed and refined. If they do not, the theory must be reformulated or rejected, no matter how appealing it originally seemed. The process is illustrated schematically in Figure E.22. This approach to investigation, combining thinking and doing—that is, theory and experiment—is known as the **scientific method**. It lies at the heart of modern science, separating science from pseudoscience, fact from fiction.

Notice that there is no end point to the process depicted in this figure. A theory can be invalidated by a single wrong prediction, but no amount of observation or experimentation can ever prove it "correct." Theories simply become more and more widely accepted as their predictions are repeatedly confirmed. The process can fail at any point in the cycle. If a theory cannot explain an experimental result or observation, or if its predictions are demonstrated to be untrue, it must be discarded or amended. And if it makes no predictions at all, then it has no real scientific value.

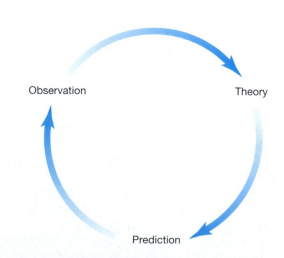

Observation

Theory

Prediction

▲ **Figure E.22 Scientific Method**   Scientific theories evolve through a combination of observation, theoretical reasoning, and prediction, which in turn suggests new observations. The process can begin at any point in the cycle, and it continues forever, or until the theory fails to explain an observation or makes a demonstrably false prediction.

As a class, modern scientific theories share several important defining characteristics:

- They must be *testable*—that is, they must admit the possibility that their underlying assumptions and their predictions can be exposed to experimental verification. This feature separates science from, for example, religion, since, ultimately, divine revelations or scriptures cannot be challenged within a religious framework—we can't design an experiment to "understand the mind of God." Testability also distinguishes science from a pseudoscience such as astrology, whose underlying assumptions and predictions have been repeatedly tested and never verified, with no apparent impact on the views of those who continue to believe in it!

E-1   DISCOVERY

Sizing up Planet Earth

In about 200 B.C. a Greek philosopher named Eratosthenes (276–194 B.C.) used simple geometric reasoning to calculate the size of our planet. He knew that at noon on the first day of summer, observers in the city of Syene (now called Aswan), in Egypt, saw the Sun pass directly overhead. This was evident from the fact that vertical objects cast no shadows and sunlight reached to the very bottoms of deep wells, as shown in the insets in the accompanying figure. However, at noon of the same day in Alexandria, a city 5000 *stadia* to the north, the Sun was seen to be displaced slightly from the vertical. The *stadium* was a Greek unit of length, roughly equal to 0.16 km—the modern town of Aswan lies about 780 (=5000 × 0.16) km south of Alexandria. Using the simple technique of measuring the length of the shadow of a vertical stick and applying elementary geometry, Eratosthenes determined the angular displacement of the Sun from the vertical at Alexandria to be 7.2°.

What could have caused this discrepancy between the two measurements? As illustrated in the figure, the explanation is simply that Earth's surface is not but, it is *curved*. Our planet is a sphere. Eratosthenes was not the first to realize that Earth is spherical—the philosopher Aristotle had done that over 100 years earlier (Section E.5)—but he was apparently the first to build on this knowledge, combining geometry with direct measurement to infer our planet's size. Here's how he did it.

Rays of light reaching Earth from a very distant object, such as the Sun, travel almost parallel to one another. Consequently, as shown in the figure, the angle measured at Alexandria between the Sun's rays and the vertical (that is, the line joining Alexandria to the center of Earth) is equal to the angle between Syene and Alexandria, as seen from Earth's center. (For the sake of clarity, this angle has been exaggerated in the drawing.) As discussed in *More Precisely E-2*, the size of this angle in turn is proportional to the fraction of Earth's circumference that lies between Syene and Alexandria:

$$\frac{7.2°}{360°} = \frac{5000 \text{ stadia}}{\text{Earth's circumference}}.$$

Sun's rays on first day of summer

Alexandria   Syene

Earth's circumference is therefore 50 × 5000, or 250,000 stadia, or about 40,000 km. Earth's radius is therefore 250,000/2π stadia, or 6366 km. The correct values for Earth's circumference and radius, now measured accurately by orbiting spacecraft, are 40,070 km and 6378 km, respectively.

Eratosthenes' reasoning was a remarkable accomplishment. More than 20 centuries ago he estimated the circumference of Earth to within 1 percent accuracy, using only simple geometry. A person making measurements on only a small portion of Earth's surface was able to compute the size of the entire planet on the basis of observation and pure logic—an early triumph of scientific reasoning.

- They must continually be *tested*, and their consequences tested, too. This is the basic circle of scientific progress depicted in Figure E.22.
- They should be *simple*. Simplicity is less a requirement than a practical outcome of centuries of scientific experience—the most successful theories tend to be the simplest ones that fit the facts. This viewpoint is often encapsulated in a principle known as *Occam's razor* (after the fourteenth-century English philosopher William of Ockham): If two competing theories both explain the facts and make the same predictions, then the simpler one is better. Put another way—"Keep it simple!" A good theory should contain no more complexity than is absolutely necessary.
- Finally, most scientists have the additional bias that a theory should in some sense be *elegant*. When a clearly stated simple principle naturally explains several phenomena previously thought to be distinct, this is widely regarded as a strong point in favor of the new theory.

You may find it instructive to apply these criteria to the many physical theories—we will encounter throughout the text.

The notion that theories must be tested and may be proven wrong sometimes leads people to dismiss their importance. We have all heard the expression, "Of course, it's only a theory," used to deride or dismiss an idea that someone finds unacceptable. Don't be fooled by this abuse of the concept! Gravity (see Section 1.4) is "only" a theory, but calculations based on it have guided human spacecraft throughout the solar system. Electromagnetism and quantum mechanics (see Chapter 2) are theories, too, yet they form the foundation for most of twentieth- (and twenty-first-) century technology.

The birth of modern science is usually associated with the Renaissance, the historical period from the late fourteenth to the mid-seventeenth century that saw a rebirth (*renaissance* in French) of artistic, literary, and scientific inquiry in European culture following the chaos of the Dark Ages. However, one of the first documented uses of the scientific method in an astronomical context was made by Aristotle (384–322 B.C.) some 17 centuries earlier. Aristotle is not normally remembered as a strong proponent of this approach—many of his best known ideas were based on pure thought, with no attempt at experimental test or verification. Still, his brilliance extended into many areas now thought of as modern science. He noted that, during a lunar eclipse (Section E.3), Earth casts a curved shadow onto the surface of the Moon. Figure E.23 shows a series of photographs taken during a recent lunar eclipse. Earth's shadow, projected onto the Moon's surface, is indeed slightly curved. This is what Aristotle must have seen and recorded so long ago.

Because the observed shadow seemed always to be an arc of the same circle, Aristotle theorized that Earth, the cause of the shadow, must be round. Don't underestimate the scope of this apparently simple statement. Aristotle also had to reason that the dark region was indeed a shadow and that Earth was its cause—facts we regard as obvious today, but far from clear 25 centuries ago. On the basis of this *hypothesis*—one possible explanation of the observed facts—he then predicted that any and all future lunar eclipses would show Earth's shadow to be curved, regardless of our planet's orientation. That prediction has been tested every time a lunar eclipse has occurred. It has yet to be proved wrong. Aristotle was not the first person to argue that Earth is round, but he was apparently the first to offer observational proof using the lunar-eclipse method.

Observation, theory, and testing—these are the cornerstones of the scientific method, a technique whose power will be demonstrated again and again throughout our text. Used properly over a period of time, this rational, methodical approach enables us to arrive at conclusions that are mostly free of the personal bias and human values of any one scientist. The scientific method is designed to yield—eventually—an objective view of the universe we inhabit.

▲ **Figure E.23 A Lunar Eclipse** This series of photographs show Earth's shadow sweeping across the Moon during an eclipse. By observing this behavior, Aristotle reasoned that Earth was the cause of the shadow and concluded that Earth must be round. His theory has yet to be disproved. (G Schneider)

◄◄◄ **CONCEPT CHECK**

Can a theory ever become a "fact," scientifically speaking?

CHAPTER REVIEW

Summary

1 Early observers grouped the stars visible to the naked eye into patterns called **constellations** (p. 4), which they imagined were attached to a vast **celestial sphere** (p. 5) centered on Earth. Constellations have no physical significance, but are still used to label regions of the sky. **Celestial coordinates** (p. 7) are a more precise way of specifying a star's location on the celestial sphere.

2 The nightly motion of the stars across the sky is the result of Earth's **rotation** (p. 6) on its axis. Because of Earth's **revolution** (p. 8) around the Sun, we see different stars at night at different times of the year. The Sun's apparent yearly path around the celestial sphere (or the plane of Earth's orbit around the Sun) is called the **ecliptic** (p. 9). We experience **seasons** (p. 10) because Earth's rotation axis is inclined to the ecliptic plane. At the **summer solstice** (p. 10), the Sun is highest in the sky and the length of the day is greatest. At the **winter solstice** (p. 10), the Sun is lowest and the day is shortest. Because of **precession** (p. 11), the orientation of Earth's axis change slowly over the course of thousands of years.

3 As the Moon orbits Earth, it keeps the same face permanently turned toward our planet. We see lunar **phases** (p. 12) as the fraction of the Moon's sunlit face visible to us varies. A **lunar eclipse** (p. 14) occurs when the Moon enters Earth's shadow. A **solar eclipse** (p. 14) occurs when the Moon passes between Earth and the Sun. An eclipse may be **total** (p. 14, 15) if the body in question (Moon or Sun) is completely obscured, or **partial** (p. 14, 15) if only a portion of the surface is affected. If the Moon happens to be too far from Earth for its disk to completely hide the Sun, an **annular eclipse** (p. 16) occurs. Because the Moon's orbit around Earth is slightly inclined to the ecliptic, solar and lunar eclipses are relatively rare events.

4 Astronomers use **triangulation** (p. 17) to measure the distances to planets and stars, forming the foundation of the **cosmic distance scale** (p. 17), the family of distance-measurement techniques used to chart the universe. **Parallax** (p. 18) is the apparent motion of a foreground object relative to a distant background as the observer's position changes. The larger the **baseline** (p. 17)—the distance between the two observation points—the greater the parallax.

5 The **scientific method** (p. 19) is a methodical approach employed by scientists to explore the universe around us in an objective manner. A **theory** (p. 19) is a framework of ideas and assumptions used to explain some set of observations and make predictions about the real world. These predictions in turn are amenable to further observational testing. In this way, the theory expands and science advances.

Review and Discussion

1. Compare the size of Earth with that of the Sun, the Milky Way galaxy, and the entire universe.

2. What is a constellation? Why are constellations useful for mapping the sky?

3. Why does the Sun rise in the east and set in the west each day? Does the Moon also rise in the east and set in the west? Why? Do stars do the same? Why?

4. How and why does a day measured by the Sun differ from a day measured by the stars?

5. How many times in your life have you orbited the Sun?

6. Why do we see different stars at different times of the year?

7. Why are there seasons on Earth?

8. What is precession, and what is its cause?

9. If one complete hemisphere of the Moon is always lit by the sun, why do we see different phases of the Moon?

10. What causes a lunar eclipse? A solar eclipse? Why aren't there lunar and solar eclipses every month?

11. Do you think an observer on another planet might see eclipses? Why or why not?

12. What is parallax? Give an everyday example.

13. Why is it necessary to have a long baseline when using triangulation to measure the distances to objects in space?

14. What two pieces of information are needed to determine the diameter of a faraway object?

15. What is the scientific method? In what ways does science differ from religion?

Conceptual Self-Test: True or False?/Multiple Choice

1. The Milky Way galaxy is about one million times larger than Earth. **(T/F)**

2. The stars in a constellation are physically close to one another. **(T/F)**

3. The solar day is longer than the sidereal day. **(T/F)**

4. The seasons are caused by the precession of Earth's axis. **(T/F)**

5. A lunar eclipse can occur only during the full phase. **(T/F)**

6. The angular diameter of an object is inversely proportional to its distance from the observer. **(T/F)**

7. If we know the distance of an object from Earth, we can determine the object's size by measuring its parallax. **(T/F)**

8. If Earth rotated twice as fast as it currently does, but its motion around the Sun stayed the same, then **(a)** the night would be twice as long; **(b)** the night would be half as long; **(c)** the year would be half as long; **(d)** the length of the day would be unchanged.

9. A long, thin cloud that stretched from directly overhead to the western horizon would have an angular size of **(a)** 45°; **(b)** 90°; **(c)** 180°; **(d)** 360°.

10. When a thin crescent of the Moon is visible just before sunrise, the Moon is in its **(a)** waxing phase; **(b)** new phase; **(c)** waning phase; **(d)** quarter phase.

11. If the Moon's orbit were a little larger, solar eclipses would be **(a)** more likely to be annular; **(b)** more likely to be total; **(c)** more frequent; **(d)** unchanged in appearance.

12. If the Moon orbited Earth twice as fast, the frequency of solar eclipses would **(a)** double; **(b)** be cut in half; **(c)** stay the same.

13. According to Figure E.8 ("The Zodiac"), in January the Sun is in the constellation **(a)** Cancer; **(b)** Gemini; **(c)** Leo; **(d)** Aquarius.

14. In Figure E.19 ("Triangulation"), using a longer baseline would result in **(a)** a less accurate distance to the tree; **(b)** a more accurate distance to the tree; **(c)** a smaller angle at point B; **(d)** a greater distance across the river.

15. In Figure E.20 ("Parallax"), a smaller Earth would result in **(a)** a smaller parallax angle; **(b)** a shorter distance measured to the object; **(c)** a larger apparent displacement; **(d)** stars appearing closer together.

Problems

Algorithmic versions of these questions are available in the Practice Problems module of the Companion Website.

The number of squares preceding each problem indicates its approximate level of difficulty.

1. ■ The vernal equinox is now just entering the constellation Aquarius. In what constellation will it lie in the year A.D. 10,000?

2. ■ Given that the distance from Earth to the Sun is 150,000,000 km, through what distance does Earth move in a second? An hour? A day?

3. ■■■ How, and by roughly how much, would the length of the solar day change if Earth's rotation were suddenly to reverse direction?

4. ■■ Through how many degrees, arc minutes, or arc seconds does the Moon move in (a) one hour of time, (b) one minute, (c) one second? How long does it take for the Moon to move a distance equal to its own diameter? Take the angular diameter of the Moon to be 0.5 degrees.

5. ■■ Given that the distance to the Moon is 384,000 km, and taking the Moon's orbit around Earth to be circular, estimate the speed (in kilometers per second) at which the Moon orbits Earth.

6. ■ At what distance is an object if its parallax, as measured from either end of a 1000-km baseline, is (a) 1°, (b) 1′, (c) 1″?

7. ■ Estimate the angular diameter of your thumb, held at arm's length.

8. ■ Given that the angular size of Venus is 55″ when the planet is 45,000,000 km from Earth, calculate Venus's diameter.

9. ■ The Moon lies 384,000 km from Earth, and the Sun lies 150,000,000 km away. If both have the same angular size, how many times larger than the Moon is the Sun?

10. ■ What angle would Eratosthenes have measured (see *Discovery E-1*) had Earth been flat?

The Companion Website at www.prenhall.com/chaisson provides algorithmically generated versions of each chapter's Problems, in addition to an Animations and Videos gallery, a Tutorials gallery and Tutorials Quiz questions, a full eBook, an Images gallery, Practice Questions, Multiple Choice, and True or False quizzes, and Labeling exercises.

1

The Copernican Revolution

The Birth of Modern Science

Living in the Space Age, we have become accustomed to the modern view of our place in the universe. Images of Earth taken from space leave little doubt that we inhabit just one planet among several, and no one seriously questions the idea that we orbit the Sun. Yet there was a time, not so long ago, when our ancestors maintained that Earth had a special role in the cosmos and lay at the center of all things. Our view of the universe—and of ourselves—has undergone a radical transformation since those early days. Humankind has been torn from its throne at the center of the cosmos and relegated to an unremarkable position on the periphery of the Milky Way Galaxy. But we have been amply compensated for our loss of prominence, for we have gained a wealth of scientific knowledge in the process. The story of how this came about is the story of the rise of science and the genesis of modern astronomy.

LEARNING GOALS

Studying this chapter will enable you to:

1 Discuss how the observed motions of the planets led to our modern view of a Sun-centered solar system.

2 Sketch the major contributions of Galileo and Kepler to the development of our understanding of the solar system.

3 State Kepler's laws of planetary motion.

4 Explain how astronomers have measured the true size of the solar system.

5 State Newton's laws of motion and his law of universal gravitation, and explain how they enable us to measure the masses of astronomical bodies.

Visit www.prenhall.com/chaisson for additional annotated images, animations, and links to related sites for this chapter.

1.1 The Motions of the Planets

Over the course of a night, the stars slide smoothly across the sky. ∞ *(Sec. E.1)* Over the course of a year, the Sun progresses along the ecliptic at an almost constant rate, varying little in brightness from day to day. ∞ *(Sec. E.2)* Over the course of a month, the Moon moves smoothly and steadily along its path on the sky relative to the stars, passing through its familiar cycle of phases. ∞ *(Sec. E.3)* In short, the behavior of the Sun, the Moon, and the stars seems fairly simple and orderly.

But ancient astronomers were also aware of five other bodies in the sky—the planets Mercury, Venus, Mars, Jupiter, and Saturn—whose behavior was not so easy to grasp. The explanation for their motion would change forever our perception of the universe and our place in it.

Wanderers in the Heavens

Planets do not behave in as regular a fashion as do the Sun, Moon, and stars. They vary in brightness and don't maintain a fixed position in the sky. Unlike the Sun and the Moon, they seem to wander irregularly around the celestial sphere—indeed, the word *planet* derives from the Greek word *planetes*, meaning "wanderer." Planets never stray far from the ecliptic and generally traverse the celestial sphere from west to east, as the Sun does. However, they seem to speed up and slow down during their journeys, and at times they even appear to loop back and forth relative to the stars, as shown in Figure 1.1. Motion in the eastward sense is usually referred to as *direct*, or *prograde*, motion. The backward (westward) loops are known as **retrograde motion**.

Unlike the Sun and stars, but like the Moon, the planets produce no visible light of their own. Instead, they shine by reflected sunlight. Ancient astronomers correctly reasoned that the apparent brightness of a planet in the night sky is related to its distance from Earth—planets appear brightest when closest to us. However, the planets Mars, Jupiter, and Saturn are always brightest during the retrograde portions of their orbits. The challenge facing astronomers was to explain the observed motions of the planets and to relate those motions to the variations in planetary brightness.

In many societies, people came to believe that there were other benefits to studying the planets. The relative positions of heavenly bodies at a person's birth were carefully studied by *astrologers*, who used the data to make predictions about that person's destiny. Thus, in a sense, astronomy and astrology arose from the same basic impulse—the desire to "see" into the future. Indeed, for a long time the disciplines were indistinguishable from one another. This chapter chronicles the period of human history when astronomy replaced astrology, and modern science was born.

Animation/Video

Retrograde Motion of Mars

◀ **Figure 1.1 Planetary Motions** Most of the time, planets move from west to east relative to the background stars. Occasionally, however, they change direction and temporarily undergo retrograde motion (east to west) before looping back. The main illustration shows an actual retrograde loop in the motion of the planet Mars. The inset depicts the movements of several planets over the course of several years, as reproduced on the inside dome of a planetarium. The motion of the planets relative to the stars (represented as unmoving points) produces continuous streaks on the planetarium "sky." (Inset: Museum of Science, Boston)

The Geocentric Universe

The earliest models of the solar system followed the teachings of the Greek philosopher Aristotle (384–322 B.C.) and were **geocentric**, meaning that Earth lay at the center of the universe and all other bodies moved around it. (Figures E.4 and E.9 illustrate the basic geocentric view.) ∞ *(Sec. E.2)* These models employed what Aristotle had taught was the perfect form: the circle. The simplest possible description—uniform motion around a circle having Earth at its center—provided a fairly good approximation to the orbits of the Sun and the Moon, but it could not account for the observed variations in planetary brightness, nor for their retrograde motion. A better model was needed to describe the planets.

In the first step toward this new model, each planet was taken to move uniformly around a small circle, called an **epicycle**, whose *center* moved uniformly around Earth on a second and larger circle, known as the **deferent** (Figure 1.2). The motion was now composed of two separate circular orbits, creating the possibility that at some times the planet's apparent motion could be retrograde. Also, the distance from the planet to Earth would vary, accounting for changes in brightness. By tinkering with the relative sizes of epicycle and deferent, with the planet's speed on the epicycle, and with the epicycle's speed along the deferent, early astronomers were able to bring this "epicyclic" motion into fairly good agreement with the observed paths of the planets in the sky. Moreover, this model had good predictive power, at least to the accuracy of observations at the time.

However, as the number and the quality of observations increased, small corrections had to be introduced into the simple epicyclic model to bring it into line with new observations. The center of the deferents had to be shifted slightly from Earth's center, and the motion of the epicycles had to be imagined uniform with respect not to Earth but to yet another point in space. Around A.D. 140 a Greek astronomer named Claudius Ptolemaeus (known today as Ptolemy) constructed perhaps the best geocentric model of all time. Illustrated in simplified form in Figure 1.3, it explained remarkably well the observed paths of the five planets then known, as well as the paths of the Sun and the Moon. However, to achieve its explanatory and predictive power, the full **Ptolemaic model** required a series of no fewer than 80 circles. To account for the paths of the Sun, the Moon, and all the nine planets (and their moons) that we know today would require a vastly more complex set.

Today our scientific training leads us to seek simplicity, because simplicity in the physical sciences has so often proved to be an indicator of truth. We would regard the intricacy of a model as complicated as the Ptolemaic system as a clear sign of a fundamentally flawed theory. (This basic principle is often referred to as *Occam's razor*—see Section E.5.) With the benefit of hindsight, we now recognize that the major error lay in the assumption of a geocentric universe, compounded by the insistence on uniform circular motion, the basis of which was largely philosophical, rather than scientific, in nature.

Actually, history records that some ancient Greek astronomers reasoned differently about the motions of heavenly bodies. Foremost among them was Aristarchus of Samos (310–230 B.C.), who proposed that all the planets, including Earth, revolve around the Sun and, furthermore, that Earth rotates on its axis once each day. This, he argued, would create an apparent motion of the sky—a simple idea that is familiar to anyone who has ridden on a merry-go-round and watched the landscape appear to move past in the opposite direction. However, Aristarchus's description of the heavens, though essentially correct, did not gain widespread acceptance during his lifetime. Aristotle's influence was too strong, his followers too numerous, his writings too comprehensive.

The Aristotelian school did present some simple and (at the time) compelling arguments in favor of their views. First, of course, Earth doesn't *feel*

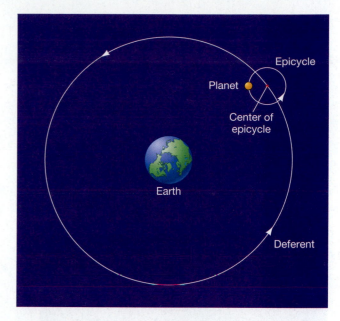

▲ **Figure 1.2 Geocentric Model** In the geocentric model of the solar system, the observed motions of the planets made it impossible to assume that they moved on simple circular paths around Earth. Instead, each planet was thought to follow a small circular orbit (the epicycle) about an imaginary point that itself traveled in a large, circular orbit (the deferent) about Earth.

Animation/Video
Geocentric Solar System

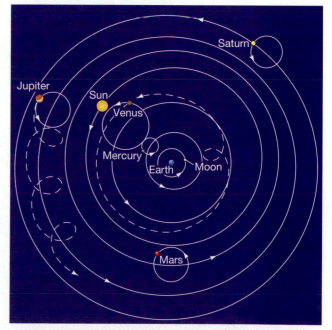

▲ **Figure 1.3 Ptolemy's Model** The basic features, drawn roughly to scale, of Ptolemy's geocentric model of the inner solar system, a model that enjoyed widespread popularity prior to the Renaissance. To avoid confusion, partial paths (dashed) of only two planets, Venus and Jupiter, are drawn here.

as if it's moving. And if it were moving, wouldn't there be a strong wind or force of some sort as we move at high speed around the Sun? For that matter, if the Sun is at the center of the universe, then why do things fall down and not up? And finally, considering that the vantage point from which we view the stars changes over the course of a year, why don't we see stellar parallax? ∞ *(Sec. E.4)*

Nowadays we might dismiss the first points as merely naïve, but the last is a valid argument and the reasoning essentially sound. We now know that there *is* stellar parallax as Earth orbits the Sun. However, because the stars are so distant, it amounts to less than 1 arc second, even for the closest stars. Early astronomers simply would not have noticed it (indeed, it was conclusively measured only in the latter half of the nineteenth century). We will encounter many other instances in astronomy where correct reasoning led to the wrong conclusions because it was based on inadequate data.

The Heliocentric Model of the Solar System

The Ptolemaic picture of the universe survived, more or less intact, for almost 13 centuries until a sixteenth-century Polish cleric, Nicholas Copernicus (Figure 1.4), rediscovered Aristarchus's **heliocentric** (Sun-centered) model. Copernicus asserted that Earth spins on its axis and, like all other planets, orbits the Sun. Not only does this model explain the observed daily and seasonal changes in the heavens, as we have seen, but it also naturally accounts for planetary retrograde motion and brightness variations. The critical realization that Earth is not at the center of the universe is now known as the **Copernican revolution**.

Figure 1.5 shows how the Copernican view explains both the changing brightness of a planet (in this case, Mars) and its apparent looping motions. If we suppose that Earth moves faster than Mars, then every so often Earth "overtakes" that planet. Each time this happens, Mars appears to move backward in the sky, in much the same way as a car we overtake on the highway seems to slip backward relative to us. Furthermore, at these times, Earth is closest to Mars, so Mars appears brightest, in agreement with observations. Notice that in the Copernican picture the planet's looping motions are only apparent—in the Ptolemaic view, they were real.

Copernicus's major motivation for introducing the heliocentric model was simplicity. Even so, he was still influenced by Greek thinking and clung to the idea of circles to model the planets' motions. To bring his theory into agreement with observations, he was forced to retain the idea of epicyclic motion, though with the deferent centered on the Sun rather than on Earth and with the epicycles being smaller than in the Ptolemaic picture. Thus, he retained unnecessary complexity and actually gained little in accuracy over the geocentric model. The heliocentric model did rectify some small discrepancies and inconsistencies in the Ptolemaic system, but for Copernicus the primary attraction of heliocentricity was its simplicity, its being "more pleasing to the mind." To this day, scientists still are guided by simplicity, symmetry, and beauty in modeling all aspects of the universe.

Copernicus's ideas were never widely accepted during his lifetime. By relegating Earth to a noncentral and undistinguished place within the solar system, heliocentricity contradicted the conventional wisdom of the time and violated the religious doctrine of the Roman Catholic Church. Copernicus surely discussed and debated his theory with his fellow scholars, but, possibly because he wished to avoid direct conflict with the Church, his book *On the Revolution of the Celestial Spheres* was not published until 1543, the year he died. Only much later, when others extended and popularized the heliocentric model—and as supporting observational evidence began to mount—did the Copernican theory gain widespread recognition.

▲ **Figure 1.4 Nicholas Copernicus** (1473–1543). (E. Lessing/Art Resource, NY)

Animation/Video
Heliocentric Solar System

CONCEPT CHECK ▶▶▶

How do the geocentric and heliocentric models of the solar system differ in their explanations of planetary retrograde motion?

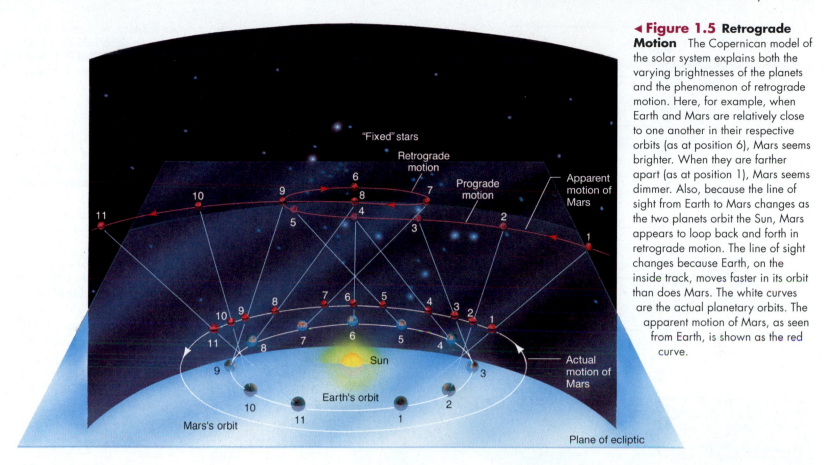

◄ **Figure 1.5** **Retrograde Motion** The Copernican model of the solar system explains both the varying brightnesses of the planets and the phenomenon of retrograde motion. Here, for example, when Earth and Mars are relatively close to one another in their respective orbits (as at position 6), Mars seems brighter. When they are farther apart (as at position 1), Mars seems dimmer. Also, because the line of sight from Earth to Mars changes as the two planets orbit the Sun, Mars appears to loop back and forth in retrograde motion. The line of sight changes because Earth, on the inside track, moves faster in its orbit than does Mars. The white curves are the actual planetary orbits. The apparent motion of Mars, as seen from Earth, is shown as the red curve.

1.2 The Birth of Modern Astronomy

In the century following the death of Copernicus and the publication of his theory of the solar system, two scientists—Galileo Galilei and Johannes Kepler—made indelible imprints on the study of astronomy. Each achieved fame for his discoveries and made great strides in popularizing the Copernican viewpoint.

Galileo's Historic Observations

Galileo Galilei (Figure 1.6) was an Italian mathematician and philosopher. By his willingness to perform experiments to test his ideas—a radical approach in those days (Section 1.5)—and by embracing the brand-new technology of the telescope, he revolutionized the way science was done, so much so that he is now widely regarded as the father of experimental science. The telescope was invented in Holland in the early seventeenth century. Having heard of the invention (but without having seen one), Galileo built a telescope for himself in 1609 and aimed it at the sky. What he saw conflicted greatly with the philosophy of Aristotle and provided much new data to support the ideas of Copernicus.

Using his telescope, Galileo discovered that:

- The Moon has mountains, valleys, and craters—terrain in many ways reminiscent of that on Earth.
- The Sun has imperfections—dark blemishes now known as *sunspots* (see Chapter 9). By noting the changing appearance of these sunspots from day to day, Galileo inferred that the Sun *rotates*, approximately once per month, around an axis roughly perpendicular to the ecliptic plane.

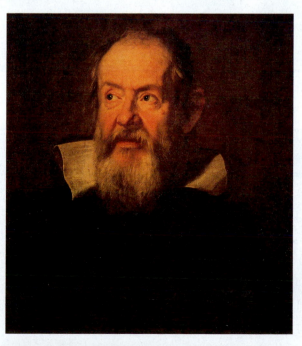

▲ **Figure 1.6 Galileo Galilei** (1564–1642). (Art Resource, NY)

◀ **Figure 1.7 Galilean Moons** The four Galilean moons of Jupiter, as sketched by Galileo in his notebook. The sketches show what Galileo saw on seven nights between 7 and 15 January, 1610. The orbits of the moons (sketched here as asterisks; now called Io, Europa, Ganymede, and Callisto) around the planet (open circle) can clearly be seen (from *Sidereus Nuncius*).

- Four small points of light, invisible to the naked eye, orbit the planet Jupiter. He realized that they were *moons* (Figure 1.7), circling that planet just as our Moon orbits Earth. To Galileo, the fact that another planet had moons provided the strongest support for the Copernican model. Clearly, Earth was not the center of all things.
- Venus shows a complete cycle of *phases* (Figure 1.8), much like the familiar monthly changes exhibited by our own Moon. This finding can be explained only by the planet's motion around the Sun.

All these observations ran directly counter to the accepted scientific beliefs of the day. They strongly supported the view that Earth is not the center of all things and that at least one planet orbited the Sun.

In 1610 Galileo published his observational findings and his controversial conclusions supporting the Copernican theory, challenging both scientific orthodoxy and the religious dogma of the day. In 1616 his ideas were judged heretical, both his works and those of Copernicus were banned by the Church, and Galileo was instructed to abandon his astronomical pursuits. This he refused to do and instead continued to amass and publish data supporting the heliocentric view. These actions brought Galileo into direct conflict with the Church. The Inquisition forced him, under threat of torture, to retract his claim that Earth orbits the Sun, and he was placed under house arrest in 1633. He remained imprisoned for the rest of his life. Not until 1992 did the Church publicly forgive Galileo's "crimes." But the damage to the orthodox view of the universe was done, and the Copernican genie was out of the bottle once and for all.

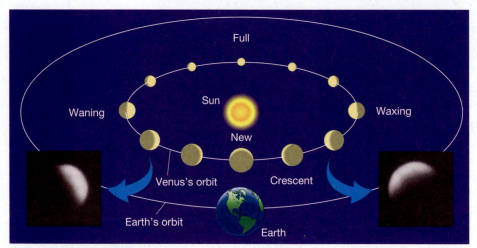

(a) Sun-centered model

R I V U X G

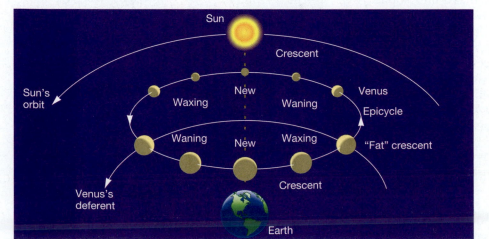

(b) Ptolemy's model

◀ **Figure 1.8 Venus Phases** Both the Ptolemaic and the Copernican models of the solar system predict that Venus should show phases as it moves in its orbit. (a) In the Copernican picture, when Venus is directly between Earth and the Sun, its unlit side faces us and the planet is invisible to us. As Venus moves in its orbit, progressively more of its illuminated face is visible from Earth. Note the connection between orbital phase and the apparent size of the planet: Venus seems much larger in its crescent phase than when it is full because it is much closer to us during its crescent phase. This is the behavior actually observed. The insets at bottom left and right are photographs of Venus taken at two of its crescent phases. (b) The Ptolemaic model (see also Figure 1.3) is unable to account for these observations. In particular, the full phase of the planet cannot be explained. Seen from Earth, Venus reaches only a "fat crescent" phase, then begins to wane as it nears the Sun. (Both these views are from a sideways perspective; from overhead, both orbits are very nearly circular, as shown in Figure 1.14.)

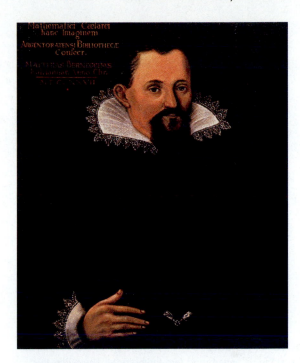

▶ **Figure 1.9** **Johannes Kepler** (1571–1630). (E. Lessing/Art Resource, NY)

The Ascendancy of the Copernican System

The Copernican revolution is a case study of how the scientific method, although affected at any given time by the subjective whims, human biases, and even sheer luck of individual researchers, does ultimately lead to a definite degree of objectivity. ∞ (Sec. E.5) Over time, many groups of scientists checking, confirming, and refining experimental tests can neutralize the subjective attitudes of individuals. Usually one generation of scientists can bring sufficient objectivity to bear on a problem, though some especially revolutionary concepts are so swamped by tradition, religion, and politics that more time is needed. In the case of heliocentricity, objective confirmation was not obtained until about three centuries after Copernicus published his work and more than 2000 years after Aristarchus had proposed the concept. Nonetheless, objectivity *did in fact* eventually prevail.

1.3 The Laws of Planetary Motion

At about the time Galileo was becoming famous for his telescopic observations, Johannes Kepler (Figure 1.9), a German mathematician and astronomer, announced his discovery of a set of simple empirical (that is, based on observation) laws that accurately described the motions of the planets. While Galileo was the first "modern" observer, who used telescopic observations of the skies to confront and refine his theories, Kepler was a pure theorist. He based his work almost entirely on the observations of another scientist (in part because of his own poor eyesight). Those observations, which predated the telescope by several decades, had been made by Kepler's employer, Tycho Brahe (1546–1601), arguably one of the greatest observational astronomers who ever lived.

Brahe's Complex Data

Tycho, as he is often called, was an eccentric aristocrat and a skillful observer. Born in Denmark, he was educated at some of the best universities in Europe, where he studied astrology, alchemy, and medicine. Most of his observations, which predated the invention of the telescope by several decades, were made at his own observatory, named Uraniborg, in Denmark (Figure 1.10). There, using instruments of his own design, Tycho maintained meticulous and accurate records of the stars, the planets, and noteworthy celestial events.

In 1597, having fallen out of favor with the Danish court, Tycho moved to Prague, which happens to be fairly close to Graz, in Austria, where Kepler lived. Kepler joined Tycho in Prague in 1600 and was put to work trying to find a theory that could explain Brahe's planetary data. When Tycho died a year later, Kepler inherited not only Brahe's position as Imperial Mathematician of the Holy Roman Empire but also his most priceless possession: the accumulated observations of the planets, spanning several decades. Tycho's observations, though made with the naked eye, were nevertheless of very high quality. Kepler set to work seeking a unifying principle to explain the motions of the planets without the need for epicycles. The effort was to occupy much of the remaining 29 years of his life.

◀◀◀ **CONCEPT CHECK**

In what ways did Galileo's observations of Venus and Jupiter conflict with the prevailing view at the time?

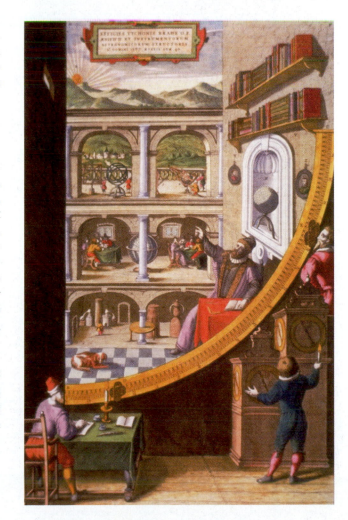

▶ **Figure 1.10** **Tycho Brahe**   The astronomer in his observatory Uraniborg, on the island of Hveen, in Denmark. (The Royal Ontario Museum)

▲ **Figure 1.11 Ellipse** An ellipse can be drawn with the aid of a string, a pencil, and two thumbtacks. The wider the separation of the foci, the more elongated, or eccentric, the ellipse. In the special case where the two foci are at the same place, the drawn curve is a circle.

CONCEPT CHECK ▶▶▶

In what ways did Galileo and Kepler differ in their approach to science? In what ways did each advance the Copernican view of the universe?

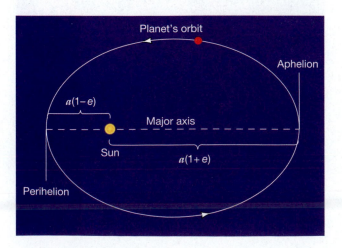

▲ **Figure 1.12 Orbital Properties** A planet's perihelion (closest approach to the Sun) and aphelion (greatest distance from the Sun) are related in a simple way to the orbital semi-major axis a and eccentricity e. Note that while the Sun resides at one focus, the other focus is empty and has no particular significance. No planet in our solar system has an orbital eccentricity as large as shown here (see Table 1.1). However, many meteoroids and all comets (see Chapter 4) have eccentricities considerably greater than this.

Kepler's goal was to find a simple description of the solar system, within the basic framework of the Copernican model, that fit Tycho's complex mass of detailed observations. In the end, he had to abandon Copernicus's original simple notion of circular planetary orbits, but even greater simplicity emerged as a result. Kepler determined the shapes and relative sizes of each planet's orbit by triangulation not from different points on Earth but from different points on Earth's orbit, using observations made at many different times of the year. ∞ *(Sec. E.4)* Noting where the planets were on successive nights, he was able to infer the speeds at which they moved. After long years working with Brahe's planetary data and after many false starts and blind alleys, Kepler succeeded in summarizing the motions of all the known planets, including Earth, in the three **laws of planetary motion** that now bear his name.

Kepler's Simple Laws

Kepler's first law has to do with the *shapes* of the planetary orbits:

> I. The orbital paths of the planets are elliptical (*not* necessarily circular), with the Sun at one focus.

An **ellipse** is simply a flattened circle. Figure 1.11 illustrates a means of constructing an ellipse using a piece of string and two thumbtacks. Each point at which the string is pinned is called a **focus** (plural: foci) of the ellipse. The long axis of the ellipse, containing the two foci, is known as the *major axis*. Half the length of this long axis is referred to as the **semi-major axis**, a conventional measure of the ellipse's size. The **eccentricity** of the ellipse is equal to the distance between the foci divided by the length of the major axis. The length of the semi-major axis and the eccentricity are all we need to describe the size and shape of a planet's orbital path. (A circle is a special kind of ellipse in which the two foci happen to coincide, so the eccentricity is zero. The semi-major axis of a circle is simply its radius. These two numbers—semi-major axis and eccentricity—are all that are needed to describe the size and shape of a planet's orbital path. From them we can derive many other useful quantities. Two of the most important (see Figure 1.12) are the planet's **perihelion** (its point of closest approach to the Sun) and its **aphelion** (greatest distance from the Sun).

In fact, no planet's elliptical orbit is nearly as elongated as the one shown in Figure 1.11. With two exceptions (the paths of Mercury and Pluto), planetary orbits have such small eccentricities that our eyes would have trouble distinguishing them from true circles. Only because the orbits are so nearly circular were the Ptolemaic and Copernican models able to come as close as they did to describing reality.

Kepler's second law, illustrated in Figure 1.13, addresses the *speed* at which a planet traverses different parts of its orbit:

> II. An imaginary line connecting the Sun to any planet sweeps out equal areas of the ellipse in equal intervals of time.

While orbiting the Sun, a planet traces the arcs labeled A, B, and C in Figure 1.13 in equal times. Notice, however, that the distance traveled along arc C is greater than the distance traveled along arc A or arc B. Because the time is the same and the distance is different, the speed must vary: when a planet is close to the Sun, as in sector C, it moves much faster than when farther away, as in sector A.

These laws are not restricted to planets. They apply to *any* orbiting object. Spy satellites, for example, move very rapidly as they swoop close to Earth's surface, not because they are propelled by powerful onboard rockets but because their highly eccentric orbits are governed by Kepler's laws.

Kepler published his first two laws in 1609, stating that he had proved them only for the orbit of Mars. Ten years later he extended them to all the known planets (Mercury, Venus, Earth, Mars, Jupiter, and Saturn) and added a third law relating the size of a planet's orbit to its sidereal orbital **period**, defined as the time needed for the planet to complete one circuit around the Sun.

Kepler's third law states:

> III. The square of a planet's orbital period is proportional to the cube of its semi-major axis.

This law becomes particularly simple when we choose the (Earth) year as our unit of time and the *astronomical unit* as our unit of length. One **astronomical unit** (AU) is the semi-major axis of Earth's orbit around the Sun—the average distance between Earth and the Sun. Like the light-year, the astronomical unit is a unit custom-made for the vast distances encountered in astronomy. Using these units for time and distance, we can conveniently write Kepler's third law for any planet in the form

$$P^2 \text{ (in Earth years)} = a^3 \text{ (in astronomical units)},$$

where P is the planet's sidereal orbital period and a is its semi-major axis.

Table 1.1 presents some basic data describing the orbits of the eight major planets of the solar system. The semi-major axis of each planet's orbit is measured in astronomical units—that is, relative to the size of Earth's orbit—and all orbital periods are measured in years. Renaissance astronomers knew these properties for the innermost six planets and used them to construct the heliocentric model of the solar system. However, Kepler's laws are obeyed by *all* the known planets orbiting the Sun, and *not just by the six on which he based his conclusions.* For purposes of verifying Kepler's third law, the rightmost column lists the ratio P^2/a^3. As we have just seen, in the units used in the table, this number should equal one in all cases. The small but significant deviations of P^2/a^3 from 1 in the cases of Uranus and Neptune are caused by the gravitational attraction between those two planets (see Chapter 7).

The laws developed by Kepler were far more than mere fits to existing data. They made definite, testable predictions about the future locations of the planets. Those predictions have been borne out to high accuracy every time they have been tested by observation—the hallmark of any credible scientific theory. ∞ *(Sec. 1.5)*

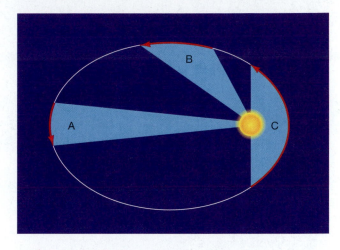

▲ **Figure 1.13 Kepler's Second Law** A line joining a planet to the Sun sweeps out equal areas in equal intervals of time. The three shaded areas A, B, and C are equal. Any object traveling along the elliptical path would take the same amount of time to cover the distance indicated by the three red arrows. Therefore, planets move faster when closer to the Sun.

◀◀◀ **CONCEPT CHECK**

Why is it significant that Kepler's laws apply to Uranus and Neptune?

TABLE 1.1 Some Planetary Properties

PLANET	ORBITAL SEMI-MAJOR AXIS, a (astronomical units)	ORBITAL PERIOD, P (Earth years)	ORBITAL ECCENTRICITY, e	P^2/a^3
Mercury	0.387	0.241	0.206	1.002
Venus	0.723	0.615	0.007	1.001
Earth	1.000	1.000	0.017	1.000
Mars	1.524	1.881	0.093	1.000
Jupiter	5.203	11.86	0.048	0.999
Saturn	9.537	29.42	0.054	0.998
Uranus	19.19	83.75	0.047	0.993
Neptune	30.07	163.7	0.009	0.986

The Dimensions of the Solar System

Kepler's laws allow us to construct a scale model of the solar system, with the correct shapes and *relative* sizes of all the planetary orbits, but they do not tell us the *actual* size of any orbit. We can express the distance to each planet only in terms of the distance from Earth to the Sun—in other words, only in astronomical units. Why is this? Because Kepler's measurements were based on triangulation using a portion of Earth's orbit as a baseline, so his distances could be expressed only relative to the size of that orbit, which was not itself determined.

Our model of the solar system would be analogous to a road map of the United States showing the *relative* positions of cities and towns but lacking the all-important scale marker indicating distances in kilometers or miles. For example, we would know that Kansas City is about three times farther from New York than it is from Chicago, but we would not know the actual mileage between any two points on the map. If we could somehow determine the value of the astronomical unit—in kilometers, say—we would be able to add the vital scale marker to our map of the solar system and compute the exact distances between the Sun and each of the planets.

The modern method for deriving the absolute scale of the solar system uses a technique called *radar ranging*. The word **radar** is an acronym for **ra**dio **d**etection **a**nd **r**anging. Radio waves are transmitted toward an astronomical body, such as a planet. Their returning echo indicates the body's direction and range, or distance, in absolute terms—in other words, in kilometers rather than in astronomical units. Multiplying the round-trip travel time of the radar signal (the time elapsed between transmission of the signal and reception of the echo) by the speed of light (300,000 km/s, which is also the speed of radio waves), we obtain twice the distance to the target planet.

We cannot use radar ranging to measure the distance to the Sun directly, because radio signals are absorbed at the solar surface and are not reflected back to Earth. Instead, the planet Venus, whose orbit periodically brings it closest to Earth, is the most common target for this technique. Figure 1.14 is an idealized diagram of the Sun–Earth–Venus orbital geometry. Neglecting for simplicity the small eccentricities of the two planets' orbits, we see from Table 1.1 that the distance from Venus to the Sun is roughly 0.7 AU. Hence (from Figure 1.14), the distance from Earth to Venus at closest approach is approximately 0.3 AU. Radar signals bounced off Venus at that instant return to a detector on Earth in about 300 seconds, indicating that Venus lies 300,000 km/s × 300 s ÷ 2 dividing by 2 (for a one-way trip) = 45,000,000 km from Earth. With this information, we can compute the magnitude of the astronomical unit: since 0.3 AU is 45,000,000 km, it follows that 1 AU is 45,000,000 km/0.3, or 150,000,000 km.

Through precise radar ranging, the astronomical unit is now known to be 149,597,870 km. In this text, we will use the rounded-off value of 1.5×10^8 km. Having determined the value of the astronomical unit, we can re-express the sizes of the other planetary orbits in more familiar units, such as miles or kilometers. The entire scale of the solar system can then be calibrated to high precision.

CONCEPT CHECK ▶▶▶

Why don't Kepler's laws tell us the value of the astronomical unit?

◀ **Figure 1.14 Astronomical Unit** Simplified geometry of the orbits of Earth and Venus as they move around the Sun. The wavy blue lines represent the paths along which radar signals might be transmitted toward Venus and received back at Earth at the moment when Venus is at its minimum distance from Earth. Because the radius of Earth's orbit is 1 AU and that of Venus is about 0.7 AU, we know that the one-way distance covered by the signal is 0.3 AU. Thus, radar measurements allow us to determine the astronomical unit in kilometers.

1.4 Newton's Laws

Kepler's three laws, which so simplified the solar system, were discovered *empirically.* In other words, they resulted solely from the analysis of observational data and were not derived from any theory or mathematical model. Indeed, Kepler did not have any appreciation for the physics underlying his laws. Nor did Copernicus understand the basic reasons *why* his heliocentric model of the solar system worked. Even Galileo, often called the father of modern physics, failed to understand why the planets orbit the Sun.

What causes the planets to revolve about the Sun, apparently endlessly? What prevents them from flying off into space or falling into the Sun? Of course, the motions of the planets obey Kepler's three laws, but only by considering something more fundamental can we truly understand planetary motion. The critical test of any complete scientific theory is its ability to explain physical phenomena, not simply describe them.

The Laws of Motion

In the seventeenth century the British physicist and mathematician Isaac Newton (Figure 1.15) developed a deeper understanding of the way *all* objects move and interact with one another. Newton's theories form the basis for what today is known as **Newtonian mechanics**. Three basic laws of motion, the law of universal gravitation, and a little calculus (which Newton also helped invent) are sufficient to explain and quantify virtually all of the complex dynamic behavior we see on Earth and throughout the universe.

Figure 1.16 illustrates *Newton's first law of motion,* which states that unless some external **force** changes its state of motion, an object at rest remains at rest and a moving object continues to move forever in a straight line with constant speed. An example of an external force would be the force exerted by, say, a brick wall when a rolling ball glances off it, or the force exerted on a pitched ball by a baseball bat. In either case, a force changes the original motion of the object. The tendency of an object to keep moving at the same speed and in the same direction unless acted upon by a force is known as **inertia**. A familiar measure of an object's inertia is its **mass**—loosely speaking, the total amount of matter the object contains. The greater an object's mass, the more inertia it has and the greater is the force needed to change its motion.

Newton's first law contrasts sharply with the view of Aristotle, who maintained (incorrectly) that the natural state of an object was to be *at rest*—most probably an opinion based on Aristotle's observations of the effect of friction. To simplify our discussion, we will neglect friction—the force that slows balls rolling along the ground, blocks sliding across tabletops, and baseballs moving through the air. In any case, this is not an issue for the planets, as there is no appreciable friction in outer space. The fallacy in Aristotle's argument was first exposed by Galileo, who conceived of the notion of inertia long before Newton formalized it into a law.

The rate of change of the velocity of an object—speeding up, slowing down, or simply changing direction—is called its **acceleration**. *Newton's second law* states that the acceleration of an object is directly proportional to the net applied force and inversely proportional to the object's mass—that is, the greater the force acting on the object, or the smaller the mass of the object, the

▲ **Figure 1.15** **Isaac Newton** (1642–1727). (The Granger Collection)

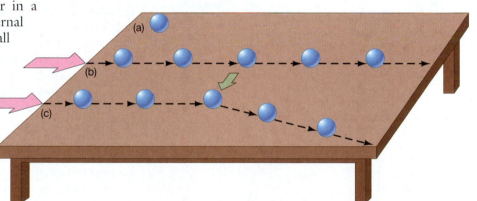

▲ **Figure 1.16** **Newton's First Law**   An object at rest will remain at rest (a) until some force acts on it. When a force (represented by the red arrow) does act (b), the object will remain in that state of uniform motion until another force acts on it. When a second force (green arrow) acts in a direction different from the first (c), the object changes its direction of motion.

Animation/Video

Thought-Experiment at Pisa

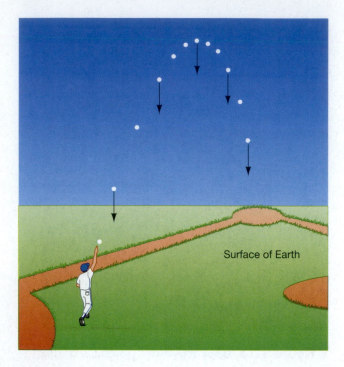

Surface of Earth

◄ Figure 1.17 Gravity A ball thrown upward from the surface of a massive object, such as a planet, is pulled continuously downward (arrows) by the gravity of that planet (and, conversely, the gravity of the ball continuously pulls the planet).

greater its acceleration. Thus, if two objects are pulled with the same force, the more massive one accelerates less. If two identical objects are pulled with different forces, the one experiencing the greater force accelerates more.

Finally, *Newton's third law* tells us that forces always occur in pairs—if body A exerts a force on body B, then body B necessarily exerts a force on body A that is equal in magnitude but oppositely directed.

Gravity

Forces can act either *instantaneously* or *continuously*. The force from a baseball bat that hits a home run can reasonably be thought of as instantaneous. A good example of a continuous force is the one that prevents the baseball from zooming off into space—**gravity**, the phenomenon that started Newton on the path to the discovery of his laws. Newton hypothesized that any object having mass exerts an attractive **gravitational force** on all other massive objects. The more massive an object, the stronger its gravitational pull.

Consider a baseball thrown upward from Earth's surface, as illustrated in Figure 1.17. In accordance with Newton's first law, the downward force of Earth's gravity steadily modifies the baseball's velocity, slowing its initial upward motion and eventually causing the ball to fall back to the ground. Of course, the baseball, having some mass of its own, also exerts a gravitational pull on Earth. By Newton's third law, this force is equal in magnitude to the weight of the ball (the weight of an object is a measure of the force with which Earth attracts that object), but oppositely directed. By Newton's second law, however, Earth has a much greater effect on the light baseball than the baseball has on the much more massive Earth. The ball and Earth each feel the same gravitational force, but Earth's *acceleration* as a result of this force is much smaller and can be safely ignored.

Now consider the trajectory of a baseball batted from the surface of the Moon. The pull of gravity is about one-sixth as great on the Moon as on Earth, so the baseball's velocity changes more slowly—a typical home run in a ballpark on Earth would travel nearly half a mile on the Moon. The Moon, less massive than Earth, has less gravitational influence on the baseball. The magnitude of the gravitational force, then, depends on the *masses* of the attracting bodies. In fact, it is directly proportional to the product of the two masses.

Studying the motions of the planets reveals a second aspect of the gravitational force. At locations equidistant from the Sun's center, the gravitational force has the same strength and is always directed toward the Sun. Furthermore, detailed calculation of the planets' accelerations as they orbit the Sun reveals that the strength of the Sun's gravitational pull decreases in proportion to the *square* of the distance from the Sun. (Newton is said to have first realized this fact by comparing the accelerations not of the planets but of the Moon and an apple falling to the ground—the basic reasoning is the same in either case.) The force of gravity is said to obey an **inverse-square law** (Figure 1.18a).

Mass m_1 Force   Mass m_2

$$F = \frac{Gm_1m_2}{r^2}$$

Distance r

(a)

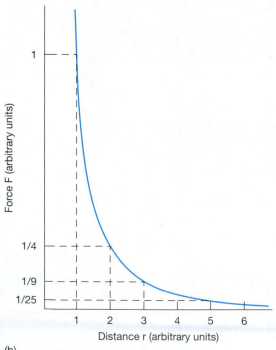

(b)

◄ Figure 1.18 Gravitational Force (a) The gravitational force between two bodies is proportional to the mass of each and is inversely proportional to the square of the distance between them. (b) Inverse-square forces weaken rapidly with distance from their source. The strength of the Sun's gravitational force decreases with the square of the distance from the Sun. The force never quite diminishes to zero, however, no matter how far away from the Sun we go.

We can combine the preceding statements about mass and distance to form a law of gravity that dictates the way in which *all* massive objects (i.e., objects having some mass) attract one another:

> Every particle of matter in the universe attracts every other particle with a force that is directly proportional to the product of the masses of the particles and inversely proportional to the square of the distance between them.

As shown in Figure 1.18(b), inverse-square forces decrease rapidly with distance from their source. For example, tripling the distance makes the force $3^2 = 9$ times weaker, while multiplying the distance by five results in a force that is $5^2 = 25$ times weaker. Despite this rapid decrease, the force never quite reaches zero. The gravitational pull of an object having some mass can never be completely extinguished.

Orbital Motion

The mutual gravitational attraction of the Sun and the planets, as expressed by Newton's law of gravity, is responsible for the observed planetary orbits. As depicted in Figure 1.19, this gravitational force continuously pulls each planet toward the Sun, deflecting its forward motion into a curved orbital path. Because the Sun is much more massive than any of the planets, it dominates the interaction. We might say that the Sun "controls" the planets, not the other way around.

The planet–Sun interaction sketched here is analogous to what occurs when you whirl a rock at the end of a string above your head. The Sun's gravitational pull is your hand and the string, and the planet is the rock at the end of that string. The tension in the string provides the force necessary for the rock to move in a circular path. If you were suddenly to release the string—which would be like eliminating the Sun's gravity—the rock would fly away along a tangent to the circle, in accordance with Newton's first law.

In the solar system, at this very moment, Earth is moving under the combined influence of these two effects: the competition between gravity and inertia. The net result is a stable orbit, despite our continuous rapid motion through space. In fact, Earth orbits the Sun at a speed of about 30 km/s, or some 70,000 mph. (Verify this for yourself by calculating how fast Earth must move to complete a circle of radius 1 AU—and hence of circumference 2π AU, or 940 million km—in one year, or 3.2×10^7 seconds.) *More Precisely 1-1* describes how astronomers can use Newtonian mechanics and the law of gravity to measure the masses of Earth, the Sun, and many other astronomical objects by studying the orbits of objects near them.

Kepler's Laws Reconsidered

Newton's laws of motion and his law of universal gravitation provided a theoretical explanation for Kepler's empirical laws of planetary motion. Just as Kepler modified the Copernican model by introducing ellipses in place of circles, so too did Newton make corrections to Kepler's first and third laws. Because the Sun and a planet feel equal and opposite gravitational forces (by Newton's third law), the Sun must also move (by Newton's first law) due to the planet's gravitational pull. As a result, the planet does not orbit the exact center of the Sun but, instead, both the planet and the Sun orbit their common **center of mass**—the "average" position of the matter comprising the two bodies (see Figure 1.20)—and Kepler's first law becomes:

> I. The orbit of a planet around the Sun is an ellipse having the center of mass of the planet–Sun system at one focus.

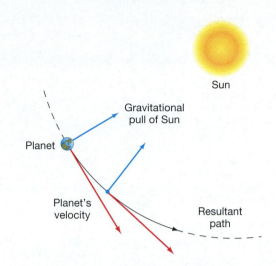

▲ **Figure 1.19 Sun's Gravity** The Sun's inward pull of gravity on a planet competes with the planet's tendency to continue moving in a straight line. These two effects combine, causing the planet to move smoothly along an intermediate path, which continuously "falls around" the Sun. This unending tug-of-war between the Sun's gravity and the planet's inertia results in a stable orbit.

Animation/Video
Earth Captures a Temporary Moon

(a) Equal masses   Center of mass

(b) Unequal masses   Center of mass

▲ **Figure 1.20 Center of Mass** (a) The center of mass of two bodies of equal mass lies midway between them. (b) As the mass of one body increases, the center of mass moves toward it. The situation is analogous to riding a see-saw—when both sides are balanced, the center of mass is at the pivot point.

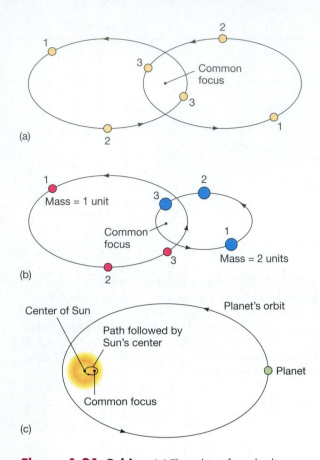

▲ Figure 1.21 Orbits (a) The orbits of two bodies (stars, for example) with equal masses, under the influence of their mutual gravity, are identical ellipses with a common focus. That focus is not at the center of either star but instead at their center of mass, located midway between them. The pairs of numbers indicate the positions of the two bodies at three times. (Note that the line joining the bodies at any given time always passes through the common focus.) (b) The orbits of two bodies when one body is twice as massive as the other. The elliptical orbits again have a common focus (at the center of mass of the two-body system), and the two ellipses have the same eccentricity, but the more massive body moves more slowly and in a smaller orbit. The larger ellipse is twice the size of the smaller one. (c) In the extreme case of a hypothetical planet orbiting the Sun, the common focus of the two orbits lies inside the Sun.

As illustrated in Figure 1.20, the center of mass of a system consisting of two objects of comparable mass does not lie within either object. Hence, for identical masses orbiting one another (Figure 1.21a) the orbits are identical ellipses with a common focus located midway between the two objects. For unequal masses (as in Figure 1.21b) the elliptical orbits still share a focus and have the same eccentricity, but the more massive object moves more slowly and on a tighter orbit. (Note that Kepler's second law, as stated earlier, continues to apply without modification to each orbit separately, but the rates at which the two orbits sweep out area are different.) In the extreme case of a planet orbiting the much more massive Sun (Figure 1.21c), the path traced out by the Sun's center lies entirely within the Sun.

The change to Kepler's third law is small in the case of a planet orbiting the Sun but may be very important in other circumstances, such as the orbits of two stars that are gravitationally bound to one another. Following through the mathematics of Newton's theory, we find that the true relationship between the semi-major axis a (measured in astronomical units) of the planet's orbit relative to the Sun and its orbital period P (in Earth years) is

$$P^2 \text{ (in Earth years)} = \frac{a^3 \text{ (in astronomical units)}}{M_{\text{total}} \text{ (in solar units)}},$$

where M_{total} is the *combined* mass of the two objects, expressed in terms of the mass of the Sun (see also *More Precisely 1-1*). Notice that this restatement of Kepler's third law preserves the proportionality between P^2 and a^3, but now the proportionality also includes M_{total}, so it is *not* quite the same for all the planets. The Sun's mass is so great, however, that the differences in M_{total} among the various combinations of the Sun and the other planets are almost unnoticeable. Therefore Kepler's third law, as originally stated, is a very good approximation. This modified form of Kepler's third law is true in *all* circumstances, inside or outside the solar system.

The Circle of Scientific Progress

The progression from the complex Ptolemaic model of the universe to the elegant simplicity of Newton's laws is a case study in the scientific method. ∞ *(Sec. E.5)* Copernicus made a radical conceptual leap away from the Ptolemaic view, gaining much in insight but little in predictive power. Kepler made critical changes to the Copernican picture and gained both accuracy and predictive power but still fell short of a true physical explanation of planetary motion within the solar system, or of orbital motion in general. Eventually, Newton showed how all known planetary motion could be explained in detail by the application of four simple, fundamental laws—the three laws of motion and the law of gravity. The process was slow, with many starts and stops and a few wrong turns, but it worked!

In a sense, then, the development of Newton's laws and their application to planetary motion represented the end of the first "loop" around the schematic diagram presented in Figure E.22. The long-standing practical and conceptual questions raised by ancient observations of retrograde motion were finally resolved, and new predictions, themselves amenable to observational testing, became possible. But Newton's laws went much further. Unlike the essentially descriptive models of Ptolemy, Copernicus, and Kepler, Newtonian mechanics is not limited to the motions of planets, or even to events occurring within our own solar system. They apply to moons, comets, spacecraft, stars, and even the most distant galaxies, extending the range of our scientific inquiries across the observable universe.

Newton's laws are still being tested today. Every time a comet appears in the night sky right on schedule or a spacecraft reaches the end of a billion-kilometer journey within meters of its target and seconds of the predicted arrival time, our confidence in the laws is further strengthened. Only in extreme circumstances do Newton's laws break down, and this fact was not realized until the twentieth century, when Albert Einstein's theories of relativity once again revolutionized our view of gravity and the universe (see Chapter 13). Most of the time, however, Newtonian mechanics provides an excellent description of the motion of planets, stars, and galaxies throughout the cosmos.

◀◀◀ | **CONCEPT CHECK**

Explain, in terms of Newton's laws of motion and gravity, why planets orbit the Sun.

1-1 | **M**ORE PRECISELY

Weighing the Sun

Mass is a fundamental property of any object, and knowing a body's mass is often the first step in understanding its appearance, structure, and even its evolution. But how can astronomers determine the mass of a faraway object—a planet, say, or a star? They can't just go out and weigh it as we would on Earth. Or can they? In fact, Newton's laws of motion and gravitation allow astronomers to do just that. Here's how.

Newton's law of gravitation states that every particle of matter in the universe attracts every other particle with a force that is directly proportional to the product of the masses of the particles and inversely proportional to the square of the distance between them. In other words, two bodies of masses m_1 and m_2, separated by a distance r, attract each other with a force F that is proportional to $(m_1 \times m_2)/r^2$. The constant of proportionality is known as the gravitational constant, or often simply as *Newton's constant*, and is always denoted by the letter G. We can thus express the law of gravity as

$$F = \frac{Gm_1 m_2}{r^2}.$$

The value of G has been measured on Earth in extremely delicate laboratory experiments. In SI (Système Internationale) units its value is 6.67×10^{-11} newton meter^2/kilogram^2 (Nm^2/kg^2). (The newton is the SI unit of force, defined as the force required to accelerate a mass of 1 kg at a rate of 1 m/s^2. One newton is approximately 0.22 pounds.)

Let's consider for a moment what happens when you step on a bathroom scale. Suppose your mass is m_1 and Earth's mass is m_2. The distance r is Earth's radius, roughly 6400 km. ∞ *(Discovery E-1)* Since G, m_2, and r are all fixed, we see that the force F is just proportional to your mass m_1, which is what you read off the scale. But notice that we can turn the reasoning around. If we already *knew* the mass m_1—say, perhaps, that we used a standard 1-kg weight—then measuring the force F and using the known values of G and r allow us to determine the value of m_2. In other words, we can use New-

ton's law to weigh planet Earth! Earth's mass is roughly 6.0×10^{24} kg, a very large number, and one about which we will have much more to say later.

Let's take this one stage further. We can use Newtonian mechanics to calculate the relationship between the distance (r) and the speed (v) of a planet moving in a circular orbit around the Sun (of mass M). By calculating the force required to keep the planet moving in a circle and comparing it with the gravitational force due to the Sun, it can be shown that the circular speed v is given by

$$v^2 = \frac{G M}{r}.$$

[Note, incidentally, that dividing this speed into the circumference of the orbit ($2\pi r$) to give the orbital period $P = 2\pi r/v$ results in Kepler's third law as stated in the text.] For our purposes, this equation may be more conveniently rewritten as

$$M = \frac{rv^2}{G}.$$

In other words, we can then use Newtonian mechanics—the combination of Newton's laws of motion and his law of gravity—to weigh the Sun. Inserting the known values of $v = 30$ km/s, $r = 1$ AU $= 1.5 \times 10^{11}$m, and G (given above), we can calculate the Sun's mass. The result is $M = 2.0 \times 10^{30}$ kg—another enormous mass, even compared with that of Earth. Similarly, knowing the distance to the Moon ($r = 384{,}000$ km) and the length of the (sidereal) month, 27.3 days, we can calculate the Moon's orbital speed ($v = 1.02$ km/s) and confirm the above measurement of Earth's mass.

In fact, this is how *all* masses are measured in astronomy. Newton's laws extend our intellectual reach far beyond the tiny fraction of the universe we can actually visit or touch. When we need to know an object's mass, we always look for its gravitational influence on something else. This principle applies to planets, stars, galaxies, and even clusters of galaxies—very different objects but all subject to the same fundamental physical laws.

CHAPTER REVIEW

Summary

1 **Geocentric** (p. 27) models of the universe, such as the **Ptolemaic model** (p. 27), have the Sun, the Moon, and all the other planets orbiting Earth. Copernicus's **heliocentric** (p. 28) view of the solar system holds that Earth, like all the other planets, orbits the Sun. This model naturally explains **retrograde motion** (p. 26) and the observed brightness variations of the planets. The widespread realization that true solar system is Sun-centered and not Earth-centered is known as the **Copernican revolution** (p. 28).

2 Galileo Galilei was the first experimental scientist. His telescopic observations of the Sun, the Moon, and the planets provided experimental evidence against the heliocentric theory and supporting Copernicus's heliocentric model. Johannes Kepler constructed a set of three simple laws describing the motions of the planets, explaining the mass of observational data accumulated by Tycho Brahe.

3 Kepler's three **laws of planetary motion** (p. 32) state that (1) planetary orbits are **ellipses** (p. 32) having the Sun as one **focus** (p. 32),

(2) a planet moves faster as its orbit takes it closer to the Sun, and (3) the **semi-major axis** (p. 32) of the orbit is related in a simple way to the planet's orbit **period** (p. 33).

4 The average distance from Earth to the Sun is one **astronomical unit** (p. 33), today precisely determined by bouncing **radar** (p. 34) signals off the planet Venus.

5 To change a body's velocity, a **force** (p. 35) must be applied. The rate of change of velocity, called the **acceleration** (p. 35), is equal to the applied force divided by the body's mass. **Gravity** (p. 36) attracts the planets to the Sun. Every object having any mass exerts a **gravitational force** (p. 36) on all other objects, and the strength of this force decreases with distance according to an **inverse-square law** (p. 36).

Review and Discussion

1. Briefly describe the geocentric model of the universe.

2. The benefit of our current knowledge lets us see flaws in the Ptolemaic model of the universe. What is its basic flaw?

3. What was the great contribution of Copernicus to our knowledge of the solar system?

4. What was the Copernican revolution?

5. What discoveries of Galileo helped confirm the views of Copernicus?

6. Why is Galileo often thought of as the first experimental scientist?

7. State Kepler's three laws of orbital motion.

8. How did Tycho Brahe contribute to Kepler's laws?

9. What is meant by the statement that Kepler's laws are empirical in nature?

10. If radio waves cannot be reflected from the Sun, how can radar be used to find the distance from Earth to the Sun?

11. List the two modifications made by Newton to Kepler's laws.

12. Why do we say that a baseball falls toward Earth, and not Earth toward the baseball?

13. Why would a baseball thrown upward from the surface of the Moon go higher than one thrown with the same velocity from the surface of Earth?

14. According to Newton, why does Earth orbit the Sun?

15. What would happen to Earth if the Sun's gravity were suddenly "turned off"?

Conceptual Self-Test: True or False?/Multiple Choice

1. Aristotle was the first to propose that all planets revolve around the Sun. **(T/F)**

2. During retrograde motion, planets actually stop and move backwards in space. **(T/F)**

3. The heliocentric model of the universe holds that Earth is at the center of the universe. **(T/F)**

4. Copernicus's theories gained widespread scientific acceptance during his own lifetime. **(T/F)**

5. Galileo's observations of the sky were made with the naked eye. **(T/F)**

6. The speed of a planet orbiting the Sun is independent of the planet's position in its orbit. **(T/F)**

7. Kepler's laws hold only for the six planets known in his time. **(T/F)**

8. You throw a baseball to someone; before the ball is caught, it is temporarily in orbit around Earth's center. **(T/F)**

9. A major flaw in Copernicus's model was that it still had **(a)** the Sun at the center; **(b)** Earth at the center; **(c)** retrograde loops; **(d)** circular orbits.

10. An accurate sketch of Mars's orbit around the Sun would show **(a)** the Sun far off center; **(b)** an oval twice as long as it is wide; **(c)** a nearly perfect circle; **(d)** phases.

11. A calculation of how long it takes a planet to orbit the Sun would be most closely related to Kepler's **(a)** first law of orbital shapes; **(b)** second law of orbital speeds; **(c)** third law of planetary distances; **(d)** first law of inertia.

12. An asteroid with an orbit lying entirely inside Earth's **(a)** has an orbital semi-major axis of less than 1 AU; **(b)** has a longer orbital period than Earth's; **(c)** moves more slowly than Earth; **(d)** has a highly eccentric orbit.

13. If Earth's orbit around the Sun were twice as large as it is now, the orbit would take **(a)** less than twice as long; **(b)** two times longer; **(c)** more than two times longer to traverse.

14. As shown in Figure 1.8 ("Venus Phases"), Galileo's observations of Venus demonstrated that Venus must be **(a)** orbiting Earth; **(b)** orbiting the Sun; **(c)** about the same diameter as Earth; **(d)** similar to the Moon.

15. Figure 1.16 ("Gravity"), showing the motion of a ball near Earth's surface, depicts how gravity **(a)** increases with altitude; **(b)** causes the ball to accelerate downward; **(c)** causes the ball to accelerate upward; **(d)** has no effect on the ball.

Problems   *Algorithmic versions of these questions are available in the Practice Problems module of the Companion Website.*

The number of squares preceding each problem indicates its approximate level of difficulty.

1. ■■ Tycho Brahe's observations of the stars and planets were accurate to about 1 arc minute (1′). To what distance does this angle correspond (a) at the distance of the Moon, (b) of the Sun, (c) of Saturn (at closest approach)?

2. ■ How long would a radar signal take to complete a round-trip between Earth and Mars when the two planets are 0.7 AU apart?

3. ■■ To an observer on Earth, through what angle will Mars appear to move relative to the stars over the course of 24 hours when the two planets are at closest approach? Assume for simplicity that Earth and Mars move on circular orbits of radii 1.0 AU and 1.5 AU, respectively, in exactly the same plane. Will the apparent motion be prograde or retrograde?

4. ■■ An asteroid has a perihelion distance of 2.0 AU and an aphelion distance of 4.0 AU. Calculate its orbital semi-major axis, eccentricity, and period (see Figure 1.12).

5. ■■ Halley's Comet has a perihelion distance of 0.6 AU and an orbital period of 76 years. What is its aphelion distance from the Sun (Figure 1.12)?

6. ■■ Using the data in Table 1.1, show that Pluto is closer to the Sun at perihelion (the point of closest approach to the Sun in its orbit) than Neptune is at any point in its orbit.

7. ■■■ Jupiter's moon Callisto orbits the planet at a distance of 1.88 million km. Callisto's orbital period about Jupiter is 16.7 days. What is the mass of Jupiter? [Assume that Callisto's mass is negligible compared with that of Jupiter, and use the modified version of Kepler's third law (Section 1.4).]

8. ■■ The acceleration due to gravity at Earth's surface is 9.80 m/s². What is the acceleration at altitudes of (a) 100 km? (b) 1000 km? (c) 10,000 km? Take Earth's radius to be 6400 km.

9. ■ Use Newton's law of gravity to calculate the force of gravity between you and Earth. Convert your answer, which will be in newtons, to pounds using the conversion 4.45 N = 1 pound. What do you normally call this force?

10. ■■ The Moon's mass is 7.3×10^{22} kg, and its radius is 1700 km. What is the speed of a spacecraft moving in a circular orbit just above the lunar surface?

 The Companion Website at www.prenhall.com/chaisson provides algorithmically generated versions of each chapter's Problems, in addition to an Animations and Videos gallery, a Tutorials gallery and Tutorials Quiz questions, a full eBook, an Images gallery, Practice Questions, Multiple Choice, and True or False quizzes, and Labeling exercises.

2

Light and Matter

The Inner Workings of the Cosmos

Stars change from birth to maturity to death, much like living things, but on vastly longer timescales. Our own star, the Sun, is about midway through its evolutionary cycle. About 5 billion years ago, it emerged from a stellar nursery much like the one shown here. The *Spitzer Space Telescope*, now orbiting the Earth, took this infrared image. By capturing radiation longer in wavelength than light, we can peer inside young star clusters like this one—a rosebud-shaped nebula called NGC 7129 about 3300 light-years away in the constellation Cepheus. (JPL)

Astronomical objects are more than just things of beauty in the night sky. Planets, stars, and galaxies are of vital significance if we are to understand fully the big picture—the grand design of the universe. Every object is a source of information about the universe—its temperature, its chemical composition, its state of motion, its past history. The starlight we see tonight began its journey to Earth decades, centuries—even millennia—ago. The faint rays from the most distant galaxies have taken billions of years to reach us. The stars and galaxies in the night sky show us not just the far away but also the long ago. In this chapter, we begin our study of how astronomers extract information from the light emitted by astronomical objects. The observational and theoretical techniques that enable researchers to determine the nature of distant atoms by the way they emit and absorb light are the indispensable foundation of modern astronomy.

LEARNING GOALS

Studying this chapter will enable you to:

1 Discuss the nature of electromagnetic radiation and tell how that radiation transfers energy and information through interstellar space.

2 Describe the major regions of the electromagnetic spectrum.

3 Explain how we can determine an object's temperature by observing the radiation it emits.

4 Describe the characteristics of continuous, emission, and absorption spectra and the conditions under which each is produced.

5 Specify the basic components of the atom and describe our modern conception of its structure.

6 Explain how electron transitions within atoms produce unique emission and absorption spectra.

7 Describe how the relative motion of a source of radiation and an observer can change the perceived wavelength of the radiation, and explain the importance of this phenomenon to astronomy.

Visit www.prenhall.com/chaisson for additional annotated images, animations, and links to related sites for this chapter.

2.1 Information from the Skies

Figure 2.1 shows our nearest large galactic neighbor, which lies in the constellation Andromeda. On a dark, clear night, far from cities or other sources of light, the Andromeda Galaxy, as it is generally called, can be seen with the naked eye as a faint, fuzzy patch on the sky, comparable in diameter to the full Moon. Yet the fact that it is visible from Earth belies this galaxy's enormous distance from us. It lies roughly 2.5 million light-years away. An object at such a distance is truly inaccessible in any realistic human sense. Even if a space probe could miraculously travel at the speed of light, it would need two and a half million years to reach this galaxy and two and a half million more to return with its findings. Considering that civilization has existed on Earth for fewer than 10,000 years (and its prospects for the next 10,000 are far from certain), even this unattainable technological feat would not provide us with a practical means of exploring other galaxies—or even the farthest reaches of our own galaxy, several tens of thousands of light-years away.

Light and Radiation

How do astronomers know anything about objects far from Earth? How can we obtain detailed information about any planet, star, or galaxy too distant for a personal visit or any kind of controlled experiment? The answer is that we use the laws of physics, as we know them here on Earth, to interpret the **electromagnetic radiation** emitted by these objects. *Radiation* is any way in which energy is transmitted through space from one point to another without the need for any physical connection between those two locations. The term *electromagnetic* means that the energy is carried in the form of rapidly changing *electric* and *magnetic fields* (see Section 2.2). Virtually all we know about the universe beyond Earth's atmosphere has been learned from analysis of electromagnetic radiation received from afar.

Visible light is the particular type of electromagnetic radiation to which the human eye happens to be sensitive. But there is also *invisible* electromagnetic radiation, which goes completely undetected by our eyes. **Radio**, **infrared**, and **ultraviolet** waves, as well as **X-rays** and **gamma rays**, all fall into this category. You should recognize that, despite the different names, the words *light, rays, electromagnetic radiation,* and *waves* really all refer to the same thing. The names are just historical accidents, reflecting the fact that it took many years for scientists to realize that these apparently very different types of radiation are in reality one and the same physical phenomenon. Throughout this text, we will use the general terms "light" and "electromagnetic radiation" more or less interchangeably.

Wave Motion

All types of electromagnetic radiation travel through space in the form of **waves.** To understand the behavior of light, then, we must know a little about this kind of motion. Simply stated, a wave is a way in which energy is transferred from place to place without physical movement of material from

◀ **Figure 2.1 Andromeda Galaxy** The pancake-shaped Anromeda Galaxy is about 2.5 million light-years away and contains a few hundred billion stars. (R. Gendler)

one location to another. In wave motion, the energy is carried by a *disturbance* of some sort that occurs in a distinctive, repeating pattern. Ripples on the surface of a pond, sound waves in air, and electromagnetic waves in space, despite their many obvious differences, all share this basic defining property.

As a familiar example, imagine a twig floating in a pond (Figure 2.2). A pebble thrown into the pond at some distance from the twig disturbs the surface of the water, setting it into up-and-down motion. This disturbance propagates outward from the point of impact in the form of waves. When the waves reach the twig, some of the pebble's energy is transferred to it, causing the twig to bob up and down. In this way, both energy and *information*—the fact that the pebble entered the water—are transferred from the place where the pebble landed to the location of the twig. We could tell just by observing the twig that a pebble (or some small object) had entered the water. With a little additional physics, we could even estimate the pebble's energy.

A wave is *not* a physical object. No water traveled from the point of impact of the pebble to the twig; at any location on the surface, the water surface simply moved up and down as the wave passed. What, then, *does* move across the pond surface? The answer is that the wave is the *pattern* of up-and-down motion, and it is this pattern that is transmitted from one point to the next as the disturbance moves across the water.

Figure 2.3 shows how wave properties are quantified. The **wave period** is the number of seconds needed for the wave to repeat itself at some point in space. The **wavelength** is the number of meters needed for the wave to repeat itself at a given moment in time. It can be measured as the distance between two adjacent wave *crests,* two adjacent wave *troughs,* or any other two similar points on adjacent wave cycles (for example, the points marked "×" in Figure 2.3). The maximum departure of the wave from the undisturbed state—still air, say, or a flat pond surface—is called its **amplitude.**

The number of wave crests passing any given point per unit time is called the wave's **frequency.** If a wave of a given wavelength moves at high speed, then many crests pass by per second and the frequency is high. Conversely, if the same wave moves slowly, then its frequency is low. The frequency of a wave is just one divided by the wave's period:

$$\text{frequency} = \frac{1}{\text{period}}.$$

Since frequency counts events per unit time, it is expressed in units of inverse time (cycles per second), called hertz (Hz) in honor of the nineteenth-century

▲ **Figure 2.2 Water Wave**   The passage of a wave across a pond causes the surface of the water to bob up and down, but there is no movement of water from one part of the pond to another. Here waves ripple out from the point where a pebble hit the water to the point where a twig is floating. The inset shows a series of "snapshots" of the pond surface as the wave passes by. The points numbered 1 through 5 represent surface locations that bob up and down with the passage of the wave.

◀ **Figure 2.3 Wave Properties**   Representation of a typical wave, showing its direction of motion, wavelength, and amplitude. In one wave period the entire pattern shown moves one wavelength to the right.

German scientist Heinrich Hertz, who studied the properties of radio waves. For example, a wave with a period of 5 seconds (5 s) has a frequency of (1/5) cycles/s = 0.2 Hz, meaning that one wave crest passes a given point in space every five seconds.

A wave moves a distance equal to one wavelength in one wave period. The product of wavelength and frequency therefore equals the *wave velocity:*

$$\text{wavelength} \times \text{frequency} = \text{velocity}.$$

Thus, if the wave in our earlier example had a wavelength of 0.5 m, its velocity is (0.5 m) × (0.2 Hz) = 0.1 m/s. Wavelength and wave frequency are *inversely* related—doubling one halves the other.

2.2 Waves in What?

Scientists know that radiation travels as a wave because light displays characteristic behavior common to *all* kinds of wave motion. For example, light waves tend to "bend around corners," just like sound waves or ocean waves passing a breakwater (Figure 2.4a). This is called **diffraction**. In addition, the crests and troughs of waves coming from different sources can reinforce or partly cancel one another (Figure 2.4b). This is known as **interference**. We are not normally aware of this behavior in the case of electromagnetic radiation because the effects are generally too small to be noticeable in everyday life. However, they are easily measured in the laboratory and are very important considerations when designing and building telescopes, as we will see in Chapter 3. Their discovery early in the nineteenth century provided strong evidence in favor of the theory that light moves as a wave. ∞ *(Sec. 1.5)*

Waves of radiation do differ in one fundamental respect from water waves, sound waves, or any other waves that travel through a material medium—radiation needs *no* such medium. When light travels from a distant cosmic object, it moves through the virtual vacuum of space. Sound waves, by contrast, cannot do this, despite what you have probably heard in almost every sci-fi movie ever made! If we were to remove all the air from a room, conversation would be impossible (even with suitable breathing apparatus to keep our test subjects alive) because sound waves cannot exist without air or some other physical medium to support them. Communication by flashlight or radio, however, would be entirely possible.

The ability of light to travel through empty space was once a great mystery. The idea that light, or any other kind of radiation, could move as a wave through nothing at all seemed to violate common sense, yet it is now a cornerstone of modern physics.

Interactions Between Charged Particles

To understand more about the nature of light, consider an *electrically charged* particle, such as an **electron** or a **proton**. Electrons and protons are elementary particles—fundamental components of matter—that carry the basic unit of charge. Electrons are said to carry a *negative* charge, while protons carry an equal and opposite *positive* charge. Just as a massive object exerts a gravitational force on any other massive object, an electrically charged particle exerts an *electrical* force on every other charged particle in the universe. ∞ *(Sec. 1.4)* Unlike gravity, however, which is always attractive, electrical forces can be either attractive or repulsive. Particles having like charges (both negative or both positive) repel one another; particles having unlike charges attract (Figure 2.5a).

Extending outward in all directions from our charged particle is an **electric field**, which determines the electric force exerted by the particle on other

CONCEPT CHECK ▶▶▶

What is a wave? What four basic properties describe a wave, and what relationships, if any, exist among them?

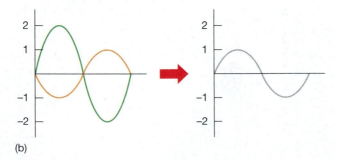

▲ **Figure 2.4 Wave Behavior** Diffraction (a) and interference (b) are two basic types of behavior shared by all waves, including electromagnetic radiation. In diffraction, a wave bends around an obstacle in its path (here a breakwater). In interference, two waves combine to produce a wave equal to the sum of the two, in this case partly canceling one another.

charged particles (Figure 2.5b). The strength of the electric field, like that of the gravitational field, decreases with increasing distance from the source according to an inverse-square law—doubling the distance reduces the strength of the field by a factor of 4. ∞ *(Sec. 1.4)* By means of the electric field, the particle's presence is "felt" by other charged particles, near and far.

Now suppose our particle begins to vibrate, perhaps because it becomes heated or collides with some other particle. Its changing position causes its associated electric field to change, and this changing field in turn causes the electrical force exerted on other charges to vary (Figure 2.5c). If we measure the changes in the forces on these other charges, we learn about our original particle. Thus, *information about our particle's motion is transmitted through space via a changing electric field*. This *disturbance* in the particle's electric field travels through space as a wave.

Electromagnetic Waves

The laws of physics tell us that a **magnetic field** must accompany every changing electric field. Magnetic fields govern the influence of *magnetized* objects on one another, much as electric fields govern interactions between charged particles. The fact that one end of a compass needle always points to magnetic north is the result of the interaction between the magnetized needle and Earth's magnetic field (Figure 2.6). Magnetic fields also exert forces on moving electric charges (that is, electric currents)—electric meters and motors rely on this basic fact. Conversely, moving charges create magnetic fields (the electromagnets found in loudspeakers and electric motors are familiar examples).

Electric and magnetic fields are inextricably linked to one another. A change in either one *necessarily* creates the other. For this reason, the disturbance produced by our moving charge actually consists of oscillating electric *and* magnetic fields, always oriented perpendicular to one another and moving together through space (Figure 2.7). These fields do not exist as independent entities. Rather, they are different aspects of a single physical phenomenon: **electromagnetism**. Together they constitute an *electromagnetic wave* that carries energy and information from one part of the universe to another.

Now consider a distant cosmic object—a star. It is made up of charged particles, mainly protons and electrons, in constant motion. As these charged contents move around, their electric fields change, and electromagnetic waves are produced. These waves travel outward into space, and eventually some reach Earth. Other charged particles, either in the molecules in our eyes or in our experimental apparatus (a telescope or radio receiver, for example), respond to the electromagnetic field changes by vibrating in tune with the received radiation. This response is how we "see" the radiation—with our eyes or with our detectors.

How quickly does one charge feel the change in the electromagnetic field when another begins to move—that is, how fast does an electromagnetic wave travel? Both theory and experiment tell us that all electromagnetic waves move at a very specific speed—the **speed of light** (always denoted by the letter c). Its value is 299,792.458 km/s in a vacuum (and somewhat less in material substances, such as air or water). In this text, we round this value off to $c = 3.00 \times 10^5$ km/s. This is an extremely high speed. In the time needed to snap your fingers—about a tenth of a second—light can travel three quarters of the way around our planet! According to the Theory of Relativity (see *More Precisely 13-1*), the speed of light is the fastest speed possible.

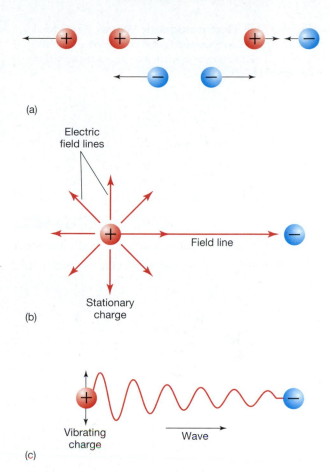

(a)

(b)

(c)

▲ **Figure 2.5 Charged Particles**   (a) Particles carrying like electrical charges repel one another; particles with unlike charges attract. (b) A charged particle is surrounded by an electric field, which determines the particle's influence on other charged particles. We represent the field by a series of field lines. (c) If a charged particle begins to vibrate, its electric field changes. The resulting disturbance travels through space as a wave, eventually interacting with distant charged particles.

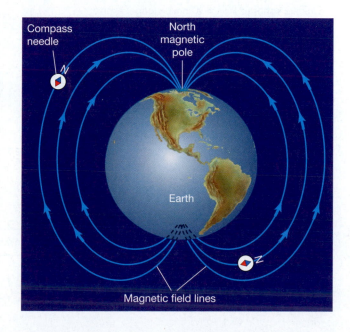

▶ **Figure 2.6 Magnetism**   Earth's magnetic field interacts with a magnetic compass needle, causing the needle to become aligned with the field—that is, to point toward Earth's north (magnetic) pole.

▶ **Figure 2.7 Electromagnetic Wave** Electric and magnetic fields vibrate perpendicular to each other. Together they form an electromagnetic wave that moves through space at the speed of light.

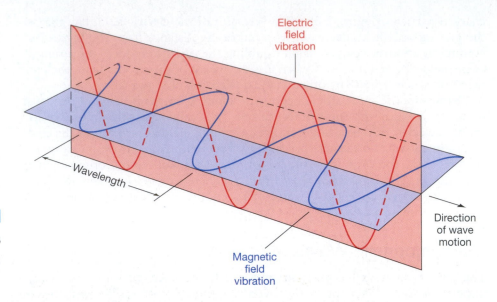

Electric field vibration

Wavelength

Direction of wave motion

Magnetic field vibration

CONCEPT CHECK ▶▶▶

What is light? List some similarities and differences between light waves and waves on water or in air.

2.3 The Electromagnetic Spectrum

White light is a mixture of colors, which we conventionally divide into six major hues—red, orange, yellow, green, blue, and violet. As shown in Figure 2.8, we can separate a beam of white light into a rainbow of these basic colors—called a *spectrum* (plural: spectra)—by passing it through a prism. This experiment was first reported by Isaac Newton more than 300 years ago. In principle, the original beam of white light could be recovered by passing the spectrum through a second prism to recombine the colored beams.

The Components of Visible Light

What determines the color of a beam of light? The answer is its frequency (or, equivalently, its wavelength)—we see different colors because our eyes react differently to electromagnetic waves of different frequencies. Red light has a frequency of roughly 4.3×10^{14} Hz, corresponding to a wavelength of about 7.0×10^{-7} m. Violet light, at the other end of the visible range, has nearly double the frequency—7.5×10^{14} Hz—and (since the speed of light is the same in either case) just over half the wavelength—4.0×10^{-7} m. The other colors we see have frequencies and wavelengths intermediate between these two extremes.

Scientists often use a unit called the *nanometer* (nm) when describing the wavelength of light (see Appendix 2). There are 10^9 nanometers in one meter. An older unit called the *angstrom* ($1 \text{ Å} = 10^{-10}$ m = 0.1 nm) is also widely used by many astronomers and atomic physicists, although the nanometer is now preferred. Thus, the

▼ **Figure 2.8 Visible Spectrum** When passed through a prism, white light splits into its component colors, spanning red to violet in the visible part of the electromagnetic spectrum. The slit narrows the beam of radiation. The image on the screen is just a series of different-colored images of the slit. Human eyes are insensitive to radiation of wavelength shorter than 400 nm or longer than 700 nm.

White light

Slit

Prism

Screen

Red Orange Yellow Green Blue Violet

Radio	Infrared	Visible	Ultraviolet	X ray	Gamma ray

4.3×10^{14} 7.5×10^{14} Frequency (Hz)

Wavelength (nm)   700   400

visible spectrum covers the wavelength range from 400 to 700 nm (4000 to 7000 Å). The radiation to which our eyes are most sensitive has a wavelength near the middle of this range, at about 550 nm (5500 Å), in the yellow-green region of the spectrum.

The Full Range of Radiation

Figure 2.9 plots the entire range of electromagnetic radiation. To the low-frequency, long-wavelength side of visible light lies radio and infrared radiation. Radio frequencies include radar, microwave radiation, and the familiar AM, FM, and TV bands. We perceive infrared radiation as heat. To the high-frequency, short-wavelength side of visible light lies ultraviolet, X-ray, and

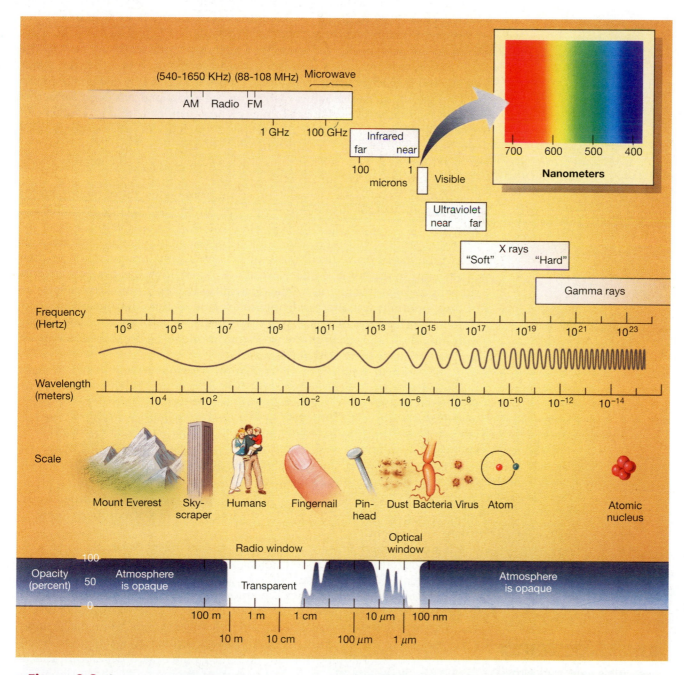

▲ **Figure 2.9 Electromagnetic Spectrum** The entire electromagnetic spectrum, showing the various regions described in the text.

gamma-ray radiation. Ultraviolet radiation, lying just beyond the violet end of the visible spectrum, is responsible for suntans and sunburns. X-rays are perhaps best known for their ability to penetrate human tissue and reveal the state of our insides without our resorting to surgery. Gamma rays are the shortest-wavelength radiation. They are often associated with radioactivity and are invariably damaging to any living cells they encounter.

All these spectral regions, including the visible, collectively make up the **electromagnetic spectrum**. Remember that despite their greatly differing wavelengths and the very different roles they play in everyday life on Earth, all types of electromagnetic radiation are basically the same and all move at the same speed—the speed of light c.

Figure 2.9 is worth studying carefully, as it contains a great deal of information. Note that wave frequency (in hertz) increases from left to right, and wavelength (in meters) increases from right to left. Scientists often disagree on the "correct" way to display wavelengths and frequencies in diagrams of this type, and it is common for astronomers working in different parts of the spectrum to adopt opposite conventions. This book consistently adheres to the convention that frequency increases toward the *right*. Notice that the wavelength and frequency scales in Figure 2.9 do not increase by equal increments of 10. Instead, successive values marked on the horizontal axis differ by factors of 10—each successive value is 10 times greater than its neighbor. This type of scale (called a *logarithmic* scale) is often used in science in order to condense a very large range of some quantity into a manageable size. Throughout the text we will often find it convenient to use such a scale in order to compress a wide range of some quantity onto a single easy-to-view plot.

Figure 2.9 shows wavelengths extending from the height of mountains for radio radiation to the diameter of an atomic nucleus for gamma-ray radiation. The box at the upper right emphasizes how small the visible portion of the electromagnetic spectrum is. Most objects in the universe emit large amounts of invisible radiation. Indeed, many objects emit only a tiny fraction of their total energy in the visible range. A wealth of extra knowledge can be gained by studying the invisible regions of the electromagnetic spectrum. To remind you of this important fact and to identify the region of the electromagnetic spectrum in which a particular observation was made, we have attached a spectrum icon—an idealized version of the wavelength scale in Figure 2.9—to every astronomical image presented in this text.

Only a small fraction of the radiation arriving at our planet actually reaches Earth's surface because of the *opacity* of Earth's atmosphere. **Opacity** is the extent to which radiation is blocked by the material through which it is passing—in this case, air. The more opaque an object is, the less radiation gets through. For example, a pane of glass has low opacity (i.e., is *transparent*) to visible light, while a sheet of paper or the air on a foggy day has high opacity. Earth's atmospheric opacity is plotted along the wavelength and frequency scales at the bottom of Figure 2.9. Where the shading is greatest, no radiation can get in or out—the energy is completely absorbed by atmospheric gases. Where there is no shading at all, our atmosphere is almost totally transparent.

Note that there are just a few *windows* (the unshaded regions) at well-defined locations in the electromagnetic spectrum, where Earth's atmosphere is transparent. In much of the radio and all of the visible portion of the spectrum, the opacity is low and we can study the universe at those wavelengths from ground level. In parts of the infrared range, the atmosphere is partially transparent, so we can make some infrared observations from the ground. Over the rest of the spectrum, however, the atmosphere is opaque. As a result, most infrared, and all ultraviolet, X-ray, and gamma-ray observations, can be made only from high-flying balloons or (more commonly) above the atmosphere, from orbiting satellites (see Section 3.5).

CONCEPT CHECK ▶▶▶

In what sense are radio waves, visible light, and X-rays one and the same phenomenon?

2.4 Thermal Radiation

All macroscopic objects—fires, ice cubes, people, stars—emit radiation at all times. They radiate because the microscopic charged particles in them are in constant random motion, and whenever charges change their state of motion, electromagnetic radiation is emitted. The **temperature** of an object is a direct measure of the amount of microscopic motion within it (see *More Precisely 2-1*). The hotter the object, the higher its temperature, the faster its constituent particles move and the more energy they radiate.

The Blackbody Spectrum

Intensity is a term often used to specify the amount or strength of radiation at any point in space. Like frequency and wavelength, intensity is a basic property of radiation. No natural object emits all of its radiation at just one frequency. Instead, the energy is often spread out over a range of frequencies. By studying the way in which the intensity of this radiation is distributed across the electromagnetic spectrum, we can learn much about the object's properties.

Figure 2.10 illustrates schematically the distribution of radiation emitted by any object. Note that the curve peaks at a single, well-defined frequency

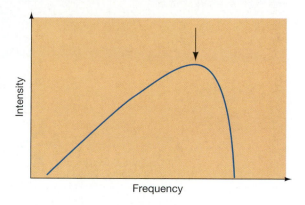

▲ **Figure 2.10 Ideal Blackbody Curve** The blackbody curve represents the spread of the intensity of the radiation emitted by any object over all possible frequencies. The arrow indicates the peak emission frequency.

2-1 MORE PRECISELY

The Kelvin Temperature Scale

The atoms and molecules that make up any piece of matter are in constant random motion. This motion represents a form of energy known as *thermal energy*. The quantity we call temperature is a direct measure of this internal motion: the higher an object's temperature, the faster the random motion of its constituent particles. More precisely, the temperature of a piece of matter specifies the average thermal energy of the particles it contains.

The temperature scale probably most familiar to you, the Fahrenheit scale, is now a peculiarity of American society. Most of the rest of the world uses the Celsius temperature scale, in which water freezes at 0 degrees (0°C) and boils at 100 degrees (100°C), as illustrated in the accompanying figure.

There are, of course, temperatures below the freezing point of water. Temperatures can in principle reach as low as −273.15°C (although we know of no matter anywhere in the universe that is actually this cold); this is the temperature at which, theoretically, all atomic and molecular motion effectively ceases. It is convenient to construct a temperature scale based on this lowest possible temperature, which is called *absolute zero*. Scientists commonly use such a scale, called the Kelvin scale in honor of the nineteenth-century British physicist Lord Kelvin. Since it takes absolute zero as its starting point, the Kelvin scale differs from the Celsius scale by 273.15°. In this book, we round off the decimal places and simply use the relationship

kelvins = degrees Celsius + 273.

Thus,

- Thermal motion of atoms and molecules ceases at 0 kelvins (0 K).
- Water freezes at 273 kelvins (273 K).
- Water boils at 373 kelvins (373 K).

Note, by the way, that the unit is "kelvins," or "K," not "degrees kelvin" or "°K."

(marked) and falls off to lesser values above and below that frequency. However, the curve is not symmetrical about the peak—the falloff is more rapid on the high-frequency side of the peak than it is toward lower frequencies. This overall shape is characteristic of the radiation emitted by *any* object, regardless of its size, shape, composition, or temperature.

The curve drawn in Figure 2.10 refers to a mathematical idealization known as a *blackbody*—an object that absorbs all radiation falling upon it. In a steady state (that is, if the temperature remains constant), a blackbody must reemit the same amount of energy as it absorbs. The **blackbody curve** shown in the figure describes the distribution of that reemitted radiation. No real object absorbs or radiates as a perfect blackbody. However, in many cases the blackbody curve is a very good approximation to reality, and the properties of blackbodies provide important insights into the behavior of real objects, including stars.

Tutorial

Continuous Spectra and Blackbody Radiation

The Radiation Laws

As illustrated in Figure 2.11, the entire blackbody curve shifts toward higher frequencies (shorter wavelengths) and greater intensities as an object's temperature increases. Even so, the *shape* of the curve remains the same. This shifting of radiation's peak frequency with temperature is familiar to us all. Very hot glowing objects, such as lightbulb filaments or stars, emit visible light because their blackbody curves peak in or near the visible range. Cooler objects, such as warm rocks or household radiators, produce invisible radiation—they are warm to the touch but are not glowing hot to the eye. These latter objects emit most of their radiation in the lower-frequency infrared portion of the electromagnetic spectrum.

There is a very simple connection between the frequency or wavelength at which most radiation is emitted and the absolute temperature (that is, temperature measured in kelvins—see *More Precisely 2-1*) of the emitting object: the peak frequency is *directly proportional* to the temperature. This relationship is more conventionally stated in terms of wavelength,

$$\text{wavelength of peak emission} \propto \frac{1}{\text{temperature}},$$

and is known as **Wien's law** (after Wilhelm Wien, the German scientist who formulated it in 1897). The symbol "$\propto$" just means "is proportional to." In essence, it tells us that the hotter the object, the bluer its radiation. For example, an object with a temperature of 6000 K (Figure 2.11c) emits most of its energy in the visible part of the spectrum, with a peak wavelength of 480 nm. At 600 K (Figure 2.11b), the object's emission would peak at 4800 nm, well into the infrared. At a temperature of 60,000 K (Figure 2.11d), the peak would move all the way through the visible range to a wavelength of 48 nm, in the ultraviolet.

It is also a matter of everyday experience that as the temperature of an object increases, the *total* amount of energy it radiates (summed over all frequencies) increases rapidly. For example, the heat given off by an electric heater increases sharply as the heater warms up and begins to emit visible light. In fact, the total amount of energy radiated per unit time is proportional to the *fourth power* of an object's temperature:

$$\text{total energy radiated per second} \propto \text{temperature}^4.$$

This relationship is called **Stefan's law**, after the nineteenth-century Austrian physicist Josef Stefan. It implies that the energy emitted by a body rises dramatically as the body's temperature increases. Doubling the temperature, for example, causes the total energy radiated to increase by a factor of 16. The radiation laws are presented in more detail in *More Precisely 2-2*.

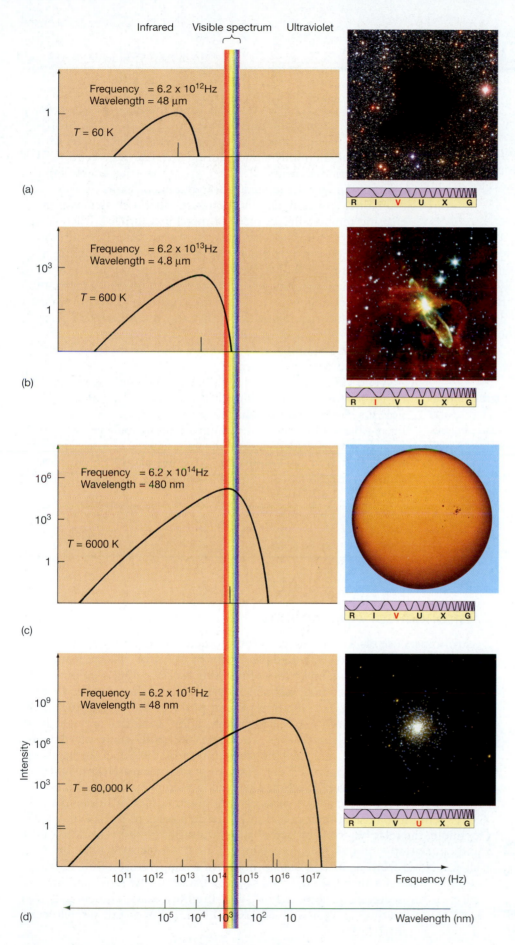

Infrared   Visible spectrum   Ultraviolet

(a)

Frequency   = 6.2 x 10^{12}Hz
Wavelength = 48 μm

T = 60 K

1

(b)

Frequency   = 6.2 x 10^{13}Hz
Wavelength = 4.8 μm

T = 600 K

10^3

1

(c)

Frequency   = 6.2 x 10^{14}Hz
Wavelength = 480 nm

T = 6000 K

10^6

10^3

1

(d)

Frequency   = 6.2 x 10^{15}Hz
Wavelength = 48 nm

T = 60,000 K

10^9

10^6

10^3

1

Intensity

10^{11} 10^{12} 10^{13} 10^{14} 10^{15} 10^{16} 10^{17} Frequency (Hz)

10^5 10^4 10^3 10^2 10   Wavelength (nm)

R I V U X G

◄ **Figure 2.11 Blackbody Curves**   Comparison of blackbody curves for four cosmic objects. (a) A cool, dark galactic cloud called Barnard 68. At a temperature of 60 K, it emits mostly radio radiation. (b) A dim, young star (shown white in the inset photograph) called Herbig-Haro 46. The star's atmosphere, at 600 K, radiates mainly in the infrared. (c) The Sun's surface, at approximately 6000 K, is brightest in the visible region of the electromagnetic spectrum. (d) Some very bright stars in a cluster called Messier 2, as observed by an orbiting space telescope above Earth's atmosphere. At a temperature of 60,000 K, these stars radiate strongly in the ultraviolet. (ESO; AURA; NASA/Spitzer/GALEX)

2-2   **MORE PRECISELY**

More on the Radiation Laws

Wien's law relates the temperature T of an object to the wavelength λ_{max} at which the object's blackbody radiation spectrum peaks. (The Greek letter λ—lambda—is conventionally used to denote wavelength.) Mathematically, if we measure T in kelvins, we find that

$$\lambda_{max} = \frac{0.29 \text{ cm}}{T}$$

Thus, at 6000 K (the approximate surface temperature of the Sun), the wavelength of maximum intensity is 0.29/6000 cm, or 480 nm, corresponding to the yellow-green part of the visible spectrum. A cooler star with a temperature of 3000 K has a peak wavelength of 970 nm, just longward of the red end of the visible spectrum, in the near infrared. The blackbody curve of a star with a temperature of 12000 K peaks at 242 nm, in the near ultraviolet, and so on.

We can also give Stefan's law a more precise mathematical formulation. With T again measured in kelvins, the total amount of energy emitted per square meter of surface per second (a quantity known as the *energy flux F*) is given by

$$F = \sigma T^4.$$

energy per unit area   temperature to the fourth power   constant

This equation is usually referred to as the *Stefan-Boltzmann* equation. (Stefan's student, Ludwig Boltzmann, was an Austrian physicist who played a central role in the development of the laws of thermodynamics during the late nineteenth and early twentieth centuries.) The constant σ (the Greek letter sigma) is known as the Stefan-Boltzmann constant.

Notice just how rapidly the energy flux increases with increasing temperature. A piece of metal in a furnace, when at a temperature of 3500 K, radiates energy at a rate of about 850 W for every square centimeter of its surface area. Doubling its temperature to 7000 K (so that it becomes yellow to white hot, by Wien's law) increases the energy emitted by a factor of 16 (four "doublings"), to 13.6 *kilo*watts (kW) (13,600 W) per square centimeter. Notice also that the law relates to energy emitted *per unit area*. The flame of a blowtorch is considerably hotter than a bonfire, but the bonfire emits far more energy *in total*, because it is much larger.

The SI unit of energy is the joule (J). Probably more familiar is the closely related unit called the watt (W), which measures power—the rate at which energy is emitted or expended by an object. One watt is the emission of one joule per second; for example, a 100-W lightbulb emits energy (mostly in the form of infrared and visible light) at a rate of 100 J/s. In these units, the Stefan-Boltzmann constant has the value $\sigma = 5.67 \times 10^{-8} \text{ W/m}^2 \cdot \text{K}^4$.

Astronomical Applications

Astronomers use blackbody curves as thermometers to determine the temperatures of distant objects. For example, study of the solar spectrum makes it possible to measure the temperature of the Sun's surface. Observations of the radiation from the Sun at many frequencies yield a curve shaped somewhat like that shown in Figure 2.10. The Sun's curve peaks in the visible part of the electromagnetic spectrum. The Sun also emits a lot of infrared and a little ultraviolet radiation. Applying Wien's law to the blackbody curve that best fits the solar spectrum, we find that the temperature of the Sun's surface is approximately 6000 K (Figure 2.11c). A more precise measurement, based on the detailed solar spectrum (Section 2.6), yields a temperature of 5800 K.

Other cosmic objects have surfaces very much cooler or hotter than the Sun's, emitting most of their radiation in invisible parts of the spectrum. For example, the relatively cool surface of a very young star might measure 600 K and emit mostly infrared radiation (Figure 2.11b). Cooler still is the interstellar gas cloud from which the star formed. At a temperature of 60 K, such a cloud would emit mainly long-wavelength radiation in the radio and infrared parts of the spectrum (Figure 2.11a). The brightest stars, by contrast, have surface temperatures as high as 60,000 K and hence emit mostly ultraviolet radiation (Figure 2.11d).

CONCEPT CHECK ▶▶▶

Describe, in terms of the radiation laws, how the appearance of an incandescent lightbulb changes as you turn a dimmer switch to increase its brightness from "off" to "maximum."

2.5 Spectroscopy

Radiation can be analyzed with an instrument known as a **spectroscope**. In its most basic form, this device consists of an opaque barrier with a slit in it (to form a narrow beam of light), a prism (to split the beam into its component colors), and either a detector or a screen (to allow the user to view the resulting spectrum). Figure 2.12 shows such an arrangement.

Emission Lines

The spectra encountered in the previous section are examples of **continuous spectra**. A lightbulb, for instance, emits radiation of all wavelengths (mostly in the visible and near-infrared ranges), with an intensity distribution that is well described by the blackbody curve corresponding to the bulb's temperature. Viewed through a spectroscope, the spectrum of the light from the bulb would show the familiar rainbow running from red to violet without interruption, as presented in Figure 2.12.

Not all spectra are continuous, however. For instance, if we took a glass jar containing pure hydrogen gas and passed an electrical discharge through it (a little like a lightning bolt arcing through Earth's atmosphere), the gas would begin to glow—that is, it would emit radiation. If we were to examine that radiation with our spectroscope, we would find that its spectrum consisted of only a few bright lines on an otherwise dark background, quite unlike the continuous spectrum described for the lightbulb. Figure 2.13 shows this schematically (the lenses have been removed for clarity), and a more detailed rendering of the spectrum of hydrogen appears in the top panel of Figure 2.14. The light produced by the hydrogen in this experiment does *not* consist of all possible colors but instead includes only a few narrow, well-defined **emission lines**, narrow "slices" of the continuous spectrum. The black background represents all the wavelengths *not* emitted by hydrogen.

After some experimentation, we would also find that although we could alter the intensity of the lines (for example, by changing the amount of hydrogen in the jar or the strength of the electrical discharge), we could not alter their color (in other words, their frequency or wavelength). The particular pattern of spectral emission lines shown in Figure 2.13 is a property of the element hydrogen—whenever we perform this experiment, the same characteristic **emission spectrum** is the result. Other elements yield different emission spectra. Depending on which element is involved, the pattern of lines can be fairly simple or very complex. Always, though, it is *unique* to that element. The emission spectrum of a gas thus provides a "fingerprint" that allows scientists to deduce its presence by spectroscopic means—in analogy with the bar codes that specify uniquely the cost of most supermarket

▲ **Figure 2.12 Spectroscope**   Diagram of a simple spectroscope. A slit in the barrier at left allows a narrow beam of light to pass. The beam continues through a prism and is split into its component colors. A lens then focuses the light into a sharp image that is either projected onto a screen, as shown here, or analyzed as it is strikes the detector.

Tutorial
Emission Spectra

► **Figure 2.13 Emission Spectrum**   Instead of a continuous spectrum, the light from excited (heated) hydrogen gas consists of a series of distinct spectral lines. (For simplicity, the focusing lenses have been omitted.) An analogy is a supermarket bar code that uniquely determines the cost of some product.

Hydrogen

Sodium

Helium

Neon

Mercury

Wavelength (nm)

▲ **Figure 2.14 Elemental Emission** The emission spectra of some well-known elements. Note that wavelengths shorter than approximately 400 nm, shown here in shades of purple, are actually in the ultraviolet part of the spectrum and are not visible. (Wabash Instrument Corporation)

products these days. Examples of the emission spectra of some common substances are shown in Figure 2.14.

Absorption Lines

When sunlight is split by a prism, at first glance it appears to produce a continuous spectrum. However, closer scrutiny shows that the solar spectrum is interrupted by a large number of narrow dark lines, as shown in Figure 2.15. These lines represent wavelengths of light that have been removed (absorbed) by gases present either in the outer layers of the Sun or in Earth's atmosphere. These gaps in the spectrum are called **absorption lines**. The absorption lines in the

◀ **Figure 2.15 Solar Spectrum** This visible spectrum of the Sun shows hundreds of vertical dark absorption lines superimposed on a bright continuous spectrum. The high-resolution spectrum is displayed in a series of 48 horizontal strips stacked vertically; each strip covers a small portion of the entire spectrum from left to right. If the strips were placed side by side, the full spectrum would be some 6 meters (20 feet) across! The scale extends from long wavelengths (red) at the upper left to short wavelengths (blue) at the lower right. (AURA)

solar spectrum are referred to collectively as *Fraunhofer lines*, after the nineteenth-century German physicist Joseph Fraunhofer, who measured and catalogued more than 600 of them.

At around the time solar absorption lines were discovered, scientists found that absorption lines could also be produced in the laboratory by passing a beam of light from a continuous source through a cool gas, as shown in Figure 2.16. They quickly observed an intriguing connection between emission and absorption lines: the absorption lines associated with a given gas occur at precisely the *same* wavelengths as the emission lines produced when the gas is heated. Both sets of lines therefore contain the *same* information about the composition of the gas.

The study of the ways in which matter emits and absorbs radiation is called **spectroscopy**. The observed relationships between the three types of spectra—continuous, emission line, and absorption line—are illustrated in Figure 2.17 and may be summarized as follows:

1. A luminous solid or liquid, or a sufficiently dense gas, emits light of all wavelengths and so produces a continuous spectrum of radiation (Figure 2.12).
2. A low-density hot gas emits light whose spectrum consists of a series of bright emission lines. These lines are characteristic of the chemical composition of the gas (Figure 2.13).
3. A low-density cool gas absorbs certain wavelengths from a continuous spectrum, leaving dark absorption lines in their place, superimposed on the continuous spectrum. These lines are characteristic of the composition of the intervening gas. They occur at precisely the same wavelengths as the emission lines produced by the gas at higher temperatures (Figure 2.16). These rules are collectively known as **Kirchhoff's laws**, after the German physicist Gustav Kirchhoff, who published them in 1859.

Astronomical Applications

Once astronomers realized that spectral lines are indicators of chemical composition, they set about identifying the observed lines in the Sun's spectrum. Almost all of the lines observed in the light received from extraterrestrial sources could be attributed to known elements (for example, many of the Fraunhofer lines in sunlight are associated with the element iron). However, some new lines also appeared in the solar spectrum. In 1868 astronomers realized that those lines must correspond to a previously unknown element. It was given the name

▲ **Figure 2.16 Absorption Spectrum** When a cool gas is placed between a source of continuous radiation (a lightbulb) and the detector/screen, the resulting color spectrum is crossed by a series of dark absorption lines. These lines are formed when the cool gas absorbs certain wavelengths (colors) from the original beam of light. The absorption lines appear at precisely the same wavelengths as the emission lines that would be produced if the gas were heated to high temperatures (see Figure 2.13).

Tutorial
Absorption Spectra

▶ **Figure 2.17 Kirchhoff's Laws** A source of continuous radiation, here represented by a lightbulb, is used to illustrate Kirchhoff's laws of spectroscopy. (a) The unimpeded beam shows the familiar continuous spectrum of colors. (b) When the source is viewed through a cloud of hydrogen gas, a series of dark hydrogen absorption lines appears in the continuous spectrum. These lines are formed when the gas absorbs some of the bulb's radiation and reemits it in random directions. Because most of the reemitted radiation does not go through the slit, the effect is to remove the absorbed radiation from the light that reaches the screen at the left. (c) When the gas is viewed from the side, a fainter hydrogen emission spectrum is seen, consisting of reemitted radiation. The absorption lines in (b) and the emission lines in (c) have the same wavelengths.

helium, after the Greek word *helios*, meaning "Sun." Only in 1895, almost three decades after its detection in sunlight, was helium discovered on Earth.

The development of spectroscopy is another example of the scientific method in action. ∞ *(Sec. E.5)* As technology evolved and experimental measurement techniques improved, scientists realized that spectra were unique fingerprints of matter. They then went on to use this knowledge as a means of determining the composition of objects that were otherwise impossible to reach. Yet for all the information that nineteenth-century astronomers could extract from observations of stellar spectra, they still lacked a theory explaining *how* those spectra arose. Despite their sophisticated spectroscopic equipment, they knew scarcely any more about the physics of stars than did Galileo or Newton. The next step in the circle of scientific progress was an explanation of Kirchhoff's empirical laws in terms of much more fundamental physical principles, just as Newtonian mechanics explained Kepler's empirical laws of planetary motion. And just as with Newton's laws, the new theory—now called *quantum mechanics*—opened the door to an explosion of scientific understanding far beyond the context in which it was originally conceived.

To understand how spectroscopy can be used to extract detailed information about astronomical objects from the light they emit, we must delve more deeply into the processes that produce line spectra.

2.6 The Formation of Spectral Lines

By the start of the twentieth century, physicists had accumulated substantial evidence that light sometimes behaves in a manner that simply cannot be explained by the wave theory of radiation. As we have just seen, the production of absorption and emission lines involves only certain, very specific wavelengths of light. This result would not be expected if light behaved only as a continuous wave and matter always obeyed the laws of Newtonian mechanics. It became clear that when light interacts with matter on very small scales, it does so not in a smooth, continuous way but in a discontinuous, stepwise manner. The challenge was to find an explanation for this unexpected behavior. The solution revolutionized our view of nature and now forms the foundation not just for physics and astronomy but for virtually all of modern science.

Atomic Structure

To explain the formation of spectral lines, we must understand not just the nature of light but also something of the structure of **atoms**—the microscopic building blocks from which all matter is constructed. Let us start with the simplest atom, hydrogen, which consists of an electron, with a negative electrical charge, orbiting a proton, which carries a positive charge. The proton forms the central **nucleus** (plural: nuclei) of the atom. Because the positive charge on the proton exactly cancels the negative charge on the electron, the hydrogen atom as a whole is electrically neutral.

How does this picture of the hydrogen atom relate to the characteristic emission and absorption lines associated with hydrogen gas? If an atom absorbs some energy in the form of radiation, that energy must cause some internal change. And if the atom emits energy, it must come from somewhere within the atom. The energy absorbed or emitted by the atom is associated with changes in the motion of the orbiting electron.

The first theory of the atom to provide an explanation of hydrogen's observed spectral lines was proposed by the Danish physicist Niels Bohr. It is now known simply as the **Bohr model** of the atom, and its essential features are as follows. First, there is a state of lowest energy—the **ground state**—which represents the "normal" condition of the electron as it orbits the nucleus. Second,

CONCEPT CHECK ▶▶▶

What are absorption and emission lines? What do they tell us about the properties of the gas producing them?

there is a maximum energy that the electron can have and still be part of the atom. Once the electron acquires more than that maximum energy, it is no longer bound to the nucleus, and the atom is said to be *ionized*. An atom having fewer (or more) than its normal complement of electrons, and hence a net electrical charge, is called an **ion**. Third, and most important (and also least intuitive), between those two energy levels, the electron can exist only in certain sharply defined energy states, often referred to as *orbitals*.

An atom is said to be in an **excited state** when an electron occupies an orbital other than the ground state. The electron then lies at a greater than normal distance from its parent nucleus, and the atom has a greater than normal amount of energy. The excited state with the lowest energy (that is, the one closest to the ground state) is called the *first excited state*, that with the second-lowest energy the *second excited state*, and so on.

In Bohr's model each electron orbital was pictured as having a specific radius, much like a planetary orbit in the solar system, as shown in Figure 2.18. However, the modern view is not so simple. Although each orbital *does* have a precise energy, the electron is now envisioned as being smeared out in an *electron cloud* surrounding the nucleus, as illustrated in Figure 2.19. It is common to speak of the average distance from the cloud to the nucleus as the radius of the electron's orbital. When a hydrogen atom is in its ground state, the radius of the orbital is about 0.05 nm (0.5 Å). As the orbital energy increases, the radius increases, too. For the sake of clarity in the diagrams that follow, we represent electron orbitals by solid lines, but bear in mind always that the fuzziness shown in part (b) of the figure is a more accurate depiction of reality.

An atom can become excited by absorbing some light energy from a source of electromagnetic radiation or by colliding with some other particle—another atom, for example. However, the atom cannot stay in that state forever. After about 10^{-8} s (second), it must return to its ground state.

The Particle Nature of Radiation

Here now is the crucial point that links atoms to radiation and allows us to interpret atomic spectra. Because electrons may exist only in orbitals having specific energies, atoms can absorb only specific amounts of energy as their electrons are boosted into excited states. Likewise, atoms can emit only specific amounts of energy as their electrons fall back to lower energy states. Thus, the amount of light energy absorbed or emitted in these processes *must correspond precisely to the energy difference between two orbitals*. This requires that light must be absorbed and emitted in the form of little "packets" of electromagnetic radiation, each carrying a very specific amount of energy. We call these packets **photons**. A photon is, in effect, a "particle" of electromagnetic radiation.

The idea that light sometimes behaves not as a continuous wave but as a stream of particles was first suggested by Albert Einstein in 1905. By that time, the experimental evidence clearly indicated that, on microscopic scales, electromagnetic radiation often displayed particle properties. Scientists understood that the notion of radiation as a wave was incomplete, but they did not know how to reconcile its two seemingly contradictory natures. Einstein's great breakthrough (for which he won the 1919 Nobel prize) came when he realized that all of the puzzling experiments could be explained by a simple, but critically important, connection between the particle and wave aspects of light. He found that the energy contained within a photon had to be proportional to the frequency of the radiation:

$$\text{photon energy} \propto \text{radiation frequency}.$$

Thus, for example, a "red" photon having a frequency of 4×10^{14} Hz (corresponding to a wavelength of about 750 nm, or 7500 Å) has 4/7 the energy of a "blue" photon, of frequency of 7×10^{14} Hz. Because it connects the

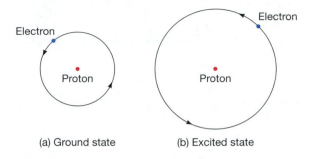

▲ **Figure 2.18 Classical Atom**   An early conception of the hydrogen atom—the Bohr model—pictured its electron orbiting the central proton in a well-defined orbital, like a planet orbiting the Sun. Two electron orbitals of different energies are shown: (a) the ground state and (b) an excited state.

▲ **Figure 2.19 Modern Atom**   The modern view of the hydrogen atom sees the electron as a "cloud" surrounding the nucleus. The same two energy states are shown as in Figure 2.18.

energy of a photon with the *color* of the light it represents, this relationship is the final piece in the puzzle of how to understand the spectra we see.

Environmental conditions ultimately determine which description—wave or stream of particles—better fits the behavior of electromagnetic radiation. As a general rule of thumb, in the macroscopic realm of everyday experience, radiation is more usefully described as a wave, and in the microscopic domain of atoms, it is best characterized as a stream of particles.

The Spectrum of Hydrogen

Absorption and emission of photons by a hydrogen atom are illustrated in Figure 2.20. Figure 2.20(a) shows the atom absorbing a photon of radiation and making a transition from the ground state to the first excited state, then emitting a photon of precisely the same energy and dropping back to the ground state. The energy difference between the two states corresponds to an ultraviolet photon, of wavelength 121.6 nm (1216 Å).

Figure 2.20(b) depicts the absorption of a more energetic (higher frequency, shorter wavelength) ultraviolet photon, this one having a wavelength

Animation/Video
Classical Hydrogen Atom I

Animation/Video
Classical Hydrogen Atom II

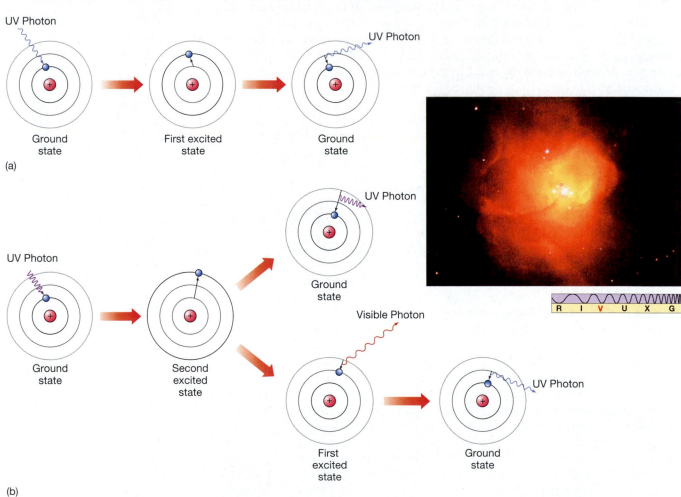

▲ **Figure 2.20 Atomic Excitation** (a) Absorption of an ultraviolet photon (left) by a hydrogen atom causes the momentary excitation of the atom into its first excited state (center). Eventually, the atom returns to its ground state (right), in the process emitting a photon having the same energy as the original photon. (b) Absorption of a higher-energy ultraviolet photon may boost the atom into a higher excited state, from which there are several possible paths back to the ground state. At the top, the electron falls immediately back to the ground state, emitting a photon identical to the one it absorbed. At the bottom, the electron first falls into the first excited state, producing visible radiation of wavelength 656.3 nm—the characteristic red glow of excited hydrogen. The object shown in the inset, designated N81, is an emission nebula: an interstellar cloud consisting largely of hydrogen gas excited by absorbing radiation emitted by some extremely hot stars (the white areas near the center). (Inset: NASA)

of 102.6 nm (1026 Å), causing the atom to jump to the *second* excited state. From that state the electron may return to the ground state via either one of two alternate paths.

1. It can proceed directly back to the ground state, in the process emitting an ultraviolet 102.6-nm photon identical to the one that excited the atom in the first place.

2. Alternatively, it can *cascade* down one orbital at a time, emitting *two* photons: one having an energy equal to the difference between the second and first excited states and the other having an energy equal to the difference between the first excited state and the ground state.

The second step in the cascade produces a 121.6-nm ultraviolet photon, just as in Figure 2.20(a). However, the first part of the cascade—the one from the second to the first excited state—produces a photon of wavelength 656.3 nm (6563 Å), which is in the visible part of the electromagnetic spectrum. This photon is seen as red light. An individual atom—if one could be isolated—would emit a momentary red flash. The inset in Figure 2.20 shows an astronomical object whose red coloration is the result of precisely this process.

   Absorption of still more energy can boost the electron to even higher orbitals within the atom. As the excited electron cascades back down to the ground state, the atom may emit many photons, each with a different energy and hence a different color, and the resulting spectrum shows many distinct spectral lines. In the case of hydrogen, all transitions ending at the ground state produce ultraviolet photons. However, downward transitions ending at the *first* excited state give rise to spectral lines in or near the visible portion of the electromagnetic spectrum (Figure 2.14). Because they form the most easily observable part of the hydrogen spectrum and were the first to be discovered, these lines (also known as *Balmer lines*) are often referred to simply as the "hydrogen series" and are denoted by the letter H. Individual transitions are labeled with Greek letters, in order of increasing energy (decreasing wavelength): the Hα (H alpha) line corresponds to the transition from the second to the first excited state and has a wavelength of 656.3 nm (red); Hβ (H beta; third to first) has wavelength 486.1 nm (green); Hγ (H gamma; fourth to first) has wavelength 434.1 nm (blue); and so on. We will use these designations (especially Hα and Hβ) frequently in later chapters.

Kirchhoff's Laws Explained

Let's reconsider our earlier discussion of emission and absorption lines in terms of the model just presented. In Figure 2.17(b) a beam of continuous radiation shines through a cloud of cool gas. The beam contains photons of all energies, but most of them cannot interact with the gas because the gas can absorb only photons having precisely the right energy to cause an electron to jump from one orbital to another. Photons having energies that cannot produce such a jump do not interact with the gas at all. They pass through it unhindered. Photons having the right energies are absorbed, excite the gas, and are removed from the beam. This is the cause of the dark absorption lines in the spectrum. These lines are direct indicators of the energy differences between orbitals in the atoms making up the gas.

   The excited gas atoms rapidly return to their original states, each emitting one or more photons in the process. Most of these reemitted photons leave at angles that do *not* take them through the slit and on to the detector. A second detector looking at the cloud from the side (Figure 2.17c) would record the reemitted energy as an emission spectrum. (This is what we are seeing in the inset to Figure 2.20.) Like the absorption spectrum, the emission spectrum is characteristic of the gas, not of the original beam.

◄◄◄ **CONCEPT CHECK**

In what ways do electron orbits in the Bohr atom differ from planetary orbits around the Sun?

(a)

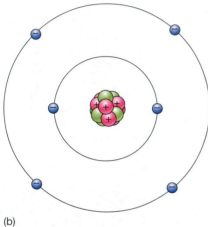

(b)

▲ **Figure 2.21 Helium and Carbon** (a) A helium atom in its ground state. Two electrons occupy the lowest-energy orbital around a nucleus containing two protons and two neutrons. (b) A carbon atom in its ground state. Six electrons orbit a six-proton, six-neutron nucleus, two of the electrons in an inner orbital, the other four at a greater distance from the center.

CONCEPT CHECK ▶▶▶

How does the structure of an atom determine the atom's emission and absorption spectra?

More Complex Spectra

All hydrogen atoms have the same structure—a single electron orbiting a single proton—but, of course, there are many other kinds of atoms, each having a unique internal structure. The number of protons in the nucleus of an atom determines the **element** that the atom represents. That is, just as all hydrogen atoms have a single proton, all oxygen atoms have eight protons, all iron atoms have 26 protons, and so on.

The next simplest element after hydrogen is helium. The central nucleus of the most common form of helium is made up of two protons and two **neutrons** (another kind of elementary particle having a mass slightly larger than that of a proton but carrying no electrical charge). About this nucleus orbit two electrons. As with hydrogen and all other atoms, the "normal" condition for helium is to be electrically neutral, with the negative charge of the orbiting electrons exactly canceling the positive charge of the nucleus (Figure 2.21a).

More complex atoms contain more protons (and neutrons) in the nucleus and have correspondingly more orbiting electrons. For example, an atom of carbon (Figure 2.21b) consists of six electrons orbiting a nucleus containing six protons and six neutrons. As we progress to heavier and heavier elements, the number of orbiting electrons increases, and consequently the number of possible electronic transitions rises rapidly. The number increases further as the temperature increases and atoms become excited or even ionized. The result is that very complicated spectra can be produced. The complexity of atomic spectra generally reflects the complexity of the source atoms. A good example is the element iron, which contributes several hundred of the Fraunhofer absorption lines seen in the solar spectrum. The many possible transitions of its 26 orbiting electrons yield an extremely rich line spectrum.

Even more complex spectra are produced by **molecules**. A molecule is a tightly bound group of atoms held together by interactions among their orbiting electrons—interactions called *chemical bonds*. Much like atoms, molecules can exist only in certain well-defined energy states, and again like atoms, molecules produce emission or absorption spectral lines when they make a transition from one state to another. Because molecules are more complex than atoms, the rules of molecular physics are also much more complex. Nevertheless, as with atomic spectral lines, painstaking experimental work over many decades has determined the precise frequencies at which millions of molecules emit and absorb radiation. These lines are molecular fingerprints, just like their atomic counterparts, enabling researchers to identify and study one kind of molecule to the exclusion of all others.

Molecular lines usually bear little resemblance to the spectral lines associated with their component atoms. For example, Figure 2.22(a) shows the emission spectrum of the simplest molecule known—molecular hydrogen. Notice how different it is from the spectrum of atomic hydrogen shown in part (b).

| 650 | 600 | 550 | 500 | 450 | 400 | 350 |

Wavelength (nm)

▲ **Figure 2.22 Hydrogen Spectra** The emission spectra of (a) molecular hydrogen and (b) atomic hydrogen. (Bausch & Lomb Inc.)

2.7 The Doppler Effect

Most of us have had the experience of hearing the pitch of a train whistle change from high shrill (high frequency, short wavelength) to low blare (low frequency, long wavelength) as the train approaches and then recedes. This motion-induced change in the observed frequency of a wave is known as the *Doppler effect*, in honor of Christian Doppler, the nineteenth-century Austrian physicist who first explained it. Applied to cosmic sources of electromagnetic radiation, it has become one of the most important observational tools in all of twentieth-century astronomy. Here's how it works.

Imagine a wave moving from the place where it is generated toward an observer who is not moving with respect to the wave source (Figure 2.23a). By noting the distances between successive wave crests, the observer can determine the wavelength of the emitted wave. Now suppose that the wave source begins to move (Figure 2.23b). Because the source moves between the times of emission of one wave crest and the next, wave crests in the direction of motion of the source are seen to be *closer together* than normal, while crests behind the source are more widely spaced. Thus, an observer in front of the source measures a *shorter* wavelength than normal, while one behind sees a *longer* wavelength.

The greater the relative speed of source and observer, the greater the observed shift. In terms of the net velocity of *recession* between source and observer, the apparent wavelength and frequency (measured by the observer) are related to the true quantities (emitted by the source) by

$$\frac{\text{apparent wavelength}}{\text{true wavelength}} = \frac{\text{true frequency}}{\text{apparent frequency}}$$

$$= 1 + \frac{\text{recession velocity}}{\text{wave speed}}.$$

A positive recession velocity means that the source and the observer are moving apart; a negative value means that they are approaching. The wave speed is the speed of light c in the case of electromagnetic radiation. For most of this text, we can assume that the recession velocity is small compared to the speed of light. Only when we come to discuss the properties of black holes (Chapter 13) and the structure of the universe on the largest scales (Chapter 17) will we have to reconsider this formula.

Note that, in the figure, the source is shown in motion. However, the same general statements hold whenever there is any *relative* motion between source and observer. Note also that only motion

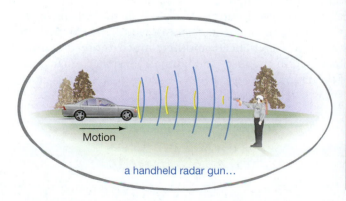

Motion

a handheld radar gun...

▼ **Figure 2.23 Doppler Effect** (a) Wave motion from a source toward an observer at rest with respect to the source. The four numbered circles represent successive wave crests emitted by the source. At the instant shown, the fifth wave crest is just about to be emitted. As seen by the observer, the source is not moving, so the wave crests are just concentric spheres (shown here as circles). (b) Waves from a moving source tend to "pile up" in the direction of motion and be "stretched out" on the other side. (The numbered points indicate the location of the source at the instant each wave crest was emitted.) As a result, an observer situated in front of the source measures a shorter-than-normal wavelength—a blueshift—while an observer behind the source sees a redshift. The inset shows a more everyday application—the hand-held radar gun used by police departments around the world to catch speeders. Here, reflected radiation (blue crests) from the oncoming car is shifted to shorter wavelengths by an amount proportional to the car's speed.

Recession 300 km/s — Hα at 657.0 nm · Hβ at 486.6 nm · Hγ at 434.5 nm

At rest 0 km/s — Hα at 656.3 nm · Hβ at 486.1 nm · Hγ at 434.1 nm

Approach 600 km/s — Hα at 655.0 nm · Hβ at 485.1 nm · Hγ at 433.3 nm

Wavelength (nm): 650 600 550 500 450 400

▲ **Figure 2.24 Doppler Shift** Because of the Doppler effect, the entire spectrum of a moving object is shifted to higher or lower frequencies, allowing specific spectral lines to be recognized and their shifts measured. The spectrum at the top shows the slight redshift of the hydrogen lines from an object moving at a speed of 300 km/s away from the observer; that at the bottom shows those same lines blueshifted while the object moves at double that speed toward the observer. The spectrum in the middle shows the unshifted spectrum of the object at rest.

Tutorial
Doppler Effect

CONCEPT CHECK ▶▶▶

How might the Doppler effect be used in determining the mass of a distant star?

along the line joining source and observer, known as *radial* motion, appears in the above equation. Motion *transverse* (perpendicular) to the line of sight has no significant effect.

In astronomical jargon, the wave measured by an observer situated in front of a moving source is said to be *blueshifted*, because blue light has a shorter wavelength than red light. Similarly, an observer situated behind the source will measure a longer-than-normal wavelength—the radiation is said to be *redshifted*. This terminology holds even for invisible radiation, for which "red" and "blue" have no meaning. Any shift toward shorter wavelengths is called a blueshift, and any shift toward longer wavelengths is called a redshift.

The Doppler effect is not normally noticeable in visible light on Earth—the speed of light is so large that the wavelength change is far too small to be noticeable for everyday terrestrial velocities. However, using spectroscopic techniques, astronomers routinely use the Doppler effect to measure the line-of-sight velocity of cosmic objects by determining the extent to which known spectral lines are shifted to longer or shorter wavelengths. For example, suppose that an astronomer observes the red Hα line in the spectrum of a star to have a wavelength of 657 nm instead of the 656.3 nm measured in the lab. (How does she know it is the same line? Because, as illustrated in Figure 2.24, she realizes that *all* the hydrogen lines are shifted by the same fractional amount—the characteristic pattern of lines identifies hydrogen as the source.) Using the above equation, she calculates that the star's radial velocity is $657/656.3 - 1 = 0.0056$ times the speed of light. In other words, the star is receding from Earth at a rate of 320 km/s.

The motions of nearby stars and distant galaxies—even the expansion of the universe itself—have all been measured in this way. Motorists stopped for speeding on the highway have experienced another, much more down-to-earth, application. As illustrated in the inset to Figure 2.23, police radar measures speed by means of the Doppler effect, as do the radar guns used to clock the velocity of a pitcher's fastball or a tennis player's serve.

2.8 Spectral-Line Analysis

Astronomers apply the laws of spectroscopy in analyzing radiation from beyond Earth. A nearby star or a distant galaxy takes the place of the lightbulb in our previous examples, an interstellar cloud or a stellar (or even planetary) atmosphere plays the role of the intervening cool gas, and a spectrograph attached to a telescope replaces our simple prism and detector. We list below a few of the properties of emitters and absorbers that can be determined by careful analysis of radiation received on (or near) Earth. We will encounter other important examples as our study of the cosmos unfolds.

1. The *composition* of an object is determined by matching its spectral lines with the laboratory spectra of known atoms and molecules.
2. The *temperature* of an object emitting a continuous spectrum can be measured by matching the overall distribution of radiation with a blackbody curve. Temperature may also be determined from detailed studies of spectral lines. In Section 10.3 we will see how stellar temperatures are most accurately measured by spectroscopic means.
3. The (*line-of-sight*) *velocity* of an object is measured by determining the Doppler shift of its spectral lines.
4. An object's *rotation rate* can be determined by measuring the *broadening* (smearing out over a range of wavelengths) produced by the Doppler effect in emitted or reflected spectral lines.
5. The *pressure* of the gas in the emitting region of an object can be measured by its tendency to broaden spectral lines. The greater the pressure, the broader the line.
6. The *magnetic field* of an object can be inferred from a characteristic splitting it produces in many spectral lines, when a single line divides into two. (This is known as the *Zeeman effect*.)

Given sufficiently sensitive equipment, there is almost no end to the wealth of data contained in starlight. However, deciphering the extent to which each of many competing factors influences a spectrum can be a very difficult task. Typically, the spectra of many elements are superimposed on one another, and several physical processes are occurring simultaneously, each modifying the spectrum in its own way. The challenge facing astronomers is to unravel the extent to which each mechanism contributes to spectral-line profiles and therefore obtain meaningful information about the source of the lines.

◄◄◄ **CONCEPT CHECK**

Why is it important for astronomers to analyze spectral lines in detail?

CHAPTER REVIEW

Summary

1 **Electromagnetic radiation** (p. 44) travels through space in the form of a **wave** (p. 44). Any electrically charged object is surrounded by an **electric field** (p. 44) that determines the force it exerts on other charged objects. When a charged particle moves, information about that motion is transmitted throughout the universe by the particle's changing **electric** and **magnetic fields** (pp. 46, 47). The information travels at the **speed of light** (p. 47) in the form of an electromagnetic wave.

2 The **electromagnetic spectrum** (p. 50) consists of (in order of increasing frequency) **radio waves, infrared radiation, visible light, ultraviolet radiation, X-rays,** and **gamma rays** (p. 44). The **opacity** (p. 50) of Earth's atmosphere—the extent to which it absorbs radiation—varies greatly with the wavelength of the radiation. Only radio waves, some infrared wavelengths, and visible light can penetrate the atmosphere and reach the ground.

3 The **temperature** (p. 51) of an object is a measure of the speed with which its constituent particles move. The **intensity** (p. 51) of radiation emitted by an object has a characteristic distribution, called a **blackbody curve** (p. 52), that depends only on the temperature of the object. **Wien's law** (p. 52) tells us that the wavelength at which the object's radiation peaks is inversely proportional to its temperature. **Stefan's law** (p. 52) states that the total amount of energy radiated is proportional to the fourth power of the temperature.

4 Many hot objects emit a **continuous spectrum** (p. 55) of radiation, containing light of all wavelengths. A hot gas may instead produce an **emission spectrum** (p. 55), consisting only of a few well-defined **emission lines** (p. 55) of specific frequencies, or colors. Passing a continuous beam of radiation through cool gas will produce **absorption lines** (p. 56) at precisely the same frequencies as would be present in the gas's emission spectrum.

5 **Atoms** (p. 58) are made up of negatively charged electrons orbiting a positively charged **nucleus** (p. 58) consisting of positively charged protons and electrically neutral **neutrons** (p. 62). The number of protons in the nucleus determines the type of **element** (p. 62) the atom repre-

sents. In the **Bohr model** (p. 58) of the hydrogen atom, the atom has a minimum-energy **ground state** (p. 58), representing its "normal" condition. When the electron has a higher-than-normal energy, the atom is in an **excited state** (p. 59). For any given atom, only certain, well-defined energies are possible. In the modern view, the electron is envisaged as being spread out in a "cloud" around the nucleus but still having a sharply defined energy.

6 When an electron moves from one energy state to another in an atom, the difference in the energy between the states is emitted or absorbed in the form of "packets" of electromagnetic radiation—**photons** (p. 59). Because the energy levels have definite energies, the photons also have definite energies that are characteristic of the type of atom involved. The energy of a photon determines the frequency, and hence the color, of the light emitted or absorbed.

7 Our perception of the wavelength of a beam of light can be altered by source's velocity relative to us. This motion-induced change in the observed frequency of a wave is called the **Doppler effect** (p. 63). Any net motion of the source away from the observer causes a redshift—a shift to lower frequencies—in the received beam. Motion toward the observer causes a blueshift. The extent of the shift is directly proportional to the source's radial velocity relative to the observer.

Review and Discussion

1. Define the following wave properties: period, wavelength, amplitude, frequency.

2. Compare the gravitational force with the electric force.

3. Describe the way in which light radiation leaves a star, travels through the vacuum of space, and finally is seen by someone on Earth.

4. What do radio waves, infrared radiation, visible light, ultraviolet radiation, X-rays, and gamma rays have in common? How do they differ?

5. In what regions of the electromagnetic spectrum is the atmosphere transparent enough to allow observations from the ground?

6. What is a blackbody? What are the characteristics of the radiation it emits?

7. If Earth were completely blanketed with clouds and we couldn't see the sky, could we learn about the realm beyond the clouds?

What forms of radiation might penetrate the clouds and reach the ground?

8. Describe how its blackbody curve changes as a red-hot glowing coal cools off.

9. What is spectroscopy? Why is it so important to astronomers?

10. What is a continuous spectrum? An absorption spectrum?

11. What is the normal condition for atoms? What is an excited atom? What are orbitals?

12. Why do excited atoms absorb and reemit radiation at characteristic frequencies?

13. Suppose a luminous cloud of gas is discovered emitting an emission spectrum. What can be learned about the cloud from this observation?

14. What is the Doppler effect, and how does it alter the way in which we perceive radiation?

15. How do astronomers use the Doppler effect to determine the velocities of astronomical objects?

Conceptual Self-Test: True or False?/Multiple Choice

1. The wavelength of green light is about the size of an atom. **(T/F)**

2. Two otherwise identical objects have temperatures of 1000 K and 1200 K, respectively. The object at 1200 K emits roughly twice as much radiation as the object at 1000 K. **(T/F)**

3. As you drive away from a radio transmitter, the radio signal you receive from the station is shifted to longer wavelengths. **(T/F)**

4. Imagine an emission spectrum produced by a container of hydrogen gas. Changing the amount of hydrogen in the container will change the colors of the lines in the spectrum. **(T/F)**

5. In the previous question, changing the gas in the container from hydrogen to helium will change the colors of the lines occurring in the spectrum. **(T/F)**

6. The energy of a photon is inversely proportional to the wavelength of the radiation. **(T/F)**

7. An electron moves to a higher energy level in an atom after absorbing a photon of a specific energy. **(T/F)**

8. Compared with ultraviolet radiation, infrared radiation has a greater **(a)** wavelength; **(b)** amplitude; **(c)** frequency; **(d)** energy.

9. An X-ray telescope located in Antarctica would not work well because of **(a)** the extreme cold; **(b)** the ozone hole; **(c)** continuous daylight; **(d)** Earth's atmosphere.

10. A star much cooler than the Sun would appear **(a)** red; **(b)** blue; **(c)** smaller; **(d)** larger.

11. The blackbody curve of a star moving toward Earth would have its peak shifted **(a)** to lower intensity; **(b)** toward higher energies; **(c)** toward longer wavelengths; **(d)** toward lower energies.

12. The visible spectrum of sunlight reflected from Saturn's cold moon Titan would be expected to be **(a)** continuous; **(b)** an emission spectrum; **(c)** an absorption spectrum.

13. Astronomers analyze starlight to determine a star's **(a)** temperature; **(b)** composition; **(c)** motion; **(d)** all of the above.

14. According to Figure 2.11 (Blackbody Curves), an object having a temperature of 1000 K emits mostly **(a)** infrared light; **(b)** visible light; **(c)** ultraviolet light; **(d)** X-rays.

15. In Figure 2.20 (Atomic Excitation), the total energy of the two photons emitted in the branch at the lower right is **(a)** greater than, **(b)** less than, **(c)** approximately equal to, **(d)** exactly equal to, the energy of the UV photon absorbed at the left.

Problems

Algorithmic versions of these questions are available in the Practice Problems module of the Companion Website.

The number of squares preceding each problem indicates its approximate level of difficulty.

1. ■ A sound wave moving through water has a frequency of 256 Hz and a wavelength of 5.77 m. What is the speed of sound in water?

2. ■ What is the wavelength of a 100-MHz (FM 100) radio signal?

3. ■ What would be the frequency of an electromagnetic wave having a wavelength equal to Earth's diameter (12,800 km)? In what part of the electromagnetic spectrum would such a wave lie?

4. ■■ The blackbody emission spectrum of object A peaks in the ultraviolet region of the electromagnetic spectrum at a wavelength of 200 nm. That of object B peaks in the red region, at 650 nm. According to Wien's law, how many times hotter than B is A? According to Stefan's law, how many times more energy per unit area does A radiate per second?

5. ■ Normal human body temperature is about 37°C. What is this temperature in kelvins? What is the peak wavelength emitted by a person with this temperature? In what part of the spectrum does this lie?

6. ■■ According to the Stefan-Boltzmann law, how much energy is radiated into space per unit time by each square meter of the Sun's surface (see *More Precisely 2-2*)? If the Sun's radius is 696,000 km, what is the total power output of the Sun?

7. ■ By what factor does the energy of a 1-nm X-ray photon exceed that of a 10-MHz radio photon?

8. ■■ How many different photons (that is, photons of different frequencies) can be emitted as a hydrogen atom in the second excited state falls back, directly or indirectly, to the ground state? What are their wavelengths? What about a hydrogen atom in the third excited state?

9. ■ The Hα line (Sec. 2.6) of a certain star is received on Earth at a wavelength of 656 nm. What is the star's radial velocity with respect to Earth?

10. ■■■ Imagine you are observing a spacecraft moving in a circular orbit of radius 100,000 km around a distant planet. You happen to be located in the plane of the spacecraft's orbit. You find that the spacecraft's radio signal varies periodically in wavelength between 2.99964 m and 3.00036 m. Assuming that the radio is broadcasting normally, at a constant wavelength, what is the mass of the planet?

 The Companion Website at www.prenhall.com/chaisson provides algorithmically generated versions of each chapter's Problems, in addition to an Animations and Videos gallery, a Tutorials gallery and Tutorials Quiz questions, a full eBook, an Images gallery, Practice Questions, Multiple Choice, and True or False quizzes, and Labeling exercises.

3

Telescopes

The Tools of Astronomy

Telescopes aid our eyes, enabling us to see to much greater distances than are accessible to the naked eye alone. They also improve other human senses, enabling us to perceive cosmic objects that emit radiation at wavelengths to which we are blind. Without telescopes, we would know only a tiny fraction of the information presented in this textbook. Here, a 2-hour time exposure captures one of the world's premier optical telescopes available to astronomers today—the Gemini North Telescope atop Mauna Kea in Hawaii, looking out toward the Milky Way. (P. Michaud, Gemini, AURA)

A t its heart, astronomy is an observational science. Painstaking observations of cosmic phenomena almost always precede any clear theoretical understanding of their nature. As a result, our detecting instruments—our telescopes—have evolved to observe as broad a range of wavelengths as possible. Until the middle of the twentieth century, telescopes were limited to visible light. Since then, however, technological advances have expanded our view of the universe to all parts of the electromagnetic spectrum. Some telescopes are sited on Earth, others must be placed in space, and design considerations vary widely from one region of the spectrum to another. Whatever the details of its construction, however, the basic purpose of a telescope is to collect electromagnetic radiation and deliver it to a detector for detailed study.

LEARNING GOALS

Studying this chapter will enable you to:

1. Sketch and describe how an optical telescope works.

2. Explain why large telescopes are needed for astronomical studies, and specify the advantages of reflecting telescopes for astronomical use.

3. Describe how Earth's atmosphere limits astronomical observations, and discuss some techniques used by ground-based astronomers to overcome these limitations.

4. Discuss some advantages and disadvantages of radio astronomy.

5. Describe interferometry, and explain how it can enhance the usefulness of radio and other observations.

6. Discuss the design, use, and placement of infrared, ultraviolet, and high-energy telescopes.

7. Explain why it is important to make astronomical observations at many different wavelengths across the electromagnetic spectrum.

Visit www.prenhall.com/chaisson for additional annotated images, animations, and links to related sites for this chapter.

▲ **Figure 3.1 Reflecting Mirror** A curved mirror can be used to focus to a single point all rays of light arriving parallel to the mirror axis. Light rays traveling along the axis are reflected back along the axis, as indicated by the arrowheads pointing in both directions. Off-axis rays are reflected through greater and greater angles the farther they are from the axis, so that they all pass through the same point—the focus. Analogously to a bucket that collects only the rain falling into it, a telescope's mirror intercepts only that light falling onto it.

3.1 Optical Telescopes

Simply put, a **telescope** is a "light bucket"—a device whose primary function is to capture as much radiation as possible from a given region of the sky and concentrate it into a focused beam for analysis. We begin our study of astronomical hardware with *optical* telescopes, designed to collect wavelengths visible to the human eye. Later we will turn our attention to telescopes designed to capture and analyze radiation in other, *invisible* regions of the electromagnetic spectrum. As we will see throughout this text, advances in our understanding of the universe have always gone hand in hand with technological improvements in both the sensitivity and the spectral range of our detectors. This coupling between technology and science is as vital today as it was when Galileo first turned his telescope toward the skies. ∞ *(Sec. 1.2)*

Reflecting and Refracting Telescopes

Optical telescopes fall into two basic categories—*reflectors* and *refractors*. Figure 3.1 shows how a **reflecting telescope** uses a curved mirror to gather and concentrate a beam of light. This mirror, usually called the *primary mirror* because telescopes generally contain more than one, is constructed so that all light rays arriving parallel to its axis (the imaginary line through the center of and perpendicular to the mirror), regardless of their distance from that axis, are reflected to pass through a single point, called the *focus*. The distance between the primary mirror and the focus is the *focal length*. In astronomical contexts, the focus of the primary mirror is referred to as the **prime focus**.

A **refracting telescope** uses a lens instead of a mirror to focus the incoming light, relying on refraction rather than reflection to achieve its purpose. **Refraction** is the bending of a beam of light as it passes from one transparent medium (for example, air) into another (such as glass). Figure 3.2(a) illustrates the process and shows how a prism can be used to change the direction of a beam of light. As illustrated in Figure 3.2(b), we can think of a lens as a series of prisms combined in such a way that all light rays striking the lens parallel to the axis are refracted to pass through the focus.

Astronomical telescopes are often used to make **images** of their fields of view. Figure 3.3 illustrates how this is accomplished, in this case by the primary mirror in a reflecting telescope. Light from a distant object reaches the telescope as nearly parallel rays. A ray of light entering the instrument parallel to the mirror axis is reflected through the focus. Light from a slightly different direction—that is, at a slight angle to the axis—is focused to a slightly different point. In this way, an image is formed near the focus. Each point in the image

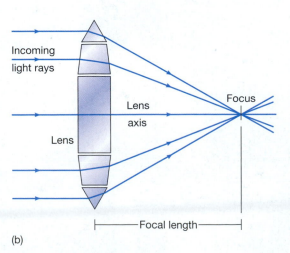

◀ **Figure 3.2 Refracting Lens** (a) Refraction by a prism changes the direction of a light ray by an amount that depends on the angle between the prism's faces. (b) A lens can be thought of as a series of prisms. A light ray traveling along the axis of a lens is unrefracted as it passes through the lens. Parallel rays arriving at progressively greater distances from the axis are refracted by increasing amounts, in such a way that all are focused to a single point.

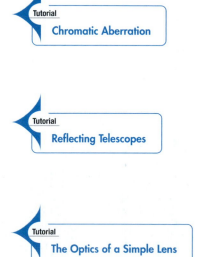

corresponds to a different angle in the field of view. Often, the image is magnified with a lens known as an *eyepiece* before being viewed by eye or recorded using a camera. Figure 3.4(a) shows the basic design of a simple reflecting telescope, illustrating how a small secondary mirror and an eyepiece are used to view the image. Figure 3.4(b) shows how a refracting telescope accomplishes the same function.

The two telescope designs shown in Figure 3.4 achieve the same result—light from a distant object is captured and focused to form an image. However, as telescope *size* has steadily increased over the years (for reasons to be discussed in Section 3.2), a number of important factors have tended to favor reflecting instruments over refractors:

1. Just as a prism separates white light into its component colors, the lens in a refracting telescope focuses red and blue light differently. This deficiency is known as *chromatic aberration*. Careful design and choice of materials can largely correct this problem, but it is difficult to eliminate it entirely, and it requires the use of very high-quality glass for the body of the lens. Mirrors do not suffer from this defect.

▲ **Figure 3.3 Image Formation**   An image can be formed by a mirror as rays of light coming from different points on a distant object are focused to slightly different locations. The result is an image of the object formed around the focus. Notice that the image is inverted (that is, upside down).

Tutorial
Chromatic Aberration

Tutorial
Reflecting Telescopes

Tutorial
The Optics of a Simple Lens

(a) Reflector

(b) Refractor

◀ **Figure 3.4 Reflectors and Refractors** Comparison of (a) reflecting and (b) refracting telescopes. Both types are used to gather and focus electromagnetic radiation to be observed by human eyes or recorded on photographs or in computers. In both cases, the image formed at the focus is viewed with a small magnifying lens called an eyepiece.

2. As light passes through the lens, some of it is absorbed by the glass. This absorption is a relatively minor problem for visible radiation, but it can be severe for infrared and ultraviolet observations because glass blocks most of the radiation coming from those regions of the electromagnetic spectrum. This problem obviously does not affect mirrors.

3. A large lens can be quite heavy. Because it can be supported only around its edge (so as not to block the incoming radiation), the lens tends to deform under its own weight. A mirror does not have this drawback because it can be supported over its entire back surface.

4. A lens has two surfaces that must be accurately machined and polished, which can be a difficult task. A mirror has only one surface.

For these reasons, *all* large modern optical telescopes are reflectors.

Telescope Design

Figure 3.5 shows some basic reflecting telescope designs. Radiation from a star enters the instrument, passes down the main tube, strikes the primary mirror, and is reflected back toward the prime focus, near the top of the tube. Sometimes astronomers place their recording instruments at the prime focus. However, it can be very inconvenient, or even impossible, to suspend bulky pieces of equipment there. More often, the light is intercepted on its path to the focus by a secondary mirror and redirected to a more convenient location, as in Figure 3.5 (b–d).

In a **Newtonian telescope** (named after Isaac Newton, who invented this design), the light is intercepted by a flat secondary mirror before it reaches the prime focus and deflected 90°, usually to an eyepiece at the side of the instrument (Figure 3.5b). This is a popular design for smaller reflecting telescopes, such as those used by amateur astronomers.

Alternatively, astronomers may choose to work on a rear platform where they can use detecting equipment too heavy or delicate to hoist to the prime focus. In this case, the light reflected by the primary mirror toward the prime focus is intercepted by a convex secondary mirror, which reflects the light back down the tube and through a small hole at the center of the primary mirror (Figure 3.5 c). This arrangement is known as a **Cassegrain telescope** (after Guillaume Cassegrain, a French lensmaker). The point behind the primary mirror where the light from the star finally converges is called the *Cassegrain focus*. A well-known example of a Cassegrain telescope is the *Hubble Space Telescope (HST; see Discovery 3-1)*, named for one of America's most notable astronomers, Edwin Hubble. The telescope's detectors all lie directly behind the primary mirror and are capable of making measurements in the optical, infrared, and ultraviolet parts of the spectrum.

A more complex observational configuration requires starlight to be reflected by several mirrors. As in the Cassegrain design, light is first reflected by the primary mirror toward the prime focus and is then reflected back down the tube by a secondary mirror. Next, a third, much smaller, mirror reflects the light

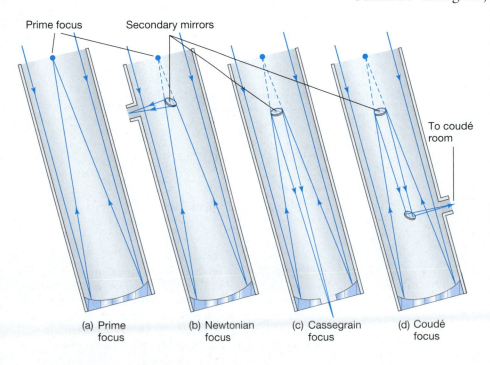

Prime focus   Secondary mirrors

To coudé room

(a) Prime focus   (b) Newtonian focus   (c) Cassegrain focus   (d) Coudé focus

◄ **Figure 3.5 Reflecting Telescopes** Four reflecting telescope designs: (a) prime focus, (b) Newtonian focus, (c) Cassegrain focus, and (d) coudé focus. Each design uses a primary mirror at the bottom of the telescope to capture radiation, which is then directed along different paths for analysis.

out of the telescope, where (depending on the details of the telescope's construction) the beam may be analyzed by a detector mounted alongside, at the *Nasmyth focus*, or it may be directed via a further series of mirrors into an environmentally controlled laboratory known as the *coudé* room (from the French word for "bent"). This laboratory is separate from the telescope itself, enabling astronomers to use very heavy and finely tuned equipment that cannot be placed at any of the other foci (all of which necessarily move with the telescope). The arrangement of mirrors is such that the light path to the coudé room does not change as the telescope tracks objects across the sky.

Figure 3.6(a) shows the twin 10-m-diameter optical/infrared telescopes of the Keck Observatory on Mauna Kea in Hawaii, operated jointly by the California Institute of Technology and the University of California. The diagram in part (b) illustrates the light paths and some of the foci. Observations may be made at the Cassegrain, Nasmyth, or coudé focus, depending on the needs of the user. As the size of the person in part (c) indicates, this is indeed a very large telescope—in fact, the two mirrors are currently among the very largest on Earth. We will see numerous examples throughout this text of Keck's many important discoveries.

Detectors and Image Processing

Computers play a vital role in observational astronomy. Most large telescopes today are controlled either by computers or by operators who rely heavily on computer assistance, and images and data are recorded in a form that can be easily read and manipulated by computer programs.

▼ **Figure 3.6 Keck Telescope** (a) The two 10-m telescopes of the Keck Observatory. (b) Artist's illustration of the telescope, the path taken by an incoming beam of starlight, and some of the locations where instruments may be placed. (c) One of the 10-m mirrors. Note the technician in orange coveralls at center. (W. M. Keck Observatory)

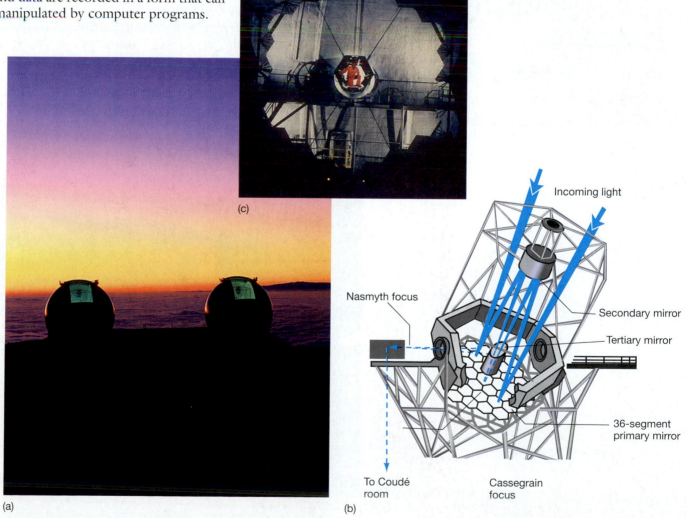

Incoming light

Nasmyth focus

Secondary mirror

Tertiary mirror

36-segment primary mirror

To Coudé room

Cassegrain focus

(a)   (b)   (c)

3-1 DISCOVERY

The Hubble Space Telescope

The *Hubble Space Telescope (HST)* is the largest, most complex, most sensitive observatory ever deployed in space. At over $7.8 billion (including the cost of several missions to service and refurbish the system), it is also the most expensive scientific instrument ever built and operated. Built jointly by NASA and the European Space Agency, *HST* was designed to allow astronomers to probe the universe with at least 10 times finer resolution and with some 30 times greater sensitivity to light than existing Earth-based devices. The first figure shows the telescope being lifted out of the cargo bay of the space shuttle *Discovery* in the spring of 1990.

(NASA)

The principal features of *HST* are displayed in the accompanying "see-through" illustration. The telescope's Cassegrain design (Section 3.1) reflects light from its 2.4-m-diameter primary mirror (the large bluish disk at center) back to a smaller, 0.3-m secondary mirror, which sends the light through a small opening in the center of the primary, from where it can be directed to any of five major instruments (shown here in various colors at left) arrayed in the aft bay of the spacecraft. The large red objects are sensors that guide the pointing of the telescope, and the huge blue panels collect light from the Sun to power everything onboard. For scale, each of the instruments is about the size of a telephone booth. They were designed to be maintained by NASA astronauts, and indeed most of the telescope's original instruments have been upgraded or replaced since *HST* was launched. The current detectors on the telescope span the visible, near-infrared, and near-ultraviolet regions of the electromagnetic spectrum, from about 100 nm (UV) to 2200 nm (IR).

Soon after launch, astronomers discovered that the telescope's primary mirror had been polished to the wrong shape. The mirror is too flat by 2 μm, about 1/50 the width of a human hair, making it impossible to focus light as well as expected. In 1993, astronauts aboard the space shuttle *Endeavour* visited *HST*. They replaced *Hubble*'s gyroscopes to help the telescope point more accurately, installed sturdier versions of the solar panels that power the telescope's electronics, and—most important—inserted an intricate set of small mirrors to compensate for the faulty primary mirror. *Hubble*'s resolution is now close to the original design specifications, and

(D. Berry)

the telescope has regained much of its lost sensitivity. Additional service missions were performed in 1997, 1999, and 2002 to replace instruments and repair faulty systems.

A good example of *Hubble*'s current scientific capabilities can be seen by comparing the two images of the spiral galaxy M100, shown here. On the left is perhaps the best ground-based photograph of this beautiful galaxy, showing rich detail and color in its spiral arms. On the right, to the same scale and orientation, is an *HST* image showing improvements in both resolution and sensitivity. (The chevron-shaped field of view is caused by the corrective optics inserted into the telescope in 1993; an additional trade-off is that *Hubble*'s field of view is smaller than those of ground-based telescopes.) The inset shows *Hubble*'s exquisite resolution of small fields of view.

During its decade and a half of operation, *Hubble* has literally revolutionized our view of the sky, helping to rewrite some theories of the universe along the way. By measuring the properties of stars and supernovae within distant galaxies, it has helped establish the size and the expansion rate of the cosmos and has provided insights into the past and future evolution of the universe. *Hubble* has studied newborn galaxies almost at the limit of the observable universe with unprecedented clarity, allowing astronomers to see the interactions and collisions that may have shaped the evolution of our own Milky Way. Turning its gaze to the hearts of galaxies closer to home, *Hubble* has provided strong evidence for supermassive black holes in their cores. Within our own Galaxy, it has given astronomers stunning new insights into the physics of star formation and the evolution of stellar systems and stars of all sizes, from superlu-

minous giants to objects barely more massive than planets. Finally, in our solar system, *Hubble* has given scientists new views both of the planets and their moons, and of the tiny fragments from which they formed long ago. Many spectacular examples of the telescope's remarkable capabilities appear throughout this book.

With *Hubble* now well into the second decade of a highly successful career, NASA scientists are planning the telescope's successor. The Next Generation Space Telescope (now known as the James Webb Space Telescope, or JWST, after the administrator who led NASA's *Apollo* program during the 1960s and 1970s), will dwarf *Hubble* in both scale and capability. Sporting a 6-m segmented mirror and containing a formidable array of detectors optimized for use at visible and infrared wavelengths, JWST will orbit at a distance of 1.5 million km from Earth, far beyond the Moon. NASA hopes to launch the instrument in 2011. The telescope's primary mission will be in understanding the formation of the first stars and galaxies, measuring the large-scale structure of the universe, and investigating the evolution of planets, stars, and galaxies.

In early 2004 NASA announced that, for budgetary and safety reasons, they had canceled the planned 2006 servicing mission that would have extended the telescope's lifetime. Not surprisingly, the decision is actively opposed by many astronomers. Without servicing, *Hubble* is expected to become inoperative by 2007.

(NASA)

(D. Malin/Anglo-Australian Telescope)

R I V U X G

It is now rare for photographic equipment to be used as the primary means of data acquisition at large observatories. Instead, electronic detectors known as **charge-coupled devices**, or **CCDs**, are in widespread use. A CCD (Figure 3.7) consists of a wafer of silicon divided into a two-dimensional array of many tiny picture elements, known as *pixels*. When light strikes a pixel, an electric charge builds up on the device. The amount of charge is directly proportional to the number of photons striking each pixel—in other words, to the intensity of the light at that point. The charge buildup is monitored electronically, and a two-dimensional image is obtained (see Figure 3.13c and d). A CCD is typically a few square centimeters in area and may contain several million pixels, generally arranged on a square grid. As technology continually improves, both the areas of CCDs and the number of pixels they contain are steadily increasing. Incidentally, the technology is not limited to astronomy—home video and digital cameras contain CCD chips similar in basic design to those in use at the great astronomical observatories of the world.

CCDs have two important advantages over photographic plates, which were the staple of astronomers for over a century. First, CCDs are much more *efficient* than photographic plates, recording as many as 75 percent of the photons striking them, compared with less than 5 percent for photographic methods. This means that a CCD instrument can image objects 10 to 20 times fainter—or the same object 10 to 20 times faster—than can a photographic plate. Second, CCDs produce a faithful representation of an image in a digital format that can be placed directly on magnetic tape or disk, or even sent across a computer network to an observer's home institution for analysis.

Using computer processing on digital CCD images, astronomers can often compensate for known instrumental defects and even partially remove unwanted background noise in the signal, allowing them to see features in their data that would otherwise remain hidden. In addition, the computer can often carry out many of the tedious and time-consuming chores that must be performed before an image or spectrum reaches its final "clean" form. Figure 3.8 illustrates how computerized image-processing techniques were used to

(a)
(b)

(c)

(d)

▲ **Figure 3.7 CCD Chip** A charge-coupled device (CCD) consists of hundreds of thousands, or even millions, of tiny light-sensitive cells called pixels, usually arranged in a square array. Light striking a pixel causes an electrical charge to build up on it. By electronically reading out the charge on each pixel, a computer can reconstruct the pattern of light—the image—falling on the chip. (a) Detail of a CCD array. (b) A CCD chip mounted for use at the focus of a telescope. (c) Typical data from the chip consist of an array of numbers, running from 0 to 9 in this simplified example. Each number represents the intensity of the radiation striking that particular pixel. (d) When interpreted as intensity levels on a computer screen, an image of the field of view results. (AURA; R. Wainscoat/ Peter Arnold)

(a)   (b)   (c)   (d)

R I V U X G

▲ **Figure 3.8 Image Processing** (a) Ground-based view of the star cluster R136, a group of stars in the Large Magellanic Cloud (a nearby galaxy). (b) The "raw" image of this same region as seen by the *Hubble Space Telescope* in 1990, before the repair mission. (c) The same image after computer processing that partly compensated for imperfections in the mirror. (d) The same region as seen by the repaired *HST* in 1994, here observed at a somewhat bluer wavelength. (AURA/NASA)

correct for known instrumental problems in the *Hubble Space Telescope*, allowing much of the planned resolution of the telescope to be recovered even before its repair in 1993.

3.2 Telescope Size

Astronomers generally prefer large telescopes over small ones for two main reasons. The first has to do with the amount of light a telescope can capture—its *light-gathering power*. The second is related to the amount of detail that can be seen—the telescope's *resolving power*.

Light-Gathering Power

Astronomers spend much of their time observing very distant—and hence very *faint*—cosmic sources. In addition, they often want to obtain *spectra*, which entail spreading the incoming light out into its component colors. ∞ *(Sec. 2.5)* In order to make such detailed observations, astronomers must collect as much light as possible for analysis by their detectors.

Assuming that we are already using the most efficient detectors available (for example, CCDs rather than photographic plates) to count as many as possible of the available photons, there are basically two ways to accomplish this. The first is simply to increase the *exposure time*—the time spent gathering light from a source. Doubling the exposure time doubles the total amount of radiation captured. In Chapter 16 (p. 438) we will see one of the most extreme examples of such a long-exposure image. Constructed in 2003 from more than 300 hours' worth of observations of the same small region of the sky, the so-called "Hubble Ultra-Deep Field" continues to provide astronomers with volumes of information on some of the most distant galaxies in the universe. However, such long exposures are rare. Access to telescopes and the sophisticated instruments attached to them is a precious commodity, and observatories are often reluctant to allocate large amounts of time to a single observation.

The second option is to improve the telescope's light-gathering ability by increasing its **collecting area**, which is the area capable of intercepting and focusing radiation. The larger the telescope's reflecting mirror (or refracting lens), the more light it collects and the easier it is to measure and study an object's radiative properties. The observed brightness of an astronomical object is

◀◀◀ **CONCEPT CHECK**

Why do the largest modern telescopes use mirrors to gather and focus light?

(a)

(b)

R I V U X G

◀ **Figure 3.9 Sensitivity**   Effect of increasing telescope size on an image of the Andromeda Galaxy. Both photographs had the same exposure time, but image (b) was taken with a telescope twice the size of that used to make image (a). Fainter detail can be seen as the diameter of the telescope mirror increases because larger telescopes are able to collect more photons per unit time. (Adapted from AURA)

directly proportional to the area of the telescope's mirror and hence to the *square* of the mirror diameter (Figure 3.9). For example, a 5-m telescope produces an image 25 times brighter than a 1-m instrument because a 5-m mirror has $5^2 = 25$ times the collecting area of a 1-m mirror. We can also think of this relationship in terms of the *time* required for a telescope to collect enough energy to create a recognizable image on a photographic plate. A 5-m telescope produces an image 25 times faster than a 1-m device because it gathers energy at a rate 25 times greater. Put another way, a one-hour time exposure with a 1-m telescope is roughly equivalent to a 2.4-minute (1/25 of an hour) time exposure with a 5-m instrument.

The twin Keck telescopes, shown in detail in Figure 3.6 and in a larger view in Figure 3.10(a), are a case in point. Each telescope combines 36 hexagonal 1.8-m mirrors into the equivalent collecting area of a single 10-m reflector. The high altitude and large size of these telescopes makes them particularly well suited for detailed spectroscopic studies of very faint objects. To the right of the Keck domes in Figure 3.10(a) is the 8.3-m Subaru (the Japanese name for the Pleiades) telescope, part of the National Astronomical Observatory of Japan. Its mirror, shown in Figure 3.10(b), is the largest single

▼ **Figure 3.10 Mauna Kea Observatory**   (a) The world's highest ground-based observatory, at Mauna Kea, Hawaii, is perched atop an extinct volcano more than 4 km (nearly 14,000 feet) above sea level. Among the domes visible in the picture are those housing the Canada–France–Hawaii 3.6-m telescope, the 8.1-m Gemini North instrument, the 2.2-m telescope of the University of Hawaii, Britain's 3.8-m infrared facility, and the twin 10-m Keck telescopes. To the right of the twin Kecks is the Japanese 8.3-m Subaru telescope. The thin air at this high-altitude site guarantees less atmospheric absorption of incoming radiation and hence a clearer view than at sea level, but the air is so thin that astronomers must occasionally wear oxygen masks while working. (b) The mirror in the Subaru telescope, currently the largest one-piece mirror in any astronomical instrument. (R. Wainscoat; R. Underwood/Keck Observatory; NAOJ)

Canada–France–
Hawaii telescope

Gemini–north
telescope

United Kingdom
infrared telescope

NASA
infrared telescope

Keck 2

Keck 1

Subaru
telescope

(a)

(b)

► **Figure 3.11** **VLT Observatory**   Located at the Paranal Observatory in
Atacama, Chile, the European Southern Observatory's Very Large Telescope (VLT) is the
world's largest optical telescope. It comprises four 8.2-m reflecting telescopes, which can
be used in tandem to create the effective area of a single 16-m mirror. (ESO)

mirror (as opposed to the segmented design used in Keck) yet built. In
terms of total available collecting area, the largest telescope system now
operating is the European Southern Observatory's Very Large Telescope
(VLT), located at Cerro Paranal, in Chile (Figure 3.11). It consists of four
separate 8.2-m mirrors that can function as a single instrument. Both
Keck and the VLT are designed to operate in the optical and near-infrared
parts of the electromagnetic spectrum.

Resolving Power

A second advantage of large telescopes is their superior **angular resolution**. In general, *resolution* refers to the ability of any device, such as a
camera or a telescope, to form distinct, separate images of objects lying
close together in the field of view. The finer the resolution, the better we
can distinguish the objects and the more detail we can see. In astronomy,
where we are always concerned with angular measurement, "close together"
means "separated by a small angle on the sky," so angular resolution is the factor that determines our ability to see fine structure (see *More Precisely E-1* and
E-2 for more detail on angular measure and its astronomical applications). In
typical situations, this often means that we need to resolve objects only a few
arc seconds (″) across. Figure 3.12 illustrates the result of increasing resolving
power with views of the Andromeda Galaxy at several different resolutions.

(a)

(b)

(c)

(d)

◄ **Figure 3.12** **Resolution**
Detail becomes clearer in the
Andromeda Galaxy as the angular
resolution is improved some 600
times, from (a) 10′, to (b) 1′,
(c) 5″, and (d) 1″. (Adapted from
AURA)

Fuzzy shadow

▲ **Figure 3.13 Diffraction**   A wave passing through a gap is diffracted through an angle that depends on the ratio of the wavelength of the wave to the size of the gap, leading to a "fuzzy" shadow on the screen. Light and dark gradations represent crests and troughs in the wave, which define the wavelength (see Figure 2.3). The longer the wavelength and/or the smaller the gap, the greater the angle through which the wave is diffracted. In the absence of diffraction, the shadow would be perfectly sharp.

One important factor limiting a telescope's resolution is *diffraction*, the tendency of light to "bend" around corners. ∞ *(Sec. 2.2)* Because of diffraction, when a parallel beam of light enters a telescope, the rays spread out slightly, making it impossible to focus the beam to a sharp point, even with a perfectly constructed mirror. Diffraction introduces a certain fuzziness, or loss of resolution, into the system (Figure 3.13). The degree of fuzziness—the minimum angular separation that can be distinguished—determines the angular resolution of the telescope. The amount of diffraction is proportional to the wavelength of the radiation divided by the diameter of the telescope mirror. As a result, we can write, in convenient units:

$$\text{angular resolution (arc seconds)} = 0.25 \frac{\text{wavelength } (\mu\text{m})}{\text{telescope diameter (m)}},$$

where $1~\mu\text{m}$ (micrometer) $= 10^{-6}~\text{m} = 1000~\text{nm}$. (Recall that the symbol μ is the Greek letter mu.)

For a given telescope size, the blurring effects of diffraction increase in proportion to the wavelength used. Observations in the infrared or radio range are often severely limited by this effect. For example, in an otherwise perfect observing environment, the best possible angular resolution of blue light (wavelength 400 nm $= 0.4~\mu\text{m}$) using a 1-m telescope is about 0.1″. This quantity is known as the *diffraction-limited resolution* of the telescope. For the same telescope used in the near infrared range, at a wavelength of 10,000 nm $= 10~\mu\text{m}$, the best resolution obtainable is only 2.5″. For light of a given wavelength, large telescopes produce less diffraction than small ones. A 5-m telescope observing in blue light has a diffraction-limited resolution of 0.02″, five times finer than that of the 1-m telescope just discussed. A 0.1-m telescope has a diffraction limit of 1″, and so on. For comparison, the angular resolution of the human eye in blue light is about 0.5′.

3.3 High-Resolution Astronomy

Even large telescopes have limitations. For example, according to the discussion in the preceding section, the 10-m Keck telescope should have an angular resolution of around 0.01″ in blue light. In practice, however, without the technological advances discussed in this section, it could not do better than about 1″. In fact, apart from instruments using special techniques developed to examine some particularly bright stars, *no* ground-based optical telescope built before 1990 can resolve astronomical objects to much better than 1″. The reason is *turbulence* in Earth's atmosphere—small-scale eddies of swirling air all along the line of sight, which blur the image of a star even before the light reaches our instruments.

Atmospheric Blurring

As we observe a star, atmospheric turbulence produces continuous small changes in the optical properties of the air between the star and our telescope (or eye). The light from the star is refracted slightly, again and again, and the stellar image dances around on the detector (or retina). This is the cause of the well-known twinkling of stars. The same basic process makes objects appear to shimmer when viewed across a hot roadway on a summer day, because turbulent hot air just above the road surface constantly deflects and distorts the light rays reaching us.

CONCEPT CHECK ▶▶▶

Give two reasons why astronomers need to build very large telescopes.

On a typical night at a good observing site, the maximum deflection produced by the atmosphere is around 1″. Consider taking a photograph of a star under such conditions. After a few minutes' exposure time (long enough for the intervening atmosphere to have undergone many small, random changes), the dancing sharp image of the star has been smeared out over a roughly circular region 1″ or so in diameter (Figure 3.14). Astronomers use the term **seeing** to describe the effects of atmospheric turbulence. The circle over which a star's light is spread is called the **seeing disk**. In this jargon, "good seeing" means that the air is relatively stable and the seeing disk is small—as little as a few tenths of an arc second across in some exceptional cases.

To achieve the best possible seeing, telescopes are sited on mountaintops (to get above as much of the atmosphere as possible) in locations where the atmosphere is known to be fairly stable and relatively free of dust, moisture, and light pollution from cities. In the continental United States, these sites tend to be in the desert Southwest. The U.S. National Observatory for optical astronomy in the Northern Hemisphere is located high on Kitt Peak near Tucson, Arizona. The site was chosen because of its many dry, clear nights, and typical seeing of around 1″. Even better conditions are found on Mauna Kea in Hawaii (Figure 3.10) and in the Andes Mountains of Chile (Figure 3.11), which is why so many large telescopes have recently been constructed at those two exceptionally clear sites.

A telescope placed above the atmosphere, in Earth orbit, can achieve resolution close to the diffraction limit, subject only to the engineering restrictions of building and placing large structures in space. The *Hubble Space Telescope* (*Discovery 3-1*) has a 2.4-m mirror and a diffraction limit of 0.05″ (in blue light).

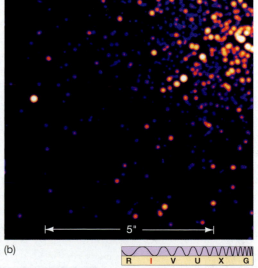

▲ **Figure 3.14 Atmospheric Turbulence** Light rays from a distant star strike a telescope detector at slightly different locations because of turbulence in Earth's atmosphere. Over time, the light covers a roughly circular region on the detector, and even the pointlike image of a star is recorded as a small disk, called the seeing disk.

Animation/Video
Hubble Space Telescope in Orbit

New Telescope Design

The latest techniques for producing ultrasharp images take these ideas of computer control and image processing several stages further. By analyzing the image *while the light is still being collected*, it is possible to adjust the telescope from moment to moment to reduce the effects of mirror distortion, temperature changes, and bad seeing. By these means, an increasing number of large telescopes can now achieve resolutions close to their theoretical (diffraction) limits.

The collection of techniques aimed at controlling environmental and mechanical fluctuations in the properties of the telescope itself is known as **active optics**. Active optics systems often include improved dome design to control airflow, precise control of the mirror temperature, and the use of actuators (pistons) behind the mirror to maintain its precise shape at all times (Figure 3.15a). Figure 3.15(b) illustrates dramatically how active optics can improve image resolution. Most recently constructed large telescopes, including Keck and the VLT, routinely employ active optics systems, improving their resolution to a few tenths of an arc second.

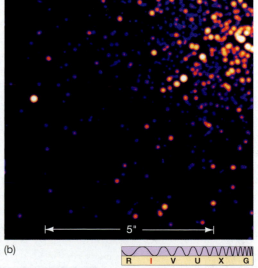

▲ **Figure 3.15 Active Optics** Infrared images of part of the star cluster R136 contrast the resolution obtained (a) without and (b) with an active-optics system. (ESO)

▶ **Figure 3.16 Adaptive Optics** (a) In this daytime photo, a test is being conducted at the Lick Observatory's 3-m Shane telescope in California, where a laser is used to create "artificial stars" (light reflected from the atmosphere back into the telescope) in the sky to improve guiding. Here, the laser beam probes the atmosphere above the telescope, allowing tiny computer-controlled changes to be made to the mirror surface thousands of times each second. (b) The improvement in seeing produced by such systems can be dramatic, as can be seen in these images acquired with the 8-m Gemini telescope atop Mauna Kea in Hawaii. The uncorrected, visible-light image (left) of the star cluster NGC 6934 is resolved to a little less than 1″. With adaptive optics applied (right), the resolution in the infrared is improved to nearly 10 times better, and more stars are seen more clearly. (Lick Observatory; NOAO)

(a)

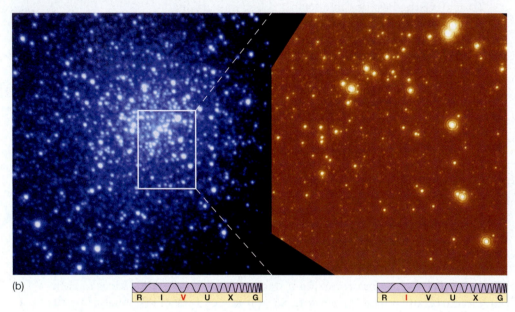

(b)

An even more ambitious undertaking is known as **adaptive optics**. This technique deforms the shape of a mirror's surface under computer control while the image is being exposed, in order to undo the effects of atmospheric turbulence. In the system shown in Figure 3.16(a), lasers probe the atmosphere above the telescope, returning information about the air's swirling motion to a computer that modifies the mirror thousands of times per second to compensate for poor seeing. Other adaptive optics systems monitor standard stars in the field of view, constantly adjusting the mirror's shape to preserve those stars' appearance. Figure 3.16(b) presents an example of the improvement in image quality that can be obtained by these means.

Many of the world's largest telescopes now incorporate sophisticated adaptive optics systems, and resolutions as fine as 0.06″ have been reported for observations at wavelengths of 1–2 μm (in the near infrared part of the spectrum). This is almost at the diffraction limit for an 8-m class telescope and significantly better than the resolution of the much smaller *HST* at the same wavelengths. Researchers anticipate that, within the next few years, adaptive-optics instrumentation will be capable of producing diffraction-limited images at near-infrared wavelengths. Remarkably, it may soon be possible to have the best of both worlds, achieving with large ground-based optical telescopes the kind of resolution once attainable only from space.

CONCEPT CHECK ▶▶▶

Why is Earth's atmosphere a problem for optical astronomers? What can they do about it?

3.4 Radio Astronomy

In addition to the visible radiation that penetrates Earth's atmosphere on a clear day, radio radiation also reaches the ground, and astronomers have built many ground-based **radio telescopes** capable of detecting cosmic radio waves. ∞ *(Sec. 2.3)* These devices have all been constructed since the 1950s—radio astronomy is a much younger field than optical astronomy.

Essentials of Radio Telescopes

Figure 3.17 shows the world's largest steerable radio telescope, the large 105-m- (340-foot-) diameter telescope located at the National Radio Astronomy Observatory in West Virginia. Conceptually, the operation of a radio telescope is the same as a prime-focus optical reflector (Figure 3.5a). It has a large, horseshoe-shaped mount supporting a huge, curved metal dish that serves as the collecting area. The dish captures cosmic radio waves and reflects them to the focus, where a receiver detects the signals and channels them to a computer.

Radio telescopes are built large in part because cosmic radio sources are extremely faint. Many sources simply don't emit many radio photons (see for example Figure 2.11), the photons don't carry much energy, and the sources themselves are often very distant. ∞ *(Sec. 2.6)* In fact, the total amount of radio energy received by Earth's entire surface is less than a trillionth of a watt. Compare this with the roughly 10 *million* watts our planet's surface receives in the form of infrared and visible light from any of the bright stars visible in the night sky. In order to capture enough radio energy to allow detailed measurements to be made, a large collecting area is essential.

Because of diffraction, the angular resolution of radio telescopes is generally quite poor compared with that of their optical counterparts. Typical wavelengths of radio waves are about a million times longer than those of visible light, and these longer wavelengths impose a corresponding crudeness in

▲ **Figure 3.17 Radio Telescope** This is the world's largest fully steerable radio telescope, the 105-m-diameter device at the National Radio Astronomy Observatory in Green Bank, West Virginia. It is 150 m tall—taller than the Statue of Liberty and nearly as tall as the Washington Monument. The schematic diagram shows the path taken by an incoming beam of radio radiation. (NRAO)

angular resolution. Even the enormous sizes of radio dishes only partly offset this effect. The 105-m telescope shown in Figure 3.17 can achieve resolution of about 1′ at a wavelength of 3 cm, although the instrument was designed to be most sensitive to wavelengths around 1 cm, where the resolution is approximately 20′. The best angular resolution obtainable with a single radio telescope is about 10″ (for the largest instruments operating at millimeter wavelengths), about 100 times coarser than the capabilities of the best optical mirrors.

Figure 3.18 shows the world's largest radio telescope, built in 1963 in Arecibo, Puerto Rico. Approximately 300 m in diameter, the telescope's reflecting surface lies in a natural depression in a hillside and spans nearly 20 acres. The receiver is strung among several limestone hills. Its enormous collecting area makes this telescope the most sensitive on Earth. The huge but fixed dish creates one distinct disadvantage, however. The Arecibo telescope cannot be pointed to follow cosmic objects across the sky. Its observations are limited to those objects that happen to pass within about 20° of overhead as Earth rotates.

The Value of Radio Astronomy

Despite the inherent disadvantage of relatively poor angular resolution, radio astronomy enjoys some advantages too. Ground-based radio telescopes can observe 24 hours a day. Unlike visible light, radio waves are not deflected or scattered by Earth's atmosphere, and the Sun itself is a relatively weak source of radio energy. As a result, radio observations can cover almost the entire sky,

▶ **Figure 3.18 Arecibo Observatory** The 300-m-diameter dish at the National Astronomy and Ionospheric Center near Arecibo, Puerto Rico. The receivers that detect the focused radiation are suspended nearly 150 m (about 45 stories) above the center of the dish. The two insets show close-ups of the receivers high above the dish (left) and technicians adjusting the dish surface to make it smoother (right). (D. Parker/T. Acevedo/NAIC; Cornell University)

even during daytime—only within a few degrees of the Sun does solar radiation swamp radio observations of more distant objects. In addition, radio observations can often be made through cloudy skies, and radio telescopes can detect the longest-wavelength radio waves even during rain or snowstorms.

However, perhaps the greatest value of radio astronomy—and, in fact, of all invisible astronomies, as discussed in the next section—is that it opens up a whole new window on the universe. There are two main reasons for this. First, just as objects that are bright in the visible part of the spectrum (the Sun, for example) are not necessarily strong radio emitters, many of the strongest radio sources in the universe emit little or no visible light. Second, visible light may be strongly absorbed by interstellar dust along the line of sight to a source. Radio waves, on the other hand, are generally unaffected by intervening matter. Many parts of the universe cannot be seen at all by optical means but are easily detectable at radio wavelengths. Radio observations therefore allow us to see whole new classes of objects that would otherwise be completely unknown.

Figure 3.19 shows a visible-light photograph of a distant galaxy called Centaurus A. Superimposed on the optical image is a radio map of the same region. The radio image is represented in *false color*, a technique commonly used for displaying images taken in nonvisible light. The colors do not represent the actual wavelength of the radiation emitted, but instead some other property of the source, in this case intensity, descending from red to yellow, green, and blue. Notice just how different the galaxy looks at radio and optical wavelengths. The visible galaxy emits little or no radio energy, while the large blobs (called *radio lobes*, and thought to have been ejected from the center of the galaxy by explosive events long ago) are completely invisible. Centaurus A (an example of a *radio galaxy*, to be studied in more detail in Chapter 15) actually emits far more energy in the form of radio waves than it does as visible light, but without radio astronomy we would be entirely ignorant of this fact, and of the galaxy's violent past.

Interferometry

Radio astronomers can sometimes overcome the problem of poor angular resolution by using a technique known as **interferometry**. This technique makes

◀ **Figure 3.20** **VLA Interferometer** (a) This large interferometer, located on the Plain of San Augustin in New Mexico, comprises 27 dishes spread along a Y-shaped pattern about 30 km across. The most sensitive radio device in the world, it is called the Very Large Array, or VLA. (b) The dishes are mounted on railroad tracks so that they can be repositioned easily. (NRAO)

it possible to produce radio images of much higher angular resolution than can be achieved with even the best optical telescopes, on Earth or in space. In interferometry, two or more radio telescopes are used in tandem to observe the same object at the same wavelength and at the same time. The combined instruments together make up an **interferometer** (Figure 3.20). By means of electronic cables or radio links, the signals received by each antenna in the array making up the interferometer are sent to a central computer that combines and stores the data as the antennas track their target.

Interferometry works by analyzing how the signals *interfere* with each other when added together. ∞ (*Sec. 2.2*) Consider an incoming wave striking two detectors (Figure 3.21). Because the detectors lie at different distances from the source, the signals they record will, in general, be out of step with one another and, when combined, will interfere destructively, partly canceling each other out. Only if the detected radio waves happen to arrive exactly in step will they combine constructively to produce a strong signal. Notice that the amount of interference depends on the *direction* in which the wave is traveling relative to the line joining the detectors. As a result, careful analysis of the strength of the combined signal can provide an accurate measurement of a source's position in the sky. As Earth rotates and the antennae track their target, a complex *interference pattern* of peaks and troughs emerges. After extensive computer processing, this pattern translates into a high-resolution image of the target object.

(b)

CONCEPT CHECK ▶▶▶

Cosmic radio waves are very weak, and the resolution of radio telescopes is often poor, so what can astronomers hope to learn from radio astronomy?

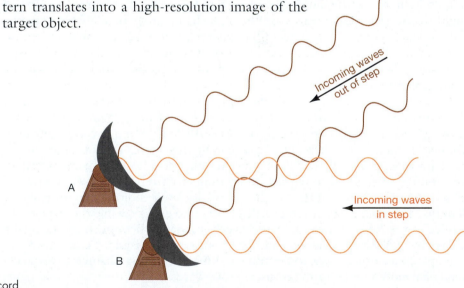

▶ **Figure 3.21** **Interferometry** Two detectors, A and B, record different signals from the same incoming wave because of the time it takes the radiation to traverse the distance between them. When the signals are combined, the amount of interference depends on the wave's direction of motion, providing a means of measuring the position of the source in the sky. Here, the brown-colored waves come from a source high in the sky and interfere destructively when captured by antennas A and B. But when the direction to the source has changed due to Earth's rotation (orange-colored waves), the interference can be constructive.

(a)

(b)

R I V U X G

R I V U X G

▲ **Figure 3.22 Radio-optical Comparison**   (a) A VLA radio image of the spiral galaxy M51 observed at radio frequencies having an angular resolution of a few arc seconds. (b) A visible-light image of M51, made with the 4-m Kitt Peak optical telescope and displayed on the same scale as (a). (NRAO; AURA)

As far as resolving power is concerned, the effective diameter of an interferometer is the distance between its outermost dishes. In other words, two small dishes can act as opposite ends of an imaginary but huge single radio telescope, dramatically improving the angular resolution. Large interferometers made up of many dishes, like the Very Large Array (VLA) shown in Figure 3.20, now routinely attain radio resolution of a few arc seconds, comparable to that of many ground-based optical instruments (without adaptive optics). Figure 3.22 compares an interferometric radio map of a nearby galaxy with a photograph of that same galaxy made using a large optical telescope. (The radio image in Figure 3.19 was also obtained by interferometric means.) The larger the distance (or *baseline*) separating the telescopes, the better the resolution attainable. Very-long-baseline interferometry (VLBI), using instruments separated by thousands of kilometers, can achieve resolution in the milliarcsecond (0.001″) range.

Although the technique was originally developed by radio astronomers, interferometry is no longer restricted to the radio domain. Radio interferometry became feasible when electronic equipment and computers achieved speeds great enough to combine and analyze radio signals from separate radio detectors without loss of data. As the technology has improved, it has become possible to apply the same methods to higher-frequency radiation. Millimeter-wavelength interferometry has already become an established and important observational technique, and infrared interferometry will become commonplace in the next few years.

Optical interferometry is the subject of intensive research. Figure 3.23 shows a recent result from an experimental system constructed at Cambridge University in England. Astronomers combined light from three small optical telescopes to produce a single, remarkably clear image, allowing them to "split" the double star Capella into its two components. The telescopes used were only 0.4 m in diameter, but with the mirrors positioned 6 m apart and operating in the near infrared, the resulting resolution was a stunning 0.01″— better by far than the resolution of the best adaptive-optics systems on the ground or of *HST* operating above Earth's atmosphere. The Keck and VLT telescopes are also designed to be used for infrared—and perhaps someday for optical—interferometric work. NASA plans to place an optical interferometer—the *Space Interferometry Mission*, or *SIM*—in Earth orbit by 2012.

0.05″

R I V U X G

▲ **Figure 3.23 Optical Interferometry**   This extremely high-resolution image of the double-star Capella was made in the near infrared at 830 nm using the Cambridge Optical Aperture Synthesis Telescope (COAST). Capella's two member stars are separated by only 0.05″, and the system is normally seen from the ground only as a slightly oblong blur. With the COAST array, the two stars are clearly resolved. (MRAO)

3.5 Other Astronomies

As just noted, the *types* of astronomical objects that can be observed may differ markedly from one wavelength range to another. Full-spectrum coverage is essential not only to see things more clearly but even to see some things at all. Because of the transmission characteristics of Earth's atmosphere, astronomers must study most wavelengths other than optical and radio from space. ∞ *(Sec. 2.3)* The rise of these "other astronomies" has therefore been closely tied to the development of the space program.

Infrared and Ultraviolet Astronomy

Infrared studies are a very important component of modern observational astronomy. Much of the gas between the stars has a temperature between a few tens and a few hundreds of kelvins, and Wien's law tells us that the infrared domain is the natural portion of the electromagnetic spectrum in which to study this material. ∞ *(Sec. 2.4)* Generally, **infrared telescopes** resemble optical telescopes, but the infrared detectors are sensitive to longer-wavelength radiation. Although most infrared radiation is absorbed by the atmosphere (primarily by water vapor), there are a few windows in the high-frequency part of the infrared spectrum where the opacity is low enough to permit ground-based observations. ∞ *(Fig. 2.9)* Indeed, as we have seen, some of the most useful infrared observing is done from the ground by large telescopes equipped with adaptive optics systems.

Astronomers can make still better infrared observations if they can place their instruments above most or all of Earth's atmosphere. Improvements in balloon-, aircraft-, rocket-, and satellite-based telescope technologies have made infrared research a powerful tool for studying the universe (Figure 3.24). However, as might be expected, the infrared telescopes that can be carried above the atmosphere are considerably smaller than the massive instruments found in ground-based observatories. The first orbiting intrared observatory was the 0.6-m British–Dutch–U.S. *Infrared Astronomy Satellite (IRAS)*, launched in 1983. During its 10-month lifetime (and long afterward—the data archives are still heavily used today), *IRAS* contributed greatly to our knowledge of comets, planets, stars, galaxies, and the scattered dust and rocky debris found in interstellar space. Its infrared view penetrated the clouds of dust and gas surrounding many cosmic objects, allowing astronomers to study regions of space that are completely obscured at visible wavelengths.

▼ **Figure 3.24 Infrared Telescopes** (a) A gondola containing a 1-m infrared telescope is readied for its balloonborne ascent to an altitude of about 30 km, where it will capture infrared radiation that cannot penetrate Earth's atmosphere. (b) An artist's conception of the *Spitzer Space Telescope*, placed in orbit in 2003. This 0.85-m telescope surveys the infrared sky at wavelengths ranging from 3 to 200 μm. During its 5-year mission, it is greatly increasing astronomers' understanding of many aspects of the universe, from the formation of stars and planets to the evolution of galaxies. (SAO; JPL)

Balloon

Gondola

(a)

(b)

Unfortunately (again by Wien's law), telescopes radiate strongly in the infrared unless they are cooled to extremely low temperatures. The end of the *IRAS* mission came not because of any equipment malfunction or unexpected mishap, but simply because its supply of liquid helium coolant ran out. *IRAS*'s own thermal emission then overwhelmed the radiation the instrument was built to detect.

In August 2003, NASA launched the 0.85-m *Space Infrared Telescope Facility* (*SIRTF*, shown in Figure 3.24b). In December of that year, when the first images from the instrument were made public, NASA renamed the spacecraft the *Spitzer Space Telescope* (*SST*), in honor of Lyman Spitzer, Jr., a renowned astrophysicist and the first person to propose (in 1946) that a large telescope be located in space. The facility's detectors are designed to operate at wavelengths in the 3-μm to 200-μm range. Unlike previous space-based observatories, the *SST* does not orbit Earth, but instead follows our planet in its orbit around the Sun, trailing millions of kilometers behind in order to minimize Earth's heating effect on the detectors. The spacecraft is currently drifting away from Earth at the rate of 0.1 AU per year. The detectors are cooled to near absolute zero in order to observe infrared signals from space without interference from the telescope's own heat. The present instrument package aboard the *Hubble Space Telescope (see Discovery 3-1)* also includes a high-resolution (0.1″) near-infrared camera and spectroscope. Figure 3.25 shows some terrestrial and astronomical examples of advantages gained by viewing the in the infrared part of the spectrum.

To the short-wavelength, higher-energy side of the visible spectrum lies the ultraviolet domain. This region of the spectrum, extending in wavelength

(a)   R I V U X G

(b)   R I V U X G

(c)   R I V U X G

(d)   R I V U X G

◄ **Figure 3.25** **Infrared Images** (a) An optical photograph taken near San Jose, California, and (b) an infrared photo of the same area taken at the same time. Infrared radiation of long wavelength can penetrate smog much better than short-wavelength visible light can. The same advantage pertains to astronomical observations. (c) An optical view of an especially dusty part of the central region of the Orion Nebula is more clearly revealed (d) in this infrared image showing a cluster of stars behind the obscuring dust. (Lick Observatory; NASA)

(a)

(b)

R   I   V   U   X   G

◄ **Figure 3.26** **Ultraviolet Images** (a) A camera on board the *Extreme Ultraviolet Explorer* satellite captured this image of the Cygnus Loop supernova remnant, the result of a massive star blowing itself virtually to smithereens. The release of energy was enormous, and the afterglow lingered for millennia. The glowing field of debris shown here within the telescope's circular field of view lies some 1500 light-years from Earth. Based on the velocity of the outflowing debris, astronomers estimate that the explosion itself must have occurred about 10,000 years ago. (b) This false-color image of the spiral galaxy M81, made by the *Galaxy Evolution Explorer* satellite launched into orbit in 2003, reveals that stars are forming in the blue arms well away from the galaxy's center. (NASA; GALEX)

from 400 nm (blue light) down to a few nanometers (the soft, or low-energy, end of the X-ray region), has only recently begun to be explored. Because Earth's atmosphere is partially opaque to radiation below 400 nm and is totally opaque to radiation below about 300 nm, astronomers cannot conduct any useful ultraviolet observations from the ground, not even from the highest mountaintop. Rockets, balloons, or satellites are therefore essential to any **ultraviolet telescope**—a device designed to capture and analyze this high-frequency radiation.

Figure 3.26(a) shows an image of a supernova remnant—the remains of a violent stellar explosion that occurred some 12,000 years ago (see Chapter 12)—obtained by the *Extreme Ultraviolet Explorer (EUVE)* satellite, launched in 1992. Since its launch, *EUVE* has mapped out our local cosmic neighborhood as it presents itself in the far ultraviolet and has radically changed astronomers' conception of interstellar space in the vicinity of the Sun. The *Hubble Space Telescope*, best known as an optical telescope, is also a superb ultraviolet instrument, as is the recently launched *Galaxy Evolution Explorer (GALEX)* satellite, one of whose images is seen in Figure 3.26(b).

High-Energy Astronomy

High-energy astronomy studies the universe as it presents itself to us in X-rays and gamma rays—the types of radiation whose photons have the highest frequencies and hence the greatest energies. How do we detect radiation of such short wavelengths? First, it must be captured high above Earth's atmosphere because none of it reaches the ground. Second, its detection requires the use of equipment basically different from that used to capture the relatively low-energy radiation discussed up to this point.

The difference in the design of **high-energy telescopes** comes about because X-rays and gamma rays cannot be reflected easily by any kind of surface. Rather, these rays tend to either pass straight through or else be absorbed by any material they strike. When X-rays barely graze a surface, however, they can be reflected from it in a way that yields an image, although the mirror design is fairly complex (Figure 3.27).

The *Einstein Observatory*, launched by NASA in 1978, was the first X-ray telescope capable of forming an image of its field of view. During its two-year lifetime, this spacecraft drove major advances in our understand-

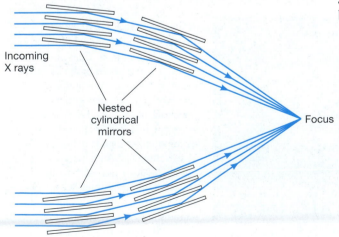

Incoming
X rays

Nested
cylindrical
mirrors

Focus

(a)

(b)

◄ **Figure 3.27** **X-Ray Telescope** (a) The arrangement of nested mirrors in an X-ray telescope allows the rays to be reflected at grazing angles and focused to form an image. (b) A cutaway 3-D rendition of the mirrors, showing their shape more clearly.

▶ **Figure 3.28 Chandra Observatory**   The *Chandra* X-ray telescope is shown here during the final stages of construction in 1998. The left end of the mirror arrangement, depicted in Figure 3.27, is at the bottom of the satellite as oriented here. *Chandra*'s effective angular resolution is 1″, allowing this spacecraft to produce images of quality comparable to that of optical photographs. *Chandra* now occupies an elliptical orbit high above Earth; its farthest point from our planet, 140,000 km out, reaches almost one-third of the way to the Moon. Such an orbit takes *Chandra* above most of the interfering radiation belts girdling Earth (see Section 5.7) and allows more efficient observations of the cosmos.

ing of high-energy phenomena throughout the universe; its observational database is still heavily used. The German *ROSAT* (short for Röntgen Satellite, after Wilhelm Röntgen, the discoverer of X-rays) was launched in 1991 and for seven years generated a wealth of high-quality observational data. It was turned off in 1999, a few months after its electronics were irreversibly damaged when the telescope was accidentally pointed too close to the Sun.

In July 1999, NASA launched the *Chandra X-Ray Observatory* (named in honor of the Indian astrophysicist Subramanyan Chandrasekhar and shown in Figure 3.28). With greater sensitivity, a wider field of view, and better resolution than either *Einstein* or *ROSAT, Chandra* is providing high-energy astronomers with new levels of observational detail. Figure 3.29 shows the first image returned by *Chandra*—a supernova remnant known as Cas A, all that now remains of a star in the constellation Cassiopeia that was observed to explode about 320 years ago. The false-color image shows 50 million-K gas in the wisps of ejected stellar material; the bright white point at the very center of the debris may be a black hole. The European *X-ray Multi-Mirror* satellite (now known as *XMM-Newton*) was launched in December 1999. *XMM-Newton* is more sensitive than *Chandra* (that is, it can detect fainter X-ray sources), but it has significantly poorer angular resolution (5′, compared with 0.5″ for *Chandra*), making the two missions complementary to one another.

Gamma-ray astronomy is the youngest entrant into the observational arena. For gamma rays, no method of producing a true image (in the sense of Section 3.1) has yet been devised—present-day gamma-ray telescopes simply point in a specified direction and count photons received. As a result, only fairly coarse (1° resolution) observations can be made. Nevertheless, even at this resolution, there is much to be learned. Cosmic gamma rays were originally detected in the 1960s by the U.S. *Vela* series of satellites, whose primary mission was to monitor illegal nuclear detonations on Earth. Since then, several X-ray telescopes have also been equipped with gamma-ray detectors.

By far the most advanced gamma-ray instrument was the *Compton Gamma-Ray Observatory (CGRO)*, launched by the space shuttle in 1991. At 17 tons, it was the largest astronomical payload ever launched by NASA at that time. *CGRO* scanned the sky and studied individual objects in much greater detail than had previously been attempted. Gamma-ray astronomy traces the most violent events in the universe, on scales ranging from stars to entire galaxies. Figure 3.30 shows a false-color *CGRO* image of a *gamma-ray blazar*, a highly energetic outburst in the nucleus of a distant galaxy. The mission ended on June 4, 2000, when, following a failure of one of the satellite's three gyroscopes, NASA opted for a controlled reentry and dropped *CGRO* into the Pacific Ocean.

Animation/Video
Chandra Light and Data Paths

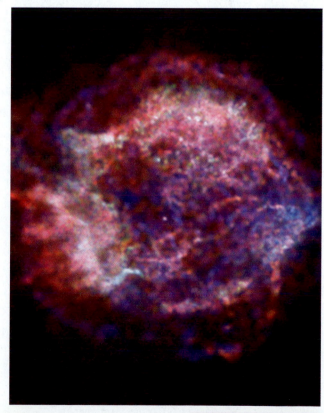

▶ **Figure 3.29 X-Ray Image**   A *Chandra* X-ray image of the supernova remnant Cassiopeia A, a debris field of scattered, glowing gases that were once part of a massive star. The red, green, and blue regions show different intensities of X-rays and also different abundances of heavy elements. Roughly 10,000 light-years from Earth and barely visible in the optical part of the spectrum, Cas A is awash in brilliantly glowing X-rays spread across some 10 light-years. (CXO/SAO)

R I V U X G

CONCEPT CHECK ▶▶▶

List two scientific benefits and two drawbacks of placing telescopes in space.

◀ **Figure 3.30 Gamma-Ray Image**   A false-color gamma-ray image from the Compton *Gamma Ray Observatory*, showing a violent event known as a gamma-ray blazar in the distant galaxy 3C279. (NASA)

Full-Spectrum Coverage

Table 3.1 lists the basic regions of the electromagnetic spectrum and describes objects typically studied in each frequency range. Bear in mind that the list is far from exhaustive and that many astronomical objects are now routinely observed at many different electromagnetic wavelengths. As we proceed through the text, we will discuss more fully the wealth of information that high-precision astronomical instruments can provide us.

As an illustration of the sort of comparison that full-spectrum coverage allows, Figure 3.31 shows a series of images of the Milky Way Galaxy. They were made by several instruments, at wavelengths ranging from radio to gamma ray, over a period of about five years. By comparing the features visible in each, we immediately see how multiwavelength observations can complement each other, greatly extending our perception of the universe around us.

TABLE 3.1 Astronomy at Many Wavelengths		
RADIATION	GENERAL CONSIDERATIONS	COMMON APPLICATIONS (CHAPTER REFERENCE)
Radio	Can penetrate dusty regions of interstellar space.	Radar studies of planets (1, 6)
	Earth's atmosphere largely transparent to radio wavelengths.	Planetary magnetic fields (7)
	Can be detected in daytime as well as at night.	Interstellar gas clouds (11)
	High resolution at long wavelengths requires very large telescopes.	Center of Milky Way Galaxy (14)
		Galactic structure (14, 15)
		Active galaxies (15)
		Cosmic background radiation (17)
Infrared	Can penetrate dusty regions of interstellar space.	Star formation (11)
	Earth's atmosphere only partially transparent to IR radiation,	Cool stars (11, 12)
	so some observations must be made from space.	Center of Milky Way Galaxy (14)
		Active galaxies (15)
		Large-scale structure of the universe (16, 17)
Visible	Earth's atmosphere transparent to visible light.	Planets (6, 7)
		Stars and stellar evolution (9, 10, 12)
		Galactic structure (14, 15)
		Large-scale structure of the universe (16, 17)
Ultraviolet	Earth's atmosphere opaque to UV radiation, so observations must be made from space.	Interstellar medium (11)
		Hot stars (12)
X-ray	Earth's atmosphere opaque to X-rays, so observations must be made from space.	Stellar atmospheres (9)
		Neutron stars and black holes (13)
	Special mirror configurations needed to form images.	Active galactic nuclei (15)
		Hot gas in galaxy clusters (16)
Gamma-ray	Earth's atmosphere opaque to gamma rays, so observations must be made from space.	Neutron stars (13)
	Cannot form images.	Active galactic nuclei (15)

▲ **Figure 3.31 Multiple Wavelengths**   The Milky Way Galaxy as it appears at (a) radio, (b) infrared, (c) visible, (d) X-ray, and (e) gamma-ray wavelengths. Each frame is a panoramic view covering the entire sky. (NRAO; NASA; Lund Observatory; MPI; NASA)

CHAPTER REVIEW

Summary

1 A **telescope** (p. 70) is a device designed to collect as much light as possible from some distant source and deliver it to a detector for detailed study. **Reflecting telescopes** (p. 70) are preferred by astronomers because large mirrors are lighter and much easier to construct than large lenses, and they also suffer from fewer optical defects. Instruments may be placed inside the telescope at the **prime focus** (p. 70), or a secondary mirror may be used to reflect the light to an external detector. Most modern telescopes use **charge-coupled devices** (p. 76) to collect and store data in digital form for later analysis.

2 The light-gathering power of a telescope depends on its **collecting area** (p. 77), which is proportional to the square of the mirror diameter. To study the faintest sources of radiation, astronomers must use large telescopes.

Large telescopes also suffer least from the effects of diffraction and hence can achieve better **angular resolution** (p. 79) once the blurring effects of Earth's atmosphere are overcome. The amount of diffraction is proportional to the wavelength of the radiation under study and inversely proportional to the size of the mirror.

3 The resolution of most ground-based optical telescopes is limited by **seeing** (p. 81)—the blurring effect of Earth's turbulent atmosphere, which smears the pointlike images of a star into a **seeing disk** (p. 81) a few arc seconds in diameter. Astronomers can improve a telescope's resolution by using **active optics** (p. 81), in which a telescope's environment and focus are carefully monitored and controlled, and **adaptive optics** (p. 82), in which the blurring effects of atmospheric turbulence are corrected for in real time.

4 **Radio telescopes** (p. 83) are conceptually similar in construction to optical reflecting telescopes, although they are generally much larger than optical instruments, in part because so little radio energy reaches Earth from space. Their main disadvantage is that diffraction of long-wavelength radio waves limits their resolution. Their principal advantage is that they allow astronomers to explore a new part of the electromagnetic spectrum and of the universe—many astronomical radio emitters are completely undetectable in visible light. In addition, radio observations are largely unaffected by Earth's atmosphere, weather, and the location of the Sun.

5 To increase the effective area of a telescope, and hence improve its resolution, several instruments may be combined into an **interferometer** (p. 86), in which the interference pattern of radiation received by two or more detectors is used to reconstruct a high-precision map of the source. Using **interferometry** (p. 85), radio telescopes can produce images much sharper than those from the best optical equipment.

6 **Infrared** (p. 90) and **ultraviolet telescopes** (p. 90) are similar in basic design to optical systems. **High-energy telescopes** (p. 90) study the X-ray and gamma-ray regions of the electromagnetic spectrum. X-ray telescopes can form images of their field of view, although the mirror design is more complex than for lower-energy instruments. Gamma-ray telescopes simply point in a certain direction and count photons received. Infrared studies in some parts of the infrared range can be carried out using large ground-based systems. However, because of atmospheric opacity, most infrared, and all ultraviolet, X-ray, and gamma-ray observations, must be made from space.

7 Different physical processes can produce very different kinds of electromagnetic radiation, and the image of a given object in long-wavelength, low-energy radio waves may bear little resemblance to its appearance in high-energy X- or gamma-rays. Observations at wavelengths spanning the electromagnetic spectrum are essential to a complete understanding of astronomical events.

Review and Discussion

1. List three advantages of reflecting telescopes over refracting telescopes.

2. What are the largest optical telescopes in use today?

3. How does Earth's atmosphere affect what is seen through an optical telescope?

4. What advantages does the *Hubble Space Telescope* have over ground-based telescopes? List some disadvantages.

5. What are the advantages of a CCD over a photographic plate?

6. Is the resolution of a 2-m telescope on Earth's surface limited more by atmospheric turbulence or by the effects of diffraction?

7. Why do radio telescopes have to be very large?

8. What kind of astronomical objects can we best study with radio techniques?

9. What is interferometry, and what problem in radio astronomy does it address?

10. Compare the highest resolution attainable with optical telescopes to the highest resolution attainable with radio telescopes (including interferometers).

11. What special conditions are required to conduct observations in the infrared?

12. In what ways do the mirrors in X-ray telescopes differ from those found in optical instruments?

13. Why was CGRO placed in space, rather than on the ground?

14. What are the main advantages of studying objects at many different wavelengths of radiation?

15. Our eyes can see light with an angular resolution of about 1′—equivalent to about a third of a millimeter at arm's length. Suppose our eyes detected only infrared radiation, with 1° angular resolution. Would we be able to make our way around on Earth's surface? To read? To sculpt? To create technology?

Conceptual Self-Test: True or False/Multiple Choice

1. The main advantage to using the *Hubble Space Telescope* is the increased amount of "night time" available to it. **(T/F)**

2. The term "seeing" is used to describe how faint an object can be detected by a telescope. **(T/F)**

3. One of the primary advantages of CCDs over photographic plates is the former's high efficiency in detecting light. **(T/F)**

4. Radio telescopes are large in part to improve their angular resolution, which is poor because of the long wavelengths at which they observe. **(T/F)**

5. Infrared astronomy can only be done from space. **(T/F)**

6. X-ray and gamma-ray telescopes employ the same basic design as optical instruments. **(T/F)**

7. Because gamma rays have very short wavelengths, gamma-ray telescopes can achieve extremely high angular resolution. **(T/F)**

8. The main reason that most professional research telescopes are reflectors is that **(a)** mirrors produce sharper images than lenses do; **(b)** their images are inverted; **(c)** they do not suffer from the effects of seeing; **(d)** large mirrors are easier to build than large lenses.

9. If telescope mirrors were made in odd sizes, the one with the *most* light-gathering power would be **(a)** a triangle with 1-m sides; **(b)** a square with 1-m sides; **(c)** a circle 1 m in diameter; **(d)** a rectangle with two 1-m sides and two 2-m sides.

10. The primary reason professional observatories are built on the highest mountaintops is to **(a)** get away from city lights; **(b)** be above the rain clouds; **(c)** reduce atmospheric blurring; **(d)** improve chromatic aberration.

11. When multiple radio telescopes are used for interferometry, resolving power is most improved by increasing **(a)** the distance between telescopes; **(b)** the number of telescopes in a given area; **(c)** the diameter of each telescope; **(d)** the electrical power supplied to each telescope.

12. The *Spitzer Space Telescope* is stationed far from Earth because **(a)** this increases the telescope's field of view; **(b)** the telescope is sensitive to electromagnetic interference from terrestrial radio stations; **(c)** doing so avoids the obscuring effects of Earth's atmosphere; **(d)** Earth is a heat source and the telescope must be kept very cool.

13. The best way to study young stars hidden behind interstellar dust clouds would be to use **(a)** X-rays; **(b)** infrared light; **(c)** ultraviolet light; **(d)** blue light.

14. The image shown in Figure 3.12 (Resolution) is sharpest when the ratio of wavelength to telescope size is **(a)** large; **(b)** small; **(c)** close to 1; **(d)** none of these.

15. Table 3.1 (Astronomy at Many Wavelengths) suggests that the best frequency range in which to study the hot (million-kelvin) gas found among the galaxies in the Virgo cluster would be **(a)** at radio frequencies; **(b)** in the infrared; **(c)** in X-rays; **(d)** in gamma-rays.

Problems

Algorithmic versions of these questions are available in the Practice Problems module of the Companion Website.

The number of squares preceding each problem indicates its approximate level of difficulty.

1. ■ A certain telescope has a $10' \times 10'$ field of view that is recorded using a CCD chip having 2048×2048 pixels. What angle on the sky corresponds to one pixel? What would be the diameter of a typical seeing disk ($1''$ radius), in pixels?

2. ■ The *SST*'s planned operating temperature is 5.5 K. At what wavelength (in micrometers, μm) does the telescope's own blackbody emission peak? How does this wavelength compare with the wavelength range in which the telescope is designed to operate? ∞ *(More Precisely 2-2)*

3. ■ A 2-m telescope can collect a given amount of light in one hour. Under the same observing conditions, how much time would be required for a 6-m telescope to perform the same task? A 12-m telescope?

4. ■ A certain space-based telescope can achieve (diffraction-limited) angular resolution of $0.05''$ for red light (wavelength 700 nm). What would its resolution be **(a)** in the infrared, at wavelength 3.5 μm, and **(b)** in the ultraviolet, at wavelength 140 nm?

5. ■■ Two identical stars are moving in a circular orbit around one another, with an orbital separation of 2 AU. The system lies 200 light-years from Earth. If we happen to view the orbit face-on, how large a telescope would we need to resolve the stars, assuming diffraction-limited optics at a wavelength of 2 μm?

6. ■■ What is the greatest distance at which *HST*, operating in blue light (400 nm), could resolve the stars in the previous question?

7. ■ The photographic equipment on a telescope is replaced by a CCD. If the photographic plate records 5 percent of the light reaching it, but the CCD records 90 percent, how much time will the new system take to collect as much information as the old detector recorded in a one-hour exposure?

8. ■ The Andromeda Galaxy lies about 2.5 million light-years away. To what distances do the angular resolutions of *SST* ($3''$), *HST* ($0.05''$), and a radio interferometer ($0.001''$) correspond at that distance?

9. ■ What would be the equivalent single-mirror diameter of a telescope constructed from two separate 10-m mirrors? Four separate 8-m mirrors?

10. ■ Estimate the angular resolutions of **(a)** a radio interferometer with a 5000-km baseline, operating at a frequency of 5 GHz, and **(b)** an infrared interferometer with a baseline of 50 m, operating at a wavelength of 1 μm.

The Companion Website at www.prenhall.com/chaisson provides algorithmically generated versions of each chapter's Problems, in addition to an Animations and Videos gallery, a Tutorials gallery and Tutorials Quiz questions, a full eBook, an Images gallery, Practice Questions, Multiple Choice, and True or False quizzes, and Labeling exercises.

Saturn ~10^8 m

Moon ~10^6 m

As our study of astronomy continues, we embark on the next in a series of increasingly larger steps that will ultimately take us to the limits of the observable universe. This journey begins at home: Planet Earth, a large ball of metal and rock surrounded by a layer of dirt, air, and water, is millions of times larger than most of the things we encounter in daily life.

Yet, as we learn about the other planets and our central Sun, one of the most striking facts is the diminutive size of Earth compared with the rest of the solar system. The dimension of our planetary family is so large that we could stack more than a million Earths side-by-side across the whole solar system. The images on this page illustrate the range of sizes and scales we will encounter in Part 2, from the scattered debris of asteroids and comets to the magnificent central Sun.

Like early Renaissance adventurers whose observations formed the basis for the Copernican revolution noted in Part 1, we are now living in another golden age of discovery—in our case, exploration of alien worlds and unearthly environments quite

Much of what we know about the planets was discovered in the last few decades.

foreign to our own. But unlike the pioneers who ventured forth in the great sailing ships to map the New World on planet Earth centuries ago, today's scientist explorers remotely crew robotic probes that trek around the solar system. Our current generation has discovered more information about the planets and their moons than all previous generations combined.

What's more, our planetary system is not alone. During just the past decade, astronomers have detected planets orbiting stars other than the Sun. These "extrasolar" planetary systems have radically changed our ideas of how our own solar system emerged and evolved to its present state. After centuries of speculation about Earth's origin, many key pieces of this grand scientific puzzle seem to be falling into place.

The next five chapters relate our present understanding of our extended, complex home in space. That understanding is still a work-in-progress, but the results thus far have revolutionized our knowledge of both our present cosmic neighborhood and Earth's rich natural history.

Sun ~10^9 m

Venus ~10^7 m

Asteroid ~10^4 m

Only within the past couple of decades have scientists taken seriously the idea that life on Earth has been disrupted over the course of billions of years by asteroid and comet impacts. Here, by contrast, is the impact of a comet by a probe sent to collide with it. This image of the oblong, 6-km-long nucleus of Comet Tempel was taken 1 minute after it obliterated a small projectile launched from the *Deep Impact* spacecraft that glided past in 2005. The objective was to gain a better understanding of comets and other nearby cosmic debris, especially given the external threat they represent to life on Earth. (JPL)

4

The Solar System

Interplanetary Matter and the Birth of the Planets

In less than a generation we have learned more about our solar system—the Sun and everything that orbits it—than in all the centuries that went before. By studying the planets, their moons, and the countless fragments of material that orbit in interplanetary space, astronomers have gained a richer outlook on our own home in space. The discoveries of the past few decades have revolutionized our understanding not only of the present state of our cosmic neighborhood but also of its history, for our solar system is filled with clues to its own origin and evolution. Paradoxically, the richest sources of information about the earliest days of the solar system are not the large bodies orbiting the Sun but rather the asteroids, comets, and meteoroids that pepper interplanetary space. These objects may seem to be only rocky and icy debris, but more than the planets themselves, they hold a record of the formative stages of our planetary system and have much to teach us about the origin of our world.

Visit www.prenhall.com/chaisson for additional annotated images, animations, and links to related sites for this chapter.

4.1 An Inventory of the Solar System

Our **solar system** contains one star (the Sun), eight planets orbiting that star, 157 moons (at last count) orbiting those planets, eight *asteroids* and more than 100 *Kuiper belt objects* larger than 300 km (200 miles) in diameter, tens of thousands of smaller (but well-studied) asteroids, myriad *comets* a few kilometers in diameter, and countless *meteoroids* less than 100 m across. This list will undoubtedly grow as we continue to explore our cosmic neighborhood. Table 4.1 lists some basic orbital and physical properties of the nine planets, with a few other solar-system objects (the Sun, the Moon, an asteroid, and a comet) included for comparison.

Animation/Video
An Astronomical Ruler

Planetary Properties

The arrangement of the major bodies in the solar system is shown in Figure 4.1. The planet closest to the Sun is Mercury. Moving outward, we encounter in turn Venus, Earth, Mars, Jupiter, Saturn, Uranus, and Neptune. Pluto is no longer classified as a major planet, but is shown here for scale. The asteroids lie mainly in a broad belt between the orbits of Mars and Jupiter.

The planets' distances from the Sun are known from their periods and Kepler's laws once the scale of the solar system is set by radar ranging on Venus. ∞ *(Sec. 1.3)* The distance from the Sun to Pluto is about 40 AU, almost a million times Earth's radius and roughly 15,000 times the distance from Earth to the Moon. Yet despite the solar system's vast extent, the planets all lie very close to the Sun, astronomically speaking. The diameter of Pluto's orbit is less than 1/1000 of a light-year, while the star nearest the solar system is several light-years distant. Note that the planetary orbits are not evenly spaced. Roughly speaking, the spacing between adjacent orbits doubles as we move outward from the Sun.

All the planets orbit the Sun counterclockwise as seen from above Earth's North Pole and in nearly the same plane as Earth (the ecliptic plane). ∞ *(Sec. 1.1)* Mercury deviates from this rule—its orbital plane lies at 7° to the ecliptic. Still, we can think of the solar system as being quite flat. If we were to view the planets' orbits from a vantage point in the ecliptic plane about 50 AU from the Sun, no planet's orbit would be noticeably tilted. Figure 4.2 is a photograph of the planets Mercury, Venus, Mars, Jupiter, and Saturn, taken during the April 2002 planetary alignment. These five planets can

(a)

10 AU

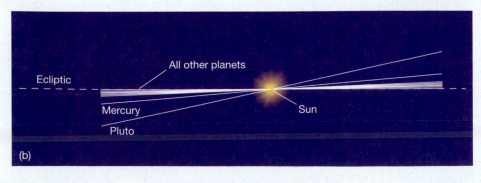

(b)

◀ **Figure 4.1 Solar System** Major bodies of the solar system: the Sun, planets, and asteroids. Part (a) shows a face-on view from directly above the ecliptic plane; part (b) shows the inclinations of the orbits in an edge-on view. Except for Mercury and Pluto (no longer considered a major planet), the orbits of the planets are almost circular and lie nearly in the same plane. As we move outward from the Sun, the distance between adjacent orbits increases. The entire solar system spans nearly 80 AU.

TABLE 4.1   Properties of Some Solar System Objects

OBJECT	ORBITAL SEMI-MAJOR AXIS (AU)	ORBITAL PERIOD (EARTH YEARS)	MASS (EARTH MASSES)	RADIUS (EARTH RADII)	NUMBER OF KNOWN MOONS	AVERAGE DENSITY (kg/m³)	AVERAGE DENSITY (Earth = 1)
Mercury	0.39	0.24	0.055	0.38	0	5400	0.98
Venus	0.72	0.62	0.82	0.95	0	5200	0.95
Earth	1.0	1.0	1.0	1.0	1	5500	1.00
Moon	—	—	0.012	0.27	—	3300	0.60
Mars	1.5	1.9	0.11	0.53	2	3900	0.71
Ceres (asteroid)	2.8	4.7	0.00015	0.073	0	2700	0.49
Jupiter	5.2	11.9	318	11.2	63	1300	0.24
Saturn	9.5	29.4	95	9.5	50	700	0.13
Uranus	19.2	84	15	4.0	27	1300	0.24
Neptune	30.1	164	17	3.9	13	1600	0.29
Pluto	39.5	249	0.002	0.2	1	2100	0.38
Comet Hale-Bopp	180	2400	1.0×10^{-9}	0.004	—	100	0.02
Sun	—	—	332,000	109	—	1400	0.25

Source: NASA

(occasionally) be found in the same region of the sky in large part because their orbits lie nearly in the same plane in space.

Earth's radius has long been known through conventional surveying techniques and, more recently, by satellite measurements. ∞ *(Discovery E-1)* Other planetary radii are found by measuring the planets' angular sizes and then employing elementary geometry. ∞ *(More Precisely E-2)* Figure 4.3 illustrates the sizes of the planets relative to the Sun.

Mass is determined by observing a planet's gravitational influence on some nearby object and applying Newton's laws of motion and gravity. ∞ *(More Precisely 1-1)* Before the Space Age, astronomers calculated planetary masses either by tracking the orbits of the planets' moons (if any) or by measuring the small but detectable distortions the planets produce in each other's orbits. Today the masses of most of the objects listed in Table 4.1 are accurately known through their gravitational effects on artificial satellites and space probes launched from Earth. Measurements of the orbits, masses, and radii of asteroids and comets are discussed in more detail in Section 4.2. With nearly 99.9 percent of the total mass, the Sun is clearly the "senior partner" in the solar system. Its gravity dominates the motion of everything else.

The final column in Table 4.1 lists a quantity called **density**, a measure of the "compactness" of an object, computed by dividing the object's mass (in kilograms) by its volume (in cubic meters). Earth's average density is 5500 kg/m³. For comparison, the density of water is 1000 kg/m³; rocks on Earth's surface have densities in the range 2000–3000 kg/m³, and iron has a density of about 8000 kg/m³. Earth's atmosphere at sea level has a density of only a few kilograms per cubic meter.

▶ **Figure 4.2 Planetary Alignment**   This image shows six planets—Mercury, Venus, Mars, Jupiter, Saturn, and Earth—during a planetary alignment in April 2002. Because the planets all orbit in nearly the same plane, it is possible for them all to appear (by chance) in the same region of the sky, as seen from Earth. The Sun and new Moon are just below the horizon. As usual, the popular press contained many sensationalized predictions of catastrophes that would occur during this rare astronomical event. Also as usual, none came true. (J. Lodriguss)

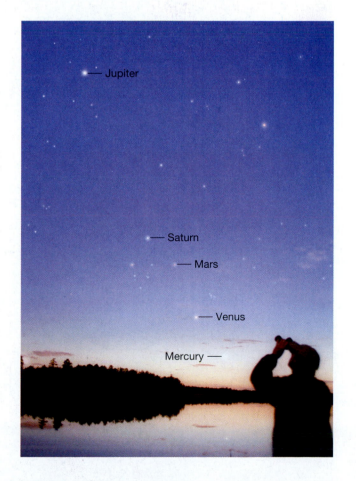

Terrestrial and Jovian Planets

Astronomers can infer a planet's overall composition from a combination of density measurements, spacecraft data, and theoretical calculations (see Chapters 5–8). Based on these measurements, we can draw a clear distinction between the inner and outer members of our planetary system. Simply put, the inner planets—Mercury, Venus, Earth, and Mars—are small, dense, and *rocky* in composition. The outer worlds—Jupiter, Saturn, Uranus, and Neptune—are large, of low density, and *gaseous*.

Because the physical and chemical properties of Mercury, Venus, and Mars are somewhat similar to Earth's, the four innermost planets are called the **terrestrial planets**. (The word *terrestrial* derives from the Latin word *terra*, meaning "land" or "earth.") The larger outer planets—Jupiter, Saturn, Uranus, and Neptune—are all similar to one another chemically and physically (and very different from the terrestrial worlds). They are labeled the **jovian planets**, after Jupiter, the largest member of the group. (The word *jovian* comes from *Jove*, another name for the Roman god Jupiter.) The jovian worlds are all much larger than the terrestrials and quite different from them in both composition and structure. Table 4.2 compares and contrasts some key properties of these two distinct planetary classes.

There are important differences among planets within each category. For example, when we take into account how the weight of the outer layers squeezes the interiors of the terrestrial planets to different extents (greatest for Earth, least for Mercury), we find that the average *uncompressed densities* of the terrestrial worlds—that is, the densities they would have in the absence of any gravitational compression—decrease steadily as we move farther from the Sun. This decrease in density indicates that, despite their other similarities, the overall compositions of these planets differ significantly one from the other. A similar trend is found among the jovian worlds. All four are thought to have large, dense "terrestrial" cores up to 10 or 15 times the mass of Earth, but these cores account for an increasing fraction of the planet's total mass as we move outward from the Sun.

Beyond the outermost jovian planet, Neptune, lie small worlds, frozen and mysterious. Notable among them, Pluto doesn't fit well into either planetary

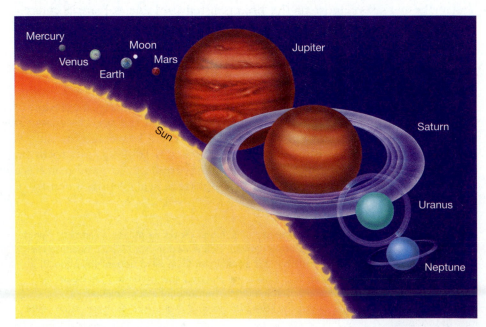

▶ **Figure 4.3 Sun and Planets** Relative sizes of the planets and our Sun. Notice that Jupiter, Saturn, Uranus, and Neptune are much larger than Earth and the other inner planets. However, even these large planets are dwarfed by the still larger Sun.

category. Indeed, recent debate among planetary scientists has caused Pluto to be removed from major planet status (we will return to this topic in Chapter 8). In both mass and composition Pluto has much more in common with the icy jovian moons than with any terrestrial or jovian planet. Many astronomers now regard it as one of the largest members of a newly recognized class of solar system objects—the Kuiper belt (see Section 4.2)—that orbit beyond the jovian worlds.

Solar System Debris

One of the principal goals of planetary science is to understand how the solar system formed and to explain the physical conditions now found on Earth and elsewhere in our planetary system. The last two sections of this chapter are devoted to this topic. Ironically, studies of the most accessible planet—Earth itself—do not help us much in our quest, because information about our planet's early stages was obliterated long ago by atmospheric erosion and geological activity, such as earthquakes and volcanoes. Much the same holds true for the other planets (with the possible exception of Mercury, where a near-airless and geologically inactive surface still retains some imprint of the distant past). The problem is that the large bodies of the solar system have all *evolved* significantly since they formed, making it difficult to decipher the circumstances of their births.

A much better place to look for hints of conditions in the early solar system is on its *small* bodies—the planetary moons and the asteroids, meteoroids Kuiper belt objects, and comets that make up interplanetary debris—for nearly all such fragments contain traces of solid and gaseous matter from the earliest times. They represent truly ancient material—many have scarcely changed since they formed along the rest of the solar system billions of years ago. Therefore, as a prologue to our study of the formation of the solar system, we first examine in more detail the contents of interplanetary space.

TABLE 4.2	Comparison Between the Terrestrial and Jovian Planets
TERRESTRIAL	**JOVIAN**
close to the Sun	far from the Sun
closely spaced orbits	widely spaced orbits
small masses	large masses
small radii	large radii
predominantly rocky	predominantly gaseous
solid surface	no solid surface
high density	low density
slower rotation	faster rotation
weak magnetic fields	strong magnetic fields
no rings	many rings
few moons	many moons

◄◄◄ **CONCEPT CHECK**

Why do astronomers draw such a clear distinction between the inner and the outer planets?

4.2 Interplanetary Matter

In the vast space among the nine known planets move countless small chunks of matter, ranging in size from a few hundred kilometers in diameter down to tiny grains of interplanetary dust. The three major constituents of this cosmic debris are asteroids, comets, and meteoroids. **Asteroids** and **meteoroids** are fragments of rocky material, somewhat similar in composition to the outer layers of the terrestrial planets. The distinction between the two is simply a matter of size—anything larger than 100 m in diameter (corresponding to a mass exceeding 10,000 tons) is conventionally called an asteroid; anything smaller is a meteoroid. **Comets** are predominantly icy rather than rocky in composition (although they do contain some rocky material) and have typical diameters in the 1–10 km range. They are quite similar in chemical makeup to some of the icy moons of the outer planets—and may very well be the progenitors of those small bodies.

Taken together, these small bodies account for a negligible fraction of the total mass of the solar system—probably only a few millionths the mass of the Sun. Their combined gravitational effect is miniscule compared with that of the Sun, Jupiter, Earth, or even our Moon, and they play no important role in the present-day workings of the planets or their moons. Why, then, do we study them in any detail? The answer is that they are of crucial importance to our studies, for they are the keys to answering some very fundamental questions about our planetary environment.

Asteroids

The name *asteroid* literally means "starlike body," but asteroids are definitely not stars. They are too small even to be classified as planets. Astronomers often refer to them as minor planets. Researchers have so far cataloged almost 100,000 asteroids with well-determined orbits. The total number of known asteroids (that is, including those whose orbits are not yet known with sufficient accuracy to make them "official") now exceeds 250,000. As sketched in Figure 4.4, the vast majority are found in a region of the solar system known as the **asteroid belt**, located between 2.1 and 3.3 AU from the Sun, roughly midway between the orbits of Mars (at 1.5 AU) and Jupiter (at 5.2 AU). All but one of the known asteroids revolve about the Sun in prograde orbits—in the same sense (direction) as Earth and the other planets. Again like the planets, most asteroids have orbits fairly close to the ecliptic plane (with inclinations less than 10°–20°). Unlike the almost circular paths of the major planets, however, asteroid orbits are generally noticeably elliptical in shape.

In addition to the main-belt asteroids, a few hundred **Trojan asteroids** share an orbit with Jupiter, remaining a constant 60° ahead of or behind that planet as it circles the Sun. This peculiar orbital behavior is not a matter of chance—the Trojan asteroids are held in place by a delicate but stable balance between the gravitational fields of Jupiter and the Sun. Calculations first performed by the eighteenth-century French mathematician Joseph Louis Lagrange show that interplanetary matter that happens to stray into one of the two regions of space now occupied by the Trojan asteroids can remain there indefinitely, perfectly synchronized with Jupiter's motion.

The orbits of most asteroids have eccentricities lying in the range 0.05–0.3, ensuring that they always remain between the orbits of Mars and Jupiter. However, a few **Earth-crossing asteroids** move on orbits that intersect the orbit of Earth, most likely because the gravitational field of Mars or Jupiter has deflected these small bodies into the inner solar system. From a human perspective, an important consequence of this fact is the very real possibility of a collision with Earth. For example, the mile-wide Earth-crossing asteroid Icarus (Figure 4.5a) missed our planet by "only" 6 million km in 1968—a close call by cosmic standards. In June 2002 the 100-m-diameter asteroid 2002 MN came much closer, missing us by a mere 120,000 km (less than one-third of the distance to the Moon). It was detected three days *after* it passed our planet!

As of mid-2005, more than 3300 Earth-crossing asteroids are known. Most have been discovered since the late 1990s, when systematic searches for such objects began. Several large, dedicated telescopes now scan the skies for faint objects in our neighborhood. More than 700 Earth crossers are officially designated "potentially hazardous," meaning that they are more than about 150 m in diameter (three times the size of the impactor responsible for the Barringer crater shown in Figure 4.17) and move in orbits that could bring them within 0.05 AU (7.5 million km) of our planet. All told, between 1995 and 2005, some 200 asteroids (that we know of!) passed within 0.05 AU of our planet. A similar number are expected to pass within this distance between 2005 and 2015. None of the currently known potentially hazardous asteroids

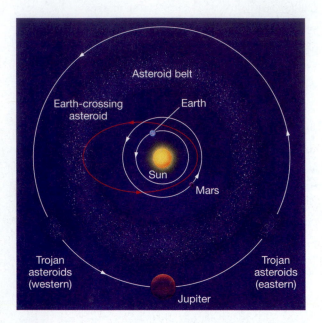

▲ **Figure 4.4 Inner Solar System** The main asteroid belt, along with the orbits of Earth, Mars, and Jupiter. Note the Trojan asteroids at two locations in Jupiter's orbit. The orbits of the main-belt asteroids are typically somewhat eccentric, but generally lie within the belt region of the figure. An Earth-crossing asteroid orbit is also shown (in red).

◄ **Figure 4.5 Asteroids, from Earth** (a) The Earth-crossing asteroid Icarus has an orbit that passes within 0.2 AU of the Sun, well within Earth's orbit. Icarus occasionally comes close to Earth, making it one of the best-studied asteroids in the solar system. Its motion relative to the stars makes it appear as a streak (marked) in this long-exposure photograph. The inset shows an ultraviolet image of the largest asteroid, Ceres, obtained by the *Hubble Space Telescope* in 2001. Little surface detail can be seen, although image processing reveals what seems to be a large impact crater some 250 m across near the center of the frame. (Palomar Observatory/Caltech; NASA/SWRI)

is expected to impact Earth during the next century—the closest predicted near-miss will occur in April 2029, when the 350-m asteroid 2004 MN4 will pass approximately 23,000 km above our planet's surface.

That said, calculations do indicate that most Earth-crossing asteroids *will* eventually collide with Earth, and that during any given million-year period, roughly three asteroids strike our planet. Several dozen large basins and eroded craters on Earth are suspected to be sites of ancient asteroid collisions. The many large-impact craters on the Moon, Venus, and Mars are direct evidence of similar events on other worlds. Such an impact could be catastrophic by human standards. Even a 1-km object carries more than 100 times more energy than all the nuclear weapons currently in existence on Earth and would devastate an area hundreds of kilometers in diameter. Should a sufficiently large asteroid hit our planet, it might even cause the extinction of an entire species. Indeed, many scientists think that the extinction of the dinosaurs was the result of just such an impact (see *Discovery 4-1*).

Asteroid Properties

Asteroids are mostly too small to be resolved by Earth-based telescopes—their sizes are generally less than the angular resolution of even the largest telescopes, and their images appear as featureless points, with no discernible surface features. ∞ *(Sec. 3.2)* However, astronomers can estimate the sizes of these tiny worlds by measuring the amount of sunlight they reflect and the heat they radiate. In addition, from time to time an asteroid happens to pass directly in front of a star, as seen from Earth, allowing astronomers to determine the asteroid's size and shape with great accuracy. The largest asteroid, Ceres (Figure 4.5b), has a diameter of 940 km. Only about two dozen are more than 200 km across, and most are far smaller. Large asteroids are roughly spherical because (as with planets) gravity is the dominant force determining their shape, but smaller bodies can be highly irregular.

Masses are known for only a few asteroids, mainly because the gravitational effect of an asteroid on its neighbors is very small and hard to measure accurately. The mass of Ceres is just 1/10,000 the mass of Earth. The total mass of all the known asteroids probably amounts to less than one-tenth the mass of the Moon.

Asteroid compositions are inferred from spectroscopic and other measurements in the infrared, visible, and ultraviolet parts of the spectrum. The darkest (least reflective) asteroids contain significant amounts of water ice and other volatile substances and are rich in organic (carbon-based) molecules. They are known as *carbonaceous* asteroids. The more reflective *silicate* asteroids are composed primarily of rocky material. The few asteroid densities known from direct measurement of masses and radii are generally consistent with compositions inferred by spectroscopic means. Silicate asteroids predominate in the inner portions of the asteroid belt, but carbonaceous asteroids are more common overall, and their fraction increases as we move outward.

The first close-up views of asteroids were provided by the Jupiter probe *Galileo*, which, on its path to the giant planet, passed twice through the asteroid belt, making close encounters with asteroid Gaspra in October 1991 and asteroid Ida in August 1993 (Figure 4.6). Both Gaspra and Ida are irregularly shaped bodies having maximum diameters of about 20 and 60 km, respectively. They are pitted with craters ranging in size from a few hundred meters to two kilometers across and are covered with layers of dust of variable thickness. Both asteroids are thought to be fragments of much larger objects that broke up following violent collisions long ago. Based on the extent of cratering on the asteroids' surfaces, astronomers estimate that the collision responsible for Ida occurred nearly a billion years ago, compared to just 200 million years for the impact that created Gaspra.

(a)

(b)

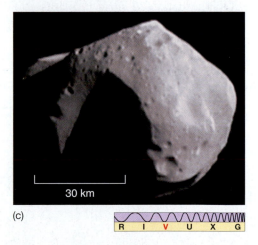

(c)

R I V U X G

▲ **Figure 4.6 Asteroids, Close-up** (a) The asteroid Gaspra as seen from a distance of 1600 km by the space probe *Galileo*. (b) The asteroid Ida photographed by *Galileo* from a distance of 3400 km. (Ida's moon, Dactyl, is visible at right.) The resolution in these photographs is about 100 m. True-color images show the surfaces of both bodies to be a fairly uniform gray. (c) Asteroid Mathilde, imaged by the *NEAR* spacecraft on its way to the asteroid Eros. The largest craters in this image are about 20 km across—much larger than those seen on Ida or Gaspra. The reason may be Mathilde's low density and rather soft composition. (NASA)

4-1 DISCOVERY

What Killed the Dinosaurs?

The name *dinosaur* derives from the Greek words *deinos* (terrible) and *sauros* (lizard), but dinosaurs were no ordinary reptiles. In their prime, roughly 100 million years ago, they were the all-powerful rulers of Earth. Their fossilized remains have been uncovered on all the world's continents. They dominated Earth for well over 100 million years (for comparison, humans have been around for a little over 2 million years). Yet according to the fossil record, these creatures vanished from Earth quite suddenly about 65 million years ago. What happened to them?

Many explanations have been offered for the extinction of the dinosaurs. Devastating plagues, magnetic field reversals (Chapter 5), increased geological activity, severe climate changes, and supernova explosions (Chapter 12) have all been proposed. In the 1980s it was suggested that a huge extraterrestrial object collided with Earth 65 million years ago, and this is now (arguably) the leading explanation for the demise of these creatures, although it is by no means universally accepted.

According to this idea, a 10- to 15-km-wide asteroid or comet struck Earth, releasing as much energy as 10 million or more of the largest hydrogen bombs humans have ever constructed and kicking huge quantities of dust (including the pulverized remnants of the asteroid itself) high into the atmosphere (see the figure). The dust may have shrouded our planet for years, virtually extinguishing the Sun's rays during this time. On the darkened surface, plants could not survive. The entire food chain was disrupted, and the dinosaurs, at the top of that chain, eventually became extinct.

Although we have no direct astronomical evidence to confirm or refute this idea, we can estimate the chances that a large asteroid or comet will strike Earth today on the basis of observations of the number of objects presently on Earth-crossing orbits. The second figure shows the likelihood of an impact as a function of the size of the impacting body. The horizontal scale indicates the energy released by the collision, measured in *megatons* of TNT. The megaton—4.2×10^{16} joules, the explosive yield of a large nuclear warhead—is the only common terrestrial measure of energy adequate to describe the violence of these occurrences. ∞ *(More Precisely 2-2)*

We see that 100 million-megaton events, like the planetwide catastrophe that supposedly wiped out the dinosaurs, are very rare, occurring only once every 10 million years or so. However, smaller impacts, equivalent to "only" a few tens of kilotons of TNT (roughly equivalent to the bomb that de-

(C. Butler/Astrostock-Sanford)

To the surprise of most astronomers, close inspection of the Ida image (Figure 4.6b) revealed the presence of a tiny moon (now named Dactyl), just 1.5 km across, orbiting the asteroid. By studying the *Galileo* images, astronomers were able to obtain some information on Dactyl's orbit around Ida and hence (using Newton's law of gravity) to estimate Ida's mass at about 5–10×10^{16} kg, and its density as 2200–2900 kg/m^3. ∞ *(More Precisely 1-1)*

In June 1997 the *Near Earth Asteroid Rendezvous (NEAR)* spacecraft visited the asteroid Mathilde on its way to the mission's main target, the asteroid Eros. Shown in Figure 4.6(c), Mathilde is some 60 km across. By sensing its gravitational pull, *NEAR* measured Mathilde's mass to be about 10^{17} kg, implying a density of just 1400 kg/m^3. To account for this low density, scientists speculate that the asteroid's interior must be quite porous. Indeed, many smaller asteroids seem to be more like loosely bound "rubble piles" than pieces of solid rock.

stroyed Hiroshima in 1945), could happen every few years—we may be long overdue for one. The most recent large impact was the Tunguska explosion in Siberia in 1908, which packed a roughly 1-megaton punch (see Figure 4.19).

The main geological evidence supporting this theory is a layer of clay enriched with the element iridium. This layer is found in 65 million-year-old rocky sediments all around our planet. Iridium on Earth's surface is rare because most of it sank into our planet's interior long ago. The abundance of iridium in this one layer of clay is about 10 times greater than in other terrestrial rocks, but it matches closely the abundance of iridium found in meteorites (and, we assume, in asteroids and comets too). The site of the catastrophic impact has also been

tentatively identified near Chicxulub, in the Yucatan Peninsula in Mexico, where evidence of a heavily eroded, but not completely obliterated, crater of just the right size and age has been found.

The theory is not without its detractors, however. Perhaps predictably, the idea of catastrophic change on Earth being precipitated by events in interplanetary space was rapidly accepted by most astronomers, but it remains controversial among (some) paleontologists and geologists. Opponents argue that the amount of iridium in the clay layer varies greatly from place to place across the globe, and there is no complete explanation of why that should be so. Perhaps, they suggest, the iridium was produced by volcanoes and had nothing to do with an extraterrestrial impact at all.

Still, in the 20 years since the idea was first suggested, the focus of the debate seems to have shifted. The reality of a major impact 65 million years ago has become widely accepted, and much of the argument now revolves around the question of whether that event actually caused the extinction of the dinosaurs or merely accelerated a process that was already under way. Thus, although the scientific issues are not fully resolved, the realization that such catastrophic events can and do occur marks an important milestone in our understanding of evolution on our planet. (This realization was bolstered by the impact of Comet Shoemaker-Levy 9 on Jupiter in 1994, as discussed in Chapter 7.) In addition, there is growing evidence for even larger impacts in the more distant past, with even more sweeping evolutionary consequences.

As a general rule, we can expect that global catastrophes are bad for the dominant species on a planet. As the dominant species on Earth, we are the ones who now stand to lose the most.

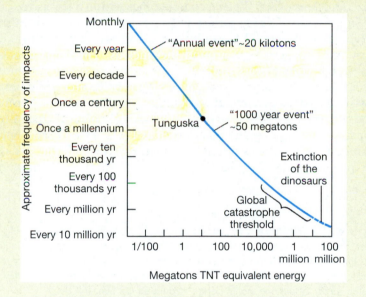

On arrival at Eros on February 14, 2000, *NEAR* (now renamed *NEAR-Shoemaker*) went into orbit around the asteroid. For one year the spacecraft sent back high-resolution images of Eros and made detailed measurements of its size, shape, gravitational and magnetic fields, composition, and structure (Figure 4.7). The craft's various sensors revealed Eros to be a heavily cratered body of mass 7×10^{15} kg and roughly uniform density of around 2700 km/m^3. These findings suggest that the asteroid's interior is solid rock—not rubble, as in the case of Mathilde—although it is extensively fractured due to innumerable impacts in the past. On February 12, 2001, *NEAR-Shoemaker* landed on Eros, sending back a series of close-up images as it descended to the surface. Remarkably, despite the lack of landing gear, the spacecraft survived the low-velocity impact. While no further images were obtained, the probe maintained radio contact with Earth for 16 more days before communication finally ceased.

◄◄◄ CONCEPT CHECK

Describe some basic similarities and differences between asteroids and the inner planets.

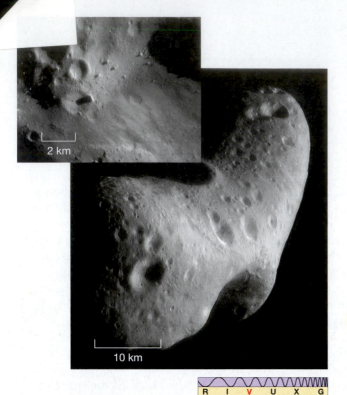

2 km

10 km

R I V U X G

◀ **Figure 4.7 Asteroid Eros**   The asteroid Eros, as seen by the *NEAR-Shoemaker* probe. Craters of all sizes, ranging from 50 m (the resolution of the image) to 5 km, pit the surface. The inset shows a close-up image of a "young" section of the surface, where loose material from recent impacts has apparently filled in and erased all trace of older craters. (JHU/NASA)

Comets

Comets are usually discovered as faint, fuzzy patches of light on the sky while still several astronomical units away from the Sun. Traveling in a highly elliptical orbit with the Sun at one focus, a comet brightens and develops an extended **tail** as it nears the Sun. (The name "comet" derives from the Greek *kome*, meaning "hair.") As the comet departs from the Sun's vicinity, its brightness and its tail diminish until it once again becomes a faint point of light receding into the distance.

The various parts of a typical comet are shown in Figure 4.8(a). Like the planets, comets emit no visible light of their own—they shine by reflected (or reemitted) sunlight. The **nucleus**, or main solid body, of a comet is only a few kilometers in diameter. During most of the comet's orbit, far from the Sun, only this frozen nucleus exists. When a comet comes within a few astronomical units of the Sun, however, its icy surface becomes too warm to remain stable. Part of it becomes gaseous and expands into space, forming a diffuse **coma** (halo) of dust and evaporated gas around the nucleus. The coma gets larger and brighter as the comet nears the Sun. At maximum size it can measure 100,000 km in diameter—almost as large as Saturn or Jupiter. Engulfing the coma, an invisible **hydrogen envelope** stretches across millions of kilometers of space. The comet's tail, most pronounced when the comet is closest to the Sun, is larger still, sometimes spanning as much as 1 AU. From Earth, only the coma and tail of a comet are visible to the naked eye. Despite the size of the tail, most of the comet's light comes from the coma. Most of the mass resides in the nucleus.

▼ **Figure 4.8 Halley's Comet**   (a) Diagram of a typical comet, showing the nucleus, coma, hydrogen envelope, and tail. The tail is not a sudden streak in time across the sky, as in the case of meteors or fireworks. Instead, it travels along with the rest of the comet as long as the comet is close enough to the Sun. (b) Halley's Comet in 1986, about one month before it rounded the Sun. (NOAO)

Hydrogen
envelope

Tail

Coma

Nucleus

To Sun

(a)

50,000 km

(b)

R I V U X G

▶ **Figure 4.9 Comet Tails** In 1997, comet Hale-Bopp displayed both an ion tail (dark blue) and a dust tail (white blue), showing the gentle curvature and inherent fuzziness of the dust. At the comet's closest approach to the Sun, its tail stretched nearly 40° across the sky. (Aaron Horowitz/Corbis)

Two types of comet tail may be distinguished. An **ion tail** is approximately straight, often made of glowing, linear streamers like those shown toward the top of Figure 4.9. Its emission spectrum indicates numerous ionized atoms and molecules that have lost some of their normal complement of electrons. ∞ *(Sec. 2.6)* A **dust tail** is usually broad, diffuse, and gently curved, as shown clearly at the center of Figure 4.9. It is rich in microscopic dust particles that reflect sunlight, making the tail visible from afar.

Both types of tails are in all cases directed away from the Sun by the **solar wind**, an invisible stream of matter and radiation escaping from the Sun. (In fact, astronomers first inferred the existence of the solar wind from observations of comet tails.) Consequently, as depicted in Figure 4.10, the tail, be it ion or dust, always lies outside a comet's orbit and actually *leads* the comet during the portion of the orbit that is outbound from the Sun. The light particles that make up the ion tails are more strongly influenced by the solar wind than by the Sun's gravity, so those tails always point directly away from the Sun. The heavier dust particles are less affected by the pressure of the solar wind and tend to follow the comet's orbit around the Sun, giving rise to the slight curvature of the dust tails.

Probably the most famous comet of all is Halley's Comet (Figure 4.8b), whose appearance at 76-year intervals has been documented at every passage since 240 B.C. Its most recent appearance was in 1986. A spectacular show, the tail of Halley's Comet can reach almost a full astronomical unit in length, stretching many tens of degrees across the sky. More recently, in 1997, comet Hale-Bopp, an unusually large and bright comet with a 40°-long tail, was almost certainly the most-watched comet in history (Figure 4.9).

Based on the best available observations, experts now consider the nucleus of a comet to be composed of dust particles plus some small rocky fragments all trapped within a loosely packed mixture of methane, ammonia, carbon dioxide and ordinary water ice, the density of the mixture being about 100 kg/m^3. Comets are often described as "dirty snowballs." Even as atoms, molecules, and dust particles boil off into space, creating a comet's coma and tail, the nucleus remains frozen at a temperature of only a few tens of kelvins. Estimates of typical cometary masses range from 10^{12} to 10^{16} kg, comparable to the masses of small asteroids. Figure 4.11 shows the nucleus of Halley's comet, as seen by the European *Giotto* spacecraft, which approached within 600 km of the nucleus in 1986.

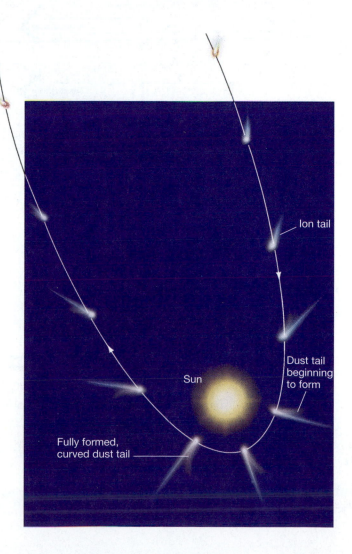

▶ **Figure 4.10 Comet Trajectory** As it approaches the Sun, a comet develops an ion tail, which is always directed away from the Sun. Closer in, a curved dust tail, also directed generally away from the Sun, may appear. Notice that the ion tail always points directly away from the Sun on both the inbound and the outgoing portion of the orbit. The dust tail has a marked curvature and tends to lag behind the ion tail. Compare this figure with comet Hale-Bopp, shown in Figure 4.9.

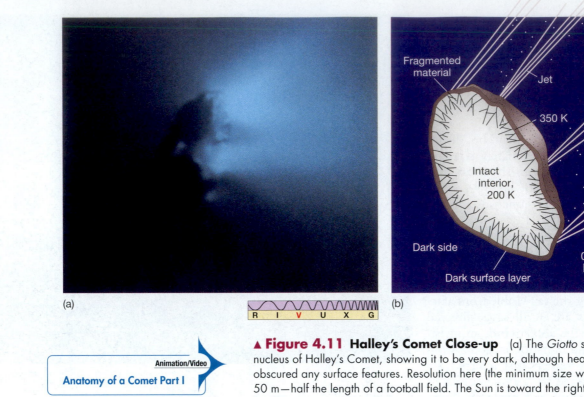

(a)

(b)

R I V U X G

▲ **Figure 4.11 Halley's Comet Close-up** (a) The *Giotto* spacecraft resolved the nucleus of Halley's Comet, showing it to be very dark, although heavy dust in the area obscured any surface features. Resolution here (the minimum size we can distinguish) is about 50 m—half the length of a football field. The Sun is toward the right in this image. The brightest areas are jets of evaporated gas and dust spewing from the comet's nucleus. (b) A diagram of Halley's nucleus, showing its size, shape, jets, and other physical and chemical properties. (European Space Agency/Max Planck Institut)

Animation/Video
Anatomy of a Comet Part I

Animation/Video
Anatomy of a Comet Part II

Animation/Video
Comet Hale-Bopp Nucleus Animation

The past few years have seen two new missions to study comets at close range. In January 2004, NASA's *Stardust* mission approached within 150 km of the nucleus of comet P/Wild 2 (*Wild* is German, pronounced "Vilt") in January 2004, collecting cometary particles in a specially designed foamlike detector (see Figure 4.12). The comet was chosen because it is a relative newcomer to the inner solar system, having been deflected onto its present orbit by an encounter with Jupiter in 1974. It therefore has not been subject to much solar heating or loss of mass by evaporation since it formed long ago. *Stardust* returned to Earth in January 2006, returning the cometary debris to mission scientists eager to study the detailed physical, chemical, and even biological properties of a body that most probably has not changed significantly since our solar system formed.

The second mission had a considerably more violent end. On July 4, 2005, a 400-kg projectile from NASA's *Deep Impact* spacecraft crashed into comet Tempel 1 at more than 10 km/s (23,000 mph), blasting gas and debris from the comet's interior out into interplanetary space, while the spacecraft itself watched from a distance of 500 km. The chapter-opening image on p. 99 shows the mothership's spectacular view of the explosion about 1 minute after impact. Spectroscopic analysis of the ejected gas has provided scientists with their first clear view of the internal composition of a comet, and hence (see Sec. 4.3) of the primordial material of the early solar system. ∞ *(Sec. 2.8)*

CONCEPT CHECK ▶▶▶

In what sense are the comets we see very unrepresentative of comets in general?

Comet Orbits

Their highly elliptical orbits take most comets far beyond Pluto, where, in accordance with Kepler's second law, they spend most of their time. ∞ *(Sec. 1.3)* The majority take hundreds of thousands, even millions, of years to complete a single orbit around the Sun. And unlike the orbits of other solar system ob-

▶ **Figure 4.12 Stardust at Wild-2**   (a) The *Stardust* spacecraft captured this image (a) of comet P/Wild-2 in January 2004, just before the craft passed through the comet's coma. (b) Onboard is a detector made of a foam-like aerogel that is 99.8% air, yet is strong enough to stop and store cometary dust particles for study upon return of the craft to Earth in 2006. (NASA)

5 km

(a)

(b)

jects, most comet orbits are *not* confined to within a few degrees of the ecliptic plane. Rather, they exhibit all inclinations and orientations, both prograde and retrograde, roughly uniformly distributed in all directions from the Sun.

A few *short-period* comets (conventionally defined as those having orbital periods of less than 200 years) are exceptions to the above general statements. According to Kepler's third law, these bodies never venture far beyond the orbit of Pluto. The orientations of their orbits also differ from those of comets in general—they tend to be prograde and lie close to the ecliptic plane, like the orbits of planets and most asteroids.

Astronomers believe that short-period comets originate beyond the orbit of Neptune, in a region of the outer solar system called the **Kuiper belt** (after Gerard Kuiper, a pioneer in infrared and planetary astronomy). Rather like the asteroids in the inner solar system, most Kuiper belt comets move in roughly circular orbits between about 30 and 100 AU from the Sun, always remaining outside the orbits of the jovian planets. Occasionally, however, either a chance encounter between two comets or (more likely) the gravitational influence of Neptune "kicks" a Kuiper belt comet into an eccentric orbit that brings it into the inner solar system and into our view. The observed orbits of these comets reflect the flattened structure of the Kuiper belt.

As of late 2005, some 900 *Kuiper belt objects* are known. Because they are so small and distant, only a tiny fraction of the total has so far been observed. Researchers estimate that the total mass of the Kuiper belt may exceed that of the asteroid belt by a factor of 100 or more (although it is probably still less than the mass of Earth). The largest known Kuiper belt object, named Quaoar (pronounced "Kwah-o-ar", and named for a native American creation god), was discovered in 2002. It orbits some 42 AU from the Sun and is roughly 1200 km across—larger than the largest asteroid, Ceres, and more than half the size of Pluto. In 2005, astronomers discovered a much *larger* object (officially designated 2003 UB313) orbiting far beyond Pluto, some 70 AU from the Sun. Preliminary estimates suggest that it may be more than 3000 km in diameter—perhaps 50 percent larger than Pluto—presenting astronomers with the problem of how to classify this strange new world. Is it a dwarf planet, or a Kuiper belt object, or a member of a brand-new class of sub-planetary bodies? Its discovery is the main reason why Pluto was recently demoted from major planet status. We will return to these thorny issues in Chapter 8.

So how do we account for the apparently random orbital orientations of the long-period comets? Only a tiny portion of a typical long-period cometary orbit lies within the inner solar system, so it follows that for every comet we see, there must be many more similar objects far from the Sun. On these general grounds, astronomers reason that there must be a huge cloud of comets lying far beyond the orbit of Pluto, completely surrounding the Sun. It is named the **Oort cloud**, after the Dutch astronomer Jan Oort, who first wrote in the 1950s of the possibility of such a vast and distant reservoir of inactive, frozen comets. The Kuiper belt and the orbits of some typical Oort cloud comets are sketched in Figure 4.13.

Based on the observed orbital properties of long-period comets, researchers believe that the Oort cloud may be up to 100,000 AU (0.5 light-years) in diameter. Like their Kuiper belt counterparts, however, most Oort cloud comets never come anywhere near the Sun. Indeed, Oort cloud comets rarely approach even the orbit of Pluto, let alone that of Earth. Only when the

▲ **Figure 4.13 Comet Reservoirs** (a) Diagram of the Oort cloud, showing a few cometary orbits. Of all the orbits shown, only the most elongated ellipse represents a comet that will enter the solar system (which, on the scale of this drawing, is smaller than the overlaid box at the center of the figure) and possibly become visible from Earth. (b) The Kuiper belt, the source of short-period comets, whose orbits hug the ecliptic plane.

gravitational field of a passing star happens to deflect a comet into an extremely eccentric orbit that passes through the inner solar system do we get to see it at all. Because the Oort cloud surrounds the Sun in all directions instead of being confined to the ecliptic plane like the Kuiper belt, the long-period comets we see can come from any direction in the sky.

Meteoroids

On a clear night it is possible to see a few *meteors*—shooting stars—every hour. A **meteor** is a sudden streak of light in the night sky caused by friction between air molecules in Earth's atmosphere and an incoming piece of asteroid, meteoroid, or comet. This friction heats and excites the air molecules, which then emit light as they return to their ground states, producing the characteristic bright streak shown in Figure 4.14. Note that the brief flash that is a meteor is in no way similar to the broad, steady swath of light associated with a comet's tail. A meteor is a fleeting event in Earth's atmosphere, whereas a comet tail exists in deep space and can remain visible in the sky for weeks or even months.

▼ **Figure 4.14 Meteor Trails** A bright streak of light called a meteor is produced when a fragment of interplanetary debris plunges into the atmosphere, heating the air to incandescence. (a) A small meteor photographed against a backdrop of stars and the Northern Lights provide the background for a brighter meteor trail. (b) These meteors (one with a red smoke trail) streaked across the sky during the height of the Leonid meteor storm in November 2001. (P. Parviainen, J. Lodriguss)

Animation/Video
Meteor Near Orion

(a)

(b)

Before encountering the atmosphere, the chunk of debris causing a meteor was almost certainly a meteoroid, simply because these small interplanetary fragments are far more common than either asteroids or comets. Any piece of interplanetary debris that survives its fiery passage through our atmosphere and finds its way to the ground is called a **meteorite**.

The smallest meteoroids are mainly the rocky remains of broken-up comets. Each time a comet passes near the Sun, some fragments dislodge from the main body. The fragments initially travel in a tightly knit group of dust or pebble-sized objects called a **meteoroid swarm**, moving in nearly the same orbit as the parent comet. Over the course of time the swarm gradually disperses along the orbit, so that eventually the **micrometeoroids**, as these small meteoroids are known, become more or less smoothly spread around the parent comet's orbit. If Earth's orbit happens to intersect the orbit of such a young cluster of meteoroids, a spectacular *meteor shower* can result. Earth's motion takes it across a given comet's orbit at most twice a year (depending on the precise orbit of each body). Because the intersection occurs at the same time each year (Figure 4.15), the appearance of certain meteor showers is a regular and (fairly) predictable event (Table 4.3).

Meteor showers are usually named for their *radiant*, the constellation from whose direction they appear to come (Figure 4.16). For example, the Perseid shower is seen to emanate from the constellation Perseus. It can last for several days but reaches its maximum every year on the morning of August 12, when upward of 50 meteors per hour can be observed. Astronomers use the speed and direction of a meteor's flight to compute its interplanetary trajectory. This is how certain meteoroid swarms have come to be identified with well-known cometary orbits.

Larger meteoroids—those more than a few centimeters in diameter—are usually *not* associated with comets. They are more likely small bodies that have strayed from the asteroid belt, possibly as the result of asteroid collisions. Their orbits can sometimes be reconstructed in a manner similar to that used to determine the orbits of meteor showers. In most cases, their computed orbits do

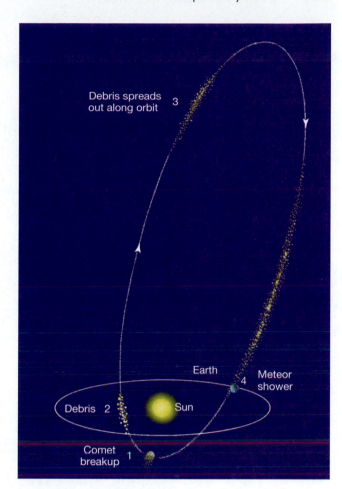

▲ **Figure 4.15 Meteor Showers** A meteoroid swarm associated with a given comet intersects Earth's orbit at specific locations, giving rise to meteor showers at specific times of the year. A portion of the comet breaks up as it rounds the Sun, at the point marked 1. The fragments continue along the comet orbit, gradually spreading out (points 2 and 3). The rate at which the debris disperses around the orbit is much slower than depicted here. It takes many orbits for the material to spread out as shown, but eventually the fragments extend all around the orbit, more or less uniformly. If the orbit happens to intersect Earth's, the result is a meteor shower each time Earth passes through the intersection (point 4).

Animation/Video
Asteroid Comet Breakup

Animation/Video
Infrared Image of Comet Fragment Impact with Jupiter

TABLE 4.3	Some Prominent Meteor Showers		
MORNING OF MAXIMUM ACTIVITY	SHOWER NAME/RADIANT	ROUGH HOURLY COUNT	PARENT COMET
Jan. 3	Quadrantid/Bootes	40	—
Apr. 21	Lyrid/Lyra	10	1861I (Thatcher)
May 4	Eta Aquarid/Aquarius	20	Halley
June 30	Beta Taurid/Taurus	25	Encke
July 30	Delta Aquarid/Aquarius/ Capricorn	20	—
Aug. 12	Perseid/Perseus	50	1862III (Swift–Tuttle)
Oct. 9	Draconid/Draco	up to 500	Giacobini–Zimmer
Oct. 20	Orionid/Orion	30	Halley
Nov. 7	Taurid/Taurus	10	Encke
Nov. 16	Leonid/Leo	12[1]	1866I (Tuttle)
Dec. 13	Geminid/Gemini	50	3200 Phaeton[2]

[1]Every 33 years, as Earth passes through the densest region of this meteoroid swarm, we see intense showers that can reach 1000 meteors per minute for brief periods of time. This occurred most recently in 1999 and 2000.

[2]Phaeton is actually an asteroid and shows no signs of cometary activity, but its orbit matches the meteoroid paths very well.

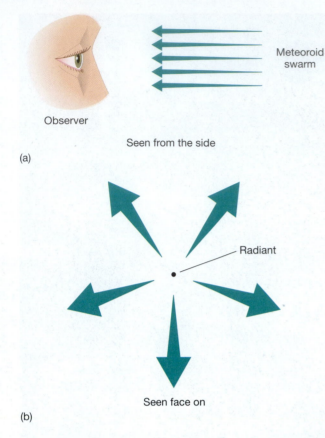

Observer

Meteoroid swarm

Seen from the side

(a)

Radiant

Seen face on

(b)

◄ **Figure 4.16 Radiant** (a) A group of meteoroids approaches an observer. All are moving in the same direction at the same speed. (b) From the observer's viewpoint, the trajectories of the meteoroids (and the meteor shower they produce) appear to spread out from a central point, called the radiant. An analogy is the way that parallel railroad tracks seem to converge in the distance (inset).

...railroad tracks converging

indeed intersect the asteroid belt, providing the strongest evidence we have that this is where they originated. These objects are responsible for most of the cratering on the surfaces of the Moon, Mercury, Venus, Mars, and some of the moons of the jovian planets. A few meteorites have been identified as originating on the Moon or Mars, blasted off the surfaces of those bodies by some impact long ago.

Meteoroids less than about a meter across (roughly a ton in mass) generally burn up in Earth's atmosphere. Larger bodies reach the surface, where they can cause significant damage, such as the kilometer-wide Barringer Crater shown in Figure 4.17. From the size of this crater, we can estimate that the meteoroid responsible must have had a mass of about 200,000 tons and a diameter of perhaps 50 m. Only 25 tons of iron meteorite fragments have been found at the crash site. The remaining mass must have been scattered by the explosion at impact, broken down by subsequent erosion, or buried in the ground. Currently, Earth is scarred with nearly 100 craters larger than 0.1 km in diameter. Most of these are so heavily eroded by weather and geological activity that they can be identified only in satellite photography, as in Figure 4.18. Fortunately, such major collisions between Earth and large meteoroids are thought to be rare events now. Researchers believe that, on average, they occur only once every few hundred thousand years (see *Discovery 4-1*).

One of the most recent documented meteoritic events occurred in central Siberia on June 30, 1908 (Figure 4.19). The presence of only a shallow depression as well as a complete lack of fragments implies that this Siberian intruder exploded several kilometers above the ground, leaving a blasted depression at ground level but no well-formed crater. Recent calculations suggest that the object in question was a rocky meteoroid about 30 m across. The explosion, estimated to have been equal in energy to a 10-megaton nuclear detonation, was heard hundreds of kilometers away and produced measurable increases in atmospheric dust levels all across the Northern Hemisphere.

Most meteorites are rocky (Figure 4.20a), although a few are composed mainly of iron and nickel (Figure 4.20b). Their basic composition is much like that of the rocky inner planets or the Moon, except that some of their lighter elements—such as hydrogen and oxygen—are depleted. Scientists believe that the rocky meteorites are associated with the rocky, silicate asteroids and that their light elements boiled away when the bodies on which they originated were wholly or partially molten. Some meteorites do show clear evidence of strong heating in the past, possibly when their parent asteroids were partially melted during the collisions that liberated the meteoritic fragments. Others, however, show no

1 km

◄ **Figure 4.17 Barringer Crater** The Barringer Meteor Crater, near Winslow, Arizona, is 1.2 km in diameter and 0.2 km deep, the result of a meteorite impact about 25,000 years ago. The meteoroid was probably about 50 m across and likely weighed around 200,000 tons. (U.S. Geological Survey)

R I V U X G

▲ **Figure 4.18** **Manicouagan Reservoir**   This photograph, taken from orbit by the *U.S. Skylab* space station, shows the ancient impact basin that forms Quebec's Manicouagan Reservoir. A large meteorite landed there about 200 million years ago. The central floor of the crater rebounded after the impact, forming an elevated central peak. The lake, 70 km in diameter, now fills the resulting ring-shaped depression. (NASA)

▲ **Figure 4.19** **Tunguska Debris**   The Tunguska event of 1908 leveled trees over a vast area. Although the impact of the blast was tremendous and its sound audible for hundreds of kilometers, the Siberian site is so remote that little was known about the event until scientific expeditions arrived to study it many years later. (Sovfoto/Eastfoto)

evidence of past heating and probably date back to the formation of the solar system. Oldest of all are the *carbonaceous* meteorites, black or dark gray in color and very likely related to the carbonaceous asteroids.

Finally, almost all meteorites are *old*. Radioactive dating shows most of them to be between 4.4 and 4.6 billion years old—roughly the age of the oldest Moon rocks brought back to Earth by *Apollo* astronauts (Chapter 5). Meteorites, along with asteroids, comets, and some lunar rocks, provide essential clues to the original state of matter in the solar neighborhood and to the birth of our planetary system.

◄◄◄ **CONCEPT CHECK**

Why are astronomers so interested in interplanetary matter?

(a)

(b)

▲ **Figure 4.20** **Meteorite Samples**   (a) A stony (silicate) meteorite often has a dark crust, created when its surface is melted by the tremendous heat generated during passage through the atmosphere. The coin at the bottom is for scale. (b) Iron meteorites are much rarer than the stony variety. Most show characteristic crystalline patterns when their surfaces are cut, polished, and etched with acid. (Science Graphics)

4.3 The Formation of the Solar System

You might be struck by the vast range of physical and chemical properties found in the solar system. Indeed, our astronomical neighborhood may seem more like a great junkyard than a smoothly running planetary system. Can we really make any sense of solar system matter? Is there some underlying principle that unifies the facts we have outlined in this chapter? Remarkably, the answer is yes. The origin of the solar system is a complex and as yet incompletely solved puzzle, but (we think!) the basic outlines are at last becoming understood.

A comprehensive theory of the formation of the solar system has been a dream of astronomers for centuries. Its development is a case study in the scientific method, on a par with the sweep of ideas presented in Chapters 1 and 2, although in this case the final form of the theory remains to be determined. ∞ (*Sec. E.5*) As we will see, competing hypotheses have risen, fallen, and in some cases risen again in response to improving observations of our planetary environment. Nevertheless, we now seem to have a coherent picture of how our solar system formed.

Until the mid-1990s, theories of planetary system formation concentrated almost exclusively on our own solar system, for the very good reason that astronomers had no other examples of planetary systems on which to test their ideas. However, all that has now changed. We currently know of more than 150 **extrasolar planets**—planets orbiting stars other than the Sun—to challenge our theories. And challenge them they do! As we will see in Section 4.4, the new planetary systems discovered so far seem to have properties quite different from our own and may require us to rethink in some radical ways our concept of how stars and planets form.

Still, we currently have only limited information on these systems—little more than orbits and mass estimates for the largest planets. Accordingly, we begin our study by outlining the comprehensive theory that accounts, in detail, for most of the observed properties of our own planetary system: the solar system. Bear in mind, though, that no part of the scenario we describe here is in any way unique to our own system. The same basic processes could have occurred during the formative stages of many of the stars in our Galaxy. Later we will see how our theories hold up in the face of the new extrasolar data.

Model Requirements

Based on the measured ages of the oldest meteorites, as well as the ages of Earth and lunar rocks, planetary scientists believe that the age of the solar system is 4.6 billion years—an enormous period of time (although still far less than the 14 billion year age of the universe—see Chapter 17). What happened so long ago to create the planetary system we see today? And what evidence survives from those early times to constrain our theories? Any theory of the origin and architecture of our planetary system must adhere to these ten known facts:

1. *Each planet is relatively isolated in space.* The planets exist as independent bodies at progressively larger distances from the central Sun; they are not bunched together.
2. *The orbits of the planets are nearly circular.* In fact, with the exception of Mercury, which we will argue is a special case, each planetary orbit closely describes a perfect circle.
3. *The orbits of the planets all lie in nearly the same plane.* The planes swept out by the planets' orbits are accurately aligned to within a few degrees. Again, Mercury is a slight exception.

4. *The direction in which the planets orbit the Sun (counterclockwise as viewed from above Earth's North Pole) is the same as the direction in which the Sun rotates on its axis.* Virtually all the large-scale motions in the solar system (other than cometary orbits) are in the same plane and in the same sense. The plane is that of the Sun's equator, and the sense is that of the Sun's rotation.

5. *The direction in which most planets rotate on their axis is roughly the same as the direction in which the Sun rotates on its axis.* This property is considerably less general than the one just described for revolution, as two planets—Venus and Uranus—do not share it.

6. *The direction in which most of the known moons revolve about their parent planet is the same as the direction in which the planet rotates on its axis.* Of the large moons in the solar system, only Neptune's moon Triton is an exception.

7. *Our planetary system is highly differentiated.* The terrestrial planets are characterized by high densities, moderate atmospheres, slow rotation rates, and few or no moons. The jovian planets have low densities, thick atmospheres, rapid rotation rates, and many moons.

8. *Asteroids are very old and exhibit a range of properties not characteristic of either the terrestrial or the jovian planets or their moons.* Asteroids share, in rough terms, the bulk orbital properties of the planets. However, they appear to be made of primitive, unevolved material, and the meteorites that strike Earth are the oldest rocks known.

9. *The Kuiper belt is a collection of asteroid-sized icy bodies orbiting beyond Neptune.* Pluto is now considered such a Kuiper belt object.

10. *The Oort cloud comets are primitive, icy fragments that do not orbit in the plane of the ecliptic and reside primarily at large distances from the Sun.* While similar to the Kuiper belt in composition, the Oort cloud is a completely distinct part of the outer solar system.

All these observed facts taken together strongly suggest a high degree of order in our solar system. The large-scale architecture is too neat, and the ages of the components too uniform, to be the result of random chaotic events. The overall organization points toward a single formation, an ancient but one-time event 4.6 billion years ago.

In the next few chapters we will see that some planetary properties (such as atmospheric composition and interior structure) have gradually *evolved* into their present states during the billions of years since the planets formed. However, *no* such evolutionary explanation exists for the items in the preceding list. For example, Newton's laws imply that the planets must move in elliptical orbits with the Sun at one focus, but they offer no explanation of why the observed orbits should be roughly circular, coplanar, and prograde. We know of no way in which the planets could have started off in random paths, then later evolved into the orbits we see today. Their basic orbital properties must have been established at the outset.

In addition to its many regularities, our solar system also has many notable *irregularities*, some of which we have already mentioned. Far from threatening our theory, however, these irregularities are important facts for us to consider in shaping our explanations. For example, any theory explaining solar system formation must not insist that *all* planets rotate in the same sense or have only prograde moons, because that is not what we observe. Instead, the theory should provide strong reasons for the observed planetary characteristics yet be flexible enough to allow for and explain the deviations, too. And, of course, the existence of the asteroids and comets that tell us so much about our past must be an integral part of the picture. That's a tall order, yet many researchers now believe we are close to that level of understanding.

◀◀◀ CONCEPT CHECK

Why is it important that a theory of solar system formation make clear statements about how planets arose yet not be too rigid in its predictions?

▲ **Figure 4.21 Angular Momentum**
(a) Conservation of angular momentum demands that a contracting, rotating cloud must spin faster as its size decreases. (b) Eventually, the solar nebula, the particular cloud destined to become the solar system, came to resemble a gigantic pancake. The large blob at the center ultimately became the Sun.

Nebular Contraction

Modern theory holds that planets are by-products of the process of star formation (Chapter 11). Imagine a large cloud of interstellar dust and gas (called a **nebula**) a light-year or so across. Now suppose that due to some external influence, such as a collision with another interstellar cloud or perhaps the explosion of a nearby star, the nebula starts to contract under the influence of its own gravity. As it contracts, it becomes denser and hotter, eventually forming a star—the Sun—at its center.

In 1796 the French mathematician-astronomer Pierre Simon de Laplace showed mathematically that conservation of angular momentum (objects spin faster as they shrink—see *More Precisely 4-1*) demands that our hypothetical nebula must spin faster as it contracts. The increase in rotation speed, in turn, causes the nebula's *shape* to change as it shrinks. Centrifugal forces (the outward "push" due to rotation) tend to oppose the contraction in directions perpendicular to the rotation axis, with the result that the nebula collapses most rapidly along the rotation axis. As shown in Figure 4.21, by the time it has shrunk to about 100 AU, the cloud has flattened into a pancake-shaped disk. This swirling mass destined to become our solar system is usually referred to as the **solar nebula**.

If we now suppose that planets form out of this spinning disk, we can begin to understand the origin of much of the architecture observed in our planetary system today, such as the near-circularity of the planets' orbits and the fact that they move in the same sense in almost the same plane. The idea that planets form from such a disk is called the **nebular theory**.

Astronomers are fairly confident that the solar nebula formed such a disk, because similar disks have been observed (or inferred) around other stars. Figure 4.22(a) shows a visible-light image of the region around the star Beta

▼ **Figure 4.22 Beta Pictoris** (a) A computer-enhanced view of a disk of warm matter surrounding the star Beta Pictoris. The images were taken at visible wavelengths, but are presented here in false color to accentuate the details. The bottom image is a close-up of the inner part of the disk, indicating the presence of a warp, possibly caused by the gravitational pull of unseen companions. In both images, the overwhelmingly bright central star has been covered to reveal the much fainter disk surrounding it. The disk is nearly edge-on to our line of sight. For scale, the dimension of Pluto's orbit (78 AU) has been drawn adjacent to the images. (b) An artist's conception of the disk of clumped matter, showing the warm disk with a young star at the center and several comet-sized or larger bodies already forming. Here the colors are thought to be accurate: at the outer edges of the disk, the temperature is low and the color is dull red. Progressing inward, the colors brighten and shift to a more yellowish tint as the temperature increases. Mottled dust is seen throughout—such protoplanetary regions are probably very "dirty." (NASA; D. Berry)

Pictoris, which lies about 50 light-years from the Sun. When the light from Beta Pictoris itself is removed and the resulting image enhanced by a computer, a faint disk of matter (viewed almost edge-on here) can be seen. This particular disk is roughly 1000 AU across—about 10 times the diameter of Pluto's orbit. Astronomers believe that Beta Pictoris is a very young star, perhaps only 100 million years old, and that we are witnessing it pass through an evolutionary stage similar to that experienced by our own Sun 4.6 billion years ago. Figure 4.22(b) shows an artist's conception of the disk.

As we saw in Chapter 1, scientific theories must continually be tested and refined as new data become available. ∞ *(Sec. 1.5)* Unfortunately for the

4-1 MORE PRECISELY

The Concept of Angular Momentum

Most celestial objects rotate. Planets, moons, stars, and galaxies all have some *angular momentum*, which we may define loosely as the tendency of a body to keep spinning or moving in a circle, or, equivalently, how much effort must be expended to stop it. Angular momentum is a fundamental property of any rotating object, as important a quantity as its mass or its energy.

Intuitively, we know that the more massive an object, or the larger it is, or the faster it spins, the harder it is to stop. In fact, angular momentum depends on the object's *mass, rotation rate* (measured in, say, revolutions per second), and *radius*, in a very specific way:

angular momentum $\propto$ mass $\times$ rotation rate $\times$ radius^2.

Recall that the symbol $\propto$ means "is proportional to." The constant of proportionality depends on the details of how the object's mass is distributed.

According to Newton's laws of motion, angular momentum is *conserved* at all times—that is, it must remain constant before, during, and after a physical change in any object, so long as no external forces act. This allows us to relate changes in size to changes in rotation rate. As illustrated in the accompanying figure, if a spherical object having some spin begins to contract, the above relationship demands that it spin faster, so that the product mass $\times$ angular speed $\times$ radius^2 remains constant. The sphere's mass does not change during the contraction, but its radius clearly decreases. The rotation speed must therefore increase in order to keep the total angular momentum unchanged. Figure skaters use this principle to spin faster by drawing in their arms and slow down by extending

them (as shown in the second accompanying figure). The mass of the human body remains the same, but its overall radius changes, causing the body's rotation rate to change in order to keep the angular momentum constant.

To take an example from the text, suppose that the interstellar gas cloud from which our solar system formed was initially one light-year across and rotating very slowly—once every 10 million years, say. The cloud's mass doesn't change (we assume) during the contraction, but its radius decreases. The rotation rate must therefore *increase* in order to keep the total angular momentum unchanged. By the time it has collapsed to a radius of 100 AU, its radius has shrunk by a factor of (one light-year/100 AU) $\approx 9.5 \times 10^{12}$ km/1.5×10^{10} km ≈ 630. Conservation of angular momentum then implies that its (average) spin rate increases by a factor of $630^2 \approx 400,000$, to roughly one revolution every 25 years, close to the orbital period of Saturn.

Incidentally, the law of conservation of angular momentum also applies to planetary orbits (where now "radius" is the distance from the planet to the Sun). In fact, Kepler's second law *is* just conservation of angular momentum, expressed another way. ∞ *(Sec. 1.3)*

Large radius

Slow rotation

Small radius

Rapid rotation

(Orban/Corbis/Sygma)

R I V U X G

▲ **Figure 4.23 Dark Cloud** Interstellar gas and dark dust lanes mark this region of star formation. The dark cloud known as Barnard 86 (left) flanks a cluster of young blue stars called NGC 6520 (right). Barnard 86 may be part of a larger interstellar cloud that gave rise to these stars. (D. Malin/Anglo-Australian Observatory)

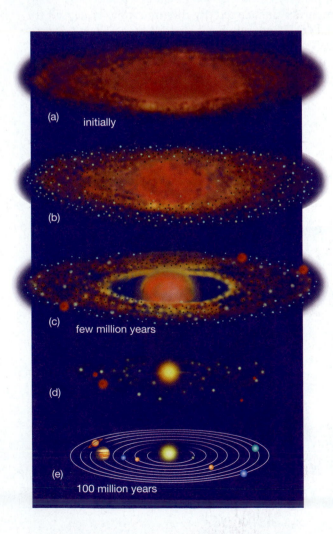

(a) initially

(b)

(c) few million years

(d)

(e) 100 million years

original nebular theory, while Laplace's description of the collapse and flattening of the solar nebula was essentially correct, we now know that a disk of gas would *not* form clumps of matter that would subsequently evolve into planets. In fact, modern computer calculations predict just the opposite: any clumps in the gas would tend to disperse, not contract further. However, the model currently favored by most astronomers, known as the **condensation theory**, rests squarely on the old nebular theory, combining its basic physical reasoning with new information about interstellar chemistry to avoid most of the original theory's problems.

The key new ingredient is *interstellar dust* in the solar nebula. Astronomers now recognize that the space between the stars is strewn with microscopic dust grains, an accumulation of the ejected matter of many long-dead stars. These dust particles probably formed in the cool atmospheres of old stars, then grew by accumulating more atoms and molecules from the interstellar gas. The end result is that interstellar space is littered with tiny chunks of icy and rocky matter, having typical diameters of about 10^{-5} m. Figure 4.23 shows one of many such dusty regions in the vicinity of the Sun.

Dust grains play an important role in the evolution of any gas cloud. Dust helps to cool warm matter by efficiently radiating heat away in the form of infrared radiation. When the cloud cools, its molecules move around more slowly, reducing the internal pressure and allowing the nebula to collapse more easily. ∞ *(More Precisely 2-1)* Furthermore, the dust grains greatly speed up the process of collecting enough atoms to form a planet. They act as **condensation nuclei**—microscopic platforms to which other atoms can attach, forming larger and larger balls of matter. This is similar to the way raindrops form in Earth's atmosphere; dust and soot in the air act as condensation nuclei around which water molecules cluster.

Planet Formation

According to the condensation theory, the planets formed from the solar nebula (Figure 4.24a) in three distinct stages. The first two apply to all planets, the third only to the giant jovian worlds.

The first stage of planet formation began when dust grains in the solar nebula formed condensation nuclei around which matter began to accumulate (Figure 4.24b). This vital step greatly hastened the critical process of forming the first small clumps of matter. Once these clumps formed, they grew rapidly by sticking to other clumps. (Imagine a snowball thrown through a fierce snowstorm, growing bigger as it encounters more snowflakes.)

As the clumps grew larger, their surface areas increased, and consequently, the rate at which they swept up new material accelerated. They gradually grew

◄ **Figure 4.24 Solar System Formation** The condensation theory of planet formation (not drawn to scale; Pluto not shown in part e). (a) The solar nebula after it has contracted and flattened to form a spinning disk (Figure 4.21b). The large blob in the center will become the Sun. Smaller blobs in the outer regions may become jovian planets (see Figure 4.26). (b) Dust grains act as condensation nuclei, forming clumps of matter that collide, stick together, and grow into moon-sized planetesimals. The composition of the grains depends on location within the nebula. (c) After a few million years, strong winds from the still-forming Sun will soon expel the nebular gas. By this time, some particularly massive planetesimals in the outer solar system have already captured large amounts of gas from the nebula. (d) With the gas ejected, planetesimals continue to collide and grow. The gas giant planets are already formed. (e) Over the course of a hundred million years or so, planetesimals are accreted or ejected, leaving a few large planets that travel in roughly circular orbits.

(a)   (b)

◄ **Figure 4.25 Newborn Solar Systems?** (a) This infrared image, taken by the *Spitzer Space Telescope*, of the bright star Fomalhaut, some 25 light-years from Earth, shows a circumstellar disk in which the process of accretion is underway. The star itself is well inside the yellowish blob at center. The outer disk, which is falsely colored orange to match the cooler dust emission, is about three times the diameter of our solar system. (b) This higher-resolution *Hubble Space Telescope* image blocks out the central (circled) parts of another such disk around a more distant star HR4796A, but shows the edges more clearly. (NASA)

Animation/Video
Orion Nebula Animation

into objects of pebble size, baseball size, basketball size, and larger. Figure 4.25 shows an infrared view of a relatively nearby star whose protostellar disk is believed to be in just this state. Eventually, this process of **accretion**—the gradual growth of small objects by collision and sticking—created objects a few hundred kilometers across.

At the end of the first stage of planet formation, the solar system was made up of hydrogen and helium gas and millions of **planetesimals**—objects the size of small moons, having gravitational fields just strong enough to affect their neighbors. By then their gravity was strong enough to sweep up material that would otherwise not have collided with them, so their rate of growth became faster still. During this second phase of planet formation, gravitational forces between planetesimals caused them to collide and merge, forming larger and larger objects. Because larger bodies exert stronger gravitational pulls, eventually almost all the planetesimal material was swept up into a few large **protoplanets**—accumulations of matter that would eventually evolve into the planets we know today.

As the protoplanets grew, a competing process became important. Their strong gravitational fields produced many high-speed collisions between planetesimals and protoplanets. These collisions led to **fragmentation** as small objects broke into still smaller chunks that were then swept up by the protoplanets. Only a relatively small number of 10- to 100-km fragments escaped capture by a planet or a moon and became the asteroids and comets.

After about 100 million years, the primitive solar system had evolved into nine protoplanets, dozens of protomoons, and a glowing **protosun** at the center. Roughly a billion more years were required to sweep the system clear of interplanetary trash. This was a period of intense meteoritic bombardment whose effects on the Moon and elsewhere are still evident today (see Chapter 5).

Animation/Video
Solar System Formation

Making the Jovian Planets

The two-stage accretion picture just described has become the accepted model for the formation of the inner solar system—the terrestrial planets. However, the origin of the giant jovian worlds is less clear. Two somewhat different views, with important consequences for our understanding of extrasolar planets, have emerged.

In the first, more conventional scenario, the four largest protoplanets in the outer solar system grew rapidly and became massive enough to enter a

▲ **Figure 4.26 Jovian Condensation** As an alternative to the growth of massive protoplanetary cores followed by accretion of nebular gas, it is possible that some or all of the giant planets formed directly by instabilities in the cool gas of the outer solar nebula. Part (a) shows the same instant as Figure 4.24(a). (b) Four gas giants have formed (red blobs), circumventing the accretion process sketched in Figure 4.24. (c) The giant planets have taken their place in the outer solar system.

third phase of planetary development—their strong gravitational fields swept up large amounts of gas directly from the solar nebula. As we will see in a moment, there was a lot more raw material available for planet building in the outer solar system, so protoplanets grew much faster there. This is the chain of events depicted in Figures 4.24(c and d). In this scenario, the cores of the jovian planets reached the point where they could capture nebular gas in less than a few million years. Compare this to the inner solar system, where the smaller, terrestrial protoplanets never reached this stage, so their growth was slow, taking 100 million years to form the planets we know today, and their masses remained relatively low.

In the second scenario the giant planets formed through instabilities in the cool outer regions of the solar nebula—not so different from Laplace's original idea—mimicking on small scales the collapse of the initial interstellar cloud. In this view, the jovian protoplanets formed directly and very rapidly, skipping the initial accretion stage and perhaps taking less than a thousand years to acquire much of their mass. These first protoplanets had gravitational fields strong enough to scoop up gas and dust from the solar nebula, allowing them to grow into the giants we see today. Figure 4.26 illustrates this alternative formation path.

Ongoing studies of planetary composition and internal structure (for example, the size of the protoplanetary core), as well as observations of extrasolar planets, may one day allow us to distinguish between these competing theories. In either case, though, once the jovian protoplanets reached the critical size at which they could capture nebular gas, they grew rapidly, their growth rate increasing as their gravitational fields intensified. Their large size today reflects the head start they received in the formation process.

Many of the jovian moons probably also formed by accretion, but on a smaller scale, in the gravitational fields of their parent planets. Once nebular gas began to flow onto the large jovian protoplanets, conditions probably resembled a miniature solar nebula, with condensation and accretion continuing to occur. The large moons of the outer planets almost certainly formed in this way. Some of the smaller moons may be captured planetesimals.

What of the gas that made up most of the original cloud? Why don't we see it today throughout the planetary system? All young stars apparently experience a highly active evolutionary stage known as the *T-Tauri phase* (see Chapter 11), during which their radiation emission and stellar winds are very intense. When our Sun entered this phase, a few million years after the solar nebula formed, any gas remaining between the planets was blown away into interstellar space by the solar wind and the Sun's radiation pressure. Afterward all that remained were protoplanets and planetesimal fragments, ready to continue their long evolution into the solar system we know today. Note that the evolution of the Sun sets the time frame for jovian planetary formation—the outer planets *must* have formed before the nebula dispersed.

The Differentiation of the Solar System

We can use the condensation theory to understand the basic composition differences in the terrestrial planets, jovian planets, and smaller bodies that constitute the solar system. Indeed, it is in this context that the term *condensation* derives its true meaning.

As the solar nebula contracted under the influence of gravity, it heated up as it flattened into a disk. The density and temperature were greatest near the central protosun and much lower in the outlying regions. In the hot inner regions, dust grains broke apart into molecules, which in turn split into atoms. Most of the original dust in the inner solar system disappeared at this stage, although the grains in the outermost parts probably remained largely intact.

The destruction of the dust in the inner solar nebula introduced an important new ingredient into the theoretical mix, one that we omitted from our earlier discussion. With the passage of time the gas radiated away its heat, and the temperature decreased at all locations except in the very core, where the Sun was forming. Everywhere beyond the protosun new dust grains began to condense out, much as raindrops, snowflakes, and hailstones condense from moist, cooling air here on Earth. It may seem strange that although there was plenty of interstellar dust early on, it was mostly destroyed, only to form again later. However, a critical change had occurred. Initially the nebular gas was uniformly peppered with dust grains. When the dust re-formed later, the distribution of grains was very different.

Figure 4.27 shows the temperature in various parts of the primitive solar system just before the onset of accretion. At any given location the only materials to condense out were those able to survive the temperature there. As marked on the figure, in the innermost regions, around Mercury's present orbit, only metallic grains could form—it was simply too hot for anything else to exist. A little farther out, at about 1 AU, it was possible for rocky, silicate grains to form, too. Beyond about 3 or 4 AU, water ice could exist, and so on. More and more material could condense out at greater and greater distances from the Sun. The composition of the grains at any given distance from the Sun determined the type of planetesimal—and ultimately planet—that formed there. The present-day structure of the asteroid belt, with rocky silicates more common in the inner regions and carbonaceous bodies, descendents of planetesimals of icier composition, most prevalent at larger radii, still reflects these early conditions.

Beyond about 5 AU from the center, the temperature was low enough to allow several abundant gases—water vapor, ammonia, and methane—to condense into solid form. (As we will see, these compounds are still important constituents of jovian atmospheres.) Consequently, the planetesimals destined to become the cores of the jovian planets were formed under cold conditions out of low-density, icy material. Because more material could condense out of the solar nebula at these radii than in the inner regions near the protosun, accretion began sooner, with more resources to draw on. If they hadn't already formed through instabilities in the cold nebular gas, the outer planets grew rapidly to the point where they could accrete not just grains but nebular gas also, and the eventual result was the hydrogen-rich jovian worlds we see today.

In the inner regions of the primitive solar system the environment was too hot for ices to survive. Many of the abundant heavier elements, such as silicon, iron, magnesium, and aluminum, combined with oxygen to produce a variety of rocky materials. Planetesimals in the inner solar system were therefore rocky or metallic, as were the protoplanets and planets they ultimately formed. Here is an additional reason why the jovian planets grew so much bigger than the terrestrial worlds. The inner regions of the nebula had to wait for the temperature to drop so that a few rocky grains could appear and begin the accretion process, whereas accretion in the outer solar system began almost as soon as the solar nebula collapsed to a disk.

The heavier materials condensed into grains in the outer solar system, too, of course. However, there they would have been vastly outnumbered by the far more abundant light elements. The outer solar system is not deficient in heavy elements—rather, the inner solar system is *underrepresented in light material.*

Asteroids and Comets

In the inner solar system, most rocky planetesimals collided with or were ejected by the growing terrestrial planets. Only a few remain today as the asteroid belt. Planetesimals beyond the orbit of Mars failed to accumulate into

(a)

(b)

▲ **Figure 4.27 Temperature in the Early Solar Nebula** (a) Theoretically computed variation of temperature across the primitive solar nebula illustrated in part (b), which shows half of the disk in Figure 4.24(b). In the hot central regions, only metals could condense out of the gaseous state to form grains. At greater distances from the central protosun, the temperature was lower, so rocky and icy grains could also form. The labels indicate the minimum radii at which grains of various types could condense out of the nebula.

a protoplanet because the huge gravitational field of nearby Jupiter continuously disturbed their motion, nudging and pulling at them, preventing them from aggregating into a planet. Many of the Trojan asteroids have likely been locked in their odd orbits by the combined gravitational pulls of Jupiter and the Sun since those earliest times.

Farther out, with the formation of the jovian worlds, the remaining planetesimals were subject to those large bodies' strong gravitational fields. Over a period of hundreds of millions of years, and after repeated gravitational "kicks" from the giant planets, particularly Jupiter and Saturn, most of the interplanetary fragments in the outer solar system were flung into orbits taking them far from the Sun (Figure 4.28). Those fragments now make up the Oort cloud. Interactions with Uranus and Neptune were gentler, and generally did not eject planetesimals to large distances. Rather, these encounters tended to deflect small bodies into eccentric orbits that carried them into the inner solar system, where they collided with a planet or were kicked into the Oort cloud by Jupiter or Saturn. Most of the original planetesimals formed beyond the orbit of Neptune are still there, making up the Kuiper belt.

The interactions that cleared the outer solar system of comets also caused significant changes in the orbits of the planets themselves. Computer simulations indicate that, during the ejection process, Jupiter moved slightly closer to the Sun, its orbital semi-major axis decreasing by a few tenths of an astronomical unit, while the other giant planets migrated outward—possibly by as much as 10 AU in the case of Neptune (Figure 4.28b).

The icy planetesimals deflected into the inner solar system played an important role in the evolution of the terrestrial planets too. A long-standing puzzle in the condensation theory's account of the formation of the inner planets was where the water and other volatile gases on Earth and elsewhere originated. At formation the inner planets' surface temperatures were far too high, and their gravity too low, to capture and retain those gases. The answer

 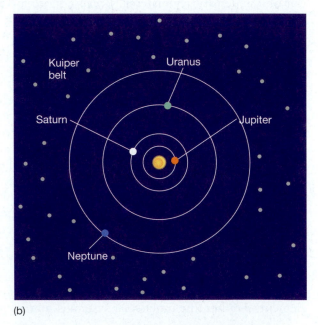

(a)   (b)

▲ **Figure 4.28 Planetesimal Ejection** The ejection of icy planetesimals to form the Oort cloud and Kuiper belt. (a) Initially, once the giant planets had formed, leftover planetesimals were found throughout the solar system. Interactions with Jupiter and Saturn apparently "kicked" planetesimals out to very large radii (the Oort cloud). Interactions with Uranus and especially Neptune tended to keep the Kuiper belt populated, but also deflected many planetesimals inward to interact with Jupiter and Saturn. (b) After hundreds of millions of years and as a result of the inward and outward "traffic," the orbits of all four giant planets were significantly modified by the time the planetesimals interior to Neptune's orbit had been ejected. As depicted here, Neptune was affected most and may have moved outward by as much as 10 AU.

seems to be that the comets from the outer solar system bombarded the newly born inner planets, supplying them with water *after* their formation.

Random Encounters in the Solar Nebula

The condensation theory accounts for the ten facts listed at the start of this section. The growth of planetesimals throughout the solar nebula, with each protoplanet ultimately sweeping up the material near it, accounts for the planets' wide spacing (point 1, although the theory does not adequately explain the regularity of the spacing). That the planets' orbits are circular (2), in the same plane (3), and in the same direction as the Sun's rotation on its axis (4) is a direct consequence of the solar nebula's shape and rotation. The rotation of the planets (5) and the orbits of the moon systems (6) are due to the tendency of smaller structures to inherit the overall sense of rotation of their parent. The heating of the nebula and the Sun's ignition resulted in the observed differentiation (7), while the debris from the accretion–fragmentation stage naturally accounts for the asteroids (8) and comets (9).

We mentioned earlier that an important aspect of any theory of solar system formation is its ability to accommodate deviations. In the condensation theory, that capacity is provided by the randomness inherent in the encounters that combined planetesimals into protoplanets. As the numbers of large bodies decreased and their masses increased, individual collisions acquired greater and greater importance. The effects of these collisions can still be seen today in many parts of the solar system.

We will discuss many solar system irregularities in the next few chapters; most can be explained as the results of random collisions. A case in point is the anomalously slow and retrograde rotation of Venus (Chapter 6), which can be explained if we assume that the last major collision in the formation history of that planet just happened to involve a near head-on encounter between two protoplanets of comparable mass. A similar encounter, this time involving Earth, may have formed our Moon (Chapter 5). Scientists usually do not like to invoke random events to explain observations. However, there seem to be many instances where pure chance has played an important role in determining the present state of the universe.

4.4 Planets Beyond the Solar System

The condensation theory was developed to explain just one planetary system— our own. However, a critical test of any scientific theory is its applicability and predictive power outside the context in which it was originally conceived. ∞ (*Sec. 1.5*) The discovery in recent years of numerous planets orbiting other stars presents astronomers with the opportunity—indeed, the scientific obligation—to confront their theories of solar system formation against a whole new body of observational data.

Detecting Extrasolar Planets

The past few years have seen enormous strides in the search for extrasolar planets. These advances have come not through any dramatic scientific or technical breakthrough but rather through steady improvements in both telescope and detector technology and computerized data analysis. The official count of extrasolar planets now stands at over 170, in more than 145 separate systems within a few hundred light years of the Sun. However, with one exception (Figure 4.29), it has not yet been possible to obtain images of any of these newly discovered worlds. The techniques used to find them are generally indirect, based on analysis of light from the parent star, not from the unseen planet.

◄◄◄ **CONCEPT CHECK**

Would you expect to find asteroids and comets orbiting other stars?

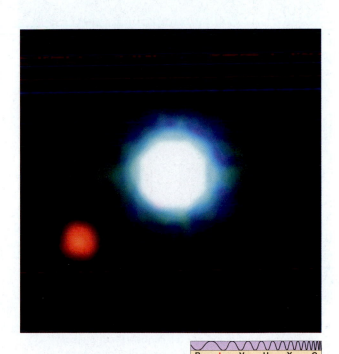

R   I   V   U   X   G

▲ **Figure 4.29 Extrasolar Planet** Most known extrasolar planets are too faint to be detectable against the glare of their parent stars. However, in this system, called 2M1207, the parent itself (centered) is very faint—a so-called *brown dwarf* (see Chapter 12)—allowing the planet (lower left) to be detected in the infrared. This planet has a mass about five times that of Jupiter and orbits 55 AU from the star, which is 230 light-years away. (ESO)

(a)

(b)

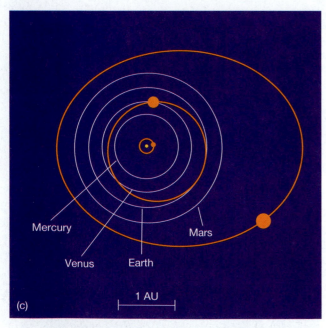

(c)

▲ **Figure 4.30** **Planets Revealed** (a) Measurements of the Doppler shift of the star 51 Pegasi reveal a clear periodic signal indicating the presence of a planetary companion of mass at least half the mass of Jupiter. (b) Radial velocity data for Upsilon Andromedae are much more complex, but are well fit (solid line) by a three-planet system orbiting the star. For reference in parts (a) and (b), the maximum possible signal produced by Jupiter orbiting the Sun is shown in blue. (c) A sketch of the inferred orbits of the three planets in the Upsilon Andromedae system (in orange), with the orbits of the terrestrial planets superimposed for comparison (in white).

As a planet orbits a star, gravitationally pulling one way and then the other, the star "wobbles" slightly. The more massive the planet or the closer its orbit to the star, the greater its gravitational pull and hence the star's movement. If the wobble happens to occur along our line of sight to the star, then we see small fluctuations in the star's radial velocity, which can be measured using the Doppler effect. ∞ (*Sec. 2.7*)

Figure 4.30 shows two sets of radial velocity data that betray the presence of planets orbiting other stars. Part (a) shows the line-of-sight velocity of the star 51 Pegasi, a near-twin to our Sun lying some 40 light-years away. These data were acquired in 1994 by Swiss astronomers using the 1.9-m telescope at Haute-Provence Observatory in France, and they were the first substantiated evidence for an extrasolar planet orbiting a Sun-like star. The regular 50 m/s fluctuations in the star's velocity have since been confirmed by several groups of astronomers and imply that a planet of at least half the mass of Jupiter orbits 51 Pegasi in a circular orbit with a period of just 4.2 days.

Figure 4.30(b) shows another set of Doppler data, this time revealing one of the most complex system of planets discovered to date—a triple-planet system orbiting another nearby Sun-like star named Upsilon Andromedae. The three planets have minimum masses of 0.7, 2.1, and 4.3 times the mass of Jupiter, and orbital semi-major axes of 0.06, 0.83, and 2.6 AU, respectively. Figure 4.30(c) sketches their orbits, with the orbits of the solar system terrestrial planets shown for scale.

The Doppler technique suffers from the limitation that the angle between the line of sight and the planet's orbital plane cannot be determined. Simply put, we cannot distinguish between a low-speed orbit seen edge-on and a high-speed orbit seen almost face-on (so only a small component of the orbital motion contributes to the line-of-sight Doppler effect). However, in one system originally discovered through Doppler measurements that is not the case. Observations of the solar-type star HD 209458, which lies some 150 light-years from Earth, reveal a small (1.7 percent) but clear drop in brightness each time its 0.6-Jupiter-mass companion, orbiting at a distance of just 7 million km (0.05 AU), passes between the star and Earth (Figure 4.31). Such planetary *transits* are rare, because they require us to see the orbit almost exactly edge-on, but when they do occur they allow an unambiguous determination of the planet's mass and radius. The resulting density of the planet is just 200 kg/m³, consistent with a high-temperature gas giant planet orbiting very close to its parent star.

Planetary Properties

Figure 4.32 shows the orbits of many of the extrasolar planets discovered, with Earth's orbit superimposed for comparison. Only a relatively small fraction (around 5 percent) of the nearby stars surveyed to date have shown evidence for extrasolar planets, but those that do present astronomers with a puzzle. Simply put, *none* of these planetary systems look anything like our own solar system! The observed planets are almost all comparable in mass to Jupiter (although a few have masses more similar to Saturn). However, their orbits are generally much smaller than those of Jupiter and Saturn and usually quite eccentric, often taking them very close (much less than 1 AU) to their parent stars. These gas-giant planets orbiting close to their parent stars have come to be known as "hot Jupiters." In addition, although about a dozen multiple (mostly double) planet systems are known, most of the planets observed appear to be the only massive objects orbiting their star.

The fact that no low-mass planets are observed is not surprising. It is what astronomers call a **selection effect**—lightweight planets, or planets far from their parent star, simply don't produce large enough velocity fluctuations to be detectable. The planets observed are precisely those most likely to be detected

◄ **Figure 4.31 An Extrasolar Transit**   If an extrasolar planet happens to pass between us and its parent star, the light from the star dims in a characteristic way. The planet orbiting a sunlike star known as HD 209458 is 200,000 km across and transits every 3.5 days, blocking about 2 percent of the star's light each time it does so.

using current techniques, although the tight eccentric orbits of the hot Jupiters discovered so far imply that these systems really do differ from the planets orbiting the Sun. As search techniques improve, astronomers are finding more and more Jupiter- (and lower) mass planets on wider and less eccentric orbits. Planet hunters are quietly confident that advances during the next decade will either bring numerous detections of extrasolar planets in orbits comparable to those in the solar system or allow astronomers to conclude that systems like our own are indeed in the minority. Either way, the consequences are profound.

The properties of the observed extrasolar planets spell disaster for any low-mass terrestrial planets in these systems. In our solar system the presence of Jupiter on a nearly circular orbit is known to have a stabilizing influence on the other planetary orbits, tending to preserve the relative tranquility of our planetary environment. In the known extrasolar systems, not only is this stabilization absent, but having a Jupiter-sized planet repeatedly plow through the inner parts of the system means that any terrestrial planets or planetesimals have almost certainly been ejected from the system.

Is Our Solar System Unusual?

Astronomers have long reasoned that the condensation scenario described earlier in this chapter was in no way unique to our own system. The same basic processes could have occurred, and perhaps *did* occur, during the formative stages of many of the stars in our galaxy, in which case planetary systems like our own should be common. Today we know that planetary systems *are* apparently fairly common, but those discovered so far don't look like our own. Do these new observations undermine our theory of solar system formation or imply that our own solar system is in some way unusual?

Current theory in fact provides a number of ways in which massive planets can end up in the orbits we observe. A Jupiter-sized planet could have been knocked onto an eccentric orbit by an interaction with another giant planet, or even by the effects of a

▶ **Figure 4.32 Extrasolar Orbits**   The orbits of most of the extrasolar planets residing more than 0.15 AU from their parent star, superimposed on a single plot, with Earth's orbit shown (dashed) for comparison. The extrasolar planets are all comparable in mass to Jupiter—these planetary systems don't look much like ours!

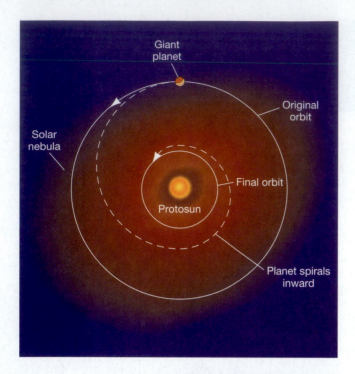

◀ **Figure 4.33** **Sinking Planet** Friction between a giant planet and the nebular disk in which it formed tends to make the planet spiral inward. The process continues until the disk is dispersed by the wind from the central star, possibly leaving the planet in a "hot Jupiter" orbit.

passing star. If it formed by gravitational instability, it might have had an eccentric orbit right from the start. And regardless of how it formed, theorists have long known that the interaction between a massive planet and the gas disk in which it moves tends to make it spiral inward and may deposit the planet in an orbit very close to the parent star (Figure 4.33). Thus, the observed planets are compatible with (if not actually predicted by) current theory. However, with no firm observational or theoretical data to tell us which conditions lead to stable planetary orbits like those we enjoy on Earth today, and which lead to the eccentric hot Jupiters observed elsewhere, we cannot yet assess whether our solar system is the norm or a rarity among the stars.

NASA has ambitious plans to search for Earth-like planets in our Galaxy. Perhaps surprisingly, the approach given the best chance of success entails searching for planets *transiting* (passing in front of) their parent stars. The effect of a transit on a star's brightness is tiny—less than one part in 10^4 for an Earth-like planet crossing the face of the Sun—and also very unlikely, as only nearly edge-on systems will show it. Nevertheless, it is measurable with existing technology. NASA's planned *Kepler* mission will monitor some 100,000 Sun-like stars for brightness fluctuations caused by Earth-like planets over a four-year period. With optimistic assumptions some 100 terrestrial transits are expected. If detected, they would imply that planets similar to our own are common in our Galaxy. *Kepler* is currently scheduled for launch in 2007.

CONCEPT CHECK ▶▶▶

Why is it not too surprising that the extrasolar planetary systems detected thus far have properties quite different from our solar system?

CHAPTER REVIEW

Summary

① The planets that make up our **solar system** (p. 100) all orbit the Sun counterclockwise, as viewed from above Earth's North Pole, on roughly circular orbits that lie close to the ecliptic plane. The orbit of Mercury is the most eccentric and also has the greatest orbital inclination. The spacing between planetary orbits increases as we move outward from the Sun. The inner **terrestrial planets** (p. 102)—Mercury, Venus, Earth, Mars—are all of comparable density and generally rocky, whereas all the outer **jovian planets** (p. 102)—Jupiter, Saturn, Uranus, and Neptune—have much lower densities and are made up mostly of gaseous or liquid hydrogen and helium.

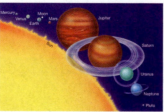

② Most **asteroids** (p. 103) orbit in a broad band called the **asteroid belt** (p. 104), lying between the orbits of Mars and Jupiter. The **Trojan asteroids** (p. 104) share Jupiter's orbit, remaining 60° ahead or behind that planet as it moves around the Sun. A few **Earth-crossing asteroids** (p. 104) have orbits that intersect Earth's orbit and will probably collide with our planet one

day. The largest asteroids are a few hundred kilometers across. Most are much smaller. Asteroids in the inner part of the asteroid belt are predominantly rocky in composition; asteroids orbiting farther out contain larger fractions of water ice and organic material.

③ **Comets** (p. 103) are fragments of icy and rocky material that normally orbit far from the Sun. We see a comet by the sunlight reflected from the dust and vapor released if its orbit happens to bring it near the Sun. Stretching behind the kilometer-size **nucleus** (p. 108) of a comet is a long **tail** (p. 108), formed by the interaction between the cometary material and the **solar wind** (p. 109). Most comets are thought to reside in the **Oort cloud** (p. 111), a vast reservoir of cometary material tens of thousands of astronomical units across, completely surrounding the Sun. Short-period comets, having orbital periods of less than about 200 years, originate in the **Kuiper belt** (p. 111), a broad band of material beyond the orbit of Neptune.

④ A **meteoroid** (p. 103) is a piece of rocky interplanetary debris smaller than 100 m across. Any meteoroid, asteroid, or comet fragment that enters Earth's atmosphere produces a **meteor** (p. 112), a bright streak of light across the sky. If part of the body causing the meteor reaches the ground, the remnant is called a **meteorite** (p. 113). Each time a comet rounds the Sun, some cometary material becomes dislodged, forming a **meteoroid swarm** (p. 113)—a group of small **micrometeoroids** (p. 113) that travel in the comet's orbit. Larger meteoroids are pieces of material chipped off asteroids following collisions in the asteroid belt. Most meteorites are between 4.4 and 4.6 billion years old.

5 The organization of the solar system, and the properties of planets, asteroids, and comets listed above, point to formation as a one-time event that occurred 4.6 billion years ago. An ideal theory of the solar system must provide strong reasons for the observed characteristics of our planetary system yet at the same time be flexible enough to allow for deviations.

6 As the **solar nebula** (p. 118) contracted under its own gravity, it began to spin faster, eventually forming a disk. **Protoplanets** (p. 121) formed in the disk and became planets, while the central **protosun** (p. 121) eventually evolved into the Sun. Particles of interstellar dust helped to cool the solar nebula and acted as **condensation nuclei** (p. 120), allowing the planet-building process to begin. Small clumps of matter grew

by **accretion** (p. 121), gradually sticking together and growing into moon-size **planetesimals** (p. 121) whose gravitational fields were strong enough to accelerate accretion. As planetesimals collided and merged, only a few planet-size objects remained.

7 Planets in the outer solar system became so large that they could capture the hydrogen and helium gas in the solar nebula, forming the jovian worlds. When the Sun became a star, its strong winds blew away any remaining gas in the solar nebula. The terrestrial planets are rocky because they formed in the hot inner regions of the solar nebula. Farther out, the nebula was cooler, so water and ammonia ice could also form. Many leftover planetesimals were ejected into the Oort cloud by the gravitational fields of the outer planets, leaving the Kuiper belt behind. The asteroid belt is a collection of planetesimals that never formed a planet, probably because of Jupiter's gravitational influence.

8 More than 150 **extrasolar planets** (p. 116) are now known. All were discovered by observing their parent star wobble back and forth as the planet orbits; one has since been observed passing in front of the star, reducing the star's brightness slightly. Most systems found so far contain a single, massive planet comparable in mass to Jupiter, on an orbit taking it close to the central star. None look much like our solar system, but this may be because current observational techniques are most sensitive only to the sorts of systems seen so far. It is currently not known whether our own solar system is unusual or normal among planetary systems.

Review and Discussion

1. Describe three differences between terrestrial and jovian planets.
2. Why are asteroids, comets, and meteoroids important to planetary scientists?
3. Do all asteroid orbits lie between Mars and Jupiter?
4. What might be the consequences of a 10-km asteroid striking Earth today?
5. What are comets like when they are far from the Sun? What happens when they enter the inner solar system?
6. Why do comets approach the Sun from any direction but asteroids generally orbit close to the ecliptic plane?
7. Explain the difference between a meteor, a meteoroid, and a meteorite.
8. What causes a meteor shower?
9. Describe the basic features of the nebular theory of solar system formation, and give three examples of how this theory explains some observed features of the present-day solar system.
10. What key ingredient in the modern condensation theory of solar system formation was missing from the original nebular theory?
11. Why are the jovian planets so much larger than the terrestrial planets?
12. How did the temperature at various locations in the solar nebula determine planetary composition?
13. Where did Earth's water come from?
14. How do astronomers detect extrasolar planets?
15. In what ways do observed extrasolar planetary systems differ from our own solar system?

Conceptual Self-Test: True or False?/Multiple Choice

1. The largest planets tend to have the lowest densities. **(T/F)**

2. The total mass of all the planets is comparable to that of the Sun. **(T/F)**

3. Asteroids were recently formed by the collision and breakup of an object orbiting in the asteroid belt. **(T/F)**

4. Most comets have short periods and orbit close to the ecliptic plane. **(T/F)**

5. The solar system is highly differentiated. **(T/F)**

6. Asteroids, meteoroids, and comets are remnants of the early solar system. **(T/F)**

7. Astronomers have no theoretical explanation for the "hot Jupiters" observed orbiting some other stars. **(T/F)**

8. If we were to construct an accurate scale model of the solar system on a football field with the Sun at one end and Pluto at the other, the planet closest to the center of the field would be **(a)** Earth; **(b)** Jupiter; **(c)** Saturn; **(d)** Uranus.

9. In the leading theory of solar system formation, the planets **(a)** were ejected from the Sun following a close encounter with another star; **(b)** formed from the same flattened, swirling gas cloud that formed the Sun; **(c)** are much younger than the Sun; **(d)** are much older than the Sun.

10. The solar system is differentiated because **(a)** all the heavy elements in the outer solar system have sunk to the center; **(b)** all the light elements in the inner solar system became part of the Sun; **(c)** all the light elements in the inner solar system were carried off in the form of comets; **(d)** only rocky and metallic particles could form close to the Sun.

11. Water on Earth was **(a)** transported here by comets; **(b)** accreted from the solar nebula; **(c)** outgassed from volcanoes in the form of steam; **(d)** created by chemical reactions involving hydrogen and oxygen shortly after Earth formed.

12. Astronomers have confirmed the existence of at least **(a)** less than 10; **(b)** roughly 50; **(c)** more than 150; **(d)** more than 1000 planets beyond our own solar system.

13. Astronomers have not yet reported any Earth-like planets orbiting other stars because **(a)** there are none; **(b)** they are not detectable with current technology; **(c)** no nearby stars are of the type expected to have Earth-like planets; **(d)** the government is preventing us from reporting their discovery.

14. According to the chart in *Discovery 4-1*, an impact resulting in global catastrophe is expected to occur roughly once per **(a)** year; **(b)** century; **(c)** millennium; **(d)** million years.

15. According to Figure 4.27, the temperature in the solar nebula at the location now the center of the asteroid belt is **(a)** 2000 K; **(b)** 900 K; **(c)** 400 K; **(d)** 100 K.

Problems

Algorithmic versions of these questions are available in the Practice Problems module of the Companion Website.

The number of squares preceding each problem indicates its approximate level of difficulty.

1. ■ Suppose the average mass of each of 20,000 asteroids in the solar system is 10^{17} kg. Compare the total mass of these asteroids to the mass of Earth. Assuming a spherical shape and a density of 3000 kg/m^3, estimate the diameter of an asteroid having this average mass.

2. ■■ The asteroid Icarus (Figure 4.5) has a perihelion distance of 0.19 AU and an orbital eccentricity of 0.83. What is its semimajor axis and aphelion distance from the Sun?

3. ■ The largest asteroid, Ceres, has a radius 0.073 times the radius of Earth and a mass of 0.0002 Earth masses. How much would an 80-kg astronaut weigh on Ceres?

4. ■■ It is observed that the number of asteroids (or meteoroids) of a given diameter is roughly inversely proportional to the square of the diameter. Assuming a density of 3000 kg/m^3 and approximating the actual distribution of asteroids and meteoroids as a single 1000-km body (Ceres), one hundred 100-km

bodies, ten thousand 10-km bodies, and so on, calculate the total mass in the form of 1000-km bodies, 100-km bodies, 10-km bodies, and 1-km bodies.

5. ■■ (a) Using the version of Kepler's laws of planetary motion from Section 1.5, calculate the orbital period of an Oort cloud comet if the semi-major axis of the comet's orbit is 50,000 AU. (b) What is the maximum possible aphelion distance for a short-period comet with an orbital period of 125 years?

6. ■■ The orbital angular momentum of a planet in a circular orbit is the product of mass times orbital speed times distance from the Sun (equivalent to the formula presented in *More Precisely 4-1*, but for a planet moving in orbit, instead of a body spinning about an axis). (a) Compare the orbital angular momenta of Jupiter, Saturn, and Earth. (b) What is the orbital angular momentum of an Oort cloud comet of mass 10^{13} kg, moving in a circular orbit 50,000 AU from the Sun?

7. ■■ Consider a planet growing by accretion of material from the solar nebula. As it grows, its density remains roughly constant. By what factor will the planet's surface gravity change as the radius doubles?

8. ■ How many 100-km-diameter rocky (3000 kg/m^3) planetesimals would have been needed to form Earth?

9. ■ A typical comet contains some 10^{13} kg of water ice. How many comets would have to strike Earth in order to account for the roughly 2×10^{21} kg of water presently found on our planet? If this amount of water accumulated over a period of 0.5 billion years, how frequently must Earth have been hit by comets during that time?

10. ■■■ Not surprisingly, as extrasolar planet masses decrease, they become harder to detect. Current Doppler techniques can reliably detect a planet having 0.25 times the mass of Jupiter orbiting a Sun-like star if the planet's *maximum* orbital speed exceeds 40 km/s. Under these conditions, what is the minimum detectable eccentricity for a planet with a five-year orbital period?

The Companion Website at www.prenhall.com/chaisson provides algorithmically generated versions of each chapter's Problems, in addition to an Animations and Videos gallery, a Tutorials gallery and Tutorials Quiz questions, a full eBook, an Images gallery, Practice Questions, Multiple Choice, and True or False quizzes, and Labeling exercises.

5

Earth and Its Moon

Our Cosmic Backyard

Photographs like this one, showing Earth hovering in space like a "blue marble," help us appreciate our place in the universe. The air, water, land, and life comprise a complex, interactive system that constantly changes. Scientists are now trying to understand not only the details of regional events, such as volcanoes, earthquakes, and weather, but also to learn more about the global changes affecting the whole planet. This image is a mosaic of many photographs taken by the GOES-7 environmental satellite. Note the hurricane in the Gulf of Mexico. (NOAA/NASA)

If we are to appreciate the universe, we must first come to know our own home. By cataloging Earth's properties and attempting to explain them, we set the stage for a comparative study of all the other planets. From an astronomical perspective, this is a compelling reason to study the structure and history of our own world. What about the Moon? It is by far our closest neighbor in space, yet despite its nearness, it is a world very different from our own. It has no air, no sound, no water. Weather is nonexistent: Clouds, rainfall, and blue sky are all absent; boulders and pulverized dust litter the landscape. Why, then, do we study it? In part, simply because it is our nearest neighbor and dominates our night sky. Beyond that, however, we study this body because it holds important clues to our own past. The very fact that it hasn't changed much since its formation means the Moon is a crucial key to unlocking the secrets of the solar system.

LEARNING GOALS

Studying this chapter will enable you to:

1 Summarize and compare the basic properties of Earth and the Moon.

2 Describe the consequences of gravitational interactions between Earth and the Moon.

3 Discuss how Earth's atmosphere helps heat us as well as protect us.

4 Outline our current model of Earth's interior structure and describe some experimental techniques used to establish this model.

5 Summarize the evidence for continental drift and discuss the physical processes that drive it.

6 Explain how dynamic events early in the Moon's history formed its major surface features.

7 Describe the nature and origin of Earth's magnetosphere.

8 Discuss the formation and evolution of Earth and the Moon.

Visit www.prenhall.com/chaisson for additional annotated images, animations, and links to related sites for this chapter.

5.1 Earth and the Moon in Bulk

In the Introduction and Chapter 1 we discussed some characteristics of Earth and the Moon known to ancient astronomers—the phases of the Moon, eclipses, and some basic geometrical inferences about the size of our planet. ∞ *(Sec. E3, E.4)* Today we have direct and much more detailed measurements of many physical properties of both Earth and the Moon. Let's start by cataloging a few facts about the Earth–Moon system.

Physical Properties

Table 5.1 presents some basic data on Earth and the Moon. The first five columns list mass, radius, and average density for Earth and the Moon, determined as described in Chapter 4. ∞ *(Sec. 4.1)* The next two columns present important measures of a body's gravitational field. **Surface gravity** is the strength of the gravitational force at the body's surface. ∞ *(More Precisely 1-1)* **Escape speed** is the speed required for any object—an atom, a baseball, or a spaceship—to escape forever from the body's gravitational pull (see *More Precisely 5-1*, p. 138). By either measure, the Moon's gravitational pull is much weaker than Earth's. As we will see, this fact has played an important role in determining the very different evolutionary paths followed by the two worlds. The final column in Table 5.1 lists (sidereal) rotation periods, which have long been accurately known from Earth-based observations.

These data will form the basis for our comparative study of our home planet and its nearest neighbor. We will expand our catalog of "planetary" characteristics and peculiarities in the next three chapters as we study in turn the other members of the solar system.

One important quantity not listed in Table 5.1 is the *distance* from Earth to the Moon. Parallax, with Earth's diameter as a baseline, has long provided astronomers with reasonably accurate measurements of the Earth–Moon distance. ∞ *(Sec. E.4)* However, radar- and laser-ranging[1] now yield far more accurate measurements of the Moon's orbit. ∞ *(Sec. 1.3)* Today the Moon's precise distance from Earth at any instant is known to within about 2 cm. The semi-major axis of the Moon's orbit around Earth is 384,000 km.

Overall Structure

Figure 5.1 compares and contrasts the main regions of these two very dissimilar worlds. As indicated in Figure 5.1(a), our planet may be divided into six main regions. In Earth's interior a thick **mantle** surrounds a smaller, two-part **core**. At the surface we have a relatively thin **crust**, comprising the solid continents and the seafloor; and the **hydrosphere**, comprising rivers, lakes, and the liquid oceans. An **atmosphere** of air lies just above the surface. At much greater altitudes a zone of charged particles trapped by our planet's magnetic field forms Earth's **magnetosphere**.

[1]This technique is conceptually similar to radar-ranging. It determines distance by measuring how long a pulse of laser light fired from Earth takes to return after reflection from the lunar surface.

TABLE 5.1	Some Properties of Earth and the Moon						
	MASS		RADIUS		AVERAGE DENSITY	SURFACE GRAVITY	ESCAPE SPEED	ROTATION
	(kg)	(Earth = 1)	(km)	(Earth = 1)	(kg/m³)	(Earth = 1)	(km/s)	PERIOD
Earth	6.0×10^{24}	1.00	6400	1.00	5500	1.00	11	23^{h}56^m
Moon	7.3×10^{22}	0.012	1700	0.27	3300	0.17	2.4	27.3 days

▶ **Figure 5.1** **Earth and Moon** The main regions of Earth and the Moon (not drawn to scale). (a) At Earth's center is an inner core, with a radius of about 1300 km. Surrounding the inner core is a liquid outer core 3500 km in radius. Most of the rest of Earth's 6400-km interior is taken up by the mantle, which is topped by a thin crust only a few tens of kilometers thick. The liquid portions of Earth's surface make up the hydrosphere. Above the hydrosphere and solid crust lies the atmosphere, most of it within 50 km of the surface. Earth's outermost region is the magnetosphere, extending thousands of kilometers into space. (b) The Moon's rocky outer mantle is about 900 km thick. The inner mantle is a semisolid layer similar to the upper regions of Earth's mantle. At the center lies the core, which may be partly molten.

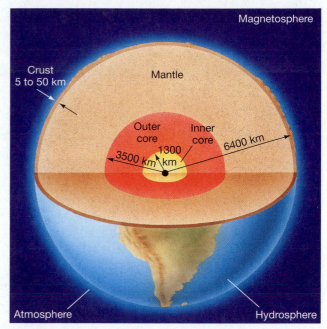

(a)

The Moon lacks a hydrosphere, an atmosphere, and a magnetosphere. Its internal structure (Figure 5.1b) is not as well studied as that of Earth, for the very good reason that the Moon is much less accessible. Nevertheless, as we will see, the basic interior regions on Earth—*crust*, *mantle*, and *core*—are also found on the Moon, although their properties differ somewhat from those of their Earthly counterparts.

5.2 The Tides

Earth is unique among the planets in that it has large quantities of liquid water on its surface. Approximately three-quarters of Earth's surface is covered by water, to an average depth of some 3.6 km. A familiar hydrospheric phenomenon is the daily fluctuation in ocean level known as the **tides**. At most coastal locations on Earth, there are two low tides and two high tides each day. The "height" of the tides—the magnitude of the variation in sea level—can range from a few centimeters to many meters, depending on the location and time of year, averaging about a meter on the open sea. An enormous amount of energy is contained in the daily tidal motion of the oceans.

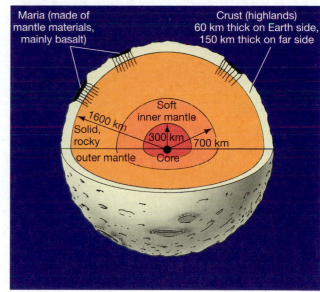

(b)

Gravitational Deformation

What causes the tides? A clue comes from the fact that they exhibit daily, monthly, and yearly cycles. In fact, the tides are a direct result of the gravitational influence of the Moon and the Sun on Earth. We have already seen how gravity keeps Earth and the Moon in orbit about one another, and both in orbit around the Sun. For simplicity, let's first consider just the interaction between Earth and the Moon.

Recall that the strength of the gravitational force depends on the inverse square of the distance separating any two objects. ∞ *(Sec. 1.4)* Thus, the Moon's gravitational attraction is greater on the side of Earth that faces the Moon than on the opposite side, some 12,800 km (Earth's diameter) farther away. This difference in the gravitational force is small, only about 3 percent, but it produces a noticeable deformation—a stretching along the line joining Earth to the Moon called a **tidal bulge**. The effect is greatest in Earth's oceans, because liquid can most easily move around on our planet's surface. As illustrated in Figure 5.2, the ocean becomes a little deeper in some places (along the Earth–Moon line) and shallower in others (in the perpendicular direction). The daily tides we see result as Earth rotates beneath this deformation.

The variation of the Moon's (or the Sun's) gravity across Earth is an example of a *differential force*, or **tidal force**. The *average* gravitational interaction between two bodies determines their orbit around one another. However, the tidal force, superimposed on that average, tends to deform the bodies. We will see many situations in this book where such forces are critically important in understanding astronomical phenomena. Note that the tidal force is not in any sense a "new" force. It is still the familiar force of gravity, just viewed in a

◄ **Figure 5.2 Lunar Tides** This exaggerated illustration shows how the Moon induces tides on both the near and far sides of Earth. The lengths of the arrows indicate the relative strengths of the Moon's gravitational pull on various parts of Earth. (a) The lunar gravitational forces acting on several locations on and inside Earth. The force is greatest on the side nearest the Moon and smallest on the opposite side. (b) The *differences* between the lunar forces experienced at the locations shown in part (a) and the force exerted by the Moon on Earth's center. Closest to the Moon, the relative force is toward the Moon because the Moon's gravitational pull is stronger at the surface than it is at the center. However, on the opposite side of Earth, the force at the surface is weaker than at the center, so the relative force is away from the Moon. The arrows thus represent the force with which the Moon tends to either pull matter away from or squeeze it toward Earth's center. Closest to the Moon, the oceans tend to be pulled away from Earth; on the far side, Earth tends to be pulled away from the oceans. The result is a tidal bulge.

different context—its effect on extended bodies such as planets and moons. Note also that we still use the word *tidal* in these other contexts, even though we are not discussing oceanic tides, and sometimes not even planets at all.

Notice in Figure 5.2 that the side of Earth opposite the Moon also experiences a tidal bulge. The different gravitational pulls—greatest on that part of Earth closest to the Moon, weaker at Earth's center, and weakest of all on Earth's opposite side—cause average tides on opposite sides of our planet to be approximately equal in height. On the side nearer the Moon, the ocean water is pulled slightly toward the Moon. On the opposite side, the ocean water is left behind as Earth is pulled closer to the Moon. Thus, high tide occurs *twice*, not once, every day.

Both the Moon and the Sun exert tidal forces on our planet. The tidal force drops off very rapidly with distance, but even though the Sun is around 375 times farther away from Earth than is the Moon, the Sun's mass is so much greater (by about a factor of 27 million) that its tidal influence is still significant—about half that of the Moon. Thus, instead of one tidal bulge being created on Earth, there are two: one pointing toward the Moon, the other toward the Sun. The interaction between them accounts for the changes in the height of the tides over the course of a month or a year. When Earth, Moon, and Sun are roughly lined up—at new or full Moon—the gravitational effects reinforce one another, and the high tides occur (Figure 5.3a). These tides are known as *spring tides*. When the Earth–Moon line is perpendicular to the Earth–Sun line (at the first and third quarters), the daily tides are smallest (Figure 5.3b). These are termed *neap tides*.

Tidal Locking

The Moon rotates once on its axis in 27.3 days—exactly the same time as it takes to complete one revolution around Earth. ∞ *(Sec. E.3)* As a result, the Moon presents the *same* face toward Earth at all times; that is, the Moon has a near side, which is always visible from Earth, and a far side, which never is (Figure 5.4a). This condition, in which the rotation period of a body is precisely equal to (or *synchronized* with) its orbital period around another body, is known as a **synchronous orbit**.

The fact that the Moon is in a synchronous orbit around Earth is no accident. It is an inevitable consequence of the tidal gravitational interaction between those two bodies. To see how this can occur, let's continue our discussion of tides on Earth. Because of Earth's rotation, the tidal bulge raised by the Moon does not point directly at the Moon, as indicated in Figure 5.2. Instead, through the effects of friction, both between the crust and the oceans and within Earth's interior, Earth's rotation tends to drag the tidal bulge

◄ **Figure 5.3 Solar and Lunar Tides** The combined effects of the Sun and the Moon produce variations in high and low tides. (a) When the Moon is either full or new, Earth, Moon, and Sun are approximately aligned, and the tidal bulges raised in Earth's oceans by the Moon and the Sun reinforce one another. (b) When the Moon is in either its first or its third quarter, the tidal effects of the Moon and the Sun partially cancel each other, and the tides are smallest. Because the Moon's tidal effect is greater than that of the Sun, the net bulge points toward the Moon.

around with it, causing the bulge to be displaced slightly ahead of the Earth–Moon line (Figure 5.4b). The Moon's asymmetrical gravitational pull on this slightly offset bulge acts to slow our planet's rotation.

According to fossil measurements, the rate of decrease is just two milliseconds every century—not much on the scale of a human lifetime, but it means that half a billion years ago Earth's day was just over 21 hours long and its year contained 410 days. At the same time, the Moon spirals slowly away from Earth, increasing its average distance from our planet by about 4 cm per century. This process will continue until (billions of years from now) Earth rotates on its axis at exactly the same rate as the Moon orbits Earth—that is, Earth's rotation will become synchronized with the Moon's motion. At that time the Moon will always be above the same point on Earth and will no longer lag behind the bulge it raises.

▶ **Figure 5.4 Tidal Locking** (a) As the Moon orbits Earth, it keeps one face permanently pointed toward our planet. To the astronaut shown here, Earth is always directly overhead. In fact, because of Earth's tides, the Moon is slightly elongated in shape, with its long axis perpetually pointing toward Earth. The elongation is highly exaggerated here. (b) The tidal bulge raised in Earth by the Moon does not point directly at the Moon. Instead, because of the effects of friction, the bulge points slightly ahead of the Moon, in the direction of Earth's rotation. (The magnitude of the effect is greatly exaggerated in this diagram.) Because the Moon's gravitational pull on the near-side part of the bulge is greater than the pull on the far side, the overall effect is to decrease Earth's rotation rate.

CONCEPT CHECK ▶▶▶

In what ways do tidal forces differ from the familiar inverse-square force of gravity?

Basically the same process is responsible for the Moon's synchronous orbit. Earth and the Moon have evolved together. Just as the Moon raises tides on Earth, Earth also produces a tidal bulge in the Moon. Indeed, because Earth is so much more massive than the Moon, the lunar tidal bulge is considerably larger, and the synchronization process correspondingly faster. The Moon's orbit became synchronous long ago—the Moon is said to have become *tidally locked* to Earth. Most of the moons in the solar system are similarly locked by the tidal forces of their parent planets.

5-1 MORE PRECISELY

Why Air Sticks Around

Why does Earth have an atmosphere but the Moon does not? Why doesn't our atmosphere disperse into space? The answer is that *gravity* holds it down. However, gravity is not the only influence acting. If it were, all of Earth's air would have fallen to the surface long ago. *Heat* competes with gravity to keep the atmosphere buoyant. Our atmosphere stays in place because the buoyancy force balances the force of gravity.

Let's explore this competition between gravity and heat in a little more detail. All gas molecules are in constant random motion. The temperature of any gas is a direct measure of this motion—the hotter the gas, the faster the molecules are moving. ∞ *(More Precisely 2-1)* The rapid movement of heated molecules creates *pressure* that tends to oppose the force of gravity, preventing our atmosphere from collapsing under its own weight.

An important measure of the strength of a body's gravity is its *escape speed*—the speed needed for any object to escape forever from its surface. This speed increases with increased mass or decreased radius of the parent body. In fact, the escape speed is proportional to the speed of a circular orbit at the body's surface. ∞ *(More Precisely 1-1)* In convenient units, it can be expressed as

$$\text{escape speed (in km/s)} = 11.2 \sqrt{\frac{\text{mass of body (in Earth masses)}}{\text{radius of body (in Earth radii)}}}.$$

In other words, you need high speed to escape the gravitational pull of a very massive or very small body, but you can escape from a less massive or larger body at lower speeds. If the *mass* of the parent body is quadrupled, the escape speed doubles. If the parent body's *radius* quadruples, then the escape speed is halved.

To determine whether or not a planet will retain an atmosphere, we must compare the planet's escape speed with the average speed of the gas particles making up the atmosphere. This speed depends not only on the temperature of the gas but also on the mass of the individual molecules—the hotter the gas or the smaller the molecular mass, the higher the average speed of the molecules:

$$\text{average molecular speed (in km/s)} = 0.157 \sqrt{\frac{\text{gas temperature(K)}}{\text{molecular mass (hydrogen atom masses)}}}.$$

Thus, increasing the absolute temperature of a sample of gas by a factor of four—from, for example, 100 K to 400 K—doubles the average speed of its constituent molecules. And at a given temperature, molecules of hydrogen in air move, on average, four times faster than molecules of oxygen, which are 16 times heavier.

At any instant, a tiny fraction of the molecules in any gas have speeds much greater than average, and some molecules are always moving fast enough to escape, even when the average molecular speed is much less than the escape speed. As a result, all planetary atmospheres slowly leak away into space. Don't be alarmed, however—the leakage is usually very gradual. As a rule of thumb, if the escape speed from a planet or moon exceeds the average speed of a given type of molecule by a factor of six or more, then molecules of that type will not have escaped from the planet's atmosphere in significant quantities since the solar system formed.

For air on Earth (with a temperature of about 300 K), the mean molecular speed of oxygen (mass = 32 hydrogen masses) and nitrogen (mass = 28) is about 0.6 km/s, comfortably below one-sixth of the escape speed (11.2 km/s). As a result, Earth is able to retain its atmosphere. However, if the Moon originally had an Earth-like atmosphere, the lunar escape speed of just 2.4 km/s means that any original atmosphere long ago dispersed into interplanetary space. The same reasoning can be used to understand atmospheric composition. For example, hydrogen molecules (mass = 2) move, on average, at about 2 km/s in Earth's atmosphere at sea level. Consequently, they have had plenty of time to escape since our planet formed, which is why we find very little hydrogen in Earth's atmosphere today.

5.3 Atmospheres

Earth's Atmosphere

Our planet's atmosphere is a mixture of gases, the most abundant of which are nitrogen (78 percent by volume), oxygen (21 percent), argon (0.9 percent), and carbon dioxide (0.03 percent). Water vapor is a variable part of the atmosphere, making up anywhere from 0.1 to 3 percent, depending on location and climate. The presence of a large amount of free oxygen makes our atmosphere unique in the solar system—Earth's oxygen is a direct consequence of the emergence of life on our planet. We will discuss the formation and evolution of planetary atmospheres in more detail in Chapter 6, when we consider the terrestrial planets as a whole.

Figure 5.5 shows a cross section of Earth's atmosphere. Compared with Earth's overall dimensions, the extent of the atmosphere is not great. Half of it lies within 5 km of the surface, and all but 1 percent of it is found below 30 km. The region below about 12 km is called the **troposphere**. Everything on Earth's surface lies within it—even Mount Everest, the tallest mountain on our planet, rises only 9 km above sea level. Above the troposphere, extending up to an altitude of 40–50 km, lies the **stratosphere**. Between 50 and 80 km from the surface lies the **mesosphere**. Above about 80 km, in the **ionosphere**, the atmosphere is kept partly ionized by solar ultraviolet radiation. These various atmospheric regions are distinguished from one another by the behavior of the temperature (decreasing or increasing with altitude) in each. Atmospheric pressure decreases steadily with increasing altitude.

The troposphere is the region of Earth's (or any other planet's) atmosphere where **convection** occurs. Convection is the constant upwelling of warm air and the concurrent downward flow of cooler air to take its place. In Figure 5.6, part of Earth's surface is heated by the Sun. The air immediately above the warmed surface is heated, expands a little, and becomes less dense. As a result, it becomes buoyant and starts to rise. At higher altitudes the opposite effect occurs: The air gradually cools, grows denser, and sinks back to the ground. The cool air at the surface rushes in to replace the hot, buoyant air that has risen. In this way, a circulation pattern is established. These *convection cells* of rising and falling air contribute to atmospheric heating and are responsible for surface winds and all the weather we experience. Above the troposphere the atmosphere is stable and the air is calm.

Within the stratosphere lies the **ozone layer**, where incoming solar ultraviolet radiation is absorbed by atmospheric oxygen, ozone, and nitrogen. (Ozone is a form of oxygen. Most atmospheric oxygen consists

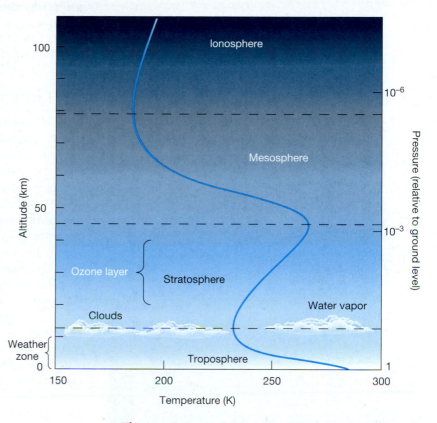

▲ **Figure 5.5 Earth's Atmosphere** Diagram of Earth's atmosphere, showing the changes in temperature (blue curve, bottom axis) and pressure (right-hand axis) from the surface to the bottom of the ionosphere. Pressure decreases steadily with increasing altitude, but the temperature may fall or rise, depending on height above the ground.

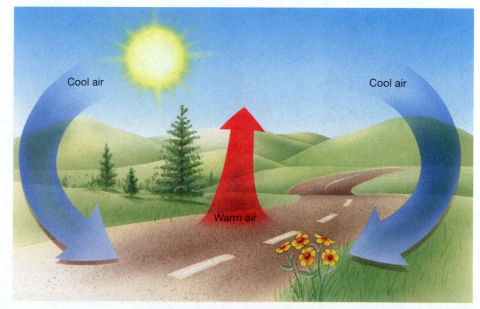

▶ **Figure 5.6 Convection** Convection occurs whenever cool matter overlies warm matter. The resulting circulation currents are familiar to us as the winds in Earth's atmosphere, caused by the solar-heated ground. Over and over, hot air rises, cools, and falls back to Earth. Eventually, steady circulation patterns are established and maintained, provided the source of heat (the Sun, in the case of Earth) remains intact.

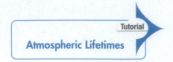

Tutorial
Atmospheric Lifetimes

of two oxygen atoms combined into a molecule; ozone consists of three oxygen atoms combined into each molecule.) The ozone concentration is greatest at an altitude of around 25 km. The ozone layer is one of the insulating layers that protect life on Earth from the harsh realities of outer space. By absorbing potentially dangerous high-frequency radiation, it acts as a planetary umbrella. Without it, advanced life (at least on Earth's surface) would be at best impaired and at worst impossible. *Discovery 5-1* discusses how human activities may now be affecting this vital part of our atmosphere.

The Greenhouse Effect

Most of the Sun's energy is emitted in the visible and near-infrared (that is, wavelengths only slightly longer than red light) regions of the electromagnetic spectrum. Because Earth's atmosphere is largely transparent to radiation of this type, almost all of the solar radiation not absorbed by or reflected from clouds in the upper atmosphere shines directly onto Earth's surface, heating it up. ∞ *(Sec. 2.3)*

Tutorial
The Greenhouse Effect

Animation/Video
Seasonal Changes on the Earth

Earth's surface reradiates the absorbed energy. As the temperature rises, the amount of radiated energy increases rapidly, according to Stefan's law, and eventually our planet radiates as much energy back into space as it receives from the Sun. In the absence of any complicating effects, this balance would be achieved at an average surface temperature of about 250 K (–23°C). At that temperature, Wien's law tells us that most of the reemitted energy is in the form of far-infrared (roughly 10 μm) radiation. ∞ *(More Precisely 2-2)*

However, there is a complication. Long-wavelength infrared radiation is partially blocked by Earth's atmosphere, mainly because of carbon dioxide and water vapor, both of which absorb very efficiently in the infrared portion of the spectrum. Even though these two gases account for only a tiny fraction of our atmosphere, they manage to absorb a large fraction of all the infrared radiation emitted from the surface. Consequently, only some of that radiation escapes back into space. The rest is radiated back to the surface, causing the temperature to rise.

This partial trapping of solar radiation is known as the **greenhouse effect**. The name comes about because a similar process operates in a greenhouse—sunlight passes relatively unhindered through glass panes, but much of the infrared radiation reemitted by the plants is blocked by the glass and cannot get out.[2] Consequently, the interior of the greenhouse heats up, and flowers, fruits, and vegetables can grow even on cold winter days. The radiative processes that determine the temperature of Earth's atmosphere are illustrated in Figure 5.7. Earth's greenhouse effect makes our planet about 40 K hotter than would otherwise be the case—a very important difference, as it raises the average temperature *above* the freezing point of water.

The magnitude of the greenhouse effect is very sensitive to the concentration of *greenhouse gases* (that is, gases that absorb infrared radiation efficiently) in the atmosphere. As just mentioned, water vapor and carbon dioxide are both important greenhouse gases. Carbon dioxide is of particular concern today, because its concentration in Earth's atmosphere is increasing, largely as a result of the burning of fossil fuels (principally oil and coal) in the industrialized world. Carbon dioxide levels have increased by more than 20 percent in the last century, and they continue to rise at a present rate of 4 percent per decade. Many scientists think that this increase, if left unchecked, may result in

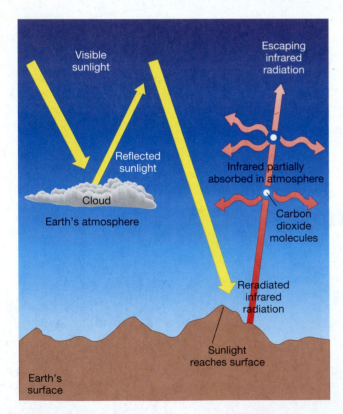

▲ **Figure 5.7 Greenhouse Effect** Sunlight that is not reflected by clouds reaches Earth's surface, warming it up. Infrared radiation reradiated from the surface is partially absorbed by carbon dioxide (and also water vapor, not shown here) in the atmosphere, causing the overall surface temperature to rise.

[2]Actually, although this process does contribute to warming the interior of a greenhouse, it is not the most important effect. A greenhouse works mainly because its glass panes prevent convection from carrying heat up and away from the interior. Nevertheless, the name "greenhouse effect" to describe the heating effect due to Earth's atmosphere has stuck.

global temperature increases of several kelvins over the next half-century, enough to melt much of the polar ice caps and cause dramatic, perhaps even catastrophic, changes in Earth's climate (see *Discovery 5-2*).

Lunar Air?

What about the Moon's atmosphere? That's easy—for all practical purposes, there is none! All of it escaped long ago. More massive objects have a better chance of retaining their atmospheres because the more massive an object, the greater the speed needed for atoms and molecules to escape (see *More Precisely 5-1*). The Moon's escape speed is only 2.4 km/s, compared with 11.2 km/s for Earth (Table 5.1). Simply put, the Moon has a lot less pulling power—any atmosphere it might once have had is gone forever.

5-1 DISCOVERY

Earth's Growing Ozone Hole

During the last two centuries, human technology has begun to produce measurable—and possibly permanent—changes in our planet. You can probably think of many instances, mostly negative, of such changes, from the threat of nuclear war to the reality of air and water pollution worldwide. One example very much in the news and relevant to the discussion in this chapter is the depletion of Earth's ozone layer. (See *Discovery 5-2* for an unrelated, but even more serious, threat to our environment.)

A particularly undesirable by-product of human technology is a group of chemicals known as chlorofluorocarbons (CFCs), relatively simple compounds once widely used for a variety of purposes—propellant in aerosol cans, dry-cleaning products, and in air conditioners and refrigerators. In the 1970s it was discovered that instead of quickly breaking down after use as had previously been thought, CFCs accumulate in the atmosphere and are carried high into the stratosphere by convection. There they are broken down by sunlight, releasing chlorine, which quickly reacts with ozone (O_3; see Section 5.3), turning it into oxygen (O_2). In chemical terms the chlorine is said to act as a *catalyst*—it is not consumed in the reaction, and so it survives to react with many more ozone molecules. A single chlorine atom can destroy up to 100,000 ozone molecules before being removed by other, less frequent chemical reactions. In this way, even a small quantity of CFCs turns out to be extraordinarily efficient at destroying atmospheric ozone.

As we saw in the text, ozone is part of the protective "blanket" of gases in our atmosphere that protect us from the harsh realities of outer space—in this case, solar ultraviolet radiation. Thus, the result of CFC emission is a substantial increase in ultraviolet radiation levels at Earth's surface, with detrimental effects to most living organisms. The accompanying figure shows the development of an ozone hole (colored magenta) over the Antarctic—a region where atmospheric circulation and low temperatures conspire each Antarctic spring to create a vast circumpolar cloud of ice crystals, which act to promote the ozone-destroying reactions, resulting in ozone levels about 50 percent below normal for the region.

Since the hole was discovered in the 1980s, its depth and area have grown each year. Its peak size is now larger than North America. Ozone depletion is not confined to the Antarctic, although the effect is greatest there. Smaller holes have been observed in the Arctic, and occasional ozone depletions of up to 20 percent have been reported at lower northern latitudes.

In the late 1980s, when the effect of CFCs on the atmosphere was realized, the world moved remarkably rapidly to curtail their production and use, with the goal of phasing them out entirely by 2030. Substantial cuts have already been made. Still, scientists think that even if all remaining CFC emissions were to stop today, it would still take several decades for CFCs to leave the atmosphere completely.

Lacking the moderating influence of an atmosphere, the Moon experiences wide variations in surface temperature. Noontime temperatures can reach 400 K, well above the boiling point of water (373 K). ∞ *(More Precisely 2-1)* At night (which lasts nearly 14 Earth days) or in the shade, temperatures fall to about 100 K, well below water's freezing point (273 K). With such high daytime temperatures and no atmosphere to help retain it, any water that ever existed on the Moon's surface has most likely evaporated and escaped. Not only is there no lunar hydrosphere, but all the lunar samples re-

5-2 DISCOVERY

The Greenhouse Effect and Global Warming

Discovery 5-1 outlined the danger to Earth's ozone layer posed by CFCs—products of modern technology with unexpected global consequences. An older, and potentially far more serious, hazard is posed by humanity's contribution to the *greenhouse effect* on our planet.

We saw the greenhouse effect in Section 5.3: so-called greenhouse gases in Earth's atmosphere, notably water vapor and carbon dioxide (CO_2), tend to trap heat leaving the surface, raising our planet's temperature by several tens of degrees Celsius. The greenhouse effect is not a bad thing—it is the reason that water exists in the liquid state on Earth's surface, and thus it is crucial to the existence and survival of life on our planet (see Chapter 18). However, if atmospheric greenhouse gas levels rise unchecked, the consequences can be catastrophic (for a particularly extreme example, see Section 6.8).

Since the Industrial Revolution in the eighteenth century, and particularly over the past few decades, human activities on Earth have steadily raised the level of carbon dioxide in our atmosphere, as illustrated in the first figure. Fossil fuels (coal, oil, and gas), still the dominant energy source of modern industry, all release CO_2 when burned. At the same time, the extensive forests that once covered much of our planet are being systematically destroyed to make room for human expansion. Forests play an important role in this problem because vegetation absorbs carbon dioxide, thus providing a natural control mechanism for atmospheric CO_2. Deforestation therefore also tends to increase the level of greenhouse gases in Earth's atmosphere.

Global warming is the slow rise in Earth's surface temperature caused by the increased greenhouse effect resulting from higher levels of atmospheric carbon dioxide. As shown in the second figure, average global temperatures have risen by about 0.5°C during the past century. This may not seem like much, but climate models predict that, if CO_2 levels continue to rise, a further increase of as much as 5°C may be possible by the end of the twenty-first century. Such a rise would be enough to cause serious climatic change on a global scale.

We saw in *Discovery 5-1* how, once the environmental impact of CFCs was identified, rapid steps were taken to curb their use. Curiously, given the more serious nature of the potential damage, a concerted response to global warming has been much slower in coming. Most scientists see the human-

enhanced greenhouse effect as a real threat to Earth's climate and urge prompt and deep reductions in CO_2 emissions, along with steps to slow and ultimately reverse deforestation. Some, however, particularly those connected with the industries most responsible for the production of greenhouse gases, argue that Earth's long-term response to increased greenhouse emissions is too complex for simple conclusions to be drawn and that immediate action is unnecessary. They suggest that the current temperature trend may be part of some much longer cycle or that natural environmental factors may in time stabilize, or even reduce, the level of CO_2 in the atmosphere without human intervention.

Given the stakes, it is perhaps not surprising that these arguments have become far more political than scientific in tone—not at all like the deliberative scientific method presented elsewhere in this text! The basic observations and much of the basic science are generally not seriously questioned, but the interpretation, long-term consequences, and proper response are all hotly debated. Separating the two sometimes is not easy, but the outcome may be of vital importance to life on Earth.

turned by the American and Soviet Moon programs were absolutely bone dry. Lunar rock doesn't even contain minerals containing water molecules locked within their crystal structure. Terrestrial rocks, conversely, are almost always 1 or 2 percent water.

The Moon is not entirely devoid of water, however. In November 1996 radar echoes from the U.S. *Clementine* spacecraft suggested the presence of water ice at the lunar poles. Since the Sun never rises more than a few degrees above the horizon, as seen from these regions, temperatures on the permanently shaded floors of craters near the poles never exceed about 100 K. Consequently, scientists theorized, ice could have remained permanently frozen there since being dumped on the Moon by comets during the very early days of the solar system, never melting or vaporizing and hence never escaping into space. ∞ *(Sec. 4.3)*

In March 1998, NASA announced that sensitive equipment on board the *Lunar Prospector* mission had confirmed *Clementine*'s findings and in fact had detected large amounts of water ice—possibly totaling trillions of tons—at both lunar poles. At first it appeared that the ice was mainly in the form of tiny crystals mixed with lunar dust, spread over many tens of thousands of square kilometers of deeply shadowed crater floors. However, subsequent analysis of the data suggests that much of the ice may exist in the form of smaller but more concentrated "lakes" of nearly pure material lying perhaps half a meter below the surface.

In an attempt to gain more information about lunar ice, NASA scientists decided to end the *Lunar Prospector* mission in a very spectacular way. As the spacecraft neared the end of its lifetime, in July 1999, it was directed to crash into one of the deep craters in which the ice was suspected to hide. The hope was that telescopes on Earth might detect spectroscopic signatures of water vapor released by the impact. No water vapor was seen, although the estimated probability of success was low—less than 10 percent—so no conclusion can be drawn from the fact that water was not directly observed. Lunar ice remains a strong possibility, but its existence has not yet been definitively verified.

5.4 Interiors

Seismology

Although we reside on Earth, we cannot easily explore our planet's interior, because drilling gear can penetrate rock only so far before breaking. No substance—not even diamond, the hardest known material—can withstand the conditions below a depth of about 10 km. That's rather shallow compared with Earth's 6400-km radius. Fortunately, geologists have developed techniques that can indirectly probe the deep recesses of our planet.

A sudden dislocation of rocky material near Earth's surface—an **earthquake**—causes the entire planet to vibrate a little. It literally rings like a giant bell (but one tuned so low that human ears cannot detect the sound). These vibrations are not random. They are systematic waves, called **seismic waves** (after the Greek word for "earthquake"), that move outward from the site of the quake. Like all waves, they carry information. This information can be detected and recorded using sensitive equipment—a *seismograph*—designed to monitor Earth tremors.

Decades of earthquake research have demonstrated the existence of many kinds of seismic waves. Two are of particular importance to the study of Earth's internal structure (Figure 5.8). First to arrive at a monitoring site after a distant earthquake are the primary waves, or *P-waves*. These are *pressure* waves, a little like ordinary sound waves in air, that alternately expand and compress the medium (the core or mantle) through which they move. Seismic

◀◀◀ CONCEPT CHECK

Why is the greenhouse effect important for life on Earth? What might be the consequences if the effect continues to strengthen?

(a) P-wave

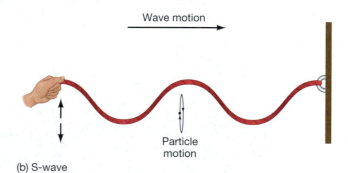

(b) S-wave

▲ **Figure 5.8 P- and S-waves**   (a) A pressure (P) wave traveling through Earth's interior causes material to vibrate in a direction parallel to the direction of motion of the wave. Material is alternately compressed and expanded. (b) A shear (S) wave produces motion perpendicular to the direction in which the wave travels, pushing material from side to side. In case (a), a typical particle oscillates forward and backward about its initial position. In (b), the particle moves up and down.

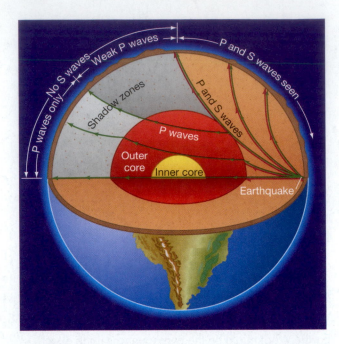

▲ Figure 5.9 **Seismic Waves** Earthquakes generate pressure (P, or primary) and shear (S, or secondary) waves that can be detected at seismographic stations around the world. The waves bend as they move through Earth's interior because of the variation in density and temperature within our planet. S-waves (colored red) are not detected by stations "shadowed" by the liquid core of Earth. P-waves (colored green) do reach the side of Earth opposite the earthquake, but their interaction with Earth's core produces another shadow zone, where almost no P-waves are seen.

P-waves usually travel at speeds ranging from 5 to 6 km/s and can travel through both liquids and solids. Some time later (the actual delay depends on the distance from the earthquake site), secondary waves, or *S-waves*, arrive. These are *shear* waves causing side-to-side motion, like waves in a guitar string. S-waves normally travel through Earth's interior at 3–4 km/s; however, they cannot travel through liquid, which absorbs them. The speed of each type depends on the density and physical state of the matter through which it is traveling. Consequently, by measuring the time taken for waves to move from the site of an earthquake to one or more monitoring stations on Earth's surface, geologists can infer the density of matter in the interior.

Modeling Earth's Interior

Because earthquakes occur often and at widespread places across the globe, geologists have accumulated a large amount of data about seismic-wave properties. They have used these data, along with direct knowledge of surface rocks, to model Earth's interior.

Figure 5.9 illustrates some paths followed by seismic waves from the site of an earthquake. Monitoring stations on the side of Earth opposite a quake never detect S-waves (colored red in the figure)—they are blocked by material within Earth's interior. Furthermore, while P-waves (green) always arrive at stations diametrically opposite the quake, other parts of Earth's surface receive almost none. Most geologists think that S-waves are absorbed by a liquid core at Earth's center and that P-waves are refracted at the core boundary, much as light is refracted by a lens. The result is the S- and P-wave "shadow zones" we observe. The fact that every earthquake exhibits these shadow zones is the best evidence that the core of our planet is molten. The radius of this **outer core**, as determined from seismic data, is about 3500 km. There is also evidence that the P-waves, which can pass through the liquid outer core, are reflected off the surface of a solid **inner core**, of radius 1300 km.

Figure 5.10 presents a model that most scientists accept. Earth's outer core is surrounded by a thick mantle and topped with a thin crust. The mantle is about 3000 km thick and accounts for the bulk (80 percent) of our planet's volume. The crust has an average thickness of only 15 km—a little less (around 8 km) under the oceans and somewhat more (20–50 km) under the continents. Note just how thin the crust is relative to the size of our planet: 15 km, about twice the height of the tallest mountains on our planet, is a mere one quarter of 1 percent of Earth's radius. The average density of crust material is around 3000 kg/m^3. Density and temperature both increase with depth. From Earth's surface to its very center, the density increases from roughly 3000 kg/m^3 to a little more than 12,000 kg/m^3, while the temperature rises from just under 300 K to well over 5000 K. Much of the mantle has a density midway between the densities of the core and crust: about 5000 kg/m^3.

The high central density suggests to geologists that the inner parts of Earth must be rich in nickel and iron. Under the heavy pressure of the overlying layers, these metals (whose densities under surface conditions are around 8000 kg/m^3) can be compressed to the high densities predicted by the model. The sharp density increase at the mantle–core boundary results from the difference in composition between the two regions. The mantle is composed of dense but rocky material—compounds of silicon and oxygen. The core consists primarily of even denser metallic elements. There is no similar jump in density or temperature at the inner core boundary—the material there simply changes from the liquid to the solid state.

Because geologists have been unable to drill deeper than about 10 km, no experiment has yet recovered a sample of the mantle. However, we are not entirely ignorant of the mantle's properties. In a *volcano*, molten rock upwells from below the crust, bringing a little of the mantle to us in the form of lava and providing some inkling of the composition of Earth's interior. The composition of the upper mantle is probably quite similar to the dark gray material known as *basalt* often found near volcanoes. With a density between 3000 and 3300 kg/m³, it contrasts with the lighter *granite* (density 2700–3000 kg/m³) that constitutes much of Earth's crust.

Differentiation

Earth, then, is not a homogeneous ball of rock. Instead, it has a layered structure, with a low-density rocky crust at the surface, intermediate-density rocky material in the mantle, and a high-density metallic core. (As we will see, the other terrestrial planets are thought to have similar internal structures.) This variation in density and composition is known as **differentiation**. Why isn't our planet just one big, rocky ball of uniform density? The answer is that at some time in the distant past, much of Earth was *molten*, allowing the higher-density matter to sink to the core, displacing lower-density material toward the surface. A remnant of this ancient heating exists today: Earth's central temperature is nearly equal to that of the Sun's surface. Two processes played important roles in heating Earth to the point where differentiation could occur. First, very early in its history, our planet experienced a violent bombardment by interplanetary debris, an essential part of the process by which Earth and the other planets formed and grew. ∞ *(Sec. 4.3)* This bombardment probably generated enough heat to melt much of our planet.

The second process that heated Earth after its formation and contributed to differentiation was **radioactivity**—the release of energy by certain unstable elements, such as uranium and thorium. These elements emit energy as their complex, heavy nuclei break up into simpler, lighter ones. Rock is such a poor conductor of heat that the energy released through radioactivity took a very long time to reach the surface and leak away into space. As a result, the heat built up in the interior, adding to the energy left there by Earth's formation. Geologists believe that enough radioactive elements were originally spread throughout the primitive planet, like raisins in a cake, that the entire planet—from crust to core—could have melted and remained molten, or at least semisolid, for about a billion years.

The Lunar Interior

The Moon's average density, about 3300 kg/m³, is quite similar to the density of the lunar surface rock obtained by U.S. and Soviet missions. This similarity all but eliminates any chance that the Moon has a large, massive, dense nickel–iron core like that within Earth. The low average lunar density suggests that the Moon contains substantially fewer heavy elements (such as iron) than does Earth.

Most of our detailed information on the Moon's interior comes from seismic data obtained from equipment left on the lunar surface by astronauts. These measurements indicate only very weak *moonquakes* deep within the lunar interior. Even if you stood directly above one of these quakes, you would not feel the vibrations. The average moonquake releases about as much energy as a firecracker, and no large quakes have ever been detected. This barely perceptible seismic activity confirms the idea that the Moon is geologically dead. Nevertheless, researchers can use these weak lunar vibrations to obtain information about the Moon's interior.

Combining all available data with mathematical models of the lunar interior indicates that the Moon is of almost uniform density, but chemically differentiated—that is, the chemical properties change from core to surface. As noted in Figure 5.1(b), the models suggest a central core perhaps 300 km in radius, surrounded by a 400-km-thick inner mantle of semisolid rock having properties similar to those of Earth's upper mantle. Above these regions lies a 900-km-thick outer mantle of solid rock, topped by a 60- to 150-km-thick crust.

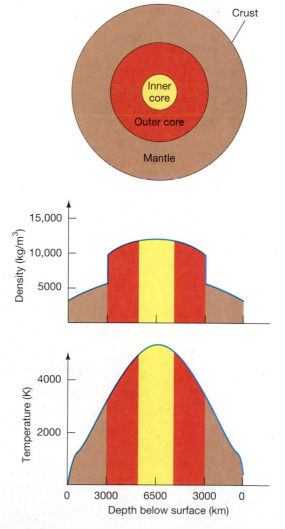

▲ **Figure 5.10 Earth's Interior**   Computer models of Earth's interior imply that the density and temperature vary considerably through the mantle and the core. Note the sharp density change between Earth's core and mantle.

The core is probably somewhat more iron-rich than the rest of the Moon, although it is still iron-poor relative to Earth's core. Data from the gravity experiment aboard the *Lunar Prospector* probe, combined with magnetic measurements made as the Moon passed through Earth's magnetic "tail" (see Figure 5.22), suggest that the Moon may have a small iron core, perhaps 300 km in radius. Near the center the models imply that the temperature may be as low as 1500 K, too cool to melt rock. However, some of the seismic data suggest that the inner core may be at least partially molten, implying a higher temperature. Our knowledge of the Moon's deep interior is still quite limited.

The crust on the Moon's far side is considerably thicker (150 km) than the crust on the side nearer Earth (60 km). Why is this? The answer is probably connected to Earth's gravitational pull. Just as heavier material tries to sink to the center of Earth, the denser far-side lunar mantle tended to sink below the lighter far-side crust in the presence of Earth's gravitational field. In other words, while the Moon was cooling and solidifying, the far-side lunar mantle was pulled a little closer to Earth than the far-side crust. In this way, the crust and mantle became slightly off-center with respect to one another. The result was the thicker crust on the Moon's far side.

CONCEPT CHECK ▶▶▶

How would our knowledge of Earth's interior be changed if our planet were geologically dead, like the Moon?

5.5 Surface Activity on Earth

Earth is geologically alive today. Its interior seethes, and its surface constantly changes. Many clear indicators of geological activity, in the form of earthquakes and volcanic eruptions, are scattered across our globe. Erosion by wind and water has obliterated much of the evidence from ancient times, but the sites of more recent activity are well documented.

Continental Drift

Figure 5.11 is a map of the currently active areas of our planet. Nearly all these sites have experienced surface activity during this century. The intriguing

▶ **Figure 5.11 Global Plates**
The red dots represent active sites where major volcanoes or earthquakes have occurred in the twentieth century. Taken together, the sites outline vast plates, indicated in dark blue, that drift around on the surface of our planet. The white arrows show the general directions and speeds of plate motions.

aspect of the figure is that the active sites are not spread evenly across our planet. Instead, they trace well-defined lines of activity, where crustal rocks shift (resulting in earthquakes) or mantle material upwells (in volcanoes).

In the mid-1960s, scientists realized that these lines are the outlines of gigantic *plates*, or slabs of Earth's surface, and that these plates are slowly drifting around the surface of our planet. These plate motions, popularly known as *continental drift*, have created the mountains, oceanic trenches, and many other large-scale features across the face of planet Earth. The technical term for the study of plate movement and the reasons for it is **plate tectonics**. Some plates are made mostly of continental landmasses, some are made of a continent plus a large part of an ocean floor, and some contain no continental land at all and are made solely of ocean floor. For the most part, the continents are just passengers riding atop much larger plates.

The plates move at an extremely slow rate. Typical speeds amount to only a few centimeters per year—about the same speed as your fingernails grow. Nevertheless, over the course of Earth history, the plates have had plenty of time to move large distances. For example, a drift rate of 2 cm per year can cause two continents (for example, Europe and North America) to separate by 4000 km—the width of the Atlantic Ocean—in 200 million years. That's a long time by human standards, but it represents only about 5 percent of Earth's age.

As the plates drift around, we might expect collisions to be routine. Indeed, plates do collide, but they are driven by enormous forces and do not stop easily. Instead, they just keep crunching into one another. Figure 5.12 shows the result of an ongoing collision as the Indian Plate thrusts northward into the Eurasian Plate. As Earth's rocky crust crumples and folds, mountains are formed—in this case, the Himalayan range, including Mount Everest, where the surface has been lifted almost 10 km by tectonic forces.

Not all plates collide head-on. As indicated by the arrows on Figure 5.11, sometimes plates slide or shear past one another. A good example is the most famous active region in North America—the San Andreas Fault in California (Figure 5.13), which forms part of the boundary between the Pacific and North American plates. Along the fault, these two plates are not moving in quite the same direction, nor at quite the same speed. Like parts in a poorly oiled machine, their motion is neither steady nor smooth. Instead, they tend to stick, then suddenly lurch forward as surface rock gives way. The resulting violent, jerky movements have been the cause of many major earthquakes along the fault line.

At still other locations, such as the seafloor under the Atlantic Ocean, the plates are moving apart. As they recede, new mantle material wells up between them, forming midocean ridges. Today hot mantle material is rising through a crack all along the Mid-Atlantic Ridge, which extends, like a seam on an enormous baseball, all the way from the North Atlantic to the southern tip of South America. Radioactive dating indicates that material has been upwelling along the ridge more or less steadily for the past 200 million years. The Atlantic seafloor is slowly growing, as the North and South American plates move away from the Eurasian and African plates.

What Drives the Plates?

What is responsible for the enormous forces that drag plates apart in some locations and ram them together in others? The answer (Figure 5.14) is *convection*—the same process we encountered earlier in our study of Earth's atmosphere (Section 5.3). Each plate is made up of crust plus a small portion of upper mantle. Below the plates, at a depth of perhaps 50 km, the temperature is sufficiently high that the mantle at that depth is soft enough to flow, very slowly, although it is not molten.

▲ **Figure 5.12 Himalayas** Mountain building results mostly from plate collisions. Here, the folding of rock is visible (in the foreground) near Mount Everest, which is part of the Himalayan mountain range at the northern end of the Indian Plate. (L. Day/Black Star)

▲ **Figure 5.13 Californian Fault** A small portion of the San Andreas fault in California. The fault is the result of the North American and Pacific plates sliding past one another. The Pacific plate, which includes a large slice of the California coast, is drifting to the northwest relative to the North American plate. The motions of the two plates are indicated by arrows whose lengths are proportional to the plate speeds. (U.S. Department of the Interior)

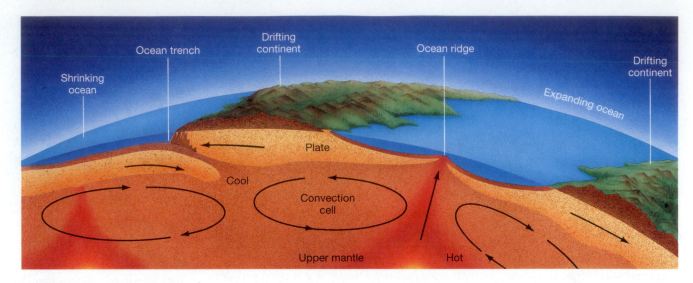

▲ **Figure 5.14 Plate Drift** Plate drift is caused by convection patterns in the upper mantle that drag the plates across Earth's surface.

This is a perfect setting for convection—warm matter underlying cool matter. The warm mantle rock rises, just as hot air rises in our atmosphere, and large circulation patterns become established. Riding atop these convection patterns are the plates. The circulation is extraordinarily sluggish. Semisolid rock takes millions of years to complete one convection cycle. Although the details are far from certain and remain controversial, many researchers believe that it is large-scale convection patterns near plate boundaries that cause the plates to move.

Figure 5.15 illustrates how all the continents nearly fit together like pieces of a puzzle. Geologists think that sometime in the past a single gargantuan landmass dominated our planet. This ancestral supercontinent, known as *Pangaea* (meaning "all lands"), is shown in Figure 5.15(a). The rest of the

▶ **Figure 5.15 Pangaea** Given the current estimated drift rates and directions of the plates, we can trace their movements back into the past. About 200 million years ago, they would have been at the approximate positions shown in (a). The continents' current positions are shown in (d).

planet was covered with water. The present locations of the continents, along with measurements of their current drift rates, suggest that Pangaea was the major land feature on Earth approximately 200 million years ago. Dinosaurs, which were then the dominant form of life, could have sauntered from Russia to Texas via Boston without getting their feet wet. The other frames in Figure 5.15 show how Pangaea split apart, its separate pieces drifting across Earth's surface, eventually becoming the familiar continents we know today. Very probably there has been a long series of "Pangaeas" stretching back in time over much of Earth's history, as tectonic forces have continually formed, destroyed, and re-formed our planet's landmasses. There probably will be many more.

Plate Tectonics on the Moon

There is no evidence for plate tectonics on the Moon today—no obvious extensive fault lines, no significant seismic activity, no ongoing mountain building. Plate tectonics requires both a relatively thin outer rocky layer, which is easily fractured into continent-sized pieces, and a soft, convective region under it, to make the pieces move. On the Moon neither of these ingredients exists. The Moon's thick crust and solid upper mantle make it impossible for pieces of the surface to move relative to one another. There simply isn't enough energy left in the lunar interior for plate tectonics to work.

◄◄◄ | **CONCEPT CHECK** ►

Describe the causes and some consequences of plate tectonics on Earth.

5.6 The Surface of the Moon

On Earth the combined actions of air, water, and geological activity erode our planet's surface and reshape its appearance almost daily. As a result, most of the ancient history of our planet's surface is lost to us. The Moon, though, has no air, no water, and no ongoing volcanic or other geological activity (although, as we will see, there is strong evidence that the young Moon underwent a period of intense geological activity before settling into its current inactive state). Consequently, features dating back almost to the formation of the Moon are still visible today. For this reason, studies of the lunar surface are of great importance to Earth geologists and have played a major role in shaping theories of the early development of our planet.

Large-Scale Features

The first observers to point their telescopes at the Moon noted large, roughly circular, dark areas that resemble (they thought) Earth's oceans. They called these regions **maria**, a Latin word meaning "seas" (singular: *mare*). The largest of them (Mare Imbrium, the Sea of Showers) is about 1100 km in diameter. Today we know that the maria are actually extensive flat plains that resulted from the spread of lava during an earlier, volcanic period of lunar evolution. In a sense, then, the maria *are* oceans—ancient seas of molten lava, now solidified.

Early observers also saw light-colored areas that resembled Earth's continents. Originally dubbed *terrae*, from the Latin word for "land," these regions are now known to be elevated several kilometers above the maria. Accordingly, they are usually called the lunar **highlands**. Both types of region are visible in Figure 5.16, a photographic mosaic of the full Moon. These light and dark surface features are also evident to the naked eye, creating the face of the familiar "man-in-the-Moon."

We have already seen numerous examples of how science progresses hand in hand with technology, and studies of the Moon are no exception. However, the Moon (for now, at least) is unique in astronomy in that manned and unmanned space probes have actually explored parts of its surface and returned

Animation/Video
One Small Step

R | I | V | U | X | G

▲ **Figure 5.16 Full Moon, Near Side**   A photographic mosaic of the full Moon, north pole at the top. Because the Moon emits no visible radiation of its own, we can see it only by the reflected light of the Sun. Some prominent maria are labeled. (UC/Lick Observatory)

R I V U X G

▲ **Figure 5.17 Full Moon, Far Side** The far side of the Moon, as seen by the *Clementine* military spacecraft. The large, dark region at center bottom outlines the South Pole–Aitken Basin, the largest and deepest impact basin known in the solar system. Only a few small maria exist on the far side. (U.S. Defense Department)

Animation/Video
Photographs of Full Rotation of the Moon

samples to Earth. Based on analyses of lunar rock brought back to Earth by *Apollo* astronauts and robot Soviet landers, geologists have found important differences in both *composition* and *age* between the highlands and the maria.

The highlands are made largely of rocks rich in aluminum, making them lighter in color and lower in density (2900 kg/m^3). The maria's basaltic matter contains more iron, giving it a darker color and greater density (3300 kg/m^3). Loosely speaking, the highlands represent the Moon's crust, while the maria are made of mantle material. Maria rock is quite similar to terrestrial basalt, and geologists believe that it arose on the Moon much as basalt did on Earth, through the upwelling of molten material through the crust. Radioactive dating indicates ages of more than four billion years for highland rocks and from 3.2 to 3.9 billion years for those from the maria.[3]

Until spacecraft flew around the Moon, no one on Earth had any idea what the Moon's hidden half looked like. To the surprise of most astronomers, when the far side of the Moon was mapped, first by Soviet and later by American spacecraft, no major maria were found. The lunar far side (Figure 5.17) is composed almost entirely of highlands.

Cratering

Because the smallest lunar features we can distinguish with the naked eye are about 200 km across, we see little more than the maria and highlands when we gaze at the Moon. Through a telescope, however (Figure 5.18), we find that the lunar surface is scarred by numerous bowl-shaped **craters** (after the Greek word for "bowl").

[3]Radioactive dating compares the rates at which different radioactive elements in a sample of rock decay into lighter elements. The age returned by this technique is the time since the rock solidified.

▼ **Figure 5.18 Moon, Close-up** (a) The Moon near third quarter. Surface features are most visible near the *terminator*, the line separating light from dark, where sunlight strikes at a sharp angle and shadows highlight the topography. (b) Magnified view of a region near the terminator, as seen from Earth through a large telescope. The central dark area is Mare Imbrium, ringed at the bottom right by the Apennine mountains. (c) Enlargement of a portion of (b). The smallest craters visible here have diameters of about 2 km, about twice the size of the Barringer Crater shown in Figure 4.17. (UC/Lick Observatory; Caltech)

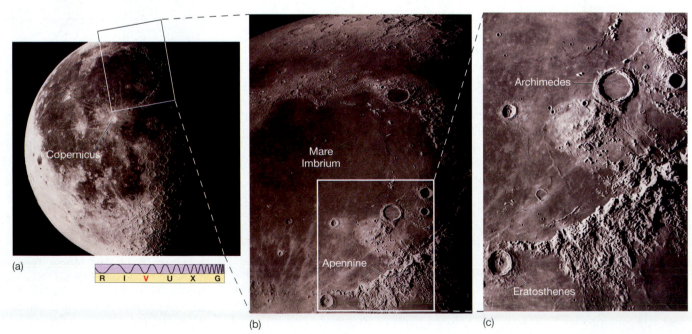

R I V U X G

Most craters formed eons ago as the result of meteoritic impact. ∞ *(Sec. 4.2)* (Compare the ancient craters in Figures 5.18c and 5.20 with the much more recent Barringer Crater in Figure 4.17.) Meteoroids generally strike the Moon at speeds of several kilometers per second. At these speeds even a small piece of matter carries an enormous amount of energy—for example, a 1-kg object hitting the Moon's surface at 10 km/s would release as much energy as the detonation of 10 kg of TNT. As illustrated in Figure 5.19, impact by a meteoroid causes sudden and tremendous pressures to build up on the lunar surface, heating the normally brittle rock and deforming the ground. The ensuing explosion pushes previously flat layers of rock up and out, forming a crater.

The material thrown out by the explosion surrounds the crater in a layer called an *ejecta blanket*, the ejected debris ranging in size from fine dust to large boulders. The larger pieces of ejected material may themselves create secondary craters. Many of the rock samples brought back by the *Apollo* astronauts show patterns of repeated shattering and melting—direct evidence of the violent shock waves and high temperatures produced in meteoritic impacts.

Lunar craters come in all sizes, reflecting the range in sizes of the impactors that create them. The largest craters are hundreds of kilometers in diameter, the smallest microscopic. Because the Moon has no protective atmosphere, even tiny interplanetary fragments can reach the lunar surface unimpeded. Figure 5.20(a) shows the result of a large meteoritic impact on the Moon, Figure 5.20(b) a crater formed by a micrometeoroid.

Craters are found everywhere on the Moon's surface, but the older highlands are much more heavily cratered than the younger maria. Knowing the ages of the highlands and maria, researchers can estimate the rate of cratering in the past. They conclude that the Moon, and presumably the entire inner solar system, experienced a sudden sharp drop in meteoritic bombardment rate about 3.9 billion years ago. The rate of cratering has been roughly constant since that time.

This time—3.9 billion years in the past—is taken to represent the end of the accretion process through which planetesimals became planets. ∞ *(Sec. 4.3)* The lunar highlands solidified and received most of their craters before that time. The great basins that formed the maria are thought to have been created during the final stages of heavy meteoritic bombardment between about 4.1 and 3.9 billion years ago. Subsequent volcanic activity filled the craters with lava, creating the formations we see today.

Lunar Erosion

Meteoritic impact is the only important source of erosion on the Moon. Over billions of years, collisions with meteoroids, large and small, have scarred, cratered, and sculpted the lunar landscape. At the present average rates, one new 10-km-diameter lunar crater is formed every 10 million years, one new 1-m-diameter crater is created about once a month, and 1-cm-diameter craters are formed every few minutes. In addition, a steady "rain" of micrometeoroids also eats away at the lunar surface (see Figure 5.21). The accumulated dust from countless impacts (called the lunar *regolith*) covers the lunar surface to an average depth of about 20 m, thinnest on the maria (about 10 m) and thickest on the highlands (more than 100 m in places).

(a)

(b)

(c)

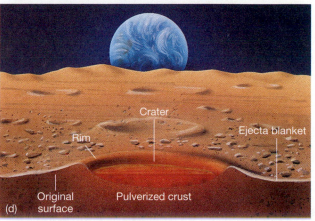
(d)

▶ **Figure 5.19 Meteoroid Impact** *Stages in the formation of a crater by meteoritic impact. (a) The meteoroid strikes the surface, releasing a large amount of energy. (b, c) The resulting explosion ejects material from the impact site and sends shock waves through the underlying surface. (d) Eventually, a characteristic crater surrounded by a blanket of ejected material results.*

▶ **Figure 5.20 Lunar Craters**
(a) Two smaller craters called Reinhold and Eddington sit amid the secondary cratering resulting from the impact that created the 90-km-wide Copernicus Crater (near the horizon) about a billion years ago. The ejecta blanket from crater Reinhold, 40 km across, in the foreground, can be seen clearly.
(b) Craters of all sizes litter the lunar landscape. Some shown here, embedded in glassy beads retrieved by *Apollo* astronauts, measure only 0.01 mm across. (The scale at the top is in millimeters.) The beads themselves were formed during the explosion following a meteoroid impact, when surface rock was melted, ejected, and rapidly cooled. (NASA)

(a)

(b)

▶ **Figure 5.21 Lunar Surface** The lunar surface is not entirely changeless. Despite the complete lack of wind and water on the airless Moon, the surface has still eroded a little under the constant "rain" of impacting meteoroids, especially micrometeoroids. Note the soft edges of the background hills of this image. In the absence of erosion, these features would be as jagged and angular today as they were when they formed. Note also the *Apollo* astronaut's boot prints in the foreground lunar dust. Those markings in the soil are only a few centimeters deep, but the low lunar erosion rate means that they will probably survive for more than a million years. (NASA)

Despite this barrage from space, the Moon's present-day erosion rate is still very low—about 10,000 times less than on Earth. For example, the Barringer Meteor Crater (Figure 4.17) in the Arizona desert, one of the largest meteor craters on Earth, is only 25,000 years old, but it is already decaying. It will probably disappear completely in a mere million years, quite a short time geologically. If a crater that size had formed on the Moon even a billion years ago, it would still be plainly visible today.

◀◀◀ **CONCEPT CHECK**

Describe three important ways in which the lunar maria differ from the highlands.

5.7 Magnetospheres

Simply put, the magnetosphere is the region around a planet that is influenced by that planet's magnetic field. It forms a buffer zone between the planet and the high-energy particles of the solar wind. It can also provide important insights into the planet's interior structure.

Earth's Magnetosphere

Shown in Figure 5.22, Earth's magnetic field extends far above the atmosphere, completely surrounding our planet. The *magnetic field lines*, which indicate the strength and direction of the field at any point in space, run from south to north, as indicated by the white arrowheads in the figure. The north and south *magnetic poles*, where the axis of an imaginary bar magnet within our planet (Figure 5.23) intersects Earth's surface, are very roughly aligned with Earth's spin axis.

Earth's inner magnetosphere contains two doughnut-shaped zones of high-energy charged particles, one located about 3000 km and the other 20,000 km above Earth's surface. These zones are named the **Van Allen belts**, after the American physicist whose instruments on board some early rocket flights during the late 1950s first detected them. We call them *belts* because they are most pronounced near Earth's equator and because they completely surround the planet. Figure 5.23 shows how these invisible regions envelop Earth except near the North and South Poles.

The particles that make up the Van Allen belts originate in the solar wind. Traveling through space, neutral particles and electromagnetic radiation are

◀◀◀ **CONCEPT CHECK**

What does the existence of a planetary magnetic field tell us about the planet's interior?

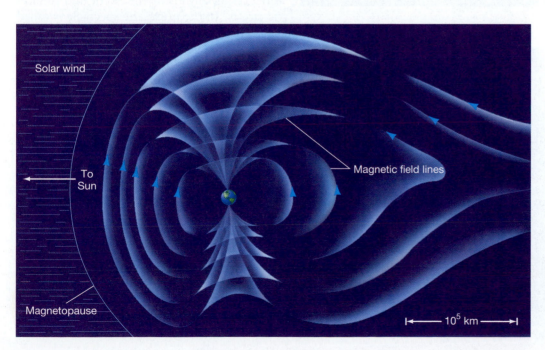

Solar wind

To Sun

Magnetic field lines

Magnetopause

|← 10⁵ km →|

◀ **Figure 5.22 Earth's Magnetosphere**
The magnetosphere is the region surrounding a planet wherein particles from the solar wind are trapped by the planet's magnetic field. Shown here is Earth's magnetosphere, with blue arrowheads on the field lines indicating the direction in which a compass needle would point. Far from Earth, the magnetosphere is greatly distorted by the solar wind, with a long "tail" extending from the nighttime side of Earth (here, at right) well into space. The magnetopause is the boundary of the magnetosphere in the sunward direction.

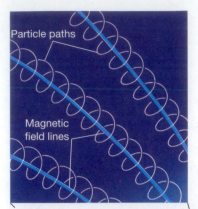

Particle paths

Magnetic
field lines

▼ Figure 5.23 Van Allen Belts Earth's magnetic field resembles somewhat the field of an enormous bar magnet buried inside our planet. High above Earth's atmosphere, the magnetosphere (light blue-green area) contains two doughnut-shaped regions (purple areas) of magnetically trapped charged particles. These are the Van Allen belts. The inset illustrates how charged particles in a magnetic field spiral around the field lines and thus tend to become trapped by Earth's magnetic field.

Van Allen belts

Magnetic
axis

unaffected by Earth's magnetism, but electrically charged particles are strongly influenced. As illustrated in the inset to Figure 5.23, a magnetic field exerts a force on a moving charged particle, causing the particle to spiral around the magnetic field lines. In this way, charged particles—mainly electrons and protons—from the solar wind can become trapped by Earth's magnetism. Earth's magnetic field exerts electromagnetic control over these particles, herding them into the Van Allen belts. The outer belt contains mostly electrons; the much heavier protons accumulate in the inner belt.

Particles from the Van Allen belts often escape from the magnetosphere near Earth's north and south magnetic poles, where the field lines intersect the atmosphere. Their collisions with air molecules create a spectacular light show called an **aurora** (Figure 5.24). This colorful display results when atmospheric atoms, excited upon collision with the charged particles, fall back to their ground states and emit visible light. ∞ *(Sec. 2.6)* Aurorae are most brilliant at high latitudes, especially inside the Arctic and Antarctic circles. In the north the spectacle is called the *aurora borealis,* or *Northern Lights.* In the south it is called the *aurora australis,* or *Southern Lights.*

Just as Earth's magnetosphere influences the charged particles of the solar wind, the stream of incoming solar wind particles also affects our magnetosphere. As shown in Figure 5.22, the sunward (daytime) side is squeezed toward Earth's surface, while the opposite side has a long "tail" often extending hundreds of thousands of kilometers into space.

Earth's magnetosphere plays an important role in controlling many of the potentially destructive charged particles that venture near our planet. Without the magnetosphere, Earth's atmosphere—and perhaps the planet surface, too—would be bombarded by harmful particles, possibly damaging many forms of life. Some researchers have even suggested that had the magnetosphere not existed in the first place, life might never have arisen on planet Earth.

◄ Figure 5.24 Aurora Borealis A colorful aurora results from the emission of light radiation after magnetospheric particles collide with atmospheric molecules. The aurora rapidly flashes across the sky, resembling huge wind-blown curtains glowing in the dark. (NCAR)

Animation/Video

Northern and
Southern Lights

Earth's magnetic field is not an intrinsic part of our planet. Instead, it is thought to be continuously generated in Earth's core and exists only because the planet is rotating. As in the dynamos that generate electrical power in automobiles and power stations, Earth's magnetism is produced by the spinning, electrically conducting, liquid metal core deep within our planet. Both rapid rotation *and* a conducting liquid core are needed for this mechanism to work. As we will see, this link between magnetism and internal structure is very important to studies of other planets, as we have few other probes of planetary interiors.

Lunar Magnetism

No Earth-based observation or spacecraft measurement has ever detected any lunar magnetic field. Based on our current understanding of how Earth's magnetic field is created, this is not surprising. As we have just seen, researchers believe that planetary magnetism requires a rapidly rotating liquid metal core. Because the Moon rotates slowly and because its core is probably neither molten nor particularly rich in metals, the absence of a lunar magnetic field is exactly what we expect.

5.8 History of the Earth–Moon System

Given all the data, can we construct a reasonably consistent history of Earth and the Moon? The answer seems to be yes. Many specifics are still debated, but a consensus now exists.

Formation

Sometime around 4.6 billion years ago Earth formed by accretion in the solar nebula. ∞ *(Sec. 4.3)* The formation of the Moon is somewhat less certain. The basic problem is that Earth and the Moon are too *dissimilar* in both density and composition for them simply to have formed together out of the same preplanetary matter (the *coformation* theory). However, there are enough *similarities*, particularly between their mantles, to make it unlikely that Earth and the Moon formed entirely independently of one another and subsequently became bound to one another, presumably after a close encounter (the *capture* theory). In addition, there are good theoretical arguments against each scenario—the former because it is hard to reconcile it with the theory of solar system formation outlined in Chapter 4, the latter because of the extreme improbability of the event.

Today many astronomers favor a scenario often called the *impact* theory, which supposes a glancing collision between a large, Mars-sized object and a youthful, molten Earth. As we have seen, such collisions were probably quite frequent in the early solar system. ∞ *(Sec. 4.3)* Computer simulations of such a catastrophic event (Figure 5.25) show that most of the bits and pieces of splattered Earth could have recombined into a stable orbit, forming the Moon. If Earth had already formed an iron core by the time the collision occurred, the Moon could indeed have ended up with a composition similar to Earth's mantle. During the collision, any iron core in the impacting object

▶ **Figure 5.25 Moon Formation**   A simulated collision between Earth and an object the size of Mars. (The sequence proceeds from top to bottom and zooms out dramatically.) The arrow in the final frame shows the newly formed Moon. Red and blue colors represent rocky and metallic regions, respectively, and the direction of motion of the blue material in frames 2, 3, 4, and 5 is toward Earth. Note how most of the impactor's metallic core becomes part of Earth, leaving the Moon composed primarily of rocky material. (W. Benz)

(a) 4 billion years ago

(b) 3 billion years ago

(c) Today

would have been left behind, eventually to become part of Earth's core. Thus, both the Moon's overall similarity to Earth's mantle and its lack of a dense central core are explained.

The quest to understand the origin of the Moon highlights the interplay between theory and observation that characterizes modern science. ∞ *(Sec. E.4)* Detailed data from generations of unmanned and manned lunar missions have allowed astronomers to discriminate between competing theories of the formation of the Moon, discarding some and modifying others. At the same time, the condensation theory of solar-system formation provides a natural context in which the currently favored impact theory can occur. ∞ *(Sec. 4.3)* Without the idea that planets formed by collisions of smaller bodies, such an impact might well have been viewed as so improbable that the theory would never have gained ground.

Evolution

The approximate age of the oldest rocks discovered in the lunar highlands is 4.4 billion years, so we know that at least part of the lunar crust must already have solidified by that time. The oldest known Earth rocks are of roughly the same age, although samples more than 4 billion years old are rare, suggesting that erosion on Earth has been quite effective at hiding the details of our planet's distant past. At formation, the Moon was already depleted in heavy metals relative to Earth.

Earth was at least partially molten during most of its first billion years of existence. Denser matter sank toward the core while lighter material rose to the surface—Earth became differentiated. The intense meteoritic bombardment that helped melt Earth at early times subsided about 3.9 billion years ago. Radioactive heating in the interior continued even after Earth's surface cooled and solidified, but it, too, diminished with time.

As our planet cooled, it did so from the outside in, because regions closest to the surface could most easily lose their excess heat to space. In this way, the surface developed a crust, and the differentiated interior attained the layered structure now implied by seismic studies. Today, radioactive heating continues throughout Earth, but there is probably not enough of it to melt any part of our planet. The high temperatures in the core are mainly the trapped remnant of a much hotter Earth that existed eons ago.

The lunar interior evolved quite differently from Earth's, in large part because of the Moon's smaller size. Small objects cool more quickly than large ones (basically because heat from the interior has less distance to go to reach the surface), and so the Moon rapidly lost its internal heat to space. During the earliest phases of the Moon's existence—roughly the first half-billion years—the meteoritic bombardment was violent enough to heat, and keep molten, most of the surface layers, perhaps to a depth of 400 km in places. However, the intense heat derived from these collisions probably did not penetrate into the deep lunar interior, because rock is a very poor conductor of heat. As on Earth, radioactivity probably heated the Moon, but (because the heat could escape more easily) not sufficiently to transform it from a warm, semisolid object to a completely liquid one. The Moon must have differentiated during this period. If it has a small iron core, that core also formed at this time.

◀ **Figure 5.26 Lunar Evolution** Paintings of the Moon (a) about 4 billion years ago, after much of the meteoritic bombardment had subsided and the surface had somewhat solidified; (b) about 3 billion years ago, after molten lava had made its way up through surface fissures to fill the low-lying impact basins and create the smooth maria; and (c) today, with much of the originally smooth maria now heavily pitted with craters formed at various times within the past 3 billion years. (U.S. Geological Survey)

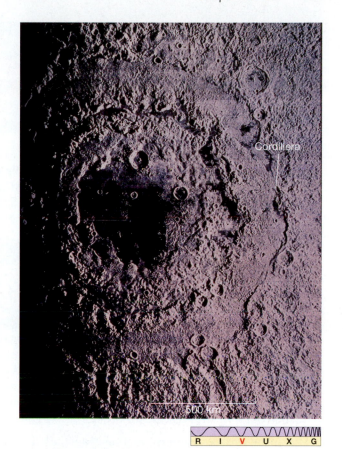

► **Figure 5.27 Large Lunar Crater** A large lunar crater, called the Orientale Basin. The meteoroid that produced this crater upthrust much surrounding matter, which can be seen as concentric rings of cliffs called the Cordillera Mountains. The outermost ring is nearly 1000 km in diameter. Notice the smaller, sharper, younger craters that have impacted this ancient basin in more recent times. (NASA)

About 3.9 billion years ago, when the heaviest bombardment ceased, the Moon was left with a solid crust dented with numerous large basins (Figure 5.26a). The crust ultimately became the highlands, and the basins soon flooded with lava and became the maria. Between 3.9 and 3.2 billion years ago, lunar volcanism filled the basins with the basaltic material we see today. The age of the youngest maria—3.2 billion years—apparently indicates the time when this volcanic activity finally subsided (Figure 5.26b). The maria are the sites of the last extensive lava flows on the Moon. Their smoothness compared with the more rugged highlands disguises their great age.

Not all these great craters became flooded with lava, however. One of the youngest is the Orientale Basin (Figure 5.27), which formed about 3.9 billion years ago. It did not undergo much subsequent volcanism, so we can recognize it as an impact crater rather than a mare. Similar "unflooded" basins can be seen on the lunar far side.

Because of Earth's gravitational pull, the lunar crust became thicker on the far side than on the near side. Therefore, lava from the interior had a shorter route through the crust to the surface on the Moon's Earth-facing side. As a result, relatively little volcanic activity occurred on the far side, and no large maria were created—the crust was simply too thick to allow that to occur.

As the Moon continued to cool, volcanic activity ended as the thickness of the solid surface layer increased. The crust is now far too thick for volcanism or plate tectonics to occur. With the exception of a few meters of surface erosion from eons of meteoritic bombardment (Figure 5.26c), the lunar landscape has remained more or less structurally frozen for the past 3 billion years. The Moon is dead now, and it has been dead for a long time.

◄◄◄ **CONCEPT CHECK**

What single factor is most responsible for the evolutionary differences between Earth and the Moon?

CHAPTER REVIEW

Summary

1 The six main regions of Earth are (from inside to outside) a central metallic **core** (p. 134), which is surrounded by a thick rocky **mantle** (p. 134) and topped with a thin **crust** (p. 134). Above the surface is the **atmosphere** (p. 134), composed primarily of nitrogen and oxygen. Surface winds and weather in the lower atmosphere, or **troposphere** (p. 139) are caused by **convection** (p. 139), whereby heat moves from one place to another by rising or sinking streams of air. Higher still lies the **magnetosphere** (p. 134), where charged particles from the Sun are trapped by Earth's magnetic field. The Moon has a partially differentiated interior consisting of a semi-solid core, a mantle and a thick crust. It has no atmosphere, because its gravity is too weak to retain one, and no magnetosphere.

2 The daily **tides** (p. 135) in Earth's oceans are caused by the gravitational effects of the Moon and the Sun, which raise **tidal bulges** (p. 135) in the oceans. Their size depends on the orientations of the Sun and the Moon relative to Earth. A differential

gravitational force is called a **tidal force** (p. 135), even when no oceans or even planets are involved. The tidal interaction between Earth and the Moon is causing Earth's spin to slow and is responsible for the Moon's **synchronous orbit** (p. 136), in which the same side of the Moon always faces our planet.

3 At high altitudes, in the **ionosphere** (p. 139), the atmosphere is kept ionized by the absorption of high-energy radiation and particles from the Sun. Between the ionosphere and the troposphere lies the **ozone layer** (p. 139), where incoming solar ultraviolet radiation is absorbed. Both these layers help protect us from dangerous radiation from space. The **greenhouse effect** (p. 140) is the absorption and trapping of infrared radiation emitted by Earth's surface by atmospheric gases (primarily carbon dioxide and water vapor). It makes our planet's surface some 40 K warmer than would otherwise be the case.

4 We study Earth's interior by observing how **seismic waves** (p. 143) produced by **earthquakes** (p. 143) travel through the mantle. Earth's iron core consists of a solid **inner core** (p. 144) surrounded by a liquid **outer core** (p. 144). The process by which dense material sinks to the center of a planet while lighter material rises to the surface is called **differentiation** (p. 145). The differentiation of Earth implies that our planet must have been at least partially molten in the past, because of bombardment by material from interplanetary space and heat released by **radioactivity** (p. 145) in Earth's interior.

5 Earth's surface is made up of enormous slabs, or plates. The slow movement of these plates across the surface is known as continental drift, or **plate tectonics** (p. 147). Earthquakes, volcanism, and mountain building are associated with plate boundaries, where plates may collide, move apart, or rub against one another. The "fit" of the continents to one another and the ages of rocks near oceanic ridges argue in favor of this theory. The motion of the plates is thought to be driven by convection in Earth's mantle. On the Moon, the crust is too thick and the mantle too cool for plate tectonics to occur.

6 The main surface features on the Moon are the dark **maria** (p. 149) and the lighter-colored **highlands** (p. 149). **Craters** (p. 150) of all sizes, caused by impacting meteoroids, are found everywhere on the lunar surface. The highlands are older than the maria and are much more heavily cratered. Meteoritic impacts are the main source of erosion on the lunar surface. There is no volcanic activity on the Moon because all volcanism was stifled by the Moon's cooling mantle shortly after extensive lava flows formed the maria more than 3 billion years ago.

7 Charged particles from the solar wind are trapped by Earth's magnetic field lines to form the **Van Allen belts** (p. 153). When particles from the Van Allen belts hit Earth's atmosphere, they heat and ionize the atoms there, causing the atoms to glow in an **aurora** (p. 154). Planetary magnetic fields are produced by the motion of rapidly rotating, electrically conducting fluid (such as molten iron) in a planet's core. The Moon rotates slowly and lacks a conducting liquid core, which accounts for the absence of a lunar magnetic field.

8 The most likely explanation for the formation of the Moon is that the newly formed Earth was struck by a Mars-sized object. The core of the impacting body remained behind as part of the core of our planet, and debris splattered into space formed the Moon.

Review and Discussion

1. Explain how the Moon produces tides in Earth's oceans.

2. What is a synchronous orbit? How did the Moon's orbit become synchronous?

3. What is convection? What effect does it have on (a) Earth's atmosphere and (b) Earth's interior?

4. In contrast to Earth, the Moon undergoes extremes in temperature. Why?

5. Use the concept of escape speed to explain why the Moon has no atmosphere.

6. What is the greenhouse effect? Is the greenhouse effect operating in Earth's atmosphere helpful or harmful? What are the consequences of an enhanced greenhouse effect?

7. The density of water in Earth's hydrosphere and the density of rocks in the crust are both lower than the average density of the planet as a whole. What does this fact tell us about Earth's interior?

8. Give two reasons geologists believe that part of Earth's core is liquid.

9. What clue does Earth's differentiation provide to our planet's history?

10. What process is responsible for the surface mountains, oceanic trenches, and other large-scale features on Earth's surface?

11. In what sense were the lunar maria once "seas"?

12. What is the primary source of erosion on the Moon? Why is the average rate of lunar erosion so much less than on Earth?

13. Name two pieces of evidence indicating that the lunar highlands are older than the maria.

14. Give a brief description of Earth's magnetosphere. Why does the Moon have no magnetosphere?

15. Describe the theory of the Moon's origin currently favored by many astronomers.

Conceptual Self-Test: True or False?/Multiple Choice

1. Because of tidal forces, the Moon is in a synchronous orbit around Earth. **(T/F)**

2. Lunar maria are extensive lava-flow regions. **(T/F)**

3. Except for the layer of air closest to Earth's surface, the ozone layer is the warmest part of the atmosphere. **(T/F)**

4. Earth's magnetic field is the result of our planet's large, permanently magnetized iron core. **(T/F)**

5. Motion of the crustal plates is driven by convection in Earth's upper mantle. **(T/F)**

6. Volcanic activity continues today on the surface of the Moon. **(T/F)**

7. Like Earth, the Moon has a molten metal core. **(T/F)**

8. If you were making a scale model of Earth, representing our planet by a 12-inch basketball, the inner core would be about the size of **(a)** a half-inch ball bearing; **(b)** a 2-inch golf ball; **(c)** a 4-inch tangerine; **(d)** a 7-inch grapefruit.

9. Earth's average density is about the same as that of **(a)** a glass of water; **(b)** a heavy iron meteorite; **(c)** an ice cube; **(d)** a chunk of black volcanic rock.

10. Sunlight absorbed by Earth's surface is reemitted in the form of **(a)** microwave, **(b)** infrared, **(c)** visible, **(d)** ultraviolet radiation.

11. The deepest that geologists have drilled into Earth is about the same as **(a)** the height of the Statue of Liberty; **(b)** the altitude most commercial jet airplanes fly; **(c)** the distance between New York and Los Angeles; **(d)** the distance between the United States and China.

12. If Earth had no Moon, then tides would **(a)** not occur; **(b)** occur more often and with more intensity; **(c)** still occur, but be barely measurable; **(d)** occur with the same frequency and slightly lower strength.

13. The most likely theory of the formation of the Moon is that it **(a)** was formed by the gravitational capture of a large asteroid; **(b)** formed simultaneously with Earth's formation; **(c)** was created from a collision scooping out the Pacific Ocean; **(d)** formed from a collision of Earth with a Mars-sized object.

14. According to Figure 5.5 (Earth's Atmosphere), commercial jet airplanes flying at 10 km are in **(a)** the troposphere; **(b)** the stratosphere; **(c)** the ozone; **(d)** the mesosphere.

15. The first figure in *Discovery 5-2* shows that CO_2 levels in Earth's atmosphere began to rise rapidly **(a)** in the middle ages; **(b)** in 1600; **(c)** in the mid-nineteenth century; **(d)** in the late twentieth century.

Problems

 Algorithmic versions of these questions are available in the Practice Problems module of the Companion Website.

The number of squares preceding each problem indicates its approximate level of difficulty.

1. ■ What would Earth's surface gravity and escape speed be if the entire planet had a density equal to that of the crust (3000 kg/m^3, say)?

2. ■ The Moon's mass is 1/80 that of Earth, and the lunar radius is 1/4 Earth's radius. Based on these figures, calculate the total weight on the Moon of a 100-kg astronaut with a 50-kg spacesuit and backpack, relative to his weight on Earth.

3. ■■ Most of Earth's ice is found in Antarctica, whose permanent ice caps cover approximately 0.5 percent of Earth's surface and are 3 km thick, on average. Earth's oceans cover roughly 71 percent of our planet to an average depth of 3.6 km. Taking water and ice to have roughly the same density, estimate by how much sea level would rise if global warming were to cause the Antarctic ice caps to melt.

4. ■■ You are standing on Earth's surface, and the full Moon is directly overhead. By what fraction is your weight decreased due to the combined tidal forces of the Sun and the Moon?

5. ■ Based on the data presented in the text, estimate the fractions of Earth's volume represented by (a) the inner core, (b) the outer core, (c) the mantle, and (d) the crust.

6. ■■ Approximating Earth's atmosphere as a layer of gas 7.5 km thick, with a uniform density of 1.3 kg/m^3, calculate the atmosphere's total mass. Compare this with Earth's mass.

7. ■■■ As discussed in the text, without the greenhouse effect, Earth's average surface temperature would be about 250 K. With the greenhouse effect, it is some 40 K higher. Use this information and Stefan's law to calculate the fraction of infrared radiation leaving Earth's surface that is absorbed by greenhouse gases in the atmosphere. ∞ *(Sec. 2.4)*

8. ■ Following an earthquake, how long would it take a seismic wave moving in a straight line with speed 5 km/s to reach the opposite side of Earth?

9. ■■ Using the rate given in the text for the formation of 10-km craters on the Moon, estimate how long it would take to cover the Moon with new craters of this size. How much higher must the cratering rate have been in the past to cover the entire lunar surface with such craters in the 4.6 billion years since the Moon formed?

10. ■ The *Hubble Space Telescope* has a resolution of about 0.05 arc second. What is the smallest object it could see on the surface of the Moon? Give your answer in meters.

The Companion Website at www.prenhall.com/chaisson provides algorithmically generated versions of each chapter's Problems, in addition to an Animations and Videos gallery, a Tutorials gallery and Tutorials Quiz questions, a full eBook, an Images gallery, Practice Questions, Multiple Choice, and True or False quizzes, and Labeling exercises.

6

The Terrestrial Planets

A Study in Contrasts

The search for life on Mars continues unabated. "Follow the water" is a good guide when prospecting for life—where there is water, there may well be life. But Mars today is as dry as any desert on Earth. However, there is growing evidence for a wetter Mars billions of years ago when the martian climate was perhaps warmer. One such piece of evidence is seen in this true-color image taken in 2004 by the *Spirit* robot on one of the Mars Exploration Rovers now exploring Mars' surface. Amid the many scattered rocks, which hold "memories" of the ancient events that formed them, we see a smooth area (at center) that might be a dried up lakebed that held shallow "puddles" of water. (JPL)

With Earth and the Moon as our guides, we now expand our field of view to study the other terrestrial planets. As we explore these worlds and seek to understand the similarities and differences among them, we begin our comparative study of the only planetary system we know. Mercury, the smallest terrestrial planet, in many ways is kin to Earth's Moon, and much can be learned by comparing Mercury with our own satellite. Venus and Mars both have properties more like Earth's, and we learn about these two terrestrial worlds by drawing parallels with our own planet. It is possible that Venus, Earth, and Mars had many similarities when they formed, yet Earth today is vibrant, teeming with life, while Venus is an uninhabitable inferno and Mars is a dry, dead world. What were the factors leading to these present conditions? In answering this question, we will discover that a planet's environment, as well as its composition, can play a critical role in determining its future.

LEARNING GOALS

Studying this chapter will enable you to:

1 Explain how Mercury's rotation has been influenced by its orbit around the Sun.

2 Describe how the atmospheres of Venus and Mars differ from one another and from Earth's.

3 Compare the surface of Mercury with that of the Moon.

4 Compare the surfaces of Venus and Mars with that of Earth.

5 Describe how we know that Mars once had running water and a thick atmosphere.

6 Discuss the similarities and differences in the internal structures and geological histories of the four terrestrial planets.

7 Explain why the atmospheres of Venus, Mars, and Earth are now so different from one another.

Visit www.prenhall.com/chaisson for additional annotated images, animations, and links to related sites for this chapter.

6.1 Orbital and Physical Properties

Mercury, the innermost planet, lies close to the Sun and is visible above the horizon for at most two hours before the Sun rises or after it sets. Orbiting somewhat farther from the Sun (but still within Earth's orbit), the next planet, Venus, is visible for a little longer—up to three hours, depending on the time of year (Figure 6.1). The interior orbits of these planets ensure that we never see them far from the Sun in the sky. Contrast this with the exterior orbit of the planet Mars. From our earthly viewpoint, Mars appears to traverse the entire sky, always keeping close to the ecliptic (and occasionally executing retrograde loops, as we saw in Chapter 1). ∞ *(Sec. 1.1)*

Venus is the third brightest object in the entire sky (only the Sun and the Moon are brighter). Like all the planets, it shines by reflected sunlight. It is so bright because almost all the sunlight reaching it is reflected from thick clouds that envelop the planet. You can even see Venus in the daytime if you know where to look. The much fainter Mercury is visible to the naked eye only when the Sun's light is blotted out—just before dawn, just after sunset, and during a total solar eclipse. Orange-red Mars is also quite easy to spot in the night sky. Because of its less reflective surface, smaller size, and greater distance from the Sun, Mars does not appear as bright as Venus, as seen from Earth. However, at its brightest—at closest approach, as marked on Figure 6.1(a)—Mars is still brighter than any star.

Table 6.1 expands Table 5.1 to include Mercury, Venus, and Mars, and adds two additional properties: surface temperature and surface atmospheric pressure. In seeking to understand the other terrestrial planets, astronomers are guided by our much more detailed knowledge of Earth and the Moon. Mercury's high average density, for instance, tells us that this planet must have a large iron core, consistent with the condensation theory's account of its formation. ∞ *(Sec. 4.3)* However, in many other respects Mercury is similar to Earth's Moon, leading us to use the Moon as a model for understanding Mercury's past. Venus and, to a lesser extent, Mars are more similar to Earth, so

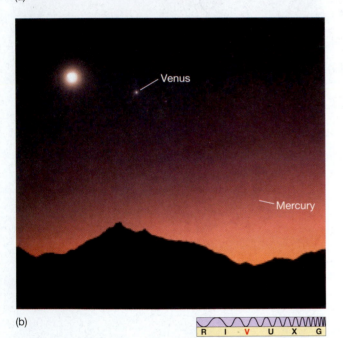

(a)

(b)

R   I · V   U   X   G

◀ **Figure 6.1 Terrestrial Planet Orbits** (a) The orbits of Mercury and Venus mean that neither planet is ever found far from the Sun, as seen from Earth. Mercury's greatest angular distance from the Sun is 28°, that of Venus 47°. Since Earth rotates through 15° per hour, this means that Mercury is generally visible only in the twilight sky within 1^{h}52^m (=28°/15° per hour) of sunrise or sunset. By the same reasoning, Venus is only visible within 3^{h}8^m of sunrise or sunset. Mars, on the other hand, lies outside Earth's orbit and can appear anywhere on the ecliptic plane, depending on the two planets' positions in their respective orbits. Mars is brightest at closest approach to Earth and progressively dimmer as the two planets separate. (b) Mercury and Venus in the evening sky. (J. Sanford/Science Photo Library)

TABLE 6.1   Some Properties of the Terrestrial Planets and Earth's Moon

	MASS (kg)	MASS (Earth = 1)	RADIUS (km)	RADIUS (Earth = 1)	AVERAGE DENSITY (kg/m³)	SURFACE GRAVITY (Earth = 1)	ESCAPE SPEED (km/s)	SURFACE ROTATION PERIOD (solar days)	SURFACE TEMPERATURE (K)	ATMOSPHERIC PRESSURE (Earth = 1)
Mercury	3.3×10^{23}	0.055	2400	0.38	5400	0.38	4.2	59	100–700	—
Venus	4.9×10^{24}	0.82	6100	0.95	5300	0.91	10	−243[1]	730	90
Earth	6.0×10^{24}	1.00	6400	1.00	5500	1.00	11	1.00	290	1.0
Mars	6.4×10^{23}	0.11	3400	0.53	3900	0.38	5.0	1.03	180–270	0.007
Moon	7.3×10^{22}	0.012	1700	0.27	3300	0.17	2.4	27.3	100–400	—

[1] The minus sign indicates retrograde rotation.

Earth provides the natural starting point for studies of those planets. For example, despite the lack of seismic data, we nevertheless assume that Venus has a metallic core and rocky mantle similar to those shown for Earth in Figure 5.1. Even the widely different atmospheres of these two worlds can be explained in familiar earthly terms.

6.2 Rotation Rates

In principle, astronomers can determine the rate at which a planet spins simply by observing the motion of some prominent surface feature across the planet's disk. Unfortunately, for Mercury and Venus this turns out not to work, because surface markings are hard or impossible to discern. Astronomers had to develop other techniques to probe the rotations of these bodies.

Mercury's Curious Spin

From Earth, even through a large telescope, we see Mercury only as a slightly pinkish, almost featureless disk. The largest ground-based telescopes can resolve features on the surface of Mercury about as well as we can perceive features on our Moon with our unaided eyes. Figure 6.2 is one of the few photographs of Mercury taken from Earth that shows any indication of surface features. In the days before close-up images were obtainable from space, astronomers could only speculate about the faint, dark markings this photograph reveals.

In the mid-nineteenth century, an Italian astronomer named Giovanni Schiaparelli attempted to measure Mercury's rotation rate by watching surface features move around the planet. He concluded that Mercury always keeps one side facing the Sun, much as our Moon always presents only one face to Earth. The explanation suggested for this synchronous rotation was the same as for the Moon: The tidal bulge raised in Mercury by the Sun had modified the planet's rotation rate until the bulge always points directly at the Sun. *(Sec. 5.2)* Although the surface features could not be seen clearly, the combination of Schiaparelli's observations and a plausible physical explanation was enough to convince most astronomers. The belief that Mercury rotated synchronously with its revolution about the Sun (once every 88 Earth days) persisted for almost a century.

In 1965 astronomers making radar observations of Mercury from the Arecibo radio telescope in Puerto Rico discovered that this long-held view was in error. *(Sec. 3.4)* The technique they used is illustrated in Figure 6.3, which shows a radar signal reflecting from the surface of a hypothetical planet. Because the planet is rotating, a pulse of outgoing radiation of a single frequency is broadened—"smeared out"—by the Doppler effect, by an amount that depends on the planet's rotation speed. *(Sec. 2.7)* The radiation reflected from the side moving toward us returns at a slightly higher frequency than does the radiation reflected from the receding side. (Think of the two hemispheres as separate sources of radiation moving at slightly different velocities, one toward us and one away from us.) By measuring the extent of this broadening, we can determine the rate at which the planet rotates.

In this way, astronomers found that the rotation period of Mercury is not 88 days, as had previously been thought, but just under 59 days—in fact, exactly two-thirds of Mercury's year. This odd state of affairs surely did not occur by chance. In fact, nineteenth-century astronomers were correct in thinking that Mercury's rotation was governed by the tidal effect of the Sun. However, the combination of the Sun's gravity and Mercury's eccentric orbit has caused the planet's rotation to be more complicated than that of the Moon. Unable to come into a state of precisely synchronous rotation (because Mercury's

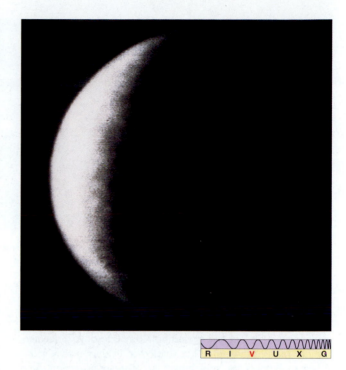

R I V U X G

▲ **Figure 6.2 Mercury**   Photograph of Mercury taken from Earth with a large ground-based optical telescope. Only a few faint surface features are discernible. (Palomar Observatory/Caltech)

Animation/Video
Transit of Mercury

▲ **Figure 6.3 Planetary Radar**   A radar beam reflected from a rotating planet yields information about both the planet's line-of-sight motion and its rotation rate.

▲ **Figure 6.4 Mercury's Rotation** Mercury's orbital and rotational motions combine to produce a solar day that is two Mercury years long. The red arrows represent an observer standing on the surface of the planet. At day 0 (center right in Year 1 drawing), it is noon for our observer and the Sun is directly overhead. By the time Mercury has completed one full orbit around the Sun and moved from day 0 to day 88 (center right in Year 2 drawing), it has rotated on its axis exactly 1.5 times, so that it is now midnight at the observer's location. After another complete orbit, it is noon once again on day 176 (center right in Year 3 drawing).

orbital speed changes significantly from place to place in its orbit), Mercury did the next best thing. It presents the same face to the Sun not every time around, but every other time. Figure 6.4 illustrates the implications of this odd rotation for a hypothetical inhabitant of Mercury. The planet's solar day—the time from one noon to the next—is two Mercury years long!

The Sun also influences the tilt of Mercury's spin axis. Because of the Sun's tides, Mercury's rotation axis is almost exactly perpendicular to its orbital plane. Thus, the noontime Sun is always directly overhead for someone standing on the equator and always on the horizon for someone standing at either pole.

Venus and Mars

The same clouds whose reflectivity make Venus so easy to see in the night sky also make it impossible for us to discern any surface features, at least in visible light. Figure 6.5, one of the best photographs of Venus taken with an Earth-based telescope, shows an almost uniform yellow-white disk, with rare hints of clouds. Because of the cloud cover, astronomers did not know Venus's rotation period until the development of radar techniques in the 1960s, when Doppler broadening of returning radar echoes indicated an unexpectedly sluggish 243-day rotation period. Furthermore, Venus's spin was found to be *retrograde*—that is, opposite that of Earth and most other solar system objects and in the direction opposite Venus's orbital motion. The planet's rotation axis is almost exactly perpendicular to its orbital plane, just as Mercury's is.

We have no "evolutionary" explanation for Venus's anomalous rotation. It is not the result of any known interaction with the Sun, Earth, or any other solar system body. At present, the best explanation astronomers can offer is that during the final stages of Venus's formation in the early solar system, the planet was struck by a large body, much like the one that may have hit Earth and formed the Moon. ∞ *(Sec. 4.3, 5.8)* That impact was sufficient to reduce the planet's spin almost to zero, leaving Venus rotating as we now observe.

In contrast to Mercury and Venus, surface markings are easily seen on Mars (Figure 6.6), allowing astronomers to track the planet's rotation. Mars rotates once on its axis every 24.6 hours—close to one Earth day. The planet's equator is inclined to the orbital plane at an angle of 24.0°, very similar to Earth's inclination of 23.5°. Thus, as Mars orbits the Sun, we find both daily and seasonal cycles, just as on Earth. However, the Martian seasons are complicated somewhat by variations in solar heating due to the planet's eccentric orbit. Figure 6.7 summarizes the rotations and orbits of the four terrestrial planets.

▲ **Figure 6.5 Venus** This photograph taken from Earth shows Venus with its creamy yellow mask of clouds. (NOAO)

R I V U X G

CONCEPT CHECK ▶▶▶

How has the Sun's gravity influenced Mercury's rotation?

▲ **Figure 6.6 Mars** A deep-red (800-nm) image of Mars taken in 1991 at Pic du Midi, an exceptionally clear site in the French Alps. One of the planet's polar caps appears at the top, and numerous other surface markings are visible. (CNRS and Université Paul Sabatier)

▲ **Figure 6.7 Terrestrial Planets' Spin** The inner planets of the solar system—Mercury, Venus, Earth, and Mars—display widely different rotational properties. Although all orbit the Sun in the same direction and in nearly the same plane, Mercury's rotation is slow and prograde, that of Venus is slow and retrograde, while those of Earth and Mars are fast and prograde. Venus rotates clockwise as seen from above the plane of the ecliptic, but Mercury, Earth, and Mars all rotate counterclockwise. This is a perspective view, roughly halfway between a flat edge-on view and a direct overhead view.

6.3 Atmospheres

In this section we outline what is known about the present-day atmospheres of the other terrestrial planets. We will return to this subject at the end of the chapter, when we discuss how the terrestrial atmospheres may have evolved into their present states.

Mercury

Mercury has no appreciable atmosphere. Although the U.S. space probe *Mariner 10*[1] did find a trace of what was at first thought to be an atmosphere, the gas is now known to be temporarily trapped hydrogen and helium from the solar wind. Mercury holds this gas for just a few weeks before it leaks away into space. The absence of an atmosphere on Mercury is readily explained by the planet's high surface temperature (up to 700 K at noon on the equator) and low mass (only 4.5 times the mass of the Moon). Any atmosphere Mercury might once have had escaped long ago. ∞ *(More Precisely 5-1)*

With no atmosphere to retain heat, Mercury's surface temperature falls to about 100 K during the planet's long night. Mercury's 600 K temperature range is the largest of any planet or moon in the solar system. Near the poles, where the Sun's light arrives almost parallel to the surface, the temperature remains low at all times. Recent Earth-based radar studies suggest that Mercury's polar temperature could be as low as 125 K and that the poles may be permanently covered with extensive thin sheets of water ice.

[1] *Mariner 10* is one of the few spacecraft to have visited Mercury. In 1973 it was placed in an eccentric orbit around the Sun that brings it close to Mercury every 176 days (two Mercury years). The three close encounters it made with Mercury in 1974 and 1975 (after which it ran out of fuel) are the source of virtually all the detailed information we have on that planet. NASA's next mission to the planet will be the *Messenger* orbiter, scheduled to arrive in 2009.

Venus

The atmosphere of Venus provides an excellent illustration of how theories are only as good as the data on which they are based. ∞ *(Discovery 1-1)* In the 1930s, scientists used spectroscopy to measure the temperature of Venus's upper atmosphere and found it to be about 240 K, not much different from that of Earth's stratosphere. ∞ *(Sec. 5.3)* Taking into account the cloud cover and the planet's nearness to the Sun, and assuming that Venus had an atmosphere much like our own, researchers concluded that Venus might have an average surface temperature only a few degrees higher than Earth's. In the 1950s, however, when radio observations of the planet penetrated the cloud layer and gave the first indication of conditions near the surface, they revealed a temperature exceeding 600 K! Almost overnight, the popular conception of Venus changed from lush tropical jungle to arid, uninhabitable desert.

Since then, spacecraft data have revealed the full extent of the differences between the atmospheres of Venus and Earth. Venus's atmosphere is much more massive than our own, and it extends to a much greater height above the planet's surface. The surface pressure on Venus is about 90 times the pressure at sea level on Earth, equivalent to an (Earth) underwater depth of about 1 km. (Unprotected humans cannot dive much below 100 m.) The surface temperature is a sizzling 730 K.

The main component (96.5 percent) of Venus's atmosphere is carbon dioxide. Almost all of the remaining 3.5 percent is nitrogen. Given Venus's similarity to Earth in mass, radius, and location in the solar system, it is commonly assumed that Venus and Earth must have started off looking somewhat alike. However, there is no sign of the large amount of water vapor that would be present if a volume of water equivalent to Earth's oceans had once existed on Venus and later evaporated. If Venus started off with Earth-like composition, something happened to its water, for the planet is now an exceedingly dry place. Even the highly reflective clouds are composed not of water vapor, as on Earth, but of sulfuric acid droplets.

Venus's atmospheric patterns are much more evident when examined with equipment capable of detecting ultraviolet radiation. Some of Venus's upper-level clouds absorb this high-frequency radiation, thereby increasing the contrast. Figure 6.8 is an ultraviolet image taken in 1979 by the U.S. *Pioneer Venus* spacecraft from a distance of 200,000 km (compare the optical image shown in Figure 6.5). The large, fast-moving cloud patterns lie between 50 and 70 km above the surface. Upper-level winds reach speeds of 400 km/h relative to the planet. Below the clouds, extending down to an altitude of 30 km, is a layer of haze. Below 30 km, the air is clear.

R I V U X G

▲ **Figure 6.8 Venus, Up Close** Venus as photographed by the *Pioneer* spacecraft's cameras 200,000 km away from the planet. This image was made by capturing solar ultraviolet radiation reflected from the planet's clouds, which are composed mostly of sulfuric acid droplets, much like the corrosive acid in a car battery. (NASA)

CONCEPT CHECK ▶▶▶

Describe some important differences between the atmospheres of Venus and Earth.

Mars

Well before the arrival of spacecraft, astronomers knew from Earth-based spectroscopy that the Martian atmosphere is quite thin and composed primarily of carbon dioxide. Spacecraft measurements confirmed these results, indicating that the atmospheric pressure is only about 1/150 the pressure of Earth's atmosphere at sea level. The Martian atmosphere is 95.3 percent carbon dioxide, 2.7 percent nitrogen, and 1.6 percent argon, plus small amounts of oxygen, carbon monoxide, and water vapor. While there is some superficial similarity in composition between the atmospheres of Mars and Venus, the two planets clearly must have had very different atmospheric histories. Average surface temperatures on Mars are about 70 K cooler than on Earth.

6.4 The Surface of Mercury

Figure 6.9 shows a picture of Mercury taken during the *Mariner 10* flyby when the spacecraft was about 200,000 km from the planet. (A "flyby," in NASA parlance, is any space mission in which a probe passes relatively close to a planet—within a few planetary radii, say—but does not go into orbit around it.) Figure 6.10 shows a higher-resolution photograph of the planet from a distance of 20,000 km. The similarities to our Moon are striking. Indeed, much of the cratered surface bears a strong resemblance to the Moon's highlands. The crater walls are generally not as high as on the Moon, and the ejected material landed closer to the impact site, exactly as we would expect on the basis of Mercury's greater surface gravity (which is a little more than twice that of the Moon). Mercury, however, shows no extensive lava flow regions akin to the lunar maria.

As on the Moon, Mercury's craters are the result of meteoritic bombardment. The craters are not so densely packed as their lunar counterparts, however, and there are extensive **intercrater plains**. One likely explanation for Mercury's relative lack of craters is that the older impact craters have been filled in by volcanic activity, in much the same way as the Moon's maria filled in craters as they formed. However, Mercury's intercrater plains do not look much like mare material. They are much lighter in color and not as flat.

Mercury has at least one type of surface feature not found on the Moon. Figure 6.11 shows a **scarp**, or cliff, that does not appear to be the result of volcanic or other familiar geological activity. The scarp cuts across several craters, which indicates that whatever produced it occurred after most of the meteoritic bombardment was over. Numerous scarps have been found in the *Mariner* images. Mercury shows no evidence of crustal motions, so the scarps could not have been formed by tectonic processes such as those responsible for fault lines on Earth. Instead, the scarps probably formed when the crust cooled, shrank, and split long ago. After the splitting, one side of the crack moved upward relative to the other side, forming the scarp cliff face. If we can apply to Mercury the cratering age estimates we use for the Moon, the scarps appeared about 4 billion years ago.

Much of the discussion in Chapter 5 about the surface of Earth's Moon applies equally well to Mercury. ∞ *(Sec. 5.6)* Figure 6.12 shows the result of what was probably the last great event in the geological history of Mercury—an immense bull's-eye crater called the Caloris Basin, formed eons ago by the impact of a large asteroid. (Because of the orientation of the planet during *Mariner 10*'s flybys, only half the basin is visible. The center is off the left-hand side of the photograph.) Compare this basin with the Orientale Basin on the Moon (Figure 5.25).

R I V U X G

▲ **Figure 6.9 Mercury, Up Close** Mercury is imaged here as a mosaic of photographs—a composite image constructed from many individual images—taken by the *Mariner 10* spacecraft in the mid-1970s during its approach to the planet. At the time, the spacecraft was 200,000 km away from Mercury. (NASA)

◀◀◀ **CONCEPT CHECK**

How do scarps on Mercury differ from geological faults on Earth?

C. Bach

100 km

R I V U X G

◀ **Figure 6.10 Mercury, Very Close** Another photograph of Mercury by *Mariner 10*, this time from a distance of about 20,000 km. The double-ringed crater at the upper left, named C. Bach and about 100 km across, exemplifies how many of the large craters on Mercury tend to form double, rather than single, rings. The reason is not yet understood. (NASA)

▲ Figure 6.11 Mercury's Surface Discovery Scarp on Mercury's surface (arrow) as photographed by *Mariner 10*. This cliff appears to have formed when the planet's crust cooled and shrank early in its history, causing a crease in the surface. Several hundred kilometers long and up to 3 km high in places, it runs diagonally across the center of the frame in this photograph. (NASA)

▲ Figure 6.12 Caloris Basin Mercury's most prominent geological feature—the Caloris Basin—measures about 1400 km across and is ringed by concentric mountain ranges (dashed lines) that reach more than 3 km high in places. This huge circular basin, only half of which can be seen in this *Mariner 10* photograph, is similar in size to the Moon's Mare Imbrium and spans more than half of Mercury's radius. (NASA)

6.5 The Surface of Venus

Although the clouds of Venus are thick and the surface is totally shrouded, we are by no means ignorant of the planet's surface. Radar astronomers have bombarded Venus with radio signals, and analysis of the radar echoes yields a map of surface features. Except for Figure 6.17, all the views of Venus in this section are "radargraphs" (not photographs) created in this way. Most recently, the U.S. *Magellan* spacecraft has provided very high-resolution radar images of the planet.

Large-Scale Topography

Figure 6.13(a) shows a relatively low-resolution map of Venus made by *Pioneer Venus* in 1979. Surface elevation above the average radius of the planet's surface is indicated by color, with white representing the highest elevations, blue the lowest. (Note that the blue has nothing to do with oceans!) For comparison, Figure 6.13(b) shows a map of Earth to the same scale and at the same spatial resolution. Figure 6.14 is a 1995 mosaic of *Magellan* images of Venus. The orange color is based on optical data returned from spacecraft that landed on the planet.

◄ Figure 6.13 Venus Radar Map (a) Radar map of the surface of Venus, based on *Pioneer Venus* data. Color represents elevation, with white the highest areas and blue the lowest. (b) A similar map of Earth, at the same spatial resolution. (NASA)

◀ **Figure 6.14 Venus Magellan Map** A planetwide mosaic of Venus made from *Magellan* images. Colors represent altitude: blue is lowest, red highest. The largest "continent" on Venus, Aphrodite Terra, is the yellow dragon-shaped area across the center of the image. (NASA)

(a)

(b)

Venus's surface appears to be mostly smooth, resembling rolling plains with modest highlands and lowlands. Only two continent-sized features, called Ishtar Terra and Aphrodite Terra, adorn the landscape, and these contain mountains comparable in height to those on Earth. The highest peaks rise some 14 km above the level of the deepest surface depressions. (The highest point on Earth, the summit of Mount Everest, lies about 20 km above the deepest section of the ocean floor.) The elevated "continents" occupy only 8 percent of Venus's total surface area whereas continents on Earth make up about 25 percent of our planet's surface.

The larger continent-sized formation, Aphrodite Terra, is located on Venus's equator. It is comparable in size to Africa. Before *Magellan*'s arrival, some researchers had speculated that Aphrodite Terra might have been the site of something equivalent to seafloor spreading on Earth, where two tectonic plates moved apart and molten rock rose to the surface in the gap between them, forming an extended ridge. This is just what is happening today at Earth's Mid-Atlantic Ridge, which is clearly visible in Figure 6.13(b). However, the Magellan images now seem to rule out any plate-tectonic activity on Venus, and the Aphrodite region shows no signs of spreading. The crust appears buckled and fractured, suggesting large compressive forces, and there seem to have been numerous periods when extensive lava flows occurred.

Volcanism and Cratering

Although erosion by the planet's atmosphere may play some part in obliterating surface features, the most important factor is volcanism (volcanic activity), which appears to resurface the planet every few hundred million years. Many areas of Venus have volcanic features. Figure 6.15(a) shows a

▶ **Figure 6.15 Venus's Surface Features** (a) A series of dome-shaped structures on Venus. They are the result of molten rock having bulged out of the ground and then retreated, leaving behind a thin, solid crust that subsequently cracked and subsided. *Magellan* found features like this in several locations on Venus. (b) A computer-generated three-dimensional representation of four of the lava domes. (c) A three-dimensional *Magellan* view of the large shield volcano known as Gula Mons. The volcanic caldera at the summit is about 100 km across. As in part (b), the vertical scale has been greatly exaggerated. The volcano is about 4 km high. Color in (b) and (c) is based on data returned by Soviet landers. (NASA)

(c)

R I V U X G

▲ **Figure 6.16 Venus Corona** This corona, called Aine, lies in the plains south of Aphrodite Terra and is about 300 km across. Coronae probably result from upwelling mantle material, causing the surface to bulge outward. Note the pancake-shaped lava domes at top, the many fractures in the crust around the corona, and the large impact craters with their surrounding white (rough) ejecta blankets that stud the region. (NASA)

Tutorial
SuperSpaceship–Voyage to Venus

CONCEPT CHECK ▶▶▶

Are the volcanoes on Venus mainly associated with the movement of tectonic plates, as on Earth?

Magellan image of seven pancake-shaped lava domes, each about 25 km across—a little bigger than Washington, D.C. They probably formed when lava oozed out of the surface, formed the dome, then withdrew, leaving the crust to crack and subside. Lava domes such as these are found in several locations on Venus. Figure 6.15(b) shows a computer-generated three-dimensional view of the domes.

The most common volcanoes on the planet are of the type known as **shield volcanoes**. Those on Earth are associated with lava welling up through a "hot spot" in the crust (like the Hawaiian Islands). They are built up over long periods of time by successive eruptions and lava flows. A characteristic of shield volcanoes is the formation of a *caldera*, or crater, at the summit when the underlying lava withdraws and the surface collapses. A large shield volcano, called Gula Mons, is shown (again as a computer-generated view) in Figure 6.15(c).

The largest volcanic structures on Venus are huge, roughly circular regions known as **coronae**. A large corona, called Aine, can be seen in Figure 6.16. Coronae are unique to Venus. They appear to have been caused by upwelling motions in the mantle that caused the surface to bulge outward, but never developed into full-fledged convection as on Earth. Coronae generally have volcanoes both in and around them, and their rims usually show evidence of extensive lava flows into the plains below.

Two pieces of indirect evidence suggest that volcanism on Venus continues today. First, the level of sulfur dioxide above Venus's clouds shows large and fairly frequent fluctuations, possibly as the result of volcanic eruptions on the surface. Second, orbiting spacecraft have observed bursts of radio energy from the planet's surface similar to those produced by the lightning discharges that often occur in the plumes of erupting volcanoes on Earth. However, while quite persuasive, these pieces of evidence are still only circumstantial. No "smoking gun" (or erupting volcano) has yet been seen, so the case for active volcanism is not yet complete.

A few Soviet *Venera* spacecraft have landed on Venus's surface. Each survived for about an hour before being destroyed by the intense heat, its electronic circuitry melting in this planetary oven. Figure 6.17 shows one of the first photographs of the surface of Venus radioed back to Earth. The flat rocks visible in this image show little evidence of erosion and are apparently quite young, supporting the idea of ongoing surface activity. Later Soviet landers performed simple chemical analyses of the surface. Some of the samples studied were found to be predominantly basaltic, again implying a volcanic past. Others resembled terrestrial granite.

Not all the craters on Venus are volcanic in origin. Some were formed by meteoritic impact. The largest impact craters on Venus are generally circular, but those less than about 15 km in diameter can be quite asymmetric. Figure 6.18(a) shows a *Magellan* image of a relatively small impact crater, about 10 km across, in Venus's southern hemisphere. Geologists think that the light-colored region is an ejecta blanket—material thrown from the crater following the impact. Its irreg-

R I V U X G

◀ **Figure 6.17 Venus in Situ** One of the first true-color views of the surface of Venus, radioed back to Earth from the Russian *Venera 14* spacecraft, which made a soft landing in 1975. The amount of sunlight penetrating Venus's cloud cover is about the same as that reaching Earth's surface on a heavily overcast day. (Russian Space Agency)

▶ **Figure 6.18 Impact Craters on Venus** (a) A *Magellan* image of an apparent multiple-impact crater in Venus's southern hemisphere. The irregular shape of the light-colored ejecta blanket seems to be the result of a meteorite that fragmented just prior to impact. The dark regions in the crater may be pools of solidified lava associated with the individual fragments. (b) Venus's largest crater, named Mead after anthropologist Margaret Mead, is about 280 km across. Bright (rough) regions clearly show its double-ringed structure. (NASA)

(a)

(b)

ular shape may be the result of a large meteorite that broke up just before impact, with the separate pieces hitting the surface near one another. This seems to be a fairly common fate for medium-sized bodies (1 km or so in diameter) that plow through Venus's dense atmosphere.

Figure 6.18(b) shows the largest-known impact feature on Venus, the 280-km-diameter crater Mead. Its double-ringed structure is in many ways similar to the Moon's Mare Orientale (Figure 5.25). Numerous impact craters, again identifiable by their ejecta blankets, can also be seen in Figure 6.16.

6.6 The Surface of Mars

The View from Earth

Earth-based observations of Mars at closest approach can distinguish surface features as small as 100 km across—about the same resolution as the unaided human eye can achieve when viewing the Moon. However, when Mars is closest to us and most easily observed, it is also full, so the angle of the Sun's rays does not permit us to see any topographical detail, such as craters or mountains. Even through a large telescope Mars appears only as a reddish disk with some light and dark patches and prominent polar caps (Figure 6.6).

Mars has been visited by far more spacecraft from Earth than any other planet, and many missions have played important roles in our understanding of the planet. Figure 6.19 shows a detailed mosaic of Mars, constructed from images made by one of the U.S. *Viking* spacecraft in orbit around the planet.

Mars's surface features undergo slow seasonal changes over the course of a Martian year—a consequence of Mars's axial tilt and somewhat eccentric orbit. The polar caps grow or shrink according to the seasons, almost disappearing during the Martian summer. The dark features also

▶ **Figure 6.19 Mars Globe** A mosaic of Mars based on images from a *Viking* spacecraft in orbit around the planet. The planet's surface is visible at a level of detail completely unattainable from Earth. Mars's Tharsis region, 5000 km across, bulges out from the equator, rising to a height of about 10 km. The two large volcanoes on the left mark the approximate peak of the Tharsis bulge. Dominating the center of the field of view is a vast "canyon" known as Valles Marineris—the Mariner Valley. (NASA)

Animation/Video
Composite Image of Mars

Chryse Planitia

Volcanoes

Tharsis bulge

Valles Marineris

Animation/Video
Mars Polar Cap

vary in size and shape. To fanciful observers around the start of the twentieth century, these changes suggested the annual growth of vegetation—and much more (see *Discovery 6-1*)—but, as with Venus, these speculations were not confirmed. The changing polar caps are mostly frozen carbon dioxide, not water ice as at Earth's North and South Poles (although smaller "residual" caps of water ice, which persist even through the warmer summer months, also exist). The dark regions are just highly cratered and eroded areas on the surface. During summer in the Martian southern hemisphere, planetwide dust storms sweep up the dry dust and carry it aloft, sometimes for months at a time, even-

6-1 DISCOVERY

Martian Canals?

The year 1877 was an important one in the human study of the planet Mars. The Red Planet came unusually close to Earth, affording astronomers an especially good view. Of particular note was the discovery, by U.S. Naval Observatory astronomer Asaph Hall, of the two moons circling Mars (see Section 6.9). But most exciting was the report of the Italian astronomer Giovanni Schiaparelli on his observation of a network of linear markings that he termed *canali*. In Italian, *canali* can simply mean "grooves" or "channels," but it can also mean "canals." As far as we know, Schiaparelli did not intend to imply that the canali were anything other than natural, but the word was translated into English as *canals*, suggesting that the grooves had been constructed by intelligent beings. Observations of these features became sensationalized in the world's press (especially in the United States), and some astronomers began drawing elaborate maps of Mars, showing oases and lakes where canals met in desert areas.

Percival Lowell (see photo), a successful Boston businessman (and brother of the poet Amy Lowell and Harvard president Abbott Lawrence Lowell), became fascinated by these reports. He abandoned his business and purchased a clear-sky site at Flagstaff, Arizona, where he built a major observatory. He devoted his life to achieving a better understanding of the

Martian "canals." In doing so, he championed the idea that Mars was drying out and that an intelligent society had constructed the canals to transport water from the wet poles to the arid equatorial deserts.

Alas, the Martian valleys and channels photographed by robot spacecraft during the 1970s are far too small to be the canali that Schiaparelli, Lowell, and others thought they saw on Mars. The entire episode represents a classic case in the history of science—a case in which well-intentioned observers, perhaps obsessed with the notion of life on other worlds, let their personal opinions and prejudices seriously affect their interpretations of reasonable data. The accompanying figures of Mars show how surface features (which were probably genuinely observed by astronomers at the turn of the century) might have been imagined to be connected. The figure on the left is a photograph of how Mars actually looked in a telescope at the end of the nineteenth century. The sketch at right is an interpretation (done at the height of the canal hoopla) of the pictured view. The human eye, under physiological stress, tends to connect dimly observed yet distinctly separated features. Humans saw patterns and canals where none in fact existed.

The chronicle of the Martian canals illustrates how the scientific method requires scientists to acquire new data to sort out sense from nonsense, fact from fiction. Rather than simply believing the claims about the Martian canals, other scientists demanded further observations to test Lowell's hypothesis. Eventually, improved observations, climaxing in the *Mariner* and *Viking* exploratory missions to the Red Planet nearly a century after all the fuss began, totally disproved the existence of canals. It often takes time, but the scientific method does eventually lead to progress in understanding reality.

(Lowell Observatory)

(Lowell Observatory)

tually depositing it elsewhere on the planet. Repeated covering and uncovering of the Martian landscape gives the impression from a distance of surface variability, but it is only the thin dust cover that changes.

Large-Scale Topography

A striking feature of the terrain of Mars is the marked difference between the northern and southern hemispheres (Figure 6.20). The northern hemisphere is made up largely of rolling volcanic plains somewhat like the lunar maria but much larger than any plains found on Earth or the Moon. They were apparently formed by eruptions involving enormous volumes of lava. The plains are strewn with blocks of volcanic rock, as well as with boulders blasted out of impact areas by infalling meteoroids. The southern hemisphere consists of heavily cratered highlands lying several kilometers above the level of the lowland north.

The northern plains are cratered much less than the southern highlands, suggesting that the northern surface is younger—perhaps 3 billion years old, compared with 4 billion in the south. In places, the boundary between the southern highlands and the northern plains is quite sharp. The surface level can drop by as much as 4 km in a distance of 100 km or so. Most scientists assume that the southern terrain is the original crust of the planet. How most of the northern hemisphere could have been lowered in elevation and flooded with lava remains a mystery.

In the late 1990s, NASA's *Mars Global Surveyor* satellite mapped the entire Martian surface to an accuracy of a few meters. Figure 6.21 shows a map of Mars based on these data, with some prominent surface features marked. The north-south asymmetry is very obvious in this map. The major geological

(a)

(b)

R I V U X G

▲ **Figure 6.20 Mars, Up Close** (a) Mars's northern hemisphere consists of rolling volcanic plains (false color image.) (b) The southern Martian highlands are heavily cratered (true color). Both photographs show roughly the same scale, nearly 1000 km across. (NASA)

▼ **Figure 6.21 Mars Map** A computer-generated map of planet Mars, based on detailed measurements made by *Mars Global Surveyor*. Color represents height above (or below) the mean planetary radius, ranging from dark blue (−8 km), through green, yellow, and red to white (over 8 km altitude), as shown in the scale at right. The Mariner valley (Figure 6.19) can be seen at the left. The opposite side of the planet is dominated by the giant Hellas impact basin. Note the great difference in elevation between the northern and southern hemispheres. Some surface features are labeled, as are the *Viking, Pathfinder,* and *Exploration Rover* robot landing sites. (NASA)

feature on the planet is the *Tharsis bulge*, visible at the left of Figure 6.19 and marked in Figure 6.21. Roughly the size of North America, Tharsis lies on the Martian equator and rises some 10 km above the rest of the Martian surface. To the east and west of Tharsis lie wide depressions, hundreds of kilometers across and up to 3 km deep. If we wished to extend the idea of "continents" from Earth and Venus to Mars, we would say that Tharsis is the only continent on the Martian surface. However, as on Venus, there is no sign of plate tectonics—the continent of Tharsis is not drifting like its Earth counterparts. ∞ *(Sec. 5.5)* Tharsis appears to be even less heavily cratered than the northern hemisphere, making it the youngest region on the planet. It is estimated to be only 2–3 billion years old.

Almost diametrically opposite Tharsis lies the Hellas Basin (labeled by its more technical name of Hellas Planitia in Figure 6.21). Located in the southern highlands, it paradoxically contains the lowest point on the Martian surface. Some 3000 km across, the floor of the basin lies nearly 9 km below its rim and over 6 km below the average level of the planet's surface. Its shape and structure identify it as an impact feature. Its formation must have caused a major redistribution of the young Martian crust—maybe even enough to account for a substantial portion of the highlands, according to some researchers. Its heavily cratered floor indicates that the impact occurred very early in Martian history—perhaps 4 billion years ago.

Volcanism

Mars contains the largest known volcanoes in the solar system. Four particularly large ones are found on the Tharsis bulge, two of them visible in Figure 6.19. Biggest of all is Olympus Mons (Figure 6.22), which lies on the northwestern slope of Tharsis, just over the left horizon of Figure 6.19, but marked on Figure 6.21. Olympus Mons is 700 km in diameter at its base—only slightly smaller than the state of Texas—and rises to a height of 25 km above the surrounding plains.

Like the volcanoes on Venus, those on Mars are not associated with plate motion, but instead are shield volcanoes, sitting atop hot spots in the Martian mantle. Spacecraft images of the Martian surface reveal many hundreds of volcanoes. Most of the largest are associated with the Tharsis bulge, but many smaller ones are also found in the northern plains. It is not known whether any of them are still active. However, based on the extent of impact cratering on their slopes, some of them apparently erupted as recently as 100 million years ago.

The great height of Martian volcanoes is a direct consequence of the planet's low surface gravity. ∞ *(Sec. 5.1)* As a shield volcano forms and lava flows and spreads, the new mountain's height depends on its ability to support its own weight. The lower the surface gravity, the less the lava weighs and the higher the mountain can be. Mars has a surface gravity only 40 percent that of Earth, and its volcanoes rise roughly 2.5 times as high.

◀ **Figure 6.22 Martian Volcano** Olympus Mons, the largest volcano known on Mars or anywhere else in the solar system, is nearly three times taller than Mount Everest on Earth, measuring about 700 km across its base and rising 25 km high at its peak. This Martian mountain seems currently inactive and may have been extinct for at least several hundred million years. By comparison, the largest volcano on Earth, Hawaii's Mauna Loa, measures a mere 120 km across and peaks just 9 km above the Pacific Ocean floor. (NASA)

300 km

R I U V X G

The Martian Grand Canyon

Associated with the Tharsis bulge is a great canyon known as Valles Marineris (the Mariner Valley). Shown in its entirety in Figure 6.21 and cutting across the center of Figure 6.19, this enormous crack in the Martian surface is not really a canyon in the terrestrial sense, because running water played no part in its formation. Geologists think it was formed by the same crustal forces that pushed the Tharsis region upward, causing the surface to split and crack. Cratering studies suggest that Valles Marineris is at least 2 billion years old. Similar (but smaller) cracks, formed in a similar way, have been found in the Aphrodite Terra region of Venus.

Valles Marineris runs for almost 4000 km along the Martian equator and extends about one-fifth of the way around the planet. At its widest, it is 120 km across, and it is as deep as 7 km in places. Earth's Grand Canyon in Arizona would easily fit into one of its side "tributary" cracks. It is so large that it can even be seen from Earth. This Martian geological feature is not a result of Martian plate tectonics, however. For some reason, the crustal forces that formed it never developed into full-fledged plate motion as on Earth.

Evidence for Past Water on Mars

Although the Valles Marineris was not formed by running water, photographic evidence reveals that liquid water once existed in great quantity on the surface of Mars. Two types of flow features were seen by the *Viking* orbiters: **runoff channels** and **outflow channels**.

The runoff channels (Figure 6.23) are found in the southern highlands. They are extensive systems—sometimes hundreds of kilometers in total length—of interconnecting, twisting channels that merge into larger, wider channels. They bear a strong resemblance to river systems on Earth, and geologists think that that is just what they are: the dried-up beds of long-gone rivers that once carried rainfall on Mars from the mountains down into the valleys. These runoff channels speak of a time four billion years ago (the age of the Martian highlands) when the atmosphere was thicker, the surface warmer, and liquid water widespread.

The outflow channels (Figure 6.24) are probably relics of catastrophic flooding on Mars long ago. They appear only in equatorial regions and generally do not form the extensive interconnected networks that characterize the runoff channels. Instead, they are probably the paths taken by huge volumes of water draining from the southern highlands into the northern plains. Judging from the width and depth of the channels, the flow rates must have been truly enormous—perhaps as much as a hundred times greater than the 10^5 tons per second carried by the Amazon River, the largest river system on Earth. Flooding shaped the outflow channels about 3 billion years ago, about the same time as the northern volcanic plains formed.

Some scientists speculate that Mars may have enjoyed an extended early period during which rivers, lakes, and perhaps even oceans adorned its surface. Figure 6.25 is a 2003 *Mars Global Surveyor* image showing what mission specialists think may be a *delta*—a fan-shaped network of channels and sediments where a river once flowed into a larger body of water, in this case a lake filling a crater in the southern highlands. Other

Tutorial
Comparative Planetology: Mars

▼ **Figure 6.23 Martian Channel**   (a) This runoff channel on Mars is about 400 km long and 5 km wide. (b) The Red River on Earth runs from Shreveport, Louisiana, to the Mississippi River. The two differ mainly in that there is currently no liquid water in this, or any other, Martian valley. (NASA)

(a)

100 km

R I V U X G

(b)

50 km

R I V U X G

| | 100 km | |

▲ **Figure 6.24 Martian Outflow** An outflow channel near the Martian equator bears witness to a catastrophic flood that occurred about 3 billion years ago. (NASA)

researchers go even farther, suggesting that the data provide evidence for large open expanses of water on the early Martian surface. Figure 6.26(a) is a computer-generated view of the Martian north polar region, showing the extent of what may have been an ancient ocean covering much of the northern lowlands. The Hellas basin (Figure 6.21) is another candidate for an ancient Martian sea.

These ideas remain controversial. Proponents point to features such as the terraced "beaches" shown in Figure 6.26(b), which could conceivably have been left behind as a lake or ocean evaporated and the shoreline receded. But detractors maintain that the terraces could also have been created by geological activity, perhaps related to the tectonic forces that depressed the northern hemisphere far below the level of the south, in which case they have nothing whatever to do with Martian water. Furthermore, *Mars Global Surveyor* data seem to indicate that the Martian surface contains too few *carbonate* rock layers—layers containing compounds of carbon and oxygen that should have been formed in abundance in an ancient ocean. Their absence supports the picture of a cold, dry Mars, which never experienced the extended mild period required to form lakes and oceans. However, data from the most recent *NASA* landers (see below) imply that at least some parts of the planet did in fact experience long periods in the past during which liquid water existed on the surface. Obviously the debate is far from over.

Where Is the Water Today?

Astronomers have no direct evidence for liquid water anywhere on the Martian surface today, and the amount of water vapor in the Martian atmosphere is tiny. Yet the extent of the outflow channels and the other evidence just described indicates that a huge total volume of water existed on Mars in the past. Where did all that water go? The most likely answer is that much of Mars's original water is now locked in a layer of **permafrost** (a layer of water ice lying just below the planet's surface, much like that found in Earth's arctic regions), with more contained in the polar caps.

Figure 6.27 shows indirect evidence for the permafrost layer in the form of a fairly typical Martian impact crater named Yuty. Unlike the lunar craters discussed in Chapter 5, Yuty's ejecta blanket gives the distinct impression of a liquid that has splashed or flowed out of the crater. Most likely, the explosive impact heated and liquefied the permafrost, resulting in the fluid appearance of the ejecta. ∞ *(Sec. 5.4)* More direct evidence for subsurface ice was obtained in 2002, when the *Mars Odyssey* orbiter detected extensive deposits of water ice crystals (actually, the hydrogen they contain) mixed with the Martian surface layers. In some locations ice appears to make up as much as 50 percent by volume of the planet's soil. Unfortunately, none of the five successful Martian landers to date (see below) has put down in a region where ice has been detected.

As noted earlier, Mars's polar caps actually consist of two distinct components. The *seasonal caps* vary in size as atmospheric carbon dioxide alternately freezes and evaporates during the winter and summer months. However, the permanently frozen *residual caps* (Figure 6.28) are now known to be composed of water ice. The presence of ice in the larger northern cap has long been known from spectroscopic observations of water vapor above them as some small fraction of the cap evaporates in the Sun's heat during the summer months. The composition of the southern residual cap was finally established

10 km		

◄ **Figure 6.25 Martian River Delta** Did this fan-shaped region of twisted streams form as a river flowed into a larger sea? If it did, the *Mars Global Surveyor* image supports the idea that Mars once had large bodies of liquid water on its surface. Not all scientists agree with this interpretation, however. (NASA) only in 2004, by spectroscopic imaging observations made by the European Space Agency's *Mars Express* orbiter. The thickness of the caps is uncertain, but it is quite possible that they represent a major storehouse for water on Mars.

Prior to the arrival of *Mars Global Surveyor*, astronomers thought that all the water on and below the Martian surface existed in the form of ice. However, in 2000, *Surveyor* mission scientists reported the discovery of numerous small-scale "gullies" in Martian cliffs and crater walls that apparently were carved by running water in the relatively recent past. These features are too small to have been resolved by *Viking* cameras. Figure 6.29 shows one such gully, found in the inner rim of a Martian impact crater in the southern highlands. Its structure has many similarities to the channels carved by flash floods on Earth. The ages of these intriguing features are uncertain, and might be as great as a million years in some cases, but the *Surveyor* team speculate that some of them may still be active today, implying that liquid water might exist in some regions of Mars at depths of less than 500 meters.

Scientists think that 4 billion years ago, as the Martian climate changed, the running water that formed the runoff channels began to freeze, forming the permafrost and drying out the river beds. Mars remained frozen for about a billion years, until volcanic (or some other) activity heated large regions of the surface, melting the permafrost and causing the flash floods that created the outflow channels. Subsequently, volcanic activity subsided, the water refroze, and Mars once again became a dry world. The present level of water vapor in the Martian atmosphere is the maximum possible given the atmosphere's present density and temperature. Estimates of the total amount of water stored as permafrost and in the polar caps are still quite uncertain, but it is likely that if all the water on Mars were to become liquid, it would cover the surface to a depth of several meters.

Exploration by Martian Landers

Remote sensing—taking images and other measurements from orbit—is very important, but in many cases there is no substitute for a close-up look. To date, five U.S spacecraft have successfully landed on the Martian surface (see Table 6.2). Their landing sites, marked on Figure 6.21, spanned a variety of Martian terrains. Their goals included detailed geological and chemical analysis of Martian surface rocks, the search for life,

▼ **Figure 6.26** **Ancient Ocean?** (a) A possible ancient Martian ocean might once have spanned the polar regions. The blue regions in this computer-generated map actually indicate elevation, but they also outline the extent of the ocean inferred from detailed study of *Mars Global Surveyor* data. (The color elevation scale is nearly the same as in Figure 6.21.) (b) This high-resolution image shows tentative evidence for erosion by standing water in the floor of Holden Crater, about 140 km across. (NASA)

(a)

Holden Crater

Uzboi Vallis

100 km

R   I   V   U   X   G

(b)

▲ **Figure 6.27 Martian Crater** The ejecta from Mars's crater Yuty (18 km in diameter) evidently was once liquid. This type of crater is sometimes called a "splosh" crater. (NASA)

▲ **Figure 6.28 Martian Polar Cap** The residual (permanent) polar caps on Mars may be major storehouses for water on the planet. This is the smaller southern cap, roughly 350 km in diameter. (ESA)

and search for water (an essential requirement for life as we know it—see Chapter 18).

Both U.S. *Viking* missions dispatched landers to the Martian surface. Figure 6.30(a) is the view from *Viking 1*, which touched down near the planet's equator just east of Tharsis. The photograph shows a windswept, gently rolling, desolate plain littered with rocks of all sizes, not unlike a high desert on Earth. This view may be quite typical of the low-latitude northern plains. The *Viking* landers performed numerous chemical analyses of the rock on Mars's surface, and performed several experiments designed specifically to test for life (see *Discovery 6-2*). One important finding of these studies was the high iron content of the planet's crust. Chemical reactions between the iron-rich surface soil and trace amounts of oxygen in the atmosphere are responsible for the iron oxide ("rust") that gives Mars its characteristic red color.

The next successful mission to the Martian surface was *Mars Pathfinder*. During the unexpectedly long life-

◀ **Figure 6.29 Running Water on Mars?** This high-resolution *Mars Global Surveyor* view (left) of a crater wall (right) near the Mariner Valley shows evidence of "gullies" apparently formed by running water in the relatively recent past. If this interpretation is correct, it suggests that liquid water may still reside below the Martian surface. (NASA)

(a)

(b)

▲ **Figure 6.30 Mars Panorama** (a) Wide-angle view from the perspective of the *Viking 1* spacecraft, now parked on the surface of Mars. The fine-grained soil and the reddish rock-strewn terrain stretching toward the horizon contain substantial amounts of iron ore; the surface of Mars is literally rusting away. The sky is a pale pink, the result of airborne dust. (b) The Mars rover *Sojourner*, seen here coming down the ramp of its mother ship, *Pathfinder*, in order to inspect a large Martian rock, nicknamed Yogi by mission scientists. For scale, the rock is about 1 m tall and lies some 10 m from *Pathfinder*. (NASA)

time of its mission in 1997 (it lasted almost three months instead of the anticipated one), the lander performed measurements of the Martian atmosphere and atmospheric dust while its robot rover *Sojourner* (Figure 6.30b) carried out chemical analyses of the soil and rocks within about 50 m of the parent craft. In addition, more than 16,000 images of the region were returned to Earth. The soil chemistry in the vicinity of the landing site was similar to that found by the *Viking* landers. Interestingly, analyses of the nearby rocks revealed chemical makeups different from that of the Martian meteorites found on Earth (see *Discovery 6-2*).

The landing site for the *Pathfinder* mission had been carefully chosen to lie near the mouth of an outflow channel, and the size distribution and composition of the many rocks and boulders surrounding the lander were consistent with their having been deposited there by flood waters. In addition, the presence of numerous rounded pebbles strongly suggested the erosive action of running water at some time in the past.

The most recent visitors to the Martian surface are the twin landers *Spirit* and *Opportunity* of the Mars Exploration Rover mission. During their operating lifetimes on the planet—now well into their second years, again far exceeding the three-month "prime mission" expectations of mission planners—these two robots have made extensive chemical and geological studies of rocks within a few kilometers of their landing sites on opposite sides of the planet, with the primary goal of finding evidence for liquid water on the surface at some time in the past. Their findings provide the best evidence yet for standing water on the ancient Martian surface and have changed the minds of many skeptical scientists on the subject of water on Mars.

Spirit's landing site was rocky and similar in many ways to the terrains encountered by earlier landers, although closer study by the science team reveals that most of the rocks in the lander's vicinity do appear to have been extensively altered by water long ago. Halfway around the planet, however, *Opportunity* appears to have hit the jackpot in its quest, finding itself surrounded by rocks showing every chemical and geological indication of having

◄◄◄ **CONCEPT CHECK**

Why do astronomers think that the Martian climate was once quite different from today's?

(a)

(b)

▲ **Figure 6.31 Opportunity Rover** (a) A panoramic view of the terrain where NASA's *Opportunity* rover landed on Mars. This is Endurance crater, roughly 130 meters across. (b) A (false-color) close-up of a clump of rocks at the bottom of Endurance crater where *Opportunity* found evidence for extensive liquid water in Mars's past. (NASA)

been very wet—possibly immersed in salt water—in the distant past (Figure 6.31). Several indicators suggest that the rocks near *Opportunity*'s landing site have been alternately underwater and dry for extended periods of time, possibly as a shallow lake alternately filled and evaporated numerous times during the course of Martian history. If life ever did exist at this site, the sorts of rocks found there might have preserved a fossil record very well, but *Opportunity* was not equipped to carry out such studies.

6.7 Internal Structure and Geological History

Mercury

Mercury's magnetic field, discovered by *Mariner 10*, is about 1/100 of Earth's field. The discovery that Mercury has a magnetic field came as a surprise to planetary scientists, who, having detected no magnetic field in the Moon, expected Mercury to have none either. In Chapter 5 we saw how a combination of liquid metal core and rapid rotation is necessary for the production of a planetary magnetic field. ∞ *(Sec. 5.7)* Mercury certainly does not rotate rapidly and it may also lack a liquid metal core, yet a magnetic field undeniably surrounds it. Although weak, the field is strong enough to deflect the solar wind and create a small magnetosphere around the planet. Scientists have no clear understanding of its origin. Possibly it is simply a "fossil remnant" dating back to the distant past when the planet's core solidified.

Mercury's magnetic field and high average density (roughly 5400 kg/m^3) taken together imply that most of the planet's interior is dominated by a large, iron-rich core having a radius of perhaps 1800 km. The ratio of core volume to total planet volume is greater for Mercury than for any other object in the solar system. Figure 6.32 illustrates the relative sizes and internal structures of Earth (and perhaps also Venus), the Moon, Mercury, and Mars.

Like the Moon, Mercury seems to have been geologically dead for roughly the past 4 billion years. Again as on the Moon, the lack of present-day geological activity on Mercury results from the mantle's being solid, preventing volcanism or tectonic motion. Largely on the basis of studies of the Moon, scientists have pieced together the following outline of Mercury's early history.

When Mercury formed 4.6 billion years ago, its location in the hot inner region of the early solar system ensured a dense, largely metallic overall composition. ∞ *(Sec. 4.3)* During the next half-billion years, Mercury melted and differentiated, just as the other terrestrial worlds did. It suffered the same intense meteoritic bombardment as the Moon. Being more massive than the

Moon, Mercury cooled more slowly, so its crust was thinner than the Moon's and volcanic activity more common. More craters were erased by lava, leading to the intercrater plains found by *Mariner 10*.

As the planet's large iron core formed and then cooled, the planet began to shrink, causing the surface to contract. This compression produced the scarps seen on Mercury's surface and may have prematurely terminated volcanic activity by squeezing shut the cracks and fissures on the surface. Thus, Mercury did not experience the subsequent extensive volcanic outflows that formed the lunar maria. Despite its larger mass and greater internal temperature, Mercury has probably been geologically inactive for even longer than the Moon.

Venus

Both U.S. and Soviet spacecraft have failed to detect any magnetosphere around Venus. Given that Venus's average density is similar to Earth's, it seems likely that Venus has an Earth-like overall composition and a partially molten iron-rich core. The lack of any detectable magnetic field, then, is almost surely the result of the planet's extremely slow rotation. ∞ *(Sec. 5.7)*

Because none of the *Venera* landers carried seismic equipment, no direct measurements of the planet's interior have ever been made, and theoretical models of the interior have very little hard data to constrain them. However, to many geologists the surface of Venus resembles that of the young Earth, at an age of perhaps a billion years. At that time, volcanic activity had already begun on Earth, but the crust was still relatively thin, and the convective processes in the mantle that drive plate tectonic motion were not yet established. The physical similarities between Venus and Earth suggest that the core/mantle structure of Venus resembles that of our own planet.

Why has Venus remained in that immature state and not developed plate tectonics as Earth did? That question remains to be answered. Some planetary geologists have speculated that the high surface temperature on Venus has inhibited evolution by slowing the planet's cooling rate. Possibly the high surface temperature has made the crust too soft for Earth-style plates to develop. Or perhaps the high temperature and soft crust led to more volcanism, tapping the energy that might otherwise have gone into convective motion.

Earth

Mars

Mercury

Moon

▶ **Figure 6.32 Terrestrial Planet Interiors** The internal structures of Earth, the Moon, Mercury, and Mars, drawn to the same scale. Note how large a fraction of Mercury's interior is core. The interior structure of Venus is largely unknown, but is thought to be similar to that of Earth.

Mars

In September 1997 the orbiting *Mars Global Surveyor* succeeded in detecting a very weak Martian field, about 1/800 that of Earth. However, this is most likely a local anomaly, not a global field. Because Mars rotates rapidly, the weakness of the magnetic field is taken to mean that the planet's core is nonmetallic, or nonliquid, or both. ∞ *(Sec. 5.7)*

6-2 DISCOVERY

Life on Mars?

Even before the *Viking* missions reached Mars in 1976, astronomers had abandoned hope of finding life on the planet. Scientists knew there were no large-scale canal systems, no surface water, almost no oxygen in the atmosphere, and no seasonal vegetation changes. The present lack of liquid water on Mars especially dims the chances for life there now. However, running water and possibly a dense atmosphere in the past may have created conditions suitable for the emergence of life long ago. In the hope that some form of microbial life might have survived to the present day, the *Viking* landers carried out experiments designed to detect biological activity. The accompanying pair of photographs show the robot arm of one of the landers digging a shallow trench.

All three *Viking* biological experiments assumed some basic similarity between hypothetical Martian bacteria and those found on Earth. A *gas-exchange* experiment offered a nutrient broth to any residents of a sample of Martian soil and looked for gases that would signal metabolic activity. A *labeled-release* experiment added compounds containing radioactive carbon to the soil, then waited for results signaling that Martian organisms had either eaten or inhaled this carbon. Finally, a *pyrolitic-release* experiment added radioactively tagged carbon dioxide to a sample of Martian soil and atmosphere, waited awhile, then removed the gas and tested the soil (by heating it) for signs that something had absorbed the tagged gas. In all cases, contamination by terrestrial bacteria was a major concern. Indeed, any release of Earth organisms would have invalidated these and all future such experiments on Martian soil. Both *Viking* landers were carefully sterilized prior to launch.

Initially, all three experiments appeared to be giving positive signals! However, subsequent careful studies showed that the results could all be explained by inorganic (that is, nonliving) chemical reactions. Thus, we have no evidence for even microbial life on the Martian surface. The *Viking* robots detected peculiar reactions that mimic in some ways the basic chemistry of living organisms, but they did not detect life itself.

One criticism of the *Viking* experiments is that they searched only for life now living. Today, Mars seems locked in an ice age—the kind of numbing cold that would prohibit sustained life as we know it. If bacterial life did arise on an Earth-like early Mars, however, then we might be able to find its fossilized remains preserved on or near the Martian surface. This is one reason that scientists are so eager to land more spacecraft on the planet, particularly at the poles, whose ice caps may offer the best environment for finding Martian life (or its remains).

Surprisingly, one alternative place to look for evidence of life on Mars is right here on Earth. The accompanying figure shows ALH84001, a blackened, 2-kg meteorite about 17 cm across, found in 1984 in Antarctica. Chemical analysis strongly indicates that it originated on Mars. It was apparently blasted off that planet long ago by a meteoritic impact of some sort, thrown into space, and eventually captured by Earth's gravity. Based on the estimated cosmic-ray exposure it received before reaching Earth, it left Mars about 16 million years ago.

In 1996 a group of scientists argued, based on all the data accumulated from studies of ALH84001, that they had discovered fossilized evidence for life on Mars. They pointed to globular structures similar to those produced by bacteria on Earth (top left inset), the presence of chemical compounds sometimes associated with Earth biology, and curved, rodlike structures (right image) resembling Earth bacteria, which the researchers interpreted as fossils of primitive Martian organisms. In 1999 the team released an analysis of a second mete-

Robot arm

Trench

(NASA)

R I V U X G

(Before)   (After)

Mars's small size means that any internal heat would have been able to escape more easily than in a larger planet like Earth or Venus. The evidence of ancient surface activity, especially volcanism, suggests that at least parts of the Martian interior must have melted at some time in the past, but the lack of current activity and the absence of any significant magnetic field indicate that the melting was never as extensive as on Earth. The latest data indicate that Mars's core has a diameter of about 2500 km, is composed largely of iron sulfide

orite also thought to have come from Mars, again reporting evidence for microbial life having similarities in size, shape, and arrangement to known nanobacteria on Earth.

These claims were and remain very controversial. Many experts categorically disagree that evidence of Martian life has been found—not even fossilized life. They maintain that, as in the case of the *Viking* experiments, the evidence could all be due to chemical reactions not requiring any kind of biology. The scale of the supposed fossil structures is also important. They are only about 0.5 μm across, 1/30 the size of ancient bacterial cells found fossilized on Earth. Furthermore, several key experiments have not yet been done, such as testing the suspected fossils for evidence of cell walls or of any internal cavities where body fluids would have resided. Nor has anyone yet found in either meteorite any amino acids, the basic building blocks of life as we know it (see Chapter 18). In addition, there is the huge problem of contamination—after all, ALH84001 was found on Earth and apparently sat in the Antarctic ice fields for 13,000 years before being picked up by meteorite hunters.

As things now stand, it's a matter of interpretation—at the frontiers of science, issues are usually not as clear-cut as we would hope. Only additional analysis and new data—perhaps in the form of samples returned directly from the Martian sur-face—will tell for sure if primitive Martian life existed long ago. Most workers in the field seem to have concluded that, taken as a whole, the results do not support the claim of ancient life on Mars. Still, even some skeptics concede that as much as 20 percent of the organic material in ALH84001 could have originated on the Martian surface—although that is a far cry from proving the existence of life there.

Whether or not ALH84001 actually turns out to contain evidence of ancient Martian life, the scientific uproar it has caused has forced scientists to think much more carefully about the kinds of life that might be able to survive in extreme environments—those with temperatures, pressures, or chemical makeup not normally regarded as hospitable to life as we know it on Earth. These studies have fueled the emerging field of *exobiology*—the search for and study of life on other planets, in our own solar system and beyond. As we will see in Chapters 9 and 18, the subsurface oceans of Jupiter's moon Europa and the hydrocarbon seas of Saturn's frigid moon Titan have joined Mars as prime candidates in the search for extraterrestrial life.

Should the claim of life on Mars hold up against the weight of healthy skepticism in the scientific community, these findings may go down in history as one of the greatest scientific discoveries of all time. We are—or at least were—not alone in the universe! Maybe. . . .

Carbonate globule

Animation/Video
Ejection of Fragments from Mars

Animation/Video
Possible Microscopic Fossils

5 cm

(NASA)

Structure inside globules

(a compound about twice as dense as surface rock), and is still at least partly molten.

Mars appears to be a planet where large-scale tectonic activity almost started but was stifled by the planet's rapidly cooling outer layers. On a larger, warmer planet, the upwelling that formed the Tharsis bulge might have developed into full-fledged plate tectonics, but the Martian mantle became too rigid, the crust too thick, for that to occur. Instead, the upwelling fired volcanic activity, perhaps even up to the present day, but geologically much of the planet died 2 billion years ago.

CONCEPT CHECK ▶▶▶

Why do Venus and Mars lack magnetic fields?

6.8 Atmospheric Evolution on Earth, Venus, and Mars

Now that we have studied the structure and history of the three terrestrial worlds that still have atmospheres today—Venus, Earth, and Mars—we can return to the question of why their atmospheres are so different from one another. Let's begin our study with Earth. The chain of events that occurred on Earth serves as a useful baseline for the histories of the other two planets.

The Development of Earth's Atmosphere

When Earth formed, any atmosphere it might have had—sometimes called the **primary atmosphere**—would have consisted of the gases most common in the early solar system: light gases such as hydrogen, helium, methane, ammonia, and water vapor, a far cry from the atmosphere we enjoy today. Almost all this light material, and especially any hydrogen or helium, escaped into space during the first half-billion or so years after Earth was formed. As discussed in Chapter 5, Earth's gravity is simply too weak to retain these light gases. ∞ *(More Precisely 5-1)*

Subsequently, Earth developed a **secondary atmosphere**, which was released from the planet's interior as a result of volcanic activity—a process called *outgassing*. Volcanic gases are rich in water vapor, carbon dioxide, sulfur dioxide, and compounds containing nitrogen. As Earth's surface temperature fell and the water vapor condensed, oceans formed. Much of the carbon dioxide and sulfur dioxide became dissolved in the oceans or combined with surface rocks. Oxygen is such a reactive gas that any free oxygen that appeared at early times was removed as quickly as it formed. As solar ultraviolet radiation liberated nitrogen from its chemical bonds with other elements, a nitrogen-rich atmosphere slowly appeared.

The final major development in the story of Earth's atmosphere was the appearance of *life* in the oceans 3.5 billion years ago. Living organisms began to produce atmospheric oxygen and eventually the ozone layer formed, shielding the surface from the Sun's harmful radiation. Eventually, life spread to the land and flourished. The fact that oxygen is a major constituent of our present-day atmosphere is a direct consequence of the evolution of life on Earth.

The Runaway Greenhouse Effect on Venus

Given the distance of Venus from the Sun, the planet was not expected to be such a pressure cooker. Why is Venus so hot? And if, as seems likely, Venus started off like Earth, then why is its atmosphere now so different from Earth's?

The answer to the first question is fairly easy. Given the present composition of its atmosphere, Venus is hot because of the greenhouse effect. Recall that *greenhouse gases* in Earth's atmosphere, particularly water vapor and carbon dioxide, tend to warm our planet. ∞ *(Sec. 5.3)* By stopping the escape of much of the infrared radiation emitted by Earth's surface, these gases increase the planet's equilibrium temperature much as an extra blanket keeps you warm on a cold night. Venus's dense atmosphere is made up almost entirely of a prime greenhouse gas, carbon dioxide (Figure 6.33). This thick blanket absorbs about 99 percent of all the infrared radiation released from the surface of Venus and is the immediate cause of the planet's sweltering 730 K surface temperature.

The answer to the second question—Why is Venus's atmosphere so different from Earth's?—is more complex. The initial stages of atmospheric development on Venus probably took place in more or less the same way as just described for our own planet. However, on Earth much of the secondary atmosphere became part of the planet surface, as carbon dioxide and sulfur dioxide dissolved in the oceans or combined with surface rocks. If all the dissolved or chemically combined carbon dioxide on Earth were released back into our present-day atmosphere, its new composition would be 98 percent carbon dioxide and 2 percent nitrogen, and it would have a pressure about 70 times its current value. In other words, apart from the presence of oxygen (which appeared on Earth only after the development of life) and water (whose absence on Venus will be explained shortly), Earth's atmosphere would look a lot like that of Venus. The real difference between Earth and Venus, then, is that Venus's greenhouse gases never left the atmosphere the way they did on Earth.

When Venus's secondary atmosphere appeared, the temperature was higher than on Earth because Venus is closer to the Sun, but the exact atmospheric temperature is uncertain. If it was so high that water vapor could not become liquid, no oceans would have formed. Consequently, outgassed water vapor and carbon dioxide would have remained in the atmosphere, and the full greenhouse effect would have gone into operation immediately. If oceans did form and most of the greenhouse gases left the atmosphere to become dissolved in the water, the temperature must still have been sufficiently high to allow a process known as the **runaway greenhouse effect** to come into play.

To understand the runaway greenhouse effect, imagine that we took Earth from its present orbit and placed it in Venus's orbit. At its new distance from the Sun, the amount of sunlight hitting Earth's surface would be almost twice its present level, and the planet would warm up. More water would evaporate from the oceans, leading to an increase in atmospheric water vapor. At the same time, the ability of both the oceans and surface rocks to hold carbon dioxide would diminish, allowing more carbon dioxide to enter the atmosphere. As a result, the greenhouse heating would increase, and the planet would warm still further, leading to a further increase in atmospheric greenhouse gases, and so on. Once started, the process would "run away," eventually leading to the complete evaporation of the oceans, restoring all the original greenhouse gases to the atmosphere. Basically, the same thing would have happened on Venus long ago, leading to the planetary inferno we see today.

The greenhouse effect on Venus was even more extreme in the past, when the atmosphere also contained water vapor. By intensifying the blanketing effect of the carbon dioxide, the water vapor helped the surface of Venus reach temperatures perhaps twice as hot as at present. At those high temperatures, the water vapor was able to rise high into the planet's upper atmosphere—so

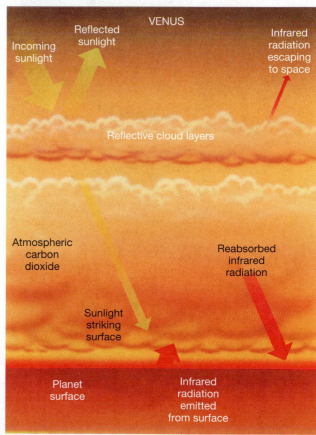

▶ **Figure 6.33 Venus's Atmosphere** *Because Venus's atmosphere is much deeper and denser than Earth's, a much smaller fraction of the infrared radiation leaving the planet's surface escapes into space. The result is a much stronger greenhouse effect than on Earth and a correspondingly hotter planet. The outgoing infrared radiation is not absorbed at a single point in the atmosphere. Instead, absorption occurs at all atmospheric levels. (The arrows indicate only that absorption occurs, not that it occurs at any specific level; the arrow thickness is proportional to amount of radiation.)*

high that it was broken up by solar ultraviolet radiation into its components, hydrogen and oxygen. The light hydrogen rapidly escaped, the reactive oxygen quickly combined with other atmospheric gases, and all water on Venus was lost forever.

Although it is highly unlikely that global warming due to human activities will ever send Earth down the path taken by Venus, this episode highlights the relative fragility of the planetary environment. ∞ *(Discovery 5-1)* No one knows how close to the Sun Earth could have formed before a runaway greenhouse effect would have occurred. But in comparing our planet with Venus, we have come to understand that there is an orbital limit, presumably between 0.7 and 1.0 AU, inside of which Earth would have suffered a similar catastrophic runaway. We must consider this "greenhouse limit" when assessing the likelihood that planets harboring life formed elsewhere in the Galaxy.

Evolution of the Martian Atmosphere

Presumably, Mars also had first a primary and then a secondary (outgassed) atmosphere early in its history. Around 4 billion years ago Mars may have had a fairly dense atmosphere rich in carbon dioxide, perhaps even with blue skies and rain. Taking into account the distance from Mars to the Sun and the fact that the Sun was about 30 percent less luminous 4 billion years ago (see Chapter 12), planetary scientists estimate that the greenhouse effect from a Martian atmosphere a few times denser than Earth's present atmosphere could have kept conditions fairly comfortable. An average surface temperature above 0°C (the freezing point of water) seems quite possible.

Sometime during the next billion years, however, most of the Martian atmosphere disappeared. Possibly some of it was expelled by impacts with large bodies in the early solar system, and a large part may have leaked away into space because of the planet's weak gravity, but most of the remainder probably became unstable and was lost in a kind of "reverse runaway greenhouse effect." ∞ *(More Precisely 5-1)* The following scenario, illustrated in Figure 6.34, is accepted by most planetary scientists (although by no means all—those who discount the evidence presented earlier for liquid water and a thick early atmosphere on Mars obviously require no explanation for their absence today).

The basic mechanism controlling the level of carbon dioxide in a planet's atmosphere is the absorption of carbon dioxide into the rocks that make up the planet's crust. The presence of liquid water (Figure 6.34a) greatly accelerates this process—carbon dioxide dissolves in the liquid water of the planet's rivers and lakes (and oceans, if any), ultimately reacting with surface material to form carbonate rocks. The result is a continual depletion of atmospheric carbon dioxide. On Earth, plate tectonics constantly recycles our planet's carbon dioxide, returning it to the atmosphere by volcanic activity. ∞ *(Sec. 5.4)* Eventually, a balance is struck between these two competing processes, and the atmospheric concentration of carbon dioxide tends to remain roughly constant. However, this is not the case on Mars. As we have just seen, Mars cooled faster than Earth and apparently never developed large-scale plate motion. As a result, even taking into consideration the large volcanoes discussed earlier in this chapter, Mars has had on average far less volcanism than Earth does, so the processes depleting carbon dioxide were much more effective than those replenishing it, creating a "one-way street" in which the level of atmospheric carbon dioxide steadily declined.

As the Martian carbon dioxide was consumed and its greenhouse effect diminished, the planet cooled, causing still more carbon dioxide to leave the at-

(a) Ancient Mars

(b) Today's Mars

(c) Terraformed Mars

◀ **Figure 6.34 Martian Evolution** (a) Mars some 4 billion years ago, with thick atmosphere and surface oceans; (b) today; and (c) after a hypothetical future terraforming mission. (Kees Veenenbos) mosphere. The reason for this is just as described above—lower temperatures allowed more carbon dioxide to be absorbed in the surface layers. The result was a runaway to lower temperatures and decreasing levels of carbon dioxide in the atmosphere. Calculations show that much of the Martian atmospheric carbon dioxide could have been depleted in this way in a relatively short period of time, perhaps as quickly as a few hundred million years. As the temperature continued to fall, water froze out of the atmosphere, lowering still further the level of atmospheric greenhouse gases and accelerating the cooling. Eventually, even carbon dioxide began to freeze out, particularly at the poles, and Mars reached the frigid state we see today (Figure 6.34b)—a cold, dry planet with most of its original complement of atmospheric gases now residing in or under the barren surface.

Mars continues to cool, slowly losing what remains of its thin atmosphere. With little internal heat remaining to drive planetwide changes, it is highly unlikely that any natural Martian phenomenon will ever restore the atmospheric conditions of long ago. However, with a human presence on Mars within the next few decades a real possibility, and with the eventual goal of human colonization of other planets in mind, some scientists are beginning to take seriously the long-standing dream of science fiction writers and are studying the issues involved in *terraforming* Mars—literally, transforming its present inhospitable climate into an Earth-like environment.

The basic idea is simple—introduce into the Martian atmosphere enough greenhouse gases to warm the planet to the point where the chain of events just described is reversed. Mars warms, the atmosphere thickens and warms some more, and eventually we end up with a habitable planet (for humans). Several possible ways of accomplishing this task have been proposed, ranging from melting parts of the polar caps using giant orbiting mirrors, to steering ammonia-rich asteroids into a collision with the planet, to building huge greenhouse gas–emitting factories on the Martian surface (as it happens, CFCs are particularly efficient greenhouse gases for this purpose). ∞ *(Discovery 5-1)* It is perhaps ironic that all of these approaches represent precisely the sorts of global environmental catastrophe currently causing such concern to scientists here on Earth! ∞ *(Discovery 4-1, 5-1, 5-2)* The value of planetary change depends very much on where one lives.

The three courses of action just listed pose substantial challenges to present technology and are not feasible today. However, they might become so within the next hundred years. Several centuries would then be required for greenhouse gases to rise to the levels needed for natural processes to take over, and perhaps one or two thousand years would be needed after that for the Martian climate to approach anything even remotely Earth-like (Figure 6.34c). Of course, humanity might instead decide that the loss to science of destroying the present Martian environment would outweigh the gain of creating a habitable world, or the practical difficulties of the task may prove too hard or expensive to overcome. We can only imagine the debate that will precede that momentous decision in human history.

◄◄◄ **CONCEPT CHECK**

If Venus and Mars had formed at Earth's distance from the Sun, what might their climates be like today?

CHAPTER REVIEW

Summary

① Mercury's rotation rate is strongly influenced by the tidal effect of the Sun, which causes the planet rotates exactly one and a half times for every one revolution around the Sun.

② Mercury has no permanent atmosphere. The atmospheres of Venus and Mars are mainly carbon dioxide. Venus's atmosphere is extremely hot and 90 times denser than Earth's, while the density of the cool Martian atmosphere is only 0.7 percent that of Earth's.

3 Mercury's surface is heavily cratered, much like the lunar highlands. Mercury lacks lunarlike maria but has extensive **intercrater plains** (p. 167) and cliff-forming cracks, or **scarps** (p. 167), in its crust. The plains were caused by lava flows early in Mercury's history. The scarps were apparently formed when the planet's core cooled and shrank, causing the surface to crack.

4 Because of its thick cloud cover, Venus's surface cannot be seen in visible light from Earth, although it has been thoroughly mapped by radar. Many lava domes and **shield volcanoes** (p. 170) have been found. No eruptions have been observed, although there is indirect evidence that Venus is still volcanically active today. The northern hemisphere of Mars consists mainly of rolling plains, while the south is rugged highlands. The reason for this asymmetry is unknown. Mars's major surface feature is the Tharsis bulge. Associated with the bulge is the largest-known volcano in the solar system and a huge crack, called the Valles Marineris, in the planet's surface.

5 There is clear evidence that water once existed in great quantity on Mars. The **runoff channels** (p. 175) are the remains of ancient Martian rivers. The **outflow channels** (p. 175) are the paths taken by floods that cascaded from the southern highlands into the northern plains. Today a large amount of that water may be locked up in the polar caps and in a layer of **permafrost** (p. 176) lying under the Martian surface. Images from orbiting spacecraft seem to show what may be ancient coastlines; others suggest that there may still be liquid water below the surface in some places. The most recent craft to land on the surface have found strong geological evidence for standing water.

6 Mercury's weak magnetic field seems to be a fossil remnant from the time when the planet's iron core solidified. Geological activity on Mercury ceased long ago. Venus does not have a detectable magnetic field even though, like Earth, it probably has a molten core. This is due to the planet's slow rotation. The surface of Venus shows no sign of plate tectonics. Features called **coronae** (p. 170) are thought to have been caused by an upwelling of mantle material that never developed into full convective motion. Mars rotates rapidly but has a very weak magnetic field, implying that its core is nonmetallic, nonliquid, or both. Mars also appears to have had some form of tectonic activity long ago, but plate tectonics never developed because the planet cooled too rapidly.

7 Any **primary atmospheres** (p. 184) of light gases that Earth, Venus, and Mars had at formation rapidly escaped. All the terrestrial planets probably developed **secondary atmospheres** (p. 184), outgassed from volcanoes, early in their lives. On Mercury the atmosphere escaped, as on the Moon. On Earth much of the outgassed material became absorbed in surface rocks or dissolved in the oceans. The development of life produced oxygen. On Venus the **runaway greenhouse effect** (p. 185) has resulted in all the planet's carbon dioxide being left in the atmosphere, leading to the extreme conditions we observe today. On Mars part of the secondary atmosphere escaped into space. Of the remainder, most of the carbon dioxide is now locked up in surface rock, and most of the water vapor is now stored in the permafrost and the polar caps.

Review and Discussion

1. In contrast to Earth, Mercury undergoes extremes in temperature. Why?

2. What do Mercury's magnetic field and large average density imply about the planet's interior?

3. How is Mercury's evolutionary history like that of the Moon? How is it different?

4. Mercury used to be called "the Moon of the Sun." Why do you think it had this name? What piece of scientific evidence proved the name inappropriate?

5. Why does Venus appear so bright to the naked eye?

6. Venus probably has a molten iron-rich core like Earth. Why doesn't it also have a magnetic field?

7. How did radio observations of Venus made in the 1950s change our conception of that planet?

8. What are the main constituents of Venus's atmosphere? What are clouds in the upper atmosphere made of?

9. What is the runaway greenhouse effect, and how might it have altered the climate of Venus?

10. Why is Mars red?

11. What is the evidence that water once flowed on Mars? Is there water on Mars today?

12. Why were Martian volcanoes able to become so large?

13. Given that Mars has an atmosphere and its composition is mostly carbon dioxide, why isn't there a significant greenhouse effect to warm its surface?

14. Do you think that sending humans to Mars in the near future is a reasonable goal? Why or why not?

15. Compare and contrast the evolution of the atmospheres of Mars, Venus, and Earth. Include a discussion of the importance of volcanoes and water.

Conceptual Self-Test: True or False/Multiple Choice

1. Mercury's solar day is longer than its solar year. **(T/F)**

2. Numerous surface features on Venus can be seen in Earth-based images made in the ultraviolet part of the spectrum. **(T/F)**

3. Lava flows are common on the surface of Venus. **(T/F)**

4. There is strong circumstantial evidence that active volcanism continues on Venus. **(T/F)**

5. Mars has the largest volcanoes in the solar system. **(T/F)**

6. There are many indications of plate tectonics on Mars. **(T/F)**

7. Valles Marineris is similar in size to Earth's Grand Canyon. **(T/F)**

8. Mercury's large overall density suggests that the planet **(a)** has an interior structure similar to that of Earth's Moon; **(b)** has a dense metal core; **(c)** has a weaker magnetic field than the Moon; **(d)** is younger than the Moon.

9. Venus's surface is permanently obscured by clouds. As a result, the surface has been studied primarily by **(a)** robotic landers; **(b)** orbiting satellites using radar; **(c)** spectroscopy; **(d)** radar signals from Earth.

10. Compared with Earth, Venus has a level of plate tectonic activity that is **(a)** much more rapid; **(b)** somewhat less rapid; **(c)** about the same; **(d)** virtually nonexistent.

11. Venus's atmospheric temperature is **(a)** about the same as Earth's; **(b)** cooler than temperatures on Mercury; **(c)** hotter than temperatures on Mercury; **(d)** high due to the presence of sulfuric acid.

12. In terms of area, the extinct Martian volcano Olympus Mons is about the size of **(a)** Mt. Everest; **(b)** Colorado; **(c)** North America; **(d)** Earth's Moon.

13. In comparison to the atmosphere of Venus, the vastly different atmospheric character of Mars is likely due to **(a)** an ineffective greenhouse effect; **(b)** a reverse greenhouse effect; **(c)** the absence of greenhouse gases in the Martian atmosphere; **(d)** Mars's greater distance from the Sun.

14. In Figure 6.3 (Planetary Radar), part (b), the highest-frequency reflected radiation is received from **(a)** the upper part of the planet; **(b)** the center of the planet; **(c)** the lower part of the planet.

15. Of the following, the best evidence for the existence of liquid water on an ancient Mars is Figure **(a)** 6.19; **(b)** 6.22; **(c)** 6.23; **(d)** 6.30.

Problems

Algorithmic versions of these questions are available in the Practice Problems module of the Companion Website.

The number of squares preceding each problem indicates its approximate level of difficulty.

1. ■ How long does it take a radar signal to travel from Earth to Mercury and back when Mercury is at its closest point to Earth?

2. ■■ Assume that a planet will have lost its initial atmosphere by the present time if the average (daytime) molecular speed exceeds one-sixth of the escape speed (see *More Precisely 5-1*). What would Mercury's mass have to be in order for it still to have a nitrogen (molecular weight 28) atmosphere?

3. ■■ Approximating Venus's atmosphere as a layer of gas 50 km thick with a uniform density of $21\ kg/m^3$, calculate its total mass. Compare your answer with the mass of Earth's atmosphere (Chapter 5, Problem 6) and with the mass of Venus.

4. ■ *Pioneer Venus* observed high-level clouds moving around Venus's equator in four days. What was their speed in kilometers per hour? In miles per hour?

5. ■■ **(a)** What is the size of the smallest feature that can be distinguished on the surface of Venus (at closest approach) by the Arecibo radio telescope at an angular resolution of 1'? **(b)** Could an infrared telescope with an angular resolution of 0.1" distinguish impact craters on the surface of Venus?

6. ■■■ Calculate the orbital period of the *Magellan* spacecraft, moving around Venus on an elliptical orbit with a minimum altitude of 294 km and a maximum altitude of 8543 km above the planet's surface. In 1993 the spacecraft's orbit was changed to have minimum and maximum altitudes of 180 km and 541 km, respectively. What was the new period?

7. ■ Verify that the surface gravity on Mars is 40 percent that of Earth. What would you weigh on Mars?

8. ■■ The outflow channel shown in Figure 6.24 is about 10 km across and 100 m deep. If it carried 10^7 metric tons (10^{10} kilograms) of water per second, as stated in the text, estimate the speed at which the water must have flowed.

9. ■■ Calculate the total mass of a uniform layer of water covering the entire Martian surface to a depth of 2 m (see Section 6.8). Compare it with the mass of Mars and with the mass of Venus's atmosphere (Problem 3).

10. ■■ Mars has two small, irregular moons, Phobos and Deimos, which orbit at distances of 9400 km and 23,500 km, respectively, from the planet. Their longest diameters are 28 km (Phobos) and 16 km (Deimos). Calculate their maximum angular diameters, as seen by an observer on the Martian surface directly under their orbits. Would the observer ever see a total solar eclipse?

The Companion Website at www.prenhall.com/chaisson provides algorithmically generated versions of each chapter's Problems, in addition to an Animations and Videos gallery, a Tutorials gallery and Tutorials Quiz questions, a full eBook, an Images gallery, Practice Questions, Multiple Choice, and True or False quizzes, and Labeling exercises.

7

Jupiter is certainly one of the most fascinating objects in the solar system. This is a true color mosaic, constructed from two dozen images taken by a camera onboard the *Cassini* spacecraft during its closest approach in 2001. It is the most detailed portrait of Jupiter ever produced, resolving features to as small as 60 kilometers. Everything seen here is a cloud, from the equatorial regions that show alternating light and dark belts, to high-latitude areas that appear more mottled. The varying structures are caused by differing cloud heights, thickness, and chemical compositions. (JPL)

The Jovian Planets

Giants of The Solar System

Beyond the orbit of Mars, the solar system is a very different place. In sharp contrast to the small, rocky, terrestrial bodies found near the Sun, the outer solar system presents us with a totally unfamiliar environment—huge gas balls, peculiar moons, planet-circling rings, and a wide variety of physical and chemical properties, many of which are only poorly understood even today. U.S. spacecraft visiting the outer planets over the past two decades have revealed these worlds in detail that was only dreamed of by centuries of Earth-bound astronomers. The jovian planets—Jupiter, Saturn, Uranus, and Neptune—differ from Earth and from each other in many ways, but they have much in common, too. As with the terrestrial planets, we learn from their differences as well as from their similarities.

LEARNING GOALS

Studying this chapter will enable you to:

1. Explain how both chance and calculation played major roles in the discovery of Uranus and Neptune.

2. Outline some basic similarities and differences among the four jovian worlds.

3. Describe some of the processes responsible for the properties of the jovian atmospheres.

4. Discuss the storm systems observed on the jovian planets, and compare them with storms on Earth.

5. Describe the internal structure and composition of the jovian planets.

6. Explain why three of the four jovian worlds radiate more energy into space than they receive from the Sun.

Visit www.prenhall.com/chaisson for additional annotated images, animations, and links to related sites for this chapter.

(a)

(b)

R I V U X G

▲ **Figure 7.1 Jupiter** (a) Photograph of Jupiter made with an Earth-based telescope, showing the planet and its four Galilean moons. (b) A *Hubble Space Telescope* image of Jupiter, in true color. Features as small as a few hundred kilometers across are resolved. (AURA/NASA)

R I V U X G

7.1 Observations of Jupiter and Saturn

The View from Earth

Named after the most powerful god of the Roman pantheon, Jupiter is the third-brightest object in the night sky (after the Moon and Venus), making it easy to locate and study from Earth. Ancient astronomers could not have known the planet's true size, but their name choice was very apt—Jupiter is by far the largest planet in the solar system. Figure 7.1(a) is a photograph of Jupiter taken through a small telescope on Earth, showing the planet's overall reddish coloration and the alternating light and dark bands paralleling its equator. Figure 7.1(b) is a *Hubble Space Telescope* image of Jupiter, showing the colored bands in much more detail and displaying numerous oval structures within them (note in particular the large red oval at lower right). These atmospheric features are quite unlike anything found on the terrestrial planets. In further contrast to the inner planets, Jupiter has many moons that vary greatly in size and other properties. The four largest, shown in Figure 7.1(a), are visible from Earth with a small telescope (and, for a few people, with the naked eye). They are known as the *Galilean moons* after Galileo Galilei, who discovered them in 1610. ∞ *(Sec. 1.2)*

Saturn (Figure 7.2), the next major body we encounter as we continue outward beyond Jupiter, was the most distant planet known to Greek astronomers. Named after the father of Jupiter in Greco–Roman mythology, Saturn orbits at almost twice Jupiter's distance from the Sun, making it considerably fainter (as seen from Earth) than either Jupiter or Mars. The planet's banded atmosphere is somewhat similar to Jupiter's, but Saturn's atmospheric bands are much less distinct. Overall, the planet has a rather uniform butterscotch hue. Again like Jupiter and unlike the inner planets, Saturn has many moons orbiting it. Saturn's best-known feature, its spectacular *ring system*, is clearly visible in Figure 7.2. The moons and rings of the jovian planets are the subject of Chapter 8.

Spacecraft Exploration

Most of our detailed knowledge of the jovian planets has come from NASA spacecraft visiting those worlds. Of particular importance are the two *Voyager* probes and the more recent *Galileo* and *Cassini* missions.

The *Voyager* spacecraft left Earth in 1977, reaching Jupiter in March (*Voyager 1*) and July (*Voyager 2*) of 1979. Both craft subsequently used Jupiter's strong gravity to send them on to Saturn in a maneuver called a *gravity assist* (see *More Precisely 7-1*). *Voyager 2* took advantage of a rare planetary configuration that enabled it to use Saturn's gravity to propel it to Uranus and then on to Neptune in a spectacularly successful "grand tour" of the outer planets. Each craft carried equipment to study planetary magnetic fields and magnetospheres, as well as radio, visible-light, and infrared sensors to analyze reflected and emitted radiation from the jovian planets and their moons. The data they returned revolutionized our knowledge of all the jovian worlds and remain our primary source of detailed information on Uranus and Neptune. The two *Voyager* craft are now headed out of the solar system, still returning data as they race toward interstellar space.

Galileo was launched in 1989 and reached Jupiter in December 1995, after a circuitous route that took it three times through the inner solar system,

◄ **Figure 7.2 Saturn** Saturn as seen by the *Hubble Space Telescope* in 2003 when the planet was at its maximum tilt (27°) toward Earth in its 30-year orbit about the Sun. The colors in this photo are natural and its resolution is about 700 km. (NASA)

7-1 MORE PRECISELY

Gravitational "Slingshots"

Sending a spacecraft to another planet requires a lot of energy—often more than can be conveniently provided by a rocket launched from Earth or safely transported in a shuttle for launch from orbit. Faced with these limitations, mission scientists make use of a "slingshot" maneuver to boost an interplanetary probe to a more energetic orbit and also aid navigation toward the target, all at no additional cost!

The first accompanying figure illustrates a gravitational slingshot, or *gravity assist*, in action. A spacecraft approaches a planet, passes close by, then escapes along a new trajectory. Obviously, the spacecraft's *direction* of motion is changed by the encounter. Less obviously, the spacecraft's *speed* is also altered as the planet's gravity propels the spacecraft in the direction of the planet's motion. By careful choice of incoming trajectory, the craft can either speed up (by passing behind the planet, as shown) *or* slow down (by passing in front), by as much as twice the planet's orbital speed. Of course, there is no free lunch—the spacecraft gains energy from, or loses it to, the planet's motion, causing its orbit to change ever so slightly. However, since planets are so much more massive than spacecraft, the effect is tiny.

Such a slingshot maneuver has been used many times in missions to both the inner and the outer planets, as illustrated in the second figure, which shows *Voyager 2*'s "grand tour" through the jovian planets. The gravitational pulls of these

giant worlds whipped the craft around at each visitation, enabling flight controllers to get considerable extra "mileage" out of the probe. Indeed, most of what we know about the outer planets has been gained from spacecraft propelled in this way. As described in the text, both the *Galileo* mission to Jupiter and the *Cassini* mission to Saturn used multiple gravity assists, both to reach the target planet and subsequently to navigate among that planet's moons. Every encounter with a moon had a slingshot effect—sometimes accelerating and sometimes slowing the probe, but each time moving it onto a different orbit—and every one was carefully calculated long before the spacecraft ever left Earth.

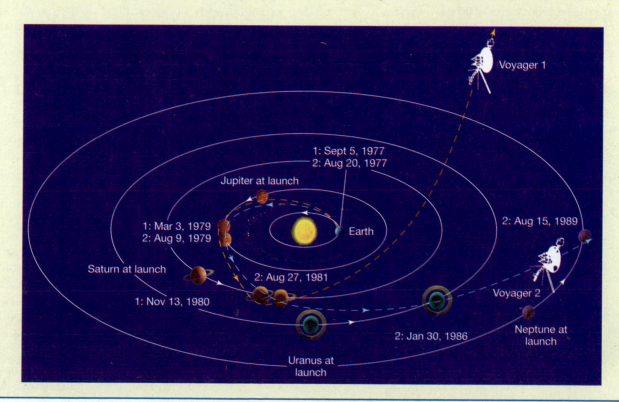

Animation/Video

Galileo Mission to Jupiter

R I V U X G

▲ **Figure 7.3** **The View from *Cassini*** This *Cassini* spacecraft image of Jupiter, taken in Decemeber 2000 from a distance of 10 million km while the vehicle was on its way to Saturn, shows intricate atmospheric clouds of different heights, thicknesses, and chemical composition. The image is approximately true color—that is, as the human eye would see it. (NASA)

receiving gravity assists from both Venus and Earth before finally reaching its target. The mission had two components: an atmospheric probe and an orbiter. The probe, slowed by a heat shield and a parachute, descended into Jupiter's atmosphere, making measurements and chemical analyses as it went. The orbiter executed a complex series of gravity-assisted maneuvers through Jupiter's moon system, returning to some moons studied by *Voyager* and visiting others for the first time. One of *Galileo*'s most important discoveries was the likely existence of an extensive saltwater ocean under the icy crust of the moon Europa, making that moon a prime candidate for the possibility of extraterrestrial life. Many more of *Galileo*'s findings are described in Chapter 8.

The mission was scheduled to end in December 1997 but was so successful that NASA extended its lifetime three times, for a total of six additional years, to obtain even more detailed data on Jupiter's inner moons. Eventually, however, the spacecraft began to run out of the fuel needed to fine-tune its orbit and keep its antenna pointed toward Earth. Knowing that they would ultimately lose control of the probe, and wishing to avoid at all costs any possible contamination of Jupiter's moon Europa (see Section 8.1), the *Galileo* team elected to end the mission by steering the spacecraft into Jupiter's atmosphere in September 2003.

In October 1997, NASA launched the *Cassini* mission to Saturn. The craft reached Saturn in July 2004 after four planetary gravity assists—two from Venus and one each from Earth and Jupiter—en route. Figure 7.3 shows *Cassini*'s close-up view of Jupiter during its December 2000 encounter with that planet. In December 2004 the craft released a probe built by the European Space Agency into the atmosphere of Titan, Saturn's largest moon (see Section 8.2). *Cassini* is now orbiting among Saturn's moons, much as *Galileo* did at Jupiter, returning detailed data on the moons, the planet, and its famous ring system. The mission was originally scheduled to last for four years after arrival at Saturn, but so accurate was the gravity-assisted maneuver that placed the probe in orbit around the planet that almost no fuel had to be used to fine-tune the trajectory. As a result, barring instrument failure, *Cassini*'s eventual lifetime may well rival that of its sister mission *Galileo*. Currently *Cassini* is extending and expanding our understanding of the Saturn system, answering some outstanding questions raised by *Voyager*, but posing many new ones, too.

7.2 The Discoveries of Uranus and Neptune

The British astronomer William Herschel discovered the planet Uranus in 1781. Because this was the first new planet discovered in recorded history, the event caused quite a stir. The story goes that Herschel's first instinct was to name the new planet *Georgium Sidus* (Latin for "George's star") after his king, George III of England. The world was saved from a planet named George by the wise advice of another astronomer, Johann Bode, who suggested instead that the tradition of using names from ancient mythology be continued and that the planet be named after Uranus, the father of Saturn.

Uranus is just barely visible to the naked eye, under the best possible observing conditions. It looks like a faint, undistinguished star. Even through a large optical telescope, Uranus appears hardly more than a tiny, pale, slightly greenish disk. Figure 7.4 shows a close-up visible-light image taken by *Voyager* 2 in 1986. Uranus's apparently featureless atmosphere contrasts sharply with the bands and spots visible in the atmospheres of the other jovian planets.

R I V U X G

◄ **Figure 7.4** **Uranus** A close-up view of Uranus sent back to Earth by *Voyager* 2 while the spacecraft was whizzing past the planet at 10 times the speed of a rifle bullet. Color is natural and nearly featureless. (NASA)

▶ **Figure 7.5 Neptune**   Neptune as seen in natural color by *Voyager 2* from a distance of 1,000,000 km. (NASA)

Once Uranus was discovered, astronomers set about charting its orbit. They quickly discovered a discrepancy between the planet's predicted and observed positions. Try as they might, they could not find an elliptical orbit that fit the planet's path. By the early nineteenth century the discrepancy had grown to a quarter of an arc minute, far too large to be explained away as observational error. Uranus seemed to be violating—slightly—Kepler's laws of planetary motion. ∞ *(Sec. 1.3)*

Astronomers realized that although the Sun's gravitational pull dominates the planet's orbital motion, the small deviation meant that some unknown body was exerting a much weaker but still measurable gravitational force on Uranus. There had to be *another* planet in the solar system influencing Uranus. In September 1845, after almost two years of work, an English mathematician named John Adams solved the problem of determining the new planet's mass and orbit. In June 1846 a French mathematician, Urbain Leverrier, independently came up with essentially the same answer. Later that year a German astronomer named Johann Galle found the new planet within one or two degrees of the predicted position. The new planet was named Neptune, and Adams and Leverrier are now jointly credited with its discovery.

Unlike Uranus, distant Neptune cannot be seen with the naked eye, although it can be seen through binoculars or a small telescope. In fact, according to his notes, Galileo may actually have seen Neptune, although he surely had no idea what it really was at the time. Through a large telescope on Earth, the planet appears only as a small, bluish disk, with few surface features visible. Even under the best observing conditions, only a few markings, suggestive of multicolored cloud bands, can be seen. With *Voyager 2*'s arrival, much more detail emerged (Figure 7.5)—in fact, the bulk of our knowledge of the two outermost jovian worlds has come from just this one spacecraft. Superficially, at least, Neptune resembles a blue-tinted Jupiter, with atmospheric bands and spots clearly evident.

◀◀◀ **CONCEPT CHECK**

How did observations of Uranus lead to the discovery of Neptune?

7.3 Bulk Properties of the Jovian Planets

Table 7.1 extends Table 6.1 to include the four jovian planets. Figure 7.6 shows them to scale, along with Earth for comparison.

Physical Characteristics

The large masses and radii and relatively low average densities of these gigantic worlds imply that they differ radically in both composition and structure from the terrestrial planets. ∞ *(Sec. 4.1)* Saturn's density is actually less than that of water—the planet would float, if only we could find a bathtub big enough! Most of the masses of Jupiter and Saturn, and about half those of Uranus and Neptune, are made up of hydrogen and helium—light gases on Earth (having densities 0.08 and 0.16 kg/m^3, respectively, at room temperature and sea level) but compressed enormously by these planets' strong gravitational fields. The abundance of hydrogen and helium on these worlds is itself a consequence of the strong jovian gravity. The jovian planets are massive enough to have retained even the lightest gas—hydrogen—and very little of their original atmospheres have escaped since the birth of the solar system 4.6 billion years ago. ∞ *(More Precisely 5-1)*

At the heart of each jovian planet lies a dense, compact core many times more massive than Earth. Three of the four (Jupiter, Saturn, and Neptune) have significant internal heating (to be discussed in Section 7.6), and this has important consequences for the behavior and appearance of their atmospheres.

TABLE 7.1 Planetary Properties

	MASS (kg)	MASS (Earth = 1)	RADIUS (km)	RADIUS (Earth = 1)	AVERAGE DENSITY (kg/m^3)	SURFACE GRAVITY (Earth = 1)	ESCAPE SPEED (km/s)	ROTATION PERIOD (solar days)	AXIAL TILT (degrees)	SURFACE TEMPERATURE (k)	SURFACE MAGNETIC FIELD (Earth = 1)
Mercury	3.3×10^{23}	0.550	2,400	0.38	5400	0.38	4.3	59	0	100–700	0.01
Venus	4.9×10^{24}	0.82	6,100	0.95	5300	0.91	10	−243[1]	179	730	0.0
Earth	6.0×10^{24}	1.00	6,400	1.00	5500	1.00	11	1.00	23	290	1.0
Mars	6.4×10^{23}	0.11	3,400	0.53	3900	0.38	5.0	1.03	24	180–270	0.0
Jupiter	1.9×10^{27}	320	71,000	11	1300	2.5	60	0.41	3	120	14
Saturn	5.7×10^{26}	95	60,000	9.5	710	1.1	36	0.43	27	97	0.7
Uranus	8.7×10^{25}	15	26,000	4.0	1200	0.91	21	−0.69[1]	98	58	0.7[2]
Neptune	1.0×10^{26}	17	25,000	3.9	1700	1.1	24	0.72	30	59	0.4[2]

[1] The minus sign indicates retrograde rotation.

[2] Average values. Because of the planet's asymmetric magnetic field geometry (see Section 7.6), the field strength varies substantially from one point on the surface to another.

None of the jovian planets has a solid surface of any kind. Their gaseous atmospheres just become hotter and denser with depth because of the pressure of the overlying layers, eventually becoming liquid in the interior. The surface we see from Earth is simply the top of the outermost layer of atmospheric clouds. Where relevant, the term *surface* in Table 7.1 and elsewhere in this chapter refers to this level in the atmosphere.

Rotation Rates

Because the jovian planets have no solid surfaces to "tie down" the gas flow, different parts of their atmospheres can and do move at different speeds. This state of affairs—when the rotation rate is not constant from one location to another—is called **differential rotation**. It is not possible in solid objects like the terrestrial worlds but is normal for fluid bodies such as the jovian planets (and the Sun—see Chapter 9).

The amount of differential rotation on Jupiter is small—the equatorial regions rotate once every $9^h 50^m$, while the higher latitudes take about 6 minutes longer. On Saturn the difference between the equatorial and polar rotation rates is a little greater (26 minutes, again slower at the poles). On Uranus and Neptune the differences are greater still—more than two hours in the case of Uranus—this time with the poles rotating more rapidly. In each case, the differen-

Jupiter

Saturn

Earth

Uranus

Neptune

◄ **Figure 7.6 Jovian Planets** Jupiter, Saturn, Uranus, and Neptune, showing their relative scales compared to Earth. Note the very different sizes and atmospheric features of these five worlds. Water in a (big!) bathtub usefully illustrates two visual analogies: Saturn is less dense than water and would therefore float, and water swirling down the drain rotates differentially— that is, faster toward the center than at the edges.

tial rotation observed in the surface layers reflects large-scale wind flows in the planet's atmosphere.

More meaningful measurements of overall planetary rotation rates are provided by observations of their magnetospheres. All four jovian worlds have strong magnetic fields and emit radiation at radio wavelengths. The strength of this radio emission varies with time and repeats itself periodically. Scientists assume that this period matches the rotation of the planet's deep interior, where (as on Earth) the magnetic field arises. ∞ *(Sec. 5.7)* The rotation periods listed in Table 7.1 were obtained in this way. There is no clear relationship between the interior and atmospheric rotation rates of the jovian planets. In Jupiter and Saturn the interior rotation rate matches the polar rotation. In Uranus the interior rotates more slowly than any part of the atmosphere. In Neptune the reverse is true.

The observations used to measure a planet's rotation period also let us determine the orientation of its rotation axis. Jupiter's rotation axis is almost perpendicular to the orbit plane, the axial tilt being only 3°, quite small compared with Earth's tilt of 23°. The tilts of Saturn and Neptune (Table 7.1) are similar to those of Earth and Mars.

Each planet in our solar system seems to have some outstanding peculiarity. In the case of Uranus, it is its rotation axis. Unlike the other major planets, which have their spin axes (very) roughly perpendicular to the ecliptic plane, Uranus's rotation axis lies almost within that plane. We might say that Uranus is "tipped over" onto its side, for its axial tilt is 98°. (Because the north pole lies below the ecliptic plane, the planet's rotation, like that of Venus, is classified as retrograde.) As a result, the north pole[1] of Uranus, at some point in the planet's orbit, points almost directly toward the Sun. Half a Uranian year later the south pole faces the Sun, as illustrated in Figure 7.7. When *Voyager 2* encountered the planet in 1986, the north pole happened to be pointing nearly

[1]We adopt here the convention that a planet's rotation is counterclockwise as seen from above the north pole (that is, planets always rotate from west to east).

Saturn would float...

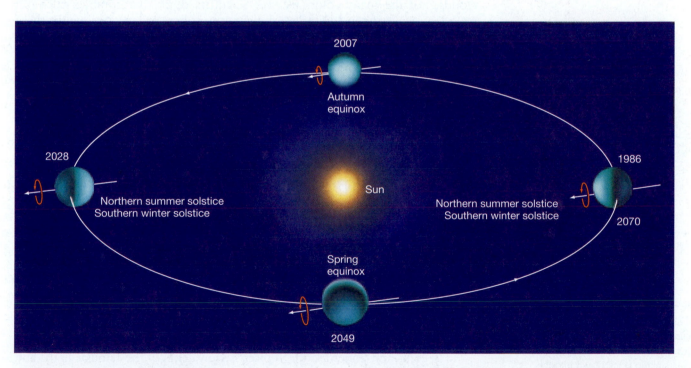

▲ **Figure 7.7 Seasons on Uranus**   Because of Uranus's axial tilt of 98°, the planet experiences the most extreme seasons known in the solar system. The equatorial regions experience two "summers" (warm seasons, at the times of the two equinoxes) and two "winters" (cool seasons, at the times of the solstices) each year, and the poles are alternately plunged into darkness for 42 years at a time.

directly at the Sun, so it was midsummer in the northern hemisphere. At the same time, the south polar regions were halfway through a 42-year period of permanent darkness.

No one knows why Uranus is tilted in this way. Astronomers speculate that a catastrophic event, such as a grazing collision between the planet and another planet-size body, might have altered the planet's spin axis. ∞ *(Sec. 4.3)* However, there is no direct evidence for such an occurrence and no theory to tell us how we should seek to confirm it.

7.4 Jupiter's Atmosphere

Just as Earth has been our yardstick for the terrestrial worlds, Jupiter will be our guide to the outer planets. We therefore begin our study of jovian atmospheres with Jupiter itself.

Overall Appearance and Composition

Visually, Jupiter is dominated by two atmospheric features (both clearly visible in Figure 7.1): a series of ever-changing atmospheric cloud bands arranged parallel to the equator and an oval atmospheric blob called the **Great Red Spot**. The cloud bands display many colors—pale yellows, light blues, deep browns, drab tans, and vivid reds, among others. The Red Spot is one of many features associated with Jupiter's weather. This egg-shaped region, the long diameter of which is approximately twice Earth's diameter, seems to be a hurricane that has persisted for hundreds of years.

The most abundant gas in Jupiter's atmosphere is molecular hydrogen (roughly 86 percent of all molecules), followed by helium (nearly 14 percent). Small amounts of atmospheric methane, ammonia, and water vapor are also found. None of these gases can, by itself, account for Jupiter's observed coloration. For example, frozen ammonia and water vapor would produce white clouds, not the many colors we see. Scientists think that complex chemical processes occurring in Jupiter's turbulent atmosphere are responsible for these colors, although the details of the chemistry are not fully understood. The trace elements sulfur and phosphorus may play important roles in influencing the cloud colors—particularly the reds, browns, and yellows. The energy that powers the reactions comes in many forms: the planet's own internal heat, solar ultraviolet radiation, aurorae in the planet's magnetosphere, and lightning discharges within the clouds. ∞ *(Sec. 5.7)*

Astronomers describe the banded structure of Jupiter's atmosphere as consisting of a series of lighter-colored *zones* and darker *belts* crossing the planet. The zones and belts vary in both latitude and intensity during the year, but the general pattern is always present. These variations appear to be the result of convective motion in the planet's atmosphere. (Recall from Chapter 5 that convection can occur whenever cool gas overlies hotter material, as is the case here.) ∞ *(Sec. 5.3)* Voyager sensors indicated that the zones lie above upward-moving convective currents, while the belts are the downward part of the cycle, where material is generally sinking, as illustrated in Figure 7.8. Thus, because of the upwelling material below them, the zones are regions of high pressure; the belts, conversely, are low-pressure regions. However, observations made during the Cassini flyby have challenged this standard view, suggesting that upward convection is actually confined to the belts. For now, planetary scientists have no clear resolution of this contradiction between the Voyager and Cassini results.

The belts and zones are the Jupiter's equivalents of the familiar high- and low-pressure systems that cause weather on Earth. A major difference from Earth is that Jupiter's rapid differential rotation causes these systems to wrap all the way around the planet, instead of forming localized circulating storms as on our own world. Because of the pressure difference between them, the belts and zones lie at slightly different heights in Jupiter's atmosphere. Cloud

CONCEPT CHECK ▶▶▶

Why do observations of a planet's magnetosphere allow astronomers to measure the rotation rate of the interior?

▲ **Figure 7.8 Jupiter's Convection** The colored bands in Jupiter's atmosphere are associated with vertical convective motion. Upwelling warm gas results in the lighter-colored zones. The darker bands lie atop lower-pressure regions, where cooler gas is sinking back down into the atmosphere. As on Earth, surface winds tend to blow in the direction from high-pressure regions to low-pressure regions. Jupiter's rotation channels these winds into an east–west flow pattern, as indicated by the three arrows drawn atop the belts and zones.

▶ **Figure 7.9 Jupiter's Atmosphere**   Models of the vertical structure of Jupiter's atmosphere indicate that the planet's clouds are arranged in three main layers, each with quite different colors and chemistry. The colors we see in photographs of the planet depend on the cloud cover. The white regions are the tops of the upper ammonia clouds. The yellows, reds, and browns are associated with the second cloud layer, which is composed of ammonium hydrosulfide ice. The lowest (bluish) cloud layer is water ice; however, the overlying layers are sufficiently thick that this level is not seen in visible light. (Compare with Figure 5.5.) The blue line shows how Jupiter's atmospheric temperature depends on altitude.

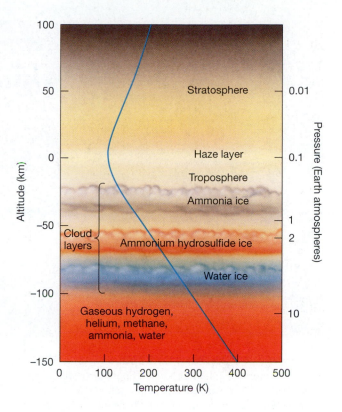

chemistry is very sensitive to temperature, and the temperature difference between the two levels is the main reason for the different colors we see.

Underlying the bands is an apparently very stable pattern of eastward and westward wind flow called Jupiter's **zonal flow.** The equatorial regions of Jupiter's atmosphere rotate faster than the planet as a whole, with an average flow speed of about 500 km/h in the easterly direction, somewhat similar to the jet stream on Earth. At higher latitudes there are alternating regions of westward and eastward flow, the alternations corresponding to the pattern of belts and zones. The flow speed diminishes toward the poles. Near the poles, where the zonal flow disappears, the band structure vanishes also.

Atmospheric Structure

Jupiter's clouds are arranged in several layers, with white ammonia clouds generally overlying the colored layers, whose composition we will discuss in a moment. Above the ammonia clouds lies a thin, faint layer of haze created by chemical reactions similar to those that cause smog on Earth. When we observe Jupiter's colors, we are looking down to many different depths in the planet's atmosphere.

Figure 7.9 is a diagram of Jupiter's atmosphere, based on observations and computer models. Because the planet lacks a solid surface to use as a reference level for measuring altitude, scientists conventionally take the top of the troposphere (the turbulent region containing the clouds we see) to lie at 0 km. With this level as the zero point, the troposphere's colored cloud layers all lie at negative altitudes in the diagram. As on other planets, weather on Jupiter is the result of convection in the troposphere. The haze layer lies at the upper edge of Jupiter's troposphere. The temperature at this level is about 110 K. Above the troposphere, as on Earth, the temperature rises as the atmosphere absorbs solar ultraviolet light.

Below the haze layer, at an altitude of −30 km, lie white, wispy clouds of ammonia ice. The temperature at this level is 125–150 K. A few tens of kilometers below the ammonia clouds, the temperature is a little warmer—above 200 K—and the clouds are made up mostly of droplets or crystals of ammonium hydrosulfide, produced by reactions between ammonia and hydrogen sulfide. At deeper levels the ammonium hydrosulfide clouds give way to clouds of water ice or water vapor. The top of this lowest cloud layer, which is not seen in visible-light images of Jupiter, lies some 80 km below the top of the troposphere.

In December 1995 the *Galileo* atmospheric probe arrived at Jupiter. The probe survived for about an hour before being crushed by atmospheric pressure at an altitude of −150 km (that is, just at the bottom of Figure 7.9). The probe's entry location (Figure 7.10) was in Jupiter's equatorial zone and, as luck would have it, coincided with an atypical hole almost devoid of upper-level clouds, and with abnormally low water content. However, once these factors were taken into account, *Galileo*'s findings on wind speed, temperature, and composition were in good agreement with the description presented above.

▶ **Figure 7.10 *Galileo's* Entry Site**   The arrow on this *Hubble Space Telescope* image shows where the *Galileo* atmospheric probe plunged into Jupiter's cloud deck on December 7, 1995. The entry location was in Jupiter's equatorial zone and apparently almost devoid of upper-level clouds. Until its demise, the probe took numerous weather measurements, transmitting those signals to the orbiting mother ship, which then relayed them to Earth. (NASA)

R   I   V   U   X   G

▲ **Figure 7.11** **Jupiter's Red Spot** *Voyager 1* took this photograph of Jupiter's Great Red Spot from a distance of about 100,000 km. Resolution is about 100 km. Note the complex flow patterns to the left of both the Red Spot and the smaller white oval below it. For scale, Earth is about the size of the white oval. The arrows indicate direction of gas flow above, below, and inside the Great Red Spot. (NASA)

Weather on Jupiter

In addition to the large-scale zonal flow, Jupiter has many smaller-scale weather patterns. The Great Red Spot (Figure 7.11) is a prime example. Observational records indicate that it has existed continuously in one form or another for more than 300 years, and it may well be much older. *Voyager* and *Galileo* observations show the Spot to be a region of swirling, circulating winds, like a whirlpool or a terrestrial hurricane, a persistent and vast atmospheric storm. The Spot varies in length, averaging about 25,000 km. It rotates around Jupiter at a rate similar to that of the planet's interior, suggesting that its roots lie far below the atmosphere. Astronomers think that the Spot is somehow sustained by Jupiter's large-scale atmospheric motion. The zonal motion north of the Spot is westward, while that to the south is eastward (Figure 7.11), supporting the idea that the Spot is confined and powered by the zonal flow. Turbulent eddies form and drift away from the spot's edge. The center, however, remains tranquil in appearance, like the eye of a hurricane on Earth.

The *Voyager* mission discovered many smaller spots that are also apparently circulating storm systems. Examples of these smaller systems are the **white ovals** seen in many images of Jupiter (see, for example, the one just south of the Great Red Spot in Figure 7.11). Their high cloud tops give these regions their white color. The white oval in Figure 7.11 is known to be at least 40 years old. Figure 7.12 shows a **brown oval**—a hole in the overlying clouds that allows us to look down into the lower atmosphere. For unknown reasons, brown ovals appear to be concentrated at latitudes around 20° north. Although not as long-lived as the Red Spot, they can persist for many years. The *Galileo* mission had been intended to continue these observations with detailed studies of Jupiter's atmosphere, but a failure of the craft's main antenna reduced its ability to transmit data, forcing mission controllers to focus their priorities on studies of the planet's moons (Section 8.1). However, many more oval storm systems—dark and light—can be seen in the *Cassini* view of Jupiter shown in Figure 7.3.

The simplified picture presented in Figure 7.8, with wind direction alternating between adjacent bands as Jupiter's rotation deflects surface winds into eastward or westward streams, is really too crude to describe the many details observed in Jupiter's atmosphere. More likely, the interaction between convection in Jupiter's atmosphere and the planet's rapid differential rotation channels the largest eddies into the observed zonal pattern, but smaller eddies—like the Red Spot and white ovals—tend to cause localized irregularities in the flow.

Although we cannot fully explain their formation, we can at least offer a partial explanation for the longevity of storm systems on Jupiter. On Earth a large storm, such as a hurricane, forms over the ocean and may survive for many days, but it dies quickly once it encounters land because the land disrupts the energy supplied and flow patterns that sustain the storm. Jupiter has no continents, so once a storm is established and has reached a size at which other storm systems cannot destroy it, apparently little affects it. The larger the system, the longer its lifetime.

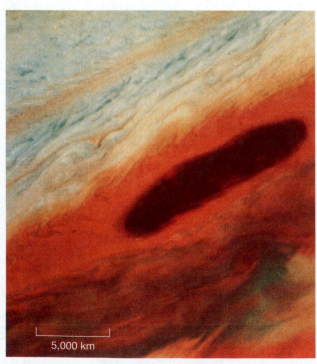

◀ **Figure 7.12** **Jupiter's Brown Oval** A brown oval in Jupiter's northern hemisphere is a break in the upper cloud layer, allowing us to see deeper into the atmosphere to where the clouds are brown. The oval's length is approximately equal to Earth's diameter. (NASA)

7.5 The Atmospheres of the Outer Jovian Worlds

Saturn's Atmospheric Composition

Saturn is not as colorful as Jupiter. Figure 7.13(a) shows yellowish and tan cloud bands that parallel the equator, but these regions display less atmospheric structure than do the bands on Jupiter. No obvious large spots or ovals adorn Saturn's cloud decks. Storms do exist, but the color changes that distinguish them on Jupiter are largely absent on Saturn. Figure 7.13(b) is the highest-resolution full-global image ever made of Saturn—a mosaic of more than a hundred photographs taken in true color—showing clarity yet subtlety in the structure of its cloud deck.

Spectroscopic studies show that Saturn's atmosphere consists of molecular hydrogen (92.4 percent) and helium (7.4 percent), with traces of methane and ammonia. As on Jupiter, hydrogen and helium dominate. However, the fraction of helium on Saturn is substantially less than on Jupiter. It is unlikely that the processes that created the outer planets preferentially stripped Saturn of nearly half its helium or that the missing helium somehow escaped from the planet while the hydrogen remained behind. Instead, astronomers theorize that at some time in Saturn's past the helium liquefied and then sank toward the center of the planet, reducing the abundance of that gas in the outer layers. The reason for this gas-to-liquid transformation is discussed in Section 7.6.

Figure 7.14 illustrates the vertical structure of Saturn's atmosphere. In many respects it is similar to Jupiter's (Figure 7.9), except that Saturn's overall atmospheric temperature is a little lower because of its greater distance from the Sun. As on Jupiter, the troposphere contains clouds arranged in three distinct layers, composed (in order of increasing depth) of ammonia ice, ammonium hydrosulfide ice, and water ice. Above the clouds lies a layer of haze.

The total thickness of the three cloud layers in Saturn's atmosphere is roughly 250 km—about three times the cloud thickness on Jupiter—and each layer is thicker than its counterpart on Jupiter. The reason for this difference is Saturn's weaker gravity. At the haze level, Jupiter's gravitational field is nearly 2.5 times stronger than Saturn's, so Jupiter's atmosphere is pulled much more powerfully toward the center of the planet and the cloud layers are squeezed more closely together.

The colors of Saturn's cloud layers, as well as the planet's overall yellowish coloration, are likely due to the same basic photochemistry as on Jupiter. However, because Saturn's clouds are thicker, there are fewer holes and gaps in the top layer, with the result that we rarely glimpse the more colorful levels below. Instead, we see only the topmost layer, which accounts for Saturn's less varied appearance.

(a)

(b)

▶ **Figure 7.13 Saturn, Up Close** (a) The banded structure of Saturn's atmosphere appears clearly in this image, taken as *Voyager 2* approached the planet. (Three of Saturn's moons appear at bottom; a fourth, not visible here, casts a black shadow on the cloud tops.) (b) This *Cassini* image is actually a mosaic of many images taken in true color; resolution is 40 km. (NASA)

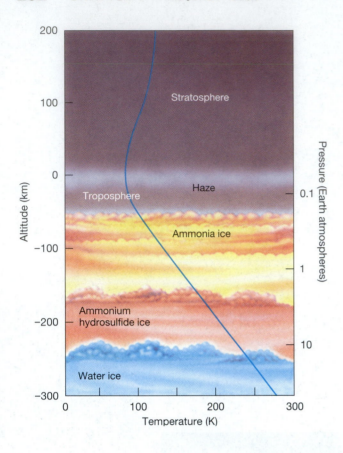

Weather on Saturn

Computer-enhanced images of Saturn clearly show the existence of bands, oval storm systems, and turbulent flow patterns, many looking very much like those on Jupiter. In addition, there is an apparently quite stable east–west zonal flow, although the wind speed is considerably greater than on Jupiter and shows fewer east–west alternations. The equatorial eastward jet stream reaches a brisk 1500 km/h. Astronomers do not fully understand the reasons for the differences between Jupiter's and Saturn's flow patterns.

Figure 7.15 shows a sequence of *Hubble Space* Telescope images of a storm system on Saturn. First detected in September 1990 as a large white spot just south of Saturn's equator, by November of that year the spot had developed into a band of clouds completely encircling the planet's equator. Astronomers think that the white coloration arose from crystals of ammonia ice that formed when an upwelling plume of warm gas penetrated the cool upper cloud layers. Because the crystals were freshly formed, they had not yet been affected by the chemical reactions that color the planet's other clouds. The turbulent flow patterns seen in and around the spot had many similarities to the flow around Jupiter's Great Red Spot. Scientists hope that routine observations of such temporary atmospheric phenomena on the outer worlds will give them greater insight into the dynamics of planetary atmospheres.

Since *Cassini*'s arrival at Saturn scientists have been able to study the planet's storm systems in much more detail. Figure 7.16 shows a particularly large and complex system that appeared in Saturn's southern hemisphere in late 2004. Dubbed the "Dragon Storm" for its twisting, convoluted shape, it is the

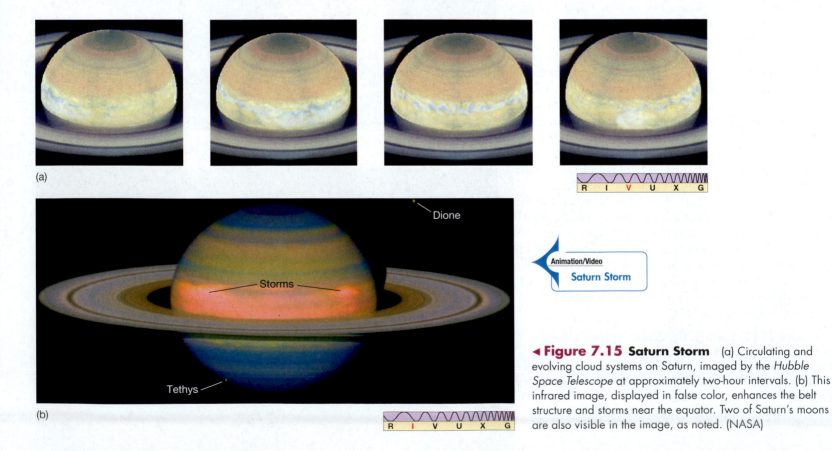

◀ **Figure 7.15** **Saturn Storm**   (a) Circulating and evolving cloud systems on Saturn, imaged by the *Hubble Space Telescope* at approximately two-hour intervals. (b) This infrared image, displayed in false color, enhances the belt structure and storms near the equator. Two of Saturn's moons are also visible in the image, as noted. (NASA)

Animation/Video
Saturn Storm

bright orange region just above and to the right of center in this (false color) image. The storm is thought to be Saturn's equivalent of a thunderstorm on Earth (although on a scale comparable to our entire planet). *Cassini*'s detectors have measured strong bursts of radio waves associated with the storm; they are most likely produced by intense lightning discharges deep below the cloud tops. The lightning is thought to be powered by convection and precipitation (water and ammonia "rain"), just as on Earth, but the bursts are millions of times stronger than anything ever witnessed here at home.

Scientists suspect that the storm is actually a long-lived phenomenon rooted deep in Saturn's atmosphere—perhaps a little like Jupiter's Great Red Spot—and normally completely hidden below the upper cloud layers. Only occasionally does the storm flare up, producing a bright plume visible from outside. *Cassini* has also detected numerous small dark storms apparently associated with this and other large systems. These smaller storms seem to be spun off from the larger systems and subsequently merge with the planet's zonal flow, perhaps providing a means for energy to flow from Saturn's warm interior into the cold atmosphere.

Atmospheric Composition of Uranus and Neptune

The atmospheres of Uranus and Neptune are very similar in composition to that of Jupiter—the most abundant gas is molecular hydrogen (84 percent), followed by helium (about 14 percent), and methane, which is more abundant on Neptune (about 3 percent) than on Uranus (2 percent). Ammonia is not observed in any significant quantity in the atmosphere of either planet, most likely because, at the low atmospheric temperatures of those planets, the ammonia exists in the form of ice crystals, which are much harder to detect spectroscopically than gaseous ammonia.

The blue color of the outer jovian planets is the result of their relatively high percentages of methane. Because methane absorbs long-wavelength red light quite efficiently, sunlight reflected from these planets' atmospheres is deficient in red and yellow photons and appears blue-green or blue. The greater the concentration of methane, the bluer the reflected light. Therefore Uranus, having less methane, looks blue-green, while Neptune, having more methane, looks distinctly blue.

Weather on Uranus and Neptune

Voyager 2 detected only a few atmospheric features on Uranus (Figure 7.17), and even they became visible only after extensive computer enhancement. Few clouds exist in Uranus's cold upper atmosphere. Clouds of the kind found on Jupiter and Saturn are found only at lower, warmer atmospheric levels. The absence of high-level clouds means we must

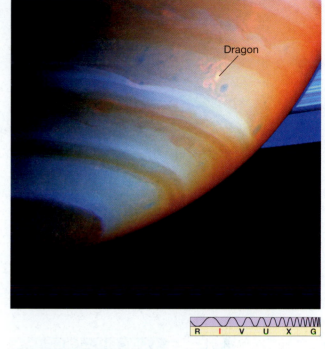

▲ **Figure 7.16 Saturn's "Dragon Storm"**   This false-color near-infrared image from *Cassini* shows what is thought to be a vast thunderstorm on Saturn. Called the "Dragon Storm," this storm also generated bursts of radio waves resembling the static created by lightning on Earth. (NASA)

◀◀◀ **CONCEPT CHECK**

List some similarities and differences between Jupiter's belts and zones and weather systems on Earth.

(a)   (b)   (c)

▲ **Figure 7.17 Uranus's Rotation**   (a), (b), and (c) These computer-enhanced *Hubble Space Telescope* images, taken at roughly four-hour intervals, show the motion of a pair of bright clouds (labeled A and B) in the planet's southern hemisphere. (The numbers at the top give the time of each photo.) (d) A near-infrared image of Uranus, also from *HST*, shows the planet's ring system, as well as a number of clouds (pink and red regions) in the upper atmosphere. The clouds circle the planet at speeds of up to 500 km/h. (NASA)

look deep into the planet's atmosphere to see any structure, and the bands and spots that characterize flow patterns on the other jovian worlds are largely washed out on Uranus by intervening stratospheric haze. Series of computer-processed images such as Figures 7.17(a–c) have shown astronomers that Uranus's atmospheric clouds and flow patterns move around the planet in the same direction as the planet's rotation, with wind speeds ranging from 200 to 500 km/h.

Although Neptune lies farther from the Sun, its internal heat (see Section 7.6) means that its upper atmosphere is actually slightly warmer than that of Uranus. This is probably the reason why Neptune's atmospheric features are more easily visible than those of Uranus (see Figure 7.18). The extra warmth causes Neptune's stratospheric haze to be thinner than the haze on Uranus and also makes the cloud layers less dense than they would otherwise be, with the result that they lie at higher levels in the atmosphere.

Voyager 2 detected numerous white methane clouds (some of them visible in Figure 7.5) lying about 50 km above the main cloud tops. Neptune's equatorial winds blow from east to west with speeds of over 2000 km/h relative to the planet's interior. Why such a cold planet should have such rapid winds, and why they should be retrograde relative to the west-to-east rotation of the interior, remains a mystery.

Neptune sports several storm systems similar in appearance to those seen on Jupiter, and probably produced and sustained by the same basic processes. The largest such storm, the **Great Dark Spot** (Figure 7.18a), was discovered by *Voyager 2* in 1989. Comparable in size to Earth, the Spot was located some 20° south of the the planet's equator—quite similar to the Great Red Spot on Jupiter. The Dark Spot also exhibited other characteristics in common with Jupiter's Red Spot. The flow around it was counterclockwise, and there appeared to be turbulence where the winds associated with the Dark Spot interacted with the zonal flow to its north and south. Astronomers did not have long to study the Dark Spot's properties, however. By the time detailed *Hubble Space Telescope* observations of the planet became available in the mid-1990s, the Spot had disappeared (Figure 7.18b). At around the same time, a new Dark Spot of comparable size became visible in Neptune's northern hemisphere. The origins and lifetimes of large storm systems on the jovian planets are currently not well understood.

CONCEPT CHECK ▶▶▶

Why are atmospheric features much less evident on Uranus than on the other jovian planets?

(a)

R I V U X G

(b)   1996   1998   2002

R I V U X G

◀ **Figure 7.18 Neptune's Dark Spot**   (a) Close-up views taken in 1989 by *Voyager 2* of the Great Dark Spot of Neptune, show a large storm system in the planet's atmosphere, possibly similar in structure to Jupiter's Great Red Spot. Resolution in the image on the right is about 50 km. The dark spot is roughly the size of Earth. (b) These *Hubble Space Telescope* views of Neptune were taken years apart (as marked), when the planet was some 4.5 billion km from Earth. The planet appears aqua because methane in its clouds has absorbed red light from the sunlight striking the planet, with the result that the reflected light we see is blue-green in color. The cloud features (mostly methane ice crystals) are tinted pink here because they were imaged in the infrared, but they are white in visible light. Notice that the Great Dark Spot has disappeared in recent years. (NASA)

7.6 Jovian Interiors

The visible cloud layers of the jovian planets are all less than a few hundred kilometers thick. How, then, do we determine the conditions beneath the clouds, in the unseen interiors of these distant worlds? We have no seismographic information on these regions (but see *Discovery 7-1*), nor do we happen to live on a more or less similar planet that we can use for comparison. Instead, we must use all available bulk data on each planet, along with our knowledge of physics and chemistry, to construct a *model* of its interior that agrees with observations. Our statements about the interiors of the jovian worlds, then, are really statements about the models that best fit the facts. As we will see throughout this book, astronomers rely heavily on model-building, often aided by powerful computers, as a powerful tool to interpret data about objects too far away for direct exploration.

Internal Structure

As illustrated in Figure 7.19, both the temperature and the pressure of Jupiter's atmosphere increase with depth. At a depth of a few thousand kilometers, the gas makes a gradual transition into the liquid state. By a depth of about 20,000 km the pressure is about 3 million times greater than atmospheric pressure on Earth. Under these conditions the hot liquid hydrogen is compressed so much that it undergoes another transition, this time to a metallic state, with properties in many ways similar to more familar liquid metals on Earth. Of particular importance for Jupiter's magnetic field, this metallic hydrogen is an excellent conductor of electricity.

The strong outward push arising from Jupiter's rotation causes a pronounced bulge at the planet's equator—Jupiter's equatorial radius (Table 7.1) exceeds its polar radius by nearly 7 percent. However, careful calculations indicate that if Jupiter were composed of hydrogen and helium alone, it should be *more* flattened than is observed. The actual degree of flattening implies a dense *core* having a mass perhaps 10 times the mass of Earth. The core's precise composition is unknown, but many scientists think it consists of materials similar to those found in the terrestrial worlds: molten, perhaps semisolid, rock. Because of Jupiter's enormous central pressure—approximately 50 million times that on Earth's surface, or 10 times that at Earth's center—the core must be compressed to very high densities (perhaps twice the core density of Earth). It is probably not much more than 20,000 km in diameter.

Saturn has the same basic internal parts as Jupiter, but their relative proportions are different: Its metallic hydrogen layer is thinner, and the central dense core (again inferred from studies of the planet's polar flattening) is larger—around 15 Earth masses. Because of its lower mass, Saturn has a less extreme core temperature, density, and pressure than Jupiter. The central pressure is around one-tenth of Jupiter's—not too different from the pressure at the center of Earth.

The interior pressures of Uranus and Neptune are sufficiently low that hydrogen stays in its molecular form all the way into the planets' cores. Currently, astronomers theorize that deep below the cloud layers, Uranus and Neptune may have high-density, "slushy" interiors containing thick layers of highly compressed water clouds. It is also possible that much of the planets' ammonia could be dissolved in the water, producing a thick, electrically conducting layer that could conceivably explain the planets' magnetic fields. At the present time, however, we don't know enough about the interiors to assess the correctness of this picture.

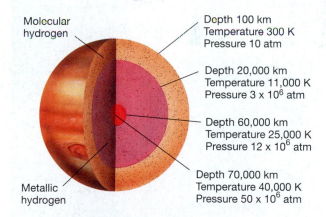

Molecular hydrogen

Metallic hydrogen

Depth 100 km
Temperature 300 K
Pressure 10 atm

Depth 20,000 km
Temperature 11,000 K
Pressure 3 x 10^6 atm

Depth 60,000 km
Temperature 25,000 K
Pressure 12 x 10^6 atm

Depth 70,000 km
Temperature 40,000 K
Pressure 50 x 10^6 atm

▲ **Figure 7.19 Jupiter's Interior** Jupiter's internal structure, as deduced from spacecraft measurements and theoretical modeling. The outermost radius shown in the diagram represents the bottom of the cloud layers, 70,000 km from the planet's center and about 100 km below the visible cloud top. Pressure and temperature increase with depth, and the atmosphere gradually liquefies over the outermost few thousand kilometers. Below 20,000 km, the hydrogen behaves like a liquid metal. At the center of the planet lies a rocky core (red region), somewhat terrestrial in composition but much larger than any of the terrestrial planets. The temperature at the center is probably about 40,000 K, and the pressure there is thought to be 50 million Earth atmospheres.

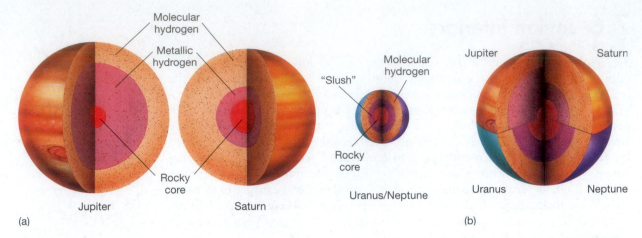

▲ **Figure 7.20 Jovian Interiors** A comparison of the interior structures of the four jovian planets. (a) The planets drawn to scale. (b) The relative proportions of the various internal zones.

Animation/Video

The Gas Giants Part II

The cores of Uranus and Neptune are inferred mainly from considerations of those planets' densities. Because of their lower masses, we would not expect the hydrogen and helium in the interiors of Uranus and Neptune to be compressed as much as in the two larger jovian worlds, yet the average densities of Uranus and Neptune are actually *greater* than the density of Saturn and similar to that of Jupiter. Astronomers interpret this to mean that Uranus and Neptune each have Earth-sized cores some 10 times more massive than our planet, having compositions similar to the cores of Jupiter and Saturn.

Our current state of knowledge is summarized in Figure 7.20, which compares the internal structures of the four jovian worlds.

Magnetospheres

The combination of rapid overall rotation and an extensive region of highly conductive fluid in its interior gives Jupiter by far the strongest planetary magnetic field in the solar system. ∞ *(Sec. 5.7)* The planet's surface field is some 14 times that of Earth, but taking the planet's radius into account implies an intrinsic field strength some 20,000 times Earth's. Jupiter is surrounded by a vast sea of energetic charged particles (mostly electrons and protons), somewhat similar to Earth's Van Allen belts but very much larger. Direct spacecraft measurements show Jupiter's magnetosphere to be almost 30 million km across (north to south), roughly a million times more voluminous than Earth's magnetosphere and far larger than the entire Sun. All four Galilean moons lie within it.

On the sunward side the boundary of Jupiter's magnetic influence on the solar wind lies about 3 million km from the planet. On the opposite side, however, the magnetosphere has a long "tail" extending away from the Sun at least as far as Saturn's orbit (some 4 AU, or 600 *million* km), as sketched in Figure 7.21. Near the planet's surface the planet's magnetic field channels particles from the magnetosphere into the upper atmosphere, forming aurorae vastly larger and more energetic than those observed on Earth (Figure 7.22). ∞ *(Sec. 5.7)*

Saturn's electrically conducting interior and rapid rotation also produce a strong magnetic field and an extensive magnetosphere. However, because of the considerably smaller mass of Saturn's metallic hydrogen zone, Saturn's surface magnetic field strength is only about 1/20 that of Jupiter. Saturn's magnetosphere extends about 1 million km toward the Sun and is large enough to contain the planet's ring system and most of its moons.

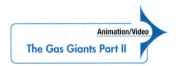

▲ **Figure 7.21 Mission** The *Pioneer 10* spacecraft (a forerunner to the *Voyager* missions) did not detect any solar particles while moving far behind Jupiter in March 1976, indicating that the tail of Jupiter's magnetosphere (drawn in red) extends beyond the orbit of Saturn.

7-1 ISCOVERY

A Cometary Impact

In 1994, astronomers were granted a rare alternative means of studying Jupiter's atmosphere and interior when a comet, called Shoemaker–Levy 9, collided with the planet. Discovered in 1993, the comet was observed to have an unusual flattened nucleus made up of several large pieces (shown in image a below), the largest no more than 1 km across. Tracing the orbit backward in time, researchers found that the comet had narrowly missed Jupiter in 1992. They realized that the objects shown in the figure were fragments produced when a previously "normal" comet was captured by Jupiter and torn apart by the planet's strong tidal forces. ∞ *(Sec. 5.2)* The data also revealed an even more remarkable fact—the comet would collide with Jupiter on its next approach!

Between July 16 and July 22, 1994, as most of the telescopes on (and in orbit around) Earth watched, the comet fragments struck Jupiter's upper atmosphere at speeds of over 60 km/s, causing a series of enormous explosions. Each impact created, for a period of a few minutes, a brilliant fireball hundreds of kilometers across and having a temperature of many thousands of kelvins. The energy released in each explosion was comparable to a billion terrestrial nuclear detonations. One of the largest pieces of the comet,

(a)   S   G   D   (NASA)

Toward Sun

Toward Jupiter

fragment G, produced the spectacular fireball shown in the second image (b). The effects on the planet's atmosphere and the vibrations produced throughout Jupiter's interior were observable for days after the impact.

As far as we can determine, none of the cometary fragments breached the jovian clouds. Only *Galileo* had a direct view of the impacts on the back side of Jupiter, and in every case the explosions seemed to occur high in the atmosphere, above the uppermost cloud layer. Spectral lines from water vapor, silicon, magnesium, and iron were observed, but all apparently came from the comet itself— which really did resemble a loosely packed snowball. ∞ *(Sec. 4.2)* Debris from the impacts, such as the "black eye" shown in image (c), spread slowly around Jupiter's bands, and after five months reached completely around the planet. It took years for all the cometary matter to settle into Jupiter's interior.

(Palomar)
(b)

(c)

The Shoemaker–Levy 9 impact has scientific significance far beyond its direct effect on Jupiter. Since the 1960s, planetary scientists have become increasingly aware that collisions between small bodies—asteroids, comets, and meteoroids—and the major planets are relatively commonplace events in the solar system. As we have seen, the consequences for life can be catastrophic. ∞ *(Discovery 4-1)* The fact that we witnessed such an event during the few decades when it has been technologically feasible for us to do so underscores just how frequent these encounters probably are and has convinced many scientists of the importance of planetary impacts to solar system evolution.

Animation/Video

Strange X-Ray Pulses from Jupiter

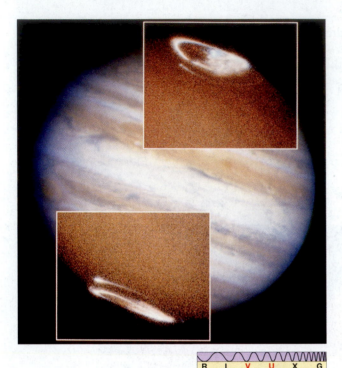

R I V U X G

▲ **Figure 7.22 Aurorae on Jupiter** Aurorae on Jupiter, as seen by the *Hubble Space Telescope*. The main image was taken in visible (true-color) light, but the two insets at the poles were taken in the ultraviolet part of the spectrum. The oval-shaped aurorae, extending hundreds of kilometers above Jupiter's limb, result from charged particles escaping the jovian magnetosphere and colliding with the atmosphere, causing the gas to glow. (See Figure 5.24.) (NASA)

Voyager 2 found that both Uranus and Neptune have fairly strong surface magnetic fields, comparable to those of Saturn and Earth. Taking the planets' radii into account, we find intrinsic field strengths a few percent that of Saturn, or 30–40 times Earth's. Both planets have substantial magnetospheres, populated largely by electrons and protons either captured from the solar wind or created from hydrogen gas escaping from the planets below. Figure 7.23 compares the magnetic field structures of Earth and the four jovian planets. The locations and orientations of the bar magnets represent the observed planetary fields, and the sizes of the bars indicate overall magnetic field strength. Note that the fields of Uranus and Neptune are not aligned with the planets' rotation axes and are significantly offset from the planets' centers. Astronomers do not yet understand why this should be so.

Internal Heating

On the basis of Jupiter's distance from the Sun, astronomers expected to find the temperature of the cloud tops to be around 105 K. At that temperature, they reasoned, Jupiter would radiate back into space the same amount of energy as it receives from the Sun. Radio and infrared observations of the planet, however, revealed a black-body spectrum corresponding to a temperature of 125 K, implying (from Stefan's law) that Jupiter emits about twice—$(125/105)^4$—as much energy as reaches the planet in the form of sunlight. ∞ *(Sec. 2.4)*

Astronomers theorize that the source of Jupiter's excess energy is heat left over from the planet's formation. ∞ *(Sec. 4.3)* In the distant past, Jupiter's interior was probably very much hotter than it is today (although not nearly hot enough for Jupiter ever to have become a star like the Sun). That early heat is slowly leaking out through the planet's heavy atmospheric blanket, resulting in the excess emission we observe today.

Infrared measurements indicate that Saturn, too, has an internal energy source. Saturn's cloud-top temperature is 97 K, substantially higher than the temperature at which Saturn would reradiate all the energy it receives from the Sun. In fact, Saturn radiates away almost *three times* more energy than it absorbs. However, the explanation for Jupiter's excess emission doesn't work for Saturn. Saturn is smaller than Jupiter and so must have cooled more rapidly—fast enough, in fact, that its original supply of heat should have been used up long ago. What then is happening inside Saturn to produce this extra heat?

The explanation for this strange state of affairs also explains Saturn's atmospheric helium deficit. At the temperatures and high pressures found in Jupiter's interior, liquid helium *dissolves* in liquid hydrogen. In Saturn, where the internal temperature is lower, the helium doesn't dissolve so easily and tends to form droplets instead. (The basic phenomenon is familiar to cooks who know that it is generally much easier to dissolve ingredients in hot liquids than in cold ones.) Saturn probably started out with a fairly uniform mix of hydrogen and helium, but the helium tended to condense out of the surrounding hydrogen, much as water vapor condenses out of Earth's atmosphere to form a mist. The amount of helium condensation was greatest in the planet's cool outer layers, where the mist turned to rain about 2 billion years ago. A light shower of liquid helium has been falling through Saturn's interior ever since. This helium precipitation is responsible for depleting the outer layers of their helium content.

▲ **Figure 7.23** **Jovian Magnetic Fields**   A comparison of the magnetic field strengths, orientations, and offsets in the four jovian planets: (a) Jupiter, (b) Saturn, (c) Uranus, (d) Neptune. The planets are drawn to scale, and in each case the magnetic field is represented as though it came from a simple bar magnet (a considerable simplification, for illustration only). The size and location of each magnet represent the strength and orientation of the planetary field. Notice that the fields of Uranus and Neptune are significantly offset from the center of the planet and are much inclined to the planet's rotation axis. Earth's magnetic field is shown for comparison. One end of each magnet is marked "N" to indicate the polarity of Earth's field.

As the helium sinks toward the center, the planet's gravitational field compresses it and heats it up. The energy thus released is the source of Saturn's internal heating. In the distant future the helium rain will stop and Saturn will cool until its outermost layers radiate only as much energy as they receive from the Sun. When that happens, the temperature at Saturn's cloud tops will be 74 K.

Uranus apparently has no internal energy source, radiating into space as much energy as reaches it from the Sun. It never experienced helium precipitation, and its initial heat was lost to space long ago. Surprisingly, given its other similarities to Uranus, Neptune *does* have an internal heat source—it emits 2.7 times more energy than it receives from the Sun. The reason is still uncertain. Some scientists think that Neptune's relatively high concentration of methane insulates the planet, helping to maintain its initially high internal temperature.

◄◄◄ **CONCEPT CHECK**

Why do all the jovian planets have strong magnetic fields?

CHAPTER REVIEW

Summary

1 Jupiter and Saturn were the outermost planets known to ancient astronomers. Uranus was discovered in the eighteenth century, by chance. Neptune was discovered after mathematical calculations of Uranus's slightly non-Keplerian orbit revealed the presence of an eighth planet.

2 The four jovian planets are all much more massive than the terrestrial worlds and are of much lower density, being composed primarily of hydrogen and helium. However, all have large, dense cores, each roughly ten times Earth's mass and of chemical composition similar to that of the ter- restrial planets. The jovian planets all rotate more rapidly than Earth. Having no solid surfaces, they display **differential rotation** (p. 196)—the rotation rate varies from place to place in the atmosphere and with depth below the surface. For unknown reasons, Uranus's spin axis lies nearly in the ecliptic plane, leading to extreme seasonal variations in solar heating on the planet as it orbits the Sun.

3 The cloud layers on all the jovian worlds are arranged into bands of light-colored zones and darker belts crossing the planets parallel to the equator. The bands are the result of convection in the planets' interiors and the planets' rapid rotation. The lighter zones are the tops of up- welling, warm currents, and the darker bands are cooler regions, where gas is sinking. Underlying them is a stable pattern of eastward or westward wind flow called a **zonal flow** (p. 199). The wind direction alternates as we move north or south away from the equator. Jupiter's atmosphere consists of three main cloud layers. The colors we see are the result of chemical reactions at varying depths below the cloud tops.

4 A major weather pattern on Jupiter is the **Great Red Spot** (p. 198), a huge hurricane that apparently has been raging for at least three centuries. Other, smaller, atmospheric features—**white ovals** (p. 200) and **brown ovals** (p. 200)—are also ob- served. Similar systems are found on the other jovian planets, although they are less distinct on Saturn and Uranus. Long-lived storms on Saturn may lie hidden below the planet's clouds. The **Great Dark Spot** (p. 204) on Neptune had many similarities to Jupiter's Red Spot. It disappeared within a few years of its discovery by *Voyager 2*. Storms in the jovian atmospheres are long-lived because there is no solid surface to dissipate their energy.

5 The atmospheres of the jovian planets become hotter and denser with depth, eventually becoming liquid. In Jupiter and Saturn the interior pressures are so high that the hydrogen near the center exists in a liquid-metallic form rather than as molecular hydrogen. In Uranus and Neptune this change from molecular hydrogen to metallic hydrogen apparently does not occur. Radio waves emitted from the planets' magnetospheres provide a measure of their interior rotation rates. The combination of rapid rotation and conductive interiors means that all four jovian planets have strong magnetic fields and extensive magnetospheres.

6 Three jovian planets radiate more energy than they receive from the Sun. On Jupiter and Neptune the source of this en- ergy is most likely heat left over from the planet's formation. On Saturn the heating is the result of helium precipitation in the interior, where helium liquefies and forms droplets, which then fall toward the center of the planet. This process is also responsible for reducing the amount of helium observed in Saturn's surface layers.

Review and Discussion

1. How was Uranus discovered?

2. Why did astronomers suspect an eighth planet beyond Uranus?

3. Why have the jovian planets retained most of their original atmospheres?

4. What is differential rotation, and what evidence do we have that it occurs on Jupiter?

5. Why does Saturn have a less varied appearance than Jupiter?

6. Compare the thicknesses of Saturn's various layers (clouds, molecular hydrogen, metallic hydrogen, and more) to the equivalent layers in Jupiter.

7. What is the Great Red Spot? What is known about the source of its energy?

8. Describe the cause of the colors in the jovian planets' atmospheres.

9. Compare and contrast the atmospheres of the jovian planets, describing how the differences affect the appearance of each.

10. Explain the theoretical model that accounts for Jupiter's internal heat source.

11. Describe the theoretical model that accounts for Saturn's missing helium and the surprising amount of heat the planet radiates.

12. What is unusual about the rotation of Uranus?

13. What is responsible for Jupiter's enormous magnetic field?

14. How are the interiors of Uranus and Neptune thought to differ from those of Jupiter and Saturn?

15. How much information do you think we might have gathered about the outer solar system without the *Voyager*, *Galileo*, and *Cassini* missions?

Conceptual Self-Test: True or False?/Multiple Choice

1. The solid surface of Jupiter lies just below the cloud layers that are visible from Earth. **(T/F)**

2. Storms in the jovian planets' atmospheres are generally much longer lived than storms on Earth. **(T/F)**

3. The element helium plays an important role in producing Saturn's atmospheric colors. **(T/F)**

4. Jupiter emits more energy than it receives from the Sun. **(T/F)**

5. Although often referred to as a gaseous planet, Jupiter is mostly liquid in its interior. **(T/F)**

6. Neptune probably does not have a rocky core. **(T/F)**

7. Around northern midsummer on Uranus, an observer near the north pole would observe the Sun high and almost stationary in the sky. **(T/F)**

8. The magnetic fields of both Uranus and Neptune are highly tilted relative to their rotation axes and significantly offset from the planets' centers. **(T/F)**

9. The main constituent of Jupiter's atmosphere is (a) hydrogen; (b) helium; (c) ammonia; (d) carbon dioxide.

10. Saturn's cloud layers are much thicker than those of Jupiter because Saturn has (a) more mass; (b) lower density; (c) a weaker magnetic field; (d) weaker surface gravity.

11. Saturn's icy–rocky core is roughly (a) half the mass of; (b) the same mass as; (c) twice as massive as; (d) 10 times more massive than Earth.

12. Uranus was discovered about the same time as (a) Columbus reached North America; (b) the U.S. Declaration of Independence; (c) the American Civil War; (d) the Great Depression in the United States.

13. Compared with Uranus, the planet Neptune is (a) much smaller; (b) much larger; (c) roughly the same size; (d) closer to the Sun.

14. According to Figure 7.9 (Jupiter's Atmosphere), if ammonia and ammonium hydrosulfide ice were transparent to visible light, Jupiter would appear (a) bluish; (b) red; (c) tawny brown; (d) exactly as it does now.

15. According to Figure 7.21, Jupiter's magnetosphere extends about (a) 1 AU; (b) 5 AU; (c) 10 AU; (d) 20 AU beyond the planet.

Problems

Algorithmic versions of these questions are available in the Practice Problems module of the Companion Website.

The number of squares preceding each problem indicates its approximate level of difficulty.

1. ■ The *Hubble Space Telescope* has an angular resolution of 0.05″. What is the size of the smallest feature visible on (a) Jupiter, at 4.2 AU, and (b) Neptune, at 29 AU?

2. ■ What is the gravitational force exerted on Uranus by Neptune, at closest approach? Compare your answer with the Sun's gravitational force on Uranus.

3. ■ What is the round-trip travel time of light from Earth to Jupiter (at the minimum Earth–Jupiter distance of 4.2 AU)? How far would a spacecraft orbiting Jupiter at a speed of 20 km/s travel in that time?

4. ■■ What would be the mass of Jupiter if it were composed entirely of hydrogen at a density of 0.08 kg/m³ (the density of hydrogen at sea level on Earth)?

5. ■ Verify the statement made in the text that the force of gravity at Jupiter's cloud tops is nearly 2.5 times that at Saturn's cloud tops.

6. ■ Given Jupiter's current atmospheric temperature, what is the smallest possible mass the planet could have and still have retained its hydrogen atmosphere? ∞ *(More Precisely 5-1)*

7. ■ How long does it take for Saturn's equatorial flow, moving at 1500 km/h, to encircle the planet?

8. ■■ If Saturn's surface temperature is 97 K and the planet radiates three times more energy than it receives from the Sun, use Stefan's law to calculate what the surface temperature would be in the absence of any internal heat source. ∞ *(More Precisely 2-2)*

9. ■■ If the core of Uranus has a radius twice that of planet Earth and an average density of 8000 kg/m³, calculate the mass of Uranus outside the core. What fraction of the planet's total mass is core?

10. ■ From Wien's law, at what wavelength does Neptune's thermal emission peak? In what part of the electromagnetic spectrum does this wavelength lie? ∞ *(More Precisely 2-2, Fig. 2.9)*

 The Companion Website at www.prenhall.com/chaisson provides algorithmically generated versions of each chapter's Problems, in addition to an Animations and Videos gallery, a Tutorials gallery and Tutorials Quiz questions, a full eBook, an Images gallery, Practice Questions, Multiple Choice, and True or False quizzes, and Labeling exercises.

8

Moons, Rings, and Pluto

Small Worlds Among Giants

The number of known moons in the solar system increased rapidly during the late 1990s. Better telescopes enabled astronomers to take a more complete inventory of objects in our cosmic neighborhood. But the more closely we examine these alien worlds, the more complex they seem to be. In this puzzling image taken by the *Cassini* spacecraft in 2005, Saturn's tiny ice-covered moon Enceladus (only 500 km in diameter) shows evidence of a youthful terrain in the south where craters are mostly absent. The long blue streaks (about 1 km wide) are likely fractures in the ice through which gas escapes to form a thin but real atmosphere. (JPL)

All four jovian planets have systems of moons and rings that exhibit fascinating variety and complexity. The six largest jovian moons have many planetary features, providing us with insight into the terrestrial worlds. Two of these moons, Jupiter's Europa and Saturn's Titan, have become prime candidates for study in the search for life elsewhere in the solar system. Like the moons, the rings of the jovian planets differ greatly from one to the next. The rings of Saturn—the best-known ring system among the jovian planets—is one of the most spectacular displays in the sky. Pluto is no longer considered a major planet; it is very much smaller than even the terrestrial worlds and in many ways is much more moonlike than planetlike. For that reason we study it here, along with the moons and rings of its giant neighbors.

LEARNING GOALS

Studying this chapter will enable you to:

1 Describe how the Galilean moons form a miniature solar system around Jupiter and exhibit a wide range of properties.

2 Discuss the composition and origin of the atmosphere on Titan, Saturn's largest moon.

3 Explain why astronomers think Neptune's moon Triton was captured by that planet.

4 Describe the properties of the medium-sized moons of the outer planets.

5 Describe the nature and detailed structure of Saturn's rings.

6 Explain why planetary rings form.

7 Discuss the differences between the rings of Saturn and those of the other jovian planets.

8 Explain why Pluto was recently demoted from major planet status.

Visit www.prenhall.com/chaisson for additional annotated images, animations, and links to related sites for this chapter.

8.1 The Galilean Moons of Jupiter

All four jovian planets have extensive moon systems—Jupiter has at least 63 natural satellites in orbit around it, Saturn has 50, Uranus 27, and Neptune 13. These moons range greatly in size and other physical properties, but they fall into three fairly natural groups.

First come the six "large" jovian moons, each comparable in size to Earth's Moon. They are listed in Table 8.1, along with our own Moon for comparison. All have diameters greater than 2500 km and are large enough to have their own distinctive geologies and histories. Second, there are the twelve "medium-sized" moons—spherical bodies with diameters ranging from about 400 to 1500 km. Their heavily cratered surfaces offer tantalizing clues to the past and present state of the environment in the outer solar system. Finally, there are the many "small" moons—irregularly shaped chunks of ice, all less than 300 km across—that exhibit a bewildering variety of complex and fascinating motions. Most are probably little more than captured interplanetary debris. ∞ *(Sec. 4.3)* Their numbers have increased rapidly in recent years (and will probably continue to rise) thanks to steadily improving ground and space-based technology.

In this chapter we will focus our attention on the large and midsize jovian moons. We begin our study of the large moons with the Galilean satellites of Jupiter. ∞ *(Sec. 1.2)*

A "Miniature Solar System"

Jupiter's four large moons travel on nearly circular, prograde orbits in Jupiter's equatorial plane. Moving outward from the planet, they are named Io, Europa, Ganymede, and Callisto, after the mythical attendants of the Roman god Jupiter. The two *Voyager* spacecraft and, more recently, the *Galileo* mission returned many remarkable images of these small worlds, allowing scientists to discern much fine surface detail on each. Figure 8.1 is a *Voyager 1* image of Io and Europa, with Jupiter providing a spectacular backdrop. Figure 8.2 shows the four Galilean moons to scale and their relative orbits sketched as well.

The Galilean moons range in size from slightly smaller than Earth's Moon (Europa) to slightly larger than Mercury (Ganymede). Their densities decrease with increasing distance from Jupiter in a manner very reminiscent of the falloff in density of the terrestrial planets with increasing distance from the Sun. ∞ *(Sec. 4.1)* Indeed, many astronomers think that the formation of Jupiter and its Galilean satellites may have mimicked on a small scale the formation of the Sun and the inner planets, with generally denser moons forming closer to the intense heat of the young Jupiter. ∞ *(Sec. 4.3)*

R I **V** U X G

▲ **Figure 8.1 Jupiter, Up Close** *Voyager 1* took this photograph of Jupiter with ruddy Io on the left and pearl-like Europa toward the right. Note the scale of objects here: Io and Europa are each comparable in size to our Moon; the Great Red Spot is roughly twice as large as Earth. (NASA)

	PARENT	DISTANCE FROM PARENT PLANET		ORBIT PERIOD	RADIUS	MASS	DENSITY
NAME	PLANET	(km)	(planet radii)	(Earth days)	(km)	(Earth's Moon = 1)	(kg/m³)
Moon	Earth	384,000	60.2	27.3	1740	1.00	3300
Io	Jupiter	422,000	5.90	1.77	1820	1.22	3500
Europa	Jupiter	671,000	9.38	3.55	1570	0.65	3000
Ganymede	Jupiter	1,070,000	15.0	7.15	2630	2.02	1900
Callisto	Jupiter	1,880,000	26.3	16.7	2400	1.46	1800
Titan	Saturn	1,220,000	20.3	16.0	2580	1.83	1900
Triton	Neptune	355,000	14.3	−5.88[1]	1350	0.29	2100

TABLE 8.1 The Large Moons of the Solar System

[1] The minus sign indicates a retrograde orbit.

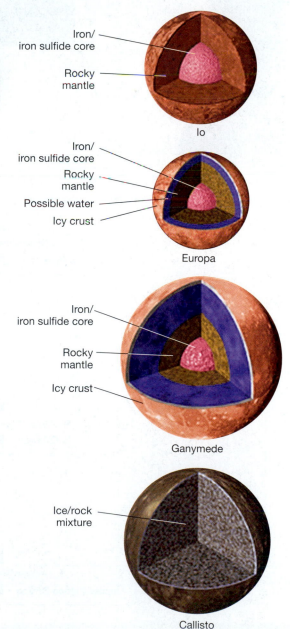

Detailed gravity measurements made by the *Galileo* orbiter support this picture of the origin of the Jupiter system. The data suggest that Io and Europa have large, iron-rich cores surrounded by thick mantles of generally rocky composition, perhaps similar to the crusts of the terrestrial planets. Europa has a water/ice outer shell between 100 and 200 km thick. Ganymede and Callisto are of more lightweight overall composition. Low-density materials, such as water ice, may account for as much as half their total mass. Ganymede has a relatively small metallic core topped by a rocky mantle and a thick icy outer shell. Callisto seems to be a largely undifferentiated mixture of rock and ice. Figure 8.3 summarizes our knowledge of the interiors of these bodies.

Io: The Most Active Moon

Io is the most geologically active object in the entire solar system. In mass and radius it is fairly similar to Earth's Moon, but there the resemblance ends. Shown in Figure 8.4, Io's uncratered surface is a collage of oranges, yellows, and blackish browns.

As *Voyager 1* sped past Io, it made an outstanding discovery: Io has active volcanoes! The spacecraft photographed eight erupting volcanoes. Six of them were still erupting when *Voyager 2* passed by four months later. Inset (a) in

▶ **Figure 8.3 Galilean Moon Interiors**   Cutaway diagrams showing the interior structure of the four Galilean satellites. Moving outward from Io to Callisto, the moons' densities steadily decrease as the composition shifts from rocky mantles and metallic cores in Io and Europa, to a thick icy crust and smaller core in Ganymede, to an almost uniform rock and ice mix in Callisto. Both Ganymede and Europa are thought to have layers of liquid water beneath their icy surfaces.

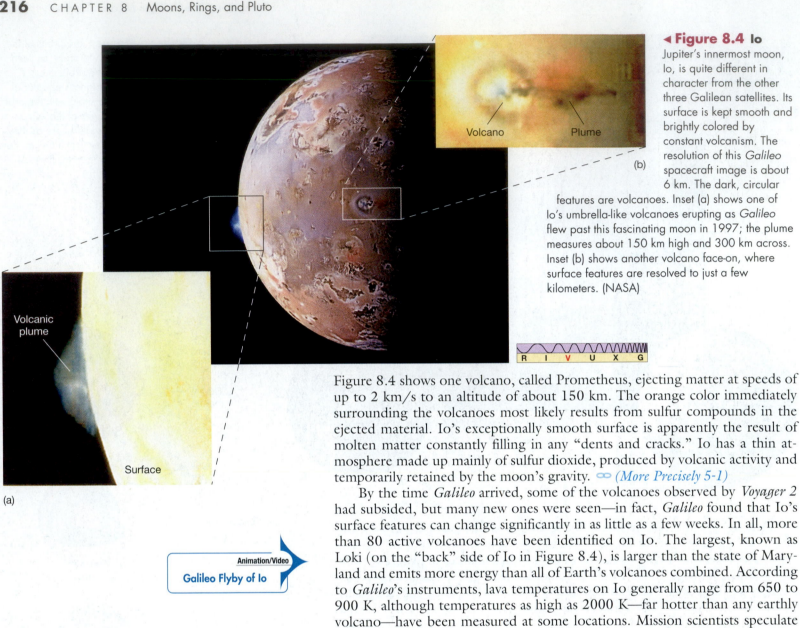

◄ Figure 8.4 Io

Jupiter's innermost moon, Io, is quite different in character from the other three Galilean satellites. Its surface is kept smooth and brightly colored by constant volcanism. The resolution of this *Galileo* spacecraft image is about 6 km. The dark, circular features are volcanoes. Inset (a) shows one of Io's umbrella-like volcanoes erupting as *Galileo* flew past this fascinating moon in 1997; the plume measures about 150 km high and 300 km across. Inset (b) shows another volcano face-on, where surface features are resolved to just a few kilometers. (NASA)

Volcano   Plume   (b)

Volcanic plume

Surface

(a)

R   I   V   U   X   G

Figure 8.4 shows one volcano, called Prometheus, ejecting matter at speeds of up to 2 km/s to an altitude of about 150 km. The orange color immediately surrounding the volcanoes most likely results from sulfur compounds in the ejected material. Io's exceptionally smooth surface is apparently the result of molten matter constantly filling in any "dents and cracks." Io has a thin atmosphere made up mainly of sulfur dioxide, produced by volcanic activity and temporarily retained by the moon's gravity. ∞ *(More Precisely 5-1)*

By the time *Galileo* arrived, some of the volcanoes observed by *Voyager 2* had subsided, but many new ones were seen—in fact, *Galileo* found that Io's surface features can change significantly in as little as a few weeks. In all, more than 80 active volcanoes have been identified on Io. The largest, known as Loki (on the "back" side of Io in Figure 8.4), is larger than the state of Maryland and emits more energy than all of Earth's volcanoes combined. According to *Galileo*'s instruments, lava temperatures on Io generally range from 650 to 900 K, although temperatures as high as 2000 K—far hotter than any earthly volcano—have been measured at some locations. Mission scientists speculate that these superhot volcanoes may be similar to those that occurred on Earth more than 3 billion years ago.

Io is far too small to have geological activity similar to that on Earth. It should be long dead, like our own Moon, its internal heat lost to space billions of years ago. Instead, the source of Io's energy is *external*—Jupiter's gravity. Because Io orbits very close to Jupiter, the planet's huge gravitational field produces strong tidal forces on the moon, resulting in a large (100 m) tidal bulge. ∞ *(Sec. 5.2)* If Io were Jupiter's only satellite, it would long ago have come into a state of synchronous rotation with the planet. In that case, Io would move in a perfectly circular orbit with one face permanently turned toward Jupiter. The tidal bulge would be stationary with respect to the moon, and there would be no internal stresses and no volcanism.

But Io is not alone. As it orbits, it is constantly tugged by the gravity of its nearest large neighbor, Europa. These tugs are small and not enough to cause any great tidal effect in and of themselves, but they are sufficient to make Io's orbit slightly noncircular, preventing the moon from settling into a precisely synchronous state. The moon's orbital speed varies from place to place as it revolves around the planet, but its rate of rotation on its axis remains constant. Thus it cannot keep one face permanently turned toward Jupiter. Instead, as seen from below, Io rocks or wobbles slightly from side to side on its axis as it moves. The

Animation/Video
Galileo Flyby of Io

Animation/Video
Io Cutaway

Animation/Video
Jupiter's Moon Europa

tidal bulge, however, always points directly toward Jupiter, so it moves back and forth across Io's surface as the moon wobbles. The conflicting forces acting on Io result in enormous tidal stresses that continually flex and squeeze its interior.

Just as repeated back-and-forth bending of a wire can produce heat through friction, Io is constantly energized by this ever-changing distortion. Researchers estimate that the total amount of heat generated within Io as a result of tidal flexing is about 100 million megawatts—more than five times the total power consumption of all the nations on Earth. The large amount of heat generated within Io causes huge jets of gas and molten rock to squirt out of the surface. It is likely that much of Io's interior is soft or molten, with only a relatively thin solid crust overlying it.

Europa: Liquid Water Locked in Ice

Europa (Figure 8.5) is a world very different from Io. Europa has relatively few craters on its surface, again suggesting that recent activity has erased the scars of ancient meteoric impacts. However, the surface displays a vast network of lines crisscrossing bright, clear fields of water ice. Many of these linear bands extend halfway around the satellite.

Based on the *Voyager* images, planetary scientists theorized that Europa is covered by an ocean of liquid water whose surface layers are frozen. More recent *Galileo* images seem to support this idea. Figure 8.5(b) is a high-resolution image of this weird moon, showing ice-filled surface cracks similar to those seen in Earth's polar ice floes. Figure 8.5(c) shows what appear to be icebergs—flat chunks of ice that have been broken apart and reassembled, perhaps by the action of water below. Mission scientists speculate that Europa's ice may be several kilometers thick and that the ocean below it may be 100 km deep. As on Io, the ultimate source of the energy that keeps the ocean liquid and also drives the internal motion leading to the cracks in the icy surface is Jupiter's gravity. The heating mechanism is essentially similar to that on Io, but the effects are less extreme on Europa because of its greater distance from the planet.

Other detailed images of the surface support this hypothesis. Figure 8.5(d) shows a region where Europa's icy crust appears to have been pulled apart and new material has filled in the gaps between the separating ice sheets. Elsewhere on the surface, *Galileo* found what appeared to be the icy equivalent of lava flows on Earth—regions where water apparently erupted through the surface and flowed for many kilometers before solidifying. The scarcity of impact craters on Europa implies that the processes responsible for these features did not stop long ago. Rather, they must be ongoing.

Further evidence comes from studies of Europa's magnetic field. Magnetic measurements made by *Galileo* on repeated flybys of the moon revealed that Europa has a weak magnetic field that constantly changes strength and direction. This finding is consistent with the idea that the field is generated by the action of Jupiter's magnetism on a shell of electrically conducting fluid about 100 km below Europa's surface—in other words, the salty liquid water layer suggested by the surface observations. These results convinced quite a few skeptical scientists of the reality of Europa's deep ocean.

(a)

(b)   (c)

(d)

R I V U X G

▶ **Figure 8.5 Europa** (a) A *Voyager 1* mosaic of Europa. Resolution is about 5 km. (b) Europa's icy surface is only lightly cratered, indicating that some ongoing process must be obliterating impact craters soon after they form. The origin of the cracks crisscrossing the surface is uncertain. (c) At 50-m resolution, this image from the *Galileo* spacecraft shows a smooth yet tangled surface resembling the huge ice floes that cover Earth's polar regions. This region is called Conamara Chaos. (d) This detailed *Galileo* image shows "pulled apart" terrain that suggests liquid water upwelling from the interior and freezing, filling in the gaps between separating surface ice sheets. (NASA)

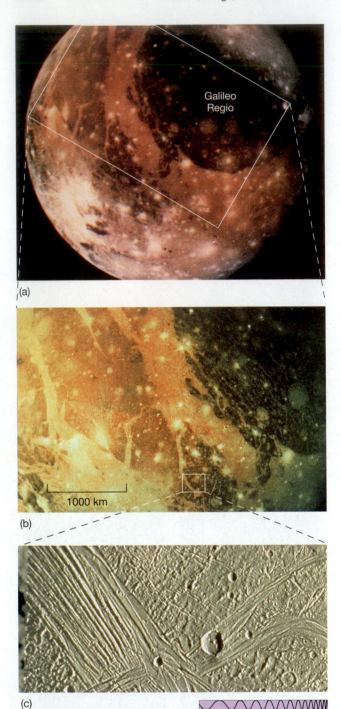

▲ **Figure 8.6 Ganymede** (a) and (b) *Voyager 2* images of Ganymede. The dark regions are the oldest parts of the moon's surface and probably represent its original icy crust. The largest dark region visible here, called Galileo Regio, spans some 3200 km. The lighter, younger regions are the result of flooding and freezing that occurred within a billion years or so of Ganymede's formation. The light-colored spots are recent impact craters. (c) Grooved terrain on Ganymede may have been caused by a process similar to plate tectonics on Earth. The area shown in this *Galileo* image spans about 50 km and reveals a multitude of ever-smaller ridges, valleys, and craters, right down to the resolution limit of the spacecraft's camera (about 300 m, or three times the length of a football field). The image suggests erosion of some sort, possibly even caused by water. (NASA)

The likelihood that Europa has an extensive layer of liquid water below its surface ice opens up many interesting avenues of speculation about the possible development of life there. In the rest of the solar system, only Earth has liquid water on or near its surface, and most scientists agree that water played a key role in the appearance of life here (see Chapter 18). Europa may well contain more liquid water than exists on our entire planet! Of course, the existence of water does not *necessarily* imply the emergence of life. Europa, even with its liquid ocean, is still a hostile environment compared to Earth. Nevertheless, the possibility, even a remote one, of life on Europa was an important motivating factor in the decision to extend the *Galileo* mission for a total of six additional years. ∞ *(Sec. 7.1)* No new visits are currently planned to the Jupiter system, although planetary scientists are lobbying hard for NASA or the European Space Agency to mount a mission dedicated to this fascinating world.

Ganymede and Callisto: Fraternal Twins

Ganymede, shown in Figure 8.6, is the largest moon in the solar system, exceeding not only Earth's Moon but also the planets Mercury and Pluto in size. It has many impact craters on its surface, along with patterns of dark and light markings reminiscent of the highlands and maria on Earth's Moon. In fact, Ganymede's history has many parallels with that of the Moon (with water ice replacing lunar rock).

As on the inner planets and Earth's Moon, we can estimate ages on Ganymede by counting craters. We learn that the darker regions are the oldest parts of Ganymede's surface. They are the original icy surface, just as the ancient highlands on our own Moon are its original crust. Ganymede's surface has darkened with age as micrometeorite dust has slowly covered it. The light-colored parts are much less heavily cratered, and so must be younger. They are Ganymede's "maria," and probably formed in a manner similar to the way the maria on the Moon were created. ∞ *(Sec. 5.6)* Intense meteoritic bombardment caused liquid water—Ganymede's counterpart to our own Moon's molten lava—to upwell from the interior, flooding the impacting regions before solidifying. Not all of Ganymede's surface features follow the lunar analogy, however. Ganymede has a system of grooves and ridges (Figure 8.6b, c) that may have resulted from crustal tectonic motion, much as Earth's surface undergoes mountain building and faulting at plate boundaries. ∞ *(Sec. 5.5)* The process apparently stopped about 3 billion years ago when the cooling crust became too thick.

Callisto, shown in Figure 8.7, is in many ways similar in appearance to Ganymede, although Callisto is more heavily cratered and has few fault lines. Callisto's most obvious feature is a huge series of concentric ridges surrounding two large basins, one clearly visible in Figure 8.7. The ridges resemble the ripples made as a stone hits water, but on Callisto they probably resulted from a cataclysmic impact with a meteorite. The upthrust ice was partially melted, but it resolidified quickly, before the ripples had a chance to subside. Today both the ridges and the rest of the crust are frigid ice and show no obvious signs of geological activity. Apparently, Callisto froze before plate tectonic or other activity could start. The density of impact craters on the large impact basins indicates that they formed long ago, perhaps 4 billion years in the past.

Ganymede's internal differentiation indicates that the moon was largely molten at some time in the past, yet Callisto is undifferentiated and hence apparently never melted (Figure 8.3). ∞ *(Sec. 5.4)* Researchers are uncertain why two such similar bodies should have evolved so differently. Complicating things further is the discovery by the *Galileo* orbiter that Ganymede has a weak magnetic field, about 1 percent that of Earth, suggesting that this moon, too, may have liquid or "slushy" water below its surface. ∞ *(Sec. 5.7)* If that is so, then Ganymede's heating and differentiation must have happened relatively

(a)

(b)

◀ Figure 8.7 Callisto
(a) Callisto is similar to Ganymede in overall composition, but is more heavily cratered. The large series of concentric ridges visible at the left is known as Valhalla. Extending nearly 1500 km from the basin center, the ridges formed when "ripples" from a large meteoritic impact froze before they could disperse completely. Resolution here is about 10 km. (b) This higher-resolution image of Callisto's equatorial region, about 300 × 200 km in area, displays more clearly its heavy cratering. (NASA)

recently—less than a billion years ago, based on recent estimates of how rapidly the moon's heat escapes into space.

Scientists have no clear explanation for how this could have occurred, although some speculate that interactions among the inner moons may have caused Ganymede's orbit to have changed significantly about 1 billion years ago and that prior tidal heating by Jupiter helped melt the moon's interior. These interactions may conceivably also be related to the near-perfect 1:2:4 ratio among the orbital periods of the three inner Galilean moons (see Table 8.1), which astronomers cannot easily explain but are unwilling to dismiss as simple coincidence.

Animation/Video
Jupiter's Moon Ganymede

◀◀◀ CONCEPT CHECK

What is the source of energy for all the activity observed on Jupiter's Galilean moons?

8.2 The Large Moons of Saturn and Neptune

Titan: A Moon with an Atmosphere

The largest and most intriguing of Saturn's moons is Titan. From Earth, it is visible only as a barely resolved orange disk. Long before the *Voyager* missions, astronomers knew from spectroscopic observations that the moon's ruddy coloration is caused by something quite special—an atmosphere. So anxious were mission planners to obtain a closer look that they programmed *Voyager 1* to pass very close to Titan, even though that meant the spacecraft could not then use Saturn's gravity to continue on to Uranus and Neptune.

Unfortunately, despite this close encounter, the moon's surface remained a mystery. A thick, uniform haze layer, in some ways similar to the smog found over many cities on Earth, envelops the moon, completely obscuring the spacecraft's view (Figure 8.8). A more detailed, close-up look at Titan is one of the principal targets of the ongoing *Cassini* mission. ∞ *(More Precisely 7-1)* The spacecraft's carefully planned trajectory includes a total of 44 visits to Titan by the end of the primary mission in 2008.

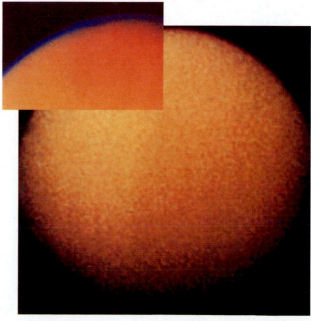

▶ Figure 8.8 Titan Titan, larger than Mercury and roughly half the size of Earth, was photographed from only 4000 km away as the *Voyager 1* spacecraft passed by in 1980. Only Titan's upper cloud deck can be seen here. The inset shows a contrast-enhanced image of the haze layer (falsely colored blue). (NASA)

▲ Figure 8.9 Titan Revealed *Cassini's* infrared telescopes revealed this infrared, false-color view of Titan's surface in late 2004. Red color corresponds to atmospheric gases and green and blue to surface features. The inset shows a surface feature thought to be an icy volcano, suggesting some geological activity on this icy moon's surface. Resolution of the larger image is 25 km; that of the inset is ten times better. (NASA/ESA)

Figure 8.9 shows how *Cassini's* infrared instruments have been able to penetrate the moon's foggy atmosphere, revealing details of the surface. The image shows light and dark regions near the center of the field of view, thought to be icy plateaus, apparently smeared with hydrocarbon tar. Ridges and cracks on the moon's surface suggest that geological activity, in the form of "titanquakes," may be common. The rather blurred boundaries between the light and dark regions, the peculiar surface markings in the light-colored region, and the absence of extensive cratering suggest that some sort of erosion is occurring, perhaps as a result of wind or volcanic activity. Radar imaging has revealed few large (10–100 km diameter) craters on the moon's surface, but there do not seem to be as many small craters as would be expected given Titan's location in Saturn's congested ring plane (see Section 8.4). The insert in Figure 8.9 shows what appears to be an icy volcano, supporting the view that the moon's surface is geologically active.

Titan's atmosphere is thicker and denser even than Earth's and is certainly more substantial than that of any other moon. Radio and infrared observations made by *Voyager 1* and *Cassini* show that the atmosphere is made up mostly of nitrogen (roughly 90 percent) and possibly argon (up to 10 percent), with a few percent of methane and trace amounts of other gases. The surface temperature is a frigid 94 K, roughly what we would expect on the basis of Titan's distance from the Sun. Under these conditions, as on Ganymede and Callisto, water ice plays the role of rock on Earth, and liquid water the role of lava. The presence of an atmosphere allows us to extend this analogy further—at the temperatures typical of the lower atmosphere, methane and ethane behave rather like water on Earth, raising the possibility of methane rain, snow, and fog, and (perhaps) even rivers and oceans!

Titan's atmosphere seems to act like a gigantic chemical factory. Powered by the energy of sunlight, it is undergoing a complex series of chemical reactions that ultimately result in the observed chemical composition. The upper atmosphere is thick with smoggy haze, and the surface may be covered with hydrocarbon sediment that has settled down from the clouds. One of the many reasons Titan is so interesting to scientists is that it may present an opportunity to study the kind of chemistry thought to have occurred billions of years ago on Earth—the prebiotic chemical reactions (preceding living organisms, but instrumental in their emergence) that eventually led to life on our own planet (see Chapter 18).

Prior to *Cassini's* arrival, speculation about what might be found on Titan's surface ran the gamut from oceans of liquid methane or ethane to icy valleys laden with hydrocarbon sludge. On January 14, 2005, the *Huygens* probe, transported to Saturn by *Cassini* and released three weeks earlier, arrived at Titan and parachuted through the thick atmosphere to the moon's surface (Figure 8.10a). Figure 8.10(b) shows an intriguing image radioed back from *Huygens* during its descent. Interpretations remain uncertain, but it appears to show a network of drainage channels leading to a shoreline. The probe landed on solid ground, and for the next hour transmitted images and instrument readings to *Cassini* as it passed overhead. The view from the surface (Figure 8.11c) reveals a hazy view of an icy landscape. The "rocks" in the foreground are a few centimeters across, and show evidence of erosion by liquid of some sort. The nature of Titan's "ocean" remains uncertain—it may well consist of slush rather than liquid.

(a)

Drainage channels

(b)

R I V U X G

(c)

Figure 8.11 shows the probable structure of Titan's atmosphere. The diagram was drawn based largely on *Voyager* measurements, but it is largely consistent with the data received from *Huygens* during its descent. Despite Titan's low mass (a little less than twice that of Earth's Moon) and hence low surface gravity (one-seventh of Earth's), the atmospheric pressure at ground level is 60 percent greater than on Earth. Titan's atmosphere contains about 10 times more gas than Earth's atmosphere and, because of Titan's weaker gravitational pull, extends 10 times farther into space than does our own. The top of the main haze layer lies about 200 km above the surface (Figure 8.8 inset), although there are additional layers, seen primarily through their absorption of ultraviolet radiation, at 300 km and 400 km. Below the haze the atmosphere is clear, although rather gloomy because so little sunlight gets through. *Cassini* has detected low-lying methane clouds at roughly the altitude (20 km) predicted by the theoretical models, but the clouds are less common than scientists had expected, with the result that the "rain" supposedly responsible for the channels in Figure 8.10(b) is not as widespread as was previously thought.

Why does Titan have such a thick atmosphere when Ganymede and Callisto do not? The answer lies largely in Titan's 94 K surface temperature, which is considerably lower than that of Jupiter's moons, making it easier for the moon to retain an atmosphere. ∞ *(More Precisely 5-2)* Also, gases may have become trapped in the surface ice when the moon formed. As Titan's internal radioactivity warmed the moon, the ice released the trapped gases, forming a thick methane–ammonia atmosphere. Sunlight split the ammonia into hydrogen, which escaped into space, and nitrogen, which remained in the atmosphere. The methane, which was less easily broken apart, survived intact. Together with argon outgassed from Titan's interior by volcanic activity, these gases form the basis of the atmosphere we see today. ∞ *(Sec. 6.8)* However, the story may still be incomplete. Detailed studies by *Cassini* scientists of the composition of Titan's atmosphere suggest that the atmosphere may have been much thicker—perhaps 20 times denser—in the distant past. How the moon could have lost so much of its atmosphere remains a mystery.

▲ **Figure 8.10 Titan's Surface** (a) Artist's conception of the *Huygens* lander parachuting through Titan's thick atmosphere. (b) This photograph of the surface was taken from an altitude of 8 km as the probe descended. It shows a network of channels reminiscent of streams or rivers draining from the light-shaded uplifted terrain (at center) into a darker, low-lying region (at bottom). Resolution is about 20 m. (c) *Huygen's* view of its landing site, in approximately true color. The foreground rocklike objects are only a few centimeters across. (D. Ducros; NASA/ESA)

▲ Figure 8.11 Titan's Atmosphere The structure of Titan's atmosphere, as deduced from *Voyager 1* observations. The solid blue line represents temperature at different altitudes.

Scientists think that Titan's overall composition must be similar to those of Ganymede and Callisto, because these three moons have such similar masses and radii and hence average densities. However, it is not known whether Titan is differentiated like Ganymede or of more uniform composition like Callisto (see Figure 8.3). As *Cassini* repeatedly passes close to Titan over the next few years, scientists hope to build up a more complete picture of the moon's internal structure.

Triton: Captured from the Kuiper Belt?

Neptune's large moon, Triton, is the smallest of the seven large moons in the solar system, with about half the mass of the next smallest, Europa. Lying 4.5 billion km from the Sun and possessing an icy surface that reflects much of the solar radiation reaching it, Triton has a temperature of just 37 K. *Voyager 2* found that this moon has an extremely thin nitrogen atmosphere and a solid, frozen surface probably composed primarily of water ice.

A *Voyager 2* mosaic of Triton's south polar region is shown in Figure 8.12. The moon's low temperatures produce a layer of nitrogen frost that forms, evaporates, and reforms seasonally over the polar caps. The frost is visible as the pinkish region at the bottom right of Figure 8.12. Overall, there is a marked lack of cratering on Triton, presumably indicating that surface activity has obliterated the evidence of most impacts. There are many other signs of an active past. Triton's face is scarred by large fissures similar to those seen on Ganymede, and the moon's odd cantaloupe-like terrain may indicate repeated faulting and deformation over the moon's lifetime. In addition, Triton has numerous frozen lakes of water ice, thought to be volcanic in origin.

Triton's surface activity is not just a thing of the past. As *Voyager 2* passed the moon, its cameras detected great jets of nitrogen gas erupting several kilometers into the sky. These geysers may result when liquid nitrogen below Triton's surface is heated and vaporized by some internal energy source, or perhaps even by the Sun's feeble light. Scientists conjecture that nitrogen geysers may be very common on Triton and are perhaps responsible for much of the moon's thin atmosphere.

Triton is unique among the large moons of the solar system in that its orbit around Neptune is *retrograde*. Also, with an orbital inclination of about 20°, Triton is the only large jovian moon not to orbit in its parent planet's equatorial plane. The event (or events) that placed Triton on a tilted, retrograde orbit are the subject of considerable speculation. Triton's peculiar orbit and surface features suggest to many astronomers that the moon did not form as part of the Neptune system but instead was captured by the planet, probably not too long ago. The surface deformations on Triton certainly suggest violent and relatively recent events in the moon's past, but the details remain uncertain. The absence of a "normal" midsize moon system around Neptune (see Section 8.3) is also often cited as evidence for some catastrophic event in the planet's history.

◄ Figure 8.12 Triton The south polar region of Triton, showing a variety of terrains ranging from deep ridges and gashes to what appear to be frozen water lakes, all indicative of past surface activity. The pinkish region at the lower right is nitrogen frost, forming the moon's polar cap. Resolution is about 4 km. The roughly circular lakelike feature shown in the inset is about 200 km across and may have been caused by the eruption of an ice volcano. The water "lava" has since solidified, leaving a smooth surface. The absence of craters indicates that the eruption was a relatively recent event. Details in this inset image are resolved to a remarkable 1 km. (NASA)

Whatever Triton's past, its future seems quite clear. Because of its retrograde orbit, the tidal bulge that Triton raises on Neptune tends to make the moon spiral *toward* the planet rather than away from it (as our Moon moves away from Earth). ∞ *(Sec. 5.2)* Calculations indicate that Triton is doomed to be torn apart by Neptune's tidal gravitational field, probably in no more than about 100 million years.

◄◄◄ **CONCEPT CHECK**

Why was Voyager 1 unable to photograph the surface of Titan? Why do we now have a much better understanding of the moon's surface?

8.3 The Medium-Sized Jovian Moons

Table 8.2 lists the 12 medium-sized satellites of the solar system, conventionally defined as those bodies having radii between about 200 and 800 km. The six that orbit Saturn are shown in Figure 8.13. Figure 8.14 shows the remainder (five orbiting Uranus, the other orbiting Neptune) to the same scale as Figure 8.13. Their densities suggest that all are composed largely of rock and water ice. All move on nearly circular paths and are tidally locked by their parent planet's gravity into synchronous rotation.

Most of these moons show no evidence for significant geological activity, although there are a few possible exceptions. The light-colored *wispy terrain* visible on Saturn's Rhea and Dione in Figure 8.13 was once thought to have been caused by events in the distant past during which water was released from the moons' interiors and condensed on their surfaces. However, *Cassini* images reveal that the markings on Dione, at least, are in fact bright ice cliffs created by *tectonic fractures*, where stresses in the moon's interior as it cools and contracts have caused the surface layers to crack and buckle. Dione also has "maria" of sorts, where flooding from ice volcanism appears to have obliterated the older craters. Saturn's moon Iapetus has a distinctly "two-faced" appearance. One hemisphere is very dark, while the other is very bright. Whether the moon is being coated by dark material falling on it from outside or being resurfaced by material from within is unknown. Both Tethys in the Saturn system and Ariel orbiting Uranus have extensive cracks on their surfaces, which may also have been caused by cooling and shrinking of the moons' outer layers, but are more likely the result of meteoritic bombardment.

		DISTANCE FROM PARENT PLANET		**ORBIT PERIOD**	**RADIUS**	**MASS**	**DENSITY**
NAME	**PARENT PLANET**	**(km)**	**(planet radii)**	**(Earth days)**	**(km)**	**(Moon = 1)**	**(kg/m^3)**
Mimas	Saturn	186,000	3.10	0.94	200	0.00051	1100
Enceladus	Saturn	238,000	3.97	1.37	250	0.00099	1000
Tethys	Saturn	295,000	4.92	1.89	530	0.0085	1000
Dione	Saturn	377,000	6.28	2.74	560	0.014	1400
Rhea	Saturn	527,000	8.78	4.52	760	0.032	1200
Iapetus	Saturn	3,560,000	59.3	79.3	720	0.022	1000
Miranda	Uranus	130,000	5.09	1.41	240	0.00090	1200
Ariel	Uranus	191,000	7.48	2.52	580	0.018	1700
Umbriel	Uranus	266,000	10.4	4.14	590	0.016	1400
Titania	Uranus	436,000	17.1	8.71	790	0.048	1700
Oberon	Uranus	583,000	22.8	13.5	760	0.041	1600
Proteus	Neptune	118,000	4.76	1.12	210	—	—

TABLE 8.2 The Medium-Sized Moons of the Solar System

R I V U X G

▲ **Figure 8.13 Saturnian Moons** Saturn's six medium-sized satellites, to scale, as imaged by the *Cassini* spacecraft. All are heavily cratered and all are shown here in natural color. Iapetus shows a clear contrast between its light-colored (icy) surface (at top in this image) and its darkly cratered hemisphere (at bottom). The light-colored wisps on Rhea are thought to be water and ice released from the moon's interior during some long-ago period of activity. Dione and Tethys arguably show evidence of ancient geological activity. Enceladus appears to still be volcanically active, but the cause of the volcanism is unexplained. Mimas's main surface feature is the large crater Herschel. For scale, part of Earth's Moon is shown at the left. (NASA)

Saturn's moon Enceladus almost certainly does have ongoing geological activity. Its surface is so bright and shiny—reflecting virtually 100 percent of the sunlight falling on it—that astronomers suspect that the moon must be completely coated with fine crystals of pure ice, possibly the icy "ash" of water volcanoes. Much of the surface is devoid of impact craters, which seem to have been erased by "lava flows" composed of water temporarily liquefied during internal upheavals and now frozen again. It is not known why there should be such activity on so small a moon.

Most of the midsized moons show heavy cratering, indicating the cluttered and violent planetary environment that existed just after the outer planets formed, as the remaining planetesimals were accreted by larger bodies or ejected from the vicinity of the jovian planets to form the Oort cloud. ∞ *(Sec. 4.3)* The innermost midsize moons, Mimas (Saturn) and Miranda (Uranus), show perhaps the clearest evidence of violent meteoritic impacts. The impact that caused the large crater visible on the image of Mimas in Figure 8.13 must have come very close to shattering the moon. Miranda (Figure 8.15) displays a wide range of surface terrains, including ridges, valleys, faults, and many other geological features. To explain why Miranda seems to combine so many types of terrain, some researchers have hypothesized that it has been catastrophically disrupted several times, with the pieces falling back together in a chaotic, jumbled way. However, it will be a long time before we can obtain more detailed information to test this theory.

CONCEPT CHECK ▶▶▶

What surface property of the medium-sized moons tells us that the outer solar system was once a very violent place?

◄**Figure 8.14 Moons of Uranus and Neptune** The five medium-sized moons of Uranus and the single medium-sized moon of Neptune, Proteus, shown to the same scale as Figure 8.13. In order of increasing distance from the planet, the moons of Uranus are Miranda, Ariel, Umbriel, Titania, and Oberon. Like their counterparts among Saturn's moons, all six moons are heavily cratered, and most show no clear evidence of any present or past geological activity. (NASA)

▼ **Figure 8.15 Miranda** The asteroid-sized innermost moon of Uranus, photographed by *Voyager 2*. Miranda has a fractured surface suggestive of a violent past, but the cause of the grooves and cracks is unknown. The resolution in the inset is about 2 km. (NASA)

30 km

8.4 Planetary Rings

All four jovian worlds have planetary ring systems girdling their equators. The properties and appearance of each planet's rings are intimately connected with the planet's small and medium-sized moons, with many of the inner jovian moons orbiting close to (or even within) the parent planet's rings. Saturn's rings are by far the best known and best observed, and we study them here in greatest detail. However, the rings of all the jovian planets have much to tell us about the environment in the outer solar system.

Saturn's Spectacular Ring System

Figure 8.16 shows the main features of Saturn's rings. The dark gap about two-thirds of the way out from the inner edge is named the **Cassini Division**. Outside the Cassini Division lies the **A ring**. The inner ring, between the Cassini Division and the planet, is in reality two rings, the **B** and **C rings**. A smaller gap, known as the **Encke gap**, is found in the outer part of the A ring. Its width is 270 km. No finer ring details are visible from Earth. Of the three main rings, the B ring is brightest, followed by the somewhat fainter A ring, and then by the almost translucent C ring. The additional rings listed in Table 8.3 cannot be seen in Figure 8.16.

Because Saturn's rings lie in the p lanet's equatorial plane and the planet's rotation axis is tilted with respect to the ecliptic, the appearance of the rings (as seen from Earth) changes seasonally, as shown in Figure 8.17. As Saturn orbits the Sun, the angle at which the rings are illuminated varies. When the planet's north or south pole is tipped toward the Sun (which means it is either summer or winter on Saturn), the highly reflective rings are at their brightest. In Saturn's spring and fall the rings are close to being edge-on both to the Sun and to observers on Earth, so they seem to disappear. One important deduction we can make from this simple observation is that the rings are very *thin*. In fact, we now know that their thickness is less than a few hundred meters, even though they are more than 200,000 km in diameter.

Saturn's rings are not solid objects. In 1857 Scottish physicist James Clerk Maxwell realized that they must be composed of a great number of small solid *particles*, all independently orbiting Saturn, like so many tiny moons. Measurements of the Doppler shift of sunlight reflected from the rings reveal that the orbital speeds decrease with increasing distance from the planet in precise accordance with Kepler's laws and Newton's law of gravity. ∞ *(Sec. 1.4)* The

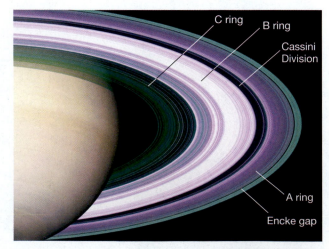

◄ **Figure 8.16 Saturn's Rings** Saturn as observed by the *Cassini* spacecraft in 2005. The main ring features are marked and are shown here in false color in order to enhance contrast. (NASA/ESA)

TABLE 8.3	The Rings of Saturn				
	INNER RADIUS		OUTER RADIUS		
RING	(km)	(planet radii)	(km)	(planet radii)	WIDTH (km)
D	60,000	1.00	74,000	1.23	14,000
C	74,000	1.23	92,000	1.53	18,000
B	92,000	1.53	117,600	1.96	25,600
A	122,200	2.04	136,800	2.28	14,600
F	140,500	2.34	140,600	2.34	100
E	210,000	3.50	300,000	5.00	90,000

highly reflective nature of the rings suggests that the particles are made of ice, and infrared observations in the 1970s confirmed that water ice is in fact a prime ring constituent. *Cassini* also detected substantial amounts of "mud"—small rocky particles and dust—mixed in. Radar observations and later *Voyager* and *Cassini* studies of scattered sunlight showed that the sizes of the icy ring particles range from fractions of a millimeter to tens of meters in diameter, with most particles being about the size (and composition) of a large and dirty snowball. The rings are thin because collisions between ring particles tend to keep them moving in circular orbits in a single plane. Any particle straying from this orderly configuration soon runs into other ring particles and is eventually forced back into the fold.

The Roche Limit

To understand why rings form, consider the fate of a small moon orbiting close to a massive planet such as Saturn (Figure 8.18). As we bring our hypothetical moon closer to the planet, the tidal force exerted on it by the planet increases, stretching it along the direction to the planet. ∞ *(Sec. 5.2)* The tidal force increases rapidly as the distance to the planet decreases. Eventually a point is reached where the planet's tidal force becomes greater than the internal forces holding the moon together, and the moon is torn apart by the planet's gravity. The pieces of the satellite move on their own individual orbits around that planet, eventually spreading all the way around it in the form of a ring.

▲ **Figure 8.17** **Saturn's Tilt** (a) Over time Saturn's rings change their appearance to terrestrial observers as the tilted ring plane orbits the Sun. The roughly true-color images (inset) span a period of several years from 1996 to 2000, showing how the rings change from our perspective, from almost edge-on to more nearly face-on. (NASA)

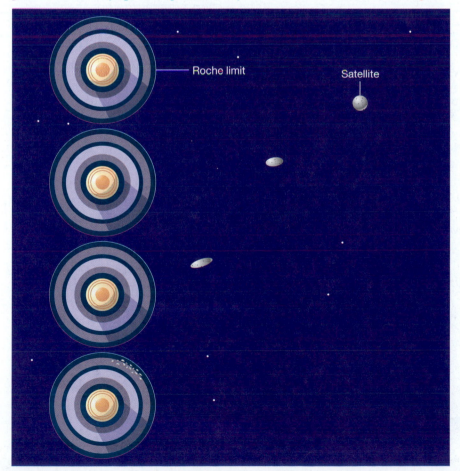

◄ **Figure 8.18** **Roche Limit** The tidal field of a planet first distorts (top) and then destroys (bottom) a moon that strays too close.

For any given planet and any given moon, this critical distance inside which the moon is destroyed is known as the **Roche limit**, after the nineteenth-century French mathematician Edouard Roche, who first calculated it. If our hypothetical moon is held together by its own gravity, and if its average density is similar to that of the parent planet (both reasonably good assumptions for Saturn's larger moons), then the Roche limit is approximately 2.5 times the radius of the planet. For example, in the case of Saturn (of radius 60,000 km), the Roche limit lies roughly 150,000 km from the planet's center—just outside the outer edge of the A Ring. The rings of Saturn occupy the region inside Saturn's Roche limit. These considerations apply equally well to the other jovian worlds. As indicated in Figure 8.19, all four ring systems in our solar system are found within (or close to) the Roche limit of their parent planet.

Note that the calculation of the Roche limit applies only to moons massive enough for their own gravity to be the dominant binding force. Space probes and sufficiently small moons can survive within the Roche limit because they are held together not by gravity but rather by interatomic (electromagnetic) forces, so the above arguments do not apply.

▲ **Figure 8.19 Jovian Ring Systems**   The rings of Jupiter, Saturn, Uranus, and Neptune. All distances are expressed in planetary radii. The red line represents the Roche limit, and all the rings lie within (or very close to) this limit of the parent planet.

Fine Structure

In 1979 the *Voyager* probes found that Saturn's main rings are actually composed of tens of thousands of narrow **ringlets**. A *Cassini* view of this fine ring structure is shown in Figure 8.20. Furthermore, this intricate structure is not fixed but instead varies both with time and with position in the rings. The ringlets are formed when the mutual gravitational attraction of the ring particles, possibly in combination with the effects of Saturn's inner moons, creates waves of matter—regions of high and low density—that move in the plane of the rings like ripples on the surface of a pond.

The narrower gaps in the rings—about 20 in all—are most likely swept clean by the action of small *moonlets* embedded in them. Because these moonlets are much larger (perhaps 10 or 20 km across)

Animation/Video
Voyager Ring Spokes Animation

▼ **Figure 8.20 Saturn's Rings, Up Close**   *Cassini* took this true-color mosaic of Saturn's dazzling ring structure just before flying through the tenuous outer rings of Saturn. Earth is superposed, to scale, for a size comparison. (NASA/ESA)

R   I   V   U   X   G

than the largest ring particles, they simply scoop up ring material as they go. Prior to *Cassini*'s arrival, only one of these moonlets had been found, despite many careful searches of the *Voyager* images. *Cassini* has now identified numerous small moonlets (a few kilometers across) in the gaps between the main rings.

The largest gap—the Cassini Division—has a different origin. It owes its existence to the gravitational influence of Saturn's innermost medium-sized moon, Mimas. Over time, particles orbiting within the Division have been deflected by Mimas's gravity into eccentric orbits that cause them to collide with other ring particles, effectively moving them into new orbits at different radii. The net result is that the number of ring particles in the Cassini Division is greatly reduced.

Outside the A ring lies perhaps the strangest ring of all: the faint, narrow **F ring** (Figure 8.21). It was discovered by *Pioneer 11* in 1979, but its full complexity became evident only when the *Voyager* spacecraft took a closer look. Its oddest feature is that it looks as though it is made up of several strands braided together. It seems as though the ring's complex structure, as well as its thinness, arises from the influence of two small moons, known as **shepherd satellites**, that orbit about 1000 km on either side of it. Their gravitational effect gently guides any particle straying too far out of the F ring back into it. *Cassini* has discovered several more braided rings within the main ring system. The braids are the result of waves created in the rings by the moons. However, the details of how the braids are produced, and why the shepherd moons are there at all, remain subjects of active research.

Voyager 2 also found a faint series of rings, now known collectively as the **D ring**, inside the inner edge of the C ring, stretching down almost to Saturn's cloud tops. The D ring contains relatively few particles and is so dark that it is completely invisible from Earth. The faint **E ring**, also a *Voyager 2* discovery, lies well outside the main ring structure. It may be the result of volcanism on the moon Enceladus.

The Rings of Jupiter, Uranus, and Neptune

One of the many remarkable findings of the 1979 *Voyager* missions was the discovery of a faint ring of matter encircling Jupiter in the plane of the planet's equator (Figure 8.22). This ring lies roughly 50,000 km above the top cloud layer, inside the orbit of the innermost moon, and is probably composed of dark fragments of rock and dust chipped off the inner moons by meteorites. The outer edge of the ring is quite sharply defined, although a thin sheet of material may extend all the way down to Jupiter's cloud tops. In

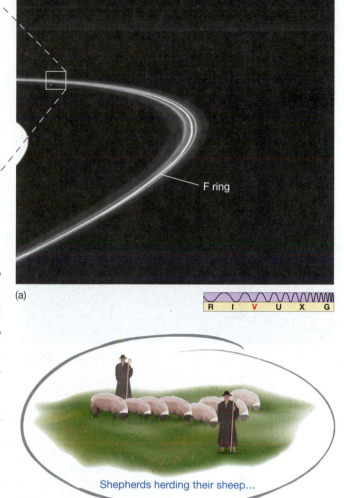

(b) Shepherd moon

(a)

R I V U X G

▲ **Figure 8.21 Shepherd Moon** (a) Saturn's narrow F ring appears to contain kinks and braids, making it unlike any of Saturn's other rings. Its thinness can be explained by the effects of two shepherd satellites that orbit the ring, one a thousand kilometers inside the ring and the other a thousand kilometers outside it. (b) One of the potato-shaped shepherding satellites, roughly 100 km across, can be seen at the right. The accompanying visual analogy depicts a couple of shepherds herding their sheep in a slightly wavy line. (NASA)

▶ **Figure 8.22 Jupiter's Faint Ring** Jupiter's faint ring as photographed by *Voyager 2*. Made up of dark fragments of rock and dust possibly chipped off the innermost moons by meteorites, the ring was unknown before the two *Voyager* spacecraft arrived at the planet. (NASA)

R I V U X G

▲ **Figure 8.23 Uranus's Rings** The main rings of Uranus, as imaged by *Voyager 2*. All nine of the rings known before *Voyager 2*'s arrival can be seen in this photo. Resolution is about 10 km, which is just about the width of most of these rings. The two rings discovered by *Voyager 2* are too faint to be seen here. The inset shows a closeup of the Epsilon ring, revealing some internal structure. (NASA)

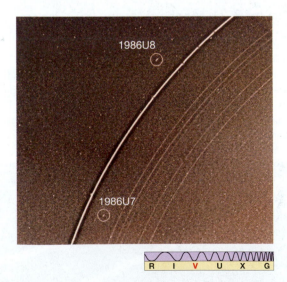

▲ **Figure 8.24 More Shepherd Moons** These two small moons, named Cordelia and Ophelia, were discovered by *Voyager 2* in 1986. They shepherd Uranus's Epsilon ring, keeping it from diffusing away. (NASA)

the direction perpendicular to the equatorial plane, the ring is only a few tens of kilometers thick.

The ring system surrounding Uranus was discovered in 1977, when astronomers observed the planet passing in front of a bright star, momentarily dimming the star's light. Such a **stellar occultation** happens a few times per decade and allows astronomers to measure planetary structures that are too small and faint to be detected directly. The ground-based observations revealed nine thin rings encircling Uranus. Two more were discovered by *Voyager 2*. The locations of all 11 rings are indicated in Figure 8.19.

The nine rings known from Earth-based observation are shown in Figure 8.23, a *Voyager 2* image that makes it readily apparent that Uranus's rings are quite different from those of Saturn. While Saturn's rings are bright and wide with relatively narrow, dark gaps between them, those of Uranus are dark, narrow, and widely spaced. However, like Saturn's rings, the rings of Uranus are all less than a few tens of meters thick (measured in the direction perpendicular to the ring plane). Like the F ring of Saturn, Uranus's narrow rings require shepherd satellites to keep them in place. During its 1986 flyby, *Voyager 2* detected the shepherds of the Epsilon ring (Figure 8.24). Many other, undetected shepherd satellites must also exist.

As shown in Figure 8.25, Neptune is surrounded by four dark rings. Three are quite narrow, like the rings of Uranus, and one is quite broad and diffuse, more like Jupiter's ring. The outermost ring is noticeably clumped in places. From Earth we see not a complete ring, but only partial arcs; the unseen parts of the ring are simply too thin (unclumped) to be detected. The connection between the rings and the planet's small inner satellites has not yet been firmly established, although many astronomers now think the clumping is caused by shepherd satellites.

The Formation of Planetary Rings

The dynamic behavior—waves, collisions, and interactions with moons—observed in the rings of Saturn and the other jovian planets suggests to many researchers that the rings are quite young—perhaps no more than 50 million years old, just 1 percent of the age of the solar system. If the rings are indeed so young, then either they are replenished from time to time, perhaps by fragments of moons chipped off by meteoritic impact, or they are the result of a relatively recent, possibly catastrophic event in the planet's system—a moon torn apart by tidal forces, or destroyed in an impact with a large comet or even with another moon.

Astronomers estimate that the total mass of Saturn's rings is enough to make a satellite about 250 km in diameter. If such a satellite strayed inside Saturn's Roche limit or was destroyed (perhaps by a collision) near that radius, a

◄ **Figure 8.25 Neptune's Faint Rings** In this long-exposure image, the planet (center) is heavily overexposed and has been artificially blotted out to make the rings easier to see. One of the two fainter rings lies between the inner bright ring and the planet. The other lies between the two bright rings. (NASA)

ring system could have resulted. While we do not know if Saturn's rings originated in this way, this is the likely fate of Neptune's large moon Triton. As mentioned earlier (in Section 8.2), Triton will be torn apart by Neptune's tidal gravitational field within 100 million years or so. By that time it is quite possible that Saturn's ring system will have disappeared, and Neptune will then be the planet in the solar system with spectacular rings!

◀◀◀ **CONCEPT CHECK**

What is the Roche limit, and what is its relevance to planetary rings?

8.5 Pluto and Its Moon

The Discovery of Pluto

In Chapter 7 we saw how irregularities in Uranus's orbit led to the discovery of Neptune in the mid-nineteenth century. ∞ *(Sec. 7.2)* By the end of that century, observations of the orbits of Uranus and Neptune suggested that Neptune's influence was not sufficient to account for all of the observed irregularities in Uranus's motion. Furthermore, it seemed that Neptune itself was being affected by some other unknown body. Astronomers hoped to pinpoint the location of this new planet using techniques similar to those that had guided them to the discovery of Neptune.

One of the most ardent searchers was Percival Lowell, a capable and persistent observer and one of the best-known astronomers of his day. ∞ *(Discovery 6-1)* Basing his investigation primarily on the motion of Uranus (Neptune's orbit was still relatively poorly determined at the time), Lowell set about calculating where the supposed ninth planet should be. He searched for it, without success, during the decade preceding his death in 1916. In 1930 the American astronomer Clyde Tombaugh, working with improved equipment and photographic techniques at the Lowell Observatory, finally succeeded in finding Lowell's ninth planet, only 6° away from Lowell's predicted position. The new object was named Pluto for the Roman god of the dead, who reigns over eternal darkness (and also because its first two letters and its astrological symbol ♇ are Lowell's initials).

Unfortunately, it now appears that the supposed irregularities in Neptune's orbit and the extra irregularities in the motion of Uranus do not really exist. Furthermore, the mass of Pluto, measured accurately only in the 1980s, is far too small to have caused them anyway. What's more, Pluto was recently reclassifed as a member of the Kuiper belt and is no longer considered a major planet at all.

At nearly 40 AU from the Sun, Pluto is often hard to distinguish from the background stars. Like Neptune, it is never visible to the naked eye. Pluto is among the largest objects in the solar system not studied at close range by unmanned spacecraft, although NASA has launched a 2006 mission to Pluto and the Kuiper belt, currently scheduled to reach the outer solar system in 2015. For now, though, every new discovery about this remote world is the result of painstaking observations either from Earth or from Earth-orbiting instruments. Figure 8.26 is a composite map of Pluto's surface obtained by combining many *Hubble Space Telescope* images. It shows surface detail at about the same level as can be seen on Mars with a small telescope. Other than the bright polar caps, none of the

Animation/Video
Pluto Viewed by HST

▼ **Figure 8.26 Pluto** (a) A surface map of Pluto—not a photograph, but rather a "modeled" view created by carefully combining 24 *Hubble Space Telescope* images with a mathematical description of the surface. (b) A computer-generated three-dimensional representation of the same data. (NASA)

(a)

(b)

R I V U X G

(a)

(b)

R   I   V   U   X   G

▲ **Figure 8.27 Pluto and Charon**   (a) The discovery photograph of Pluto's moon Charon. The moon is the small bump on the top right portion of the image. (b) The Pluto–Charon system, as seen by the *Hubble Space Telescope*. The angular separation of the planet and its moon is about 0.9 arc seconds. (U.S. Naval Observatory; NASA)

Animation/Video
Historical Observations of Pluto

Animation/Video
Orbits of Neptune and Pluto

CONCEPT CHECK ▶▶▶

What do Pluto and Triton have in common?

---

surface features has been conclusively identified. Some of the dark regions may be craters or impact basins, as on Earth's Moon.

The Pluto–Charon System

In 1978 astronomers at the U.S. Naval Observatory discovered that Pluto has a satellite. It is now named Charon, after the mythical boatman who ferried the dead across the river Styx into Hades, Pluto's domain. The discovery photograph of Charon is shown in Figure 8.27(a). Charon is the large white bump at the top right of the image. Figure 8.27(b) shows a 2005 *Hubble Space Telescope* image that clearly resolves the two bodies. Based on studies of Charon's orbit, astronomers have measured the mass of Pluto to great accuracy. It is just 0.0021 Earth masses ($\sim 10^{22}$ kg)—or about 0.17 times the mass of Earth's Moon.

By pure chance, Charon's orbit over the six-year period from 1985 to 1991 produced a series of *eclipses* as Pluto and Charon repeatedly passed in front of one another, as seen from Earth. With more good fortune, these eclipses took place while Pluto was closest to the Sun, making for the best possible Earth-based observations. Basing their calculations on the variations in light as Pluto and Charon periodically hid each other, astronomers computed the masses and radii of both bodies and determined their orbit plane. The mass of Charon is about 0.12 that of Pluto, giving the Pluto–Charon system by far the largest satellite-to-object mass ratio in the solar system. Pluto's radius is 1150 km, about one-fifth that of Earth. Charon is about 600 km in radius and orbits 19,600 km from Pluto.

By studying sunlight reflected from their surfaces, astronomers also found that Pluto and Charon are tidally locked as they orbit each other. Their orbital period, and the rotation period of each, is 6.4 days. As shown in Figure 8.28, Charon's orbit and the spin axes of both objects are inclined at an angle of 118° to the plane of the ecliptic.

In late 2005, scientists using *HST* reported the discovery of two additional small moons, perhaps 100-200 km in diameter, orbiting Pluto at about twice the distance of Charon. As of early 2006, they have yet to be officially confirmed (see Figure 8.27).

Pluto's Origin

Pluto's average density of 2000 kg/m³ is too low for a terrestrial planet but far too high for a jovian one. Instead, the mass, radius, and density of Pluto are just what we expect for one of the icy moons of a jovian planet. Indeed, Pluto is comparable in both mass and radius to Neptune's large moon, Triton.

This similarity has fueled much speculation, both about Pluto's origin and about the strange state of Neptune's moon system. However, despite considerable effort, theorists have never been able to construct a plausible sequence of events that simultaneously accounts for Pluto's eccentric, inclined orbit and Triton's unusual retrograde path around Neptune. Instead, it now seems much more likely that the similarities between the two bodies stem from the fact that both formed and grew by accretion in the same environment—the Kuiper belt. ∞ *(Sec. 4.3)*

Many researchers accept that there may have been thousands of Pluto-size objects initially present in the outer solar system. Most have long since been ejected from the Kuiper belt—into the inner solar system or out to the Oort cloud—or, like Triton, have been captured by or collided with one of the outer planets. In this view, like Ceres in the inner asteroid belt, Pluto is simply one of the largest surviving members of the class (see Section 8.6). Charon

▲ **Figure 8.28 Pluto–Charon Orbit**
Charon's path around Pluto is circular, synchronous, and inclined at 118° to the orbit plane of the Pluto–Charon system about the Sun. The Pluto–Charon orbit plane is itself inclined at 17.2° to the plane of the ecliptic.

was captured by Pluto late in the formation process, following a collision (or near-miss) between the two, in a manner reminiscent of the formation of Earth's own Moon. ∞ *(Sec. 5.8)*

8.6 Pluto's Demotion from Major Planet Status

Let's review what we have learned about Pluto and its environment. Pluto was discovered in the 1930s after a search for a hypothetical perturber of Neptune's orbit. We now know that Pluto is much smaller and less massive than was originally supposed and could not possibly perturb Neptune's orbit in the way imagined. Nevertheless, and despite the misgivings of some astronomers of the day, Pluto became established as the ninth planet of our solar system.

In the 1950s astronomer Gerard Kuiper proposed that the source of short-period comets was a belt—now known as the Kuiper belt—of small bodies orbiting beyond the orbit of Neptune, some 40–50 AU from the Sun. ∞ *(Sec. 4.2)* These Kuiper belt objects, or KBOs, are small and dark, and very hard to see. Only in the 1990s were the first KBOs actually observed (see Figure 8.29). More than 700 are now known; probably more than 100,000 exist in total.

As the numbers of observed KBOs have increased, astronomers have built up a better understanding of their sizes and orbits. We now know, for example,

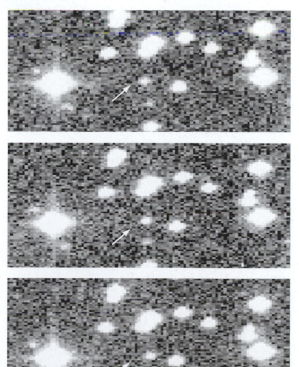

▶ **Figure 8.29 Kuiper Belt Object**   Some of the best available images of a Kuiper belt object. Known as Pholus, the object itself is the fuzzy blob (marked with an arrow) that changes position between one frame and the next. It may be almost 1000 km across, and lies more than 40 AU from Earth. (LPL)

R I V U X G

Varuna

Ixian Sedna

Quaoar

2000 km

2003 UB313

Pluto

Moon

Earth

▲ **Figure 8.30 KBOs Compared** Some large Kuiper belt objects, including the largest known, called Quaoar, with Pluto, Earth, the Moon, and the possible Oort cloud object Sedna added for scale. All KBO diameters are approximate, as is that of Sedna; they are estimated from the object's observed brightness. (NASA; Caltech)

that about 25 percent of all KBOs have the *same* orbital period as Pluto. This is not a coincidence, of course—it is caused by gravitational interactions between these bodies and the planet Neptune, maintaining all their orbital periods at precisely 1.5 times that of Neptune. And it is also becoming clear as the details are filled in that there is no real distinction between the Pluto and the largest KBOs.

Figure 8.30 shows the diameters of a few large members of the Kuiper belt, as well as Earth and the Moon for scale. The figure also includes two intriguing new objects. One, called Sedna, was discovered in 2003 and is thought to have a diameter of about 1500 km. Although it currently lies within the Kuiper belt, it is probably not really a KBO, as its highly elliptical orbit takes it out almost to the (theoretical) inner edge of the Oort cloud at a distance of some 1000 AU. The other is 2003UB313, a KBO apparently larger than Pluto and currently the most distant solar-system object known. ∞ *(Sec. 4.2)* Its confirmation in 2005 marked the beginning of the end of Pluto's tenure as a planet. Looking at the figure (or the underlying data), it is hard to avoid the conclusion that these bodies, along with Pluto, are simply the largest known members of the KBO class—the "Kings of the Kuiper belt" perhaps, but not true planets. And even larger objects at greater distances from the Sun are not ruled out observationally. Indeed, most professional astronomers now view Pluto as a *dwarf planet*, playing much the same role in the Kuiper belt as Ceres (or perhaps Pallas) does among the asteroids. ∞ *(Sec. 4.2)*

Some people are disturbed at Pluto's recent demotion from the solar system "A list." After all, Pluto orbits the Sun, has enough mass to be spherical due to its own gravity, and even has its own moon. So what is the problem with calling it a planet? The answer is that Pluto *is* still a minor planet, in the same sense that all KBOs are minor planets—all are remnants of a long-gone protoplanetary disk, and all orbit the Sun as independent bodies. ∞ *(Sec. 4.3)* However, KBOs are not *major* planets, and Pluto is no longer sufficiently different from them to warrant inclusion in a separate category. Few astronomers doubt that if Pluto were discovered today, it would be immediately classified as a member of the Kuiper belt.

Pluto's recent reclassification illustrates the way in which science evolves. Our conception of the cosmos has undergone many changes—some radical, others more gradual—since the time of Copernicus. ∞ *(Sec. 1.2)* We have seen several examples already, and we will see many more later in this book. As our understanding grows, our terminology and classifications change. The situation with Pluto has a close parallel in the discovery of the first asteroids in the early nineteenth century. They, too, were initially classified as planets—indeed, in the 1840s, leading astronomy texts listed no fewer than 11 planets in our solar system, including as numbers 5 through 8 the asteroids Vesta, Juno, Ceres, and Pallas. Within a couple of decades, however, the discovery of several dozen more asteroids had made it clear that these small bodies represented a whole new class of solar system objects, separate from the major planets, and the number of major planets fell to eight (including the newly discovered Neptune).

Much of the observational work on the Kuiper belt began as a search for a tenth planet. It is ironic that the end result of these efforts has been a reduction in the number of "true" (that is, major) solar system planets back to eight.

◄◄◄ **CONCEPT CHECK**

Why do many astronomers now feel that Pluto is not a major planet?

CHAPTER REVIEW

Summary

1 There are six large moons in the outer solar system, each comparable to or larger than Earth's Moon in size and mass. In addition, there are 12 medium-sized moons and many smaller ones. The four Galilean moons of Jupiter have densities that decrease with increasing distance from the planet. The innermost, Io, has active volcanoes powered by the constant flexing of the moon by Jupiter's tidal forces. Europa has a cracked, icy surface that may well conceal an ocean of liquid water. Ganymede and Callisto have heavily cratered surfaces. Ganymede's magnetic field implies relatively recent geological activity, but Callisto shows none.

2 Saturn's large moon Titan has a thick atmosphere that obscures the moon's surface and may be the site of complex cloud and surface chemistry. The moon's surface is so cold that water behaves like rock and liquid methane flows like water. Sensors aboard *Cassini* have allowed mission scientists to map the moon's surface for the first time, revealing evidence for ongoing erosion and volcanic activity. The *Huygens* probe landed on the icy surface and photographed what may be channels carved by flowing methane.

3 Neptune's moon Triton has a fractured surface of water ice and a thin atmosphere of nitrogen, probably produced by nitrogen "geysers" on its surface. Triton is the only large moon in the solar system to have a retrograde orbit around its parent planet. This orbit is unstable and will eventually cause Triton to be torn apart by Neptune's gravity.

4 The medium-sized moons of Saturn and Uranus are made up predominantly of rock and water ice. Many of them are heavily cratered and in some cases must have come close to being destroyed by the meteoritic impacts whose craters we now see. Some show evidence of recent or ongoing geological activity, although in most cases the cause of the activity is not known.

5 From Earth the main visible features of Saturn's rings are the **A, B,** and **C rings,** the **Cassini Division,** and the **Encke gap**

(p. 226). The Cassini Division is a dark region between the A and B rings. The Encke gap lies near the outer edge of the A ring. The rings are made up of trillions of individual particles, ranging in size from dust grains to boulders. Their total mass is comparable to that of a small moon. Saturn's rings are made up of tens of thousands of narrow **ringlets** (p. 228). Interactions between the ring particles and the planet's inner moons are responsible for much of the fine structure observed. Saturn's narrow **F ring** (p. 229) lies just outside the A ring. It has a kinked, braided structure, apparently caused by two small **shepherd satellites** (p. 228) that orbit close to it and prevent it from breaking up.

6 The **Roche limit** (p. 228) of a planet is the distance within which the planet's tidal field overwhelms the internal gravity of an orbiting moon, tearing the moon apart and forming a ring. All the rings of the four jovian planets lie inside (or close to) their parent planets' Roche limits.

7 Jupiter has a faint, dark ring extending down to the planet's cloud tops. Uranus has a series of dark, narrow rings first detected from Earth by **stellar occultation** (p. 230), which occurs when a body passing between Earth and some distant star obscures the light received from that star. Shepherd satellites keep Uranus's rings from breaking apart. Neptune has three narrow rings like Uranus's and one broad ring, like Jupiter's.

8 Pluto was discovered in 1930 after a laborious search for a planet that was supposedly affecting Uranus's orbital motion. We now know that Pluto is far too small to have had any detectable influence on Uranus's path. Pluto has a moon, Charon, the mass of which is about one-eighth that of Pluto. Studies of Charon's orbit around Pluto have allowed the masses and radii of both bodies to be accurately determined. Pluto's properties are far more moonlike than planetlike. Triton, Pluto, Charon, and the newly discovered 2003UB313 are all examples of large Kuiper belt objects.

Review and Discussion

1. How does the density of the Galilean moons vary with increasing distance from Jupiter? Is there a trend to this variation? If so, explain the reason for it.

2. What is unusual about Jupiter's moon Io?

3. Why is there speculation that the Galilean moon Europa might be an abode for life?

4. Why do astronomers think that Ganymede may have experienced a relatively recent episode of internal heating?

5. What property of Saturn's largest moon, Titan, makes it of particular interest to astronomers?

6. What is the predicted fate of Triton?

7. What is unique about Miranda? Give a possible explanation for this uniqueness.

8. Seen from Earth, Saturn's rings sometimes appear broad and bright but at other times seem to disappear. Why?

9. Why do many astronomers think Saturn's rings formed quite recently?

10. What effect does Mimas have on the rings of Saturn?

11. Describe the behavior of shepherd satellites.

12. How do the rings of Neptune differ from those of Uranus and Saturn?

13. Why did astronomers recently demote Pluto as a major planet?

14. In what respect is Pluto more like a moon than a jovian or terrestrial planet?

15. How was the mass of Pluto determined?

Conceptual Self-Test True or False?/Multiple Choice

1. Scientists speculate that Europa may have liquid water below its frozen surface. **(T/F)**

2. The composition of Saturn's ring particles is predominantly water ice. **(T/F)**

3. Saturn's rings lie within the planet's Roche limit. **(T/F)**

4. Two small shepherd satellites are responsible for the unusually complex form of Saturn's F ring. **(T/F)**

5. Titan's surface is obscured by thick clouds of ammonia ice. **(T/F)**

6. Uranus has no large moons. **(T/F)**

7. Triton's orbit is unusual because it is retrograde. **(T/F)**

8. Pluto was discovered via its gravitational effect on Neptune. **(T/F)**

9. Io's surface appears very smooth because it **(a)** is continually resurfaced by volcanic activity; **(b)** is covered with ice; **(c)** has been shielded by Jupiter from meteorite impacts; **(d)** is liquid.

10. The Galilean moons of Jupiter are sometimes described as a miniature inner solar system because **(a)** there are the same number of Galilean moons as there are terrestrial planets; **(b)** the moons have generally "terrestrial" composition; **(c)** the moons' densities decrease with increasing distance from Jupiter; **(d)** the moons all move on circular, synchronous orbits.

11. The atmospheric pressure at the surface of Titan is **(a)** less than; **(b)** about the same as; **(c)** about 1.5 times greater than; **(d)** about 16 times greater than the atmospheric pressure at Earth's surface.

12. The rings of Uranus **(a)** are broad and bright; **(b)** are narrow and dark; **(c)** are narrow and bright; **(d)** cannot be detected from Earth.

13. The solar system object most similar to Pluto is **(a)** Mercury; **(b)** the Moon; **(c)** Titan; **(d)** Triton.

14. In Figure 8.11 (Titan's Atmosphere), the atmospheric temperature at the top of the main haze layer is approximately **(a)** 50 K; **(b)** 100 K; **(c)** 150 K; **(d)** 200 K.

15. From Figure 8.17 (Saturn's Tilt), the next time Saturn's rings will appear roughly edge-on as seen from Earth will be around **(a)** 2004; **(b)** 2010; **(c)** 2020; **(d)** 2035.

Problems

 Algorithmic versions of these questions are available in the Practice Problems module of the Companion Website.

The number of squares preceding each problem indicates its approximate level of difficulty.

1. ■ What is the orbital speed, in kilometers per second, of ring particles at the inner edge of Saturn's B ring? Compare with the speed of a satellite in low Earth orbit (500 km altitude, say). Why are these speeds so different? ∞ *(More Precisely 1-1)*

2. ■■ Io orbits Jupiter at a distance of six planetary radii from the planet's center once every 42 hours. If Jupiter completes one rotation every 10 hours, use Kepler's third law to calculate how far from the center a satellite must orbit in order to appear stationary above the planet. ∞ *(Sec. 1.3)*

3. ■■■ Calculate the difference between the acceleration due to Jupiter's gravity at Io's center and the acceleration on Io's surface at the point closest to the planet. This is the tidal acceleration exerted by Jupiter on the moon. ∞ *(Sec. 5.2)* Compare this with Io's surface gravity. Repeat the comparison for Earth and Earth's Moon.

4. ■ Show that Titan's surface gravity is about one-seventh that of Earth. What is Titan's escape speed? ∞ *(More Precisely 5-2)*

5. ■■ Compare the apparent sizes of the Galilean moons, as seen from Jupiter's cloud tops, with the angular diameter of the Sun at Jupiter's distance. Would you expect ever to see a total solar eclipse from Jupiter's cloud tops? ∞ *(More Precisely E-2)*

6. ■■ Assuming a spherical shape and a uniform density of 2000 kg/m³, calculate how small an icy moon of one of the outer planets would have to be before a fastball pitched at 40 m/s (about 90 mph) could escape.

7. ■ The total mass of material in Saturn's rings is about 10^{15} tons (10^{18} kg). Suppose the average ring particle is 6 cm in radius (a large snowball) and has a density of 1000 kg/m³. How many ring particles are there?

8. ■ How close is Charon to Pluto's Roche limit?

9. ■ What would be your weight on Pluto? On Charon?

10. ■■■ What is the round-trip travel time of light from Earth to Pluto (at a distance of 40 AU)? How far would a spacecraft moving in a circular orbit 500 km above Pluto's surface travel during that time? To how many orbits does this correspond?

 The Companion Website at www.prenhall.com/chaisson provides algorithmically generated versions of each chapter's Problems, in addition to an Animations and Videos gallery, a Tutorials gallery and Tutorials Quiz questions, a full eBook, an Images gallery, Practice Questions, Multiple Choice, and True or False quizzes, and Labeling exercises.

PART 3 The Stars

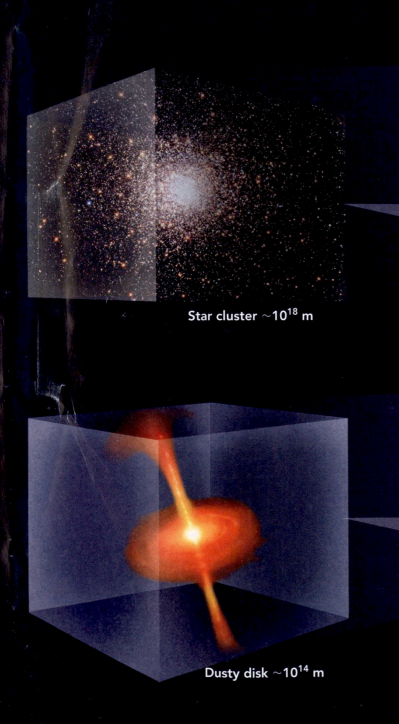

Star cluster ~10^{18} m

Dusty disk ~10^{14} m

The Sun—comprising 99.9% of all the matter in our solar system—is an entirely different kind of astronomical object from any planet or moon. The Sun is not even remotely similar to any of the bodies we have studied thus far, for the Sun is a star. Its size, mass, and energy output simply defy terrestrial comparison.

Yet despite these extreme properties, astronomers regard the Sun as a typical star, having average qualities among the family of all stars seen in the universe. And despite its fundamentally different internal structure from any planet, the Sun is a relatively simple ball of gas that astronomers have come to understand remarkably well. Partly because the Sun is so relatively near and we can study it up close (well almost, as it still resides 8 light-minutes away, or 150 million kilometers), it is often said that we know more about the Sun than we do about the Earth.

All the other stars we see in the nighttime sky as small, twinkling points of light are much, much farther away—a realization that became clear only in the twentieth century. There, among the vast swath of billions of

Nearly a trillion trillion stars inhabit our universe; the Sun is but one of these stars.

Galaxy ~10^{21} m

stars known collectively as the Milky Way, astronomers have also found giant stars and dwarf stars, dusty disks and gassy nebulae, and not least probable, black holes. Stars together do resemble a family, or perhaps a zoo—young and old members, bright and dim ones, indeed some more temperamental than others.

Most remarkable of all, stars have evolutionary cycles—they are born, mature, and die. Stars originate in huge clouds of gas and dust, ignite much like the Sun, and eventually terminate by returning to space some of the heavy elements created within their cores. The poetic notion that we are made of stardust is a bit romantic—but it also happens to be true!

Part 3 describes many diverse stellar and interstellar regions that often resemble a "galactic ecosystem"—an evolutionary balance nearly as intricate and delicate as life in a tide pool or a tropical forest. With the theory of stellar evolution, astronomers have discovered a powerful, unifying theme with which to study all stars.

Nebula ~10^{16} m

Red giant ~10^{11} m

9

The Sun

Our Parent Star

The Sun is our star—the main source of energy that powers weather, climate, and life on Earth. Humans simply would not exist without the Sun. Although we take it for granted each and every day, the Sun is of great importance to us in the cosmic scheme of things. This spectacular image shows a small piece of the Sun "up close"—a high-resolution photograph taken from Earth, revealing its granulated surface and dark spots, all of which are gas. The region shown is 50,000 kilometers on a side, equal to about four times the size of Earth, but only a few percent of the Sun's complete surface. (Royal Swedish Academy of Sciences)

Living in the solar system, we have the chance to study, at close range, perhaps the most common type of cosmic object—a star. Our Sun is a star, and a fairly average star at that, but with one unique feature: It is very close to us—some 300,000 times closer than our next nearest neighbor, Alpha Centauri. Whereas Alpha Centauri is 4.3 light-years distant, the Sun is only 8 light-minutes away from us. Consequently, astronomers know far more about the properties of the Sun than about any of the other distant points of light in the universe. Indeed, a good fraction of all our astronomical knowledge is based on modern studies of the Sun. Just as we studied our parent planet, Earth, to set the stage for our exploration of the solar system, we now study our parent star, the Sun, as the next step in our exploration of the universe.

LEARNING GOALS

Studying this chapter will enable you to:

1 Summarize the overall properties and internal structure of the Sun.

2 Describe the concept of luminosity and how it is measured.

3 List and describe the outer layers of the Sun.

4 Explain how studies of the solar surface tell us about the Sun's interior.

5 Discuss the nature of the Sun's magnetic field and its relationship to the various types of solar activity.

6 Outline the process by which energy is produced in the Sun's interior.

7 Explain how observations of the Sun's core have challenged our understanding of solar physics.

Visit www.prenhall.com/chaisson for additional annotated images, animations, and links to related sites for this chapter.

Tutorial
SuperSpaceship—Voyage to the Sun

▼ Figure 9.1 The Sun The inner part of this composite, filtered image of the Sun shows a sharp edge, although our star, like all stars, is made of a gradually thinning gas. The edge appears sharp because the solar photosphere is so thin. The outer portion of the image is the solar corona, normally too faint to be seen, but visible during an eclipse, when the light from the solar disk is blotted out. (Note the blemishes; they are sunspots.) (NOAO)

R I V U X G

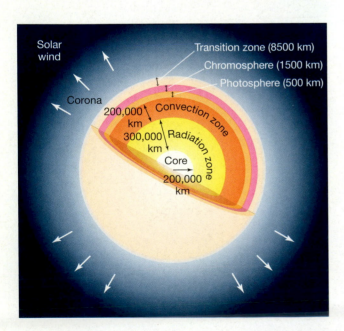

9.1 The Sun in Bulk

The Sun is the sole source of light and heat for the maintenance of life on Earth. It is a **star**, a glowing ball of gas held together by its own gravity and powered by nuclear fusion (where two or more atomic nuclei combine, releasing energy in the process—see Section 9.5) at its center. In its physical and chemical properties, the Sun is very similar to most other stars, regardless of when and where they formed. Far from detracting from our interest in the Sun, this very mediocrity is one of the main reasons we study it—most of what we know about the Sun also applies to many other stars in the sky. The Sun is the best-studied star in the universe, and it is no exaggeration to say that our knowledge of the cosmos rests squarely on our understanding of solar physics.

Overall Structure

Table 9.1 lists some basic solar data. Having a radius of more than 100 Earth radii, a mass of more than 300,000 Earth masses, and a surface temperature well above the melting point of any known material, the Sun is clearly a body very different from any other we have encountered so far.

The Sun has a surface of sorts—not a solid surface (the Sun contains no solid material), but rather that part of the brilliant gas ball we perceive with our eyes or view through a heavily filtered telescope. The part of the Sun that emits the radiation we see is called the **photosphere** (after the Greek word *photos*, meaning "light"). Its radius (listed as the radius of the Sun in Table 9.1) is about 700,000 km. However, for reasons to be discussed in a moment, its thickness is probably no more than 500 km—less than 0.1 percent of the radius—which is why we perceive the Sun as having a well-defined, sharp edge (Figure 9.1).

The main regions of the Sun are illustrated in Figure 9.2. Just above the photosphere is the Sun's lower atmosphere, called the **chromosphere**, about 1500 km thick. Above that lies a region called the **transition zone**, where the temperature rises dramatically. Above 10,000 km, and stretching far beyond, is a thin, hot upper atmosphere, the solar **corona**. At still greater distances the corona turns into the **solar wind**, which flows away from the Sun and permeates the entire solar system. ∞ *(Sec. 4.2)*

Extending down some 200,000 km below the photosphere is the **convection zone**, a region where the material of the Sun is in constant convective motion. ∞ *(Sec. 5.3)* Below the convection zone lies the **radiation zone**, where solar energy is transported toward the surface by radiation rather than by convection. The term *solar interior* is often used to mean both the radiation and convection zones. The central **core**, roughly 200,000 km in radius, is the site of powerful nuclear fusion reactions that generate the Sun's enormous energy output.

Solar rotation can be measured by timing sunspots and other surface features as they cross the solar disk. ∞ *(Sec. 1.2)* These observations indicate that the Sun rotates in about a month, but it does not do so as a solid body. Instead, it spins differentially—faster at the equator and slower at the poles, like Jupiter and Saturn. The equatorial rotation period is about 25 days. Sunspots are never seen above latitude 60° (north or south), but at that latitude they indicate a 31-day period. Other measurement techniques, such as those described in Section 9.2, reveal that the rotation period continues to increase as we approach the poles. The polar period may be as long as 36 days.

◀ Figure 9.2 Solar Structure The main regions of the Sun, not drawn to scale, with some physical dimensions labeled. The photosphere is the visible "surface" of the Sun. Below it lie the convection zone, the radiation zone, and the core. Above the photosphere, the solar atmosphere consists of the chromosphere, the transition zone, and the corona.

Luminosity

The Sun radiates an enormous amount of energy into space. We can measure the Sun's total energy output in two stages, as follows. First, by holding a light-sensitive device—a solar cell, perhaps—perpendicular to the Sun's rays above Earth's atmosphere, we can measure how much solar energy is received per square meter of area every second at Earth's distance from the Sun. This quantity, known as the **solar constant**, is approximately 1400 watts per square meter (W/m^2).[1] About 50–70 percent of this energy reaches Earth's surface; the rest is intercepted by the atmosphere (30 percent) or reflected away by clouds (0–20 percent). Thus, for example, a sunbather's body having a total surface area of about 0.5 m^2 receives solar energy at a rate of roughly 500 watts, approximately equivalent to the output of a small electric room heater or five 100-W lightbulbs.

Knowing the solar constant, we can calculate the *total* amount of energy radiated in all directions from the Sun. Imagine a three-dimensional sphere that is centered on the Sun and is just large enough that its surface intersects Earth's center (Figure 9.3). The sphere's radius is one AU, and its surface area is therefore $4\pi \times (1\ \text{AU})^2$, or approximately 2.8×10^{23} m^2. Assuming that the Sun radiates energy equally in all directions, we can determine the total rate at which energy leaves the Sun's surface simply by multiplying the solar constant by the total surface area of our imaginary sphere. The result is just under 4×10^{26} W. This quantity is called the **luminosity** of the Sun. As we will see, it is quite typical of the luminosities of many of the stars in the sky.

Take a moment to consider the magnitude of the solar luminosity. The Sun is an enormously powerful source of energy. *Every second* it produces an amount of energy equivalent to the detonation of about 100 billion 1-megaton nuclear bombs. There is simply nothing on Earth comparable to a star!

9.2 The Solar Interior

How have we learned about the inside of the Sun? After all, the sunlight we see comes from the photosphere; none of the radiation from the interior reaches us directly. In this section we discuss in a little more detail how we know what we know about conditions below the solar surface.

Lacking any direct measurements of the Sun's interior, researchers must use indirect means to probe in more detail the inner workings of our parent star. To accomplish this, they construct *mathematical models* of the Sun, combining all available observations with theoretical insight into solar physics. The result is the **standard solar model**, which has gained widespread acceptance among astronomers. We can construct similar models of other stars, but because the Sun is so close and well studied, the standard solar model is by far the best tested.

Modeling the Structure of the Sun

The Sun's bulk properties—its mass, radius, temperature, and luminosity—do not vary much from day to day or from year to year. Although we will see in Chapter 12 that stars like the Sun do change significantly over periods of *billions* of years, for our purposes here this slow evolution may be ignored. On "human" time scales, the Sun may reasonably be thought of as unchanging.

---

[1] The SI unit of energy is the *joule* (J). Probably more familiar to most readers is the closely related unit *watt* (W), which measures *power*, defined as the rate at which energy is emitted or expended by an object. One watt is equal to one joule per second. For example, a lightbulb with a power rating of 100 W radiates 100 J of energy each second.

TABLE 9.1	Some Solar Properties
Radius	696,000 km
Mass	1.99×10^{30} kg
Average density	1410 kg/m^3
Rotation period	25.1 days (equator); 30.8 days (60° latitude) 36 days (poles) 26.9 days (interior)
Surface temperature	5780 K
Luminosity	3.86×10^{26} W

◀◀◀ **CONCEPT CHECK**

Why must we assume that the Sun radiates equally in all directions when computing the solar luminosity from the solar constant?

▲ **Figure 9.3 Solar Luminosity** We can draw an imaginary sphere around the Sun so that the sphere's surface passes through Earth's center. The radius of this imaginary sphere is one AU. The "solar constant" is the amount of power striking a 1 m^2 detector at Earth's distance, as shown in the inset. By multiplying the sphere's surface area by the solar constant, we can measure the Sun's luminosity, the amount of energy it emits each second.

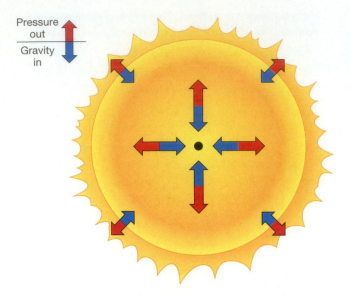

Pressure out / Gravity in

▲ **Figure 9.4 Stellar Balance** In the interior of a star such as the Sun, the outward pressure of hot gas balances the inward pull of gravity. This is true at every point within the star, guaranteeing its stability.

Accordingly, as illustrated in Figure 9.4, theoretical models generally begin by assuming that the Sun is in a state of **hydrostatic equilibrium**, in which pressure's outward push exactly counteracts gravity's inward pull. This stable balance between opposing forces is the basic reason that the Sun neither collapses under its own weight nor explodes into interstellar space. ∞ *(More Precisely 5-1)* It also has an important consequence for the solar interior. Because the Sun is very massive, its gravitational pull is very strong, so very high internal pressure is needed to maintain hydrostatic equilibrium. This high pressure in turn requires a very high central temperature, a fact crucial to our understanding of solar energy generation (Section 9.5). In this way the assumption of hydrostatic equilibrium lets us determine the density and temperature inside the Sun. This information, in turn, allows the model to make predictions about other observable solar properties—luminosity, radius, spectrum, and so on—and the internal details of the model are fine-tuned until the predictions agree with observations. This is the scientific method at work; the standard solar model is the result. ∞ *(Sec. E.5)*

To test and refine the standard solar model, astronomers are eager to obtain information about the solar interior. However, with so little direct information about conditions below the photosphere, we must rely on more indirect techniques. In the 1960s, measurements of the Doppler shifts of solar spectral lines revealed that the surface of the Sun oscillates, or vibrates, like a complex set of bells. ∞ *(Sec. 2.7)* These vibrations, illustrated in Figure 9.5, are the result of internal pressure ("sound") waves that reflect off the photosphere and repeatedly cross the solar interior. These waves penetrate deep inside the Sun, and analysis of their surface patterns allows scientists to study conditions far below the Sun's surface. This process is similar to the way in which seismologists study Earth's interior by observing seismic waves produced by earthquakes. ∞ *(Sec. 5.4)* For this reason, study of solar surface patterns is usually called **helioseismology**, even though solar pressure waves have nothing whatever to do with solar seismic activity (which doesn't exist, as the Sun is entirely gaseous).

The most extensive study of solar vibrations is the ongoing GONG (short for Global Oscillations Network Group) project. By making continuous observations of the Sun from many clear sites around Earth, solar astronomers have obtained uninterrupted high-quality solar data spanning weeks at a time. The *Solar and Heliospheric Observatory* (*SOHO;* see *Discovery 9-1*), launched by the European Space Agency in 1995 and now permanently stationed between Earth and the Sun some 1.5 million km from our planet, provides continuous monitoring of the Sun's surface and atmosphere. Analysis of the data from all these sources provides important detailed information about the temperature, density, rotation, and convective state of the solar interior. Direct comparison with theory is possible throughout a large portion of the Sun, and the agreement between model and observations is spectacular—the frequencies and wavelengths of observed solar oscillations are within 0.1 percent of the predictions of the standard solar model.

Figure 9.6 shows the solar density and temperature according to the standard solar model, plotted as functions of distance from the Sun's center. Notice

◄ **Figure 9.5 Solar Oscillations** (a) The Sun has been found to vibrate in a very complex way. By observing the motion of the solar surface, scientists can determine the wavelengths and frequencies of the individual waves and deduce information about the Sun not obtainable by other means. The alternating patches represent gas moving down (red) and up (blue). (See also *Discovery 9-1*.) (b) Depending on their initial directions, the waves contributing to the observed oscillations may travel deep inside the Sun, providing vital information about the solar interior. (National Solar Observatory)

(a)   (b)

how the density drops sharply at first, then more slowly near the photosphere. The variation in density is large, ranging from a core value of about 150,000 kg/m³, 20 times the density of iron, to an extremely small photospheric value of 2×10^{-4} kg/m³, about 10,000 times less dense than air at Earth's surface. The *average* density of the Sun (Table 9.1) is 1400 kg/m³, about the same as the density of Jupiter.

As shown in Figure 9.6, the solar temperature also decreases with increasing radius, but not as rapidly as the density. Computer models indicate a central temperature of about 15 million K. The temperature decreases steadily, reaching the observed value of 5800 K at the photosphere.

Energy Transport

The very hot solar interior ensures violent and frequent collisions among gas particles. In and near the core, the extremely high temperatures guarantee that the gas is completely ionized. ∞ *(Sec. 2.6)* With no electrons left on atoms to capture photons and move into more excited states, the deep solar interior is quite transparent to radiation. Only occasionally does a photon scatter off a free electron. The energy produced by nuclear reactions in the core travels outward toward the surface in the form of radiation with relative ease.

As we move outward from the core, the temperature falls, and eventually some electrons can remain bound to nuclei. With more and more atoms retaining electrons that can absorb the outgoing radiation, the gas in the interior changes from being relatively transparent to being almost totally opaque. By the outer edge of the radiation zone, 500,000 km from the center, *all* of the photons produced in the Sun's core have been absorbed. Not one of them reaches the surface. But what happens to the energy they carry?

The photon's energy must travel beyond the solar interior—the fact that we see sunlight proves that the energy escapes. In fact, that energy is carried to the solar surface by *convection*—the same basic physical process we saw in our study of Earth's atmosphere, although operating in a very different environment in the Sun. ∞ *(Sec. 5.3)* Convection can occur whenever cooler material overlies warmer material, and this is just what happens in the outer part of the Sun's interior. Hot solar gas physically moves outward, while cooler gas above it sinks, creating a characteristic pattern of convection cells. All through the convection zone, energy is transported to the surface by physical motion of the solar gas. (Note that this actually represents a departure from hydrostatic equilibrium, as defined above, but it can still be handled within the standard solar model.) Remember that there is no physical movement of material when radiation is the energy-transport mechanism; convection and radiation are *fundamentally different* ways in which energy can be transported from one place to another.

Figure 9.7 is a schematic diagram of the solar convection zone. There is a hierarchy of convection cells, organized in tiers at different depths. The deepest tier, about 200,000 km below the photosphere, is thought to contain cells tens of

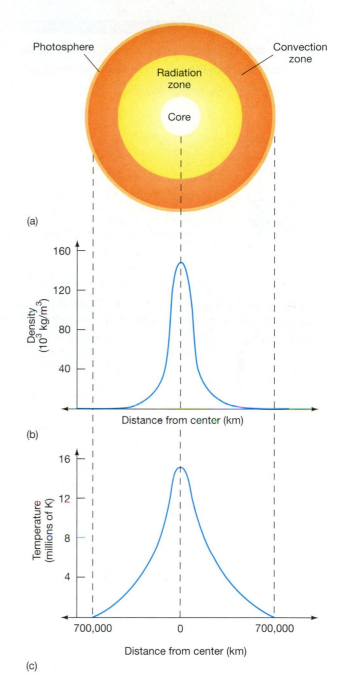

(a)

(b)

(c)

▲ **Figure 9.6 Solar Interior** Theoretical density and temperature distributions in the interior of the Sun. (a) A cross-sectional cut through the center of the Sun. (b) The variation of density across the cut. (c) The corresponding run of temperature.

◄ **Figure 9.7 Solar Convection** Physical transport of energy in the Sun's convection zone. This region is visualized as a boiling, seething sea of gas. Each convective cell at the top of the convection zone is about 1000 km across. The convective cell sizes become progressively smaller closer to the surface. (This is a highly simplified diagram; there are many different cell sizes, and they are not so neatly arranged.)

CONCEPT CHECK ▶▶▶

Describe the two distinct ways in which energy moves outward from the solar core to the photosphere.

thousands of kilometers in diameter. Energy is carried upward through a series of progressively smaller cells, stacked one upon another, until, at a depth of about 1000 km below the photosphere, the individual cells are about 1000 km across. The top of this uppermost tier of convection lies just below the solar photosphere.

Convection does not proceed into the solar atmosphere. In and above the photosphere the density is so low that the gas is transparent and radiation once again becomes the mechanism of energy transport. Photons reaching the photosphere escape freely into space. The solar density falls off so rapidly in this region of the Sun that the gas makes the transition from completely opaque to completely transparent over a very small distance—just a few hundred kilometers—accounting for the thinness of the photosphere and the observed sharp "edge" of the Sun.

9-1 DISCOVERY

SOHO: Eavesdropping on the Sun

Throughout the few decades of the Space Age, various nations, led by the United States, have sent spacecraft to most of the major bodies in the solar system. One of the as-yet-unexplored bodies is Pluto, the most notable member of the Kuiper belt, which has never been visited by a robot orbiter or even a flyby craft—although this may change soon. ∞ *(Sec. 8.5)* The other unexplored body is the Sun. Currently, the next best thing to a dedicated reconnoitering spacecraft is the *Solar and Heliospheric Observatory (SOHO)*, launched in 1995 and (as of 2005) still operating, which has radioed back to Earth volumes of new data—and a few new puzzles—about our parent star.

SOHO is a billion-dollar mission operated primarily by the European Space Agency. The 2-ton robot is now on-station about 1.5 million km sunward of Earth—about 1 percent of the distance from Earth to the Sun. This is the so-called L_1 Lagrangian point, where the gravitational pull of the Sun and Earth are precisely equal—a good place to park a monitoring platform. There *SOHO* looks unblinkingly at our star 24 hours a day, eavesdropping on the Sun's surface, atmosphere, and interior. This automated vehicle carries a dozen instruments, capable of measuring almost everything from the Sun's corona and magnetic field to its solar wind and internal vibrations. The accompanying figure shows a false-color image of the Sun's lower corona, obtained by combining *SOHO* data obtained at three different extreme ultraviolet wavelengths.

SOHO is positioned just beyond Earth's magnetosphere, so its instruments can study cleanly the charged particles of the solar wind—high-speed matter escaping from the Sun, flowing outward from the corona. Coordinating these on-site measurements with *SOHO* images of the Sun itself, astronomers now think they can follow solar magnetic field loops expanding and breaking as the Sun prepares itself for mass ejections several days before they actually occur. Given that such coronal storms can endanger pilots and astronauts and play havoc with communications, power grids, satellite electronics, and other human activity, the prospect of having accurate forecasts of disruptive solar events is a welcome development.

In fact, there do seem to be some correlations between the 22-year solar cycle (two sunspot cycles, with oppositely direct-

ed magnetic fields) and periods of climatic dryness here on Earth. For example, near the start of the past eight cycles, there have been droughts in North America—at least within the middle and western plains from South Dakota to New Mexico. The most recent of these droughts, which typically last three to six years, came in the late 1950s. The one expected in the 1980s, however, did not occur as clearly as anticipated.

Other possible Sun–Earth connections include a link between solar activity and increased atmospheric circulation on our planet. As circulation increases, terrestrial storm systems deepen, extend over wider ranges of latitude, and carry more moisture. The relationship is complex, and the subject controversial, because no one has yet shown any physical mechanism (other than the Sun's heat, which does not vary much during the solar cycle) that would allow solar activity to stir our terrestrial atmosphere. Without a better understanding of the physical mechanism involved, none of these effects can be incorporated into our weather-forecasting models.

(NASA/ESA)

R I V U X G

Evidence for Solar Convection

In part, our knowledge of solar convection is derived indirectly, from computer models of the solar interior. However, astronomers also have some direct evidence of conditions in the convection zone. Figure 9.8 is a high-resolution photograph of the solar surface. The visible surface is highly mottled with regions of bright and dark gas known as *granules*. This **granulation** of the solar surface is a direct reflection of motion in the convection zone. Each bright granule measures about 1000 km across—comparable to one of the larger states in the United States—and has a lifetime of between 5 and 10 minutes.

Each granule forms the topmost part of a solar convection cell. Spectral lines detected in the bright granules are slightly blueshifted, indicating Doppler-shifted matter approaching us—that is, moving upward—at about 1 km/s. Similarly, material in the dark granules is moving down into the solar interior, exactly as we would expect for the topmost tier of convection in Figure 9.7. ∞ *(Sec. 2.7)* The brightness variations of the granules result from differences in temperature. The upwelling gas is hotter and therefore (by Stefan's law) emits more radiation than the cooler downwelling gas. ∞ *(Sec. 2.4)* The adjacent bright and dark gases appear to contrast considerably, but in reality their temperature difference is less than about 500 K.

Careful measurements also reveal a much larger-scale flow beneath the solar surface. **Supergranulation** is a flow pattern quite similar to granulation except that supergranulation cells measure some 30,000 km across. As with granulation, material upwells at the center of the cells, flows across the surface, then sinks down again at the edges. Scientists think that supergranules are the imprint on the photosphere of the deepest tier of large convective cells depicted in Figure 9.7.

9.3 The Solar Atmosphere

Astronomers can obtain an enormous amount of information about the Sun (or any star) from analysis of absorption lines in its spectrum. ∞ *(Sec. 2.5)* Figure 9.9 (see also Figure 2.14) is a detailed visible spectrum of the Sun, spanning the range of wavelengths from 360 to 690 nm. Based on spectral analysis, Table 9.2 lists the 10 most common elements in the Sun. This distribution is very similar to what we saw on the jovian planets, and it is what we will find for the universe as a whole. Hydrogen is by far the most abundant element, followed by helium.

Strictly speaking, analysis of spectral lines allows us to draw conclusions only about the part of the Sun where the lines form—the photosphere and chromosphere. However, the data in Table 9.2 are thought to be representative of the entire Sun (with the exception of the solar core, where nuclear reactions are steadily changing the composition—see Section 9.5).

▲ **Figure 9.8 Solar Granulation** A photograph of the granulated solar photosphere, taken from the *Skylab* space station looking directly down on the Sun's surface. Typical solar granules are comparable in size to Earth's continents. The bright portions of the image are regions where hot material is upwelling from below, as shown in Figure 9.7. The dark regions correspond to cooler gas that is sinking back down into the interior. The inset drawing shows a perpendicular cut through the surface. (NASA)

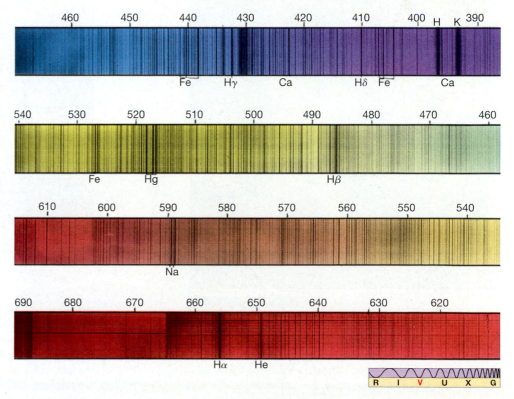

▶ **Figure 9.9 Solar Spectrum** A detailed visible spectrum of our Sun shows thousands of dark absorption lines, indicating the presence of 67 elements in various stages of excitation and ionization in the lower solar atmosphere. The numbers give wavelengths, in nanometers. (Palomar Observatory/Caltech)

TABLE 9.2	The Composition of the Sun	
ELEMENT	PERCENTAGE OF TOTAL NUMBER OF ATOMS	PERCENTAGE OF TOTAL MASS
Hydrogen	91.2	71.0
Helium	8.7	27.1
Oxygen	0.078	0.97
Carbon	0.043	0.40
Nitrogen	0.0088	0.096
Silicon	0.0045	0.099
Magnesium	0.0038	0.076
Neon	0.0035	0.058
Iron	0.0030	0.14
Sulfur	0.0015	0.040

The Chromosphere

Density continues to decrease rapidly as we move outward through the solar atmosphere. The low density of the chromosphere means that it emits very little light of its own and cannot be observed visually under normal conditions. The photosphere is just too bright, dominating the chromosphere's radiation. Still, although it is not normally seen, astronomers have long been aware of the chromosphere's existence. Figure 9.10 shows the Sun during an eclipse in which the photosphere was obscured by the Moon but the chromosphere was not. The chromosphere's characteristic pinkish hue, the result of the red Hα emission line of hydrogen, is plainly visible. ∞ *(Sec. 2.6)*

The chromosphere is far from tranquil. Every few minutes, small solar storms erupt there, expelling jets of hot matter known as *spicules* into the Sun's upper atmosphere (Figure 9.11). These long, thin spikes of matter leave the Sun's surface at typical speeds of about 100 km/s, reaching several thousand kilometers above the photosphere. Spicules tend to accumulate around the edges of supergranules. The Sun's magnetic field is also known to be somewhat stronger than average in those regions. Scientists speculate that the downwelling material there tends to strengthen the solar magnetic field and that spicules are the result of magnetic disturbances in the Sun's churning outer layers.

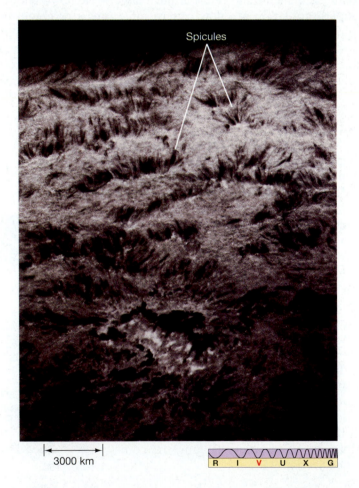

Spicules

3000 km

R I V U X G

▲ **Figure 9.11 Solar Spicules** Short-lived, narrow jets of gas that typically last mere minutes can be seen sprouting up from the chromosphere in this Hα image of the Sun. The spicules are the thin, dark, spikelike regions. They appear dark against the face of the Sun because they are cooler than the photosphere. (NOAO)

R I V U X G

▲ **Figure 9.10 Solar Chromosphere** This photograph of a total solar eclipse shows the solar chromosphere a few thousand kilometers above the Sun's surface. Note the prominence at left. (G. Schneider)

▶ **Figure 9.12 Solar Corona** When both the photosphere and the chromosphere are obscured by the Moon during a solar eclipse, the faint solar corona becomes visible. This photograph shows clearly the emission of radiation from a relatively inactive corona. (Bencho Angelov)

The Transition Zone and Corona

During the brief moments of a solar eclipse, if the Moon's angular size is large enough that both the photosphere and the chromosphere are blocked, the ghostly solar corona can be seen, as in Figure 9.12. With the photospheric light removed, the pattern of spectral lines changes dramatically. The spectrum shifts from absorption to emission, and an entirely new set of spectral lines suddenly appears.

The shift from absorption to emission is entirely in accordance with Kirchhoff's laws, because we see the corona against the blackness of space, not against the bright continuous spectrum from the photosphere below. ∞ *(Sec. 2.5)* The new lines arise because atoms in the corona are much more highly *ionized* than atoms in the photosphere or chromosphere. ∞ *(Sec. 2.6)* Their internal electronic structure and hence their spectra are therefore quite different from those of atoms at lower atmospheric levels. The cause of this extensive electron stripping is the high coronal temperature. The degree of ionization inferred from spectra observed during solar eclipses tells us that the temperature of the upper chromosphere exceeds that of the photosphere. Furthermore, the temperature of the solar corona, where even more ionization is seen, is higher still.

Figure 9.13 shows how the temperature varies with altitude above the photosphere. The temperature decreases to a minimum of about 4500 K some 500 km above the photosphere, after which it rises steadily. About 1500 km above the photosphere, in the transition zone, the gas temperature increases sharply, reaching more than one million K at an altitude of 10,000 km. Thereafter, in the corona, the temperature remains roughly constant at around 3 million K, although *SOHO* and other orbiting instruments have detected coronal "hot spots" having temperatures many times higher than this average value. The rapid temperature rise in the transition zone is still not fully understood. The temperature profile runs contrary to intuition: Moving away from a heat source, we would normally expect the heat to diminish, but this is not the case in the lower atmosphere of the Sun. The corona must have another energy source. Astronomers now think that magnetic disturbances in the solar photosphere (see Section 9.4) are ultimately responsible for heating the corona.

The Solar Wind

Electromagnetic radiation and fast-moving particles—mostly protons and electrons—escape from the Sun all the time. The radiation moves away from the photosphere at the speed of light, taking eight minutes to reach Earth. The particles travel much more slowly, although at the still considerable speed of about 500 km/s, reaching Earth in a few days. This constant stream of escaping solar particles is the solar wind.

The solar wind is a direct consequence of the high coronal temperature. About 10 million km above the photosphere, the coronal gas is hot enough to escape the Sun's gravity, and it begins to flow outward into space. At the same time, the solar atmosphere is continuously replenished from below. If that were not the case, the corona would disappear in about a day. The Sun is, in effect, "evaporating"—constantly shedding mass through the solar wind. The wind is an extremely thin medium, however. Although it carries away about a million tons of solar matter each second, less than 0.1 percent of the Sun's mass has been lost since the solar system formed 4.6 billion years ago.

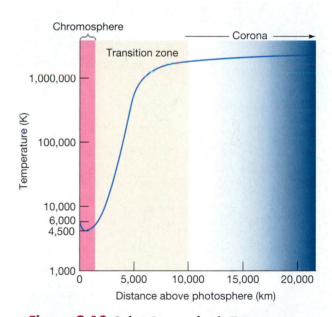

▲ **Figure 9.13 Solar Atmospheric Temperature** The change in gas temperature in the lower solar atmosphere is dramatic. The temperature, indicated by the blue line, reaches a minimum of 4500 K in the chromosphere and then rises sharply in the transition zone, finally leveling off at around 3 million K in the corona.

◀◀◀ **CONCEPT CHECK**

Describe two ways in which the spectrum of the solar corona differs from that of the photosphere.

9.4 The Active Sun

Most of the Sun's luminosity results from continuous emission from the photosphere. However, superimposed on this steady, predictable aspect of our star's energy output is a much more irregular component, characterized by explosive and unpredictable surface activity. Solar activity contributes little to the Sun's total luminosity and has little effect on the evolution of the Sun as a star, but it does affect us here on Earth. The size and duration of coronal holes are strongly influenced by the level of solar activity and hence so is the strength of the solar wind. This, in turn, directly affects Earth's magnetosphere.

Sunspots

Figure 9.14 is an optical photograph of the entire Sun showing numerous dark **sunspots** on the surface. ∞ *(Sec. 1.2)* They typically measure about 10,000 km across—about the size of Earth. At any given instant, the Sun may have hundreds of sunspots, or it may have none at all.

Studies of sunspots show an *umbra*, or dark center, surrounded by a grayish *penumbra*. The close-up view of a pair of sunspots in Figure 9.15(a) shows both of these dark areas against the bright background of the undisturbed photosphere. This gradation in darkness indicates a gradual change in photospheric temperature. In other words, sunspots are simply relatively *cooler* regions of the photospheric gas. The temperature of the umbra is about 4500 K, that of the penumbra 5500 K. The spots, then, are certainly composed of hot gases. They seem dark only because they appear against an even brighter background (the 5800 K photosphere).

▼ Figure 9.14 Sunspots This photograph of the entire Sun, taken during a period of maximum solar activity, shows several groups of sunspots. The largest spots in this image are more than 20,000 km across, nearly twice the diameter of Earth. Typical sunspots are only about half this size. (Palomar Observatory/Caltech)

R I V U X G

(a) |←— 50,000 km —→|

(b) |←— 10,000 km —→|

Penumbra

Umbra

R I V U X G

► Figure 9.15 Sunspots, Up Close (a) The largest pair of sunspots in Figure 9.14, each consisting of a cool, dark inner region called the umbra surrounded by a warmer, brighter region called the penumbra. The spots appear dark because they are slightly cooler than the surrounding photosphere. (b) A high-resolution, true-color image of a single sunspot shows details of its structure as well as the granules surrounding it. This spot is about the size of Earth. (Palomar Observatory/Caltech; National Solar Observatory)

As mentioned in Sec. 9.1 (see Table 9.1), observations of sunspots indicate that the Sun does not rotate as a solid body. Instead, it spins *differentially*—faster at the equator and more slowly at the poles, like Jupiter and Saturn. The combination of differential rotation and convection radically affects the character of the Sun's magnetic field, which in turn plays a major role in determining the numbers and location of sunspots.

Solar Magnetism

Spectroscopic studies indicate that the magnetic field in a typical sunspot is about 1000 times greater than the field in the surrounding, undisturbed photosphere (which is itself several times stronger than Earth's magnetic field). Furthermore, the field lines are not randomly oriented, but instead are directed roughly perpendicular to (out of or into) the Sun's surface. Scientists think that sunspots are cooler than their surroundings because these abnormally strong fields interfere with the normal convective flow of hot gas toward the solar surface.

The **polarity** of a sunspot simply indicates which way its magnetic field is directed. We will conventionally label spots where the field lines emerge from the interior as S and those where the lines dive below the photosphere as N (so that field lines above the surface run from S to N, as on Earth). Sunspots almost always come in pairs whose members lie at roughly the same latitude and have opposite magnetic polarities. As illustrated in Figure 9.16(a), magnetic field lines emerge from the solar interior through one member (S) of a sunspot pair, loop through the solar atmosphere, then reenter the photosphere through the other member (N). As in Earth's magnetosphere, charged particles tend to follow the solar magnetic field lines. ∞ *(Sec. 5.7)* Figure 9.16(b) shows an actual image of solar magnetic loops; the light we see in this image is emitted by charged particles trapped by the solar field.

▼ **Figure 9.16 Sunspot Magnetism** (a) Sunspot pairs are linked by magnetic field lines. The Sun's magnetic field lines emerge from the surface through one member of a pair and reenter the Sun through the other member. The leading members of all sunspot pairs in the solar northern hemisphere have the same polarity (labeled N or S, as described in the text). If the magnetic field lines are directed into the Sun in one leading spot, they are inwardly directed in all other leading spots in that hemisphere. The same is true in the southern hemisphere, except that the polarities are always opposite those in the north. The entire magnetic field pattern reverses itself roughly every 11 years. (b) A far-ultraviolet image taken by NASA's *Transition Region and Coronal Explorer (TRACE)* satellite in 1999, showing magnetic field lines arching between two sunspot groups. Note the complex structure of the field lines, which are seen here via the radiation emitted by superheated gas flowing along them. Resolution here is about 700 km. In this negative image (which shows the lines more clearly), the darkest regions have temperatures of about 2 million K. (NASA)

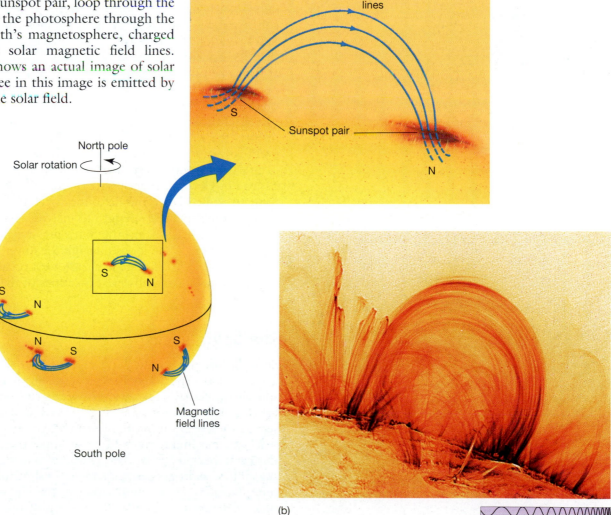

(a)

(b)

▶ **Figure 9.17 Solar Rotation** (a, b) The Sun's differential rotation wraps and distorts the solar magnetic field. (c) Occasionally, the field lines burst out of the surface and loop through the lower atmosphere, thereby creating a sunspot pair. The underlying pattern of the solar field lines explains the observed pattern of sunspot polarities. (If the loop happens to occur near the edge of the Sun and is seen against the blackness of space, we see a phenomenon called a prominence, see Figure 9.20.) A reasonable analogy might be the loops and kinks that form in a tangled garden hose.

Despite the irregular appearance of the sunspots themselves, there is a great deal of order in the underlying solar field. *All* the sunspot pairs in the same solar hemisphere (north or south) at any instant have the *same* magnetic configuration. That is, if the leading spot (measured in the direction of the Sun's rotation) has N polarity, as shown in the figure, then all leading spots in that hemisphere have the same polarity. What's more, in the other hemisphere at the same time, all sunspot pairs have the *opposite* magnetic configuration (S polarity leading). To understand these regularities in sunspot polarities, we must look in more detail at the Sun's magnetic field.

As illustrated in Figure 9.17 (a–c), the Sun's differential rotation greatly distorts the solar magnetic field, "wrapping" it around the solar equator and eventually causing any originally north—south magnetic field to reorient itself in an east–west direction. At the same time, convection causes the magnetized gas to upwell toward the surface, twisting and tangling the magnetic field pattern. In some places, the field lines becomes kinked like a knot in a garden hose, causing the field strength to increase. Occasionally, the field becomes so strong that it overwhelms the Sun's gravity, and a "tube" of field lines bursts out of the surface and loops through the lower atmosphere, forming a sunspot pair (Figure 9.17c). The general east–west organization of the underlying solar field accounts for the observed polarities of the resulting sunspot pairs in each hemisphere.

The Solar Cycle

Sunspots are not steady. Most change their size and shape, and all come and go. Individual spots may last anywhere from 1 to 100 days. A large group of spots typically lasts 50 days. However, centuries of observations have established a clear **sunspot cycle**. Figure 9.18(a) shows the number of sunspots observed each year during the twentieth century. The average number of spots reaches a maximum every 11 or so years, then falls off almost to zero before the cycle begins afresh. The latitudes at which sunspots appear vary as the sunspot cycle progresses. Individual sunspots do not move up or down in latitude, but new spots appear closer to the equator as older ones at higher latitudes fade away. Figure 9.18(b) is a plot of observed sunspot latitudes as a function of time.

▶ **Figure 9.18** **Sunspot Cycle** (a) Annual number of sunspots throughout the twentieth century, showing five-year averages of annual data to make long-term trends more evident. The (roughly) 11-year solar cycle is clearly visible. At the time of minimum solar activity, hardly any sunspots are seen. About four years later, at the time of maximum solar activity, as many as 200 spots are observed per year. (b) Sunspots cluster at high latitudes when solar activity is at a minimum. They appear at lower and lower latitudes as the number of sunspots peaks. They are again prominent near the Sun's equator as solar minimum is again approached. The most recent solar maximum occurred in 2001. The blue lines in the upper plot indicate the average latitude of the spots at any given time.

In fact, the 11-year sunspot cycle is only half of a 22-year **solar cycle**. For the first 11 years of the solar cycle, the leading spots of all sunspot pairs in the same solar hemisphere have one polarity, and spots in the other hemisphere have the opposite polarity (Figure 9.16). These polarities then reverse for the next 11 years.

Astronomers theorize that the Sun's magnetic field is both generated and amplified by the constant stretching, twisting, and folding of magnetic field lines that results from the combined effects of differential rotation and convection. The theory is similar to the "dynamo" theory that accounts for the magnetic fields of Earth and the jovian planets, except that the solar dynamo operates much faster and on a much larger scale. ∞ *(Sec. 5.7)* One important prediction of this theory is that the Sun's magnetic field should rise to a maximum, then fall to zero and reverse itself in a roughly periodic way, just as observed. Solar surface activity, such as the sunspot cycle, simply follows the changes in the underlying field.

Figure 9.19 plots all sunspot data recorded since the invention of the telescope. As can be seen, the 11-year "periodicity" of the solar sunspot cycle is far from perfect. Not only does the period vary from 7 to 15 years, but the sunspot cycle has disappeared entirely in the relatively recent past. The lengthy period of solar inactivity that extended from 1645 to 1715 is called the *Maunder minimum*, after the British astronomer who drew attention to these historical records. (Interestingly, the period also seems to correspond fairly well to the coldest years of the so-called "Little Ice Age" that chilled northern Europe during the late seventeenth century, suggesting a link between solar activity and climate on Earth.) The cause of the Sun's century-long variations remains a mystery. Lacking a complete understanding of the solar cycle, we cannot easily explain how it could virtually shut down for such an extended period of time.

Active Regions

Sunspots are relatively gentle aspects of solar activity. However, the photosphere surrounding them occasionally erupts violently, spewing forth into the

▲ **Figure 9.19** **Maunder Minimum** Number of sunspots occurring each year over the past four centuries. Note the absence of spots during the late seventeenth century.

(a)

(b)

R I V U X G

▲ **Figure 9.20 Solar Prominences** (a) This particularly large solar prominence was observed by ultraviolet detectors aboard the *SOHO* spacecraft in June 2002. (b) Like a phoenix rising from the solar surface, this filament of hot gas measures more than 100,000 km in length. Earth could easily fit between its outstretched "arms." Dark regions in this *TRACE* image have temperatures less than 20,000 K; the brightest regions are about 1 million K. The ionized gas follows the solar magnetic field lines away from the Sun. Most of the gas will subsequently cool and fall back into the photosphere. (NASA)

Animation/Video
May 12, 1997 Solar Flare Event

Animation/Video
Eruption of a Solar Active Region

R I V U X G

corona large quantities of energetic particles. The sites of these explosive events are known simply as **active regions**. Most pairs or groups of sunspots have active regions associated with them. Like all other aspects of solar activity, these phenomena tend to follow the solar cycle and are most frequent and violent around the time of sunspot maximum.

Figure 9.20 shows two solar **prominences**—loops or sheets of glowing gas ejected from an active region on the solar surface. Prominences move through the inner parts of the corona under the influence of the Sun's magnetic field. Magnetic instabilities in the strong fields found in and near sunspot groups may cause the prominences, although the details are still not completely understood. A typical solar prominence measures some 100,000 km in extent, nearly 10 times the diameter of planet Earth. Some may persist for days or even weeks. Prominences as large as that shown in Figure 9.20(b) (which spanned almost half a million kilometers of the solar surface—half the Sun's diameter) are less common and usually appear only at times of greatest solar activity.

Flares are another type of solar activity observed near active regions (Figure 9.21). Also the result of magnetic instabilities, flares are much more violent (and even less well understood) than prominences. They flash across a region of the Sun in minutes, releasing enormous amounts of energy as they go. Temperatures in the extremely compact hearts of flares can reach 100 million K (roughly six times hotter than the solar core). So energetic are these cataclysmic explosions that some researchers have likened them to bombs exploding in the lower regions of the Sun's atmosphere. Unlike the trapped gas that makes up the characteristic loop of a prominence, the particles produced by a

◀ **Figure 9.21 Solar Flare** Much more violent than a prominence, a solar flare is an explosion on the Sun's surface that sweeps across an active region in a matter of minutes, accelerating solar material to high speeds and blasting it into space. (USAF)

▶ **Figure 9.22 Coronal Mass Ejection**   (a) A few times per week, on average, a giant magnetized "bubble" of solar material detaches itself from the Sun and rapidly escapes into space, as shown in this *SOHO* image taken in 2002. The circles are artifacts of an imaging system designed to block out the light from the Sun itself and exaggerate faint features at larger radii. (b) Should such a coronal mass ejection encounter Earth with its magnetic field oriented opposite to our own, as shown, the field lines can join together, allowing high-energy particles to enter and possibly severely disrupt our planet's magnetosphere. If the fields are oriented differently, the coronal mass ejection simply slides by Earth with little effect. (NASA/ESA)

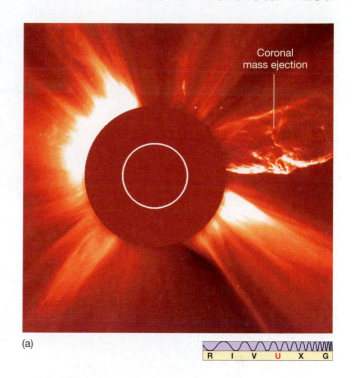

Coronal mass ejection

(a)

R I V U X G

flare are so energetic that the Sun's magnetic field is unable to hold them and shepherd them back to the surface. Instead, the particles are simply blasted into space by the violence of the explosion. Flares may also be responsible for many of the internal pressure waves that give rise to solar surface oscillations.

Figure 9.22(a) shows a **coronal mass ejection** from the Sun. Sometimes (but not always) associated with flares and prominences, these phenomena are giant magnetic "bubbles" of ionized gas that separate from the rest of the solar atmosphere and escape into interplanetary space. Such ejections occur about once per week at times of sunspot minimum, but up to two or three times per day at solar maximum. Carrying an enormous amount of energy, they can—if their fields are properly oriented—connect with Earth's magnetic field via a process known as *reconnection*, dumping some of their energy into the magnetosphere and potentially causing widespread communications and power disruptions on our planet (Figure 9.22b).

The Sun in X Rays

Unlike the 5800-K photosphere, which emits most strongly in the visible part of the electromagnetic spectrum, the million-kelvin coronal gas radiates at much higher frequencies—primarily in X rays. ∞ *(Sec. 2.4)* For this reason, X-ray telescopes have become important tools in studying the solar corona. Figure 9.23 shows a sequence of X-ray images of the Sun. The full corona extends well beyond the regions shown, but the density of coronal particles emitting the radiation diminishes rapidly with distance from the Sun. The intensity of X-ray radiation farther out is too dim to be seen here.

In the mid-1970s instruments aboard NASA's *Skylab* space station revealed that the solar wind escapes mostly through solar "windows" called **coronal holes**. The dark area moving from left to right in Figure 9.23(a) (which shows more recent data from the Japanese *Yohkoh* X-ray solar observatory) is a coronal hole. Not really holes, such structures are simply deficient in matter—vast regions of the Sun's atmosphere where the density is about 10 times lower than the already tenuous, normal corona. Note that the underlying solar photosphere looks black in these images because it is far too cool to emit X rays in any significant quantity. The largest coronal holes can be hundreds of thousands of kilometers across. Structures of this size are seen only a few times each decade. Smaller holes—perhaps only a few tens of thousand kilometers in size—are much more common, appearing every few hours.

Coronal holes are lacking in matter because the gas there is able to stream freely into space at high speeds, driven by disturbances in the Sun's atmosphere and magnetic field. Figure 9.23(b) illustrates how, in coronal holes, the solar magnetic field lines extend from the surface far out into interplanetary space. Because charged particles tend to follow the field lines, they can escape, particularly from the Sun's polar regions, according to recent findings from *SOHO* and NASA's *Ulysses* spacecraft, which flew high above the ecliptic plane to explore the Sun's polar regions. (Note, however, that the coronal hole shown in Figure 9.23a spans the Sun's equator.) In other regions of the corona, the solar magnetic field lines stay close to the Sun, keeping charged

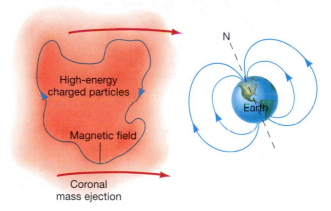

High-energy charged particles

Magnetic field

Coronal mass ejection

N

Earth

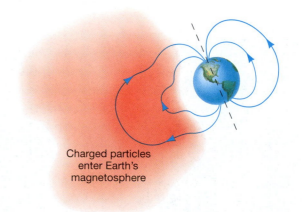

Charged particles enter Earth's magnetosphere

(b)

Animation/Video

Coronal Mass Ejection

(a)

Field lines loop back to Sun—particles trapped

Field lines extend into interplanetary space—particles escape

(b)

▶ **Figure 9.23 Coronal Hole** (a) Images of X-ray emission from the Sun observed by the *Yohkoh* satellite. These frames were taken at roughly two-day intervals, starting at the left. Note the dark, V-shaped coronal hole traveling from left to right, where the X-ray observations outline in dramatic detail the abnormally thin regions through which the high-speed solar wind streams forth. (b) Charged particles follow magnetic field lines that compete with gravity. When the field is trapped and loops back toward the photosphere, the particles are also trapped; otherwise, they can escape as part of the solar wind. (ISAS/Lockheed Martin)

Animation/Video
Ejection of Material from the Sun

particles near the surface and inhibiting the outward flow of the solar wind (just as Earth's magnetic field tends to prevent the incoming solar wind from striking Earth), so the density remains relatively high. ∞ *(Sec. 5.7)*

The Changing Solar Corona

The solar corona also varies in step with the sunspot cycle. The photograph of the corona in Figure 9.12 shows the Sun at sunspot minimum. The corona is fairly regular in appearance and surrounds the Sun more or less uniformly. Compare this image with Figure 9.24, which was taken in 1991, close to a peak of the sunspot cycle. The active corona here is much more irregular than the one in Figure 9.12 and extends farther from the solar surface. The streamers of coronal material pointing away from the Sun are characteristic of this phase.

Astronomers think that the corona is heated primarily by solar surface activity, which can inject large amounts of energy into the upper solar atmosphere. *SOHO* observations have implicated a "magnetic carpet" as the source of much of the heating, where small magnetic loops continually sprout up, then disappear, dumping vast amounts of energy into the lower solar atmosphere. These small-scale magnetic loops probably provide most of the energy needed to heat the corona. In addition, more extensive disturbances often move through the corona above an active site in the photosphere, distributing the energy throughout the coronal gas. Given this connection, it is hardly surprising that both the appearance of the corona and the strength of the solar wind are closely correlated with the solar cycle.

CONCEPT CHECK ▶▶▶

What do observations of sunspot polarities tell us about the solar magnetic field?

◀ **Figure 9.24 Active Corona** Photograph of the solar corona during the July 1991 eclipse, at the peak of the sunspot cycle. At these times, the corona is much less regular and much more extended than at sunspot minimum (compare Figure 9.12). Astronomers think that coronal heating is caused by surface activity on the Sun. The changing shape and size of the corona are the direct result of variations in prominence and flare activity over the course of the solar cycle. (National Solar Observatory)

9.5 The Heart of the Sun

By what process does the Sun shine, day after day, year after year, eon after eon? The answer to this question is central to all of astronomy. Without it we can understand neither the evolution of stars and galaxies in the universe nor the existence of life on Earth.

Nuclear Fusion

Only one known energy-generation mechanism can conceivably account for the Sun's enormous energy output. That process is **nuclear fusion**—the combining of light nuclei into heavier ones. We can represent a typical fusion reaction symbolically as

nucleus 1 + nucleus 2 → nucleus 3 + energy.

For powering the Sun and other stars, the most important piece of this equation is the energy produced.

The essential point here is that, during a fusion reaction, the total mass *decreases*—the mass of nucleus 3 is less than the combined masses of nuclei 1 and 2. Where does this mass go? It is converted to energy in accordance with Einstein's famous equation

$$E = mc^2,$$

or

energy = mass × (speed of light)^2.

This equation says that to determine the amount of energy corresponding to a given mass, simply multiply the mass by the square of the speed of light (c in the equation). The speed of light is so large that even a small amount of mass translates into an enormous amount of energy.

The production of energy by a nuclear fusion reaction is an example of the **law of conservation of mass and energy**, which states that the sum of mass and energy (properly converted to the same units, using Einstein's equation) must always remain constant in any physical process. There are no known exceptions. The light we see coming from the Sun means that the Sun's mass must be slowly, but steadily, decreasing with time.

The Proton–Proton Chain

The lightest and most common element in the universe is hydrogen, and it is the fusion of hydrogen nuclei (protons) into nuclei of the next lightest element (helium) that powers the Sun. But all atomic nuclei are positively charged, so they repel one another. Furthermore, by the inverse-square law, the closer two nuclei come to one another, the greater is the repulsive force between them (Figure 9.25a). ∞ *(Sec. 2.2)* How then do nuclei—two protons, say—ever manage to fuse into anything heavier? The answer is that if they collide at high enough speeds, one proton can momentarily plow deep into the other, eventually coming within the exceedingly short range of the **strong nuclear force**, which binds nuclei together (see *More Precisely 9-1*). At distances less than about 10^{-15} m, the attraction of the nuclear force overwhelms the electromagnetic repulsion, and fusion can occur (Figure 9.25b). Speeds in excess of a few hundred kilometers per second, corresponding to a gas temperature of at least 10 million K, are needed to slam protons together fast enough to initiate fusion. ∞ *(More Precisely 5-1)* Such conditions are found in the core of the Sun and at the centers of all stars. The radius at which the temperature drops below this value defines the outer edge of the solar core (Figure 9.6).

(a)

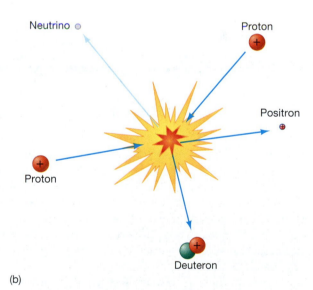

(b)

▲ **Figure 9.25 Proton Interactions**   (a) Since like charges repel, two low-speed protons veer away from one another, never coming close enough for fusion to occur. (b) Sufficiently high-speed protons may succeed in overcoming their mutual repulsion, approaching close enough for the strong force to bind them together—in which case they collide violently, triggering nuclear fusion that ultimately powers the Sun.

Several new particles are created in the reaction shown in Figure 9.25(b). The **deuteron** is a nucleus of *deuterium*, or "heavy hydrogen," containing a proton and an extra neutron. The **positron** is the positively charged *antiparticle* of an electron. Its properties are identical to those of a normal, negatively charged electron, except for the positive charge. Particles and antiparticles annihilate (destroy) one another when they meet, producing pure energy in the form of gamma-ray photons. The **neutrino** is a chargeless and virtually massless elementary particle. (The name derives from the Italian for "little neutral one.") Neutrinos move at nearly the speed of light and interact with hardly anything. They can penetrate, without stopping, several light-years of lead. Their interactions with matter are governed by the **weak nuclear force,** described in *More Precisely 9-1*. Yet despite their elusiveness, neutrinos can be detected with carefully constructed instruments. In the final section we discuss some rudimentary neutrino "telescopes" and the important contribution they have made to solar astronomy.

The basic set of nuclear reactions powering the Sun (and the vast majority of all stars) is sketched in Figure 9.26. In fact it is not a single reaction, but rather a sequence of reactions called the **proton–proton chain**. The following stages are shown in the figure.

(I) First two protons combine, as in Figure 9.25.

(II) The resulting positrons annihilate with electrons, producing gamma rays, while the deuterons combine with protons to create a type of helium called helium-3, containing only one neutron. Two of each of these sets of reactions are shown in Figure 9.26.

(III) Finally, two helium-3 nuclei combine to form helium-4 and two protons.

Setting aside the temporary intermediate nuclei produced, the net effect is that four hydrogen nuclei (protons) combine to create one nucleus of the next lightest element, helium-4 (containing 2 protons and 2 neutrons, for a total mass of 4), creating two neutrinos, and releasing energy in the form of gamma-ray radiation: ∞ *(Sec. 2.6)*

$$4 \text{ protons} \rightarrow \text{helium-4} + 2 \text{ neutrinos} + \text{energy.}$$

9-1 MORE PRECISELY

The Strong and Weak Nuclear Forces

As far as we can tell, the behavior of all matter in the universe, from elementary particles to clusters of galaxies, is ruled by just four (or fewer) basic forces, which are fundamental to everything in the universe. We have already encountered two of them in this text—the *gravitational force*, which binds planets, stars, and solar systems together, and the much stronger *electromagnetic force*, which governs the properties of atoms and molecules, as well as magnetic phenomena on planets and stars. ∞ *(Secs. 1.5, 2.2)*

Both of these forces are governed by inverse-square laws and are long range. In contrast, the other two fundamental forces, known as the *weak nuclear force* and the *strong nuclear force* are of extremely short range—about 10^{-18} m and 10^{-15} m, respectively, less than the size of an atomic nucleus.

The weak force governs the emission of radiation during some radioactive decays, as well as interactions between neutrinos and other particles. In fact, according to current theory, the electromagnetic and weak force are not really distinct entities, but are actually just different aspects of a single "electroweak" force.

The strong force holds atomic nuclei together. It binds particles with enormous strength, but only when they approach one another very closely. Two protons must come within about 10^{-15} m of one another before the attractive strong force can overcome their electromagnetic repulsion.

Very loosely speaking, we can say that the strong force is about 100 times stronger than electromagnetism, 100,000 times stronger than the weak force and 10^{39} times stronger than gravity. But not all particles are subject to all types of force. All particles interact through gravity because all have mass (or equivalently, energy). However, only charged particles interact electromagnetically. Protons and neutrons are affected by the strong force, but electrons are not. Under the right circumstances, the weak force can affect any type of subatomic particle, regardless of its charge.

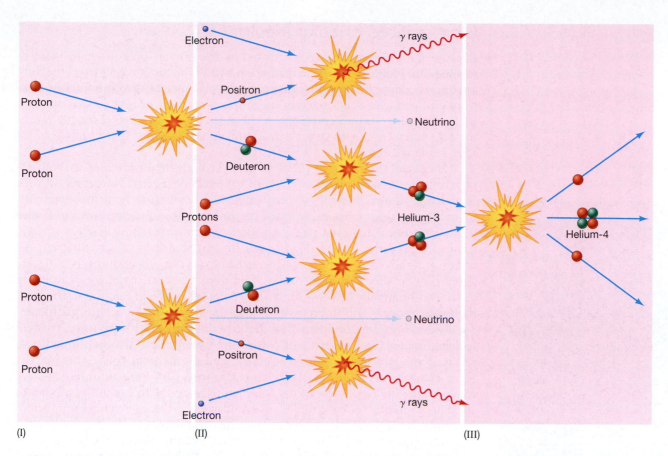

▲ **Figure 9.26 Solar Fusion**   In the proton–proton chain, a total of six protons (and two electrons) are converted to two protons, one helium-4 nucleus, and two neutrinos. The two leftover protons are available as fuel for new proton–proton reactions, so the net effect is that four protons are fused to form one helium-4 nucleus. Energy, in the form of gamma rays, is produced at each stage.

Other reaction sequences, involving the more massive nuclei of carbon, nitrogen, and oxygen, can produce the same end result—four protons fuse into helium-4 and energy is produced in the form of gamma rays. However, the sequence shown in Figure 9.26 is the simplest, and is responsible for almost 90 percent of the Sun's luminosity. Notice how, at each stage of the chain, more complex nuclei are created from simpler ones. We will see later that not all stars are powered by hydrogen fusion; nevertheless, this basic idea of simpler elements combining to form more complex ones, releasing energy in the process, continues to apply.

Gargantuan quantities of protons are fused into helium by the proton–proton chain in the core of the Sun every second. As discussed in more detail in *More Precisely 9-2*, to fuel the Sun's present energy output, hydrogen must be fused into helium in the core at a rate of 600 million tons per second—a lot of mass, but only a tiny fraction of the total amount available. As we will see in Chapter 12, the Sun will be able to sustain this rate of core burning for about another 5 billion years.

The Sun's nuclear energy is produced in the core in the form of gamma rays. However, as the energy passes through the cooler overlying layers of the solar interior and photons are repeatedly absorbed and reemitted, the radiation's blackbody spectrum steadily shifts toward lower and lower temperatures and, by Wien's law, the characteristic wavelength of the radiation increases. ∞ *(Sec. 2.4)* The energy eventually leaves the solar photosphere mainly in the form of visible and infrared radiation. A comparable amount of energy is carried off by the neutrinos, which escape unhindered into space at almost the speed of light.

Animation/Video

Tritium-Helium Fusion

Observations of Solar Neutrinos

Because the gamma-ray energy created in the proton–proton chain is transformed into visible and infrared radiation by the time it emerges from the Sun, astronomers have no direct electromagnetic evidence of the core nuclear reactions. Instead, the *neutrinos* created in the proton–proton chain are our best bet for learning about conditions in the solar core. They travel cleanly out of the Sun, interacting with virtually nothing, and escape into space a few seconds after being created.

Of course, the fact that they can pass through the entire Sun without interacting also makes neutrinos difficult to detect on Earth! Nevertheless, with knowledge of neutrino physics it is possible to construct neutrino detectors. Over the past three decades, several experiments (two of them shown in Figure 9.27) have been designed to detect solar neutrinos reaching Earth's surface. Their designs differ widely, they are sensitive to neutrinos of very different energies, and they disagree somewhat on the details of their findings, but they all agree on one very important point: The number of solar neutrinos reaching Earth is substantially less (by 30 to 50 percent) than the prediction of the standard solar model. This discrepancy is known as the **solar neutrino problem**.

Researchers are confident that the experimental results are to be trusted. Indeed, the lead investigators of two of the experiments just mentioned have received the strongest possible scientific endorsement of their work, in the form of the 2002 Nobel Prize! However, theorists have long considered it very unlikely that the resolution of the solar neutrino problem lies in the physics of the Sun's interior. For example, we could reduce the theoretical number of neutrinos by postulating a lower temperature in the solar core, but the nuclear reactions just described are too well known and the agreement between the standard solar model and helioseismological observations (Section 9.2) is far too close to permit conditions in the core to deviate much from the predictions of the model.

How, then, do we explain the clear disagreement between theory and observation? The answer involves the properties of the neutrinos themselves and has caused scientists to rethink some very fundamental concepts in particle physics. If neutrinos have even a minute amount of mass, theory indicates that it should be possible for them to change their properties—even to transform into other particles—during their eight-minute flight from the solar core to Earth, through a process known as **neutrino oscillations**. In this picture, neutrinos are produced in the Sun at the rate required by the standard solar model, but some turn into something else—actually, other types of neutrinos—on their way to Earth and hence go undetected in the experiments just described. (In the jargon of the field, the neutrinos are said to "oscillate" into other particles.)

(a)

(b)

◀ **Figure 9.27 Neutrino Telescope** (a) This swimming pool–sized "neutrino telescope" is buried beneath a mountain near Tokyo, Japan. Called Super Kamiokande, it is filled (in operation) with 50,000 tons of purified water, and contains 13,000 individual light detectors (some shown here being inspected by technicians) to sense the telltale signature—a brief burst of light—of a neutrino passing through the apparatus. The instrument was badly damaged by a freak accident in November 2001, when one of the detectors apparently imploded, creating a shock wave and chain reaction that destroyed more than half the detectors. It is currently under repair. (b) The Sudbury Neutrino Observatory (SNO), situated 2 km underground in Ontario, Canada. The SNO detector is similar in design to the Kamiokande device, but, by using "heavy" water (with hydrogen replaced by deuterium) instead of ordinary water (H_2O), and adding two tons of salt, it also becomes sensitive to other neutrino types. The device contains 10,000 light-sensitive detectors, arranged on the inside of the large sphere shown here. (ICRR, SNO)

In 1998 the Japanese group operating the Super Kamiokande detector shown in Figure 9.27(a) reported the first experimental evidence of neutrino oscillations (and hence of nonzero neutrino masses), although the observed oscillations did not involve neutrinos of the type produced in the Sun. More recently, in 2001, measurements made at the Sudbury Neutrino Observatory (SNO) in Ontario, Canada (Figure 9.27b), revealed strong evidence for the "other" neutrinos into which the Sun's neutrinos have been transformed. Further SNO observations with a modified detector, reported in 2002, confirmed the result. The total numbers of neutrinos observed are completely consistent with the standard solar model. The solar neutrino problem has at last been solved—and neutrino astronomy can claim its first major triumph!

◄◄◄ **CONCEPT CHECK**

Why does the fact that we see sunlight imply that the Sun's mass is slowly decreasing?

9-2 | **M**ORE PRECISELY

Energy Generation in the Proton–Proton Chain

Let's look in a little more detail at the energy produced by fusion in the solar core and compare it with the energy needed to account for the Sun's luminosity. As discussed in the text and illustrated below, the net effect of the fusion process is that four protons combine to produce a nucleus of helium-4, in the process creating two neutrinos and two positrons (which are quickly converted into energy by annihilation with electrons).

We can calculate the total amount of energy released by accounting carefully for the total masses of the nuclei involved and applying Einstein's famous formula $E = mc^2$. This in turn allows us to relate the Sun's total luminosity to the consumption of hydrogen fuel in the core. Careful laboratory experiments have determined the masses of all the particles involved in the above reaction: The total mass of the protons is 6.6943×10^{-27} kg, the mass of the helium-4 nucleus is 6.6466×10^{-27} kg, and the neutrinos are virtually massless. We omit the positrons here—their masses will end up being counted as part of the total energy released. The difference between the total mass of the four protons and that of the final helium-4 nucleus, 0.0477×10^{-27} kg, is not great, but it is easily measurable.

Multiplying the vanished mass by the square of the speed of light yields 0.0477×10^{-27} kg $\times (3.00 \times 10^8$ m/s$)^2 = 4.28 \times 10^{-12}$ J. This is the energy produced in the form of radiation when 6.69×10^{-27} kg (the rounded-off mass of the four protons) of hydrogen fuses to helium. It follows that fusion of 1 kg of hydrogen generates $4.28 \times 10^{-12}/6.69 \times 10^{-27} = 6.40 \times 10^{14}$ J. Put another way, the process converts about 0.71 percent of the original mass into pure energy.

Thus we have established a direct connection between the Sun's energy output and the consumption of hydrogen in the core. The Sun's luminosity of 3.86×10^{26} W (see Table 9.1), or 3.86×10^{26} J/s (joules per second), implies a mass consumption rate of 3.86×10^{26} J/s $\div 6.40 \times 10^{14}$ J/kg $= 6.03 \times 10^{11}$ kg/s—roughly 600 million tons of hydrogen every second. That sounds like a lot—the mass of a small mountain, and 600 times the mass loss rate in the solar wind—but it represents only a few million million millionths of the total mass of the Sun. Our parent star will be able to sustain this loss for a very long time—this is the subject of Chapter 12.

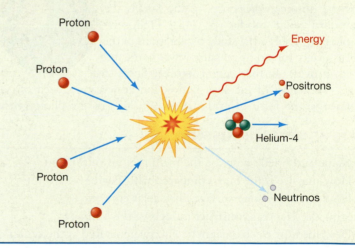

CHAPTER REVIEW

Summary

① Our Sun is a **star** (p. 242), a glowing ball of gas held together by its own gravity and powered by nuclear fusion at its center. The **photosphere** (p. 242) is the region at the Sun's surface from which virtually all the visible light is emitted. The main interior regions of the Sun are the **core** (p. 242), where nuclear reactions generate energy; the **radiation zone** (p. 242), where the energy travels outward in the form of electromagnetic radiation; and the **convection zone** (p. 242), where the Sun's matter is in constant convective motion.

② The amount of solar energy reaching a one square meter area at the top of Earth's atmosphere each second is a quantity known as the **solar constant** (p. 243). The Sun's **luminosity** (p. 250) is the total amount of energy radiated from the solar surface per second.

③ Above the photosphere lies the **chromosphere** (p. 242), the Sun's lower atmosphere. In the **transition zone** (p. 242) above the chromosphere, the temperature increases from a few thousand to around a million kelvins. Above the transition zone is the Sun's thin, hot upper atmosphere, the solar **corona** (p. 242). At a distance of about 15 solar radii, the gas in the corona is hot enough to escape the Sun's gravity, and the corona begins to flow outward as the **solar wind** (p. 242).

④ Much of our knowledge of the solar interior comes from mathematical models. The model that best fits the observed properties of the Sun is the **standard solar model** (p. 243). **Helioseismology** (p. 244)—the study of vibrations of the solar surface caused by pressure waves in the interior—provides further insight into the Sun's structure. The effect of the solar convection zone can be seen on the surface in the form of **granulation** (p. 247) of the photosphere.

Lower levels in the convection zone also leave their mark on the photosphere in the form of larger transient patterns called **supergranulation** (p. 247).

⑤ **Sunspots** (p. 250) are Earth-sized regions on the solar surface that are a little cooler than the surrounding photosphere. They are regions of intense magnetism. Both the numbers and locations of sunspots vary in a roughly 11-year **sunspot cycle** (p. 252) as the Sun's magnetic field rises and falls. The overall direction of the field reverses from one sunspot cycle to the next. The 22-year cycle that results when the direction of the field is taken into account is called the **solar cycle** (p. 253). Solar activity tends to be concentrated in **active regions** (p. 254) associated with sunspot groups; this activity plays an important role in heating the corona and driving the solar wind, which flows along open field lines from low-density regions of the corona called **coronal holes** (p. 255).

⑥ The Sun generates energy by converting hydrogen to helium in its core by the process of **nuclear fusion** (p. 257). When four protons are converted to a helium nucleus in the **proton–proton chain** (p. 258), some mass is lost. The **law of conservation of mass and energy** (p. 257) requires that this mass appear as energy, eventually resulting in the light we see.

⑦ **Neutrinos** (p. 258) are nearly massless particles that are produced in the proton–proton chain and escape from the Sun. Despite their elusiveness, it is possible to detect a small fraction of the neutrinos streaming from the Sun. Observations over several decades led to the **solar neutrino problem** (p. 260)—substantially fewer neutrinos are observed than are predicted by theory. The accepted explanation, supported by recent observational evidence, is that **neutrino oscillations** (p. 260) convert some neutrinos to other (undetected) particles en route from the Sun to Earth.

Review and Discussion

1. Name and briefly describe the main regions of the Sun.
2. How massive is the Sun, compared with Earth?
3. How hot is the solar surface? The solar core?
4. How do scientists construct models of the Sun?
5. Describe how electromagnetic energy generated at the center of the Sun reaches Earth.
6. Why does the Sun appear to have a sharp edge?
7. What evidence do we have for solar convection?
8. What is the solar wind?
9. What is the cause of sunspots, flares, and prominences?
10. What is the Maunder minimum?

11. What fuels the Sun's enormous energy output?
12. What is the law of conservation of mass and energy? How is it relevant to nuclear fusion in the Sun?
13. What are the ingredients and the end result of the proton–proton chain? Why is energy released in the process?
14. Why have scientists been trying so hard to detect solar neutrinos?
15. What would we observe on Earth if the Sun's internal energy source suddenly shut off? Would the Sun darken instantaneously? If not, how long do you think it might take—minutes, days, years, millions of years—for the Sun's light to begin to fade? Repeat the question for solar neutrinos.

Conceptual Self-Test: True or False?/Multiple Choice

1. The most abundant element in the Sun is hydrogen. **(T/F)**

2. There are as many absorption lines in the solar spectrum as there are elements present in the Sun. **(T/F)**

3. The density and temperature in the solar corona are much higher than in the photosphere. **(T/F)**

4. Sunspots appear dark because they are hotter than the surrounding gas of the photosphere. **(T/F)**

5. Flares are caused by magnetic disturbances in the lower atmosphere of the Sun. **(T/F)**

6. The proton–proton chain releases energy because mass is created in the process. **(T/F)**

7. Neutrinos have never been detected experimentally. **(T/F)**

8. Compared with Earth's diameter, the Sun's diameter is about **(a)** the same; **(b)** ten times larger; **(c)** one hundred times larger; **(d)** one million times larger.

9. The Sun spins on its axis roughly once each **(a)** hour; **(b)** day; **(c)** month; **(d)** year.

10. If astronomers lived on Venus instead of on Earth, the solar constant they measure would be **(a)** larger; **(b)** smaller; **(c)** the same.

11. A typical solar granule is about the size of **(a)** a U.S. city; **(b)** a large U.S. state; **(c)** the Moon; **(d)** Earth.

12. The time between successive sunspot maxima is about **(a)** a month; **(b)** a year; **(c)** a decade; **(d)** a century.

13. The primary source of the Sun's energy is **(a)** fusion of light nuclei to make heavier ones; **(b)** fission of heavy nuclei into lighter ones; **(c)** the slow release of heat left over from the Sun's formation; **(d)** the solar magnetic field.

14. According to the standard model of the Sun (Figure 9.6, Solar Interior), as the distance from the center increases, **(a)** the density decreases at about the same rate as the temperature decreases; **(b)** the density decreases faster than the temperature decreases; **(c)** the density decreases more slowly than the temperature decreases; **(d)** the density decreases but the temperature increases.

15. Comparing Figures 9.6 (Solar Interior) and 9.13 (Solar Atmospheric Temperature), the temperature in the solar corona is about the same as **(a)** in the core; **(b)** near the bottom of the radiation zone; **(c)** near the bottom of the convection zone; **(d)** at the photosphere.

Problems

Algorithmic versions of these questions are available in the Practice Problems module of the Companion Website.

The number of squares preceding each problem indicates its approximate level of difficulty.

1. ■ Use the reasoning presented in Section 9.1 to calculate the value of the "solar constant" (a) on Mercury, at perihelion, and (b) on Jupiter.

2. ■■ The largest-amplitude solar pressure waves have periods of about five minutes and move at the speed of sound in the outer layers of the Sun, about 10 km/s. (a) How far does the wave move during one wave period? (b) Approximately how many wavelengths are needed to completely encircle the Sun? (c) Compare the wave period with the orbital period of an object moving just above the solar photosphere. ∞ *(More Precisely E-2)*

3. ■ Given that the solar spectrum corresponds to a temperature of 5800 K and peaks at a wavelength of 500 nm, use Wien's law to determine the wavelength corresponding to the peak of the blackbody curve (a) in the core of the Sun, where the temperature is 10^7 K, (b) in the solar convection zone (10^5 K), and (c) just below the solar photosphere (10^4 K). ∞ *(More Precisely 2-2)* What form (visible, infrared, X-ray, etc.) does the radiation take in each case?

4. ■ If convected solar material moves at 1 km/s, how long does it take to flow across the 1000-km expanse of a typical granule? Compare this length of time with the roughly 10-minute lifetimes observed for most solar granules.

5. ■■ Use Stefan's law to calculate how much less energy is emitted per unit area of a 4500-K sunspot than is emitted per unit area of the surrounding 5800-K photosphere. ∞ *(More Precisely 2-2)* Give your answer as a percentage.

6. ■■ The solar wind carries mass away from the Sun at a rate of about 2 million ton/s (where 1 ton = 1000 kg). (a) Compare this rate with the rate at which the Sun loses mass in the form of radiation. (b) At this rate, how long would it take for all of the Sun's mass to escape?

7. ■■ The Sun's differential rotation is responsible for wrapping the solar magnetic field around the Sun (Figure 9.17). Using the figures presented in Table 9.1, estimate how long it takes for material at the solar equator to "lap" the material near the poles—that is, to complete one extra trip around the Sun's rotation axis.

8. ■■ How long does it take the Sun to convert one Earth mass of hydrogen to helium?

9. ■■ Assuming that (1) the solar luminosity has been constant since the Sun formed, and (2) the Sun was initially of uniform composition throughout, as described by Table 9.2, estimate how long it would take the Sun to convert all of its original hydrogen into helium.

10. ■■■ The entire reaction sequence shown in Figure 9.26 generates 4.3×10^{-12} joules of electromagnetic energy and releases two neutrinos. Assume that all of the Sun's energy comes from this sequence and that neutrino oscillations transform half of the neutrinos produced into other particles by the time they travel one AU from the Sun. Estimate the total number of solar neutrinos that pass through Earth each second.

The Companion Website at www.prenhall.com/chaisson provides algorithmically generated versions of each chapter's Problems, in addition to an Animations and Videos gallery, a Tutorials gallery and Tutorials Quiz questions, a full eBook, an Images gallery, Practice Questions, Multiple Choice, and True or False quizzes, and Labeling exercises.

10

Measuring the Stars

Giants, Dwarfs, and the Main Sequences

We have now studied Earth, the Moon, the solar system, and the Sun. To continue our inventory of the contents of the universe, we must move away from our local environment into the depths of space. In this chapter we take a great leap in distance and consider stars in general. Our primary goal is to comprehend the nature of the stars that make up the constellations, as well as the myriad more distant stars we cannot perceive with our unaided eyes. Rather than studying their individual peculiarities, however, we will concentrate on determining the physical and chemical properties they share. There is order in the legions of stars scattered across the sky. Like comparative planetology in the solar system, comparing and cataloging the stars play vital roles in furthering our understanding of the galaxy and the universe we inhabit.

The total number of stars, even in our local neighborhood, is virtually beyond our ability to count. Relatively few of them have been studied in detail. Yet astronomers have learned a great deal about stars in general, and their ranges of properties—their masses, their luminosities, even their ages and (as we shall see in later chapters) their destinies. Here the Very Large Telescope in Chile has imaged the magnificent star cluster NGC 3603, a rich group of more than a thousand stars roughly 20,000 light-years away and probably less than a million years old. (European Southern Observatory)

LEARNING GOALS

Studying this chapter will enable you to:

1 Explain how stellar distances are determined.

2 Distinguish between luminosity and apparent brightness, and explain how stellar luminosity is determined.

3 Explain how stars are classified according to their colors, surface temperatures, and spectral characteristics.

4 Discuss how physical laws are used to estimate stellar sizes.

5 Describe how a Hertzsprung–Russell diagram is constructed and used to identify stellar properties.

6 Explain how knowledge of a star's spectroscopic properties can lead to an estimate of its distance.

7 Explain how stellar masses are measured and how mass is related to other stellar properties.

Visit www.prenhall.com/chaisson for additional annotated images, animations, and links to related sites for this chapter.

10.1 The Solar Neighborhood

Looking deep into space, astronomers can observe and study millions of individual stars. By observing distant galaxies, we can infer statistically the properties of trillions more. All told, the observable universe probably contains several tens of sextillions (1 sextillion = 10^{21}) stars. Yet, remarkably, despite their incredible numbers, the essential properties of stars—their appearance in the sky, their births, lives, deaths, even their interactions with their environment—can be understood in terms of just a few basic physical stellar quantities: luminosity (brightness), temperature (color), chemical composition, size, and mass. In this chapter we examine these basic stellar properties, describe how we measure them, and show how they lead to a simple but powerful means of classifying stars of all kinds. With this foundation we can then discuss the life cycles of stars in the next three chapters.

As with the planets, knowing the *distances* to the stars is essential to understanding many of their other properties. We therefore begin our study of stellar astronomy by reviewing the use of simple geometry in determining the distances to our neighbors.

Stellar Parallax

The distances to the nearest stars can be measured using *parallax*—the apparent shift of an object relative to some distant background as the observer's point of view changes. ∞ *(Sec. E.4)* However, even the closest stars are so far away that no baseline on Earth is adequate to measure their parallaxes. But, by comparing observations made of a star *at different times of the year*, as shown in Figure 10.1 (compare Figure E.20), we can effectively extend the baseline to the diameter of Earth's orbit around the Sun, two AU. The parallax is determined by comparing photographs made from the two ends of the baseline. As indicated on the figure, a star's parallactic angle—or, more commonly, just its "parallax"—is conventionally defined to be *half* of its apparent shift relative to the background as we move from one side of Earth's orbit to the other.

Because the stars are so far away, stellar parallaxes are always very small, and astronomers find it convenient to measure them in arc seconds rather than in degrees. ∞ *(More Precisely E-1)* If we ask at what distance from the Sun a star must lie in order for its observed parallax to be exactly 1″, we get an answer of 206,265 AU, or 3.1×10^{16} m. ∞ *(More Precisely E-1)* Astronomers call this distance one **parsec** (1 pc), from "*par*allax in arc *sec*onds." Because parallax decreases as distance increases, we can relate a star's parallax to its distance from the Sun by the following simple formula:

$$\text{distance (in parsecs)} = \frac{1}{\text{parallax (in arcseconds)}}.$$

Thus, a star with a measured parallax of 1″ lies at a distance of 1 pc from the Sun. The parsec is defined so as to make the conversion between parallactic angle and distance easy. An object with a parallax of 0.5″ lies at a distance of $1/0.5 = 2$ pc, an object with a parallax of 0.1″ lies at $1/0.1 = 10$ pc, and so on. One parsec is approximately equal to 3.3 light-years.

◄ **Figure 10.1 Stellar Parallax** (a) The geometry of stellar parallax. For observations made six months apart, the baseline is twice the Earth–Sun distance, or two AU. Compare Figure E.20, which shows exactly the same geometry, but on a much smaller scale. (b) The parallactic shift (greatly exaggerated by the white arrow in this drawing) is usually measured photographically. Images of the same region of the sky made at different times of the year are used to determine a star's apparent movement relative to background stars.

Our Nearest Neighbors

The closest star to the Sun is called Proxima Centauri. It is a member of a triple-star system (three stars orbiting one another, bound together by gravity) known as the Alpha Centauri complex. Proxima Centauri displays the largest known stellar parallax, 0.77″, which means that it is about $1/0.77 = 1.3$ pc away—about 270,000 AU, or 4.3 light-years. That's the *nearest* star to Earth—at almost 300,000 times the Earth–Sun distance! This is a fairly typical interstellar distance in the Milky Way Galaxy.

Vast distances can sometimes be grasped by means of analogies. Imagine Earth as a grain of sand orbiting a golfball-sized Sun at a distance of 1 m. The nearest star, also a golfball-sized object, is then 270 *kilo*meters away. Except for the planets in our solar system, ranging in size from grains of sand to small marbles within 50 m of the "Sun," and the Kuiper belt and Oort cloud, trillions of tiny dust grains scattered over a volume 100 km across, nothing of consequence exists in the 270 km separating the Sun and the other star.

The next nearest neighbor to the Sun beyond the Alpha Centauri system is called Barnard's Star. Its parallax is 0.55″, so it lies at a distance of 1.8 pc, or 6.0 light-years—370 km in our model. All told, fewer than 100 stars lie within 5 pc (1000 km in our model) of the Sun. Such is the void of interstellar space. Figure 10.2 is a map of our nearest galactic neighbors—the 30 or so stars lying within 4 pc of Earth. *Discovery 10-1* (p. 277) describes how some of the stars in this figure got their peculiar names.

Ground-based images of stars are generally smeared out into a seeing disk of radius 1″ or so by turbulence in Earth's atmosphere. ∞ *(Sec. 3.3)* However, astronomers have special techniques that can routinely measure stellar parallaxes of 0.03″ or less, corresponding to stars within about 30 pc (100 light-years) of Earth. Several thousand stars lie within this range, most of them of much lower luminosity than the Sun and invisible to the naked eye. Adaptive optics systems allow even more accurate measurements of stellar positions, extending the parallax range to over 100 pc. ∞ *(Sec. 3.3)* In the 1990s, data from the European *Hipparcos* satellite have extended this range to well over 200 pc, encompassing nearly a million stars. Even so, almost all of the stars in our Galaxy are far more distant. Proposed next-generation space missions will have the incredible measurement range of up to 25,000 pc, spanning our entire Galaxy and almost certainly revolutionizing our knowledge of its structure.

Stellar Motion

In addition to the apparent motion caused by parallax, stars have real spatial motion, too. A star's *radial* velocity—along the line of sight—can be measured using the Doppler effect. ∞ *(Sec. 2.7)* For many nearby stars the *transverse* velocity—perpendicular to our line of sight—can also be determined by careful monitoring of the star's position on the sky.

Figure 10.3(a) shows two photographs of the sky around Barnard's Star, made on the same day of the year, but 22 years apart. Note that the star (marked by the arrow) has moved during the 22-year interval shown. Because Earth was at the same point in its orbit when the photographs were taken, the observed displacement is *not* the result of parallax. Instead, it indicates *real* space motion of Barnard's Star relative to the Sun. This annual movement of a star

◄◄◄ **CONCEPT CHECK**

Why can't astronomers use simultaneous observations from different parts of Earth's surface to determine stellar distances?

▼ **Figure 10.2 Sun's Neighborhood** A plot of the 30 closest stars to the Sun, projected so as to reveal their three-dimensional relationships. Notice that many are members of multiple-star systems. All lie within 4 pc (about 13 light-years) of Earth. The gridlines represent distances in the Galactic plane.

(a)

◄ **Figure 10.3 Real Space Motion**
(a) Comparison of two photographs taken 22 years apart shows evidence of real space motion for Barnard's Star (denoted by an arrow). (b) The motion of the Alpha Centauri star system drawn relative to our solar system. The radial component of the velocity is measured by using the Doppler shift of lines in Alpha Centauri's spectrum. The transverse component has been determined by observing the system's proper motion. The true spatial velocity, indicated by the red arrow, results from the combination of the two. (Harvard College Observatory)

(b)

across the sky, as seen from Earth and corrected for parallax, is called **proper motion**. Barnard's Star moved 228″ in 22 years. Its proper motion is therefore 228″/22 years, or 10.3″/yr.

A star's transverse velocity is easily calculated once its proper motion and its distance are known. At the distance of Barnard's Star (1.8 pc), an angle of 10.3″ corresponds to a distance of 0.00009 pc, or about 2.8 billion km. Barnard's Star takes a year to travel this distance, so its transverse velocity is 2.8 billion km/3.2×10^7 s, or 88 km/s. Even though stars' transverse velocities are often quite large—tens or even hundreds of kilometers per second—their great distances from the Sun mean that it usually takes many years for us to discern their movement across the sky. In fact, Barnard's Star has the largest-known proper motion of any star. Only a few hundred stars have proper motions greater than 1″/yr.

A star's total space velocity is the combination of its radial and transverse components. Figure 10.3(b) illustrates the relationship among these quantities for another near neighbor, Alpha Centauri. Spectral lines from Alpha Centauri are blueshifted by a tiny amount—about 0.0067 percent—allowing astronomers to measure the star system's radial velocity (relative to the Sun) as 300,000 km/s $\times$ 6.7×10^{-5} = 20 km/s toward us. ∞ *(Sec. 2.7)* Alpha Centauri's proper motion has been measured to be 3.7″/yr. At Alpha Centauri's distance of 1.35 pc (corresponding to a parallax of 0.74″), that measurement implies a transverse velocity of 24 km/s. We can combine the transverse (24 km/s) and radial (20 km/s) velocities according to the Pythagorean theorem, as indicated in the figure. The total velocity is $\sqrt{24^2 + 20^2}$, or about 31 km/s, in the direction shown by the horizontal red arrow.

10.2 Luminosity and Apparent Brightness

Luminosity is an *intrinsic* property of a star—it does not depend in any way on the location or motion of the observer. It is sometimes referred to as the star's *absolute brightness*. However, when we look at a star, we see not its luminosity but rather its **apparent brightness**—the amount of energy striking unit area of some light-sensitive surface or device, such as a human eye or a CCD chip, per unit time. (Notice that this is precisely the same definition as we saw in Chapter 9 for the solar constant—in our new terminology, the solar constant is just the apparent brightness of the Sun.) ∞ *(Sec. 9.1)* In this section we discuss how these important quantities are related to one another.

Another Inverse-Square Law

Figure 10.4 shows light leaving a star and traveling through space. Moving outward, the radiation passes through imaginary spheres of increasing radius surrounding the source. The amount of radiation leaving the star per unit

▲ **Figure 10.4 Inverse-Square Law** As light moves away from a source such as a star, it is steadily diluted while spreading over progressively larger surface areas (depicted here as sections of spherical shells). Thus, the amount of radiation received by a detector (the source's apparent brightness) varies inversely as the square of its distance from the source.

time—the star's luminosity—is constant, so the farther the light travels from the source, the less energy passes through each unit of area. Think of the energy as being spread out over an ever larger area and therefore spread more thinly, or "diluted," as it expands into space. Because the area of a sphere grows as the square of the radius, the energy per unit area—the star's apparent brightness—is inversely proportional to the square of the distance from the star. Doubling the distance from a star makes it appear 2^2, or four, times fainter. Tripling the distance reduces the apparent brightness by a factor of 3^2, or nine, and so on.

Of course, the star's luminosity also affects its apparent brightness. Doubling the luminosity doubles the energy crossing any spherical shell surrounding the star and hence doubles the apparent brightness. The apparent brightness of a star is therefore directly proportional to the star's luminosity and inversely proportional to the square of its distance:

$$\text{apparent brightness} \propto \frac{\text{luminosity}}{\text{distance}^2}.$$

(Reminder: $\propto$ is the symbol representing proportionality.) Thus, two identical stars can have the same apparent brightness if (and only if) they lie at the same

10-1 DISCOVERY

Naming the Stars

Looking at Figure 10.2, you might well wonder how the stars shown there got their names. In fact, there is no coherent convention for naming stars—or rather, there are many, often leading to many names for any given object! Here is a brief (and roughly chronological) description of some common naming schemes:

1. The brightest stars in the sky usually have ancient names. Many of the names are of Arabic origin, dating back to around the tenth century A.D., when Muslim astronomy flourished, preserving the scientific knowledge of Greek and other early cultures while European science floundered in the Dark Ages. Examples are the well-known stars Betelgeuse (from *Ibt al Jauzah*, Armpit of the Central One, corrupted to Bed Elgueze), Rigel (from *Rijl Jauzah al Yusra*, Left Foot of the Giant), Aldebaran (from *Al Dabaran*, the Follower), and Deneb (from *Al Dhanab al Dulfin*, the Dolphin's Tail). The names Procyon and Sirius (see Figure 10.2) are Greek.

2. A more systematic scheme was introduced in 1603 by German lawyer Johann Bayer. We rank the stars in a given constellation (Orion, say) by brightness, labeling them with Greek letters: Alpha Orionis (Betelgeuse), Beta Orionis (Rigel), or Epsilon Eridani (Figure 10.2). ∞ *(Sec. E.1)*

3. There are only 24 Greek letters, so eventually the Bayer scheme breaks down. In the early eighteenth century, British Astronomer Royal John Flamsteed suggested simply numbering the stars from west to east within a constellation. In the Flamsteed scheme, Betelgeuse is 58 Ori, Rigel is 19 Ori, and so on (see also 61 Cygni in Figure

10.2). Sometimes the Bayer and Flamsteed designations are combined, as in "58 α Ori."

4. As telescopes improved and more and more stars were discovered, astronomers started compiling their own catalogues, each with their own numbering system, and the confusion really started! Some catalogues listed stars by celestial coordinates—see, for example, star BD + 5°1668 in Figure 10.2, which appears in the nineteenth-century German Bonner Durchmusterung (BD) catalogue, and lies just over 5° above the celestial equator. ∞ *(Sec. E.1)* Other catalogues simply named stars by the order in which they were discovered (or added to the list)—the star Lalande 21185 in Figure 10.2 is the 21,185th entry in a catalogue compiled by eighteenth-century French astronomer Joseph de Lalande. Unfortunately, the catalogues often overlap, so any given star probably appears in many different catalogues with very different names. For example, Lalande 21185 is also known as HD 95735 (in the Henry Draper catalogue), while Betelgeuse is also known as HD 39801 and SAO 113271 (in the Smithsonian Astrophysical Observatory catalogue), among many other aliases.

5. Today newly discovered stars are defined simply by their coordinates in the sky, and not given any special name. If you like, you can (for a fee, of course) register your own star name, but don't expect any astronomer ever to cite that name, or to take the registry...seriously—only "official" names approved by *International Astronomical Union* are used in astronomy.

Many other naming schemes exist (e.g., UV Ceti in Figure 10.2, which is a variable star), and still other catalogues exist for galaxies, but we won't go into them here. By now you are probably completely perplexed—our job as astronomers is done!

◄ **Figure 10.5 Luminosity** Two stars A and B of different luminosities can appear equally bright to an observer on Earth if the brighter star B is more distant than the fainter star A.

distance from Earth. However, as illustrated in Figure 10.5, two nonidentical stars can also have the same apparent brightness if the more luminous one lies farther away. A bright star (that is, one having large apparent brightness) is a powerful emitter of radiation (high luminosity), is near Earth, or both. A faint star (small apparent brightness) is a weak emitter (low luminosity), is far from Earth, or both.

Determining a star's luminosity is a twofold task. First, the astronomer must determine the star's apparent brightness by measuring the amount of energy detected through a telescope in a given amount of time. Second, the star's distance must be measured—by parallax for nearby stars and by other means (to be discussed later) for more distant stars. The luminosity can then be found using the inverse-square law. This is basically the same reasoning we used earlier in our discussion of how astronomers measure the luminosity of the Sun. ∞ *(Sec. 9.1)*

The Magnitude Scale

Instead of measuring apparent brightness in SI units (for example, watts per square meter, W/m², the unit in which we expressed the solar constant in Section 9.1), optical astronomers find it more convenient to work in terms of a construct called the **magnitude scale**. This scale dates back to the second century B.C., when the Greek astronomer Hipparchus ranked the naked-eye stars into six groups. The brightest stars were categorized as first magnitude. The next brightest stars were labeled second magnitude, and so on, down to the faintest stars visible to the naked eye, which were classified as sixth magnitude. The range one (brightest) through six (faintest) spanned all the stars known to the ancients. Notice that a *large* magnitude means a *faint* star.

When astronomers began using telescopes with sophisticated detectors to measure the light received from stars, they quickly discovered two important facts about the magnitude scale. First, the one through six magnitude range defined by Hipparchus spans about a factor of 100 in apparent brightness—a first-magnitude star is approximately 100 times brighter than a sixth-magnitude star. Second, the characteristics of the human eye are such that a change of one magnitude corresponds to a *factor* of about 2.5 in apparent brightness. In other words, to the human eye a first-magnitude star is roughly 2.5 times brighter than a second-magnitude star, which in turn is roughly 2.5 times brighter than a third-magnitude star, and so on. (By combining factors of 2.5, we confirm that a first-magnitude star is indeed $(2.5)^5 \approx 100$ times brighter than a sixth-magnitude star.)

In the modern version of the magnitude scale, astronomers *define* a change of five in the magnitude of an object to correspond to *exactly* a factor of 100 in apparent brightness. Because we are really talking about apparent (rather than absolute) brightnesses, the numbers in Hipparchus's ranking system are now called **apparent magnitudes**. In addition, the scale is no longer

Apparent magnitude

30	*Hubble, Keck* limit (30)
	Hale telescope limit (28)
	4-meter telescope (CCD detector) limit (26)
20	
	1-meter telescope limit (18-20)
	25-cm (10-inch) telescope limit (14)
	Binocular limit (10)
10	Barnard's Star (9.5)
	Naked-eye limit (6)
	Polaris (2.5)
	Betelgeuse (0.8)
0	Alpha Centauri (0)
	Sirius (–1.5)
	Venus (at brightest –4.4)
–10	
	Full moon (–12.5)
–20	
	Sun (–26.7)
–30	

◄ **Figure 10.6 Apparent Magnitude** This graph illustrates the apparent magnitudes of some astronomical objects, and the limiting magnitudes (that is, the faintest magnitudes attainable) of some telescopes used to observe them. The original magnitude scale was defined so that the brightest stars in the night sky had magnitude 1 while the faintest stars visible to the naked eye had magnitude 6. It has since been extended to cover much brighter and much fainter objects. An increase of 1 in apparent magnitude corresponds to a decrease in apparent brightness by a factor of approximately 2.5.

limited to whole numbers, and magnitudes outside the original range 1–6 are allowed—very bright objects can have apparent magnitudes less than 1, and very faint objects can have apparent magnitudes much greater than 6. Figure 10.6 illustrates the apparent magnitudes of some astronomical objects, ranging from the Sun, at −26.8, to the faintest object detectable by the *Hubble* or *Keck* telescopes, at an apparent magnitude of +30—about as faint as a firefly seen from a distance equal to Earth's diameter.

Apparent magnitude measures a star's apparent brightness when seen at the star's actual distance from us. To compare intrinsic, or absolute, properties of stars, however, astronomers imagine looking at all stars from a standard distance of 10 pc. (There is no particular reason to use 10 pc—it is simply convenient.) A star's **absolute magnitude** is its apparent magnitude when viewed from a distance of 10 pc. Because distance is fixed in this definition, absolute magnitude is a measure of a star's absolute brightness, or luminosity. Despite the Sun's large negative (that is, very bright) apparent magnitude, our star's absolute magnitude is 4.8. In other words, if the Sun were moved to a distance of 10 pc from Earth, it would be only a little brighter than the faintest stars visible in the night sky. As discussed further in *More Precisely 10-1*, the numerical difference between a star's absolute and apparent magnitudes is a measure of the distance to the star.

◄◄◄ CONCEPT CHECK

Two stars are observed to have the same apparent magnitude. Based on this information, what, if anything, can be said about their luminosities?

10-1 MORE PRECISELY

More on the Magnitude Scale

Let's restate our discussion of two important topics—stellar luminosity and the inverse-square law—in terms of magnitudes.

Absolute magnitude is equivalent to luminosity—an intrinsic property of a star. Given that the Sun's absolute magnitude is 4.8, we can construct a conversion chart (shown at right) relating these two quantities. Since an increase in brightness by a factor of 100 corresponds to a decrease in magnitude by five units, it follows that a star with luminosity 100 times that of the Sun has absolute magnitude $4.8 - 5 = -0.2$, while a 0.01-solar luminosity star has absolute magnitude $4.8 + 5 = 9.8$. We can fill in the gaps by noting that one magnitude corresponds to a factor of $100^{1/5} \approx 2.512$, two magnitudes to $100^{2/5} \approx 6.310$, and so on. A factor of 10 in brightness corresponds to 2.5 magnitudes. You can use this chart to convert between solar luminosities and absolute magnitudes in many of the figures in this and later chapters.

To cast the *inverse-square law* in these terms, recall that increasing the distance to a star by a factor of 10 decreases its apparent brightness by a factor of 100 (by the inverse-square law) and hence increases its apparent magnitude by 5 units. Increasing the distance by a factor of 100 increases the apparent magnitude by 10, and so on. Every increase in distance by a factor of 10 increases the apparent magnitude by 5.

Since absolute magnitude is simply apparent magnitude at a distance of 10 pc, we can write:

apparent magnitude − absolute magnitude

$$= 5 \log_{10}\left(\frac{\text{distance}}{10 \text{ pc}}\right).$$

(The *logarithm* function—the LOG key on your calculator—is defined by the property that, if $a = \log_{10}(b)$, then $b = 10^a$.) Even though it doesn't look much like it, this equation is precisely equivalent to the inverse-square law presented in the text!

For stars more than 10 pc from Earth, the apparent magnitude is greater than the absolute magnitude, while the reverse is true for stars closer than 10 pc. Thus, for example, the Sun, with an absolute magnitude of 4.8, seen from a distance of 100 pc, would have an apparent magnitude of $4.8 + 5 \log_{10}(100) = 14.8$, since $\log_{10}(100) = 2$. This is well below the threshold of visibility for binoculars or even a large amateur telescope (see Figure 10.6). We can also turn this around to illustrate how knowledge of a star's absolute and apparent magnitudes tells us its distance. The star Rigel Kentaurus (also known as Alpha Centauri) has absolute magnitude +4.34 and is observed to have apparent magnitude −0.01. Its magnitude difference is −4.35, and its distance must therefore be $10 \text{ pc} \times 10^{-4.34/5} = 1.35 \text{ pc}$, in agreement with the result (obtained by parallax) given in the text.

10.3 Stellar Temperatures

Color and the Blackbody Curve

Looking at the night sky, you can tell at a glance which stars are hot and which are cool. In Figure 10.7(a), which shows the constellation Orion as it appears through a small telescope, the colors of the cool red star Betelgeuse (α) and the hot blue star Rigel (β) are clearly evident. ∞ *(Sec. E.2)* Astronomers can determine a star's surface temperature by measuring its apparent brightness (radiation intensity) at several frequencies, then matching the observations to the appropriate blackbody curve. ∞ *(Sec. 2.4)* In the case of the Sun, the theoretical curve that best fits the emission describes a 5800-K emitter. The same technique works for any star, regardless of its distance from Earth.

Because the basic shape of the blackbody curve is so well understood, astronomers can estimate a star's temperature using as few as *two* measurements at selected wavelengths. This is accomplished through the use of telescope filters that block out all radiation except that within specific wavelength ranges. For example, a B (blue) filter rejects all radiation except for a certain range of violet to blue light. Similarly, a V (visual) filter passes only radiation in the green to yellow range (the part of the spectrum to which human eyes happen to be particularly sensitive).

Figure 10.8 shows how these filters admit different amounts of light for objects of different temperatures. In curve (a), corresponding to a very hot 30,000-K emitter, considerably more radiation is received through the B filter than through the V filter. In curve (b) the temperature is 10,000 K and the B and V intensities are about the same. In the cool 3000-K curve (c) far more energy is received in the V range than in the B range. In each case it is possible to reconstruct the entire blackbody curve on the basis of only those two measurements because no other blackbody curve can be drawn through both measured points.

To the extent that a star's spectrum is well approximated as a blackbody, measurements of the B and V intensities are enough to specify the star's blackbody curve and thus yield its surface temperature. Astronomers often refer to the ratio of a star's B to V intensities (or, equivalently, the difference between the star's apparent magnitudes measured through the B and V filters) as the star's *color index* (or just its color). Table 10.1 lists surface temperatures and colors for a few well-known stars.

Stellar Spectra

Color is a useful way to describe a star, but astronomers often use a more detailed scheme to classify stellar properties, incorporating additional knowledge of stellar physics obtained through spectroscopy. ∞ *(Sec. 2.5)* Figure 10.9 compares the spectra of seven stars, arranged in order of decreasing surface temperature. All the spectra extend from 400 to 650 nm. Like the solar spectrum, each shows a series of dark absorption lines superimposed on a background of continuous color. ∞ *(Sec. 9.3)* However, the line patterns differ

(a)

(b)

R I V U X G

◄ **Figure 10.7 Star Colors** (a) The different colors of the stars making up the constellation Orion are easily distinguished in this photograph taken by a wide-field camera attached to a small telescope. The bright red star at the upper left (α) is Betelgeuse; the bright blue-white star at the lower right (β) is Rigel. (Compare with Figure E.2.) The scale of this photograph is about 20° across. (b) An incredibly rich field of colorful stars, this time in the direction of the center of the Milky Way. Here, the field of view is just 2 arc minutes across—much smaller than in (a). The image is heavily populated with stars of many different temperatures. (J Sanford/Astrostock-Sanford; NASA)

TABLE 10.1	Stellar Colors and Temperatures	
SURFACE TEMPERATURE (K)	COLOR	FAMILIAR EXAMPLES
30,000	electric blue	Mintaka (δ Orionis)
20,000	blue	Rigel
10,000	white	Vega, Sirius
7,000	yellow-white	Canopus
6,000	yellow	Sun, Alpha Centauri
4,000	orange	Arcturus, Aldebaran
3,000	red	Betelgeuse, Barnard's Star

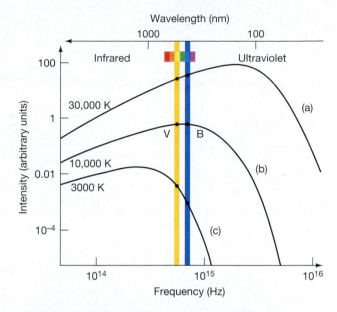

▲ **Figure 10.8 Blackbody Curves** Blackbody curves for three temperatures, along with the locations of the B (blue) and V (visual) filters. Star (a) is very hot—30,000 K—so its B intensity is greater than its V intensity. Star (b) has roughly equal B and V readings and so appears white. Its temperature is about 10,000 K. Star (c) is red; its V intensity greatly exceeds the B value, and its temperature is 3000 K.

greatly from one star to another. Some stars display strong lines in the long-wavelength part of the spectrum, while others have their strongest lines at short wavelengths. Still others show strong absorption lines spread across the whole visible spectrum. What do these differences tell us?

Although spectral lines of many elements are present at widely varying strengths, the differences among the spectra in Figure 10.9 are not due to differences in chemical composition—all seven stars are, in fact, more or less solar in their makeup. Instead, the differences are due almost entirely to the stars' *temperatures*. The spectrum at the top of Figure 10.9 is exactly what we expect from a star having solar composition and a surface temperature of about 30,000 K, the second from a 20,000-K star, and so on, down to the 3000-K star at the bottom.

The main differences among the spectra in Figure 10.9 are as follows:

- Spectra of stars having surface temperatures exceeding 25,000 K usually show *strong* absorption lines of singly ionized helium (i.e., helium atoms that have lost one orbiting electron) and multiply ionized heavier elements, such as oxygen, nitrogen, and silicon (the latter lines are not shown in the figure). These strong lines are not seen in the spectra of cooler stars because only very hot stars can excite and ionize such tightly bound atoms.
- In contrast, the hydrogen absorption lines in the spectra of very hot stars are relatively *weak*. The reason is not a lack of hydrogen, which is by far the most abundant element in all stars. At these high temperatures, however, much of the hydrogen is ionized, so there are few intact hydrogen atoms to produce strong spectral lines.
- Hydrogen lines are strongest in stars having intermediate surface temperatures of around 10,000 K. This temperature is just right for electrons to move frequently between hydrogen's second and higher orbitals, producing the characteristic visible hydrogen spectrum. ∞ *(Sec. 2.6)*
- Hydrogen lines are again weak in stars with surface temperatures below about 4000 K, but now because the temperature is too low to boost many electrons out of the ground state. ∞ *(Sec. 2.6)* The most intense spectral lines in these stars are due to weakly excited heavy atoms and from some molecules, as noted at the bottom of Figure 10.9.

▶ **Figure 10.9 Stellar Spectra** Comparison of spectra observed for seven stars having a range of surface temperatures. These are not actual spectra, which are messy and complex, but simplified artists' renderings illustrating a few spectral features. The spectra of the hottest stars, at the top, show lines of helium and multiply ionized heavy elements. In the coolest stars, at the bottom, there are no lines for helium, but lines of neutral atoms and molecules are plentiful. At intermediate temperatures, hydrogen lines are strongest. All seven stars have about the same chemical composition.

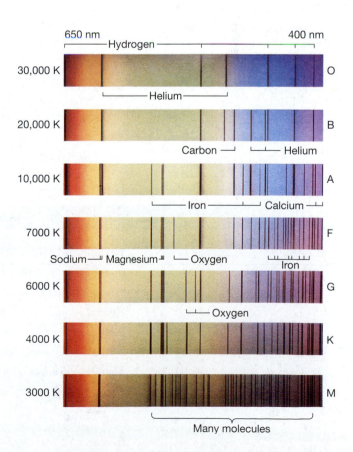

Stellar spectra are the source of all the detailed information we have on stellar composition, and they do in fact reveal significant differences in composition among stars, particularly in the abundances of carbon, nitrogen, oxygen, and heavier elements. However, as we have just seen, these differences are *not* the primary reason for the different spectra that are observed. Instead, the main determinant of a star's spectral appearance is its temperature, and stellar spectroscopy is a powerful and precise tool for measuring this important stellar property.

Spectral Classification

Stellar spectra like those shown in Figure 10.9 were obtained for numerous stars around the start of the twentieth century, as observatories around the world amassed stellar spectra from both hemispheres of the sky. Between 1880 and 1920, researchers correctly identified many of the observed spectral lines by comparing them with lines obtained in the laboratory. However, modern atomic theory had not yet been developed, so the correct interpretation of the line strengths, as just described, was impossible at the time. Lacking full understanding of how atoms produce spectra, astronomers organized their data by classifying stars according to their most prominent spectral features—especially their hydrogen-line intensities. They adopted an A, B, C, D,... scheme in which A stars, with the strongest hydrogen lines, were thought to have more hydrogen than did B stars, and so on. The classification extended as far as the letter P.

In the 1920s scientists began to understand the intricacies of atomic structure and the causes of spectral lines. Astronomers quickly realized that stars could be more meaningfully classified according to surface temperature. Instead of adopting an entirely new scheme, however, they chose to shuffle the existing alphabetical categories—those based on the strength of hydrogen lines—into a new sequence based on temperature. In the modern scheme, in order of decreasing temperature, the letters now run O, B, A, F, G, K, M. (The other letter classes have been dropped.) These stellar designations are called **spectral classes** (or *spectral types*). Use the mnemonic "**O**h, **B**e **A** **F**ine **G**uy/**G**irl, **K**iss **M**e" to remember them in the correct order.[1]

---

[1]Recently, some astronomers have proposed the addition of two new spectral classes—L and T—for low-mass, low-temperature stars whose odd spectra distinguish them from the M-class stars in the current scheme. For now, at least, the new classification is not in widespread use. These new objects are probably not "true" stars, fusing hydrogen into helium in their cores—rather, they are brown dwarfs (see Section 11.4) that never achieved high enough central temperatures for fusion to begin.

TABLE 10.2 Spectral Classes			
SPECTRAL CLASS	TEMPERATURE (K)	PROMINENT ABSORPTION LINES	FAMILIAR EXAMPLES
O	30,000	Ionized helium strong; multiply ionized heavy elements; hydrogen faint	Mintaka (O9)
B	20,000	Neutral helium moderate; singly ionized heavy elements; hydrogen moderate	Rigel (B8)
A	10,000	Neutral helium very faint; singly ionized heavy elements; hydrogen strong	Vega (A0), Sirius (A1)
F	7000	Singly ionized heavy elements; neutral metals; hydrogen moderate	Canopus (F0)
G	6000	Singly ionized heavy elements; neutral metals; hydrogen relatively faint	Sun (G2), Alpha Centauri (G2)
K	4000	Singly ionized heavy elements; neutral metals strong; hydrogen faint	Arcturus (K2), Aldebaran (K5)
M	3000	Neutral atoms strong; molecules moderate; hydrogen very faint	Betelgeuse (M2)
			Barnard's Star (M5)

Astronomers further divide each lettered spectral class into 10 subdivisions, denoted by the numbers zero to nine. By convention, the lower the number, the hotter the star. Thus, for example, our Sun is classified as a G2 star (a little cooler than G1 and a little hotter than G3), Vega is A0, Barnard's Star is M5, Betelgeuse is M2, and so on. Table 10.2 lists the main properties of each stellar spectral class for the stars presented in Table 10.1.

We should not underestimate the importance of the early work in classifying stellar spectra. Even though the original classification was based on erroneous assumptions, the painstaking accumulation of large quantities of accurate and well-categorized data paved the way for rapid improvements in understanding once a theory came along that explained the observations.

10.4 Stellar Sizes

Direct and Indirect Measurements

Virtually all stars are unresolvable points of light in the sky, even when viewed through the largest telescopes. Still, a few are big enough, bright enough, and close enough to allow us to measure their sizes *directly*. In some cases, the results are even detailed enough to allow a few surface features to be distinguished (Figure 10.10). By measuring a star's angular size and knowing its distance from the Sun, astronomers can determine its radius by simple geometry. ∞ *(More Precisely E-1)* The sizes of a few dozen stars have been measured in this way.

Most stars are too distant or too small for such direct measurements to be possible. Their sizes must be inferred by more indirect means, using the radiation laws. ∞ *(Sec. 2.4)* According to the Stefan–Boltzmann law, the rate at which a star emits energy into space—the star's *luminosity*—is proportional to the fourth power of the star's *surface temperature*. ∞ *(More Precisely 2-2)* However, the luminosity also depends on the star's *surface area*, because large bodies radiate more energy than do small bodies having the same surface temperature. Because the surface area of a star is proportional to the square of its radius, we can combine these proportionalities to say

$$\text{luminosity} \propto \text{radius}^2 \times \text{temperature}^4.$$

This **radius–luminosity–temperature relationship** is important because it demonstrates that knowledge of a star's luminosity and temperature can yield an estimate of its radius—an *indirect* determination of stellar size.

Giants and Dwarfs

Let's consider some examples to clarify these ideas. *More Precisely 10-2* (p. 286) presents a little more detail on the calculations involved. The star known as Aldebaran (the orange-red "eye of the bull" in the constellation Taurus) has a surface temperature of about 4000 K and a luminosity of 1.3×10^{29} W. Thus, its surface temperature is 0.7 times, and its luminosity about 330 times, the corresponding quantities for our Sun. The radius–luminosity–temperature

◀◀◀ **CONCEPT CHECK**

Why does a star's spectrum depend on its temperature?

(a)

R I V U X G

(b)

Size of Earth's orbit

R I V U X G

▶ **Figure 10.10 Betelgeuse** (a) The swollen star Betelgeuse (shown here in false color) is close enough for us to directly resolve its size, along with some surface features thought to be storms similar to those that occur on the Sun. Betelgeuse is such a huge star (600 times the size of the Sun) that its photosphere exceeds the size of Mars's orbit. Most of the surface features discernible here are larger than the entire Sun. (The dark lines are added contours of constant brightness, not part of the star itself.) (b) An ultraviolet view of Betelgeuse, in false color, as seen by a European camera onboard the *Hubble Space Telescope*, shows more clearly just how large this huge star is. (NOAO; NASA/ESA)

CONCEPT CHECK ▶▶▶

Can we measure the radius of a star without knowing its distance?

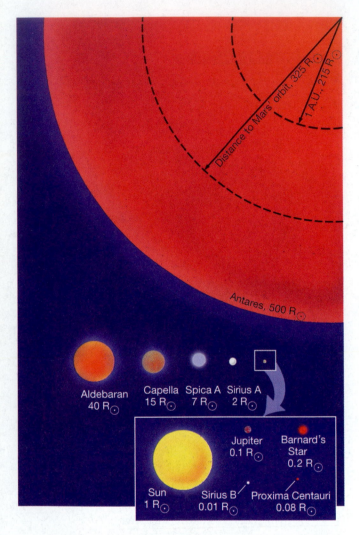

▲ **Figure 10.11 Stellar Sizes** Star sizes vary greatly. Shown here are the estimated sizes of several well-known stars, including a few of those discussed in this chapter. Only part of the red-giant star Antares can be shown on this scale; the supergiant Betelgeuse would fill the entire page. (Here and in other figures, the symbol "⊙" stands for the Sun, so the symbol "R⊙" here means "solar radius.")

relationship *(More Precisely 10-2)* then implies that the star's radius is almost 40 times the solar value. If our Sun were that large, its photosphere would extend halfway to the orbit of Mercury and, seen from Earth, would cover more than 20° on the sky.

A star as large as Aldebaran is known as a *giant*. More precisely, **giants** are stars having radii between 10 and 100 times that of the Sun. Since any 4000-K object is reddish in color, Aldebaran is known as a **red giant**. Even larger stars, ranging up to 1000 solar radii in size, are known as **supergiants**. Betelgeuse is a prime example of a **red supergiant**.

Now consider Sirius B—a faint companion to Sirius A, the brightest star in the night sky. Sirius B's surface temperature is roughly 24,000 K, four times that of the Sun. Its luminosity is 10^{25} W, about 0.025 times the solar value. Substituting these quantities into our radius–luminosity–temperature relationship we obtain a radius of 0.01 times the solar radius—roughly the size of Earth. Sirius B is much hotter but smaller and fainter than our Sun. Such a star is known as a *dwarf*. In astronomical parlance, the term **dwarf** refers to any star of radius comparable to or smaller than the radius of the Sun (including the Sun itself). Because any 24,000 K object glows bluish-white, Sirius B is an example of a **white dwarf**.

Stellar radii range from less than 0.01 to more than 100 times the radius of the Sun. Figure 10.11 illustrates the sizes of a few well-known stars.

10.5 The Hertzsprung–Russell Diagram

Figure 10.12 plots luminosity versus temperature for a few well-known stars. A figure of this sort is called a *Hertzsprung–Russell diagram*, or **H–R diagram**, after Danish astronomer Ejnar Hertzsprung and U.S. astronomer Henry Norris Russell, who independently pioneered the use of such plots in the second decade of the twentieth century. The vertical luminosity scale, expressed in units of solar luminosity (3.9×10^{26} W), extends over a large range, from 10^{-4} to 10^4. The Sun appears right in the middle of the luminosity range, at a luminosity of 1. Surface temperature is plotted on the horizontal axis, although in the unconventional sense of temperature increasing to the *left*, so that the spectral sequence O, B, A, . . . reads left to right. (Consult also Overlay 1 of the acetate insert near the middle of this book to see how sizes and colors of the data points can be used to represent stellar radii and temperatures, respectively.)

The Main Sequence

The few stars plotted in Figure 10.12 give little indication of any particular connection between stellar properties. However, as Hertzsprung and Russell plotted more and more stellar temperatures and luminosities, they found that a relationship does in fact exist. Stars are *not* uniformly scattered across the H–R diagram. Instead, most are confined to a fairly well-defined band stretching diagonally from top left (high-temperature, high-luminosity) to bottom right (low-temperature, low-luminosity). In other words, cool stars tend to be faint (less luminous) and hot stars tend to be bright (more luminous). This band of stars spanning the H–R diagram is known as the **main sequence**. Figure 10.13 is an H–R diagram for stars lying within 5 pc of the Sun. Note that most stars in the solar neighborhood lie on the main sequence.

A useful analogy involves human beings. Our height and weight are also correlated along a "main sequence." That is, as shown near Figure 10.13, the majority of the human population falls within a diagonal area (bounded by the shaded area). This is true since tall people usually weigh more than shorter ones. Some exceptions—including dwarfs and basketball players—lie outside this well-defined range.

▶ **Figure 10.12 H–R Diagram of Well-Known Stars** A plot of luminosity against surface temperature (or spectral class), known as an H–R diagram, is a useful way to compare stars. Plotted here are the data for some stars mentioned earlier in the text. The Sun, of course, has a luminosity of 1 solar unit. Its temperature, read off the bottom scale, is 5800 K—a G-type star. Similarly, the B-type star Rigel, at top left, has a temperature of about 11,000 K and a luminosity more than 10,000 times that of the Sun. The M-type star Proxima Centauri, at bottom right, has a temperature of 3000 K and a luminosity less than 1/10,000 that of the Sun. (See also Overlay 1 of the acetate insert.)

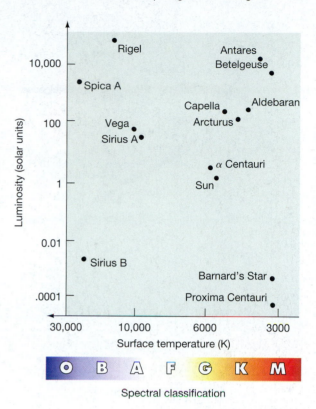

The surface temperatures of main-sequence stars range from about 3000 K (spectral class M) to more than 30,000 K (spectral class O). This temperature range is relatively small—only a factor of 10. In contrast, the observed range in luminosities is very large, covering eight orders of magnitude (that is, a factor of 100 million), from 10^{-4} to 10^4 times the luminosity of the Sun.

Using the radius–luminosity–temperature relationship (Section 10.4), astronomers find that stellar radii also vary along the main sequence. The faint, red M-type stars in the bottom right of the H–R diagram are only about 1/10 the size of the Sun, whereas the bright, blue O-type stars in the upper left are about 10 times larger than the Sun. The dashed lines in Figure 10.13 represent constant stellar radius, meaning that any star lying on a given line has the same radius, regardless of its luminosity or temperature. Along a constant-radius line, the radius–luminosity–temperature relationship implies

$$\text{luminosity} \propto \text{temperature}^4.$$

By including such lines on our H–R diagrams, we can indicate stellar temperatures, luminosities, and radii on a single plot.

At the top end of the main sequence, the stars are large, hot, and very luminous. Because of their size and color, they are referred to as **blue giants**. The very largest are called **blue supergiants**. At the other end, stars are small, cool, and faint. They are known as **red dwarfs**. Our Sun lies right in the middle.

Figure 10.14 shows an H–R diagram for a different group of stars—the 100 stars of known distance having the greatest apparent brightness, as seen from Earth. Notice that here there are many more stars at the upper end of the main sequence than at the lower end. The reason for this excess of blue giants is simple—we can see very luminous stars a long way off. The stars shown in Figure 10.14 are scattered through a much greater volume of space than those in Figure 10.13, and the Figure 10.14 sample is heavily biased toward the most luminous objects. In fact, of the 20 brightest stars in the sky, only six lie within 10 pc of us. The rest are visible, despite their great distances, because of their high luminosities.

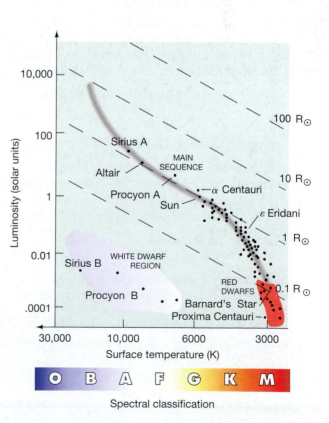

▶ **Figure 10.13 H–R Diagram of Nearby Stars** Most stars have properties within the long, thin shaded region of the H–R diagram known as the main sequence. The points plotted here are for stars lying within about 5 pc of the Sun. Each dashed diagonal line corresponds to a constant stellar radius, so that stellar size can be indicated on the same diagram as stellar luminosity and temperature. (Recall that the symbol "$R_\odot$" means "solar radius.") As noted in the analogy, most people fall along a "main sequence" as well.

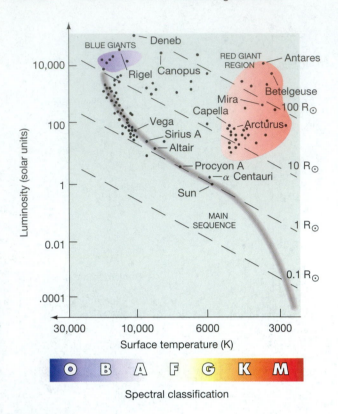

An H–R diagram for the 100 brightest stars in the sky is biased in favor of the most luminous stars—which appear toward the upper left—because we can see them more easily than we can the faintest stars. (Compare with Figure 10.13, which shows only the closest stars.)

If very luminous blue giants are overrepresented in Figure 10.14, low-luminosity red dwarfs are surely underrepresented. In fact, no dwarfs appear on this diagram. This absence is not surprising, because low-luminosity stars are difficult to observe from Earth. In the 1970s, astronomers began to realize that they had greatly underestimated the number of red dwarfs in our galaxy. As hinted at by the H–R diagram in Figure 10.13, which shows an unbiased sample of stars in the solar neighborhood, red dwarfs are actually the most common type of star in the sky. They probably account for upward of 80 percent of all stars in the universe. In contrast, O- and B-type supergiants are extremely rare—only about one star in 10,000 falls into this category.

The White-Dwarf and Red-Giant Regions

Some of the points plotted in Figures 10.12–10.14 clearly do not lie on the main sequence. One such point in Figure 10.12 represents Sirius B, the white dwarf we met in Section 10.4 (see *More Precisely 10-2*). Its surface temperature (24,000 K) is about four times greater than that of the Sun, but its luminosity is hundreds of times less than the solar value. A few more such faint bluish-white stars can be seen in the bottom left-hand corner of Figure 10.13, in the **white-dwarf region**.

10-2 MORE PRECISELY

Estimating Stellar Radii

We can combine the Stefan–Boltzmann law $F = \sigma T^4$ with the formula for the area of a sphere, $A = 4\pi r^2$, to obtain the relationship between radius (r), luminosity (L), and temperature (T) described in the text ∞ *(More Precisely 2–2)*:

$$L = FA = 4\pi\sigma R^2 T^4$$

or

$$\text{luminosity} \propto \text{radius}^2 \times \text{temperature}^4.$$

If we adopt convenient solar units, in which L is measured in solar luminosities (3.9×10^{26} W), r is measured in solar radii (696,000 km), and T in units of the solar temperature (5800 K), we can remove the constant $4\pi\sigma$ and write this equation as

$$L\text{(in solar luminosities)} = R^2 \text{ (in solar radii)} \times T^4 \text{ (in units of 5800 K)}.$$

As illustrated in the accompanying figure, both the radius and the temperature are important in determining the star's luminosity.

To compute the radius of a star from its luminosity and temperature, we rearrange this equation to read (in the same units)

$$R = \frac{\sqrt{L}}{T^2}.$$

This simple application of the radiation laws is the basis for almost every estimate of stellar size made in this book.

Let's take as examples the two stars discussed in the text. The red giant star Aldebaran has luminosity $L = 330$ units and temperature $T = 4000/5800 = 0.69$ unit. According to the foregoing equation for R, its radius is therefore $\sqrt{330}/0.69^2 = 18/0.48 = 39$ times the radius of the Sun. At the opposite extreme, the white dwarf Sirius B has $L = 0.025$ and $T = 4.7$, so its radius is $\sqrt{0.025}/4.7^2 = 0.16/22 \approx 0.007$ solar radii.

Blue giant	Red giant	Red dwarf
10^4 $L_\odot$	80 $L_\odot$	0.05 $L_\odot$
20 $R_\odot$	20 $R_\odot$	0.5 $R_\odot$
13,000 K	4000 K	4000 K

▶ **Figure 10.15** *Hipparcos* **H–R Diagram** This simplified version of the most complete H–R diagram ever compiled represents more than 20,000 data points, as measured by the European *Hipparcos* spacecraft for stars within a few hundred parsecs of the Sun.

Also shown in Figure 10.12 is Aldebaran, with a surface temperature (4000 K) about half that of the Sun, but its luminosity is more than 300 times greater than the Sun's. Another point represents Betelgeuse, the ninth brightest star in the sky, a little cooler than Aldebaran, but nearly 100 times yet more luminous. The upper right-hand corner of the H–R diagram (marked on Figure 10.14), where these stars are found, is called the **red-giant region**. No red giants are found within 5 pc of the Sun (Figure 10.13), but many of the brightest stars seen in the sky are in fact red giants (Figure 10.14). Red giants are relatively rare, but they are so bright that they are visible to very great distances. They form a third distinct class of stars in the H–R diagram, very different in their properties from both main-sequence stars and white dwarfs.

The recent *Hipparcos* mission (Section 10.1), in addition to determining hundreds of thousands of stellar parallaxes with unprecedented accuracy, also measured the colors and luminosities of more than 2 million stars. Figure 10.15 shows an H–R diagram based on a tiny portion of the enormous *Hipparcos* dataset. The main sequence and red giant regions are clearly evident. Few white dwarfs appear, however, simply because the telescope was limited to observations of relatively bright objects—brighter than apparent magnitude 12. Almost no white dwarfs lie close enough to Earth that their magnitudes fall below this limit.

About 90 percent of all stars in our solar neighborhood, and presumably a similar percentage elsewhere in the universe, are main-sequence stars. About 9 percent of stars are white dwarfs, and 1 percent are red giants.

10.6 Extending the Cosmic Distance Scale

We have already discussed the connections between luminosity, apparent brightness, and distance. Knowledge of a star's apparent brightness and its distance allows us to determine its luminosity using the inverse-square law. But we can also turn the problem around. If we somehow knew a star's luminosity and then measured its apparent brightness, the inverse-square law would tell us its distance from the Sun.

Spectroscopic Parallax

Most of us have a rough idea of the approximate intrinsic brightness (that is, the luminosity) of a typical traffic signal. Suppose you are driving down an unfamiliar street and see a red traffic light in the distance. Your knowledge of the light's luminosity enables you immediately to make a mental estimate of its distance. A light that appears relatively dim (low apparent brightness) must be quite distant (assuming it's not just dirty). A bright one must be relatively close. Thus, *measurement of the apparent brightness of a light source, combined with some knowledge of its luminosity, can yield an estimate of its distance.*

For a star the trick is to find an independent measure of luminosity without knowing the distance. The H–R diagram can provide just that. For example, suppose we observe a star and determine its apparent magnitude to be 10. By itself, that doesn't tell us much–the star could equally be faint and close, or bright and distant (Figure 10.5). But suppose we have some additional information: The star lies on the main sequence and has spectral type A0. Then we can read the star's luminosity off a graph such as Figure 10.13 or 10.14. A main-sequence A0 star has a luminosity of approximately 100 solar units.

Spectral classification

◀◀◀ **CONCEPT CHECK**

Only a tiny fraction of all stars are giants. Why, then, do giants account for so many of the stars visible to the naked eye in the night sky?

CONCEPT CHECK ▶▶▶

Suppose astronomers discover that due to a calibration error, all distances measured by geometric parallax are 10 percent larger than currently thought. What effect would this have on the "standard" main sequence used in spectroscopic parallax?

According to *More Precisely 10-1*, this corresponds to an absolute magnitude of 0 and hence to a distance of 1000 pc.

This process of using stellar spectra to infer distances is called **spectroscopic parallax**.[2] The key steps are

1. we measure the star's apparent brightness and spectral type *without* knowing how far away it is,
2. we use the spectral type to estimate the star's luminosity, assuming that it lies on the main sequence, and
3. finally, we apply the inverse-square law to determine the distance to the star.

The existence of the main sequence allows us to make a connection between an easily measured quantity (spectral type) and the star's luminosity, which would otherwise be unknown. As we will see in upcoming chapters, this essential logic (with a variety of different techniques replacing step 2) is used again and again as a means of distance measurement in astronomy. In practice, the "fuzziness" of the main sequence translates into a small (10–20 percent) uncertainty in the distance, but the basic idea remains valid.

In Chapter 1 we introduced the first rung on a ladder of distance-measurement techniques that will ultimately carry us to the edge of the observable universe. That rung is radar ranging on the inner planets. ∞ *(Sec. 1.3)* It establishes the scale of the solar system and defines the astronomical unit. At the beginning of this chapter we discussed a second rung on the cosmic distance ladder—stellar parallax—which is based on the first, since Earth's orbit is the baseline. Now, having used the first two rungs to determine the distances and other physical properties of many nearby stars, we use that knowledge to construct the third rung—spectroscopic parallax. Figure 10.16 illustrates schematically how this third rung builds on the first two and expands our cosmic field of view still deeper into space.

Spectroscopic parallax can be used to determine stellar distances out to several thousand parsecs. Beyond that, spectra and colors of individual stars are difficult to obtain. Note that, in using this method, we are assuming (without proof) that distant stars are basically similar to those nearby—in particular, that main-sequence stars *fall on the same main sequence*. Only by making this assumption can we use spectroscopic parallax to expand the boundaries of our distance-measurement techniques.

Each rung in the distance ladder is calibrated using data from the lower rungs, so changes made at any level will affect measurements made on *all* larger scales. As a result, the impact of the enormous dataset of high-quality observations returned by the *Hipparcos* mission (Section 10.1) extends far beyond the volume of space actually surveyed by the satellite. By recalibrating the local foundations of the cosmic distance scale, *Hipparcos* has caused astronomers to revise their estimates of distances on *all* scales—up to and including the scale of the universe itself. All distances quoted throughout this text reflect updated values based on *Hipparcos* data.

▲ **Figure 10.16 Stellar Distances** Knowledge of a star's luminosity and apparent brightness can yield an estimate of its distance. Astronomers use this third rung on our distance ladder, called spectroscopic parallax, to measure distances as far out as individual stars can be clearly discerned—several thousand parsecs. We do much the same in everyday life: Traffic light standards that are dim are usually judged to be distant, while those that are bright are often closer.

traffic lights in perspective...

Luminosity Class

What if the star in question happens to be a red giant or a white dwarf and does not lie on the main sequence? As mentioned in Chapter 2, detailed analysis of spectral *line widths* can provide information on the pressure, and hence the density, of the gas where the line formed. ∞ *(Sec. 2.6)* The atmosphere of a red giant is much less dense than that of a main-sequence star, and this in

[2]This unfortunate name is very misleading, as the method has nothing in common with stellar (geometric) parallax other than its use as a means of determining stellar distances.

TABLE 10.3	Stellar Luminosity Classes
CLASS	DESCRIPTION
Ia	Bright supergiants
Ib	Supergiants
II	Bright giants
III	Giants
IV	Subgiants
V	Main-sequence stars/dwarfs

TABLE 10.4	Variation in Stellar Properties within a Spectral Class			
SURFACE TEMPERATURE (K)	LUMINOSITY (solar luminosities)	RADIUS (solar radii)	OBJECT	EXAMPLE
4900	0.3	0.8	K2V main-sequence star	ε Eridani
4500	110	21	K2III red giant	Arcturus
4300	4000	140	K2Ib red supergiant	ε Pegasi

turn is much less dense than the atmosphere of a white dwarf. Figures 10.17(a) and (b) illustrate the difference between the spectra of a main-sequence star and a red giant of the same spectral type.

Over the years, astronomers have developed a system for classifying stars according to the widths of their spectral lines. Because line width depends on pressure in the stellar photosphere, and because this pressure in turn is well correlated with luminosity, this stellar property has come to be known as **luminosity class**. The standard luminosity classes are listed in Table 10.3 and shown on the H–R diagram in Figure 10.17(c). By determining a star's luminosity class, astronomers can usually tell with a high degree of confidence what sort of object it is, and the full specification of a star's spectral properties includes its luminosity class. For example, the Sun, a G2 main-sequence star, is of class G2V, the B8 blue supergiant Rigel is B8Ia, the red dwarf Barnard's Star is M5V, the red supergiant Betelgeuse is M2Ia, and so on.

Consider, for example, a K2-type star (Table 10.4) with a surface temperature of approximately 4500 K. If the widths of the star's spectral lines tell us that it lies on the main sequence (i.e., it is a K2V star), then its luminosity is about 0.3 times the solar value. If the star's spectral lines are observed to be narrower than lines normally found in main-sequence stars, the star may be recognized as a K2III giant, with a luminosity 100 times that of the Sun (Figure 10.17c). If the lines are very narrow, the star might instead be classified as a K2Ib supergiant, brighter by a further factor of 40, at 4000 solar luminosities. In each case, knowledge of luminosity classes allows astronomers to identify the object and make an appropriate estimate of its luminosity and hence its distance.

10.7 Stellar Masses

As with all other objects, we measure a star's mass by observing its gravitational influence on some nearby body—another star, perhaps, or a planet. If we know the distance between the two bodies, then we can use Newton's laws to calculate their masses. *(More Precisely 1-1)*

Binary Stars

Most stars are members of *multiple-star systems*—groups of two or more stars in orbit around one another. The majority are found in **binary-star systems**, which consist of two stars in orbit about their common center of mass, held together by their mutual gravitational attraction. (The Sun is not part of a multiplestar system; if it has anything at all uncommon about it, it is this lack of stellar companions.)

Astronomers classify binary-star systems (or simply *binaries*) according to their appearance from Earth and the ease with which they can be observed.

(a)

(b)

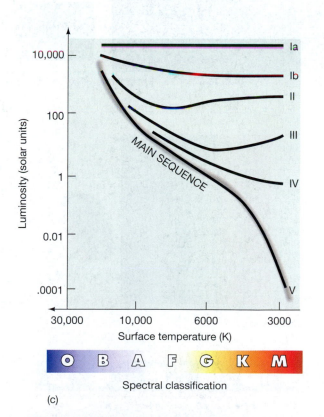

(c)

▲ **Figure 10.17 Stellar Luminosities** The widths of absorption lines provide information on the density of a star's atmosphere. (a) The denser atmosphere of a main-sequence K-type star has broader lines than a giant star of the same spectral class (b). (c) Approximate locations of the standard stellar luminosity classes in the H–R diagram.

Binary Stars—Radial Velocity Curves

Tutorial

Eclipsing Binary Stars—Light Curves

Tutorial

Visual binaries have widely separated members bright enough to be observed and monitored separately (Figure 10.18a). The more common **spectroscopic binaries** are too distant from us to be resolved into separate stars, but they can be indirectly perceived by monitoring the back-and-forth Doppler shifts of their spectral lines as the stars orbit one another and their line-of-sight velocities vary periodically. ∞ *(Sec. 2.7)*

In a *double-line* spectroscopic binary, two distinct sets of spectral lines—one for each component star—shift back and forth as the stars move. Because we see particular lines alternately approaching and receding, we know that the objects emitting the lines are in orbit. In the more common *single-line* systems (Figure 10.18b), one star is too faint for its spectrum to be distinguished, so we see only one set of lines shifting back and forth. This shifting means that the detected star must be in orbit around another star, even though the companion cannot be observed directly. (If this idea sounds familiar, it should—all the extrasolar planetary systems discovered to date are extreme examples of single-line spectroscopic binaries.) ∞ *(Sec. 4.4)*

In the much rarer **eclipsing binaries**, the orbital plane of the pair of stars is almost edge-on to our line of sight. In this situation, we observe a periodic decrease of starlight intensity as one member of the binary passes in front of the other (Figure 10.18c). By studying the variation of the light from an eclipsing binary system—called the binary's **light curve**—astronomers can derive detailed information not only about the stars' orbits and masses but also about their radii.

Note that these categories are not mutually exclusive. For example, an eclipsing binary may also be (and, in fact, often is) a spectroscopic binary system.

Mass Determination

The period of a binary can be measured by observing the orbits of the component stars, the back-and-forth motion of the stellar spectral lines, or the dips in the light curve—whatever information is available. Observed binary periods span a broad range—from hours to centuries. How much additional information can be extracted depends on the type of binary involved.

If the distance to a *visual* binary is known, then the orbits of each component can be individually tracked and the masses of the components can be determined. ∞ *(Sec. 1.4)*

For *spectroscopic* binaries, Doppler-shift measurements give us information only on the radial velocities of the component stars, and this limits the information we can obtain. Simply put, we cannot distinguish between a slow-moving binary seen edge-on and a fast-moving binary seen almost face-on (so that only a small component of the orbital motion is along the line of sight). For a double-line system, only lower limits on the individual masses can be obtained. For single-line systems, even less information is available, and only a rather complicated relation between the component masses (known as the mass function) can be derived.

(a) Visual

R I V U X G

| Lab spectrum (reference) |
| Starlight redshifted at time 1 |
| Starlight blueshifted at time 2 |
| Lab spectrum (reference) |

(b) Spectroscopic

Motion away from observer causes redshift

(1)

Motion toward observer causes blueshift

(2)

#1 #2 #3 #4 #5 #6

Intensity

Time

(c) Eclipsing

◄ **Figure 10.18 Binary Stars** (a) The period and separation of a binary system (here the visual binary Kruger 60) can be observed directly if each star is clearly seen. (b) Binary properties may be determined indirectly by measuring the periodic Doppler shift of one star relative to the other as they move in their orbits. This is a single-line system, in which only one spectrum (from the brighter component) is visible. (c) If the two stars happen to eclipse one another, additional information can be obtained by observing the periodic decrease in starlight as one star passes in front of the other. (Harvard College Observatory)

If a spectroscopic binary happens also to be an *eclipsing* system, then the uncertainty in the orbital inclination is removed, as the binary is known to be edge-on (or very nearly so). In that case, both masses can be determined for a double-line binary. For a single-line system, the mass function is simplified to the point where the mass of the unseen component is known if the mass of the brighter component can be obtained by other means (for example, if it is recognized as a main-sequence star of a certain spectral class—see Figure 10.19).

Despite all these qualifications, individual component masses have been obtained for many nearby binary systems. A simple example is presented in *More Precisely 10-3*. Virtually all we know about the masses of stars is based on such observations.

Mass and Other Stellar Properties

Table 10.5 summarizes the various observational and theoretical techniques used to measure the stars, listing the quantities that are assumed known (usually as a result of the application of other techniques listed in the table), those that are measured, and the theory that is applied to turn the observations into the desired result. We end our introduction to the stars with a brief look at how mass is correlated with the other stellar properties discussed in this chapter.

Figure 10.19 is a schematic H–R diagram showing how stellar mass varies along the main sequence. There is a clear progression from low-mass red dwarfs to high-mass blue giants. With few exceptions, main-sequence stars range in mass from about 0.1 to 20 times the mass of the Sun. The hot O- and B-type stars are generally about 10 to 20 times more massive than our Sun. The coolest K- and M-type stars contain only a few tenths of a solar mass. The mass of a star at the time of its formation determines its location on the main sequence. Based on observations of stars within a few hundred light-years of the Sun, Figure 10.20 illustrates how the masses of main-sequence stars are

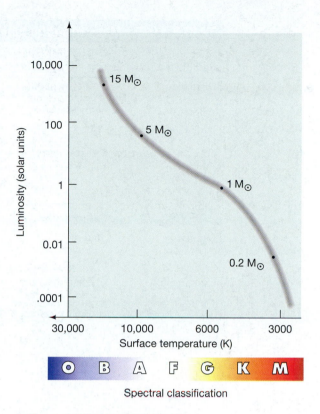

▲ **Figure 10.19 Stellar Masses**   More than any other stellar property, mass determines a star's position on the main sequence. Low-mass stars are cool and faint; they lie at the bottom of the main sequence. Very massive stars are hot and bright; they lie at the top of the main sequence. (The symbol "$M_\odot$" means "solar mass.")

10-3 MORE PRECISELY

Measuring Stellar Masses in Binary Stars

As discussed in the text, most stars are members of binary systems—where two stars orbit one another, bound together by gravity. Here we describe how—in an idealized case where the relevant orbital parameters are known—we can use the observed orbital data, together with our knowledge of basic physics, to determine the masses of the component stars.

Consider the nearby visual binary system made up of the bright star Sirius A and its faint companion Sirius B, sketched in the accompanying figure. The binary's orbital period can be measured simply by watching the stars orbit one another, or alternatively by following the back-and-forth velocity wobbles of Sirius A due to its faint companion. It is almost exactly 50 years. The orbital semi-major axis can also be obtained by direct observation of the orbit, although in this case we must use some additional knowledge of Kepler's laws to correct for the binary's 46° inclination to the line of sight. ∞ *(Sec. 1.3)* It is 20 AU—an angular size of 7.5″ at a distance of 2.7 pc. ∞ *(Sec. 1.3)* Once we know these two key orbital parameters, we can use the modified version of Kepler's third law to calculate the sum of the masses of the two stars. The result is $20^3/50^2 = 3.2$ times the mass of the Sun. ∞ *(Sec. 1.4)*

Further study of the orbit allows us to determine the individual stellar masses. Doppler observations show that Sirius A moves at approximately half the speed of its companion relative to their center of mass. ∞ *(Sec. 1.4, 2.7)* This implies that Sirius A must have twice the mass of Sirius B. It then follows that the masses of Sirius A and Sirius B are 2.1 and 1.1 solar masses, respectively.

Often the calculation of the masses of binary components is complicated by the fact that only partial information is available—we might only be able to see one star, or perhaps only spectroscopic velocity information is available (see Section 10.7). Nevertheless, this technique of combining elementary physical principles with detailed observations is how virtually every stellar mass quoted in this text has been determined.

TABLE 10.5 Measuring the Stars					
STELLAR PROPERTY	MEASUREMENT TECHNIQUE	"KNOWN" QUANTITY	MEASURED QUANTITY	THEORY APPLIED	SECTION
Distance	stellar parallax spectroscopic parallax	astronomical unit main sequence	parallactic angle spectral type apparent magnitude	elementary geometry inverse-square law	10.1 10.6
Radial velocity		speed of light atomic spectra	spectral lines	Doppler effect	10.1
Transverse velocity	astrometry	distance	proper motion	elementary geometry	10.1
Luminosity		distance main sequence	apparent magnitude spectral type color	inverse square law	10.2 10.6
Temperature	photometry spectroscopy		spectral type	blackbody law atomic physics	10.3 10.3
Radius	direct indirect	distance	angular size luminosity temperature	elementary geometry radius–luminosity– temperature relationship	10.4 10.4
Composition	spectroscopy		spectrum	atomic physics	10.3
Mass	observations of binary stars	(distance)	binary period binary orbit orbital velocity	Newtonian gravity and dynamics	10.7

CONCEPT CHECK ▶▶▶

How do we know about the masses of stars that aren't members of binaries?

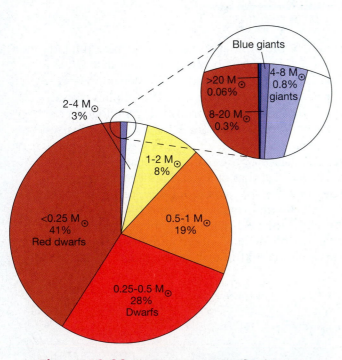

▲ **Figure 10.20 Stellar Mass Distribution** The distribution of masses of main-sequence stars, as determined from careful measurement of stars in the solar neighborhood.

distributed. Notice the huge fraction of low-mass stars, as well as the tiny fraction contributed by stars of more than a few solar masses.

Figure 10.21 illustrates how the radii and luminosities of main-sequence stars depend on mass. The two plots are called the *mass–radius* (part a) and *mass–luminosity* (part b) relationships. Along the main sequence, both radius and luminosity increase with mass. The radius rises slightly more slowly than the mass, whereas luminosity increases much faster—more like the *fourth power* of the mass (indicated by the straight line in Figure 10.21b). For example, a 2-solar mass main-sequence star has a radius roughly twice that of the Sun and a luminosity of 16 (2^4) solar luminosities; a 0.2-solar mass main-sequence star has a radius of about 0.2 solar radii and a luminosity of 0.0016 (0.2^4) solar luminosities.

Table 10.6 compares key properties of some well-known main-sequence stars, arranged in order of decreasing mass. Notice that the central temperature (obtained from mathematical models similar to those discussed in Chapter 9) differs relatively little from star to star, compared to the large spread in stellar luminosities. The final column in the table presents an estimate of each star's *lifetime*, obtained by dividing the amount of fuel available (that is, the star's mass) by the rate at which the fuel is consumed (the luminosity),

$$\text{stellar lifetime} \propto \frac{\text{stellar mass}}{\text{stellar luminosity}},$$

and noting that the lifetime of the Sun (see Chapter 12) is about 10 billion years.

Because luminosity increases so rapidly with mass, the most massive stars are by far the shortest lived. For example, according to the mass–luminosity relationship, the lifetime of a 10-solar mass O-type star is roughly $10/10^4 = 1/1000$ that of the Sun, or about 10 million years. We can be sure that all the O- and B-type stars we now observe are quite young—less than a few tens of millions of years old. Their nuclear reactions proceed so rapidly that their fuel is quickly depleted despite their large masses. At the opposite end of the main sequence, the low core density and temperature of an 0.1-solar mass M-type star mean that its proton–proton reactions churn away much more sluggishly than in the Sun's core, leading to a very low luminosity and a correspondingly long lifetime. ∞ *(Sec. 9.5)* Many of the K- and M-type stars now visible in the sky will shine on for at least another trillion years. The evolution of stars—large and small—is the subject of the next two chapters.

STAR	SPECTRAL TYPE/ LUMINOSITY CLASS[1]	MASS (solar masses)	CENTRAL TEMPERATURE (10^6 K)	LUMINOSITY (solar luminosities)	ESTIMATED LIFETIME (millions of years)
Spica B*	B2V	6.8	25	800	90
Vega	A0V	2.6	21	50	500
Sirius	A1V	2.1	20	22	1000
Alpha Centauri	G2V	1.1	17	1.6	7000
Sun	G2V	1.0	15	1.0	10,000
Proxima Centauri	M5V	0.1	0.6	0.00006	16,000,000

TABLE 10.6 Key Properties of Some Well-Known Main-Sequence Stars

* The "star" Spica is, in fact, a binary system comprising a B1III giant primary (Spica A) and a B2V main-sequence secondary (Spica B).

(a)

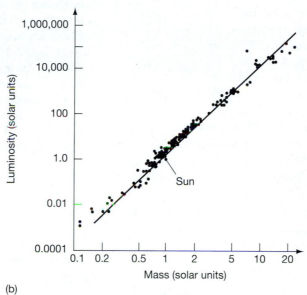

(b)

◄ **Figure 10.21 Stellar Radii and Luminosities** (a) Dependence of stellar radius on mass for main-sequence stars; actual measurements are plotted here. The radius increases almost in proportion to the mass over much of the range. (b) Dependence of luminosity on mass. The luminosity increases roughly as the fourth power of the mass (indicated by the straight line).

CHAPTER REVIEW

Summary

1 The distances to the nearest stars can be measured by stellar parallax. A star with a parallax of 1 arc second is 1 **parsec** (p. 266)—about 3.3 light-years—away from the Sun. A star's **proper motion** (p. 268), which is its true motion across the sky, is a measure of the star's velocity perpendicular to our line of sight.

2 The **apparent brightness** (p. 268) of a star is the rate at which energy from the star reaches a detector. Apparent brightness falls off as the inverse square of the distance. Optical astronomers use the **magnitude scale** (p. 270) to express and compare stellar brightnesses. The greater the magnitude, the fainter the star; a difference of five magnitudes corresponds to a factor of 100 in brightness. **Apparent magnitude** (p. 270) is a measure of apparent bright-

ness. The **absolute magnitude** (p. 271) of a star is the apparent magnitude it would have if placed at a standard distance of 10 pc from the viewer. It is a measure of the star's luminosity.

3 Astronomers often measure the temperatures of stars by measuring their brightnesses through two or more optical filters, then fitting a blackbody curve to the results. Spectroscopic observations of stars provide an accurate means of determining both stellar temperatures and stellar composition. Astronomers classify stars according to the absorption lines in their spectra. The standard stellar **spectral classes** (p. 274), in order of decreasing temperature, are O B A F G K M.

4 Only a few stars are large enough and close enough to Earth that their radii can be measured directly. The sizes of most stars are estimated indirectly through the **radius–luminosity–temperature relationship** (p. 275). Stars are categorized as **dwarfs** (p. 276) comparable in size to, or smaller than, the Sun; **giants** (p. 276) up to 100 times larger than the Sun, and **supergiants** (p. 276) more than 100 times larger than the Sun. In addition to "normal" stars such as the Sun, two other important classes of star are **red giants** (p. 276), which are large, cool, and luminous, and **white dwarfs** (p. 276), which are small, hot, and faint.

5 A plot of stellar luminosity versus stellar spectral class (or temperature) is called an **H–R diagram** (p. 276). About 90 percent of all stars plotted on an H–R diagram lie on the **main sequence** (p. 276), which stretches from hot, bright **blue supergiants** (p. 277) and **blue giants** (p. 277), through intermediate stars such as the Sun, to cool, faint **red dwarfs** (p. 277). Most main-sequence stars are red dwarfs; blue giants are quite rare. About 9 percent of stars lie in the **white-dwarf region** (p. 278) and the remaining 1 percent lie in the **red-giant region** (p. 279).

6 If a star is known to lie on the main sequence, measurement of its spectral type allows its luminosity to be estimated and its distance

to be measured. This method of distance determination, which is valid for stars up to several thousand parsecs from Earth, is called **spectroscopic parallax** (p. 280). A star's **luminosity class** (p. 281) allows astronomers to distinguish main-sequence stars from giants and supergiants of the same spectral type.

7 Most stars are not isolated in space but instead orbit other stars in **binary-star systems** (p. 281). In a **visual binary** (p. 282), both stars can be seen and their orbits charted. In a **spectroscopic binary** (p. 282), the stars cannot be resolved, but their orbital motion can be detected spectroscopically. In an **eclipsing binary** (p. 282), the orbits are oriented in such a way that one star periodically passes in front of the other as seen from Earth and dims the light we receive. Studies of binary stars often allow stellar masses to be measured. The mass of a star determines its size, temperature, and brightness. Hot blue giants are much more massive than the Sun; cool red dwarfs are much less massive. High-mass stars burn their fuel rapidly and have much shorter lifetimes than the Sun. Low-mass stars consume their fuel slowly and may remain on the main sequence for trillions of years.

Review and Discussion

1. How is stellar parallax used to measure the distances to stars?

2. What is a parsec? Compare it to the astronomical unit.

3. Explain how a star's real motion through space translates into motion observable from Earth.

4. Describe some characteristics of red giants and white dwarfs.

5. What is the difference between absolute and apparent magnitude?

6. How do astronomers measure star temperatures?

7. Briefly describe how stars are classified according to their spectral characteristics.

8. What information is needed to plot a star on the Hertzsprung–Russell diagram?

9. What is the main sequence? What basic property of a star determines where it lies on the main sequence?

10. How are distances determined using spectroscopic parallax?

11. Which stars are most common in the Milky Way Galaxy? Why don't we see many of them in H–R diagrams?

12. How can stellar masses be determined by observing binary-star systems?

13. If a high-mass star starts off with much more fuel than a low-mass star, why doesn't the high-mass star live longer?

14. In general, is it possible to determine the life span of an individual star simply by noting its position on an H–R diagram?

15. Visual binaries and eclipsing binaries are relatively rare compared to spectroscopic binaries. Why is this?

Conceptual Self-Test: True or False?/Multiple Choice

1. Star A appears brighter than star B, as seen from Earth. Therefore, star A must be closer to Earth than star B. **(T/F)**

2. A star of apparent magnitude +5 looks brighter than one of apparent magnitude +2. **(T/F)**

3. Red giants are very bright because they are extremely hot. **(T/F)**

4. The radius of a star can be determined if the star's distance and luminosity are known. **(T/F)**

5. Astronomers can distinguish between main-sequence and giant stars by purely spectroscopic means. **(T/F)**

6. In a spectroscopic binary, the orbital motion of the component stars appears as variations in the overall apparent brightness of the system. **(T/F)**

7. It is possible to have a 1-billion-year-old O-type main-sequence star. **(T/F)**

8. From a distance of 1 parsec, the angular size of Earth's orbit would be **(a)** 1 degree; **(b)** 2 degrees; **(c)** 1 arc minute; **(d)** 2 arc seconds.

9. Compared with a star of absolute magnitude −2 at a distance of 100 pc, a star of absolute magnitude 5 at a distance of 10 pc will appear **(a)** brighter; **(b)** fainter; **(c)** to have the same brightness; **(d)** bluer.

10. Stars of spectral class M do not show strong lines of hydrogen in their spectra because **(a)** they contain very little hydrogen; **(b)** their surfaces are so cool that most hydrogen is in the ground state; **(c)** their surfaces are so hot that most hydrogen is ionized; **(d)** the hydrogen lines are swamped by even stronger lines of other elements.

11. Cool stars can be very luminous if they are very **(a)** small; **(b)** hot; **(c)** large; **(d)** close to our solar system.

12. The mass of a star may be determined **(a)** by measuring its luminosity; **(b)** by determining its composition; **(c)** by measuring its Doppler shift; **(d)** by studying its orbit around a binary companion.

13. Pluto's apparent magnitude is approximately 14. According to Figure 10.6 (Apparent Magnitude), Pluto can be seen **(a)** with the naked eye on a dark night; **(b)** using binoculars; **(c)** using a 1-meter telescope; **(d)** only with the *Hubble Space Telescope*.

14. According to Figure 10.12 (H-R Diagram of Well-known Stars), Barnard's Star must be **(a)** hotter; **(b)** larger; **(c)** closer to us; **(d)** bluer than Proxima Centauri.

15. According to Figures 10.19 (Stellar Masses) and 10.20 (Stellar Mass Distribution), most stars are **(a)** much hotter; **(b)** much cooler; **(c)** much larger; **(d)** much brighter than the Sun.

Problems

Algorithmic versions of these questions are available in the Practice Problems module of the Companion Website.

The number of squares preceding each problem indicates its approximate level of difficulty.

1. ■ How far away is the star Spica, whose parallax is 0.013″? What would Spica's parallax be if it were measured from an observatory on Neptune's moon Triton as Neptune orbited the Sun?

2. ■■ A star lying 20 pc from the Sun has proper motion of 0.5″/yr. What is its transverse velocity? If the star's spectral lines are observed to be redshifted by 0.01 percent, calculate the magnitude of its three-dimensional velocity relative to the Sun.

3. ■■ (a) What is the luminosity of a star having three times the radius of the Sun and a surface temperature of 10,000 K? (b) A certain star has a temperature twice that of the Sun and a luminosity 64 times greater than the solar value. What is its radius, in solar units?

4. ■■ Two stars—A and B, of luminosities 0.5 and 4.5 times the luminosity of the Sun, respectively—are observed to have the same apparent brightness. Which one is more distant, and how much farther away is it than the other?

5. ■■ Two stars—A and B, of absolute magnitudes 3 and 8, respectively—are observed to have the same apparent magnitude. Which one is more distant, and how much farther away is it than the other?

6. ■■ Calculate the amount of energy received per unit area per unit time from the Sun, as seen from a distance of 10 pc. Compare this with the solar constant at Earth. ∞ *(Sec. 9.1)*

7. ■ Astronomical objects visible to the naked eye range in apparent brightness from faint 6th-magnitude stars to the Sun, with magnitude −27. What is the range in apparent brightness corresponding to this magnitude range?

8. ■ A star has apparent magnitude 10.0 and absolute magnitude 2.5. How far away is it?

9. ■■■ Two stars in an eclipsing spectroscopic binary system are observed to have an orbital period of 25 days. Further observations reveal that the orbit is circular, with a separation of 0.3 AU, and that one star is 1.5 times the mass of the other. What are the masses of the stars? ∞ *(More Precisely 1-1)*

10. ■■ Given that the Sun's lifetime is about 10 billion years, estimate the life expectancy of (a) a 0.2-solar mass, 0.01-solar luminosity red dwarf; (b) a 3.0-solar mass, 30-solar luminosity star; (c) a 10-solar mass, 1000-solar luminosity blue giant.

The Companion Website at www.prenhall.com/chaisson provides algorithmically generated versions of each chapter's Problems, in addition to an Animations and Videos gallery, a Tutorials gallery and Tutorials Quiz questions, a full eBook, an Images gallery, Practice Questions, Multiple Choice, and True or False quizzes, and Labeling exercises.

11

The Interstellar Medium

Star Formation in the Milky Way

Stars and planets are not the only inhabitants of our Galaxy. The space around us harbors invisible matter throughout the dark voids between the stars. The density of this interstellar matter is extremely low—approximately a trillion trillion times less dense than matter in either stars or planets, far thinner than the best vacuum attainable on Earth. Only because the volume of interstellar space is so vast does the mass of interstellar matter amount to anything at all. Why bother to study this near-perfect vacuum? We do so for three important reasons. First, there is as much mass in the "voids" between the stars as there is in the stars themselves. Second, interstellar space is the region out of which new stars are born. Third, it is the region into which old stars expel their matter when they die. It is one of the most significant crossroads through which matter passes in the history of our universe.

Interstellar space is the place both where stars are "born" and to which they return at "death." Rich in gas and dust, yet extraordinarily thinly distributed throughout the vast, dark regions among the stars, interstellar matter occasionally glows as nebulae or contracts to form new stars (and sometimes planets). Here, we see a billowing tower of cold, loose matter extending for nearly 10 light-years from a stellar nursery known as the Eagle Nebula. Radiation from its young stars excites the surrounding gas, which then re-emits colorful light characteristic of abundant elements (red is hydrogen, blue is oxygen). (STScI)

LEARNING GOALS

Studying this chapter will enable you to:

1. Summarize the composition and physical properties of the interstellar medium.

2. Describe the characteristics of emission nebulae, and explain their significance in the life cycle of stars.

3. Discuss the properties of dark interstellar dust clouds.

4. Specify some of the radio techniques used to probe the nature of interstellar matter.

5. Summarize the sequence of events leading to the formation of a star like our Sun.

6. Describe some of the observational evidence supporting the modern theory of star formation.

7. Explain how the formation of a star is affected by its mass.

8. Discuss the properties of star clusters, and explain how they form.

Visit www.prenhall.com/chaisson for additional annotated images, animations, and links to related sites for this chapter.

288

11.1 Interstellar Matter

Figure 11.1 is a mosaic of photographs covering a much greater expanse of universal "real estate" than anything we have studied thus far. From our vantage point on Earth, the panoramic view shown here stretches all the way across the sky. On a clear night it is visible to the naked eye as the *Milky Way*. In Chapter 14 we will come to recognize this band as the flattened disk, or *plane*, of our own Galaxy.

The bright regions in this image are made up of innumerable unresolved distant stars, merging together into a continuous blur at the resolution of the telescope. However, the dark areas are *not* simply "holes" in the stellar distribution. They are regions of space where *interstellar matter* obscures the light from stars beyond, blocking from our view what would otherwise be a smooth distribution of bright starlight. In this chapter we focus not on the stars themselves but on the vast reaches of interstellar space in which they form.

Gas and Dust

The matter between the stars is collectively termed the **interstellar medium**. It is made up of two components—*gas* and *dust*—intermixed throughout all of space. The gas is made up mainly of individual atoms and small molecules. The dust consists of clumps of atoms and molecules—not unlike the microscopic particles that make up smoke or soot.

Apart from numerous narrow atomic and molecular absorption lines, the gas alone does not block electromagnetic radiation to any great extent. The patchy obscuration evident in Figure 11.1 is caused by dust. Light from distant stars cannot penetrate the densest accumulations of interstellar dust any more than a car's headlights can illuminate the road ahead in a thick fog. As a rule of thumb, a beam of light can be absorbed or scattered only by particles having diameters comparable to or larger than the wavelength of the radiation involved. The amount of obscuration (absorption or scattering) produced by particles of a given size increases with decreasing wavelength. The typical diameter of an interstellar dust particle—or **dust grain**—is about 10^{-7} m, comparable in size to the wavelength of visible light. Consequently, dusty regions of interstellar space are transparent to long-wavelength radio and infrared radiation but opaque to shorter-wavelength optical, ultraviolet, and X-ray radiation.

Because the opacity of the interstellar medium increases with decreasing wavelength, light from distant stars is preferentially robbed of its higher-frequency ("blue") components. Hence, in addition to being generally diminished in overall brightness, stars also appear redder than they really are. Illustrated in Figure 11.2, this effect, known as **reddening**, is similar to the

▼ **Figure 11.1** **Milky Way Mosaic** The Milky Way photographed almost from horizon to horizon, spanning nearly 180°. This band contains high concentrations of stars as well as interstellar gas and dust. The grey box shows the field of view of Figure 11.4. (Axel Mellinger)

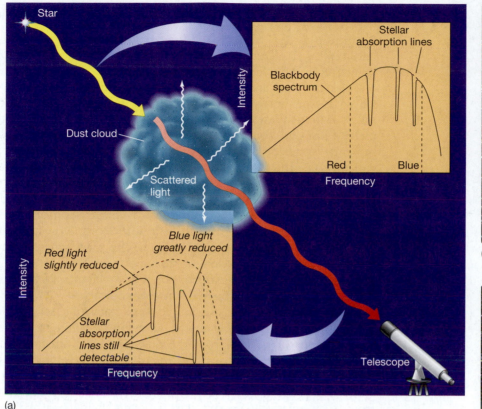

(a)

▲ **Figure 11.2 Reddening** (a) Starlight passing through a dusty region of space is both dimmed and reddened, but spectral lines are still recognizable in the light that reaches Earth. Note that this reddening has *nothing* to do with the Doppler effect—the frequencies of the lines are unchanged, although their intensities may be substantially reduced. (b) This dusty interstellar cloud, called Barnard 68, is opaque to visible light, except near the edges, where some light from background stars can be seen. Because blue light is more easily scattered or absorbed by dust than is red light, stars seen through the cloud appear red. The cloud spans about 0.2 pc and lies about 160 pc away. Frame (c) illustrates (in false color) how infrared radiation can penetrate Barnard 68, although it too is preferentially stripped of its shorter wavelengths. (ESO)

process that produces spectacular red sunsets here on Earth (see Figure 11.3). Reddening can be seen very clearly in Figure 11.2(b), which shows a type of compact, dusty interstellar cloud called a *globule*. (We will discuss such dark interstellar clouds in more detail in Section 11.3.) The center of this cloud, called Barnard 68, is opaque to all optical wavelengths, so starlight cannot pass through it. However, near the edges, where there is less intervening cloud matter, some light does make it through; in the infrared image of Figure 11.2(c), even more radiation gets through. Notice how stars seen through the cloud are both dimmed and reddened relative to those seen directly.

As indicated in Figure 11.2(a), absorption lines in a star's spectrum are still recognizable in the radiation reaching Earth, allowing the star's spectral class, and hence its true luminosity and color, to be determined. ∞ *(Sec. 10.6)* Astronomers can then measure the degree to which the star's light has been dimmed and reddened en route to Earth. This in turn allows them to estimate the amount of dust along the line of sight.

▶ **Figure 11.3 Reddening in Earth's Atmosphere** The phenomenon of reddening is especially evident as sunlight passes through Earth's atmosphere at the end of a long, hot summer day. Airborne dust particles and evaporated seawater molecules most easily scatter the Sun's blue light, leaving only its dimmed, red light—a spectacular sunset. (NCAR)

Density and Composition of the Interstellar Medium

By measuring its effect on the light from many different stars, astronomers have built up a picture of the distribution and chemical properties of interstellar matter in the solar neighborhood.

Gas and dust are found everywhere in interstellar space. No part of our Galaxy is truly devoid of matter, although the density of the interstellar medium is extremely low. Overall, the gas averages roughly 10^6 atoms per cubic meter—just one atom per cubic centimeter. (For comparison, the best vacuum presently attainable in laboratories on Earth contains about 10^9 atoms per cubic meter.) Interstellar dust is even rarer—about one dust particle for every trillion or so atoms. The space between the stars is populated with matter so thin that harvesting all the gas and dust in an interstellar region the size of Earth would yield barely enough matter to make a pair of dice.

Despite such low densities, over sufficiently large distances interstellar matter accumulates slowly but surely, to the point where it can block visible light and other short-wavelength radiation from distant sources. All told, space in the vicinity of the Sun contains about as much mass in the form of interstellar gas and dust as exists there in the form of stars.

Interstellar matter is distributed very unevenly (see *Discovery 11-1*). In some directions it is largely absent, allowing astronomers to study objects literally billions of parsecs from the Sun. In other directions there are small amounts of interstellar matter, so the obscuration is moderate, preventing us from seeing objects more than a few thousand parsecs away, but allowing us to study nearby stars. Still other regions are so heavily obscured that starlight from even relatively nearby stars is completely absorbed before reaching Earth. Even in regions of dense obscuration, however, there are often "windows" through which we can see to great distances—several can be seen in Figure 11.1.

The composition of interstellar gas is reasonably well known from spectroscopic studies of interstellar absorption lines. ∞ *(Sec. 2.5)* Generally speaking, it mirrors the composition of other astronomical objects, such as the Sun, the stars, and the Jovian planets: Most of the gas—about 90 percent—is atomic or molecular hydrogen, 9 percent is helium, and the remaining 1 percent consists of heavier elements. The gas is deficient in some heavy elements, such as carbon, oxygen, silicon, magnesium, and iron, most likely because these elements have gone to form the interstellar dust.

In contrast, the composition of the dust is not well known, although there is some infrared evidence for silicates, carbon, and iron, supporting the theory that interstellar dust forms out of interstellar gas. The dust probably also contains some "dirty ice," a frozen mixture of water ice contaminated with trace amounts of ammonia, methane, and other compounds, much like cometary nuclei in our own solar system. ∞ *(Sec. 4.2)*

CONCEPT CHECK ▶▶▶

If space is a near-perfect vacuum, how can there be enough dust in it to block starlight?

11.2 Star-Forming Regions

Figure 11.4 shows a magnified view of the central part of Figure 11.1 (the region indicated by the rectangle in the earlier figure), in the general direction of the constellation Sagittarius. The field of view is mottled with stars and dark interstellar matter. In addition, several fuzzy patches of light are clearly visible. Labeled M8, M16, M17, and M20, they correspond to the 8th, 16th, 17th, and 20th entries in a catalog compiled by Charles Messier, an eighteenth-century French astronomer. Today these four objects are known as examples of **emission nebulae**—glowing clouds of hot interstellar gas. Figure 11.5 enlarges the left side of Figure 11.4, showing the nebulae more clearly.

◀ **Figure 11.4** **Milky Way in Sagittarius**
Enlargement of the central portion of Figure 11.1, showing regions of brightness (vast fields of stars) as well as regions of darkness (where interstellar matter obscures the light from more distant stars). The field of view is roughly 30° across. The four red emission nebulae discussed in the text are labeled. (Palomar Observatory/Caltech)

11-1 DISCOVERY

Ultraviolet Astronomy and the "Local Bubble"

The ultraviolet is that part of the spectrum where we expect to witness events and see objects involving temperatures in the hundreds of thousands, even millions, of kelvins—hot regions like the seething atmospheres or eruptive flares of stars; violent events, such as massive stars caught in the act of exploding; and active galaxies whose spinning hearts may harbor black holes. Yet ultraviolet astronomy has also contributed greatly to the study of the interstellar medium, by allowing a unique mapping of our local cosmic neighborhood.

Only 30 years ago astronomers had presumed that the gases filling the spaces among the stars would absorb virtually all short-wavelength UV radiation before it had a chance to reach Earth. But in 1975, during a historic linkup of the *Apollo* (U.S.) and *Soyuz* (USSR) space capsules, the onboard astronauts and cosmonauts performed a key experiment—they used a small telescope to detect extreme ultraviolet radiation from a few nearby, very hot stars. Shortly thereafter, theoretical ideas changed as astronomers began to realize that interstellar gas is distributed very unevenly in space, in cool dense clumps interspersed by irregularly shaped regions of hot, low-density gas.

Since then, space-based ultraviolet observations have shown that some regions of interstellar space are much thinner (5000 atoms/m³) and hotter (500,000 K) than previously expected. Part of the space among the dust clouds and the emission nebulae seems to contain extremely dilute yet seething plasma, probably the result of the concussion and expanding debris from stars that exploded long ago. These superheated interstellar "bubbles," or *intercloud medium*, may extend far into interstellar space beyond our local neighborhood and conceivably into the even vaster spaces among the galaxies.

The Sun seems to reside in one such low-density region— a huge cavity called the *Local Bubble*, sketched in the accompa-

nying figure. Only because we live within this vast, low-density bubble can we detect so many stars in the far UV; the hot, tenuous interstellar gas is virtually transparent to this radiation. The Local Bubble contains about 200,000 stars and extends for hundreds of trillions of kilometers, or nearly 100 pc. It was probably carved out by a supernova explosion that occurred in these parts roughly 300,000 years ago. ∞ *(Sec. 12.4)* Our distant ancestors must have seen it—a stellar catastrophe as bright as the full Moon.

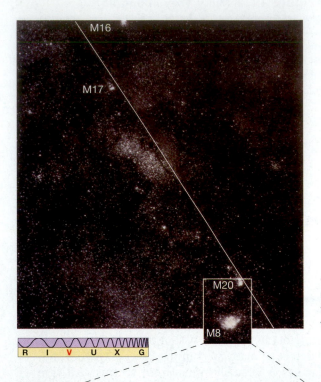

Historically, astronomers have used the term **nebula** to refer to any "fuzzy" patch (bright or dark) on the sky—any region of space that is clearly distinguishable through a telescope, but not sharply defined, unlike a star or a planet. We now know that many (although not all) nebulae are clouds of interstellar dust and gas. If a cloud happens to obscure stars lying behind it, we see it as a dark patch on a bright background, as in Figure 11.2(b)—a *dark nebula*. But if something within the cloud—a group of hot young stars, for example—causes it to glow, then we see a bright emission nebula instead. The method of spectroscopic parallax applied to stars visible within the emission nebulae shown in Figure 11.5 indicates that their distances from Earth range from 1200 pc (M8) to 1800 pc (M16). ∞ *(Sec. 10.6)* M16, at the top left, is approximately 1000 pc from M20, near the bottom.

We can gain a better appreciation of these nebulae by examining progressively smaller fields of view. Figure 11.6 is an enlargement of the region near the bottom of Figure 11.5, showing M20 at the top and M8 at the bottom.

◄ **Figure 11.5 Galactic Plane** A photograph of a smaller portion (about 12° across) of the region of the sky shown in Figure 11.4, showing stars, gas, and dust as well as several distinct fuzzy patches of light known as emission nebulae. The plane of the Milky Way is marked with a white diagonal line. (Harvard College Observatory)

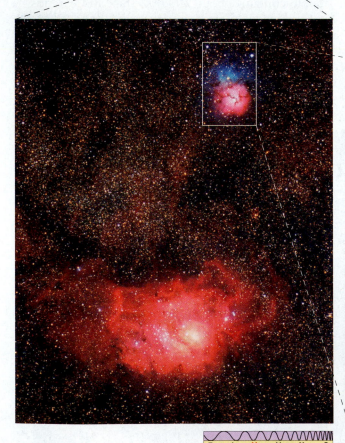

▲ **Figure 11.6 M20–M8 Region** An enlargement of the bottom of Figure 11.5, showing M20 (top) and M8 (bottom) more clearly. The two nebulae are only a few degrees apart. (P. Perkins)

▲ **Figure 11.7 Trifid Nebula** (a) Further enlargement of the top of Figure 11.6, showing only M20 and its interstellar environment. The nebula itself (in red) is about 6 pc in diameter. It is often called the Trifid Nebula because of the dust lanes (in black) that trisect its midsection. (b) A false-color infrared image taken by the *Spitzer Space Telescope* reveals bright regions of star-forming activity mostly in those lanes of dust. (AURA; NASA)

Figure 11.7 further enlarges the top of Figure 11.6, presenting a close-up of M20 and its immediate environment in both the visible and infrared parts of the spectrum. The total area displayed measures some 12 pc across. Emission nebulae are among the most spectacular objects in the universe, yet they appear only as small, undistinguished patches of light when viewed in the larger context of the Milky Way. Perspective is crucial in astronomy.

The nebulae shown in Figures 11.4–11.7 are regions of glowing, ionized gas. At or near the center of each is at least one newly formed hot O- or B-type star producing copious amounts of ultraviolet light. As ultraviolet photons travel outward from the star, they heat and ionize the surrounding gas. As electrons recombine with nuclei, they emit visible radiation, causing the gas to glow. The predominant red coloration is the result of hydrogen atoms emitting Hα light in the red part of the visible spectrum. ∞ *(Sec. 2.6)*

Woven through the glowing nebular gas, and plainly visible in Figures 11.5–11.7, are dark **dust lanes** obscuring the nebular light. These dust lanes are part of the nebulae and are not just unrelated dust clouds that happen to lie along our line of sight. The bluish region visible in Figure 11.7 immediately above M20 is another type of nebula unrelated to the red emission nebula itself. Called a *reflection nebula*, it is caused by starlight reflecting off intervening dust particles along the line of sight. The blue coloration occurs because short-wavelength blue light is more easily scattered by interstellar matter back toward Earth and into our detectors. Figure 11.8 sketches some of the key features of emission nebulae, illustrating the connection between the central stars, the nebula itself, and the surrounding interstellar medium.

Figure 11.9 shows enlargements of two of the other nebulae visible in Figure 11.5. Notice again the hot bright stars embedded within the glowing nebular gas and the overall red hue of the reemitted radiation in parts (a) and (c). (The colors in the *Hubble* images b and d accentuate observations made at different wavelengths.) The relationship between the nebula and the dust lanes is again evident in Figure 11.9(b) and (d), where regions of gas and dust are simultaneously silhouetted against background nebular emission and illuminated by foreground nebular stars. The interaction between stars and gas is particularly striking in Figure 11.9(b). The three dark "pillars" visible in this spectacular image are part of the interstellar cloud from which the stars formed. The rest of the cloud in the vicinity of the new stars has already been dispersed by their radiation. The fuzz around the edges of the pillars, especially at top right and center, is the result of this ongoing process, where intense

▼ **Figure 11.8 Nebular Structure**   An emission nebula results when ultraviolet radiation from one or more hot stars ionizes part of an interstellar cloud. The nebula's reddish color is produced as electrons and protons recombine to form hydrogen atoms. Dust lanes may be seen if part of the parent cloud happens to obscure the emitting region. If some starlight happens to encounter another dusty cloud (or perhaps another part of the cloud harboring the emission nebula), some of the radiation, particularly at the shorter wavelength blue end of the spectrum, may be scattered back toward Earth, forming a reflection nebula.

CONCEPT CHECK ▶▶▶

If emission nebulae are powered by ultraviolet light from hot stars, why do they appear red?

stellar radiation continues to "eat away" the cloud, heating and dispersing the dense molecular gas. The process removes the less dense material first, leaving behind delicate sculptures composed of the denser parts of the original cloud, much as wind and water create spectacular structures in Earth's deserts and shores by eroding away the softest rock. The pillars will eventually be destroyed, but probably not for another hundred thousand or so years.

Table 11.1 lists some vital statistics for the nebulae shown in Figure 11.5. Unlike stars, nebulae are large enough for their sizes to be measurable by simple geometry. Coupling this size information with estimates of the amount of matter along our line of sight (as revealed by a nebula's total light emission), we can find the nebula's density. Generally, emission nebulae have densities of a few hundred particles, mostly protons and electrons, per cubic centimeter

▼ **Figure 11.9** **Emission Nebulae** Enlargements of selected portions of Figure 11.5. (a) M16, the Eagle Nebula. (b) A *Hubble Space Telescope* image of huge pillars of cold gas and dust inside M16 shows delicate sculptures created by the action of stellar ultraviolet radiation on the original cloud. (c) M8, the Lagoon Nebula. (d) A high-resolution view of the core of M8, a region known as the Hourglass. Notice the irregular shape of the emitting regions, the characteristic red color of the light, the bright stars within the gas, and the patches of obscuring dust. The varied colors of the insets result from observations at different wavelengths. (AURA; NASA)

Animation/Video
M16 The Eagle Nebula

Animation/Video
Gaseous Pillars of Star Birth

(a)

(b)

(c)

R I V U X G

(d)

TABLE 11.1	Nebular Properties				
OBJECT	APPROX DISTANCE (pc)	AVERAGE DIAMETER (pc)	DENSITY (10^6 particles/m^3)	MASS (SOLAR MASSES)	TEMPERATURE (K)
M8	1200	14	80	2600	7500
M16	1800	8	90	600	8000
M17	1500	7	120	500	8700
M20	1600	6	100	250	8200

(10^8 per cubic meter). Figure 11.10 shows a typical emission nebula, along with its spectrum, spanning part of the visible and near-ultraviolet wavelength interval. ∞ *(Sec. 2.5)* Numerous emission lines can be seen, and information on the nebula can be extracted from all of them. Analyses of such spectra reveal compositions similar to those found in the Sun and other stars and elsewhere in the interstellar medium. Spectral-line widths imply that the gas atoms and ions in this nebula have temperatures around 8000 K. ∞ *(Sec. 2.8)*

A note on terminology: Spectroscopists describe the *ionization state* of an atom by attaching a roman numeral to the chemical symbol for the atom—I for the neutral (that is, not ionized) atom, II for a singly ionized atom (an atom missing one electron), and so on. ∞ *(Sec. 2.6)* Because emission nebulae are composed mainly of ionized hydrogen, astronomers often refer to them as *HII regions*. Regions of space containing primarily neutral (atomic) hydrogen are known as *HI regions*.

R I V U X G

▲ **Figure 11.10 Nebular Spectrum** The visible spectrum of the hot gases in a nearby star-forming region known as the Omega nebula (M17). Shining by the light of several very hot stars, the nebula produces a complex spectrum of bright and dark lines (bottom), also shown here as an intensity trace from red to blue (center). (ESO)

11.3 Dark Dust Clouds

Emission nebulae are only one small component of interstellar space. Most of space—in fact, more than 99 percent of it—is devoid of such regions and contains no stars. It is simply dark. The average temperature of a typical dark region of interstellar space is about 100 K. Compare this with 273 K, at which water freezes, and 0 K, at which atomic and molecular motions cease. ∞ *(More Precisely 2-1)*

Within the dark voids among the nebulae and the stars lurks another distinct type of astronomical object, the **dark dust cloud**. These clouds are cooler than their surroundings (with temperatures as low as a few tens of kelvins), and thousands or even millions of times denser. In some regions, densities exceeding 10^9 atoms/m^3 (1000 atoms/cm^3) are found. Dark dust clouds are often called *dense* interstellar clouds by researchers, but recognize that even these densest interstellar regions are barely denser than the best vacuum achievable in terrestrial laboratories. Still, it is because their density is much larger than the average value of 10^6 atoms/m^3 in interstellar space that we can distinguish these clouds from the surrounding expanse of the interstellar medium.

Obscuration of Visible Light

Dark dust clouds bear little resemblance to terrestrial clouds. Most are bigger than our solar system, and some are many parsecs across (yet, even so, they make up no more than a few percent of the entire volume of interstellar space). Despite their name, these clouds are made up primarily of gas, just like the rest

(a)

R I V U X G

(b)

R I V U X G

▲ **Figure 11.11** **Obscuration and Emission** (a) At optical wavelengths, this dark dust cloud (known as L977) can be seen only by its obscuration of background stars. (b) At radio wavelengths, it emits strongly in the CO molecular line, with the most intense radiation coming from its densest part. (C. and E. Lada)

of the interstellar medium. However, their absorption of starlight is due almost entirely to the dust they contain.

Figure 11.11(a) shows a region called L977, in the constellation Cygnus. It is a classic example of a dark dust cloud. Some early (eighteenth-century) observers thought that these dark patches on the sky were simply empty regions of space that happened to contain no bright stars. However, by the late nineteenth century, astronomers had discounted this idea. They realized that seeing clear spaces among the stars would be like seeing clear tunnels between the trees in a huge forest, and it was extremely unlikely that as many tunnels would lead away from Earth as would be required to explain the observed dark regions.

Despite this realization, however, before the advent of radio astronomy astronomers had no direct means of studying clouds like L977. Emitting no visible light, they are generally undetectable to the eye except by the degree to which they dim starlight. However, as shown in Figure 11.11(b), the cloud's radio emission—in this case from carbon monoxide (CO) molecules contained within its volume (see p. 300)—outlines it clearly at radio wavelengths, providing an indispensable tool for the study of such objects.

Figures 11.12(a) and (b) are optical photographs of another dark dust cloud. Taking its name from a neighboring star, Rho Ophiuchi, it resides relatively nearby—about 300 pc from the Sun. Pockets of intense blackness mark regions where the dust and gas are especially concentrated and the light from background stars is completely obscured. Note the long "streamers" of dust and gas, making this dark dust cloud (as for most) very irregularly shaped. The

▼ **Figure 11.12** **Dark Dust Cloud** (a) The dark dust cloud Rho Ophiuchi is "visible" only because it blocks light coming from stars behind it. The dashed line indicates the cloud's approximate outline. (b) Another view of the region on a slightly enlarged scale, showing fainter foreground objects and more subtle colors. To orient (a) and (b), note the "pentagon" of bright objects visible in the lower half of each image. The bright star Antares is at the bottom. Up and to its right, near the edge of the image, is a star cluster called M4. Rho Ophiuchi itself is the bright object near the top, surrounded by a blue reflection nebula. (c) An infrared map of the same region, to roughly the same scale as parts (a) and (b). The very bright source near the top of the cloud is a hot emission nebula, also visible in the optical images. The "streamers" at left are the dark dust lanes evident in parts (a) and (b). (The black diagonal streak at right is an instrumental effect.) (Harvard Observatory; D. Malin; NASA)

(a) |← 1 pc →|

R I V U X G

(b)

R I V U X G

(c)

R I V U X G

bright patches within the dark region are emission nebulae in the foreground. Some of them are part of the cloud, where newly formed stars near the surface have created a "hot spot" in the cold, dark gas. Others have no connection to the cloud and just happen to lie along our line of sight. Like all dark dust clouds, this cloud is too cold to emit any visible light. However, it does radiate strongly at longer wavelengths. Figure 11.12(c) shows an infrared image of the same region.

Figure 11.13 shows a particularly striking and well-known example of a dark dust cloud—the Horsehead Nebula in Orion. This curiously shaped finger of gas and dust projects out from the much larger dark cloud in the bottom half of the image and stands out clearly against the red glow of a background emission nebula.

(a)

(b)

▲ **Figure 11.13 Horsehead Nebula** (a) The Horsehead Nebula in Orion is a striking example of a dark dust cloud, silhouetted against the bright background of an emission nebula. (b) A stunning image of the Horsehead, taken at highest resolution by the Very Large Telescope (VLT) in Chile. ∞ *(Sec. 3.2)* The "neck" of the horse is about 0.25 pc across. The nebular region lies roughly 1500 pc from Earth. (Royal Observatory of Belgium; ESO)

21-Centimeter Radiation

In many cases, dark dust clouds absorb so much light from background stars that they cannot easily be studied by the optical techniques described in Section 11.1. Fortunately, astronomers have an important alternative means of probing their structure that relies on low-energy radio emissions produced by the interstellar gas itself.

Much of the gas in interstellar space consists of atomic hydrogen. Recall that a hydrogen atom consists of one electron orbiting a single-proton nucleus. ∞ *(Sec. 2.6)* Besides its orbital motion around the central proton, the electron also has some rotational motion—that is, *spin*—about its own axis. The proton also spins. This model parallels a planetary system, in which, in addition to the orbital motion of a planet about a central star, both the planet (electron) and the star (proton) rotate about their axes.

The laws of physics dictate that there are just two possible spin configurations for a hydrogen atom in its ground state. As illustrated in Figure 11.14, the electron and proton can either rotate in the *same* direction, with their spin axes parallel, or they can rotate in *opposite* directions, with their axes parallel but oppositely oriented. The former configuration has slightly higher energy than the latter. When a slightly excited hydrogen atom having the electron and proton spinning in the same direction drops down to the less-energetic, opposite-spin state, the transition releases a photon with energy equal to the energy difference between the two states.

Because the energy difference between the two states is very small, the energy of the emitted photon is very low. Consequently, the wavelength of the radiation is long—about 21 cm. That wavelength lies in the radio portion of the electromagnetic spectrum. Researchers refer to the spectral emission line that results from this hydrogen–spin-flip process as the **21-centimeter line**. It provides a vital probe into any region of the universe containing atomic hydrogen gas. Needing no visible starlight to help calibrate their signals, radio astronomers can observe any interstellar region that contains enough hydrogen to produce a detectable signal. Even the low-density regions between dark dust clouds can be studied.

Of great importance is the fact that the wavelength of 21-cm radiation is much larger than the typical size of interstellar dust particles. Accordingly, this radio radiation reaches Earth completely unaffected by interstellar debris. The

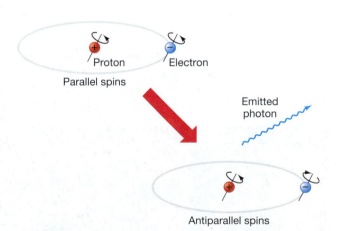

▲ **Figure 11.14 Hydrogen 21-cm Emission** A ground-level hydrogen atom changing from a higher-energy state (electron and proton spinning in the same direction) to a lower-energy state (spinning in opposite directions). The emitted photon carries away an energy equal to the energy difference between the two spin states.

opportunity to observe interstellar space well beyond a few thousand parsecs, and in directions lacking detectable background stars, makes 21-cm observations among the most important and useful in all of astronomy. As we will see in Chapters 14 and 15, it has been indispensable in allowing astronomers to map out the large-scale structure of our own and other galaxies.

Molecular Gas

In certain interstellar regions of cold (typically 20 K) neutral gas, densities can reach as high as 10^{12} particles/m^3. The gas particles in these regions are not atoms but *molecules*. Because of the predominance of molecules in these dense interstellar regions, they are known as **molecular clouds**.

Only long-wavelength radio radiation can escape from these dense, dusty parts of interstellar space. Molecular hydrogen (H_2) is by far the most common constituent of molecular clouds, but unfortunately this molecule does not emit or absorb radio radiation, so it cannot easily be used as a probe of cloud structure. Nor are 21-cm observations helpful—they are sensitive only to *atomic* hydrogen, not to the *molecular* form of the gas. Instead, astronomers use radio observations of other molecules, such as carbon monoxide (CO), hydrogen cyanide (HCN), ammonia (NH$_3$), water (H$_2$O), and formaldehyde (H$_2$CO), to study the dark interiors of these dusty regions. These molecules are produced by chemical reactions within molecular clouds and have emission lines in the radio part of the spectrum. They are found only in very small quantities—perhaps one part per billion—but when we observe them we know that the regions under study must also contain high densities of molecular hydrogen, dust, and other important constituents. Figure 11.15 shows a contour map of the distribution of formaldehyde molecules in the immediate vicinity of M20. Notice that the amount of formaldehyde (and, we assume, the amount molecular hydrogen) peaks well away from the visible nebula.

Molecular observations such as this reveal that the spectacular star-forming regions we see are actually parts of much larger molecular clouds. The bright emission nebula "bursts out" of its dark parent cloud when the radiation from a group of hot, newborn stars heats and ionizes the cold molecular gas. Radio maps of interstellar gas and infrared maps of interstellar dust reveal that molecular clouds do not exist as distinct and separate objects in space. Rather, they make up huge **molecular cloud complexes**, some spanning as much as 50 pc across and containing enough gas to make millions of stars like our Sun. About 1000 such complexes are known in our Galaxy. Figure 11.16 is a radio map of a large (330 square degree) region of the sky, made using certain spectral lines

▲ Figure 11.15 Molecules Near M20 Contour map of the amount of formaldehyde near the M20 nebula, demonstrating how formaldehyde is especially abundant in the darkest interstellar regions. Other kinds of molecules have been found to be similarly distributed. The contour values increase from the outside to the inside, so the maximum density of formaldehyde lies just to the bottom right of the visible nebula. The green and red contours outline the intensity of formaldehyde lines at different frequencies. (AURA; contours by the authors)

▼ Figure 11.16 Molecular Cloud Complexes This radio map shows the outer portion of the Milky Way (that is, looking in the opposite direction from Figure 11.1) as it appears in CO emission. The bright regions are molecular cloud complexes, dense regions of interstellar space where molecules abound and, apparently, stars are forming. The map extends over a full quadrant of the Galaxy, or 90°, along the Galactic plane and was made from 1,696,800 observations of CO spectra. (Five College Radio Astronomy Observatory)

of the CO molecule, showing numerous molecular cloud complexes stretching across the entire field of view.

Why are molecules found only in the densest and darkest of interstellar clouds? One possible reason is that the dust protects the fragile molecules from the harsh interstellar environment—the same absorption that prevents high-frequency radiation from getting out to our detectors also prevents it from getting in to destroy the molecules. Another possibility is that the dust acts as a catalyst that helps form the molecules. The dust grains provide both a place where atoms can stick and react and a means of dissipating any heat associated with the reaction, which might otherwise destroy the newly formed molecules. Probably the dust plays both roles, although the details are still debated.

11.4 The Formation of Stars Like the Sun

Let us now turn our attention to the connection between the interstellar medium and the stars in our Galaxy. How do stars form? What factors determine their masses, luminosities, and spatial distribution? In short, what basic processes are responsible for the appearance of our night sky?

Gravity and Heat

Simply stated, star formation begins when part of the interstellar medium—one of the cold, dark clouds discussed in the previous section—collapses under its own weight. An interstellar cloud is maintained in hydrostatic equilibrium by a balance between two basic opposing influences: gravity (which is always directed inward) and heat (in the form of outwardly directed pressure). ∞ (Sec. 9.2) However, if gravity somehow begins to dominate over heat, then the cloud can lose its equilibrium and start to contract.

Consider a small portion of a large interstellar cloud. Concentrate first on just a few atoms, as shown in Figure 11.17(a). Even though the cloud's temperature is very low, each atom has some random motion. ∞ (More Precisely 2-1) Each atom is also influenced by the gravitational attraction of all its neighbors. The gravitational force is not large, however, because the mass of each atom is so small. When a few atoms accidentally cluster for an instant, as sketched in Figure 11.17(b), their combined gravity is insufficient to bind them into a lasting, distinct clump of matter. This accidental cluster disperses as quickly as it formed (Figure 11.17c). The effect of heat—the random motion of the atoms—is much stronger than the effect of gravity.

As the number of atoms increases, their gravitational attraction increases too, and eventually the collective gravity of the clump is strong enough to prevent it from dispersing back into interstellar space. How many atoms are required for this to be the case? The answer, for a typical cool (100 K) cloud, is about 10^{57}—in other words, the clump must have mass comparable to that of the Sun. If our hypothetical clump is more massive than that, then it will not disperse. Instead, its gravity will cause it to contract, ultimately to form a star.

Of course, 10^{57} atoms don't just clump together by random chance. Rather, star formation is *triggered* when a sufficiently massive pocket of gas is squeezed by some external event. Perhaps it is compressed by the *shock wave* produced when a nearby group of O- or B-type stars form and heat their surroundings, creating an emission nebula, or when a neighboring star explodes as a supernova (see Section 12.4). Or possibly part of an interstellar cloud simply becomes too cold for its internal pressure to support it against its own gravity. Whatever the cause, theory suggests that once the collapse begins, star formation inevitably follows.

◄◄◄ **CONCEPT CHECK**

In mapping molecular clouds, why do astronomers use observations of "minority" molecules such as carbon monoxide and formaldehyde when these molecules constitute only a tiny fraction of the total number of molecules in interstellar space?

▼ **Figure 11.17 Atomic Motions**   The motions of a few atoms within an interstellar cloud are influenced by gravity so slightly that the atoms' paths are hardly changed (a) before, (b) during, and (c) after an accidental, random encounter.

TABLE 11.2 Prestellar Evolution of a Sun-like Star

STAGE	APPROXIMATE TIME TO NEXT STAGE (yr)	CENTRAL TEMPERATURE (K)	SURFACE TEMPERATURE (K)	CENTRAL DENSITY (particles/m³)	DIAMETER[1] (km)	OBJECT
1	2×10^6	10	10	10^9	10^{14}	Interstellar cloud
2	3×10^4	100	10	10^{12}	10^{12}	Cloud fragment
3	10^5	10,000	100	10^{18}	10^{10}	Cloud fragment/protostar
4	10^6	1,000,000	3000	10^{24}	10^8	Protostar
5	10^7	5,000,000	4000	10^{28}	10^7	Protostar
6	3×10^7	10,000,000	4500	10^{31}	2×10^6	Star
7	10^{10}	15,000,000	6000	10^{32}	1.5×10^6	Main-sequence star

[1]For comparison, recall that the diameter of the Sun is 1.4×10^6 km and that of the solar system roughly 1.5×10^{10} km.

Table 11.2 lists seven evolutionary stages that an interstellar cloud goes through before becoming a main-sequence star like our Sun. These stages are characterized by different central temperatures, surface temperatures, central densities, and radii of the prestellar object. They trace its progress from a quiescent interstellar cloud to a genuine star. The numbers given in Table 11.2 and the following discussion are valid *only* for stars of approximately the same mass as the Sun. In the next section we will relax this restriction and consider the formation of stars of other masses.

Stage 1—An Interstellar Cloud

The first stage in the star-formation process is a dense interstellar cloud—the core of a dark dust cloud or perhaps a molecular cloud. These clouds are truly vast, sometimes spanning tens of parsecs (10^{14}–10^{15} km) across. Typical temperatures are about 10 K throughout, with a density of perhaps 10^9 particles/m³. Stage-1 clouds contain thousands of times the mass of the Sun, mainly in the form of cold atomic and molecular gas. (The dust they contain is important for cooling the cloud as it contracts and also plays a crucial role in planet formation, but it constitutes a negligible fraction of the total mass.) ∞ *(Sec. 4.3)*

The dark region outlined by the red and green radio contours in Figure 11.15 (*not* the emission nebula itself, where stars have already formed) probably represents a stage-1 cloud just starting to contract. Doppler shifts of the observed formaldehyde lines indicate that it is probably infalling. Less than a light year across, this region has a total mass over 1000 times the mass of the Sun—considerably greater than the mass of M20 itself.

Once the collapse begins, theory indicates that fragmentation into smaller and smaller clumps of matter naturally follows due to gravitational instabilities in the gas. As illustrated in Figure 11.18, a typical cloud can break up into tens, hundreds, even thousands, of fragments, each imitating the shrinking behavior

▶ **Figure 11.18 Cloud Fragmentation** As an interstellar cloud contracts, gravitational instabilities cause it to fragment into smaller pieces. The pieces themselves continue to contract and fragment, eventually forming many tens or hundreds of individual stars.

of the parent cloud and contracting ever faster. The whole process, from a single quiescent cloud to many collapsing fragments, takes a few million years. Depending on the precise conditions under which fragmentation takes place, an interstellar cloud can produce either a few dozen stars, each much larger than our Sun, or a collection of hundreds of stars comparable to or smaller than our Sun. There is little evidence for stars born in isolation, one star from one cloud. Most stars—perhaps all stars—appear to originate as members of multiple systems.

The process of continuing fragmentation is eventually stopped by the increasing density within the shrinking cloud. As fragments continue to contract, they eventually become so dense that radiation cannot get out easily. The trapped radiation causes the temperature to rise, the pressure to increase, and the fragmentation to stop. However, the contraction continues.

Stages 2 and 3—A Contracting Cloud Fragment

As it enters stage 2, a fragment destined to form a star like the Sun—the end product of the process sketched in Figure 11.18—contains between one and two solar masses of material. Estimated to span a few hundredths of a parsec across, this fuzzy, gaseous blob is still about 100 times the size of our solar system. Its central density is about 10^{12} particles/m^3.

Even though the fragment has shrunk substantially, its average temperature is not much different from that of its parent cloud. The reason is that the gas constantly radiates large amounts of energy into space. The material of the fragment is so thin that photons produced anywhere within it easily escape without being reabsorbed, so virtually all the energy released in the contraction is radiated away and does not cause any significant increase in temperature. Only at the center, where the radiation must traverse the greatest amount of material in order to escape, is there any appreciable temperature increase. The gas there might be as warm as 100 K by this stage. For the most part, however, the fragment stays cold as it shrinks.

Several tens of thousands of years after it first began contracting, the start of stage 3 occurs when a stage-2 fragment has shrunk to a gaseous sphere with a diameter roughly the size of our solar system (still 10,000 times the size of our Sun). The inner regions have become opaque to their own radiation and have started to heat up considerably, as noted in Table 11.2. The central temperature has reached about 10,000 K—hotter than the hottest steel furnace on Earth. However, the gas near the edge is still able to radiate its energy into space and so remains cool. The central density by this time is approximately 10^{18} particles/m^3 (still only 10^{-9} kg/m^3 or so).

For the first time, our fragment is beginning to resemble a star. The dense, opaque region at the center is called a **protostar**—an embryonic object perched at the dawn of star birth. Its mass increases as more and more material rains down on it from outside, although its radius continues to shrink because its pressure is still unable to overcome the relentless pull of gravity. By the end of stage 3, we can distinguish a "surface" on the protostar—its *photosphere*. Inside the photosphere, the protostellar material is opaque to the radiation it emits. (Note that this is the same operational definition of *surface* we used for the Sun.) ∞ *(Sec. 9.1)* From here on, the surface temperatures listed in Table 11.2 refer to the photosphere and not to the edge of the collapsing fragment, where the temperature remains low.

Figure 11.19 shows a star-forming region in Orion. Lit from within by several O-type stars, the bright Orion Nebula is partly surrounded by a vast molecular cloud that extends well beyond the roughly 5×10 pc region bounded by the photograph in Figure 11.19(b). The Orion complex harbors several smaller sites of intense radio emission from molecules deep within its core. Shown in Figure 11.19(c), they measure about 10^{10} km, about the diameter of our solar system. Their density is about 10^{15} particles/m^3, much higher than the density of the surrounding cloud. Although the temperatures of these regions cannot be estimated reliably, many researchers regard the regions as objects between stages 2 and 3, on the threshold of becoming protostars. Figures 11.15(d) and (e) are visible-light images of part of the nebula showing other evidence for protostars.

Stage 4—A Protostar

As the protostar evolves, it shrinks, its density increases, and its temperature rises, both in the core and at the photosphere. Some 100,000 years after the fragment formed, it reaches stage 4, where its center seethes at about 1,000,000 K. The electrons and protons ripped from atoms whiz around at hundreds of kilometers per second, but the temperature is still short of the 10^7 K needed to ignite the proton–proton nuclear reactions that fuse hydrogen into helium. Still much larger than the Sun, our gassy heap is now about the size of Mercury's orbit. Its surface temperature has risen to a few thousand kelvins.

▼ **Figure 11.19 Orion Nebula, up Close**   (a) The constellation Orion, with the region around its famous emission nebula marked by a rectangle. The Orion Nebula is the middle "star" of Orion's sword. (b) Enlargement of the framed region of part (a), suggesting how the nebula is partly surrounded by a vast molecular cloud. Various parts of this cloud are probably fragmenting and contracting, with even smaller sites forming protostars. The three frames at the right show some of the evidence for those protostars: (c) false-color radio image of some intensely emitting molecular sites, (d) real-color visible image of embedded nebular "knots" thought to harbor protostars, and (e) high-resolution image of several young stars surrounded by disks of gas and dust where planets might ultimately form. (J. Sanford/Astrostock-Sanford; AURA; Harvard-Smithsonian Center for Astrophysics; NASA)

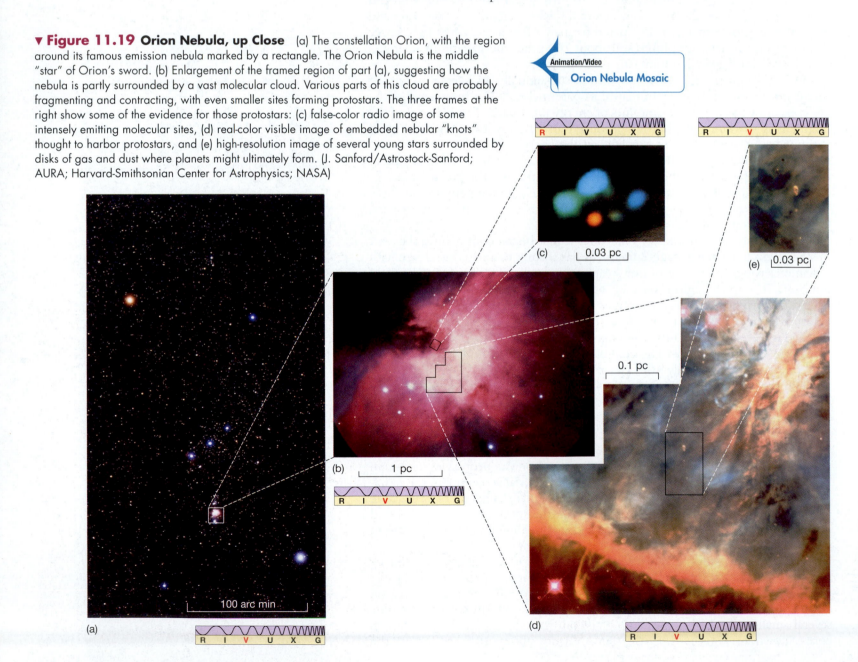

▶ **Figure 11.20 Protostar on the H–R Diagram** The red arrow indicates the approximate evolutionary track followed by an interstellar cloud fragment before becoming a stage-4 protostar. The boldface numbers on this and subsequent H–R plots refer to the prestellar evolutionary stages listed in Table 11.2 and described in the text. Recall from Chapter 10 that $R_\odot$ means the radius of the Sun.

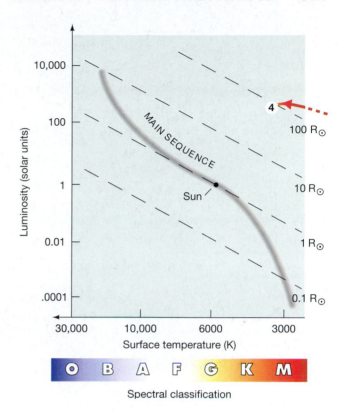

Knowing the protostar's radius and surface temperature, we can calculate its luminosity using the radius–luminosity–temperature relationship. ∞ *(Sec. 10.4)* Remarkably, it turns out to be around 1000 times the luminosity of the Sun. Because nuclear reactions have not yet begun in the protostar's core, this luminosity is due entirely to the release of gravitational energy as the protostar continues to shrink and material from the surrounding fragment (which we called the solar nebula back in Chapter 4) rains down on its surface. ∞ *(Sec. 4.3)*

In the hunt for objects at more advanced stages of star formation, radio techniques become less useful because stages 4, 5, and 6 have increasingly higher temperatures. By Wien's law, their emission shifts toward shorter wavelengths, and so they shine most strongly in the infrared. ∞ *(Sec. 2.4)* One particularly bright infrared emitter, known as the Becklin–Neugebauer object, was detected in the core of the Orion molecular cloud in the 1970s. Its luminosity is around 1000 times the luminosity of the Sun. Most astronomers agree that this warm, dense blob is a high-mass protostar, probably around stage 4.

By the time stage 4 is reached, our protostar's physical properties can be plotted on a Hertzsprung–Russell (H–R) diagram. ∞ *(Sec. 10.5)* At each phase of a star's evolution, its surface temperature and luminosity can be represented by a single point on the diagram. The motion of that point around the diagram as the star evolves is known as the star's **evolutionary track**. It is a graphical representation of a star's life. Note that it has nothing to do with any real spatial motion of the star. The red track on Figure 11.20 depicts the approximate path followed by our interstellar cloud fragment since it became a protostar at the end of stage 3 (which itself lies off the right-hand edge of the figure). Figure 11.21 is an artist's sketch of an interstellar gas cloud proceeding along the evolutionary path outlined so far.

Stage 5—Protostellar Evolution

Our protostar is still not in equilibrium. Even though its temperature is now so high that outward-directed pressure has become a powerful countervailing influence against gravity's inward pull, the balance is not yet perfect. The protostar's internal heat gradually diffuses out from the hot center to the cooler surface, where it is radiated away into space. As a result, the contraction slows, but it does not stop completely.

After stage 4, the protostar moves downward on the H–R diagram (toward lower luminosity) and slightly to the left (toward higher temperature), as shown in Figure 11.22. By stage 5, the protostar has shrunk to about 10 times the size of the Sun, its surface temperature is about 4000 K, and its luminosity has fallen to about 10 times the solar value. The central temperature has risen to about 5,000,000 K. The gas is completely ionized by now, but the protons still do not have enough thermal energy for nuclear fusion to begin.

Protostars often exhibit violent surface activity during this phase of their evolution, resulting in extremely strong protostellar winds, much denser than that of our

▼ **Figure 11.21 Interstellar Cloud Evolution** Artist's conception of the changes in an interstellar cloud during the early evolutionary stages outlined in Table 11.2. (Not drawn to scale.) The duration of each stage, in years, is indicated.

▲ Figure 11.22 Newborn Star on the H–R Diagram The changes in a protostar's observed properties are shown by the path of decreasing luminosity, from stage 4 to stage 6. At stage 7, the newborn star has arrived on the main sequence.

Sun. This portion of the evolutionary track is often called the **T-Tauri phase**, after T-Tauri, the first "star" (actually protostar) to be observed in this stage of prestellar development. The interaction between the strong winds and the nebular disk out of which the protostar (and possibly planets) is still forming often results in a *bipolar flow*—expelling two "jets" of matter perpendicular to the disk. Figure 11.23(a) shows a particularly striking example.

These outflows can be very energetic. Figure 11.23(b) shows a portion of the Orion molecular cloud, south of the Orion Nebula, where a newborn star is seen still surrounded by a bright nebula, its turbulent wind spreading out into the interstellar medium. Below it (enlarged in the inset) are twin jets known as HH1 and HH2. (HH stands for Herbig–Haro, after the investigators who first catalogued such objects.) Formed in another (unseen) protostellar disk, these jets have traveled outward for almost half a light-year before colliding with interstellar matter. More Herbig–Haro objects can be seen in the upper right portion of the figure.

Events proceed more slowly as the protostar approaches the main sequence. The initial contraction and fragmentation of the interstellar cloud occurred quite rapidly, but by stage 5, as the protostar nears the status of a full-fledged star, its evolution slows. Its contraction is governed largely by the rate at which it can radiate its internal energy into space. As the luminosity decreases, so, too, does the contraction rate.

Until the *Infrared Astronomy Satellite (IRAS)* was launched in the early 1980s, astronomers were aware only of very massive stars forming in clouds far away. But *IRAS* showed that stars are forming much closer to home, and some of these protostars have masses comparable to that of our Sun. Figure 11.24 shows two examples of low-mass protostars—both spotted in a rich star-forming region in Orion. Their infrared heat signatures are those expected of a stage-5 object.

Stages 6 and 7—A Newborn Star

Some 10 million years after its first appearance, the protostar finally becomes a true star. By stage 6, when our roughly one-solar-mass object has shrunk to a radius of about 1,000,000 km, the contraction has raised the central temperature to 10,000,000 K—enough to ignite nuclear burning. Protons begin fusing into helium nuclei in the core, and a star is born. As shown in Figure 11.22, the star's surface temperature at this point is about 4500 K, still a little cooler than the Sun. Even though the radius of the newly formed star is slightly larger than that of the Sun, its lower temperature means that its luminosity is slightly less than (actually, about two-thirds of) the present solar value.

Indirect evidence for stage-6 stars comes from infrared observations of objects that seem to be luminous hot stars hidden from optical view by surrounding dark clouds. Their radiation is mostly absorbed by a "cocoon" of dust, then reemitted by the dust as infrared radiation. Two considerations support the idea that the hot stars responsible for heating the clouds have only recently ignited: (1) dust cocoons are predicted to disperse quite rapidly once their central stars form, and (2) these objects are invariably found in the dense cores of molecular clouds, consistent with the star-formation sequence just outlined.

Over the next 30 million years or so, the stage-6 star contracts a little more. Its central density rises to about 10^{32} particles/m^3 (more conveniently expressed as 10^5 kg/m^3), the central temperature increases to 15,000,000 K, and the surface temperature reaches 6000 K. By stage 7 the star finally reaches the main sequence just about where our Sun now resides. Pressure and gravity are finally balanced in the stellar interior, and the rate at which nuclear energy is generated in the core exactly matches the rate at which energy is radiated from the surface.

CONCEPT CHECK ▶▶▶

What distinguishes a collapsing cloud from a protostar, and a protostar from a star?

▼ **Figure 11.23** **Protostellar Outflow** (a) This remarkable image shows two jets emanating from the young star system HH30, the result of matter accreting onto an embryonic star near the center. The system is viewed roughly edge-on to the disk. (b) This view of the Orion molecular cloud shows the outflow from a newborn star, still surrounded by nebular gas. The inset shows a pair of jets called HH1/HH2, formed when matter falling onto another protostar (still obscured by the dusty cloud fragment from which it formed) creates a pair of high-speed jets of gas perpendicular to the flattened protostellar disk. The jets are nearly 1 light-year across. Several more Herbig–Haro objects can be seen at top right. One of them, the oddly shaped "waterfall" at top right, may be due to an earlier outflow from the same protostar responsible for HH1 and HH2. (AURA; NASA)

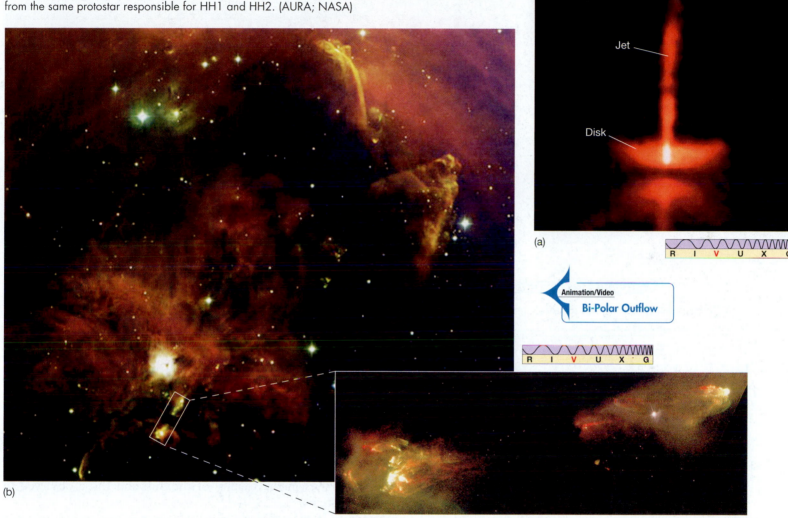

(a)

R I V U X G

Animation/Video
Bi-Polar Outflow

R I V U X G

(b)

(a) R I V U X G

(b) R I V U X G

◄ **Figure 11.24** **Protostars** (a) An infrared image of a planetary system–sized dusty disk in the Orion region, showing heat and light emerging from its center. On the basis of its temperature and luminosity, this unnamed source appears to be a low-mass protostar around Stage 5 in the H–R diagram. (b) An optical image of a slightly more advanced circumstellar disk surrounding an embedded protostar in Orion. (NASA)

The evolutionary events just described occur over the course of 40–50 million years. Although this is a long time by human standards, it is still less than 1 percent of the Sun's lifetime on the main sequence. Once an object begins fusing hydrogen and establishes a "gravity-in/pressure-out" hydrostatic equilibrium, it burns steadily for a very long time. The star's location on the H–R diagram—that is, its surface temperature and luminosity—will remain almost unchanged for the next 10 billion years.

11.5 Stars of Other Masses

The numerical values and evolutionary track just described are valid only for 1-solar-mass stars. The temperatures, densities, and radii of prestellar objects of other masses exhibit similar trends, but the details differ, in some cases quite considerably. Perhaps not surprisingly, the most massive fragments formed within interstellar clouds tend to produce the most massive protostars and eventually the most massive stars. Similarly, low-mass fragments give rise to low-mass stars.

The Zero-Age Main Sequence

Figure 11.25 compares the theoretical pre-main-sequence evolutionary track taken by our Sun with the corresponding tracks of a 0.3-solar-mass star and a 3.0-solar-mass star. All three tracks traverse the H–R diagram in the same general manner, but cloud fragments that eventually form stars more massive than the Sun approach the main sequence along a higher track on the diagram, while those destined to form less massive stars take a lower track. The *time* required for an interstellar cloud to become a main-sequence star also depends strongly on its mass. ∞ *(Sec. 10.7)* The most massive fragments contract into O-type stars in a mere million years, roughly 1/50 the time taken by the Sun. The opposite is the case for prestellar objects having masses much less than our Sun. A typical M-type star, for example, requires nearly a billion years to form.

Whatever the mass, the endpoint of the prestellar evolutionary track is the main sequence. A star is considered to have reached the main sequence when hydrogen burning begins in its core and the star's properties settle down to stable values. The main-sequence band predicted by theory is usually called the **zero-age main sequence** (ZAMS). It agrees quite well with the main sequences observed for stars in the vicinity of the Sun and those observed in more distant stellar systems.

It is important to realize that *the main sequence is not an evolutionary track—stars do not evolve along it.* Rather, it is a "waystation" on the H–R diagram where stars stop and spend most of their lives—low-mass stars at the bottom, high-mass stars at the top. Once on the main sequence, a star stays in roughly the same location in the H–R diagram during its whole time as a stage-7 object. In other words, a star that arrives on the main sequence as, say, a G-type star can never "work its way up" to become a B- or an O-type main-sequence star, nor move down to become an M-type red dwarf. As we will see in Chapter 12, the next stage of stellar evolution occurs when a star leaves the main sequence. When this occurs, the star will have pretty much the same surface temperature and luminosity it had when it arrived on the main sequence millions (or billions) of years earlier.

"Failed" Stars

Some cloud fragments are too small ever to become stars. Pressure and gravity come into equilibrium before the central temperature becomes hot enough to fuse hydrogen, so they never evolve beyond the protostar stage. Rather than

CONCEPT CHECK ▶▶▶

Do stars evolve along the main sequence?

▲ **Figure 11.25 Prestellar Evolutionary Tracks**
Prestellar evolutionary paths for stars more massive and less massive than our Sun.

(a)   R I V U X G

(b)   R I V U X G

(c)   R I V U X G

turning into stars, such low-mass fragments continue to cool, eventually becoming compact, dark "clinkers"—cold fragments of unburned matter—in interstellar space. Small, faint, and cool (and growing ever colder), these objects are known collectively as **brown dwarfs**. On the basis of theoretical modeling, astronomers calculate that the minimum mass of gas needed to generate core temperatures high enough to begin nuclear fusion is about 0.08 solar masses—80 times the mass of Jupiter.

Vast numbers of brown dwarfs may well be scattered throughout the universe—fragments frozen in time somewhere in the cloud-contraction phase. With our current technology we have great difficulty in detecting them, be they planets associated with stars or interstellar cloud fragments alone in space. We can telescopically detect stars and spectroscopically infer atoms and molecules, but low-mass astronomical objects outside our solar system are very hard to see. Based on current observations, it seems that up to 100 billion cold, dark, brown dwarfs may lurk in the depths of interstellar space—comparable to the total number of "real" stars in our Galaxy.

Recent advances in observational hardware and image-processing techniques have identified several likely brown-dwarf candidates, often using techniques similar to those employed in the search for extrasolar planets. ∞ *(Sec. 4.4)* (Recall from Section 10.7 that most stars are found in binaries; the same may very well be true of brown dwarfs.) Figure 11.26(a) shows Gliese 623, a binary system containing a brown dwarf candidate originally identified by radial velocity measurements. Figures 11.26(b) and (c) show the binary star system Gliese 229, first identified as a possible brown dwarf by ground-based infrared observations, and more recently imaged from space.

▲ **Figure 11.26 Brown Dwarfs** (a) This *Hubble Space Telescope* image shows Gliese 623, a binary system that may contain a brown dwarf (marked by an arrow). Astronomers hope that continued observations of this system will allow the companion's mass to be measured with sufficient accuracy to determine whether or not it really is a brown dwarf. The "rings" in the image are instrumental artifacts. (b) A ground-based image of the binary-star system Gliese 229. The two objects are only 7″ apart; the fainter "star" (arrow) has a luminosity only a few millionths that of the Sun and an estimated mass about 50 times that of Jupiter. (c) A *Hubble* image of the same system. (The bright diagonal streak in the latter image is caused by a hardware problem in the CCD chip used to record it.) (NASA; S. Kulkarni; NASA)

11.6 Star Clusters

The end result of cloud collapse is a group of stars, all formed from the same parent cloud and lying in the same region of space. Such a collection of stars is called a **star cluster**. Figure 11.27 shows a spectacular view of a newborn star cluster and (part of) the interstellar cloud from which it came. Because all the stars formed at the same time out of the same cloud of interstellar gas and under the same environmental conditions, clusters are near-ideal "laboratories" for stellar studies—not in the sense that astronomers can perform experiments on them, but because the properties of the stars are very tightly constrained. The only factor distinguishing one star from another in the same cluster is mass, so theoretical models of star formation and evolution can be compared with reality without the complications introduced by broad spreads in age, chemical composition, and place of origin.

R I V U X G

▲ **Figure 11.27 Newborn Cluster** The star cluster NGC 3603 and a portion of the larger molecular cloud in which it formed. The cluster contains about 2000 bright stars and lies some 6000 pc from Earth. The field of view shown here spans about 20 light-years. Radiation from the cluster has cleared a cavity in the cloud several light-years across. The inset shows the central area more clearly, including the most massive star in the region (called Sher 25, above and to the left of the cluster), which is already near the end of its lifetime, having ejected part of its outer layers and formed a ring of gas. Many low-mass stars, less massive than the Sun, can also be seen. (ESO; NASA)

Clusters and Associations

Figure 11.28(a) shows a small star cluster called the Pleiades, or Seven Sisters, a well-known naked-eye object in the constellation Taurus, lying about 120 pc from Earth. This type of loose, irregular cluster, found mainly in the plane of the Milky Way, is called an **open cluster**. Open clusters typically contain from a few hundred to a few tens of thousands of stars and are a few parsecs across. Figure 11.28(b) shows the H–R diagram for stars in the Pleiades. The cluster contains stars in all parts of the main sequence. The blue stars must be relatively young, for, as we saw in Chapter 10, they burn their fuel very rapidly. ∞ *(Sec. 10.7)* Thus, even though we have no direct evidence of the cluster's birth, we can estimate its age as less than 20 million years, the lifetime of an O-type star. The wisps of leftover gas evident in the photograph are further evidence of the cluster's youth.

Less massive, but more extended, clusters are known as **associations**. These typically contain no more than a few hundred stars, but may span many tens of parsecs. Associations tend to be rich in very young stars. They are very loosely bound and may survive for only a few tens of millions of years before dissolving in the Galactic tidal field. Open clusters are also eventually destroyed, but generally live much longer—hundreds of millions or billions of years. It is quite likely that the main difference between associations and open clusters is simply the efficiency with which stars formed—that is, how large a fraction of the original cloud ended up in the form of stars.

Figure 11.29(a) shows a very different type of star cluster, called a **globular cluster**. All globular clusters are roughly spherical (which accounts for their name), are generally found away from the Milky Way plane, and contain hundreds of thousands, and sometimes millions, of stars spread out over about 50 pc. The H–R diagram for this cluster (called Omega Centauri) is shown in Figure 11.29(b).

The most outstanding feature of globular clusters is their lack of upper main-sequence stars. In fact, globular clusters contain no main-sequence stars with masses greater than about 0.8 times the mass of the Sun. (The A-type

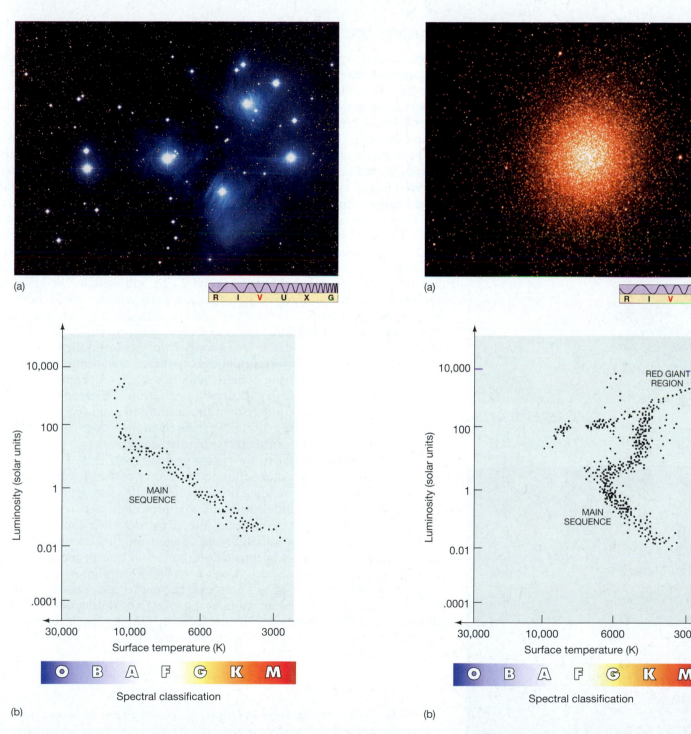

▲ **Figure 11.28** **Open Cluster** (a) The Pleiades cluster (also known as the Seven Sisters or M45) lies about 120 pc from the Sun. (b) An H–R diagram for the stars of this well-known open cluster. (NOAO)

▲ **Figure 11.29** **Globular Cluster** (a) The globular cluster Omega Centauri is approximately 5000 pc from Earth and some 40 pc in diameter. (b) An H–R diagram for some of its stars. (J. Lodriguss)

CONCEPT CHECK ▶▶▶

If stars in a cluster all start to form at the same time, how can some influence the formation of others?

stars in this plot are stars at a much later evolutionary stage that happen to be passing through the location of the upper main sequence.) Their more massive O- through F-type stars have long since exhausted their nuclear fuel and disappeared from the main sequence. On the basis of these and other observations, astronomers estimate that all globular clusters are at least 10 billion years old. They contain the oldest known stars in our Galaxy. Astronomers speculate that the 150 or so globular clusters observed today are just the survivors of a much larger population of clusters that formed long ago.

Clusters and Nebulae

Until relatively recently, the existence of star clusters within the emission nebulae surrounding newborn stars was largely conjecture—most of the stars cannot be seen optically because they are obscured by dust. However, infrared observations now clearly demonstrate that star clusters really are found there. Figure 11.30 compares optical and infrared views of the central regions of the Orion Nebula. The optical image in Figure 11.30(a) is centered on the Trapezium, the group of four bright stars responsible for ionizing the nebula, but shows little evidence for the stars in the system. However, the false-color infrared image in Figure 11.30(b) reveals an extensive cluster of stars, with a broad range of masses, within and behind the visible nebula.

How many stars form in a cluster, and of what type are they? How much gas is left over? What does the collapsed cloud look like once star formation has run its course? At present, although the main stages in the formation of individual stars (stages 3–7) seem fairly well established, the answers to these broader questions (involving stages 1 and 2) are still sketchy. We know from the H–R diagrams of observed stars that low-mass stars are much more common than high-mass ones, but an explanation of exactly why this is so awaits a more thorough understanding of the star-formation process. ∞ *(Sec. 10.7)* We saw earlier how, when one or more O- or B-type stars form, their intense radiation and winds cause the surrounding gas to disperse, shutting down the star-formation process and leaving behind a young star cluster (Figures 11.19 and 11.27). Recent computer simulations of star-forming clouds (see Figure 11.31) suggest that the sequence of events leading to main-sequence stars can also be strongly influenced by physical interactions—close encounters and even collisions—between protostars within the cluster.

The simulations reveal that the strong gravitational fields of the most massive protostars give them a competitive advantage over their smaller rivals in attracting gas from the surrounding nebula, causing the giant protostars to grow even faster. Close encounters with massive protostars also tend to disrupt smaller protostellar disks, terminating the growth of their central protostars and ejecting planets and low-mass brown dwarfs from the disk into intracluster space. In dense clusters these interactions may even lead to mergers and further growth of massive objects. Thus, even before the intense radiation of newborn O and B stars begins to disperse the cluster gas, the formation of a few large bodies can significantly inhibit the growth of smaller ones. (Compare this picture of large objects dominating the accretion process at the expense of the smaller bodies around them with the standard view of planet formation presented in Chapter 4.) ∞ *(Sec. 4.3)*

(a)

R I V U X G

(b)

R I V U X G

◀ **Figure 11.30 Young Stars in Orion** (a) Visible-light and (b) infrared *Hubble* views of the central part of the Orion nebula. Note the four bright stars of the Trapezium. The visible image is dominated by the emission nebula and shows few stars. However, the infrared image shows an extensive star cluster containing stars of many masses, possibly including many brown dwarfs. (NASA)

▶ **Figure 11.31 Protostellar Collisions** In the congested environment of a young cluster, star formation is a competitive and violent process. Large protostars may grow by "stealing" gas from smaller ones, and the extended disks surrounding most protostars can lead to collisions and even mergers. The cluster environment can play a crucial role in determining the types of stars that form. This frame from a supercomputer simulation shows a small star cluster emerging from an interstellar cloud that originally contained some 50 solar masses of material, distributed over a volume 1 light-year across. Remnants of the cloud are shown here in red. (M. Bate, I. Bonnell, and V. Bromm)

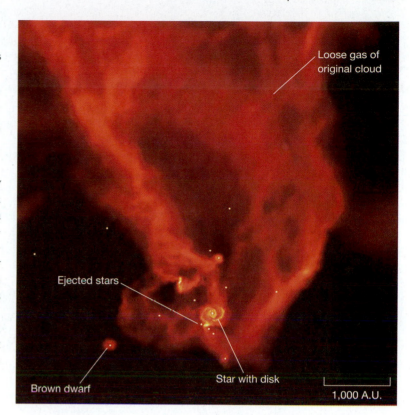

All these considerations provide important insight into why low-mass stars are so much more common than high-mass stars. The first few massive bodies to form tend to prevent the formation of additional high-mass stars by stealing their "raw material" and ultimately disrupting the environment in which other stars are growing. This also helps explain the existence of brown dwarfs, by providing at least two natural ways (disk destruction and gas dispersion) in which star formation can stop before nuclear fusion begins in a growing stellar core.

Take another look at the sky one clear, dark evening. Ponder all of the cosmic activity you have learned about as you peer upward at the stars. After studying this chapter, you may find that you have to modify your view of the night sky. Even the seemingly quiet nighttime darkness is dominated by continual change.

CHAPTER REVIEW

Summary

1 The **interstellar medium** (p. 290) occupies the space between stars. It is made up of cold (less than 100 K) gas, mostly atomic or molecular hydrogen and helium, and **dust grains** (p. 290). Interstellar dust is very effective at blocking our view of distant stars, even though the density of the interstellar medium is very low. The spatial distribution of interstellar matter is very patchy. The dust preferentially absorbs short-wavelength radiation, leading to **reddening** (p. 290) of light passing through interstellar clouds.

2 **Emission nebulae** (p. 292) are extended clouds of hot, glowing interstellar matter. Those associated with star formation are caused by hot O- and B-type stars heating and ionizing their surroundings. They are often crossed by dark **dust lanes** (p. 295)—part of the larger molecular cloud from which they formed.

3 **Dark dust clouds** (p. 297) are cold, irregularly shaped regions in the interstellar medium that diminish or completely obscure the light from background stars. The interstellar medium also contains many cold, dark **molecular clouds** (p. 300). Dust within these clouds probably both protects the molecules and acts as a catalyst to help them form. Molecular clouds are likely sites of future star formation. Often, several molecular clouds are found close to one another, forming a **molecular cloud complex** (p. 300) millions of times more massive than the Sun.

4 Astronomers can study dark interstellar clouds by observing their effect on the light from more distant stars. Another way to observe these regions of interstellar space is through spectral analysis of the **21-centimeter line** (p. 299) produced whenever the electron in a hydrogen atom reverses its spin, changing its energy very slightly in the process. Astronomers usually study these clouds through observations of other molecules that are less common than hydrogen but much easier to detect.

5 Stars form when an interstellar cloud collapses under its own gravity and breaks up into smaller pieces. The evolution of the contracting cloud can be represented as an **evolutionary track** (p. 305) on the Hertzsprung–Russell diagram. A cold interstellar cloud (stage 1)

containing a few thousand solar masses of gas can fragment into tens or hundreds of smaller clumps of matter, from which stars eventually form. As a collapsing prestellar fragment heats up and becomes denser (stages 2 and 3), it eventually becomes a **protostar** (stage 4; p. 303)—a warm, very luminous object that emits radiation mainly in the infrared portion of the electromagnetic spectrum. The protostar contracts as it radiates its internal energy into space (stage 5). Eventually, the central temperature becomes high enough for hydrogen fusion to begin (stage 6), and the protostar becomes a star (stage 7).

6 Dark dust clouds and globules are examples of stage-1 clouds and stage-2 collapsing fragments. Examples of stage-3 objects have been observed in some star-forming regions, such as Orion. Stars in the **T-Tauri phase** (p. 306) are examples of stage-4/5 protostars. Low-mass stage-6 stars on their final approach to the main sequence are found in star-forming regions.

7 Stars of different masses go through similar formation stages, but end up at different locations along the main sequence, high-mass stars at the top, low-mass stars at the bottom. The **zero-age main sequence** (p. 308) is the main-sequence predicted by stellar evo-

lutionary theory. It agrees very well with observed main sequences. The most massive stars have the shortest formation times and the shortest main-sequence lifetimes. At the other extreme, some low-mass fragments never reach the point of nuclear ignition. Objects not massive enough to fuse hydrogen to helium are called **brown dwarfs** (p. 309). They may be very common in the universe.

8 A single collapsing and fragmenting cloud can give rise to hundreds or thousands of stars—a **star cluster** (p. 309). **Open clusters** (p. 310), which are loose, irregular clusters that typically contain from a few tens to a few thousands of stars, are found mostly in the Milky Way plane.

They typically contain many bright blue stars, indicating that they formed relatively recently. **Globular clusters** (p. 310) are roughly spherical and may contain millions of stars. They include no main-sequence stars more massive than the Sun, indicating that they formed long ago. Infrared observations have revealed young star clusters within several emission nebulae. Eventually, star clusters break up into individual stars, although the process may take billions of years to complete.

Review and Discussion

1. What is the composition of interstellar gas? Of interstellar dust?

2. How is interstellar matter distributed through space?

3. What are some methods that astronomers use to study interstellar dust?

4. What is an emission nebula?

5. What is 21-centimeter radiation, and why is it useful to astronomers?

6. If our Sun were surrounded by a cloud of gas, would this cloud be an emission nebula? Why or why not?

7. Briefly describe the basic chain of events leading to the formation of a star like the Sun.

8. What is an evolutionary track?

9. Why do stars tend to form in groups?

10. What critical event must occur in order for a protostar to become a star?

11. What are brown dwarfs?

12. Because stars live much longer than we do, how do astronomers test the accuracy of theories of star formation?

13. At what evolutionary stages must astronomers use radio and infrared radiation to study prestellar objects? Why can't they use visible light?

14. Explain the usefulness of the Hertzsprung–Russell diagram in studying the evolution of stars. Why can't evolutionary stages 1–3 be plotted on the diagram?

15. Compare and contrast the properties of open and globular star clusters.

Conceptual Self-Test: True or False?/Multiple Choice

1. Interstellar matter is quite evenly distributed throughout the Milky Way Galaxy. **(T/F)**

2. Emission nebulae radiate mainly in the ultraviolet part of the electromagnetic spectrum. **(T/F)**

3. Twenty-one-centimeter radiation can be used to probe the interiors of molecular clouds. **(T/F)**

4. The most massive stars form most rapidly. **(T/F)**

5. Brown dwarfs take a long time to form, but will eventually become visible as stars on the lower main sequence. **(T/F)**

6. The formation of the first high-mass stars in a collapsing cloud tends to inhibit further star formation within that cloud. **(T/F)**

7. Most stars form as members of groups or clusters. **(T/F)**

8. The dark interstellar globule shown in Figure 11.2(b) is about the same size as **(a)** a cloud in Earth's atmosphere; **(b)** the entire planet Earth; **(c)** a star like the Sun; **(d)** the Oort cloud.

9. The red glow of an emission nebula **(a)** is emitted by warm gas falling onto the still-forming stars at the center; **(b)** is produced by hydrogen gas heated to high temperatures by massive stars

within the nebula; **(c)** is a reflection of light from stars near the nebula; **(d)** is the unresolved light of many faint red stars within the nebula.

10. The Rho Ophiuchi cloud, shown in Figure 11.12(a) (Dark Dust Cloud), is dark because **(a)** there are no stars in this region; **(b)** the stars in this region are young and faint; **(c)** starlight from behind the cloud does not penetrate the cloud; **(d)** the region is too cold to sustain stellar fusion.

11. Of the following telescopes, the one best suited to observing dark dust clouds is **(a)** an X-ray telescope; **(b)** a large visible-light telescope; **(c)** an orbiting ultraviolet telescope; **(d)** a radio telescope.

12. A protostar that will eventually turn into a star like the Sun is significantly **(a)** smaller; **(b)** more luminous; **(c)** fainter; **(d)** less massive than the Sun.

13. One of the primary differences between the Pleiades cluster (Figure 11.28a) and Omega Centauri (Figure 11.29a) is that the Pleiades cluster is much **(a)** larger; **(b)** younger; **(c)** farther away; **(d)** denser.

14. In Figure 11.2 (Reddening), the effect of interstellar absorption on spectral lines is **(a)** to shift them to longer wavelengths; **(b)** to shift them to shorter wavelengths; **(c)** to leave their wavelengths unchanged; **(d)** to erase them.

15. In Figure 11.22 (Newborn Star on the H–R Diagram), as the star moves from stage 6 to stage 7 it becomes **(a)** cooler and fainter; **(b)** hotter and smaller; **(c)** redder and brighter; **(d)** larger and cooler.

Problems

 Algorithmic versions of these questions are available in the Practice Problems module of the Companion Website.

The number of squares preceding each problem indicates its approximate level of difficulty.

1. ■ The average density of interstellar gas within about 100 pc of the Sun is much lower than the value mentioned in the text—in fact, it is just 10^3 hydrogen atoms/m³. Given that the mass of a hydrogen atom is 1.7×10^{-27} kg, calculate the total mass of interstellar matter contained within a local volume equal to that of Earth.

2. ■■ Calculate the frequency of 21-cm radiation. If interstellar clouds along our line of sight have radial velocities in the range 75 km/s (receding) to 50 km/s (approaching), calculate the range of frequencies and wavelengths over which the 21-cm line will be observed. ∞ *(More Precisely 2-3)*

3. ■ Calculate the radius of a spherical molecular cloud whose total mass equals the mass of the Sun. Assume a cloud density of 10^{12} hydrogen molecules per cubic meter. Give your answer in AU.

4. ■■ A beam of light shining through a dense molecular cloud is diminished in intensity by a factor of 2 for every 5 pc it travels. By what factor is the light from a background star diminished in intensity, if the total thickness of the cloud is 60 pc? What is this change, in magnitudes?

5. ■■ Estimate the escape speeds near the edges of the four emission nebulae listed in Table 11.1, and compare them with the average speeds of hydrogen nuclei in those nebulae. ∞ *(More Precisely 5-1)* Do you think it is likely that the nebulae are held together by their own gravity?

6. ■■ For an interstellar gas cloud to contract, the average speed of its constituent atoms must be less than half the cloud's escape speed. Will a molecular hydrogen cloud of mass 1000 solar masses, radius 10 pc, and temperature of 10 K begin to collapse, and why? ∞ *(More Precisely 5-1)*

7. ■ A protostar evolves from a temperature $T = 3500$ K and a luminosity $L = 5000$ times that of the Sun to $T = 5000$ K and $L = 3$ solar units. What is its radius **(a)** at the start, and **(b)** at the end of the evolution? ∞ *(Sec. 10.4)*

8. ■ Through approximately how many magnitudes does a three-solar-mass star decrease in brightness as it evolves from stage 4 to stage 6?

9. ■■ Use the radius–luminosity–temperature relation to calculate the luminosity, in solar units, of a brown dwarf whose radius is 0.1 solar radii and whose surface temperature is 600 K (0.1 times that of the Sun). ∞ *(Sec. 10.4)*

10. ■■ Approximating the gravitational field of our Galaxy as a mass of 10^{11} solar masses at a distance of 8000 pc (see Chapter 14), estimate the "tidal radius" of a 20,000-solar-mass star cluster—that is, the distance from the cluster center outside of which the tidal force due to the Galaxy overwhelms the cluster's gravity. ∞ *(Sec. 5.2)*

 The Companion Website at www.prenhall.com/chaisson provides algorithmically generated versions of each chapter's Problems, in addition to an Animations and Videos gallery, a Tutorials gallery and Tutorials Quiz questions, a full eBook, an Images gallery, Practice Questions, Multiple Choice, and True or False quizzes, and Labeling exercises.

12

Stellar Evolution

The Lives And Deaths of Stars

Although no one has seen a single star move through all of its evolutionary paces, the theory of stellar evolution is well understood. Like archaeologists who study bones and artifacts of different ages to understand more about ourselves, astronomers observe stars of different ages and then try to assemble the pieces of data into a consistent picture of how stars evolve over billions of years. In this striking infrared image downlinked to Earth by the *Spitzer Space Telescope*, we can note exquisite detail in the Helix Nebula, whose aged central star is shedding its outer envelope. Compare this infrared image with its optical equivalent on the cover of this book. (G. Fazio, IRAC)

After reaching the main sequence, a newborn star changes little in outward appearance for more than 90 percent of its lifetime. However, at the end of this period, as the star begins to run out of fuel and die, its properties once again change greatly. Aging stars travel along evolutionary tracks that take them far from the main sequence. The fate of a star depends primarily on its mass. Low-mass stars are destined to end their lives quiescently, their outer layers eventually escaping into interstellar space, while very massive stars die violently, in spectacular explosions of almost unimaginable fury. In either case, the deaths of stars enrich the Galaxy with newly created heavy elements. By continually comparing theoretical calculations with detailed observations of stars of all types, astronomers have refined the theory of stellar evolution into a precise and powerful tool for understanding the universe.

LEARNING GOALS

Studying this chapter will enable you to:

1 Explain why stars evolve off the main sequence.

2 Summarize the evolutionary stages followed by a Sun-like star once it leaves the main sequence, and describe the resulting remnant.

3 Explain how white dwarfs in binary systems can become explosively active.

4 Contrast the evolutionary histories of high-mass with those of low-mass stars.

5 Describe the two types of supernova, and explain how each is produced.

6 Explain the origin of elements heavier than helium, and discuss the significance of these elements for the study of stellar evolution.

7 Discuss the observations that help verify the theory of stellar evolution.

Visit www.prenhall.com/chaisson for additional annotated images, animations, and links to related sites for this chapter.

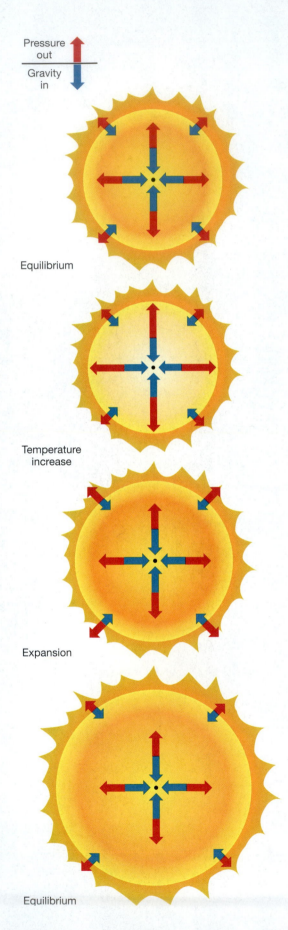

Pressure out

Gravity in

Equilibrium

Temperature increase

Expansion

Equilibrium

12.1 Leaving the Main Sequence

Most stars spend most of their lives on the main sequence. A star like the Sun, for example, after spending a few tens of millions of years in formation (stages 1–6 in Chapter 11), will reside on or near the main sequence (stage 7) for roughly 10 billion years before evolving into something else. ∞ *(Sec. 11.3)* That "something else" is the topic of this chapter.

Stars and the Scientific Method

No one has ever witnessed the complete evolution of any star, from birth to death. Stars take an enormously long time—millions, billions, even trillions of years—to evolve. ∞ *(Sec. 10.7)* Yet in less than a century, astronomers have developed a comprehensive theory of stellar evolution and now even go so far as to claim that this theory is one of the best-tested in all of astronomy. How can this be? How can we can talk so confidently about what took place billions of years in the past, and what will happen billions of years in the future? The answer is that we can observe billions of stars in the universe, enough to see examples of every possible stage of stellar development, allowing us to test and refine our theoretical ideas. ∞ *(Discovery 1-1)* Just as we might piece together a picture of the human life cycle by studying a snapshot of all the residents of a large city, so we can construct a picture of stellar evolution by studying the myriad stars we see in the night sky.

The modern theory of the lives and deaths of stars is another example of the scientific method in action. ∞ *(Sec. E.5)* Faced with a huge volume of observational data, with little or no theory to organize or explain it, astronomers in the late 19th and early 20th centuries painstakingly classified and categorized the properties of the stars they observed. ∞ *(Sec. 10.5)* During the first half of the 20th century, as quantum mechanics began to yield detailed explanations of the behavior of light and matter on subatomic scales, theoretical explanations of many key stellar properties fell into place. ∞ *(Sec. 2.6)* Since the 1950s, a truly comprehensive theory has emerged, tying together the basic disciplines of atomic and nuclear physics, electromagnetism, thermodynamics, and gravitation into a coherent whole. Today, theory and observation proceed hand in hand, each refining and validating the details of the other as astronomers continue to hone their understanding of stellar evolution.

Note, by the way, that astronomers always use the term "evolution" in this context to mean change *during the lifetime of an individual star.* Contrast this usage with the meaning of the term in biology, where it refers to changes in the characteristics of a *population* of plants or animals over many generations. In fact, as we will see at the end of this chapter, populations of stars do also evolve much as in the latter "biological" sense, as the overall composition of the interstellar medium (and hence of each new stellar generation) changes slowly over time due to nuclear fusion in stars. However, in astronomical parlance, "stellar evolution" normally refers to changes that occur during a single stellar lifetime.

Structural Change

On the main sequence, a star slowly fuses hydrogen into helium in its core. This process is usually called **core-hydrogen burning**, but realize that here is

◀ **Figure 12.1 Hydrostatic Equilibrium**   In a steadily burning star on the main sequence, the outward pressure exerted by hot gas balances the inward pull of gravity. This is true at every point within the star, guaranteeing its stability. From top to bottom, the frames show how a star alters its size to maintain equilibrium.

another instance where astronomers use a familiar term in an unfamiliar way: To astronomers, "burning" always means nuclear fusion in a star's core, and not the chemical reaction (such as the combustion of wood or gasoline in air) we would normally think of in everyday speech. Chemical burning does not directly affect atomic nuclei.

As illustrated schematically in Figure 12.1, a main sequence star is in a state of hydrostatic equilibrium, in which pressure's outward push exactly counteracts gravity's inward pull. ∞ *(Sec. 9.2)* This is a stable balance between gravity and pressure, in which a small change in one always results in a small compensating change in the other. For example, as shown in the figure, a small increase in the star's central temperature leads to an increase in pressure, causing the star to expand and cool, thus recovering its equilibrium. Conversely, a small temperature decrease leads to a slight pressure decrease, and gravity causes the star to contract and heat up; again, equilibrium is restored. You should keep Figure 12.1 in mind as you study the various stages of stellar evolution described next. Much of a star's complex behavior can be understood in these simple terms.

As the hydrogen in the star's core is consumed, the balance between these opposing forces starts to shift. Both the star's internal structure and its outward appearance begin to change. The star leaves the main sequence. The post-main-sequence stages of stellar evolution—the end of the star's life—depend critically on the star's mass. As a rule of thumb, we can say that low-mass stars die gently, while high-mass stars die catastrophically. The dividing line between these two very different outcomes lies around eight times the mass of the Sun, and in this chapter we will refer to stars of more than 8 solar masses as "high-mass" stars. However, within both the "high-mass" and the "low-mass" (i.e., less than 8 solar masses) categories, there are substantial variations, some of which we point out as we proceed.

Rather than dwelling on the many details, we will concentrate on a few representative evolutionary sequences. We begin by considering the evolution of a fairly low-mass star like the Sun. Later, we will broaden our discussion to include all stars, large and small.

12.2 Evolution of a Sun-like Star

Figure 12.2 illustrates how the composition of a main-sequence star's interior changes as the star ages. The star's helium content increases fastest at the center, where temperatures are highest and the burning (creating helium from hydrogen) is fastest. ∞ *(Sec. 9.5)* The helium content also increases near the edge of the core, but more slowly because the burning rate is less rapid there. The inner, helium-rich region becomes larger and more hydrogen deficient as the star continues to shine.

Stages 8 and 9—the Subgiant and Red Giant Branches

Recall from Chapter 9 that a temperature of 10^7 K is needed to fuse hydrogen into helium. Only above that temperature do colliding hydrogen nuclei (that is, protons) have enough speed to overwhelm their mutual electromagnetic repulsion. ∞ *(Sec. 9.5)* Because helium nuclei (with two protons each, compared to one for hydrogen) carry a greater positive charge, their electromagnetic repulsion is larger, and so higher temperatures are needed to cause them to fuse. The core temperature at this stage is too low for helium fusion to begin. Eventually, hydrogen becomes depleted at the center, the nuclear fires there subside, and the location of principal burning moves to higher layers in the core. An inner core of nonburning pure helium starts to grow.

◄◄◄ **CONCEPT CHECK** ►

In what sense is the Sun stable?

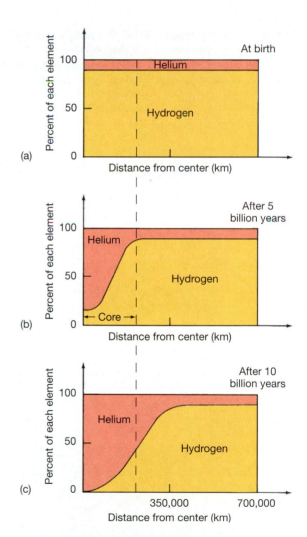

▲ **Figure 12.2 Solar Composition Change**
Theoretical estimates of the changes in a Sun-like star's composition. Hydrogen (yellow) and helium (orange) abundances are shown (a) at birth, just as the star arrives on the main sequence; (b) after 5 billion years; and (c) after 10 billion years. At stage (b) only a few percent of the star's total mass has been converted from hydrogen to helium. The change speeds up as the nuclear burning rate increases with time.

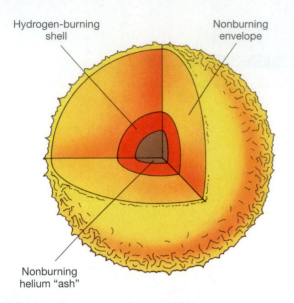

Hydrogen-burning shell

Nonburning envelope

Nonburning helium "ash"

▲ **Figure 12.3 Hydrogen Shell Burning** As a star's core converts more and more of its hydrogen into helium, the hydrogen in the shell surrounding the nonburning helium "ash" burns ever more violently. Note that the diagram is not to scale. By the time shown here (a little after stage 8 in Table 12.1) the core has shrunk to a few tens of thousands of kilometers in diameter, while the star's surface is ten times the star's original size.

Without nuclear burning to maintain it, the outward-pushing gas pressure weakens in the helium inner core; however, the inward pull of gravity does not. Once the pressure is relaxed—even a little—structural changes in the star become inevitable. As soon as hydrogen becomes substantially depleted, about 10 billion years after the star arrived on the main sequence, the helium core begins to contract.

The shrinkage of the helium core releases gravitational energy, driving up the central temperature and heating the overlying layers, causing the hydrogen there to fuse even more rapidly than before. Figure 12.3 depicts this **hydrogen shell-burning** stage, in which hydrogen is burning at a furious rate in a relatively thin layer surrounding the nonburning inner core of helium "ash." The hydrogen shell generates energy faster than did the original main-sequence star's hydrogen-burning core, and the shell's energy production continues to increase as the helium core contracts. Strange as it may seem, the star's response to the disappearance of the nuclear fire at its center is to get brighter!

The pressure exerted by this enhanced hydrogen burning causes the star's nonburning outer layers to increase in radius. Not even gravity can stop them. Even while the core is shrinking and heating up, the overlying layers are expanding and cooling. The star, aged and unbalanced, is on its way to becoming a red giant. The change from normal main-sequence star to elderly red giant takes about 100 million years.

We can trace these large-scale changes on an H–R diagram. Figure 12.4 shows the star's path away from the main sequence, labeled as stage 7. ∞ *(Sec. 11.3)* The star first evolves to the right on the diagram, its surface temperature dropping while its luminosity increases only slightly. The star's roughly horizontal track from its main-sequence location (stage 7) to stage 8 on the figure is called the **subgiant branch**. By stage 8 the star's radius has increased to about three times the radius of the Sun. The nearly vertical (constant temperature) path followed by the star between stages 8 and 9 is known as the **red giant branch** of the H–R diagram. (Figure 12.4 corresponds to a point just after stage 8, as the star begins to ascend the giant branch.)

The red giant is huge. By stage 9 its luminosity is many hundreds of times the solar value. Its radius is around 100 solar radii—about the size of Mercury's orbit. In contrast, its helium core is surprisingly small, less than 1/1000 the size of the entire star, or just a few times larger than Earth. Continued shrinkage of the red giant's core has compacted its helium gas to approximately 10^8 kg/m³. Contrast this with the 10^{-3} kg/m³ in the giant's outermost layers, with the 5500 kg/m³ average density of Earth, and with the 150,000 kg/m³ in the present core of the Sun. About 25 percent of the mass of the entire star is packed into its planet-sized core.

A familiar example of a low-mass star in the red giant phase is the KIII giant Arcturus (see Figure 10.14), one of the brightest stars in our night sky. Its mass is about 1.5 times that of the Sun. Currently in the hydrogen shell-burning stage and ascending the red giant branch, its radius is some 21 times the solar value. ∞ *(Table 10.4)* It emits more than 160 times more energy than the Sun, much of it in the infrared part of the spectrum.

Stage 10—Helium Fusion

This simultaneous contraction of the red giant's core and expansion of its envelope—the nonburning layers surrounding the core—does not continue in-

◀ **Figure 12.4 Red Giant on the H–R Diagram** As its helium core shrinks and its outer envelope expands, the star leaves the main sequence (stage 7). At stage 8, the star is well on its way to becoming a red giant. The star continues to brighten and grow as it ascends the red giant branch to stage 9. As noted in Chapter 10, the dashed diagonal lines are lines of constant radius, allowing us to gauge the changes in the size of the star.

definitely. A few hundred million years after a solar-mass star leaves the main sequence, the central temperature reaches the 10^8 K needed for helium to fuse into carbon, and the nuclear fires reignite.

For stars comparable in mass to the Sun, the high densities and pressures found in the stage-9 core mean that the onset of helium fusion is a very violent event. Once the burning starts, the core cannot respond quickly enough to the rapidly changing conditions within it, and its temperature rises sharply in a runaway explosion called the **helium flash**. For a few hours, the helium burns ferociously, like an uncontrolled bomb. Eventually, the star's structure "catches up" with the flood of energy dumped into it by helium burning. The core expands, its density drops, and equilibrium is restored as the inward pull of gravity and the outward push of gas pressure come back into balance. The core, now stable, begins to burn helium into carbon at temperatures well above 10^8 K.

The helium flash terminates the star's ascent of the red giant branch of the H–R diagram at stage 9 in Figure 12.4. Yet despite the explosive detonation of helium in the core, the flash does *not* increase the star's luminosity. On the contrary, the helium flash produces an expansion and cooling of the core that ultimately results in a *reduction* in the energy output as the star jumps from stage 9 to stage 10. As indicated in Figure 12.5, the surface temperature is now higher than it was on the red giant branch, although the luminosity is considerably less than at the helium flash. This adjustment in the star's properties occurs quite quickly—in about 100,000 years.

At stage 10 our star is now stably burning helium in its inner core and fusing hydrogen in a shell surrounding that core. It resides in a well-defined region of the H–R diagram known as the **horizontal branch**, where core helium-burning stars remain for a time before resuming their journey around the H–R diagram. The star's specific position within this region is determined mostly by its mass—not its original mass, but whatever mass remains after its ascent of the red giant branch. The two masses differ because, during the red giant stage, strong stellar winds eject large amounts of matter—as much as 20–30 percent of the initial mass—from a star's surface. The more massive stars have thicker envelopes and lower surface temperatures at this stage, but all stars have roughly the same luminosity after the helium flash. As a result, stage-10 stars tend to lie along a horizontal line on the H–R diagram, with more massive stars to the right and less massive ones to the left.

Stage 11—Back to the Giant Branch

Nuclear reactions in stars proceed at rates that increase very rapidly with temperature. At the extremely high temperatures found in the core of a horizontal-branch star, the helium fuel doesn't last long—no more than a few tens of million years after the initial flash.

As helium fuses to carbon, a new carbon-rich inner core begins to form and phenomena similar to the earlier buildup of helium occur. Helium becomes depleted at the center, and eventually fusion ceases there. The nonburning carbon core shrinks in size—even as its mass increases due to helium fusion—and heats up as gravity pulls it inward, causing the hydrogen- and helium-burning rates in the overlying layers to increase. The star now contains a contracting carbon-ash inner core surrounded by a helium-burning shell, which is in turn surrounded by a hydrogen-burning shell (Figure 12.6). The outer envelope of the star expands, much as it did earlier in the first red giant

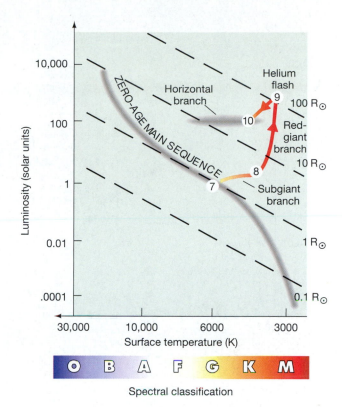

▲ **Figure 12.5 Horizontal Branch**   A large increase in luminosity occurs as a star ascends the red-giant branch, ending in the helium flash. The star then settles down into another equilibrium state at stage 10, on the horizontal branch.

◀◀◀ **CONCEPT CHECK**

Why does a star get brighter as it runs out of fuel in its core?

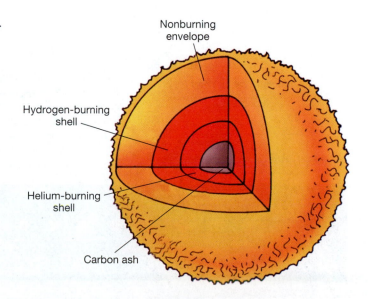

▶ **Figure 12.6 Helium Shell Burning**   Within a few million years after the onset of helium burning (stage 9), carbon ash accumulates in the star's inner core (not to scale). Above this core, hydrogen and helium are still burning in concentric shells.

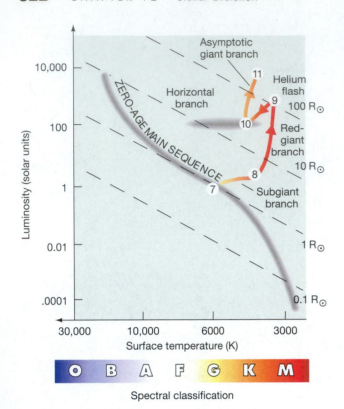

Spectral classification

◄ **Figure 12.7 Reascending the Giant Branch** A carbon-core star reenters the giant region of the H–R diagram (stage 11) for the same reason it evolved there the first time around: Lack of nuclear burning at the center causes the core to contract and the overlying layers to expand. The star here is midway between stages 10 and 11 in Table 12.1.

stage. By the time it reaches stage 11 in Figure 12.7, the star has become a swollen red giant for a second time.

The burning rates in the shells surrounding the inner core are much fiercer during the star's second trip into the red giant region, and the radius and luminosity increase to values even greater than those reached during the first visit (stage 9). Its carbon core continues to shrink, driving the hydrogen- and helium-burning shells to higher and higher temperatures and luminosities. To distinguish this second ascent of the red-giant branch from the first, the star's track during this stage is often referred to as the *asymptotic giant branch.*[1]

Table 12.1 summarizes the key stages through which our solar-mass star evolves. It is a continuation of Table 11.2, except that the density units are now the more convenient kilograms per cubic meter and we now express radii in units of the solar radius. As in Chapter 11, the numbers in the "Stage" column refer to the evolutionary stages noted in the figures and discussed in the text. Figure 12.8 illustrates the stages through which a G-type star like our Sun will evolve over the course of its lifetime.

12.3 The Death of a Low-Mass Star

As our star moves from stage 10 to stage 11, its envelope swells while its inner carbon core, too cool for further nuclear burning, continues to contract. If the central temperature could become high enough for carbon fusion to occur, still heavier elements could be synthesized, and the newly generated energy might again support the star, restoring for a time the equilibrium between gravity and pressure. For solar-mass stars, however, this does not occur—a lot more mass is needed, as we will see in the next section. The temperature never reaches the 600 million K needed for new nuclear reactions to occur. The red giant is now very close to the end of its nuclear-burning lifetime.

[1]The term is borrowed from mathematics. An asymptote to a curve is a second curve that approaches ever closer to the first as the two are extended to infinity. Theoretically, if the star remained intact, the asymptotic-giant branch would approach the red giant branch from the left as the luminosity increased and would effectively merge with the red giant branch near the top of Figure 12.10. However, a Sun-like star will not live long enough for that to occur.

Death of the Sun Part I Animation/Video

Death of the Sun Part II Animation/Video

▼ **Figure 12.8 G-type Star Evolution** Relative sizes and colors of a normal G-type star (such as our Sun) during its formative stages, on the main sequence, and while passing through the red giant and white dwarf stages. At maximum swelling, the red giant is approximately 70 times the size of its main-sequence parent; the core of the giant is about 1/15 the main-sequence size and would be barely discernible if this figure were drawn to scale. The duration of time spent in the various stages—protostar, main-sequence star, red giant, and white dwarf—is roughly proportional to the lengths shown in this imaginary trek through space. The star's brief stays on the horizontal and asymptotic giant branches are not shown.

Protostar   Main-sequence G-type star

Stage 4   Stage 7

TABLE 12.1	Evolution of a Sun-like Star						
STAGE	APPROX. TIME TO NEXT STAGE (yr)	CENTRAL TEMPERATURE (K)	SURFACE TEMPERATURE (K)	CENTRAL DENSITY (kg/m³)	RADIUS (km)	RADIUS (solar radii)	OBJECT
7	10^{10}	1.5×10^7	6,000	10^5	7×10^5	1	Main-sequence star
8	10^8	5×10^7	4,000	10^7	2×10^6	3	Subgiant
9	10^5	10^8	4,000	10^8	7×10^7	100	Red giant/Helium flash
10	5×10^7	2×10^8	5,000	10^7	7×10^6	10	Horizontal branch
11	10^4	2.5×10^8	4,000	10^8	4×10^8	500	Red giant (AGB)
	10^5	3×10^8	100,000	10^{10}	10^4	0.01	Carbon core
12	—	—	3,000	10^{-17}	7×10^8	1,000	Planetary nebula*
13	—	10^8	50,000	10^{10}	10^4	0.01	White dwarf
14	—	Close to 0	Close to 0	10^{10}	10^4	0.01	Black dwarf

*Values in columns 2–7 refer to the envelope.

Stage 12—A Planetary Nebula

Our aging star is now in quite a predicament. Its carbon core is, for all practical purposes, dead, while the outer hydrogen- and helium-burning shells consume fuel at a furious and increasing rate. As it reascends the giant branch, the star begins to fall apart. Driven by the intense radiation from within, its outer layers begin to drift away into interstellar space. Slowly at first, then more rapidly as the core luminosity increases, the star loses virtually its entire envelope in less than a million years. The former red giant now consists of two distinct pieces: the exposed core, very hot and still very luminous, surrounded by a cloud of dust and cool gas—the escaping envelope—expanding away at a typical speed of a few tens of kilometers per second.

As the core exhausts its last remaining fuel, it contracts and heats up, moving to the left in the H–R diagram. Eventually it becomes so hot that its ultraviolet radiation ionizes the inner parts of the surrounding cloud, producing a spectacular display called a **planetary nebula** (see Figure 12.9). The word *planetary* here is misleading, for these objects have no connection whatever with planets. The name originated in the eighteenth century when, viewed at poor resolution through small telescopes, these shells of gas looked to some astronomers like the circular disks of the planets in our solar system.

Note that the mechanism by which planetary nebulae shine is very similar to that powering the emission nebulae we studied earlier—ionizing radiation from a hot star embedded in a cool gas cloud. ∞ *(Sec. 11.2)* However, recognize that these two classes of object have very different origins and represent completely separate phases of stellar evolution. The emission nebulae in Chapter 11 were the signposts of recent stellar birth. Planetary nebulae indicate impending stellar death.

Astronomers once thought that the escaping giant envelope would be more or less spherical in shape, completely surrounding the core in three dimensions, just as it had while still part of the star. Figure 12.9(a) shows an example in which this may well, in fact, be the case. The nebula looks brighter

Animation/Video
Helix Nebula Animation

Animation/Video
Bi-Polar Planetary Nebula

Red giant

White dwarf

Core

Stage 8

Stage 9

(a)

(b)

(c)

(d)

R I V U X G

▲ **Figure 12.9 Planetary Nebulae** A planetary nebula is an extended region of glowing gas surrounding a small, dense core (visible in each case shown here). (a) Abell 39, some 2100 pc away, is a classic planetary nebula shedding a spherical shell about 1.5 pc across; (b) Ring Nebula, perhaps the most famous of all planetaries at 1500 pc distant and 0.5 pc across, is probably a genuine ring, yet too small and dim to be seen with the naked eye; (c) Cat's Eye Nebula, about 1000 pc away and 0.1 pc across, is an example of a much more complex planetary, possibly produced by a pair of binary stars (unresolved at center) that have both shed envelopes; and (d) M2–9, some 600 pc away and 0.5 pc end-to-end, shows surprising twin lobes (or jets) of glowing gas emanating from a central, dying star and racing out at speeds of about 300 km/s. The white superposed arrows in each frame denote the white dwarf remnants. (AURA; NASA)

near the edges simply because there is more emitting gas along the line of sight there, creating the illusion of a bright ring. However, such cases now seem to be in the minority. There is growing evidence that, for reasons not yet fully understood, the final stages of red giant mass loss are often decidedly *non*spherical. The Ring nebula shown in Figure 12.9(b) may very well *be* a ring and not just our view of a glowing shell of gas as was once thought, and many planetary nebulae are much more complex than this. As illustrated in Figures 12.9(c) and (d), some exhibit jetlike and other irregular structures, and the star's environment (such as the existence of a binary companion) also seems to play an important role in determining the nebula's shape and appearance.

The central star fades and eventually cools, and the expanding gas cloud becomes more and more diffuse, gradually dispersing into interstellar space. In doing so it plays a vital role in the evolution of our Galaxy. During the final stages of the red giant's life, nuclear reactions between carbon and unburned helium in the core create oxygen, and in some cases even heavier elements, such as neon and magnesium. Some of these reactions also release neutrons that, carrying no electrical charge, have no electrostatic repulsion to overcome and hence can interact with existing nuclei to form still heavier elements. ∞ *(Sec. 2.6, 9.5)* All of these elements—helium, carbon, oxygen, and heavier—are "dredged up" from the depths of the core into the envelope by convection during the star's final years and enrich the interstellar medium when the giant envelope escapes. ∞ *(Sec. 9.2)* The evolution of low-mass stars is the source of virtually all of the carbon-rich dust observed throughout the plane of our Galaxy. ∞ *(Sec. 11.1)*

Dense Matter

Before the shrinking carbon core can become hot enough for carbon to fuse, the rising central pressure stops the contraction and stabilizes the temperature. However, this pressure is not the "normal" thermal pressure of high-speed particles in a very hot gas. ∞ *(More Precisely 5-1)* Instead, the core enters a state in which the free electrons in the core—a vast sea of charged particles stripped from their parent nuclei by the ferocious heat in the stellar interior—play a critical role in determining the star's future.

The core density at this stage (stage 12 in Table 12.1) is enormous—about 10^{10} kg/m³. This density is much higher than anything we have seen so far in our study of the cosmos. A single cubic centimeter of core matter would weigh 1000 kg on Earth—a ton of matter compressed into a volume about the size of a grape! Under these extreme conditions, a law of quantum physics known as the *Pauli Exclusion Principle* comes into play, effectively preventing the electrons in the core from being crushed any closer together. (In fact, the core was in a similar high-density state at the moment of helium ignition, causing the helium flash.)

In essence, the electrons behave like tiny rigid spheres that can be squeezed relatively easily up to the point of contact, but become virtually incompressible thereafter. By stage 12 the pressure in the carbon core resisting

▶ **Figure 12.10 White Dwarf on H–R Diagram**   A star's passage from the horizontal branch (stage 10) to the white dwarf stage (stage 13) creates an evolutionary path that cuts across the H–R diagram.

> **Tutorial**
> **Evolution of a 1-Solar-Mass Star**

> **Animation/Video**
> **H-R Diagram Tracks Stellar Evolution**

> **Animation/Video**
> **Evolution of a 1-Solar-Mass Star**

the force of gravity is supplied almost entirely by tightly packed electrons, and this pressure brings the core back into equilibrium at a central temperature of "only" about 300 million K—too cool to fuse carbon into any of the heavier elements. This stage represents the maximum compression that the star can achieve—there is simply not enough matter in the overlying layers to bear down any harder. Supported by the resistance of its electrons to further compression, the core contraction stops.

Stage 13—A White Dwarf

Formerly concealed by the atmosphere of the red giant star, the carbon core becomes visible as a *white dwarf* as the envelope recedes. ∞ *(Sec. 10.4)* The core is very small, about the size of Earth, with a mass about half that of the Sun. Shining only by stored heat, not by nuclear reactions, this small star has a white-hot surface when it first becomes visible, although it appears dim because of its small size. This is stage 13 of Table 12.1. The approximate path followed by the star on the H–R diagram as it evolves from a stage 11 red giant to a stage 13 white dwarf is shown in Figure 12.10.

Not all white dwarfs are seen as the cores of planetary nebulae. Several hundred have been discovered "naked," their envelopes expelled to invisibility long ago. Figure 12.11 shows an example of a white dwarf, Sirius B, that happens to lie particularly close to Earth; it is the faint binary companion of the much brighter Sirius A. ∞ *(Sec. 10.7)* Some of its properties are listed in Table 12.2. With more than the mass of the Sun packed into a volume smaller than Earth, Sirius B's density is about a million times greater than anything familiar to us in the solar system. Sirius B has an unusually high mass for a white dwarf. It is thought to be the evolutionary product of a star roughly four times the mass of the Sun.

Hubble Space Telescope observations of nearby globular clusters have revealed the white-dwarf sequences long predicted by theory but previously too faint to detect at such large distances. Figure 12.12(a) shows a ground-based view of the globular cluster M4, lying 2100 pc from Earth. Part (b) of the figure shows a *Hubble* close-up of a small portion of the cluster, revealing dozens of white dwarfs (some marked) among the cluster's much brighter main-sequence, red giant, and horizontal-branch stars. When plotted on an H–R diagram (see Figure 12.26), the white dwarfs fall nicely along the line indicated on Figure 12.10.

The white dwarf continues to cool and dim with time, following the white–yellow–red line near the bottom of Figure 12.10, eventually becoming a

TABLE 12.2	Sirius B—A Nearby White Dwarf
Mass	1.1 Solar masses
Radius	0.008 Solar radii (5500 km)
Luminosity (total)	0.04 Solar luminosities
Surface temperature	24,000 K
Average density	3×10^9 kg/m³

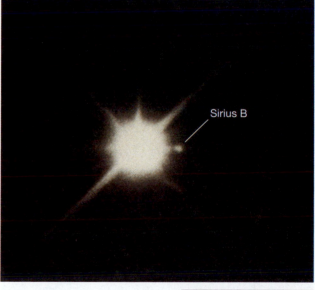

R I V U X G

▲ **Figure 12.11 Sirius Binary System**   Sirius B (the speck of light to the right of the much larger and brighter Sirius A) is a white dwarf star, a companion to Sirius A. The "spikes" on the image of Sirius A are not real; they are caused by the support struts of the telescope. (Palomar Observatory)

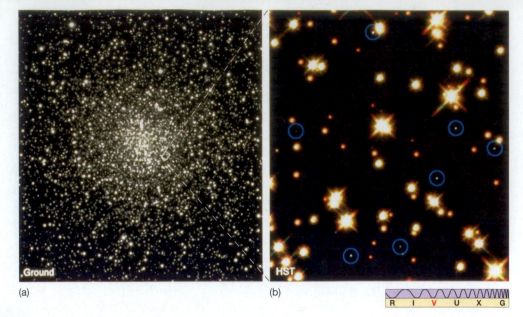

▶ **Figure 12.12 Distant White Dwarfs**
(a) The globular cluster M4, as seen through a large ground-based telescope at Kitt Peak National Observatory in Arizona. M4 is the closest globular cluster to us, at 2100 pc away; it spans some 16 pc. (b) A peek at M4's suburbs by the *Hubble Space Telescope* shows nearly a hundred white dwarfs within a small 0.2 square parsec region. Some of the brightest ones are circled in blue. (AURA; NASA)

(a)

(b)

(c)

black dwarf—a cold, dense, burned-out ember in space. This is stage 14 of Table 12.1, the graveyard of stars. The cooling dwarf does not shrink much as it fades away, however. Even though its heat is leaking away into space, gravity does not compress it further. Its tightly packed electrons will support the star even as its temperature drops (after trillions of years) almost to absolute zero. As the dwarf cools, it remains about the size of Earth.

Most white dwarfs are composed of carbon and oxygen, but theory predicts that very low-mass stars (less than about one-quarter the mass of the Sun) will never reach the point of helium fusion. Instead, the core of such a star will become supported by the resistance of its electrons to further compression before its central temperature reaches the 100 million K needed for helium to fuse. Eventually, the star's envelope will be ejected in a manner similar to that of a more massive star, forming a *helium white dwarf*. The time needed for this kind of transformation to occur is very long—hundreds of billions of years—so no helium white dwarfs have ever actually formed in this way. ∞ *(Sec. 10.7)* However, if a solar-mass star is a member of a binary system, it is possible for its envelope to be stripped away during the red giant stage by the gravitational pull of its companion, exposing the helium core and terminating the star's evolution before helium fusion can begin. Several such low-mass helium white dwarfs have in fact been detected in binary systems.

In stars much more massive than the Sun (close to the 8-solar mass limit on "low-mass" stars at the time the carbon core forms—see Section 12.4), temperatures in the core may become high enough that oxygen can combine with helium to form neon, eventually leading to the formation of a rare *neon–oxygen white dwarf* when fusion finally stops.

Novae

In some cases, the white dwarf stage does not represent the end of the road for a Sun-like star. Given the right circumstances, it is possible for a white

◀ **Figure 12.13 Nova** A nova is a star that suddenly increases enormously in brightness, then slowly fades back to its original luminosity. Novae are the result of explosions on the surfaces of faint white dwarf stars, caused by matter falling onto their surfaces from the atmosphere of larger binary companions. Shown is Nova Herculis 1934 in (a) March 1935 and (b) May 1935, after brightening by a factor of 60,000. (c) The light curve of a typical nova. The rapid rise and slow decline in the light received from the star, as well as the maximum brightness attained, are in good agreement with the explanation of the nova as a nuclear flash on a white dwarf's surface. (UC/Lick Observatory)

▶ **Figure 12.14 Close Binary System** If a white dwarf in a binary system is close enough to its companion, its gravitational field can tear matter from the companion's surface. Notice that the matter does not fall directly onto the white dwarf's surface. Instead, it forms an accretion disk of gas spiraling down onto the dwarf.

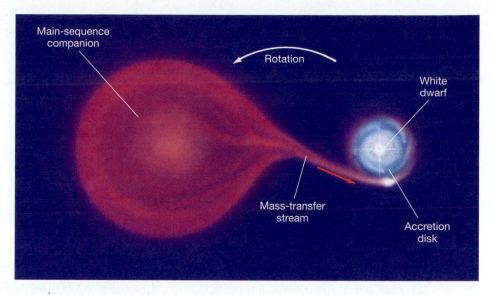

dwarf to become explosively active, in the form of a highly luminous **nova** (plural: novae). The word *nova* means "new" in Latin, and to early observers, these stars did indeed seem new as they suddenly appeared in the night sky. Astronomers now recognize that a nova is what we see when a white dwarf undergoes a violent explosion on its surface, resulting in a rapid, temporary increase in luminosity. Figures 12.13(a) and (b) illustrate the brightening of a typical nova over a period of three days. Figure 12.13(c) shows the nova's light curve, demonstrating how the luminosity rises dramatically in a matter of days, then fades slowly back to normal over the course of several months. On average, two or three novae are observed each year.

What could cause such an explosion on a faint, dead star? The energy involved is far too great to be explained by flares or other surface activity, and as we have just seen, there is no nuclear activity in the dwarf's interior. The answer to this question lies in the white dwarf's surroundings. If the white dwarf is isolated, then it will indeed cool and ultimately become a black dwarf, as just described. However, should the white dwarf be part of a binary system in which the other star is either a main-sequence star or a giant, an important new possibility exists.

If the distance between the dwarf and the other star is small enough, the dwarf's gravitational field can pull matter—primarily hydrogen and helium—away from the surface of the companion, as illustrated in Figure 12.14. Because of the binary's rotation, material leaving the companion does not fall directly onto the white dwarf. Instead, it "misses" the dwarf, loops around behind it, and goes into orbit around it, forming a swirling, flattened disk of matter called an **accretion disk**, as shown in the figure. As it builds up on the white dwarf's surface, the "stolen" gas becomes hotter and denser. Eventually its temperature exceeds 10^7 K, and the hydrogen ignites, fusing into helium at a furious rate. This surface-burning stage is as brief as it is violent. The star suddenly flares up in luminosity, then fades away as some of the fuel is exhausted and the rest is blown off into space (Figure 12.15). If the event happens to be visible from Earth, we see a nova.

Once the nova explosion is over and the binary has returned to normal, the mass-transfer process can begin again. Astronomers know of many *recurrent novae*—stars that have been observed to "go nova" several times over the course of a few decades. Such systems can, in principle, repeat their violent outbursts many dozens, if not hundreds, of times.

Animation/Video
Recurrent Nova

◀◀◀ **CONCEPT CHECK**

Will the Sun ever become a nova?

(a)

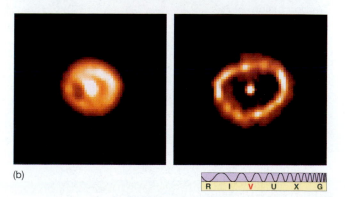

(b)

R I V U X G

▶ **Figure 12.15 Nova Matter Ejection** (a) The ejection of material from the star's surface can clearly be seen in this image of Nova Persei, taken some 50 years after it suddenly brightened by a factor of 40,000 in 1901. (b) Nova Cygni, imaged here with a European camera on the *Hubble Space Telescope*, erupted in 1992. At left, more than a year after the blast, a rapidly billowing bubble is seen; at right, seven months after that, the shell continued to expand and distort. The images are fuzzy because the object is more than 10,000 light-years away. (Palomar Observatory; ESA)

◄ Figure 12.16 High-Mass Evolutionary Tracks
Evolutionary tracks for stars of 1, 4, and 10 solar masses (shown only up to helium ignition in the 1-solar mass case). Stars with masses comparable to the mass of the Sun ascend the giant branch almost vertically, whereas higher-mass stars move roughly horizontally across the H–R diagram from the main sequence into the red giant region. The most massive stars experience smooth transitions into each new burning stage. Some points are labeled with the element that has just started to fuse in the inner core.

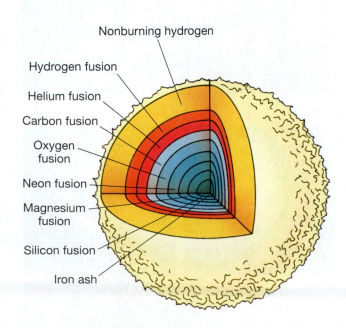

12.4 Evolution of Stars More Massive than the Sun

All stars leaving the main sequence on the journey toward the red giant region of the H–R diagram have similar internal structure. Thereafter, however, their evolutionary tracks diverge.

Heavy Element Fusion

Figure 12.16 compares the post-main-sequence evolution of three stars having masses 1, 4, and 10 times the mass of the Sun. Note that while the solar-mass star ascends the red giant branch almost vertically, the higher-mass stars tend to move nearly horizontally across the H–R diagram after leaving the upper main sequence. Their luminosities stay roughly constant as their radii increase and their surface temperatures drop.

The 4- and 10-solar mass stars loop back and forth across the diagram as they evolve. They do not experience a helium flash when helium fusion begins, and they reach temperatures high enough for helium and carbon to fuse into oxygen. In the 4-solar mass star, nuclear burning subsequently ceases, eventually leading to a carbon-oxygen white dwarf, much as described previously. The fate of the 10-solar mass star is a little less certain. Depending on how much mass it loses as it evolves (see below), it may also end up as a white dwarf, or it might retain enough mass for fusion to continue to more massive nuclei. Even higher-mass stars can fuse not just hydrogen, helium, and carbon but also oxygen, neon, magnesium, silicon, and even heavier elements as its inner core continues to contract and its central temperature continues to rise.

Evolution proceeds so rapidly for stars bigger than 10 solar masses that it doesn't even reach the red giant region before helium fusion begins. Such stars achieve a central temperatures of 10^8 K while still quite close to the main sequence. As each element is burned to depletion at the center, the core contracts, heats up, and fusion starts again. A new inner core forms, contracts again, heats again, and so on. The star's evolutionary track continues smoothly across the H–R diagram, seemingly unaffected by each new phase of burning. The star's luminosity stays roughly constant as its radius increases and its surface temperature drops. It swells to become a **red supergiant**.

Figure 12.17 is a cutaway diagram of an evolved high-mass star nearing the end of its red supergiant stage. Note the numerous layers where various nuclei burn. As the temperature increases with depth, the product of each burning stage becomes the fuel for the next. At the relatively cool periphery of the core, hydrogen fuses into helium. In the intermediate layers, shells of helium, carbon, and oxygen burn to form heavier nuclei. Deeper down reside neon, magnesium, silicon, and other heavy nuclei, all produced by nuclear fusion in the layers overlying the nonburning inner core. The inner core itself is composed of iron.

Observations of Supergiants

A good example of a post-main-sequence blue supergiant is the bright star Rigel in the constellation Orion. With a radius some 70 times that of the Sun

◄ Figure 12.17 Heavy-Element Fusion Cutaway diagram of the interior of a highly evolved high-mass star. The interior resembles the layers of an onion, with shells of progressively heavier elements burning at smaller and smaller radii and at higher and higher temperatures. The core is only a few times larger than planet Earth, while the star is hundreds of times larger than the Sun.

and a luminosity of around 60,000 solar luminosities, Rigel is thought to have had an original mass about 17 times that of the Sun, although a strong stellar wind has probably carried away a significant fraction of the star's mass since it formed. Although still near the main sequence, Rigel is probably already fusing helium into carbon in its core.

Astronomers have discovered that, while stars of all spectral types have stellar winds, the highly luminous supergiants have by far the strongest. ∞ *(Sec. 9.3)* Satellite and rocket observations have shown that the wind speeds of some blue supergiants may reach 3000 km/s and carry off more than 10^{-6} solar masses of material per year. Over the relatively short span of a few million years, these stars can blow a tenth or more of their total mass—a Sun's worth of matter—off into space. During violent outbursts near the end of the star's life, the mass loss rate may exceed 10^{-3} solar masses per year. Figure 12.18(a) shows a luminous blue variable star called Eta Carinae that is currently displaying this type of behavior. The glowing nebula surrounding the star itself probably contains some 2–3 solar masses of gas shed during this and previous outbursts over the past few hundred years.

The three-part *Hubble* image shown in Figure 12.18(b) captures another stellar outburst, in which a red supergiant suddenly brightened in 2002 to more than a half-million times our Sun's luminosity. Astronomers don't fully understand what caused the star to brighten in this way. Note that what we are seeing here is not matter being expelled outward as fast as the images suggest. Rather, a burst of light—often called a "light echo"—is illuminating shells of gas and dust shed long ago and now surrounding the star. By studying the gas revealed by the expanding light echo, we can learn about the star's mass loss history.

Perhaps the best-known red supergiant is Betelgeuse (shown in Figures 10.7 and 10.10), also in Orion and Rigel's rival for the title of brightest star in the constellation. Its luminosity is 10,000 times that of the Sun in visible light and perhaps four times that in the infrared. Astronomers think that Betelgeuse is currently fusing helium into carbon and oxygen in its core, but its eventual fate is uncertain. As best we can tell, the star's mass at formation was between 12 and 17 times the mass of the Sun. However, like Rigel and many other supergiants, Betelgeuse has a strong stellar wind and is known to be surrounded by a huge shell of dust of its own making. This suggests that Betelgeuse has lost a lot of mass since it formed, but just how much remains uncertain.

(b)

▲ **Figure 12.18 Mass Loss from Supergiants**   (a) With an estimated mass of around 100 times the mass of the Sun and a luminosity of 5 million times the solar value, Eta Carinae is one of the most massive and most luminous stars known. This *HST* image shows blobs of ejected material racing away from the star at hundreds of kilometers per second, following a series of explosive outbursts that ejected 2 or 3 solar masses of material into space. (b) The star V838 Monocerotis is a highly variable (and poorly understood) red supergiant about 20,000 light-years distant. After the star brightened suddenly in 2002, a pulse of light spread outward illuminating shells of gas and dust expelled long ago. For scale, the rightmost image is about 7 light-years across. (NASA)

CONCEPT CHECK ▶▶▶

Why does the iron core of a high-mass star collapse?

The End of the Road

Through each period of stability and instability and each new burning stage, the star's central temperature increases, the nuclear reactions speed up, and the newly released energy supports the star for ever shorter periods of time. For example, in round numbers a star 20 times more massive than the Sun burns hydrogen for 10 million years, helium for 1 million years, carbon for 1000 years, oxygen for one year, and silicon for a week. Its iron core grows for less than a day.

Once the inner core begins to change to iron, our high-mass star is in trouble. Nuclear fusion involving iron does not produce energy, because iron nuclei are so tightly bound that energy cannot be extracted by combining them into heavier elements. In effect, iron plays the role of a fire extinguisher, damping the inferno in the stellar core. With the appearance of substantial quantities of iron, the central fires cease for the last time, and the star's internal support begins to dwindle. The star's foundation is destroyed, and its equilibrium is gone forever. Even though the temperature in the iron core has reached several billion kelvins by this stage, the enormous inward gravitational pull of matter ensures catastrophe in the very near future. Gravity overwhelms the pressure of the hot gas, and the star implodes, falling in on itself.

The core temperature rises to nearly 10 billion K. At these temperatures individual photons are energetic enough to split iron into lighter nuclei and then break those lighter nuclei apart until only protons and neutrons remain. This process is known as *photodisintegration*. In less than a second the collapsing core undoes all the effects of nuclear fusion that occurred during the previous 10 million years. But to split iron and lighter nuclei into smaller pieces requires a lot of energy. After all, this splitting is the opposite of the fusion reactions that generated the star's energy during earlier times. The process thus *absorbs* some of the core's heat energy, reducing the pressure and accelerating the collapse.

Now the core consists entirely of electrons, protons, neutrons, and photons at enormously high densities, and it is still shrinking. As the density continues to rise, the protons and electrons are crushed together, combining to form more neutrons and releasing neutrinos. Even though the central density by this time may have reached 10^{12} kg/m^3 or more, most of the neutrinos pass through the core as if it weren't there. ∞ *(Sec. 9.5)* They escape into space, carrying away still more energy as they go.

There is nothing to prevent the collapse from continuing all the way to the point at which the neutrons themselves come into contact with each other, at the incredible density of about 10^{15} kg/m^3. At this density, the neutrons in the shrinking core play a role similar in many ways to that of the electrons in a white dwarf. When far apart, they offer little resistance to compression, but when brought into contact, they produce enormous pressures that strongly oppose further contraction. The collapse finally begins to slow. By the time it is actually halted, however, the core has overshot its point of equilibrium and may reach a density as high as 10^{18} kg/m^3 before beginning to reexpand. Like a fast-moving ball hitting a brick wall, the core becomes compressed, stops, then rebounds—with a vengeance!

The events just described do not take long. Only about a second elapses from the start of the collapse to the "bounce" at nuclear densities. Driven by the rebounding core, an enormously

▼ **Figure 12.19 Supernova 1987A** A supernova called SN1987A (arrow) was exploding near this nebula (30 Doradus) at the moment the photograph at right was taken. The photograph at left is the normal appearance of the star field. (See *Discovery 12-1*.) (AURA)

R I V U X G

energetic shock wave then sweeps outward through the star at high speed, blasting all the overlying layers—including the heavy elements outside the iron inner core—into space. Although the details of how the shock reaches the surface and destroys the star are still uncertain, the end result is not: The star explodes in one of the most energetic events known in the universe (Figure 12.19). This spectacular death rattle of a high-mass star is known as a **core-collapse supernova**.

Our earlier dividing line of 8 solar masses between "low mass" and "high mass" really refers to the mass at the time the carbon core forms. Since very luminous stars often have strong stellar winds, it is possible that main-sequence stars as massive as 10 to 12 times the mass of the Sun may still manage to avoid going supernova. Unfortunately, we do not know exactly how much mass either Rigel or Betelgeuse has lost, so we cannot yet tell whether they are above or below the supernova threshold. Either might explode or instead become a (neon–oxygen) white dwarf, but for now we can't say which. We may just have to wait and see—in a few million years we will know for sure!

Table 12.3 summarizes the possible outcomes of stellar evolution for stars of different masses. For completeness, brown dwarfs—the end product of low-mass protostars unable even to fuse hydrogen in their cores—are included in the list. ∞ *(Sec. 11.4)*

TABLE 12.3	End Points of Evolution for Stars of Different Masses
INITIAL MASS (SOLAR MASSES)	FINAL STATE
Less than 0.08	(Hydrogen) brown dwarf
0.08–0.25	Helium white dwarf
0.25–8	Carbon–oxygen white dwarf
8–12 (approx.)*	Neon–oxygen white dwarf
Greater than 12*	Supernova

*Precise numbers depend on the (poorly known) amount of mass lost while the star is on, and after it leaves, the main sequence.

12.5 Supernova Explosions

Novae and Supernovae

Observationally, a *supernova*, like a nova, is a "star" that suddenly increases dramatically in brightness, then slowly dims again, eventually fading from view. However, despite some similarities in their light curves, novae and supernovae are very different phenomena. As we saw in Section 12.2, a nova is a violent explosion on the surface of a white dwarf in a binary system.[2] Supernovae are much more energetic—about a million times brighter than novae—and are driven by very different underlying physical processes. A supernova produces a burst of light billions of times brighter than the Sun, reaching that brightness within just a few hours of the start of the outburst. The total amount of electromagnetic energy radiated by a supernova during the few months it takes to brighten and fade away is roughly the same as the Sun will radiate during its *entire* 10^{10}-year lifetime!

Astronomers divide supernovae into two classes. **Type I supernovae** contain very little hydrogen, according to their spectra, and have light curves (see Figure 12.20) somewhat similar in shape to those of typical novae—a sharp rise in intensity followed by steady, gradual decline. **Type II supernovae** are hydrogen-rich, and usually have a characteristic "plateau" in the light curve a few months after the maximum. Observed supernovae are divided roughly evenly between these two categories.

Type I and Type II Supernovae Explained

How do we explain the two types of supernova just described? In fact, we already have half of the answer—the characteristics of Type II supernovae are exactly consistent with the core-collapse supernovae discussed in the previous

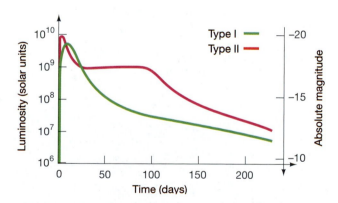

▲ **Figure 12.20 Supernova Light Curves** The light curves of typical Type I and Type II supernovae. In both cases, the maximum luminosity can sometimes reach that of a billion suns, but there are characteristic differences in the falloff of the luminosity after the initial peak. Type I light curves somewhat resemble those of novae (see Figure 12.13c), but the total energy released is much larger. Type II curves have a characteristic plateau during the declining phase.

[2]Note that when discussing novae and supernovae, astronomers tend to blur the distinction between the observed event (the sudden appearance and brightening of an object in the sky) and the process responsible for it (a violent explosion in or on a star). The two terms can have either meaning, depending on context.

section. The Type II light curve is in good agreement with computer simulations of a stellar envelope expanding and cooling as it is blown into space by a shock wave sweeping up from below. In addition, since the expanding material consists mainly of unburned hydrogen and helium, it is not surprising that those elements are strongly represented in the supernova's spectrum.

What of Type I supernovae? Is there more than one way for a supernova explosion to occur? The answer is yes. To understand the alternative supernova mechanism, we must reconsider the long-term consequences of the accretion–explosion cycle that causes a nova. A nova explosion ejects matter from a white dwarf's surface, but it does not necessarily expel or burn all the material that has accumulated since the last outburst. In other words, there is a tendency for the dwarf's mass to increase slowly with each new nova cycle. As its mass grows and the internal pressure required to support its weight rises, the white dwarf can enter into a new period of instability—with disastrous consequences.

Recall that a white dwarf is held up by the pressure of electrons that have been squeezed so close together that they have effectively come into contact with one another. However, there is a limit to the mass of a white dwarf, beyond which the electrons cannot support the star against its own gravity. Detailed calculations show that the maximum mass of a white dwarf is about 1.4 solar masses, a mass often called the *Chandrasekhar mass*, after the Indian-American astronomer Subramanyan Chandrasekhar, whose work in theoretical astrophysics earned him a Nobel Prize in physics in 1983. More recently, in 1999, NASA's latest X-ray telescope was named *Chandra* in his honor. ∞ *(Sec. 3.5)*

If an accreting white dwarf exceeds the Chandrasekhar mass, it immediately starts to collapse. When that occurs, its internal temperature rapidly rises to the point at which carbon can (at last) fuse into heavier elements. Carbon fusion begins everywhere throughout the white dwarf almost simultaneously, and the entire star explodes in a **carbon-detonation supernova**—an event comparable in violence to the core-collapse supernova associated with the death of a high-mass star. In an alternative and (many astronomers think) possibly more common scenario, two white dwarfs in a binary system may collide and merge to form a massive, unstable star. The end result is the same—a carbon-detonation supernova.

The detonation of a carbon white dwarf, the descendant of a *low-mass* star, is a Type I supernova. Because the explosion occurs in a system containing virtually no hydrogen, we can readily see why the supernova spectrum shows little evidence of that element. The appearance of the light curve results almost entirely from the radioactive decay of unstable heavy elements produced during the explosion. These explosions have one other important property: Because they all originate in essentially similar systems—white dwarfs whose masses have just exceeded the Chandrasekhar limit—Type I supernovae show remarkably little variation in their properties from one supernova to the next, no matter where (or when) they occur. As we will see in Chapters 15 and 17, this fact, coupled with the huge luminosities involved, make Type I supernovae invaluable tools for mapping out the structure and evolution of the universe on the largest scales.

Figure 12.21 summarizes the processes responsible for the two types of supernovae. We emphasize that, despite the similarity in their peak luminosities, Type I and Type II supernovae are actually unrelated to one another. They occur in stars of very different types, under very different circumstances. All high-mass stars become Type II (core-collapse) supernovae, but only a tiny fraction of low-mass stars evolve into white dwarfs that ultimately explode as Type I (carbon-detonation) supernovae. However, there are far more low-mass stars than high-mass stars, resulting in the remarkable coincidence that the two types of supernova occur at roughly the same rate.

(a) Type I Supernova

White dwarf

Planetary nebula

Binary star system

Accretion disk

Red giant

Growing white dwarf

Detonation

Time

(b) Type II Supernova

Helium, carbon

Hydrogen

Normal star fusion

Heavy elements

Hydrogen

Iron core

Massive star imploding

Hydrogen

Core rebound

Remnant core

Shock wave

Explosion

▲ **Figure 12.21 Two Types of Supernova**   Type I and Type II supernovae have different causes. These sequences depict the evolutionary history of each type. (a) A Type I supernova usually results when a carbon-rich white dwarf pulls matter onto itself from a nearby red giant or main-sequence companion. (b) A Type II supernova occurs when the core of a high-mass star collapses and then rebounds in a catastrophic explosion.

Supernova Remnants

We have plenty of evidence that supernovae have occurred in our Galaxy. Occasionally, the supernova explosions are visible from Earth. In other cases, we can detect their glowing remains, or **supernova remnants**. One of the best-studied supernova remnants is the Crab Nebula (Figure 12.22a). Its brightness has greatly dimmed now, but the original explosion in the year A.D. 1054 was so brilliant that it is prominently recorded in the manuscripts of Chinese and Middle Eastern astronomers. For nearly a month, this exploded star reportedly could be seen in broad daylight. Even today, the knots and filaments give a strong indication of past violence. The nebula—the envelope of the high-mass star that exploded to create this Type II supernova—is still expanding into space at several thousand kilometers per second.

Figure 12.22(b) shows another example (also of Type II). The expansion velocity of the Vela supernova remnant implies that its central star exploded around 9000 B.C. This remnant lies only 500 pc from Earth. Given its proximity, it may have been as bright as the Moon for several months. See also Figure 3.29 for an X-ray image of a nearby supernova remnant.

Although hundreds of supernovae have been observed in other galaxies during the 20th century, no one using modern equipment has ever observed one in our own Galaxy (see *Discovery 12-1* for a discussion of a recent supernova in a galaxy very close to our own). A viewable Milky Way star has not exploded since Galileo first turned his telescope to the heavens almost four centuries ago. Based on stellar evolutionary theory, astronomers estimate that

Animation/Video

Supernova Explosions

(a)

(b)

R I V U X G

◀ **Figure 12.22** **Supernova Remnants**   (a) This remnant of an ancient Type II supernova is called the Crab Nebula (M1 in the Messier catalog). It resides about 1800 pc from Earth and has an angular diameter about one-fifth that of the full Moon. Its debris is scattered over a region about 2 pc across. In A.D. 1054 Chinese astronomers observed this supernova explosion. The main image was taken with the Very Large Telescope of the European Southern Observatory in Chile, the inset by the *Hubble* telescope in orbit. (b) The glowing gases of the Vela supernova remnant are spread across 6° of the sky. Note the extensive filamentary structure. (ESO; NASA; D. Malin/Anglo-Australian Observatory)

an observable supernova ought to occur in our Galaxy every 100 years or so. Our local neighborhood seems long overdue for one. Unless stars explode much less frequently than predicted by theory, we should be treated to a (relatively) nearby version of nature's most spectacular cosmic event any day now.

The Formation of the Heavy Elements

From the point of view of life on Earth, probably the most important aspect of stellar evolution is its role in creating, and then dispersing, the heavy elements out of which both our planet and our bodies are made. Since the 1950s, astronomers have come to realize that all of the hydrogen and most of the helium in the universe are *primordial*—that is, they date back to the very earliest times, long before the first stars formed. ∞ *(Sec. 17.6)* All other elements (and, in particular, virtually everything we see around us on Earth) formed later, in stars.

CONCEPT CHECK ▶▶▶

How did astronomers know, even before the mechanisms were understood, that there were at least two distinct physical processes at work in creating supernovae?

We have already seen something of how heavy elements are created from light ones by nuclear fusion. This is the basic process that powers all stars. Hydrogen fuses to helium, then helium to carbon and oxygen. Subsequently, in high-mass stars, carbon and oxygen can fuse to form still heavier elements. Neon, magnesium, sulfur, silicon—in fact, elements up to and including iron—are created in turn by fusion reactions in the cores of the most massive stars.

However, fusion stops at iron. The fact that iron nuclei will not fuse to release energy and create more massive nuclei is the basic underlying cause of Type II supernovae. How then were even heavier elements, such as copper, lead, gold, and uranium, formed? As mentioned in Section 12.3, some of these elements were formed during the late red-giant stages of low-mass stars via reactions involving neutrons. Similar reactions occur in supernovae too, where neutrons and protons, produced when some nuclei are ripped apart by the almost unimaginable violence of the blast, are crammed into other nuclei, creating heavy elements that cannot have formed by any other means. Many of the heaviest elements were formed *after* their parent stars had already died, even as the debris from the explosion signaling the stars' deaths was hurled into interstellar space. In this way stellar evolution steadily enriches the composition of each new generation of stars.

Although no one has ever observed directly the formation of heavy nuclei in stars, astronomers are confident that the chain of events just described really occurs. When nuclear reaction rates (determined from laboratory experiments) are incorporated into detailed computer models of stars and supernovae, the results agree very well, point by point, with the composition of the universe inferred from analysis of meteorites and spectroscopic studies of planets, stars, and the interstellar medium. The reasoning is indirect, but the agreement between theory and observation is so striking that most astronomers regard it as strong evidence supporting the entire theory of stellar evolution.

12.6 Observing Stellar Evolution in Star Clusters

Star clusters provide excellent test sites for the theory of stellar evolution. Every star in a given cluster formed at the same time, from the same interstellar cloud, with virtually the same composition. ∞ *(Sec. 11.5)* Only the mass varies from one star to another. This uniformity allows us to check the accuracy of our theoretical models in a very straightforward way. Having studied in some detail the evolutionary tracks of individual stars, let us now consider how their collective appearance changes in time.

We begin our study shortly after the cluster's formation, with the high-mass stars already fully formed and burning steadily on the upper main sequence and lower-mass stars just beginning to arrive on the main sequence (Figure 12.23a). The appearance of the cluster at this early stage is dominated by its most massive stars—the bright blue supergiants.

▶ **Figure 12.23 Cluster Evolution on the H–R Diagram**   The changing H–R diagram of a hypothetical star cluster. (a) Initially, stars on the upper main sequence are already burning steadily while the lower main sequence is still forming. (b) At 10^7 years, O-type stars have already left the main sequence (as indicated by the arrows), and a few red giants are visible. (c) By 10^8 years, stars of spectral type B have evolved off the main sequence. More red giants are visible, and the lower main sequence is almost fully formed. (d) At 10^9 years, the main sequence is cut off at about spectral type A. The subgiant and red giant branches are just becoming evident, and the formation of the lower main sequence is complete. A few white dwarfs may be present. (e) At 10^{10} years, only stars less massive than the Sun still remain on the main sequence. The cluster's subgiant, red giant, horizontal, and asymptotic-giant branches are all discernible. Many white dwarfs have now formed.

(a)

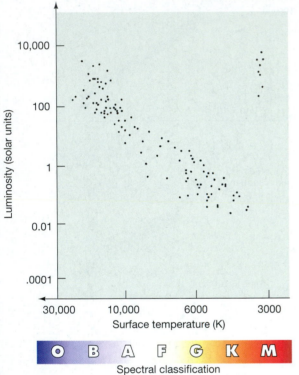

(b)

◀ **Figure 12.24 Newborn Cluster H–R Diagram** (a) The "double cluster" h and χ Persei. (b) The H–R diagram of the pair indicates that the stars are very young—probably only about 10 million years old. Even so, the most massive stars have already left the main sequence. (NOAO)

Figure 12.23(b) shows our cluster's H–R diagram after 10 million years. The most massive O-type stars have evolved off the main sequence. Most have already exploded and vanished, but one or two may still be visible as supergiants traversing the top of the diagram. The remaining cluster stars are largely unchanged in appearance. The cluster's H–R diagram has a slightly shortened main sequence. Figure 12.24 shows the twin open clusters h and χ (the Greek letter chi) Persei, along with their observed H–R diagram. Comparing Figure 12.24(b) with Figure 12.23(b), astronomers estimate the age of this double cluster to be about 10 million years.

After 100 million years (Figure 12.23c), stars brighter than type B5 or so (about 4 to 5 solar masses) have left the main sequence, and a few more supergiants are visible. By this time most of the cluster's low-mass stars have finally arrived on the main sequence, although the dimmest M-type stars may still be in their contraction phase. The appearance of the cluster is now dominated by bright B-type stars and brighter supergiants.

At any time during the evolution, the cluster's original main sequence is intact up to some well-defined stellar mass, corresponding to the stars that are just leaving the main sequence at that instant. We can imagine the main sequence being "peeled away" from the top down, with fainter and fainter stars turning off and heading for the giant branch as time goes on. Astronomers refer to the high-luminosity end of the observed main sequence as the **main-sequence turnoff**. The mass of the star that is just evolving off the main sequence at any moment is known as the *turnoff mass*. Knowing the turnoff mass is equivalent to knowing the age of the cluster—stars less massive than the turnoff are still on the main sequence, while more massive stars have already evolved into something else.

At 1 billion years (Figure 12.23d) the main-sequence turnoff mass is around 2 solar masses, corresponding roughly to spectral type A2. The subgiant and giant branches associated with the evolution of low-mass stars are just becoming apparent, and the formation of the lower main sequence is now complete. In addition, the first white dwarfs have just appeared, although they are often too faint to be observed at the distances of most clusters. Figure 12.25 shows the Hyades open cluster and its H–R diagram. The H–R diagram appears to lie between Figures 12.22(c) and 12.22(d). More careful measurements yield a cluster age of about 600 million years.

At 10 billion years, the turnoff point has reached solar-mass stars, of spectral type G2. The subgiant and giant branches are now clearly discernible in the H–R diagram (Figure 12.23e), and the horizontal branch appears as a distinct region. Many white dwarfs are also present in the cluster. Figure 12.26 shows a composite H–R diagram for several globular clusters in our Galaxy. The various evolutionary stages predicted by theory are all clearly visible in this figure, although the individual points are shifted somewhat to the left relative to our previous H–R diagrams because of differences in composition between stars like the Sun and stars in globular clusters. (Globular cluster stars tend to be hotter than solar-type stars of the same mass.) The clusters represented in Figure 12.26(b) are quite deficient in heavy elements compared with the Sun. Figure 12.26(a) shows a recent (*Hubble*) image of the globular cluster M80, which has a similarly low concentration of heavy elements—only about 2 percent the solar value—and whose H–R diagram looks qualitatively very similar to Figure 12.26(b).

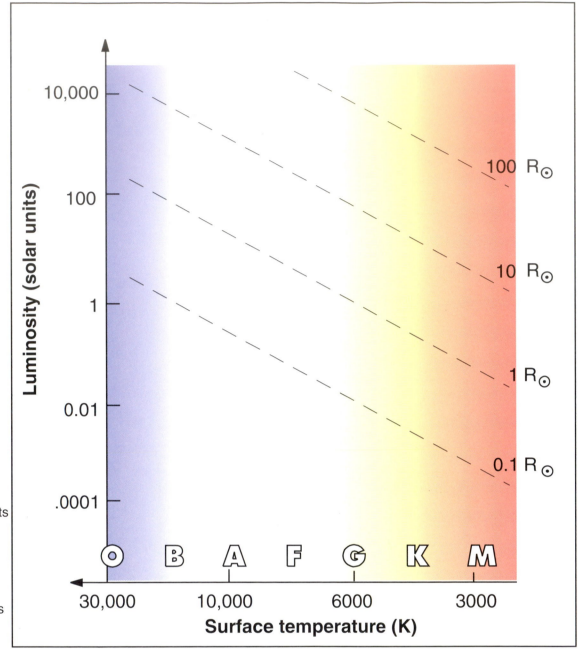

The H-R diagram plots stars by luminosity (vertical axis) and temperature, or spectral class (horizontal axis). The dashed diagonal lines are lines of constant radius.

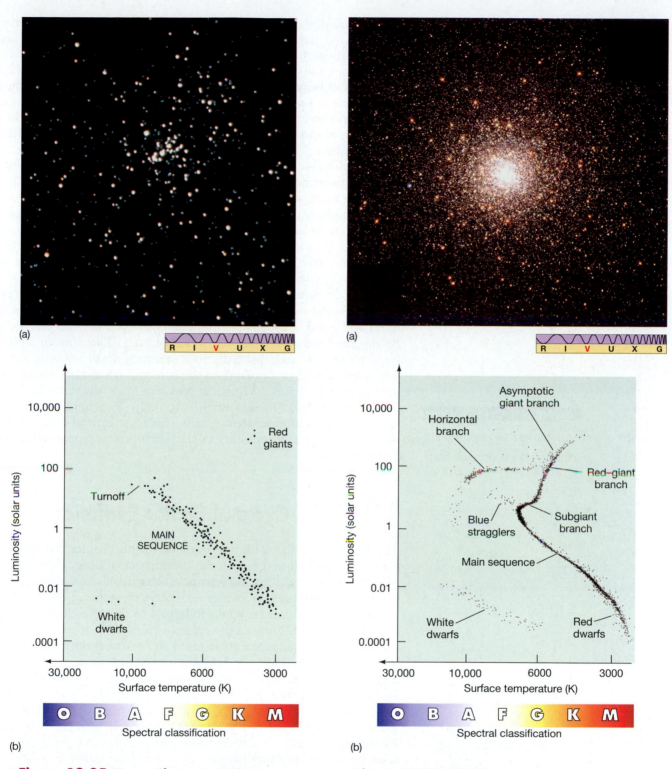

▲ Figure 12.25 Young Cluster H–R Diagram
(a) The Hyades cluster, a relatively young group of stars visible to the naked eye. (b) The H–R diagram for this cluster is cut off at about spectral type A, implying an age of about 600 million years. A few massive stars have already become white dwarfs. (NOAO)

▲ Figure 12.26 Old Cluster H–R Diagram (a) The globular cluster M80, which lies some 8 kpc from Earth. (b) Combined H–R diagram, based on ground- and space-based observations, for several globular clusters similar in overall composition to M80. Fitting the main-sequence turnoff and the giant and horizontal branches to theoretical models implies an age of about 12 billion years, making these clusters among the oldest known objects in the Milky Way Galaxy. This is consistent with the low concentrations of heavy elements observed in their spectra. (NASA; data courtesy of William E. Harris)

By carefully adjusting their theoretical models until the main sequence, subgiant, red giant, and horizontal branches are all well matched, astronomers have determined that this diagram corresponds to a cluster age of roughly 12 billion years, a little older than the hypothetical cluster in Figure 12.23(e). The deficiency of heavy elements is consistent with these clusters being among the first objects to form in our Galaxy (see Section 12.7). In fact, globular cluster ages determined this way show a remarkably small spread. Most of the globular clusters in our Galaxy appear to have formed between about 10 and 12 billion years ago.

The objects labeled as *blue stragglers* in Figure 12.26(b) appear at first sight to contradict the theory just described. They lie on the main sequence, but in locations suggesting that they should have evolved into white dwarfs long ago, given the cluster's age of 12 billion years. Blue stragglers are observed in many star clusters. They are main-sequence stars, but they did not form when the cluster did. Instead, they formed much more recently, through *mergers* of lower mass stars—so recently, in fact, that they have not yet had time to evolve into giants.

How did these stars merge? In some cases, such mergers were probably the result of stellar evolution in binary systems, as the component stars evolved, grew, and came into contact. In others, however, the mergers are thought to be the result of actual *collisions* between stars. The core of M80 contains a huge number of stars packed into a relatively small volume. For example, a sphere of radius 2 pc centered on the Sun contains exactly four stars, including the Sun itself. ∞ *(Sec. 10.1)* At the center of M80, the same 2-pc sphere would contain more than 1 *million* stars—our night sky would be ablaze with thousands of objects brighter than Venus! The dense central cores of globular clusters are among the few places in the entire universe where stellar collisions are likely to occur.

CONCEPT CHECK ▶▶▶

Why are observations of star clusters so important to the theory of stellar evolution?

12.7 The Cycle of Stellar Evolution

The theory of stellar evolution outlined in this chapter can naturally account for the observed differences in the abundances of heavy elements between the old globular-cluster stars and stars now forming in our Galaxy. Even though an evolved star continuously creates new heavy elements in its interior, changes in the star's composition are confined largely to the core, and the star's spectrum may give little indication of the events within. Only at the end of the star's life are its newly created elements released and scattered into space.

Thus the spectra of the *youngest* stars show the *most* heavy elements, because each new generation of stars increases the concentration of these elements in the interstellar clouds from which the next generation forms. Accordingly, the photosphere of a recently formed star contains a much greater abundance of heavy elements than that of a star formed long ago. Knowledge of stellar evolution allows astronomers to estimate the ages of stars from purely spectroscopic studies, even when the stars are isolated and are not members of any cluster.

We have now seen all the ingredients that make up the complete cycle of star formation and evolution. Let us now briefly summarize that process, which is illustrated in Figure 12.27.

1. Stars form when part of an interstellar cloud is compressed beyond the point at which it can support itself against its own gravity. The cloud collapses and fragments, forming a cluster of stars. The hottest stars

▶ **Figure 12.27 Stellar Recycling** The cycle of star formation and evolution continuously replenishes the Galaxy with new heavy elements and provides the driving force for the creation of new generations of stars. Clockwise from the top are an interstellar cloud (Barnard 68), a star-forming region in our Galaxy (RCW 38), a massive star ejecting a "bubble" and about to explode (NGC 7635), and a supernova remnant and its heavy-element debris (N49). (ESO; NASA)

Interstellar medium

Star formation

Stellar evolution

Supernova + heavy elements

heat and ionize the surrounding gas, sending shock waves through the surrounding cloud, modifying the formation of lower-mass stars and possibly triggering new rounds of star formation. ∞ *(Secs. 11.2, 11.3)*

2. Within the cluster, stars evolve. The most massive stars evolve fastest, creating heavy elements in their cores and spewing them forth into the interstellar medium in supernova explosions. Low-mass stars take longer to evolve, but they too can create heavy elements and contribute significantly to the "seeding" of interstellar space when they shed their envelopes as planetary nebulae. Roughly speaking, low-mass stars are responsible for most of the carbon, nitrogen, and oxygen that make life on Earth possible. High-mass stars produced the iron and silicon that make up Earth itself, as well as the heavier elements on which much of our technology is based.

3. The creation and explosive dispersal of new heavy elements are accompanied by further shock waves. The passage of these shock waves through the interstellar medium simultaneously enriches the medium and compresses it into further star formation. Each generation of stars increases the concentration of heavy elements in the interstellar clouds from which the next generation forms. As a result, recently formed stars contain a much greater abundance of heavy elements than stars that formed long ago.

In this way, although some material is used up in each cycle—turned into energy or locked up in stars less massive than the Sun (most of which have not yet left the main sequence)—the Galaxy continuously recycles its matter. Each new round of formation creates stars containing more heavy elements than the preceding generation had. From the old globular clusters, which are observed to be deficient in heavy elements relative to the Sun, to the young open clusters, which contain much larger amounts of these elements, we observe this enrichment process in action. Our Sun is the product of many such cycles. We ourselves are another. Without the heavy elements synthesized in the hearts of stars, life on Earth would not exist.

Stellar evolution is one of the great success stories of astrophysics. Like all good scientific theories, it makes definite, testable predictions about the universe, at the same time remaining flexible enough to incorporate new discoveries as they occur. Theory and observation have advanced hand in hand. At the start of the 20th century, many scientists despaired of ever knowing even the compositions of the stars, let alone why they shine and how they change. Today, the theory of stellar evolution is a cornerstone of modern astronomy.

◀◀◀ **CONCEPT CHECK**

Why is stellar evolution important to life on Earth?

12-1 DISCOVERY

Supernova 1987A

Although there have been no observable supernovae in our own Galaxy since the invention of the telescope, astronomers were treated to the next best thing in 1987 when a 15-solar mass B-type supergiant exploded in the Large Magellanic Cloud (LMC), a small satellite galaxy orbiting our own (see Sec. 15.1). For a few weeks the Type II supernova, designated SN1987A and shown in Figure 12.19, outshone all the other stars in the LMC combined. Because the LMC is relatively close to Earth and the explosion was detected soon after it occurred, SN1987A has provided astronomers with a wealth of detailed information on supernovae, allowing them to make key comparisons between theoretical models and observational reality. By and large, the theory of stellar evolution has held up very well. Still, SN1987A did hold a few surprises.

The light curve of SN1987A, shown at right, differed somewhat from the "standard" Type II shape (Figure 12.20). The peak brightness was only about 1/10 the standard value and occurred much later than expected. These differences are mainly the result of the (relatively) small size of SN1987A's parent star, which was a blue supergiant at the time of the explosion and not a red supergiant, as indicated by the end of the topmost evolutionary track in Figure 12.16. The reason for this is that the parent star's envelope was deficient in heavy elements, significantly altering the star's evolutionary track. Having looped back toward the main sequence after helium ignition, the star had just turned back to the right on the H–R diagram following the ignition of carbon, with a surface temperature of around 20,000 K, when the rapid chain of events leading to the supernova occurred.

Because the parent star was small and quite tightly bound by gravity, a lot of the energy produced in the form of electromagnetic radiation was used in expanding SN1987A's stellar

envelope, so far less was left over to be radiated into space. Thus, SN1987A's luminosity during the first few months was therefore lower than expected, and the early peak evident in Figure 12.20 did not occur. The peak in the SN1987A light curve at about 80 days actually corresponds to the "plateau" in the Type II light curve in Figure 12.20.

About 20 hours before the supernova was detected optically, underground detectors in Japan and the United States recorded a brief (13-second) burst of neutrinos. The neutrinos preceded the light because they escaped during the collapse, whereas the first light of the explosion was emitted only after the supernova shock had plowed through the body of the star to the surface. Theoretical models, consistent with these observations, suggest that many tens of thousands of times more energy was emitted in the form of neutrinos than in any other form. Detection of this neutrino pulse is considered to be a brilliant confirmation of theory and may well herald a new age of astronomy. For the first time, astronomers have received information from beyond the solar system by radiation outside the electromagnetic spectrum.

Theory predicts that the expanding remnant of SN1987A is now on the verge of being resolvable by optical telescopes. The accompanying photographs show the barely resolved rem-

CHAPTER REVIEW

Summary

1 Stars spend most of their lives on the main sequence, in the **core-hydrogen burning** (stage 7; p. 318) phase of stellar evolution. Stars leave the main sequence when the hydrogen in their cores is exhausted. With no internal energy source, the star's helium core is unable to support itself against its own gravity and begins to shrink. The star at this stage is in the **hydrogen shell-burning** (p. 320) phase, with nonburning helium at the center surrounded by a layer of burning hydrogen. The energy released by the contracting helium core heats the hydro-

gen-burning shell, greatly increasing the nuclear reaction rates there. As a result, the star becomes much brighter, while the envelope expands and cools. A solar-mass star moves off the main sequence on the H–R diagram first along the **subgiant branch** (stage 8; p. 320), then almost vertically up the **red giant branch** (stage 9; p. 320).

2 As the helium core contracts, it heats up. Eventually, helium begins to fuse into carbon. In a star like the Sun, helium burning begins explosively, in the **helium flash** (p. 321). The flash expands the core and reduces the star's luminosity, sending it onto nant (at left below) surrounded by a much larger shell of glowing gas (in yellow). Scientists reason that the progenitor star expelled this shell during its red giant phase, some 40,000 years before the explosion. The image we see results from the initial flash of ultraviolet light from the supernova hitting the ring and causing it to glow brightly. As the debris from the explosion itself strikes the ring, it has become a temporary but intense source of X-rays. As shown in the *Chandra* X-ray image

in (false-color) blue at bottom right, and taken in 2000, the fastest-moving ejecta have already reached it, forming the small (1000 AU diameter), glowing regions on the left side of the ring. The multi-colored illustration to its right is an artist's interpretation of this fascinating piece of galactic real estate.

These images also show core debris moving outward toward the ring. The six insets at bottom (also the sickle-shaped region in the bottom right inset) show material expanding at nearly 3000 km/s. The main image also revealed, to everyone's surprise, two additional faint rings that might be caused by radiation sweeping across an hourglass-shaped bubble of gas, itself perhaps the result of a nonspherical "bipolar" stellar wind from the progenitor star before the supernova occurred.

Buoyed by the success of stellar-evolution theory and armed with firm theoretical predictions of what should happen next, astronomers eagerly await future developments in the story of this remarkable object.

(NASA)

Expanding debris

(NASA)   Illuminated ring

R I V U X G

Animation/Video

Composition and Structure of the Ring Around Supernova 1987A

Animation/Video

Shockwaves Hit the Ring Around Supernova 1987A

1994   1996   1998   2000   2002   2004

R I V U X G

---

the **horizontal branch** (stage 10; p. 321) of the H–R diagram. The star now has a core of burning helium surrounded by a shell of burning hydrogen. An inner core of nonburning

carbon forms, shrinks and heats the overlying burning layers, and the star once again becomes a red giant, even more luminous than before (stage 11). The core of a solar-mass star never becomes hot enough to fuse carbon. Such a star continues to brighten and expand until its envelope is ejected into space, forming a **planetary nebula** (stage 12; p. 323). At that point the core becomes visible as a hot, faint, and extremely dense white dwarf (stage 13). The

white dwarf cools and fades, eventually becoming a cold **black dwarf** (stage 14; p. 326).

3 A **nova** (p. 327) is a star that suddenly increases greatly in brightness, then slowly fades back to its normal appearance over a period of months. It is the result of a white dwarf in a binary system drawing hydrogen-rich material from its companion. The gas spirals inward in an **accretion disk** (p. 327) and builds up on the white dwarf's surface, eventually becoming hot and dense enough for the hydrogen to burn explosively, temporarily causing a large increase in the dwarf's luminosity.

4 Stars more massive than about 8 solar masses form heavier and heavier elements in their cores, at a more and more rapid pace. As they evolve into **red supergiants** (p. 328) their cores   form a layered structure consisting of burning shells of successively heavier elements. The process stops at iron, whose nuclei can neither be fused together nor split to produce energy. As a star's iron core grows in mass, it eventually becomes unable to support itself against gravity and begins to collapse. At the high temperatures produced during the collapse, iron nuclei are broken down into protons and neutrons. The protons combine with electrons to form more neutrons. Eventually, when the core becomes so dense that the neutrons are effectively brought into physical contact with one another, their resistance to further squeezing stops the collapse and the core rebounds, sending a violent shock wave out through the rest of the star. The star explodes in a **core-collapse supernova** (p. 331).

5 **Type I supernovae** (p. 331) are hydrogen poor and have a light curve similar in shape to that of a nova. **Type II supernovae** (p. 331) are hydrogen rich and have a characteristic bump in the light curve a few months after

maximum. A Type II supernova is a core-collapse supernova. A Type I supernova is a **carbon-detonation supernova** (p. 332), which occurs when a white dwarf in a binary system collapses and then explodes as its carbon ignites. We can see evidence for a past supernova in the form of a **supernova remnant** (p. 333), a shell of exploded debris surrounding the site of the explosion and expanding into space at a speed of thousands of kilometers per second.

6 All elements heavier than helium formed in evolved stars or in supernova explosions. With each new generation of stars the fraction of heavy elements in the universe increases. Comparisons between theoretical predictions of element production and observations of element abundances in the Galaxy provide strong support for the theory of stellar evolution.

7 The theory of stellar evolution can be tested by observing star clusters. At any instant, stars with masses above the cluster's **main-sequence turnoff** (p. 336) have evolved off the main sequence. By comparing a cluster's main-sequence turnoff mass with theoretical predictions, astronomers can determine the cluster's age.

Review and Discussion

1. For how long can a star like the Sun keep burning hydrogen in its core?

2. Why is the depletion of hydrogen in the core of a star such an important event?

3. What makes an ordinary star become a red giant?

4. How big (in AU) will the Sun become when it enters the red giant phase?

5. How long does it take for a star like the Sun to evolve from the main sequence to the top of the red giant branch?

6. What is the helium flash?

7. How do stars of low mass die? How do stars of high mass die?

8. What is a planetary nebula? With what stage of stellar evolution is it associated?

9. What are white dwarfs? What is their ultimate fate?

10. Under what circumstances will a binary star produce a nova?

11. What occurs in a massive star to cause it to explode?

12. What are the observational differences between Type I and Type II supernovae?

13. How do the mechanisms that cause Type I and Type II supernovae explain their observed differences?

14. What evidence do we have that many supernovae have occurred in our Galaxy?

15. Why do the cores of massive stars evolve into iron and not heavier elements?

Conceptual Self-Test: True or False?/Multiple Choice

1. All of the single red-dwarf stars that ever formed are still on the main sequence today. **(T/F)**

2. The Sun will get brighter as it begins to run out of fuel in its core. **(T/F)**

3. A planetary nebula is the disk of matter around a star that will eventually form a planetary system. **(T/F)**

4. The various stages of stellar evolution predicted by theory can be tested by observations of stars in clusters. **(T/F)**

5. A nova is a sudden outburst of light coming from an old main sequence star. **(T/F)**

6. It takes less and less time to fuse heavier and heavier elements inside a high-mass star. **(T/F)**

7. In a core-collapse supernova, the outer part of the core rebounds from the inner, high-density core, destroying the entire outer part of the star. **(T/F)**

8. Because of stellar nucleosynthesis, the spectra of old stars show more heavy elements than those of young stars. **(T/F)**

9. A white dwarf is supported by the pressure of tightly packed (a) electrons; (b) protons; (c) neutrons; (d) photons.

10. A star like the Sun will end up as a (a) blue giant; (b) white dwarf; (c) binary star; (d) red dwarf.

11. A white dwarf can dramatically increase in brightness only if it (a) has another star nearby; (b) can avoid nuclear fusion in its core; (c) is spinning very rapidly; (d) is descended from a very massive star.

12. Nuclear fusion in the Sun will (a) never create elements heavier than helium; (b) create elements up to and including oxygen; (c) create all elements up to and including iron; (d) create some elements heavier than iron.

13. Most of the carbon in our bodies originated in (a) the core of the Sun; (b) the core of a red giant star; (c) a supernova; (d) a nearby galaxy.

14. If the evolutionary track in *Overlay 3*, showing a Sun-like star, were instead illustrating a significantly more massive star, its starting point (stage 7) would be (a) up and to the right; (b) down and to the left; (c) up and to the left; (d) down and to the right.

15. Figure 12.20 ("Supernova Light Curves") indicates that a supernova whose luminosity declines steadily in time is most likely associated with a star that is (a) without a binary companion; (b) more than eight times the mass of the Sun; (c) on the main sequence; (d) comparable in mass to the Sun.

Problems *Algorithmic versions of these questions are available in the Practice Problems module of the Companion Website.*

The number of squares preceding each problem indicates its approximate level of difficulty.

1. ■ Calculate the average density of a red giant core of mass 0.25 solar masses and radius 15,000 km. Compare this with the average density of the giant's envelope, if the mass of the envelope is 0.5 solar masses and its radius is 0.5 AU. Compare each with the central density of the Sun. ∞ *(Sec. 9.2)*

2. ■■■ A main sequence star at a distance of 20 pc is barely visible through a certain telescope. The star subsequently ascends the giant branch, during which time its temperature drops by a factor of three and its radius increases 100-fold. What is the new maximum distance at which the star would still be visible using the same telescope?

3. ■■ If the Sun will reside on the main sequence for 10^{10} years and the luminosity of a main-sequence star is proportional to the fourth power of the star's mass, what mass star is just now leaving the main sequence in a cluster that formed (a) 400 million, (b) 2 billion years ago?

4. ■ How long will it take the Sun's planetary nebula, expanding at a speed of 20 km/s, to reach the orbit of Neptune? How long to reach the nearest star?

5. ■ What are the escape speed (in km/s) and surface gravity (relative to Earth's gravity) of Sirius B (Table 12.2)? ∞ *(More Precisely 5-1)*

6. ■■■ A certain telescope could just detect the Sun at a distance of 10,000 pc. What is the apparent magnitude of the Sun at this distance? (For convenience, take the Sun's absolute magnitude to be 5.) What is the maximum distance at which the telescope could detect a nova having a peak luminosity of 10^5 solar luminosities? Repeat the calculation for a supernova having a peak luminosity 10^{10} times that of the Sun.

7. ■■ At what distance would the supernova in the previous question look as bright as the Sun (apparent magnitude –27)? Would you expect a supernova to occur that close to us?

8. ■■ A (hypothetical) supernova at a distance of 150 pc has an absolute magnitude of –20. Compare its apparent magnitude with that of (a) the full Moon; (b) Venus at its brightest (see Figure 10.6). Would you expect a supernova to occur so close to us?

9. ■ The Crab Nebula is now about 1 pc in radius. If it was observed to explode in A.D. 1054, roughly how fast is it expanding? (Assume constant expansion velocity. Is that a reasonable assumption?)

10. ■■ A supernova's energy is often compared to the total energy output of the Sun over its lifetime. Using the Sun's current luminosity, calculate the total solar energy output, assuming a 10^{10} year main-sequence lifetime. Using Einstein's formula $E = mc^2$, calculate the equivalent amount of mass, in Earth masses. ∞ *(Sec. 9.5)*

 The Companion Website at www.prenhall.com/chaisson provides algorithmically generated versions of each chapter's Problems, in addition to an Animations and Videos gallery, a Tutorials gallery and Tutorials Quiz questions, a full eBook, an Images gallery, Practice Questions, Multiple Choice, and True or False quizzes, and Labeling exercises.

13

Neutron Stars and Black Holes

Strange States of Matter

Neutron stars and black holes are among the more exotic members of the vast population of stars throughout the universe. These objects represent the end states of stellar systems, yet despite their bizarre nature they do seem to fit quite well within our models of stellar evolution. This stunning image is actually a composite of three images taken by telescopes in Earth orbit: optical light (yellow) observed with *Hubble*, X-ray radiation (blue and green) with *Chandra* and infrared radiation (red) with *Spitzer*. The object is Cassiopeia A, the remnant of a supernova that exploded about 300 years ago: the small turquoise dot at the center may be a neutron star. (NASA)

Our study of stellar evolution has led us to some very unusual and unexpected objects. Red giants, white dwarfs, and supernova explosions surely represent extreme states of matter completely unfamiliar to us here on Earth. Yet stellar evolution can have even more bizarre consequences. The strangest states of all result from the catastrophic implosion–explosion of stars much more massive than our Sun. The almost unimaginable violence of a supernova explosion may bring into being objects so extreme in their behavior that they require us to reconsider some of our most hallowed laws of physics. They open up a science-fiction writer's dream of fantastic phenomena. They may even one day force scientists to construct a whole new theory of the universe. Yet incredibly, as far as we can tell, they may actually exist!

LEARNING GOALS

Studying this chapter will enable you to:

1 Describe the properties of neutron stars and explain how these strange objects are formed.

2 Explain the nature and origin of pulsars and account for their characteristic radiation.

3 List and explain some of the observable properties of neutron star binary systems.

4 Discuss the basic characteristics of gamma-ray bursts and some theoretical attempts to explain them.

5 Describe Einstein's theories of relativity, and discuss how they relate to neutron stars and black holes.

6 Describe how black holes are formed, and discuss their effects on matter and radiation in their vicinity.

7 Relate the phenomena that occur near black holes to the warping of space around them.

8 Discuss the difficulties in observing black holes, and explain some of the ways in which the presence of a black hole might be detected.

Visit www.prenhall.com/chaisson for additional annotated images, animations, and links to related sites for this chapter.

13.1 Neutron Stars

What remains after a supernova explosion? Is the original star blown to bits and dispersed into interstellar space, or does some portion of it survive? For a Type I (carbon-detonation) supernova, most astronomers regard it as unlikely that anything is left behind after the explosion. ∞ *(Sec. 12.5)* The entire star is shattered by the blast. However, for a Type II (core-collapse) supernova, theoretical calculations indicate that part of the star survives the explosion.

Recall that in a Type II supernova, the iron core of a massive star collapses until its neutrons effectively come into contact with one another. At that point the central portion of the core rebounds, creating a powerful shock wave that races outward through the star, violently expelling matter into space. ∞ *(Sec. 12.4)* The key point here is that the shock wave does not start at the very center of the collapsing core. The innermost part of the core—the region that rebounds—remains intact as the shock wave it produces destroys the rest of the star. After the violence of the supernova has subsided, this ultracompressed ball of neutrons is all that is left. Researchers call this core remnant a **neutron star**.

Neutron stars are extremely small and very massive. Composed purely of neutrons packed together in a tight ball about 20 km across, a typical neutron star is not much bigger than a small asteroid or a terrestrial city (Figure 13.1), yet its mass is greater than that of the Sun. With so much mass squeezed into such a small volume, neutron stars are incredibly dense. Their average density can reach 10^{17} or even 10^{18} kg/m^3, nearly a billion times denser than a white dwarf. ∞ *(Sec. 12.3)* A single thimbleful of neutron-star material would weigh 100 million tons—about as much as a good-sized terrestrial mountain.

Neutron stars are solid objects. Provided that a sufficiently cool one could be found, you might even imagine standing on it. However, this would not be easy, as its gravity is extremely powerful. A 70-kg human would weigh the Earth equivalent of about 1 billion kg (one million tons). The severe pull of a neutron star's gravity would flatten you much thinner than this piece of paper.

In addition to large mass and small size, newly formed neutron stars have two other very important properties. First, they *rotate* extremely rapidly, with periods measured in fractions of a second. This is a direct result of the law of conservation of angular momentum (Chapter 4), which tells us that any rotating body must spin faster as it shrinks, and the core of the parent star almost certainly had some rotation before it began to collapse. ∞ *(More Precisely 4-1)* Second, they have very strong *magnetic fields*. The original field of the parent star is amplified as the collapsing core squeezes the magnetic field lines closer together, creating a magnetic field trillions of times stronger than Earth's.

In time, theory indicates, a neutron star will spin more and more slowly as it radiates its energy into space, and its magnetic field will diminish. However, for a few million years after its birth, these two properties combine to provide the primary means by which this strange object can be detected and studied, as we now discuss.

13.2 Pulsars

The first observation of a neutron star occurred in 1967 when Jocelyn Bell, a graduate student at Cambridge University, observed an astronomical object emitting radio radiation in the form of rapid *pulses* (Figure 13.2). Each pulse consisted of a roughly 0.01-s burst of radiation, separated by precisely 1.34 s from the next.

More than 1500 of these pulsating objects are now known. They are called **pulsars**. Each has its own characteristic pulse shape and period. Observed pulsar periods are generally quite short, ranging from about a few milliseconds to

CONCEPT CHECK ▶▶▶

Are all supernovae expected to lead to neutron stars?

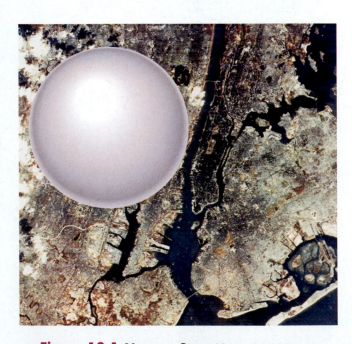

▲ **Figure 13.1 Neutron Star** Neutron stars are not much larger than many of Earth's major cities. In this fanciful comparison, a typical neutron star sits alongside Manhattan Island. (NASA)

about a second, corresponding to flashing rates of between one and several hundred times per second. In some cases, the periods are extremely stable—in fact, constant to within a few seconds in a million years—making pulsars the most accurate natural clocks known in the universe, more accurate even than the best atomic clocks on Earth.

A Pulsar Model

When Bell made her discovery in 1967, she did not know what she was looking at. Indeed, no one at the time knew what a pulsar was. The explanation won Bell's thesis advisor, Anthony Hewish, the 1974 Nobel Prize in physics. Hewish reasoned that the only physical mechanism consistent with such precisely timed pulses is a small, rotating source of radiation. Only rotation can cause the high degree of regularity of the observed pulses, and only a small object can account for the sharpness of each pulse (see Section 13.4, p. 352). The best current model describes a pulsar as a compact, spinning neutron star that periodically flashes radiation toward Earth.

...a lighthouse beacon

Figure 13.3 outlines the main features of this pulsar model. Two "hot spots," either on the surface of a neutron star or in the magnetosphere just above it, continuously emit radiation in a narrow beam. These spots are most likely localized regions near the neutron star's magnetic poles, where charged particles, accelerated to extremely high energies by the star's rotating magnetic field, emit radiation along the star's magnetic axis. The hot spots radiate more or less steadily, and the resulting beams sweep through space like a revolving lighthouse beacon as the neutron star rotates. Indeed, this pulsar model is often known as the **lighthouse model**. If the neutron star happens to be oriented such that one of the rotating beams sweeps across Earth, then we see the pulses. The period of the pulses is the star's rotation period.

► **Figure 13.3 Pulsar Model**
This diagram of the lighthouse model of neutron star emission accounts for many of the observed properties of pulsars. Charged particles, accelerated by the magnetism of the neutron star, flow along the magnetic field lines, producing radio radiation that beams outward. At greater distances from the star, the field lines channel these particles into a high-speed outflow in the star's equatorial plane, forming a pulsar wind. The beam sweeps across the sky as the neutron star rotates. If it happens to sweep across Earth, we see a pulsar—much like a lighthouse beacon.

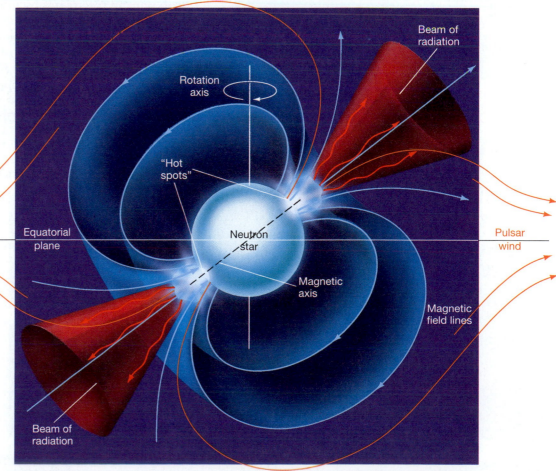

CONCEPT CHECK ▶▶▶

Why don't we see pulsars at the centers of all supernova remnants?

A few pulsars are definitely associated with supernova remnants, clearly establishing those pulsars' explosive origin, although not all such remnants have a detectable pulsar within them. Figure 13.4(c) shows optical images of the Crab pulsar, at the center of the Crab supernova remnant (Figure 13.4a). ∞ *(Sec. 12.5)* By observing the speed and direction of the Crab's ejected matter, astronomers can work backward to pinpoint the location in space at which the explosion must have occurred. It corresponds to the location of the pulsar. This is all that remains of the once massive star whose supernova was observed in 1054.

As indicated in Figure 13.3, the neutron star's strong magnetic field and rapid rotation channel high-energy particles from near the star's surface into the surrounding nebula. The result is an energetic *pulsar wind* that flows outward at almost the speed of light, primarily in the star's equatorial plane. Figure 13.4(b) shows this process in action in the Crab—this combined *Hubble/Chandra* image reveals rings of hot X-ray-emitting gas moving rapidly away from the pulsar. Eventually, the energy from the pulsar wind is deposited in the surrounding Crab nebula, heating the gas to very high temperatures and powering the spectacular display we see from Earth.

Most pulsars emit pulses in the form of radio radiation. However, some also pulse in the visible, X-ray, and gamma-ray parts of the spectrum. Figure 13.5 shows the Crab and nearby Geminga pulsars in gamma rays. Geminga is unusual in that, while it pulsates strongly in gamma rays, it is barely detectable in visible light and not at all at radio wavelengths. Whatever types of radiation are produced, pulsar flashes at different frequencies all occur at regular, repeated time intervals—as we would expect since they arise from the same object—although the pulses may not all occur at the same instant.

Most known pulsars are observed (usually by Doppler measurements) to have high speeds—much greater than the typical speeds of stars in the Galaxy. The most likely explanation for these anomalously high speeds is that a neutron star can receive a substantial "kick" due to asymmetries in the supernova in which it forms. If the supernova's enormous energy is channeled even slightly in one direction, the newborn neutron star/pulsar can recoil in the opposite direction with a speed of many tens or even hundreds of kilometers per second. Thus observations of pulsar velocities give theorists additional insight into the detailed physics of supernovae.

Neutron Stars and Pulsars

All pulsars are neutron stars, but not all neutron stars are pulsars, for two reasons. First, the two ingredients that make the neutron star pulse—rapid rotation and strong magnetic field—both diminish with time, so the pulses

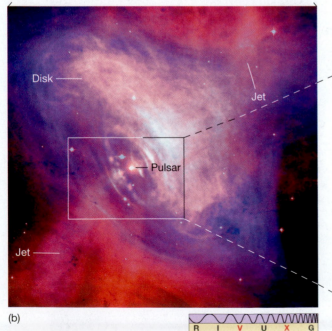

(a)

R I V U X G

Disk ——

Jet

—— Pulsar

Jet ——

(b)

R I V U X G

ON

OFF

(c)

R I V U X G

◀ **Figure 13.4 Crab Pulsar** In the core of the Crab Nebula (a), the Crab pulsar (c) blinks on and off about 30 times each second. In this pair of optical images, the pulsing can be clearly seen. In the top frame, the pulsar (arrow) is on; in the bottom frame, it is off. (b) This more recent *Chandra* X-ray image of the Crab, superimposed on a *Hubble* optical image, shows the central pulsar, as well as rings of hot X-ray-emitting gas in the equatorial plane, driven by the pulsar wind and moving rapidly outward. Also visible in the image is a jet of hot gas (*not* the beam of radiation from the pulsar) escaping perpendicular to the equatorial plane. (ESO; NASA; UC/Lick Observatory)

(b) |← —————————— 0.24 second —————————— →|

gradually weaken and become less frequent. Theory indicates that within a few tens of millions of years, the pulses have all but stopped. Second, even a young, bright pulsar is not necessarily visible from Earth. The pulsar beam depicted in Figure 13.3 is relatively narrow—perhaps as little as a few degrees across. Only if the neutron star happens to be oriented in just the right way do we see pulses. When we see the pulses from Earth, we call the object a pulsar. Note that we are using the term "pulsar" here to mean the pulsing object we observe if the beam crosses Earth. However, many astronomers use the word more generically to mean *any* young neutron star producing beams of radiation as in Figure 13.3. Such an object will be a pulsar as seen from some directions—just not necessarily ours!

Figure 13.6 shows *Hubble* images of a neutron star that is not a pulsar. First identified by X-ray observations, this object is just 30 km across and has a surface temperature of about 700,000 K. Despite this enormous temperature, its small size means that it is very faint—just 25th magnitude in visible light. ∞ *(Sec. 10.4)* The star is thought to be about one million years old, lies about 60 pc away, and is moving at a speed of 110 km/s across our line of sight. Such faint objects are very difficult to detect, however, and a detection of a "bare" neutron star like this is a rare event. Most cataloged neutron stars have been detected either as pulsars or by their interaction with a "normal" stellar companion in a binary system (see Section 13.3).

Given our current knowledge of star formation, stellar evolution, and neutron stars, observations of pulsars are consistent with the ideas that (1) *every* high-mass star dies in a supernova explosion, (2) most supernovae leave a neutron star behind (a few result in black holes, as discussed in a moment), and (3) *all* young neutron stars emit beams of radiation, just like the pulsars we actually detect. On the basis of estimates of the rate at which massive stars have formed over the lifetime of the Milky Way, astronomers reason that, for every pulsar we know of, there must be several hundred thousand more neutron stars moving unseen somewhere in the Galaxy. Some formed relatively recently—less than a few million years ago—and simply happen not to be beaming their energy toward Earth. However, the vast majority are old, their youthful pulsar phase long past.

Neutron stars (and black holes too) were predicted by theory long before they were actually observed, although their extreme properties made many scientists doubt that they would ever be found in nature. The fact that we now have strong observational evidence not just for their existence but for the vitally important roles they play in many areas of high-energy astrophysics is yet another testament to the fundamental soundness of the theory of stellar evolution.

13.3 Neutron Star Binaries

We noted in Chapter 10 that most stars are not single but instead are members of binary systems. ∞ *(Sec. 10.8)* Although many pulsars are known to be isolated (that is, not part of any binary), at least some do have binary companions, and the same is true of neutron stars in general (that is, even the ones not seen as pulsars). One important consequence of this is that some neutron star masses have been determined very accurately. All the measured masses are fairly close to 1.4 times the mass of the Sun—the Chandrasekhar mass of the stellar core that collapsed to form the neutron star remnant.

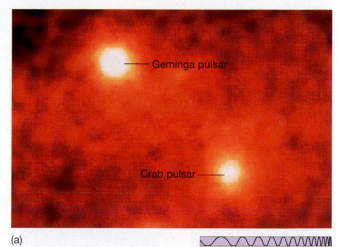

(a)

R I V U X G

▲ **Figure 13.5 Gamma-Ray Pulsars** (a) The Crab and Geminga pulsars, which happen to lie fairly close (about 4.5 degrees) to one another in the sky. Unlike the Crab, Geminga is barely visible at optical wavelengths, and undetectable in the radio. (b) Sequence of *Compton Gamma-Ray Observatory* images showing Geminga's 0.24-s pulse period. (The Crab's 33-millisecond period is too rapid to be resolved by the detector.) (NASA)

R I V U X G

▲ **Figure 13.6 Isolated Neutron Star** This lone neutron star was first detected by its X-ray emission, and subsequently imaged by *Hubble*. It lies about 60 pc from Earth and is thought to be about 1 million years old. This triple exposure shows it streaking across the sky at more than 100 km/s. (NASA)

X-Ray Sources

With the launch of the first orbiting X-ray telescopes in the 1970s, numerous X-ray sources were found near the central regions of our Galaxy and also near the centers of a few globular star clusters. ∞ *(Sec. 11.6)* Some of these sources, known as **X-ray bursters**, emit much of their energy in violent eruptions, each thousands of times more luminous than our Sun but lasting only seconds. A typical burst is shown in Figure 13.7.

The X-ray emission is thought to arise on or near neutron stars that are members of binary systems. Matter torn from the surface of the (main-sequence or giant) companion by the neutron star's strong gravitational pull accumulates on the neutron star's surface. (Figure 12.14 shows the white dwarf equivalent.) ∞ *(Sec. 12.3)* The gas forms an accretion disk around the neutron star, then slowly spirals inward. The inner portions of the disk become extremely hot, releasing a steady stream of X-rays.

As gas builds up on the neutron star's surface, its temperature rises due to the pressure of overlying material. Eventually, it becomes hot enough to fuse hydrogen. The result is a sudden period of rapid nuclear burning that releases a huge amount of energy in a brief but intense X-ray burst. After several hours of renewed accumulation, a fresh layer of matter produces the next burst. Thus, the mechanism is very similar to that responsible for a nova explosion on a white dwarf, except that it occurs on a far more violent scale because of the neutron star's much stronger surface gravity.

Millisecond Pulsars

In the mid-1980s astronomers discovered an important new category of pulsars—a class of very rapidly rotating objects called **millisecond pulsars**. These objects spin hundreds of times per second (that is, their pulse periods are a few milliseconds, 0.001 s). This speed is about as fast as a typical neutron star can spin without flying apart. In some cases the star's equator is moving at more than 20 percent of the speed of light. This suggests a phenomenon bordering on the incredible—a cosmic object just 20 km across, more massive than our Sun, spinning almost at breakup speed, making nearly 1000 complete revolutions *every second*. Yet the observations and their interpretation leave little room for doubt. More than 150 millisecond pulsars are now known.

The story of these remarkable objects is further complicated by the fact that many of them are found in globular clusters. This is odd because globular clusters are known to be at least 10 billion years old, yet Type II supernovae (the kind that create neutron stars) are associated with massive stars that explode within a few tens of *millions* of years after their formation, and no stars have formed in any globular cluster since the cluster came into being. ∞ *(Sec. 11.6, 12.4)* Thus, no new neutron star has been produced in a globular cluster in a very long time. Furthermore, as mentioned earlier, the pulsar produced in a supernova explosion is expected to slow down and fade away in only a few tens of millions of years—after 10 billion years, its rotation rate should be very slow. The rapid rotation of the pulsars found in globular clusters therefore cannot be a relic of their birth. These pulsars must have been "spun up"—that is, had their rotation rates increased—by some other, much more recent, mechanism.

▲ **Figure 13.7 X-Ray Burster** An X-ray burster produces a sudden, intense flash of X-rays followed by a period of relative inactivity lasting as long as several hours. Then another burst occurs. The bursts are thought to be caused by explosive nuclear burning on the surface of an accreting neutron star, similar to the explosions on a white dwarf that give rise to novae. (a) An optical photograph of the globular star cluster Terzan 2, showing a 2″ dot at the center where the X-ray bursts originate. (b) X-ray images taken before and during the outburst. The most intense X-rays correspond to the position of the dot shown in (a). (SAO)

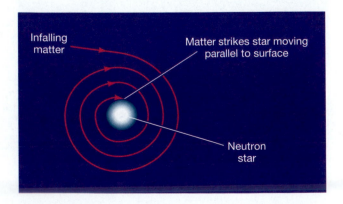

◄ **Figure 13.8 Millisecond Pulsar** Gas from a companion star spirals down onto the surface of a neutron star. As the infalling matter strikes the star, it moves almost parallel to the surface, so it tends to make the star spin faster. Eventually, this process can result in a millisecond pulsar—a neutron star spinning at the incredible rate of hundreds of revolutions per second.

▶ **Figure 13.9 Cluster X-Ray Binaries**
The dense core of the old globular cluster *47 Tucanae* harbors more than 100 separate X-ray sources (shown in the *Chandra* image at right). More than half of these are thought to be binary millisecond pulsars, still accreting small amounts of gas from their companions after an earlier period of mass transfer spun them up to millisecond speeds. (ESO; NASA)

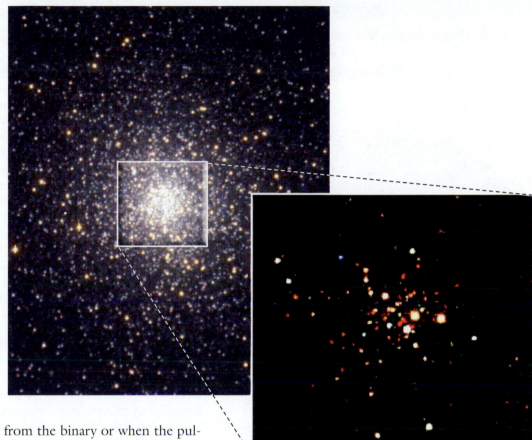

The most likely explanation for the high rotation rate of millisecond pulsars is that they have been spun up by drawing in matter from a companion star. As matter spirals down onto the neutron star's surface in an accretion disk, it provides the "push" needed to make the neutron star spin faster (Figure 13.8). Theoretical calculations indicate that this process can spin the star up to breakup speed in about a hundred million years. This general picture is supported by the finding that, of the 100 or so millisecond pulsars observed in globular clusters, roughly half are known currently to be members of binary systems. The remaining solo millisecond pulsars were probably formed when an encounter with another star ejected the pulsar from the binary or when the pulsar's own intense radiation destroyed its companion.

Notice that the scenario of accretion onto a neutron star from a binary companion is the same scenario that we just used to explain the existence of X-ray binaries. In fact, the two phenomena are very closely linked. Many X-ray binaries may be on their way to becoming millisecond pulsars, and many millisecond pulsars are X-ray sources, powered by the trickle of material still falling onto them from their binary companions. Figure 13.9 shows the globular cluster 47 Tucanae, together with a remarkable *Chandra* image of its core showing no fewer than 108 X-ray sources—about 10 times the number that had been known in the cluster prior to *Chandra*'s launch. Roughly half of these sources are millisecond pulsars; the cluster also contains two or three "conventional" neutron star binaries. Most of the remaining sources are white dwarf binaries, similar to those discussed in Chapter 12. ∞ *(Sec. 12.3)*

Pulsar Planets

Radio astronomers can capitalize on the precision with which pulsar signals repeat themselves to make extremely accurate measurements of pulsar motion. In 1992 radio astronomers at the Arecibo Observatory found that the pulse period of a millisecond pulsar lying some 500 pc from Earth exhibits tiny but regular fluctuations (at the level of about one part in 10^7) on two distinct time scales—67 and 98 days.

The most likely explanation for these fluctuations is that they are caused by the Doppler effect as the pulsar wobbles back and forth in space under the combined gravitational pulls of two small bodies, each about three times the mass of Earth. One orbits the pulsar at a distance of 0.4 AU, the other at a distance of 0.5 AU, with orbital periods of 67 and 98 days, respectively. Further observations not only confirmed these findings, but also revealed the presence of a third body, with a mass comparable to that of Earth's Moon, orbiting just 0.2 AU from the pulsar.

CONCEPT CHECK ▶▶▶

What is the connection between X-ray sources and millisecond pulsars?

April 21, 1991

April 25, 1991

Time (seconds)

May 18, 1991

(a)

2704 BATSE gamma-ray bursts

(b)

These remarkable results constituted the first definite evidence of planet-sized bodies outside our solar system. A few other millisecond pulsars have since been found with similar behavior. However, it is unlikely that any of these planets formed in the same way as our own. Any planetary system orbiting the pulsar's parent star was almost certainly destroyed in the supernova explosion that created the pulsar. As a result, scientists are uncertain about how these planets came into being. One possibility involves the binary companion that provided the matter necessary to spin the pulsar up to millisecond speeds. Possibly the pulsar's intense radiation and strong gravity destroyed the companion, then spread its matter out into a disk (a little like the solar nebula) in whose cool outer regions the planets might have condensed.

Astronomers searched for decades for planets orbiting main-sequence stars like our Sun, on the assumption that planets are a natural by-product of star formation. ∞ *(Sec. 4.3)* As we have seen, these searches have now identified many extrasolar planets, although the planetary systems discovered to date generally don't look much like our own—just one planet even comparable in mass to Earth has so far been detected. ∞ *(Sec. 4.4)* It is ironic that the first Earth-sized planets found outside the solar system orbit a dead star, and have little or nothing in common with our own world.

13.4 Gamma-Ray Bursts

Discovered serendipitously in the late 1960s by military satellites looking for violations of the Nuclear Test Ban Treaty, and first made public in the 1970s, **gamma-ray bursts** have developed into one of the deepest mysteries in astronomy today. The bursts consist of bright, irregular flashes of gamma rays typically lasting only a few seconds (Figure 13.10a). Until the 1990s it was thought that gamma-ray bursts were basically "scaled-up" versions of X-ray bursters, in which even more violent nuclear burning resulted in the release of more energetic gamma rays. However, it now appears that this is not the case.

Distances and Luminosities

Figure 13.10(b) shows an all-sky plot of the positions of 2704 bursts detected by the *Compton Gamma-Ray Observatory (CGRO)* during its nine-year operational lifetime. ∞ *(Sec. 3.5)* On average, *CGRO* detected gamma-ray bursts at the rate of about one per day. Note that the bursts are distributed uniformly across the sky, rather than being confined to the relatively narrow band of the Milky Way (compare Figure 3.30). This widespread distribution has convinced most astronomers that the bursts do not originate within our own Galaxy, as had once been assumed, but instead are produced at much greater distances.

Actually measuring the distance to a gamma-ray burst is no easy task. The gamma-ray observations do not provide enough information in and of themselves to tell us how far away the burst is, so astronomers must instead associate the burst with some other object in the sky—called the burst *counterpart*—whose distance can be measured by other means. The techniques for studying counterparts generally involve observations in the optical

◀ **Figure 13.10 Gamma-Ray Bursts** (a) Plots of intensity versus time (in seconds) for three gamma-ray bursts. Note the substantial differences between them. Some bursts are irregular and spiky, whereas others are much more smoothly varying. Whether this wide variation in burst appearance means that more than one physical process is at work is presently unknown. (b) Positions on the sky of all the gamma-ray bursts detected by the *Compton Gamma-Ray Observatory.* (BATSE stands for Burst and Transient Source Experiment, one of the instruments on *CGRO.*) The bursts appear to be distributed uniformly across the entire sky. (NASA) or X-ray parts of the electromagnetic spectrum. The problem is that the resolution of a gamma-ray telescope is quite poor, so the burst positions are uncertain by about a degree and a relatively large region of the sky must be scanned in search of a counterpart. ∞ *(Sec. 3.5)* Furthermore, the "afterglow" of a burst at X-ray or optical wavelengths fades quite rapidly, severely limiting the time available to complete the search.

The most successful searches for burst counterparts have been carried out by satellites combining gamma-ray detectors with X-ray and/or optical telescopes. A recent example is NASA's *Swift* mission, which became operational in 2005. *Swift* combines a wide-angle gamma-ray detector (to monitor as much of the sky as possible) with two telescopes: one X-ray and one optical/ultraviolet instrument. The gamma-ray detector system pinpoints the direction of the burst to an accuracy of about 4 arc minutes, and within seconds the onboard computer automatically repositions the satellite to point the X-ray and optical telescopes in that direction. At the same time, the craft relays the burst position to other instruments in space and on the ground. *Swift* detects burst counterparts at the rate of about one per week, and is playing a pivotal role in advancing our understanding of these violent phenomena.

The first direct measurement of the distance to a gamma-ray burst was made in 1997 when astronomers detected the visible afterglow created as the burst faded and cooled. Using spectroscopic distance-measurement techniques (to be discussed in more detail in Section 16.1), they found that the event occurred more than 2 *billion* parsecs from Earth. Figure 13.11 shows an optical image of another gamma-ray burst, this one almost 5 billion parsecs distant, along with what may be its host galaxy—this is one of the first images to associate a burst with a possible galactic host. In all, as of early 2006, more than 100 afterglows of gamma-ray bursts have been detected, and about 30 distances are known. All the distances are very large, implying that the bursts must be extremely energetic, since otherwise they wouldn't be detectable by our equipment.

If we assume that the gamma rays are emitted equally in all directions (a big assumption!), then we can calculate the total energy emitted, using the inverse-square law. ∞ *(Sec. 10.2)* We find that each burst generates more energy—in some cases hundreds of times more energy—than a typical supernova explosion, all in a matter of seconds!

What Causes the Bursts?

Not only are gamma-ray burst sources extremely energetic, they are also very *small*. The millisecond flickering in the bursting gamma rays detected by *GRO* implies that whatever their origin, all of their energy must come from a volume no larger than a few hundred kilometers across. The reasoning is as follows: If the emitting region were, say, 300,000 km—one light-second—across, even an instantaneous change in intensity at the source would be smeared out over a time interval of 1 s as seen from Earth, because light from the far side of the object would take 1 s longer to reach us than light from the near side. For the gamma-ray variation not to be blurred by the light-travel time, the source cannot be more than one light-millisecond, or 300 km, in diameter.

Most theoretical attempts to explain gamma ray bursts picture the burst as a *relativistic fireball*—an expanding region, possibly a jet, of superhot gas radiating furiously in the gamma-ray part of the spectrum. (The term "relativistic" here means that the gas particles are moving at speeds comparable to the speed of light, and Einstein's theory of relativity is needed to describe

▼ **Figure 13.11 Gamma-Ray Burst Counterpart** Optical images of the gamma-ray burst GRB 971214 (the numbers simply mean December 14, 1997). Image (a) at left, taken by the Keck telescope, shows the visible afterglow of the gamma-ray source (arrow) to be quite bright, comparable to two other prominent sources in the overlaid box. Its distance is almost five billion parsecs. By the time the *Hubble* image (b) was taken (about two months after the Keck image and four months after the initial burst of gamma rays was detected by the Italian–Dutch *BeppoSAX* satellite), the afterglow had faded, but a faint image of a host galaxy remains. The different colors result from the use of different filters to observe the light. (Keck; NASA)

(a)   (b)

R I V U X G

them—see Section 13.6.) The complex burst structure and afterglows we see are produced as the fireball expands, cools, and interacts with its surroundings. If the energy is emitted as a jet, then the overall energy of the burst may be reduced to more "manageable" levels, as the luminosity we see is representative of only a small fraction of the sky. This may help make the huge luminosity of the most energetic bursts easier to explain. As an analogy, consider a hand-held laser pointer of the sort commonly used in talks and lectures. It radiates only a few milliwatts of power, much less than a household lightbulb, but it appears enormously bright if you happen to look directly into the beam. (*Don't* do this, by the way!) The laser beam is bright because all of its energy is concentrated in almost a single direction, instead of being radiated in all directions into space.

Two leading models for the energy source have emerged, as sketched in Figure 13.12. The first (Figure 13.12a) is the "true" end point of a binary star system. Suppose that both members of the binary evolve to become neutron stars. As the system continues to evolve, gravitational radiation (see *Discovery 13-1*) is released, and the two ultradense stars spiral in toward each other. Once they are within a few kilometers of one another, coalescence is inevitable. Such a merger will likely produce a violent explosion comparable in energy to a supernova and perhaps energetic enough to explain the flashes of gamma rays we observe. The overall rotation of the binary system may channel the energy into a high-speed, high-temperature jet.

The second model (Figure 13.12b), sometimes called a *hypernova*, is a "failed" supernova—but what a failure! In this picture a very massive star undergoes core collapse just as described earlier for a type-II supernova, but instead of forming a neutron star, the core collapses in on itself, forming a black hole (see Section 13.5). ∞ *(Sec. 12.4)* Instead of being blown to pieces, the star implodes onto the black hole, forming an accretion disk and again generating a relativistic jet. The jet punches its way out of the star, producing a gamma-ray burst as it slams into the surrounding shells of gas expelled from the star during the final stages of its nuclear-burning lifetime. At the same

▼ **Figure 13.12 Gamma-Ray Burst Models** Two models have been proposed to explain gamma-ray bursts. Part (a) depicts the merger of two neutron stars; part (b) shows the collapse of a single star. Both models predict (at right) a relativistic fireball, perhaps in the form of jets, as shown.

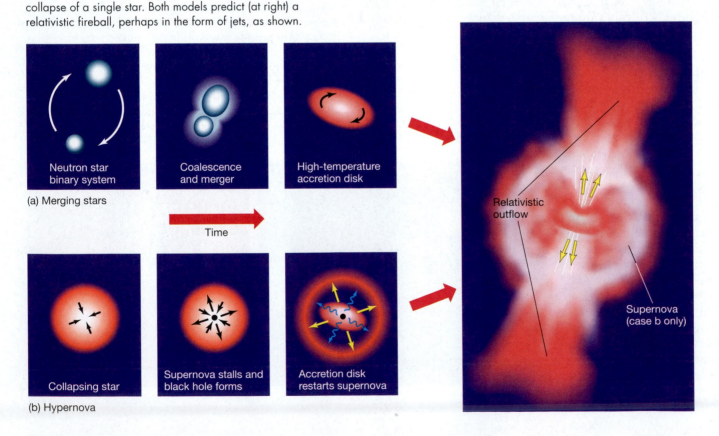

(a) Merging stars

Neutron star binary system

Coalescence and merger

High-temperature accretion disk

Time

(b) Hypernova

Collapsing star

Supernova stalls and black hole forms

Accretion disk restarts supernova

Relativistic outflow

Supernova (case b only)

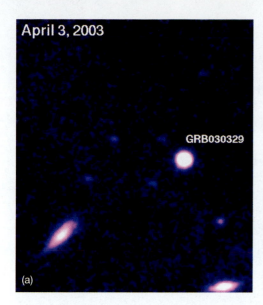

April 3, 2003

GRB030329

(a)

May 1, 2003

GRB030329

(b)

R   I   V   U   X   G

◄ **Figure 13.13** **Hypernova?**   The gamma-ray burst GRB 030329 may prove crucial to theorists' understanding of the physical processes underlying these violent phenomena. First detected by NASA's *High Energy Transient Explorer 2 (HETE-2)* satellite and subsequently observed at radio, optical, and X-ray wavelengths, the burst counterpart has all the hallmarks of a high-mass supernova, lending strong support to the hypernova model. Here, the counterpart is shown as seen through the 8.2-m European VLT telescope in Chile, (a) near the moment of the burst and (b) fading a month later. (ESO)

time, intense radiation from the accretion disk may restart the stalled supernova, blasting what remains of the star into space.

The idea of a relativistic fireball has become widely accepted among workers in this fast-changing branch of astrophysics. Many researchers regard it as likely that the energy is released as a jet and that the most intense gamma-ray bursts occur when the jet is directed toward us ("looking into the laser"). There is still no consensus among experts on which (if either) of the two models just described is correct, although the apparent association of some recent supernovae with earlier gamma-ray bursts argues in favor of the hypernova scenario. The strongest evidence of this connection was obtained in 2003, when NASA's *HETE-2* satellite detected a particularly bright burst known as GRB 030329 (Figure 13.13). Within 24 hours, the 8.2-m *VLT* in Chile had observed the counterpart and measured its spectrum and continued to monitor the afterglow as it faded over the next month. ∞ *(Sec. 3.2)* Both the spectrum and the light curve of the counterpart are precisely what astronomers would expect from the supernova produced by a very massive (25-solar mass) star. ∞ *(Sec. 12.4)*

The hypernova model predicts bursts of relatively long duration and is now the leading explanation of the "long" gamma-ray bursts lasting more than about two seconds, such as the second event shown in Figure 13.10a. However, the model has difficulty explaining the many "short" bursts that are also observed (for example, the other two bursts shown in that figure). Detailed calculations indicate that the neutron star merger model can naturally account for the short bursts. In mid-2005, *Swift* detected a rapidly fading X-ray afterglow from a short burst that appears to be consistent with the theoretical predictions of the merger scenario. It is quite possible that there are, in fact, two distinct classes of gamma-ray bursts and that both of the models depicted in Figure 13.12 contribute to the total. However, the final word on these enigmatic objects clearly has yet to be written.

◄◄◄ **CONCEPT CHECK**

What are gamma-ray bursts, and why do they pose such a challenge to current theory?

13.5 Black Holes

Neutron stars are supported by the resistance of tightly packed neutrons to further compression. Squeezed together, these elementary particles form a hard ball of ultradense matter that not even gravity can compress further. Or do they? Is it possible that, given enough matter packed into a small enough volume, the collective pull of gravity can eventually crush even a neutron star? Can gravity continue to compress a massive star into an object the size of a planet, a city, a pinhead—even smaller? The answer, apparently, is yes.

13-1 DISCOVERY

Gravity Waves—A New Window on the Universe

Electromagnetic waves are common everyday phenomena. They involve periodic changes in the strengths of electric and magnetic fields. ∞ *(Fig. 2.6)* They move through space and transport energy. Any accelerating charged particle, such as an electron in a broadcasting antenna or on the surface of a star, generates electromagnetic waves.

The modern theory of gravity—Einstein's theory of relativity—also predicts waves that move through space. A *gravity wave* is the gravitational counterpart of an electromagnetic wave. *Gravitational radiation* results from changes in the strength of a gravitational field. In theory, any time an object having any mass accelerates, a gravity wave is emitted at the speed of light. The passage of a gravity wave should produce small distortions in the space through which it passes. Gravity is an exceedingly weak force compared with electromagnetism, so these distortions are expected to be very small—in fact, much smaller than the diameter of an atomic nucleus for waves that might be produced by any known astrophysical source. Yet many researchers think that these tiny distortions should be measurable. No experiment has yet succeeded in detecting gravity waves, but their detection would provide such strong support for the theory of relativity that scientists are eager to find them.

The leading candidates for systems likely to produce gravity waves detectable on Earth include (1) close binary systems containing black holes, neutron stars, or white dwarfs, and (2) the collapse of a star into a black hole. Each of these possibilities involves the acceleration of huge masses, resulting in rapidly changing gravitational fields. The first probably presents the best chance to detect gravity waves. Binary star systems should emit gravitational radiation as the component stars orbit one another. As energy escapes in the form of gravity waves, the two stars spiral toward one another, orbiting more rapidly and emitting even more gravitational radiation. For sufficiently close systems, the stars merge in a few tens or hundreds of millions of years (although most of the radiation is emitted during the last few seconds). As we saw in Section 13.4, neutron star mergers may well also be the origin of some gamma-ray bursts, so gravitational radiation might provide an alternative means of studying these violent and mysterious phenomena.

Such a slow but steady decay in the orbit of a binary system has in fact been detected. In 1974, radio astronomer Joseph Taylor and graduate student Russell Hulse at the University of Massachusetts discovered a very unusual binary system. Both components are neutron stars, and one is observable from Earth as a pulsar. This system has become known as the *binary pulsar*. Measurements of the periodic Doppler shift of the pulsar's radiation show that its orbit is slowly shrinking at exactly the rate predicted by relativity theory if the energy were being carried off by gravity waves. Even though the gravity waves themselves have not been detected, the binary pulsar is regarded by most astronomers as very strong evidence supporting general relativity. Taylor and Hulse received the 1993 Nobel Prize in physics for their discovery.

In January 2004, radio astronomers announced the discovery of a *double-pulsar* binary system with an even shorter period than the binary pulsar, implying stronger relativistic effects and a shorter merger time—less than 100 million years. Because both components are pulsars, and since the system, by pure luck, also happens to be seen almost exactly edge on by observers on Earth, leading to eclipses, we can expect this system to provide a wealth of information on both neutron stars and gravitational physics in coming years.

The figure below shows part of an ambitious gravity-wave observatory called LIGO—short for Laser Interferometric Gravity-Wave Observatory—which became operational in 2003. Twin detectors, one (shown here) in Hanford, Washington, the other in Livingston, Louisiana, use laser beams to measure the extremely small (less than one-thousandth the diameter of an atomic nucleus) distortions of space produced in the lengths of the 4-km-long arms should a gravity wave pass by. The instrument is in theory capable of detecting waves from many Galactic and extragalactic sources, although, so far, no gravity waves have actually been detected. However, the system is scheduled to be upgraded and its sensitivity greatly increased in 2007. Much more sensitive space-based instruments, with detecting arms literally a million times larger than those in LIGO, are planned to come on-line around 2010.

If successful, the discovery of gravity waves could herald a new age in astronomy, in much the same way that invisible electromagnetic waves, virtually unknown a century ago, revolutionized classical astronomy and led to the field of modern astrophysics.

The Final Stage of Stellar Evolution

Although the precise figure is uncertain, mainly because the behavior of matter at very high densities is not well understood, most researchers concur that the mass of a neutron star cannot exceed about three times the mass of the Sun. This is the neutron-star equivalent of the white dwarf mass limit discussed in the previous chapter. ∞ *(Sec. 12.5)* Above this mass, not even tightly packed neutrons can withstand the star's gravitational pull. In fact, we know of *no* force that can counteract gravity beyond this point. Thus, if enough material is left behind after a supernova explosion that the central core exceeds this limit, or if enough matter falls back onto the neutron star after the supernova has occurred, gravity wins the battle with pressure once and for all, and the central core collapses forever. Stellar evolution theory indicates that this is the fate of any star whose main-sequence mass exceeds about 25 times the mass of the Sun.

As the core shrinks, the gravitational pull in its vicinity eventually becomes so great that nothing—not even light—can escape. The resultant object therefore emits no light, no other form of radiation, no information whatsoever. Astronomers call this bizarre endpoint of stellar evolution, in which the core of a very-high-mass star collapses in on itself and vanishes forever, a **black hole**.

Escape Speed

Newtonian mechanics—up to now our reliable and indispensable tool in understanding the universe—cannot adequately describe conditions in or near black holes. ∞ *(Sec. 1.4)* To comprehend these collapsed objects, we must turn instead to the modern theory of gravity, Einstein's **general theory of relativity**, discussed in Section 13.6. Still, we can still usefully discuss some aspects of these strange bodies in more or less Newtonian terms. Let's reconsider the familiar Newtonian concept of escape speed—the speed needed for one object to escape from the gravitational pull of another—supplemented by two key facts from relativity: (1) nothing can travel faster than the speed of light, and (2) all things, *including light*, are attracted by gravity.

A body's escape speed is proportional to the square root of the body's mass divided by the square root of its radius. ∞ *(More Precisely 5-1)* Earth's radius is 6400 km, and the escape speed from Earth's surface is just over 11 km/s. Now consider a hypothetical experiment in which Earth is squeezed on all sides by a gigantic vise. As our planet shrinks under the pressure, the escape speed increases because the radius is decreasing. Suppose Earth were compressed to one-fourth its present size. The escape speed would double (because $1/\sqrt{1/4} = 2$). Any object escaping from the surface of this hypothetically compressed Earth would need a speed of at least 22 km/s.

Imagine compressing Earth by an additional factor of 1000, making its radius hardly more than a kilometer. Now a speed of about 700 km/s—more than the escape speed of the Sun—would be needed to escape. Compress Earth still further, and the escape speed continues to rise. If our hypothetical vise were to squeeze Earth hard enough to crush its radius to about a centimeter, the speed needed to escape its surface would reach 300,000 km/s. But this is no ordinary speed. It is the speed of light, the fastest speed allowed by the laws of physics as we currently know them.

Thus, if by some fantastic means the entire planet Earth could be compressed to less than the size of a grape, the escape speed would exceed the speed of light. And because nothing can exceed that speed, the compelling conclusion is that nothing—absolutely nothing—could escape from the surface of such a compressed body. Even radiation—radio waves, visible light, X-rays, photons of all wavelengths—would be unable to escape its intense gravity. Our planet would be invisible and uncommunicative—no signal of any sort could be sent to the universe outside. The origin of the term *black hole* becomes clear. For all practical purposes, Earth could be said to have disappeared from the universe. Only its gravitational field would remain behind, betraying the presence of its mass, now shrunk to a point.[1]

The Event Horizon

Astronomers have a special name for the critical radius at which the escape speed from an object would equal the speed of light and within which the object could no longer be seen. It is the **Schwarzschild radius**, after Karl Schwarzschild, the German scientist who first studied its properties. The Schwarzschild radius of any object is proportional to its mass. For Earth, it is 1 cm. For Jupiter, at about 300 Earth masses, it is about 3 m. For the Sun, at 300,000 Earth masses, it is 3 km. For a 3-solar mass stellar core, the Schwarzschild radius is about 9 km. As a convenient rule of thumb, the Schwarzschild radius of an object is simply 3 km multiplied by the object's mass measured in solar masses.

Every object has a Schwarzschild radius—it is the radius to which the object would have to be compressed for it to become a black hole. Put another way, a black hole is an object that happens to lie within its own Schwarzschild radius. The surface of an imaginary sphere having a radius equal to the Schwarzschild radius and centered on a collapsing star is called the **event horizon**. No event occurring within this region can ever be seen, heard, or known by anyone outside. Even though there is no matter of any sort associated with it, we can think of the event horizon as the "surface" of a black hole.

Theory indicates that, if more than 3 solar masses of material remain behind after a supernova explosion, the remnant core will collapse catastrophically, diving below the event horizon in less than a second. The event horizon is not a physical boundary of any kind—just a communications barrier. The core would shrink right past the Schwarzschild radius to ever-diminishing size on its way to being crushed to a point. It simply "winks out," disappearing and becoming a small dark region from which nothing can escape—a literal black hole in space. This is the predicted fate of stars having more than about 20–25 times the mass of the Sun.

Tutorial

Escape Speed and Black Hole Event Horizons

CONCEPT CHECK ▶▶▶

What happens if you compress an object within its own Schwarzschild radius?

13.6 Einstein's Theories of Relativity

By the latter part of the 19th century, physicists were well aware of the special status of the speed of light, *c*. It was, they knew, the speed at which all electromagnetic waves traveled, and, as best they could tell, it represented an upper limit on the speeds of *all* known particles. Scientists struggled without success to construct a theory of mechanics and radiation in which *c* was a natural speed limit.

Special Relativity

In 1887, a fundamental experiment carried out by the American physicists A. A. Michelson and E. W. Morley compounded theorists' problems further by demonstrating another important and unique aspect of light: The measured speed of a beam of light is *independent* of the motion of either the observer or the source (see *More Precisely 13-1*). No matter what our motion may be rela-

---

[1]In fact, regardless of the composition or condition of the object that formed the black hole, only three physical properties can be measured from the outside—the hole's mass, charge, and angular momentum. All other information is lost once matter falls into the black hole. Only these three numbers are required to describe completely a black hole's outward appearance. In this chapter we consider only holes that formed from nonrotating, electrically neutral matter. Such objects are completely specified once their masses are known.

tive to the source of the radiation, we always measure precisely the same value for c: 299,792.458 km/s.

A moment's thought tells us that this is a decidedly nonintuitive statement. For example, if we were traveling in a car moving at 100 km/h and we fired a bullet forward with a speed of 1000 km/h relative to the car, an observer standing at the side of the road would see the bullet pass by at $100 + 1000 = 1100$ km/h, as illustrated in Figure 13.14(a). However, the Michelson–Morley experiment tells us that if we were traveling in a rocket ship at one-tenth the speed of light, $0.1c$, and we shone a searchlight beam ahead of us (Figure 13.14b), an outside observer would measure the speed of the beam not as $1.1c$, as the example of the bullet would suggest, but as c. The rules that apply to particles moving at or near the speed of light are different from those we are used to in everyday life.

The *special theory of relativity* (or just *special relativity*) was proposed by Einstein in 1905 to deal with the preferred status of the speed of light. The theory is the mathematical framework that allows us to extend the familiar laws of physics from low speeds (i.e., speeds much less than c, which are often referred to as *nonrelativistic*) to very high (or *relativistic*) speeds, comparable to c.

The essential features of special relativity are as follows:

1. The speed of light, c, is the maximum possible speed in the universe, and all observers measure the *same* value for c, regardless of their motion. Einstein broadened this statement into the *principle of relativity*: The basic laws of physics are *the same* to all observers.
2. There is no absolute frame of reference in the universe; that is, there is no "preferred" observer relative to whom all other velocities can be measured. Instead, only relative velocities between observers matter (hence the term "relativity").
3. Neither space nor time can be considered independently of one another. Rather, they are each components of a single entity: *spacetime*. There is no absolute, universal time—observers' clocks tick at different rates, depending on the observers' motions relative to one another.

Special relativity is equivalent to Newtonian mechanics in describing objects that move much more slowly than the speed of light, but it differs greatly in its predictions at relativistic velocities (see *More Precisely 13-1*). Yet, despite their often nonintuitive nature, all of the theory's predictions have been repeatedly verified to a high degree of accuracy. Today, special relativity lies at the heart of modern science. No scientist seriously doubts its validity.

General Relativity

In constructing his special theory of relativity, Einstein rewrote the laws of motion expounded by Newton more than two centuries previously. ∞ *(Sec. 1.4)* Fitting Newton's other great legacy—the theory of gravitation—into the framework of relativity was a much more complex mathematical problem, which took Einstein another decade to complete. The result once again overturned scientists' conception of the universe.

In 1915, Einstein illustrated the connection between special relativity and gravity with the following famous "thought experiment." Imagine that you are enclosed in an elevator with no windows, so that you cannot directly observe the outside world, and the elevator is floating in space. You are weightless.

▲ Figure 13.14 Speed of Light (a) A bullet fired from a speeding car would be measured by an outside observer to have a speed equal to the sum of the speeds of the car and of the bullet. (b) A beam of light shining forward from a high-speed spacecraft would still be observed to have speed c, regardless of the speed of the spacecraft. The speed of light is thus independent of the speed of the source or of the observer.

Now suppose that you begin to feel the floor press up against your feet. Weight has apparently returned. There are two possible explanations for this, as shown in Figure 13.15. A large mass could have come nearby, and you are feeling its downward gravitational attraction (Figure 13.15a), *or* the elevator has begun to accelerate upward, and the force you feel is that exerted by the elevator as it accelerates you at the same rate (Figure 13.15b). The crux of Einstein's argument is this: There is *no* experiment that you can perform within the elevator (without looking outside) that will let you distinguish between these two possibilities.

Thus, Einstein reasoned, there is no way to tell the difference between a gravitational field and an accelerated frame of reference (such as the rising elevator in the thought experiment). This statement is known more formally as the *equivalence principle*. Using it, Einstein succeeded in incorporating gravity into special relativity as a general acceleration of all particles. The resulting the-

13-1 MORE PRECISELY

Special Relativity

Einstein's theory of special relativity was constructed in large part to solve the puzzle of the 1887 Michelson–Morley experiment, which had demonstrated that the observed speed of light is *independent* of the observer's motion through space. Michelson and Morley attempted to determine Earth's motion relative to the "absolute" space through which light supposedly moved by measuring the speed of light at different times of the day (so that the orientation of their equipment would change as Earth rotated) and on different days of the year (so that Earth's velocity would vary as our planet orbited the Sun). As illustrated in the first figure (see also Figure 13.14), they expected to measure a faster-moving beam of light when it was aligned with Earth's motion (to the right in the figure) and a slower beam when opposed to Earth's motion (to the left). But in fact they measured precisely the *same* velocity for any orientation of their apparatus. Far from measuring the properties of absolute space, the Michelson–Morley experiment ultimately demolished the entire concept—and with it, the 19th-century view of the universe.

With special relativity, Einstein elevated the speed of light to the status of a constant of nature, rewrote the laws of mechanics to reflect that new fact, and opened the door to a flood of new physics and a much deeper understanding of the universe. But many commonsense ideas had to be abandoned in the process and replaced with some decidedly less-intuitive concepts. We describe here some of those odd consequences of Einstein's theory. Imagine that you are an observer watching a rocket ship fly past at relative velocity v and that the craft is close enough for you to make detailed observations inside its cabin. If v is much less than the speed of light, c, you would see nothing out of the ordinary—special relativity is consistent with familiar Newtonian mechanics at low velocities. However, if v is comparable to c, unexpected things start to happen.

As the ship's velocity increases, you begin to notice that the ship appears to *contract* in the direction in which it is mov-

ing. A meterstick on board, identical at launch to the one you keep in your laboratory, is now shorter than its twin. This is called *Lorentz contraction* (or sometimes Lorentz–Fitzgerald contraction). The second figure shows the stick's measured length aboard the moving ship: At low velocities (bottom) the meterstick measures 1 m, but at high velocities (top), the stick is shortened considerably. Because of Lorentz contraction, a meterstick moving at 90 percent of the velocity of light would shrink to a little less than half a meter. (This is not an optical illusion.)

At the same time, the ship's clock, synchronized prior to launch with your own, now ticks *more slowly*. This phenomenon, known as *time dilation*, has been observed many times in laboratory experiments in which fast-moving radioactive particles are observed to decay more slowly than if they were at rest in the lab. Their internal clocks are slowed by their rapid motion. Although no material particle can actually reach the speed

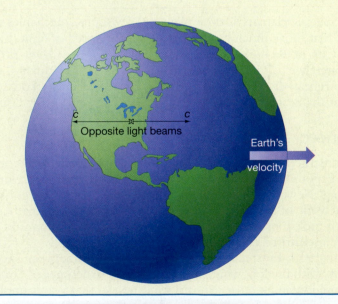

Opposite light beams

Earth's velocity

ory is called *general relativity*. However, Einstein found that another huge conceptual leap was needed. To include the effects of gravity, the mathematics forced Einstein to the unavoidable conclusion that spacetime—the single entity combining space and time, central to special relativity—has to be *curved*.

The central concept of general relativity is this: Matter—all matter—tends to "warp" or curve space in its vicinity. Objects such as planets and stars react to this warping by changing their paths. In the Newtonian view of gravity, particles move on curved trajectories because they are acted upon by a gravitational force. ∞ *(Sec. 1.4)* In Einsteinian relativity, those same particles move on curved trajectories because they are falling freely through space, following the curvature of spacetime produced by some nearby massive object. The more the mass, the greater is the warping. Thus, in general relativity, there is no such thing as a "gravitational force" in the Newtonian sense. Objects move as they do because they follow the curvature of spacetime, which is determined

of light, Einstein's theory implies that, as v approaches c, the measured length of the meterstick will shrink to nearly zero and the clock will slow to a virtual stop.

The effects of Lorentz contraction and time dilation are described by the same quantity, usually called the *Lorentz factor* and denoted by the Greek letter γ (gamma). For a particle moving with velocity v, this factor is given by

$$\gamma = \frac{1}{\sqrt{1 - v^2/c^2}}$$

As v increases, lengths are reduced and clocks slow down by this factor (which is always greater than or equal to 1). For Earth, orbiting the Sun at 30 km/s, $v = 0.0001c$ and $\gamma = 1/\sqrt{0.99999999} = 1.000000005$, so the effects of relativity are negligibly small. However, for a charged particle in a pulsar wind (Section 13.2) traveling at 90 percent of the speed

of light, $v/c = 0.9$, and now $\gamma = 1/\sqrt{0.19} = 2.3$. Speeds very close to that of light are needed before the effect becomes large—even at one-tenth the speed of light, 30,000 km/s, γ is still only 1.005, so relativistic effects are barely noticeable.

Of course, from the point of view of an astronaut on board the spaceship, *you* are the one moving rapidly. Hence, as seen from the ship, *you* appear to be compressed in the direction of motion, and *your* clock runs slowly! How can both observations be correct? After all, you both know that nothing has happened to your mersticks or clocks. You feel no force squeezing your bodies. What is going on?

When measuring the length of the moving meterstick, you do so by noting the positions of the two ends *at the same time, according to your clock.* But those two events—the two measurements you make—do *not* occur at the same time as seen by the astronaut on the spaceship. In relativity, time is relative, and simultaneity (the idea that two events happen "at the same time") is no longer a well-defined concept. From the astronaut's viewpoint, your measurement of the leading end of the meterstick occurs *before* your measurement of the trailing end. This time discrepancy ultimately results in the Lorentz contraction you observe. A similar argument applies to measurements of time, such as the period between two clock ticks. Time dilation occurs because the measurements occur at the same location and different times in one frame, but at *different places and times* in the other.

Einstein's ideas were revolutionary at the time, requiring physicists to abandon some long-held, cherished, and "obvious" facts about the universe. Reading the above discussion, you can perhaps begin to understand why special relativity initially encountered such resistance among some scientists (and why it still confuses nonscientists today!). Nevertheless, the gain in scientific understanding ultimately overcame the price in unfamiliarity. Within just a few years, special relativity had become almost universally accepted (and Einstein was well on his way to becoming the best-known scientist on the planet).

◀ **Figure 13.15 Einstein's Elevator** A passenger in a closed elevator floating in space feels weight return—he feels a force exerted on his feet by the elevator floor. Einstein reasoned that no experiment conducted entirely within the elevator can tell the passenger whether the force is (a) due to the gravity of a nearby massive object or (b) caused by the acceleration of the elevator itself.

by the amount of matter present. Stated more loosely, "Spacetime tells matter how to move, and matter tells spacetime how to curve."[2]

Some props may help you visualize these ideas. Bear in mind, however, that these props are not real, but only tools to help you grasp some exceedingly strange concepts. Imagine a pool table with the tabletop made of a thin rubber sheet rather than the usual hard felt. As Figure 13.16 suggests, such a rubber sheet becomes distorted when a heavy weight (such as a rock) is placed on it. The heavier the rock (Figure 13.16b), the larger is the distortion.

Trying to play pool on this table, you would quickly find that balls passing near the rock were deflected by the curvature of the tabletop (Figure 13.16b). The pool balls are not attracted to the rock in any way; rather, they respond to the curvature of the sheet produced by the rock's presence. In much the same way, anything that moves through space—matter *or radiation*—is deflected by the curvature of spacetime near a star. (See *More Precisely 13-2* on p. 381 for a solar system measurement, and Sections 14.6 and 16.3 for examples on much larger scales.) Earth's orbital path is the trajectory that results as our planet falls freely in the relatively gentle curvature of space created by our Sun. When the curvature is small (i.e., gravity is weak), both Einstein and Newton predict the same orbit—the one we observe. However, as the gravitating mass increases, the two theories begin to diverge.

Curved Space and Black Holes

Modern notions about black holes rest squarely on the general theory of relativity. Although white dwarfs and (to a lesser extent) neutron stars can be adequately described by the classical Newtonian theory of gravity, only the modern Einsteinian theory of relativity can properly account for the bizarre physical properties of black holes.

Let's consider another analogy. Imagine a large extended family of people living on a huge rubber sheet—a sort of gigantic trampoline. Deciding to hold a reunion, they converge on a given place at a given time. As shown in Figure 13.17, one person remains behind, not wishing to attend. She keeps in touch with her relatives by means of "message balls" rolled out to her (and back from her) along the surface of the sheet. These message balls are the analog of radiation carrying information through space.

As the people converge, the rubber sheet sags more and more. Their accumulating mass creates an increasing amount of space curvature. The message balls can still reach the lone person far away in nearly flat space, but they arrive less frequently as the sheet becomes more and more warped and stretched—as shown in Figures 13.14(b) and (c)—and the balls have to climb out of a deeper and deeper well. Finally, when enough people have arrived at the appointed

(a)

Deflection of pool ball

(b)

[2]This characterization of gravity is due to the renowned relativist John Archibald Wheeler.

◀ **Figure 13.16 Curved Space** (a) A pool table made of a thin rubber sheet sags when a weight is placed on it. Likewise, space is bent, or warped, in the vicinity of any massive object. (b) As the weight increases, so does the warping of space. A ball shown rolling across the table is deflected by the curvature of the surface, in much the same way as a planet's curved orbit is determined by the curvature of spacetime produced by the Sun.

▲ **Figure 13.17 Space Warping**   Any mass causes a rubber sheet (space) to be curved. As people assemble at the appointed spot on the sheet (marked by an "×"), the curvature grows progressively larger, as shown in (a)–(c). The blue arrows represent some directions in which information can be transmitted from place to place. (d) The people are eventually sealed inside the bubble, forever trapped and cut off from the outside world.

spot, the mass becomes too great for the rubber to support them. As illustrated in Figure 13.17(d), the sheet pinches off into a "bubble," compressing the people into oblivion and severing their communications with the lone survivor outside. This final stage represents the formation of an event horizon around the reunion party.

Right up to the end—the pinching off of the bubble—two-way communication is possible. Message balls can reach the outside from within (but at a slower and slower rate as the rubber stretches), and messages from outside can get in without difficulty. However, once the event horizon (the bubble) forms, balls from the outside can still fall in, but they can no longer be sent back out to the person left behind, no matter how fast they are rolled. They cannot make it past the "lip" of the bubble in Figure 13.17(d). This analogy (very) roughly depicts how a black hole warps space completely around on itself, isolating its interior from the rest of the universe.

The essential ideas—the slowing down and eventual cessation of outward-going signals and the one-way nature of the event horizon once it forms—all have parallels in the case of stellar black holes. In Einsteinian terms, a *black hole* is a region of space where the gravitational field becomes overwhelming and the curvature of space extreme. At the event horizon itself, the curvature is so great that space "folds over" on itself, causing objects within to become trapped and disappear.

◀◀◀ | **CONCEPT CHECK** ▶

How do Newton's and Einstein's theories differ in their descriptions of gravity?

13.7 Space Travel Near Black Holes

Black holes are *not* cosmic vacuum cleaners. They don't cruise around interstellar space, sucking up everything in sight. The orbit of an object near a black hole is essentially the same as its orbit near a "normal" star of the same mass as the black hole. Only if the object happened to pass within a few Schwarzschild radii (perhaps 50 or 100 km for a typical 5- to 10-solar mass black hole formed in a supernova explosion) of the event horizon would there be any significant difference between its actual orbit and the one predicted by Newtonian gravity and described by Kepler's laws. Of course, if some matter did fall into the black hole—that is, if its orbit happened to take it too close to the event horizon—it would be unable to get out. Black holes are like turnstiles, permitting matter to flow in only one direction—inward.

Tidal Forces

Matter flowing into a black hole is subject to enormous gravitational stress. An unfortunate explorer falling feet first into a solar mass black hole would find that the pull of gravity is much stronger at her feet (which are closer to the hole) than at her head. This force difference is a *tidal force*—precisely the

▲ **Figure 13.18 Black Hole Heating** Any matter falling into the clutches of a black hole will become severely distorted and heated. This sketch shows an imaginary planet being pulled apart by a black hole's gravitational tides.

▲ **Figure 13.19 Robot–Astronaut** A hypothetical robot–astronaut can travel toward a black hole while performing experiments that humans, farther away, can monitor to learn something about the nature of space near the event horizon.

same basic phenomenon that is responsible for ocean tides on Earth and the spectacular volcanoes on Io—except that the tidal forces at work in and near a black hole are far more powerful than any we are ever likely to find in the solar system. ∞ *(Sec. 5.2, 8.1)* The tidal force due to the black hole is so great that our explorer would be stretched enormously in height and squeezed unmercifully laterally. She would be torn apart long before she reached the event horizon.

The net result of all this stretching and squeezing is numerous and violent collisions among the resulting debris, causing a great deal of frictional heating of any infalling matter. As illustrated (with some artistic license) in Figure 13.18, material is simultaneously torn apart and heated to high temperatures as it plunges into the hole. So efficient is this heating that prior to reaching the hole's event horizon, matter falling into the hole emits radiation of its own accord. For a black hole of roughly solar mass, the energy is expected to be emitted in the form of X-rays. Thus, contrary to what we might expect from an object whose defining property is that nothing can escape from it, the region surrounding a black hole is expected to be a *source* of energy. Of course, once the hot matter falls below the event horizon, its radiation is no longer detectable—it can never leave the hole.

Approaching the Event Horizon

One safe way to study a black hole would be to go into orbit around it, safely beyond the disruptive influence of the hole's strong tidal forces. After all, Earth and the other planets of our solar system all orbit the Sun without falling into it and without being torn apart. The gravity field around a black hole is basically no different. Let's imagine sending a less fragile (and more expendable) observer—a robot probe—toward the center of the hole, as illustrated in Figure 13.19. If our robot is small enough (to minimize the tidal force) and of sufficiently strong construction, it might conceivably survive to reach the event horizon, even if a human could not. Watching from a safe distance in our orbiting spacecraft, we can then examine the nature of space and time, at least down to the event horizon. After that boundary is crossed, there is no way for the probe to return any information about its findings.

Suppose our robot has an accurate clock and a light source of known frequency mounted on it. From our safe vantage point far outside the event horizon, we could use telescopes to read the clock and measure the frequency of the light we receive. What might we discover? We would find that the light from the robot would become more and more redshifted—shifted toward longer wavelengths—as the robot neared the event horizon (Figure 13.20). Even if the robot used rocket engines to remain motionless, the redshift would still be detected. The redshift is not caused by motion of the light source—it is not the result of the Doppler effect. ∞ *(Sec. 2.7)* Rather, it is a result of the black hole's gravitational field, clearly predicted by Einstein's general theory of relativity and known as **gravitational redshift**.

We can explain the gravitational redshift as follows. Photons are attracted by gravity. As a result, in order to escape from a source of gravity, photons must expend some energy—they have to work to get out of the gravitational field. They don't slow down (photons always move at the speed of light); they just lose energy. Because a photon's energy is proportional to the frequency of its radiation, light that loses energy must have its frequency reduced (or, conversely, its wavelength increased).

As photons traveled from the robot's light source to our orbiting spacecraft, they would become gravitationally redshifted. From our standpoint, a yellow light would appear orange, then red as the robot astronaut approached the black hole. As the robot neared the event horizon, the light would become undetectable with an optical telescope. First infrared and then radio telescopes would be needed to detect it. Theoretically, light emitted *from the event horizon itself* would be gravitationally redshifted to infinitely long wavelengths.

What about the robot's clock? What time does it tell? Is there any observable change in the rate at which the clock ticks while moving deeper into the hole's gravitational field? We would find, from our safely orbiting spacecraft, that any clock close to the hole would appear to tick more *slowly* than an equivalent clock on board our ship. The closer the clock came to the hole, the slower it would appear to run. Upon reaching the event horizon, the clock would seem to stop altogether. All action would become frozen in time. Consequently, we would never actually witness the robot cross the event horizon. The process would appear to take forever.

This apparent slowing down of the robot's clock is known as **time dilation** and is closely related to gravitational redshift. To see why this is so, think of the robot's light source as a clock, with the passage of a wave crest (say) constituting a "tick," so the clock ticks at the frequency of the radiation. As the wave is redshifted, the frequency drops, and fewer wave crests pass us each second—the clock appears to slow down. This thought experiment demonstrates that the redshift of the radiation and the slowing of the clock are one and the same thing.

From the point of view of the robot, relativity theory predicts no strange effects at all. The light source hasn't reddened and the clock keeps perfect time. In the robot's frame of reference, everything is normal. Nothing prohibits it from approaching within the Schwarzschild radius of the hole (so long as the probe is strong enough to avoid being torn apart by tidal forces)—no law of physics constrains an object from passing through an event horizon. There is no barrier at the event horizon and no sudden lurch as it is crossed. It is just a boundary in space, but one that can be crossed in only one direction.

Perhaps counterintuitively, the tidal force at the event horizon is *smaller* for more massive black holes (see problems 8 and 9 at the end of this chapter). Although, as we saw, an intrepid explorer would be shredded long before reaching the event horizon of a 1-solar mass black hole, travelers passing through the event horizon of a multimillion solar mass hole (such as might lurk in the heart of our own Galaxy, as we will see in the next Chapter) might not feel any particular discomfort. They would remain able to contemplate their surroundings—and the all-important fact that they couldn't reverse course—for a few more seconds before the rapidly rising tidal forces in the black hole's interior finally finished them off.

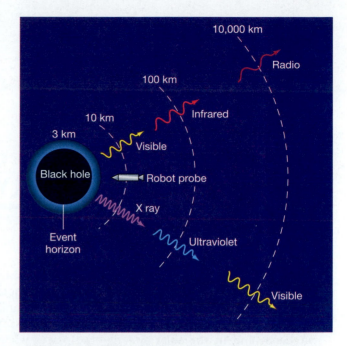

▲ **Figure 13.20 Gravitational Redshift**   As a photon escapes from the strong gravitational field close to a black hole, it must expend energy to overcome the hole's gravity. This energy does not come from a change in the speed at which the photon travels (that speed is always 300,000 km/s, even under these extreme conditions). Rather, the photon "gives up" energy by increasing its wavelength. Thus, the photon's color changes. This figure shows the effect on two beams of radiation, one of visible light and one of X-rays, emitted from a space probe as it nears the event horizon of a 1-solar mass black hole. The beams are shifted to longer and longer wavelengths as they move farther from the black hole.

Deep Down Inside

No doubt, you are wondering what lies within the event horizon of a black hole. The answer is simple: No one knows. However, the question raises some very fundamental issues that lie at the forefront of modern physics.

General relativity predicts that without some agent to compete with gravity, the core remnant of a high-mass star will collapse all the way to a point at which both its density and its gravitational field become infinite—a so-called **singularity**. The same applies to any infalling matter. However, we should not take this prediction of infinite density too literally. Singularities always signal the breakdown of the theory producing them. The present laws of physics are simply inadequate to describe the final moments of a star's collapse.

As it stands today, the theory of gravity is incomplete because it does not incorporate a proper description of matter on very small scales. As our collapsing stellar core shrinks to smaller and smaller radii, we eventually lose our ability even to describe, let alone predict, its behavior. Perhaps matter trapped in a black hole never actually reaches a singularity. Perhaps it just approaches this bizarre state, in a manner that we will someday understand as the subject of *quantum gravity*—the merger of general relativity (the theory of the universe on the largest scales) with quantum mechanics (the theory of matter on atomic and subatomic scales)—develops.

Having said that, we can at least estimate how small the core can get before current theory fails. It turns out that by the time that stage is reached, the core is already much smaller than any elementary particle. Thus, although a complete description of the endpoint of stellar collapse may well require a major overhaul of the laws of physics, for all practical purposes the prediction of collapse to a point is valid. Even if a new theory somehow succeeds in doing away with the central singularity, it is very unlikely that the external appearance

13-2 MORE PRECISELY

Tests of General Relativity

Special relativity is the most thoroughly tested and most accurately verified theory in the history of science. General relativity, however, is on somewhat less firm experimental ground. The problem with verifying general relativity is that its effects on Earth and in the solar system—the places where we can most easily perform tests—are very small. Just as special relativity produces major departures from Newtonian mechanics only when speeds approach the speed of light, general relativity predicts large departures from Newtonian gravity only when extremely strong gravitational fields are involved.

Here we consider just two "classical" tests of the theory—solar system experiments that helped ensure acceptance of Einstein's theory. Bear in mind, however, that there are no known tests of general relativity in the "strong-field" regime—that part of the theory that predicts black holes, for example—so the full theory has never been experimentally tested.

At the heart of general relativity is the premise that everything, including light, is affected by gravity because of the curvature of spacetime. Shortly after he published his theory in 1915, Einstein noted that light from a star should be deflected by a measurable amount as it passes the Sun. The closer to the Sun the light comes, the more it is deflected. Thus, the maximum deflection should occur for a ray that just grazes the solar surface. Einstein calculated that the deflection angle should be 1.75″, a small but detectable amount.

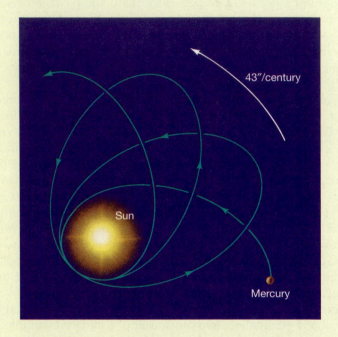

Of course, it is normally impossible to see stars so close to the Sun. During a solar eclipse, however, when the Moon blocks the Sun's light, the observation does become possible, as illustrated in the accompanying figure. In 1919 a team of observers led by the British astronomer Sir Arthur Eddington succeeded in measuring the deflection of starlight during an eclipse. The results were in excellent agreement with the prediction of general relativity.

Another prediction of general relativity is that planetary orbits should deviate slightly from the perfect ellipses of Kepler's laws. Again, the effect is greatest where gravity is strongest—that is, closest to the Sun. Thus, the largest relativistic effects are found in the orbit of Mercury. Relativity predicts that Mercury's orbit is not exactly a closed ellipse. Instead, its orbit should rotate slowly, as shown in the second (highly exaggerated) diagram above. The amount of rotation is very small—only 43″ per century—but Mercury's orbit is so well charted that even this tiny effect is measurable.

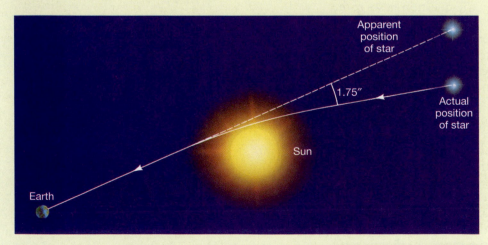

of the hole, or the existence of its event horizon, will change. Any modifications to general relativity are expected to occur only on submicroscopic scales, not on the macroscopic (kilometer-sized) scale of the Schwarzschild radius.

Singularities are places where the rules break down, and some very strange things may occur near them. Many possibilities have been suggested—gateways to other universes, time travel, or the creation of new states of matter, for instance—but none has been proved, and certainly none has been observed. Because these regions are places where science fails, their presence causes serious problems for many of our cherished laws of physics, including causality (the idea that cause should precede effect, which runs into immediate problems if time travel is possible) and energy conservation (which is violated if material can hop from one universe to another through a black hole). It is unclear whether the removal of the singularity by some future, all-encompassing theory would necessarily also eliminate all of these problematic side effects.

◄◄◄ **CONCEPT CHECK**

Why would you never actually witness an infalling object crossing the event horizon of a black hole?

13.8 Observational Evidence for Black Holes

Theoretical ideas aside, is there any observational evidence for black holes? What proof do we have that these strange invisible objects really do exist?

Black Holes in Binary Systems

Perhaps the most promising way to find black holes is to look for their effects on other astronomical objects. Our Galaxy harbors many binary star systems in which only one object can be seen, but we need observe the motion of only one star to infer the existence of an unseen companion and measure some of its properties. ∞ *(Sec. 10.7)* In the majority of cases, the invisible companion is simply small and dim, nothing more than an M-type star hidden in the glare of an O- or B-type partner, or perhaps shrouded by dust or other debris, making it invisible to even the best available equipment. In either case, the invisible object is not a black hole.

A few close binary systems, however, have peculiarities suggesting that one of their members may be a black hole. Figure 13.21(a) shows the area of the sky in the constellation Cygnus, where the evidence is particularly strong. The rectangle outlines the celestial system of interest, some 2000 pc from Earth. The black-hole candidate is an X-ray source called Cygnus X-1, discovered by the *Uhuru* satellite in the early 1970s. Its visible companion is a blue B-type giant. Assuming that it lies on the main sequence, its mass must be around 25 times the mass of the Sun. Spectroscopic observations indicate that the binary system has an orbital period of 5.6 days. Combining this information with spectroscopic measurements of the visible component's orbital speed, astronomers estimate the total mass of the binary system to be around 35 solar masses, implying that Cygnus X-1 has a mass of about 10 times that of the Sun.

X-ray radiation emitted from the immediate neighborhood of Cygnus X-1 indicates the presence of high-temperature gas, perhaps as hot as several million kelvins. Furthermore, the X-ray emission has been observed to vary in intensity on time scales as short as a millisecond. As discussed earlier in the context of gamma-ray bursts, for this variation not to be blurred by the travel time of light across the source, Cygnus X-1 cannot be more than 1 light-millisecond, or 300 km, in diameter.

These properties suggest that Cygnus X-1 could be a black hole. Figure 13.22 is an artist's conception of this intriguing object. The X-ray-emitting region is likely an accretion disk formed as matter drawn from the visible star spirals down onto the unseen component. The rapid variability of the X-ray

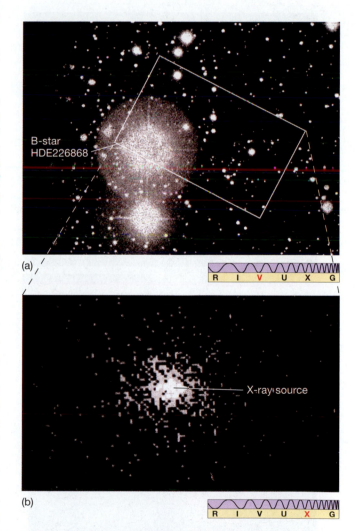

▲ **Figure 13.21 Cygnus X-1** (a) The brightest star in this photograph is a member of a binary system whose unseen companion, called Cygnus X-1, is a leading black hole candidate. (b) An X-ray image of the field of view outlined by the rectangle in part (a). (Harvard-Smithsonian Center for Astrophysics; NASA)

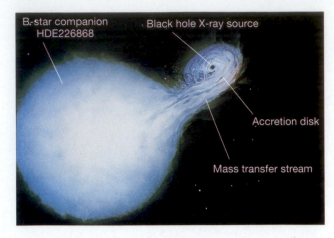

B-star companion HDE226868
Black hole X-ray source
Accretion disk
Mass transfer stream

Animation/Video
Black Hole Geometry

◄ **Figure 13.22 Black Hole** Artist's conception of a binary system containing a large, bright, visible star and an invisible, X-ray-emitting black hole. This painting is based on data obtained from detailed observations of Cygnus X-1. (L. Chaisson)

emission indicates that the unseen component must be very compact—a neutron star or a black hole. Its large mass, well above the neutron star mass limit discussed earlier, implies the latter.

A few other black hole candidates are known. For example, LMC X-3—the third X-ray source ever discovered in the Large Magellanic Cloud, a small galaxy orbiting our own (see Section 15.2)—is an invisible object that, like Cygnus X-1, orbits a bright companion star. LMC X-3's visible companion seems to be distorted into the shape of an egg by the unseen object's intense gravitational pull. Reasoning similar to that applied to Cygnus X-1 leads to the conclusion that LMC X-3 has a mass nearly 10 times that of the Sun, making it too massive to be anything but a black hole. The X-ray binary system A0620-00 has been found to contain an invisible compact object of mass 3.8 times the mass of the Sun. In total, about 20 known objects may turn out to be black holes; Cygnus X-1, LMC X-3, and A0620-00 probably have the strongest claims.

Black Holes in Galaxies

Perhaps the strongest evidence for black holes comes not from binary systems in our own galaxy but from observations of the centers of many galaxies (including our own), where astronomers have found that stars and gas are moving extremely rapidly, orbiting some very massive, unseen object. Masses inferred from Newton's laws range from millions to billions of times the mass of the Sun. ∞ *(More Precisely 1-1)* The leading (and at present, the only) explanation is that these objects are black holes. We will return to these observations, and the question of how such *supermassive black holes* might have formed, in Chapters 14 and 16.

In 2000, X-ray astronomers reported evidence for a long-sought but elusive missing link between "stellar-mass" black holes in binaries and the supermassive black holes in the hearts of galaxies. Figure 13.23 shows an unusual looking galaxy called M82, currently the site of an intense and widespread burst of star formation (see Chapter 15). The inset shows a *Chandra* image of the innermost few thousand parsecs of M82, revealing a number of bright X-ray sources close to—but *not* at—the center of the galaxy. Their spectra and X-ray luminosities suggest that they may be accreting compact objects with masses ranging from 100 to almost 1000 times the mass of the Sun. If confirmed, they will be the first *intermediate-mass black holes* ever observed.

Too large to be the remnant of a normal star, and too small to warrant the "supermassive" label, these objects present a puzzle to astronomers. One possible origin is suggested by an apparent association between many of the X-ray emitters in M82 and elsewhere, and dense, young star clusters. Theorists speculate that collisions between high-mass stars in the congested cores of such clusters might lead to the runaway growth of an extremely massive and very unstable star, which then collapses to form an intermediate-mass black hole. However, both the observations and the explanation remain controversial—further systematic studies of many other galaxies will be needed to resolve this issue.

R I V U X G

Suspected black hole

R I V U X G

◄ **Figure 13.23 Intermediate-Mass Black Holes?** X-ray observations (inset) of the center of the starburst galaxy M82 reveal a collection of bright sources that may be the result of matter accreting onto intermediate-mass black holes. The black holes are probably young, have masses between 100 and 1000 times the mass of the Sun, and lie relatively far from the center of M82. (Subaru; NASA)

Do Black Holes Exist?

So have black holes really been discovered? The answer is probably yes. Skepticism is healthy in science, but only the most stubborn astronomers (and some do exist!) would take strong issue with the reasoning that supports the case for black holes. Can we guarantee that future modifications to the theory of compact objects will not invalidate our arguments? No, but similar statements could be made in many other areas of astronomy—indeed, about almost any theory in any area of science. We conclude that, strange as they are, black holes have been detected in our Galaxy and beyond. Perhaps someday future generations of space travelers will visit Cygnus X-1 or the center of our Galaxy and (carefully!) test these conclusions firsthand. Until then, we will have to continue to rely on improving theoretical models and observational techniques to guide our discussions of these mysterious and intriguing objects.

◀◀◀ **CONCEPT CHECK**

How do astronomers "see" black holes?

CHAPTER REVIEW
Summary

1 A core-collapse supernova may leave behind an ultracompressed ball of material called a **neutron star** (p. 346). This is the remnant of the inner core that rebounded and blew the rest of the star apart. Neutron stars are extremely dense and, at formation, are predicted to be extremely hot, strongly magnetized, and rapidly rotating. They cool down, lose much of their magnetism, and slow down as they age.

2 According to the **lighthouse model** (p. 347), neutron stars, because they are magnetized and rotating, send regular bursts of electromagnetic energy into space. The beams are produced by charged particles confined by the strong magnetic fields. When we can see the beams from Earth, we call the source neutron star a **pulsar** (p. 346). The pulse period is the rotation period of the neutron star.

3 A neutron star that is a member of a binary system can draw matter from its companion, forming an accretion disk, which is usually a strong source of X-rays. As gas builds up on the star's surface, it eventually becomes hot enough to fuse hydrogen. When hydrogen burning starts on a neutron star, it does so explosively, and an **X-ray burster** (p. 350) results. The rapid rotation of the inner part of the accretion disk causes the neutron star to spin faster as new gas arrives on its surface. The eventual result is a very rapidly rotating neutron star—a **millisecond pulsar** (p. 350). Many millisecond pulsars are found in the hearts of old globular clusters. They cannot have formed recently, and must have been spun up by interactions with other stars. Careful analysis of the radiation received has shown that some millisecond pulsars are orbited by planet-sized objects.

4 **Gamma-ray bursts** (p. 352) are very energetic flashes of gamma rays that occur about once a day and are distributed uniformly over the entire sky. In some cases, their distances have been measured, placing them at very large distances and implying that they are extremely luminous. The leading theoretical models for these explosions involve the violent merger of neutron stars in a distant binary system, or the recollapse and subsequent violent explosion following a "failed" supernova in a very massive star.

5 Einstein's special theory of relativity deals with the behavior of particles moving at speeds comparable to the speed of light. It agrees with Newton's theory at low velocities, but makes many very different predictions for high-speed motion. All of its predictions have been repeatedly verified by experiment. The modern replacement for Newtonian gravity is Einstein's **general theory of relativity** (p. 357), which describes gravity in terms of the warping, or bending, of space by the presence of mass. The more mass, the greater the warping. All particles—including photons—respond to that warping by moving along curved paths.

6 The upper limit on the mass of a neutron star is about three solar masses. Beyond that mass, the star can no longer support itself against its own gravity, and it collapses to form a **black hole** (p. 357), a region of space from which nothing can escape. The most massive stars, after exploding in a supernova, form black holes rather than neutron stars. Conditions in and near black holes can only be described by general relativity. The radius at which the escape speed from a collapsing star equals the speed of light is called the **Schwarzschild radius** (p. 358). The surface of an imaginary sphere centered on the collapsing star and having a radius equal to the star's Schwarzschild radius is called the **event horizon** (p. 358).

7 To a distant observer, light leaving a space-ship that is falling into a black hole would be subject to **gravitational redshift** (p. 364) as the light climbed out of the hole's intense gravitational field. At the same time, a clock on the spaceship would show **time dilation** (p. 365)—the clock would appear to slow down as the ship approached the event horizon. The observer would never see the ship reach the surface of the hole. Once within the event horizon, no known force can prevent a collapsing star from contracting all the

way to a point-like **singularity** (p. 365), at which point both the density and the gravitational field of the star become infinite. This prediction of relativity theory has yet to be proved. Singularities are places where the known laws of physics break down.

8 Once matter falls into a black hole, it can no longer communicate with the outside. However, on its way in, it can form an accretion disk and emit X-rays. The best place to look for a black hole is in a binary system in which one component is a compact X-ray source. Cygnus X-1, a well-studied X-ray source in the constellation Cygnus, is a long-standing black-hole candidate. Studies of orbital motions imply that some binaries contain compact objects too massive to be neutron stars, leaving black holes as the only alternative. There is also substantial evidence for more massive black holes residing in or near the centers of many galaxies, including our own.

Review and Discussion

1. How does the way in which a neutron star forms determine some of its most basic properties?

2. What would happen to a person standing on the surface of a neutron star?

3. Why aren't all neutron stars seen as pulsars?

4. What are X-ray bursters?

5. Why do astronomers think that gamma-ray bursts are very energetic?

6. What is the favored explanation for the rapid spin rates of millisecond pulsars?

7. Why do you think astronomers were surprised to find a pulsar with a planetary system?

8. What does it mean to say that the measured speed of a light beam is independent of the motion of the observer?

9. Use your knowledge of escape speed to explain why black holes are said to be "black."

10. What is an event horizon?

11. Why is it so difficult to test the predictions of the theory of general relativity? Describe two tests of the theory.

12. What would happen to someone falling into a black hole?

13. What makes Cygnus X-1 a good black-hole candidate?

14. Imagine that you had the ability to travel at will through the Milky Way Galaxy. Explain why you would discover many more neutron stars than those known to observers on Earth. Where would you be most likely to find these objects?

15. Do you think that planet-sized objects discovered in orbit around a pulsar should be called planets? Why or why not?

Conceptual Self-Test: True or False?/Multiple Choice

1. Newly formed neutron stars have extremely strong magnetic fields. **(T/F)**

2. All millisecond pulsars are now, or once were, members of binary-star systems. **(T/F)**

3. The fact that gamma-ray bursts are so distant means that they must be very energetic events. **(T/F)**

4. All things, except light, are attracted by gravity. **(T/F)**

5. According to general relativity, space is warped, or curved, by matter. **(T/F)**

6. Although visible light cannot escape from a black hole, high-energy radiation, like gamma rays, can. **(T/F)**

7. Thousands of black holes have now been identified in our Galaxy. **(T/F)**

8. A neutron star is about the same size as (a) a school bus; (b) a U.S. city; (c) the Moon; (d) Earth.

9. The most rapidly "blinking" pulsars are those that (a) spin fastest; (b) are oldest; (c) are most massive; (d) are hottest.

10. The X-ray emission from a neutron star in a binary system comes mainly from (a) the hot surface of the neutron star itself; (b) heated material in an accretion disk around the neutron star; (c) the neutron star's magnetic field; (d) the surface of the companion star.

11. Gamma-ray bursts are observed to occur (a) mainly near the Sun; (b) throughout the Milky Way galaxy; (c) approximately uniformly over the entire sky; (d) near pulsars.

12. If the Sun were magically to turn into a black hole of the same mass, (a) Earth would start to spiral inward; (b) Earth's orbit would remain unchanged; (c) Earth would fly off into space; (d) Earth would be torn apart by the black hole's gravity.

13. The best place to search for black holes is in a region of space that (a) is dark and empty; (b) has recently lost some stars; (c) has strong X-ray emission; (d) is cooler than its surroundings.

14. According to Figure 13.3 (Pulsar Model), the beam of a pulsar is emitted (a) along the rotation axis; (b) in the equatorial plane; (c) from the magnetic poles; (d) in a single direction in space.

15. Figure 13.10 (Gamma-Ray Bursts) shows that gamma-ray bursts (a) are all very similar; (b) sometimes last for less than a second; (c) are confined to the Galactic plane; (d) come from great distances.

Problems

Algorithmic versions of these questions are available in the Practice Problems module of the Companion Website.

The number of squares preceding each problem indicates its approximate level of difficulty.

1. ■ The angular momentum of a spherical body is proportional to the body's angular speed times the square of its radius. ∞ *(More Precisely 4-1)* Using the law of conservation of angular momentum, estimate how fast a collapsed stellar core would spin if its initial spin rate was one revolution per day and its radius decreased from 10,000 km to 10 km.

2. ■ What would your mass be if you were composed entirely of neutron star material of density 3×10^{17} kg/m^3? (Assume that your average density is 1000 kg/m^3.) Compare this with the mass of (a) the Moon; (b) a typical 1-km diameter asteroid.

3. ■ Calculate the surface gravity (relative to Earth's gravity of 9.8 m/s^2) and the escape speed of a 1.4-solar mass neutron star with a radius of 10 km. What would be the escape speed of a 1-solar mass object with a radius of 3 km? ∞ *(More Precisely 5-1)*

4. ■■ Use the radius–luminosity–temperature relation to calculate the luminosity of a 10-km-radius neutron star for temperatures of 10^5 K, 10^7 K, and 10^9 K. What do you conclude about the visibility of neutron stars? Could the coolest of them be plotted on our H–R diagram?

5. ■■ A gamma-ray detector of area 0.5 m^2 observing a gamma-ray burst records photons having total energy 10^{-8} joules. If the burst occurred 1000 Mpc away, calculate the total amount of energy it released (assuming that the energy was emitted equally in all directions). How would this figure change if the burst occurred 10,000 pc away instead, in the halo of our galaxy? What if it occurred within the Oort cloud of our own solar system, at a distance of 50,000 AU?

6. ■■■ An unstable elementary particle is known to decay into other particles in 2 μs, as measured in a laboratory where the particle are at rest. A beam of such particles is accelerated to a speed of 99.99 percent of the speed of light. How long do the particles in the beam take to decay, in the laboratory frame of reference?

7. ■ Supermassive black holes are thought to exist in the centers of some galaxies. What would be the Schwarzschild radii of black holes of one million and one billion solar masses? How does the first black hole compare in size with the Sun? How does the second compare in size with the solar system?

8. ■■ Calculate the tidal acceleration on a 2-m-tall human falling feet first into a one-solar mass black hole—that is, compute the difference in the accelerations (forces per unit mass) on his head and his feet just before his feet cross the event horizon. Repeat the calculation for a one million-solar mass black hole and for a one billion-solar mass black hole (see the previous question). Compare these accelerations with the acceleration due to gravity on Earth (9.8 m/s^2).

9. ■■■ Endurance tests suggest that the human body cannot withstand stress greater than about 10 times the acceleration due to gravity on Earth's surface. At what distance from a one-solar mass black hole would the human in the previous question be torn apart? Calculate the minimum mass of a black hole for which an infalling human could just reach the event horizon intact.

10. ■■ Using the data given in the text (assume the upper limit on the stated range for the black hole mass), calculate the orbital separation of Cygnus X-1 and its B-type stellar companion. ∞ *(More Precisely 1-1)*

The Companion Website at www.prenhall.com/chaisson provides algorithmically generated versions of each chapter's Problems, in addition to an Animations and Videos gallery, a Tutorials gallery and Tutorials Quiz questions, a full eBook, an Images gallery, Practice Questions, Multiple Choice, and True or False quizzes, and Labeling exercises.

PART 4 Galaxies and the Universe

Galaxy cluster ~10^{23} m

P arts 1, 2, and 3 of our text have explored the inner workings of atoms, the origin and evolution of our solar system, and the births and deaths of stars. Our studies have spanned a truly vast range in scale—a factor of a trillion (10^{12}) from atomic nuclei to humans, and another factor of a trillion out to the size of Jupiter's orbit. Even with this jump of a trillion trillion—24 powers of ten—we are still well within the confines of our solar system.

Roughly another factor of a million is needed to reach the dimensions of some of the largest structures within our Milky Way Galaxy, such as the rich star clusters. And yet it is nearly *another* factor of a million in size before encountering notable groups of galaxies well beyond our own. Yet Part 4 goes beyond that: Here we reach the limits of the observable universe, a hundred times yet again larger than the gargantuan galaxy clusters—a total, incredible range of about 10^{38} in size and scale, from subatomic elementary particles to the biggest structures known in the universe.

The Milky Way itself—our home Galaxy—is altogether a thousand times larger than

A single galaxy has more stars in it than the number of people who have ever lived on Earth.

Large-scale structure ~10^{25} m

anything we have thus far considered, such as giant molecular clouds about 100 light-years across. Thought to be the center of immensities less than a century ago, our Galaxy is now known to be but one of billions of others strewn throughout deep space, each galaxy spanning some 100,000 light-years across.

These are the basic "building blocks" of the ultimate "big picture" presented in this book. The galaxies, which are huge collections of roughly a hundred billion (10^{11}) stars, themselves number an estimated hundred billion, and many of these galaxies clustered into much, much greater patterns that top the hierarchy of material structures anywhere. It is here, on the largest scales of all, that we enter the realm of fundamental inquiry that has preoccupied sages and philosophers since the dawn of recorded time.

How big is the universe? What is it made of? Where did it come from? How will it end? And are we the only intelligent life forms in all this vast universe of space and time, or do we share it with alien beings as yet unknown?

Active galaxy ~10^{22} m

Dwarf galaxy ~10^{20} m

14

The Milky Way Galaxy

A Grand Design

Looking up on a dark, clear night, we are struck by two aspects of the night sky. The first is that the individual stars we see are roughly uniformly distributed in all directions. They all lie relatively close to us, mapping out the local galactic neighborhood within a few hundred parsecs of the Sun. But this is only a local impression. Ours is a rather provincial view. Beyond those nearby stars, the second thing we notice is a fuzzy band of light—the Milky Way—stretching across the heavens. From the Northern Hemisphere, this band is most easily visible in the summertime, arcing high above the horizon. Its full extent forms a great circle that encompasses the entire celestial sphere. This is the insider's view of the galaxy in which we live, the blended light of countless distant stars. As we consider much larger volumes of space, on scales far, far greater than the distances between neighboring stars, a new level of organization becomes apparent as the large-scale structure of the Milky Way Galaxy is revealed.

Stars cluster into gargantuan assemblages called galaxies, of which our Milky Way Galaxy is just one among more than a hundred billion others. Our Sun sits in the suburbs of the Milky Way, nearly 30,000 light-years from its central regions, which appear to be a good deal more violent than the part of the Galaxy in which we live. Here displayed is the magnificent sprial galaxy, M81, which resides about 12 million light-years away, and whose size, shape, and mass resemble those of our own Galaxy. This is a false-color infrared view, taken by the *Spitzer Space Telescope* and highlighting (in red) glowing dust in the arms. (JPL) ▷

LEARNING GOALS

Studying this chapter will enable you to:

1. Describe the overall structure of the Milky Way Galaxy, and specify how the various regions differ from one another.

2. Explain the importance of variable stars in determining the size and shape of our Galaxy.

3. Describe the orbital paths of stars in different regions of the Galaxy, and explain how these motions are accounted for by our understanding of how our Galaxy formed.

4. Discuss some possible explanations for the existence of the spiral arms observed in our own and many other galaxies.

5. Explain what studies of Galactic rotation reveal about the size and mass of our Galaxy, and discuss the possible nature of dark matter.

6. Describe some of the phenomena observed at the center of our Galaxy.

Visit www.prenhall.com/chaisson for additional annotated images, animations, and links to related sites for this chapter.

14.1 Our Parent Galaxy

A **galaxy** is a gargantuan collection of stellar and interstellar matter—stars, gas, dust, neutron stars, black holes—isolated in space and held together by its own gravity. Astronomers are aware of literally billions of galaxies beyond our own. The particular galaxy we happen to inhabit is known as the **Milky Way Galaxy**, or just "the Galaxy," with a capital "G."

Our Sun lies in a part of the Galaxy known as the *Galactic disk*, an immense, circular, flattened region containing most of our Galaxy's luminous stars and interstellar matter (and virtually everything we have studied so far in this book). Figure 14.1 illustrates how, viewed from within, the Galactic disk appears as a band of light—the *Milky Way*—stretching across our nighttime sky at an angle of about 60° to the celestial equator. Seen from Earth, the center of the disk (indeed, of the entire Galaxy) lies in the direction of the constellation Sagittarius. As indicated in the figure, if we look in a direction away from the Galactic disk (red arrows), relatively few stars lie in our field of view. However, if our line of sight happens to lie within the disk (white and blue arrows), we see so many stars that their light merges into a continuous blur.

Paradoxically, although we can study individual stars and interstellar clouds near the Sun in great detail, our location within the Galactic disk makes deciphering our Galaxy's large-scale structure from Earth a very difficult task—a little like trying to unravel the layout of paths, bushes, and trees in a city park without being able to leave one particular park bench. In some directions, the interpretation of what we see is ambiguous and inconclusive. In others, foreground objects completely obscure our view of what lies beyond. As a result, astronomers who study the Milky Way Galaxy are often guided in their efforts by comparisons with more distant, but much more easily observable, galaxies. Conversely, there is every reason to suppose that *all* the features of our Galaxy described in this chapter are shared by billions of other galaxies across the universe.

Figure 14.2 shows three **spiral galaxies** thought to resemble our own in overall structure. Figure 14.2(a) is the Andromeda Galaxy, the nearest major galaxy to the Milky Way Galaxy, lying about 800 kpc (about 2.5 million light-years) away. Andromeda's apparent elongated shape is a consequence of the angle at which we happen to view it. In fact, this galaxy, like our own, consists of a thin circular **galactic disk** of matter that fattens to a **galactic bulge**

(a) Artist's view of Milky Way from afar

(b) Real image of Milky Way from inside

◀ **Figure 14.1 Galactic Plane** (a) Gazing from Earth toward the Galactic center (white arrow) in this artist's conception, we see myriad stars stacked up within the thin band of light known as the Milky Way. Looking in the opposite direction (blue arrow), we still see the Milky Way band, but now it is fainter because our Sun lies far from the Galactic center, so we see more stars when looking toward the center than when looking in the opposite direction. Looking perpendicular to the disk (red arrows), we see far fewer stars. (b) This is an optical view of the sky all around us; the white band is the disk of our Milky Way Galaxy—dimly seen with the naked eye from very dark locations on Earth. This is what one sees looking toward the Galactic center—in the direction of the white arrow in part (a). (Lund Observatory)

(b)

(c)

▲ **Figure 14.2 Spiral Galaxies** (a) The Andromeda Galaxy probably resembles fairly closely the overall layout of our own Milky Way Galaxy. The disk and bulge are clearly visible in this image, which is about 30,000 pc across. The faint halo stars cannot be seen. The white stars sprinkled all across this image are not part of Andromeda's halo. They are foreground stars in our own Galaxy, lying in the same region of the sky as Andromeda but about a thousand times closer. (b) This galaxy, catalogued as M101 (the 101st object in the Messier Catalog of astronomical objects) and seen nearly face-on, is somewhat similar in its overall structure to our own Milky Way Galaxy and Andromeda. (c) The galaxy NGC 4565 (part of the New General Catalog of objects) happens to be oriented in such a way that we see it edge-on, allowing us to see clearly its disk and central bulge. (R. Gendler)

at the center. The disk and bulge are embedded in a roughly spherical ball of faint old stars known as the **galactic halo**. These three basic galactic regions are indicated on the figure (the halo stars are so faint that they cannot be discerned here; see Figure 14.9). Figures 14.2(b) and (c) show views of two other galaxies—one seen face-on, the other edge-on—that illustrate these points more clearly.

◀◀◀ CONCEPT CHECK

Why do we see the Milky Way as a band of light across the sky?

14.2 Measuring the Milky Way

Before the twentieth century, astronomers' conception of the cosmos differed markedly from the modern view. The growth in our knowledge of our Galaxy, as well as the realization that there are many other distant galaxies similar to our own, has gone hand in hand with the development of the cosmic distance scale.

Star Counts

In the late 18th century, long before the distances to any stars were known, the English astronomer William Herschel tried to estimate the size and shape of our Galaxy simply by counting how many stars he could see in different directions in the sky. Assuming that all stars were of about equal brightness, he concluded

◄ **Figure 14.3 Herschel's Galaxy Model** Eighteenth-century English astronomer William Herschel constructed this "map" of the Galaxy by counting the numbers of stars he saw in different directions in the sky. Our Sun (marked by the large yellow dot) appears to lie near the center of the distribution, and the long axis of the diagram lies in the plane of the Galactic disk.

that the Galaxy was a somewhat flattened, roughly disk-shaped collection of stars lying in the plane of the Milky Way, with the Sun at its *center* (Figure 14.3). Subsequent refinements to this approach led to essentially the same picture. Early in the twentieth century some researchers went so far as to estimate the dimensions of this "Galaxy" as about 10 kpc in diameter by 2 kpc thick.

In fact, the Milky Way Galaxy is several tens of kiloparsecs across, and the Sun lies far from the center. The flaw in the above reasoning is that the observations were made in the visible part of the electromagnetic spectrum, and astronomers failed to take into account the absorption of visible light by the then unknown interstellar gas and dust. ∞ *(Sec. 11.1)* Only in the 1930s did astronomers begin to realize the true extent and importance of the interstellar medium.

The apparent falloff in the density of stars with distance in the plane of the Milky Way is not a real thinning of their numbers in space, but simply a consequence of the murky environment in the Galactic disk. Objects in the disk lying more than a few kiloparsecs away are hidden from our view by the effects of interstellar absorption. The long "fingers" in Herschel's map are directions where the obscuration happens to be a little less severe than in others.

The falloff in density perpendicular to the disk is real. Radiation coming to us from above or below the plane of the Galaxy, where there is less gas and dust along the line of sight, arrives on Earth relatively unscathed. There is still some patchy obscuration, but the Sun happens to lie in a location where the view out of the disk is largely unimpeded by nearby interstellar clouds.

Observations of Variable Stars

An important by-product of the laborious effort to catalog stars around the turn of the twentieth century was the systematic study of **variable stars**. These are stars whose luminosities change significantly over relatively short periods of time—some quite erratically, others more regularly. Only a small fraction of stars fall into this category, but those that do are of great astronomical significance, as we will see in a moment.

We have encountered several examples of variable stars in earlier chapters. Often, the variability is the result of membership in a binary system. Eclipsing binaries and novae are cases in point. However, sometimes the variability is an intrinsic property of the star itself, and is not dependent on its being a part of a binary. Particularly important to Galactic astronomy are the **pulsating variable stars**,[1] which vary cyclically in luminosity in very characteristic ways. Two types of pulsating variable stars that have played central roles in revealing both the true extent of our Galaxy and the distances to our galactic neighbors are the **RR Lyrae** and **Cepheid** variables. (Following long-standing astronomical practice, the names come from the first star of each class to be discovered—in this case the variable star labeled RR in the constellation Lyra and the variable star Delta Cephei, the fourth brightest star in the constellation Cepheus.)

RR Lyrae and Cepheid variable stars are recognizable by the characteristic shapes of their light curves. RR Lyrae stars all pulsate in essentially similar ways (Figure 14.4a), with only small differences in period (the time from peak to

▲ **Figure 14.4 Variable Stars** (a) Light curve of the pulsating variable star RR Lyrae. All RR Lyrae-type variables have essentially similar light curves, with periods of less than a day. (b) The light curve of a Cepheid variable star called WW Cygni. (c) This Cepheid is shown here (boxed) on successive nights, near its maximum and minimum brightness; two photos, one from each night, were superposed and then slightly displaced. (Harvard College Observatory)

---

[1]Note that these stars have *nothing* whatsoever to do with pulsars. ∞ *(Sec. 13.2)* Pulsars are rapidly rotating ultracompact neutron stars, while pulsating variable stars vary in luminosity due to changes in their internal structure.

▶ **Figure 14.5 Variable Stars on the H–R Diagram** Pulsating variable stars are found in the instability strip of the H–R diagram. As a high-mass star evolves through the strip, it becomes a Cepheid variable. Low-mass horizontal-branch stars in the instability strip are RR Lyrae variables.

peak) between one RR Lyrae variable and another. Observed periods range from about 0.5 to 1 day. Cepheid variables also pulsate in distinctive ways (the regular "sawtooth" pattern in Figure 14.4b), but different Cepheids can have very different pulsation periods, ranging from about 1 to 100 days. In either case, the stars can be recognized and identified *just by observing the variations in the light they emit.*

Pulsating variable stars are normal stars experiencing a brief (perhaps a few million years) period of instability as a natural part of stellar evolution. The conditions necessary to cause pulsations are not found in main-sequence stars. They occur in post-main-sequence stars as they evolve through a region of the Hertzsprung–Russell diagram known as the *instability strip* (Figure 14.5). When a star's temperature and luminosity place it in this strip, the star becomes internally unstable, and both its temperature and its radius vary in a regular way, causing the pulsations we observe. As the star brightens, its surface becomes hotter and its radius shrinks; as its luminosity decreases, the star expands and cools. Cepheid variables are high-mass stars evolving across the upper part of the H–R diagram. ∞ *(Sec. 12.4)* The less luminous RR Lyrae variables are lower-mass horizontal-branch stars that happen to lie within the lower portion of the instability strip. ∞ *(Sec. 12.2)*

A New Yardstick

The importance of pulsating variable stars to Galactic astronomy lies in the fact that once we recognize a star as being of the RR Lyrae or Cepheid type, we can infer its luminosity, and that in turn allows us to measure its distance. The distance calculation is the same as that presented in Chapter 10 during our discussion of spectroscopic parallax. ∞ *(Sec. 10.6)* Comparing the star's (known) luminosity with its (observed) apparent brightness yields an estimate of its distance, by the inverse-square law: ∞ *(Sec. 10.2)*

$$\text{apparent brightness} \propto \frac{\text{luminosity}}{\text{distance}^2}.$$

In this way, astronomers can use pulsating variables as a means of determining distances, both within our own Galaxy and far beyond.

How do we infer a variable star's luminosity? For RR Lyrae variables, this is simple. All such stars have basically the same luminosity (averaged over a complete pulsation cycle)—about 100 times that of the Sun. For Cepheids, we make use of a close correlation between average luminosity and pulsation period, discovered in 1908 by Henrietta Leavitt of Harvard University (see *Discovery 14-1*, p. 382) and known simply as the **period–luminosity relationship.** Cepheids that vary slowly—that is, have long periods—have high luminosities, while short-period Cepheids have low luminosities.

Figure 14.6 illustrates the period–luminosity relationship for Cepheids found within a thousand parsecs or so of Earth. Astronomers can plot such a diagram for relatively nearby Cepheid variables because they can measure the stars' distances, and hence their luminosities, using stellar or spectroscopic parallax. Thus, a simple measurement of a Cepheid variable's pulsation period immediately tells us its luminosity—we just read it off the graph in Figure 14.6. (The roughly constant luminosities of the RR Lyrae variables are also indicated in the figure.)

Animation/Video
Cepheid Star in Distant Galaxy

▲ **Figure 14.6 Period–Luminosity Plot** A plot of pulsation period versus average absolute brightness (that is, luminosity) for a group of Cepheid variable stars. The two properties are quite tightly correlated. The pulsation periods of some RR Lyrae variables are also shown.

This distance-measurement technique works well provided the variable star can be clearly identified and its pulsation period measured. With Cepheids, this method allows astronomers to estimate distances out to about 25 million parsecs, more than enough to take us to the nearest galaxies. Indeed, the existence of galaxies beyond our own was first established in the late 1920s, when American astronomer Edwin Hubble observed Cepheids in the Andromeda Galaxy and thereby succeeded in measuring its distance. The less luminous RR Lyrae stars are not as easily detectable as Cepheids, so their usable range is not as great. However, they are much more common, so within this limited range, they are actually more useful than Cepheids.

Figure 14.7 extends our cosmic distance ladder, begun in Chapter 1 with radar ranging in the solar system and expanded in Chapter 10 to include stellar and spectroscopic parallax, by adding variable stars as a fourth method of determining distance.

CONCEPT CHECK ▶▶▶

Can variable stars be used to map out the structure of the Galactic disk?

The Size and Shape of Our Galaxy

Many RR Lyrae variables are found in globular clusters, those tightly bound swarms of old, reddish stars we first met in Chapter 11. ∞ *(Sec. 11.6)* Early in the twentieth century, the American astronomer Harlow Shapley used observations of RR Lyrae stars to make two very important discoveries about the Galactic globular cluster system. First, he showed that most globular clusters reside at great distances—many thousands of parsecs—from the Sun. Second, by measuring the direction and distance of each cluster, he was able to determine their three-dimensional distribution in space (Figures 14.8 and 14.9). In this way, Shapley demonstrated that the globular clusters map out a truly gigantic, and roughly *spherical*, volume of space, now known to be about 30 kpc across.[2] However, the center of the distribution lies nowhere near our Sun. It is located nearly 8 kpc away from us, in the constellation Sagittarius.

In a brilliant intellectual leap, Shapley realized that the distribution of globular clusters maps out the true extent of stars in the Milky Way Galaxy—the region that we now call the Galactic halo. The hub of this vast collection of matter, 8 kpc from the Sun, is the **Galactic center**. As illustrated in Figure 14.9, we live in the suburbs of this huge ensemble, in the Galactic disk—the thin sheet of young stars, gas, and dust that cuts through the center of the halo. Since Shapley's time, astronomers have identified many individual stars—that is, stars not belonging to any globular cluster—within the Galactic halo.

Shapley's bold interpretation of the globular clusters as defining the overall distribution of stars in our Galaxy was an enormous step forward in human understanding of our place in the universe. Five hundred years ago Earth was considered the center of all things. Copernicus argued otherwise, demoting our planet to an undistinguished place removed from the center of the solar system. In Shapley's time the prevailing view was that our Sun was the center not only of the Galaxy but also of the universe. Shapley showed otherwise. With his observations of globular clusters, he simultaneously increased the size of our Galaxy by almost a factor of 10 over earlier estimates and banished our parent Sun to its periphery, virtually overnight!

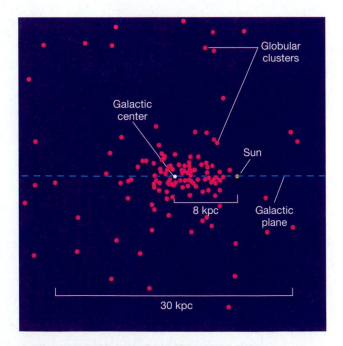

▲ **Figure 14.8 Globular Cluster Distribution**
Our Sun does not coincide with the center of the very large collection of globular clusters (indicated by the pink dots). Instead, more globular clusters are found in one direction than in any other. The Sun resides closer to the edge of the collection, which measures roughly 30 kpc across. We now know that the globular clusters outline the true distribution of stars in the Galactic halo.

[2]The Galactic globular cluster system and the Galactic halo of which it is a part are somewhat flattened in the direction perpendicular to the disk, but the degree of flattening is uncertain. The halo is certainly much less flattened than the disk, however.

◀ **Figure 14.9 Stellar Populations in Our Galaxy**
Artist's conception of a (nearly) edge-on view of the Milky Way Galaxy, showing schematically the distributions of young blue stars, open clusters, old red stars, and globular clusters. (The brightness and size of our Sun are greatly exaggerated for clarity.)

14.3 Galactic Structure

Mapping the Galaxy

Based on optical, infrared, and radio studies of our Galaxy, the very different spatial distributions of the disk, bulge, and halo components of the Milky Way Galaxy are shown in Figure 14.9. The extent of the halo is based largely on optical observations of globular clusters and other halo stars. However, as we have seen, optical techniques can cover only a small portion of the dusty Galactic disk. Much of our knowledge of the structure of the disk on larger scales is based on radio observations, particularly of the 21-cm radio emission line produced by atomic hydrogen. ∞ *(Sec. 11.3)* Because long-wavelength radio waves are largely unaffected by interstellar dust, and hydrogen is by far the most abundant element in interstellar space, the 21-cm signals are strong enough that virtually the entire disk can be observed in this way.

According to the radio studies, the center of the gas distribution coincides roughly with the center of the globular cluster system, lying about 8 kpc from the Sun. In fact, this figure is derived most accurately from radio observations of Galactic gas. The densities of both stars and gas in the disk decline quite rapidly beyond about 15 kpc from the Galactic center (although some radio-emitting gas has been observed out to at least 50 kpc).

Perpendicular to the Galactic plane, the disk in the vicinity of the Sun is "only" about 300 pc thick, or about 1/100 of the 30-kpc Galactic diameter. Don't be fooled, though. Even if you could travel at the speed of light, it would take you 1000 years to traverse the thickness of the Galactic disk. The disk may be thin compared with the Galactic diameter, but it is huge by human standards.

Also shown in Figure 14.9 is our Galaxy's central bulge, measuring roughly 6 kpc across in the plane of the Galactic disk by 4 kpc perpendicular to that plane. Obscuration by interstellar dust makes it difficult to study the bulge at optical wavelengths. However, at longer wavelengths, which are less affected by interstellar matter, a much clearer picture emerges (Figure 14.10; compare Figures 14.1b and 14.2c). Detailed measurements of the motion of gas and

▼ **Figure 14.10 Infrared View of the Milky Way Galaxy**   A wide-angle infrared image of the disk and bulge of the Milky Way Galaxy, as observed by the Two Micron All Sky Survey. Compare Figure 14.2(c). (UMass/Caltech)

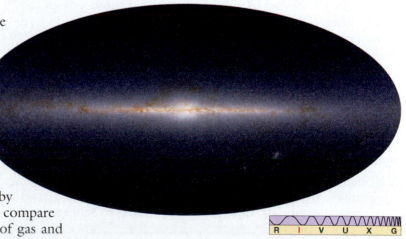

R I V U X G

stars in and near the bulge imply that it is actually football shaped, about half as wide as it is long, with the long axis of the football lying in the Galactic plane. (In this case, our Galaxy is probably of the "barred-spiral" type—see Section 15.1.)

Stellar Populations

Aside from their shapes, the three components of the Galaxy—disk, bulge, and halo—have other properties that distinguish them from one another. First, the halo contains essentially *no* gas or dust—just the opposite of the disk and bulge, in which interstellar matter is common. Second, there are clear differences in *color* between disk, bulge, and halo stars. Stars in the Galactic bulge and halo are distinctly redder than those found in the disk. Observations of other spiral galaxies also show this trend—the blue-white tint of the disk and the yellowish coloration of the bulge are evident in Figures 14.2(a) and (c).

All the bright, blue stars visible in our night sky are part of the Galactic disk, as are the young open star clusters and star-forming regions. In contrast, the cooler, redder stars—including those found in the old globular clusters—are more uniformly distributed throughout the disk, bulge, and halo. Galactic disks appear bluish because main-sequence O- and B-type blue supergiants are very much brighter than G-, K-, and M-type dwarfs, even though the dwarfs are present in far greater numbers.

The explanation for the marked difference in stellar content between disk and halo is that, whereas the gas-rich Galactic disk is the site of ongoing star formation and thus contains stars of all ages, all the stars in the Galactic halo

14-1 ISCOVERY

Early Computers

A large portion of the early research in observational astronomy focused on monitoring stellar luminosities and analyzing stellar spectra. Much of this pioneering work was done using photographic methods. What is not so well known is that most of the labor was accomplished by women. Around the turn of the twentieth century, a few dozen dedicated women—assistants at the Harvard College Observatory—created an enormous database by observing, sorting, measuring, and cataloging photographic information that helped form the foundation of modern astronomy. Some of them went far beyond their duties in the lab to make many of the basic astronomical discoveries often taken for granted today.

This 1910 photograph shows several of those women carefully examining star images and measuring variations in luminosity or wavelengths of spectral lines. In the cramped quarters of the Harvard Observatory these women inspected image after image to collect a vast body of data on hundreds of thousands of stars. Note the plot of stellar luminosity changes pasted on the wall at the left. The cyclical pattern is so regular that it likely belongs to a Cepheid variable. Known as "computers" (there were no electronic devices then), these women were paid 25 cents an hour.

In 1880 these workers began a survey of the skies that would continue for half a century. Their first major accomplishment was a catalog of the brightnesses and spectra of tens of thousands of stars, published in 1890 under the direction of

Williamina Fleming. On the basis of this compilation, several of these women made fundamental contributions to astronomy. In 1897 Antonia Maury undertook the most detailed study of stellar spectra to that time, enabling Hertzsprung and Russell independently to develop what is now called the H–R diagram. In 1898 Annie Cannon proposed the spectral classification system (described in Chapter 10) that is now the international standard for categorizing stars. ∞ *(Sec. 10.5)* In 1908 Henrietta Leavitt discovered the period–luminosity relationship for Cepheid variable stars, which later allowed astronomers to recognize our Sun's true position in our Galaxy, as well as our Galaxy's true place in the universe.

▶ **Figure 14.11 Orbital Motion in the Galactic Disk** Stars and interstellar clouds in the neighborhood of the Sun show systematic Doppler motions, implying that the disk of the Galaxy is spinning in a well-ordered way. These four Galactic quadrants are drawn to intersect not at the Galactic center, but at the Sun, the location from which our observations are made. Because the Sun orbits faster than stars and gas at larger radii, we are drawing away from material at top left and gaining on that at top right, resulting in the Doppler shifts indicated. Similarly, stars and gas in the bottom left quadrant are gaining on us, while material at bottom right is pulling away. The curved arrows represent the angular speed of the disk material.

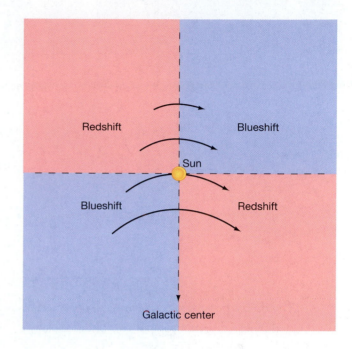

are *old*. The absence of dust and gas in the halo means that no new stars are forming there, and star formation apparently ceased long ago—at least 10 billion years in the past, judging from the ages of the globular clusters. ∞ *(Sec. 12.6)* The gas density is very high in the inner part of the Galactic bulge, making this region the site of vigorous ongoing star formation, and both very old and very young stars mingle there. The bulge's gas-poor outer regions have properties more similar to those of the halo. As we will see in Chapter 15, all these statements are equally true for other spiral galaxies—young, bright stars are always found in the disk and inner bulge because the interstellar medium is densest there.

Spectroscopic studies indicate that halo stars are far less abundant in heavy elements (that is, elements heavier than helium) than are nearby stars in the disk. Each successive cycle of star formation and evolution enriches the interstellar medium with the products of stellar evolution, leading to a steady increase in heavy elements with time. ∞ *(Sec. 12.7)* Thus, the scarcity of these elements in halo stars is consistent with the view that the halo formed long ago.

Astronomers often refer to young disk stars as *Population I* stars and old halo stars as *Population II* stars. The idea of two stellar "populations" dates back to the 1930s, when the differences between disk and halo stars first became clear. It represents something of an oversimplification, as there is actually a continuous variation in stellar ages throughout the Milky Way Galaxy and not a simple division of stars into two distinct "young" and "old" categories. Nevertheless, the terminology is still widely used.

Orbital Motion

Are the internal motions of our Galaxy's members chaotic and random, or are they part of some gigantic "traffic pattern"? The answer depends on our perspective. The motion of stars and clouds we see on small scales (within a few tens of parsecs of the Sun) seems random, but on larger scales (hundreds or thousands of parsecs) the motion is much more orderly.

As we look around the Galactic disk in different directions, a clear pattern of motion emerges (see Figure 14.11). Radiation received from stars and interstellar gas clouds in the upper-right and the lower-left quadrants of Figure 14.11 is generally *blueshifted*. At the same time, radiation from stars and gas sampled in the upper-left and lower-right quadrants tends to be *redshifted*. In other words, some regions of the Galaxy (in the blueshifted directions) are approaching the Sun, while others (the redshifted ones) are receding from us. Careful study of the positions and velocities of stars and gas clouds near the Sun leads us to the conclusion that the entire Galactic disk is *rotating* about the Galactic center. In the vicinity of the Sun, 8 kpc from the center, the orbital speed is about 220 km/s, so material takes about 225 million years to complete one circuit. At other distances from the center, the rotation period is different—shorter closer to the center, longer at greater distances—that is, the Galactic disk rotates not as a solid object but *differentially*.

This picture of orderly circular orbital motion about the Galactic center applies only to the Galactic disk. The old globular clusters in the halo and the faint, reddish individual stars in both the halo and the bulge do *not* share the

▲ **Figure 14.12 Stellar Orbits in Our Galaxy** Stars in the Galactic disk (blue curves) move in orderly, circular orbits about the Galactic center. In contrast, halo stars (orange curves) have orbits with largely random orientations and eccentricities. The orbit of a typical halo star takes it high above the Galactic disk, then down through the disk plane, then out the other side and far below the disk. The orbital properties of bulge stars are intermediate between those of disk stars and those of halo stars.

disk's well-defined rotation. Instead, as shown in Figure 14.12, their orbits are oriented randomly.[3] Although they do orbit the Galactic center, these stars move in all directions, their paths filling an entire three-dimensional volume rather than a nearly two-dimensional disk.

At any given distance from the Galactic center, bulge or halo stars move in their orbits at speeds comparable to the disk's rotational speed at that radius. However, unlike stars in the disk, bulge and halo stars orbit in *all* directions, not just in a roughly circular path confined to a narrowly defined plane. Their orbits carry these stars repeatedly through the disk and out the other side. (They don't collide with stars in the disk because interstellar distances are huge compared with the diameters of individual stars—a star or even an entire star cluster passes through the disk almost as though it wasn't there— see *Discovery 15-1*.)

Recently, astronomers have detected numerous *tidal streams* in the Galactic halo, groups of stars thought to be the remnants of globular clusters and even small satellite galaxies (see Section 15.1) torn apart by our Galaxy's tidal field. Just as micrometeoroid swarms in our solar system follow the orbit of their disrupted parent comet long after the comet itself is gone, stars in a tidal stream are now spread out around the entire original orbit of their parent cluster or galaxy. ∞ *(Sec. 4.2)*

CONCEPT CHECK ▶▶▶

Why do astronomers regard the disk and halo as distinctly different components of our Galaxy?

14.4 The Formation of the Milky Way

Table 14.1 compares some key properties of the three basic components of the Galaxy. Is there some evolutionary scenario that can naturally account for the structure we see today? The answer is yes, and it takes us all the way back to the birth of our Galaxy, more than 10 billion years ago. Not all the details are agreed upon by all astronomers, but the overall picture is now fairly widely accepted. For simplicity we confine our discussion here to the Galactic disk and halo. In many ways the bulge is intermediate in its properties between these two extremes.

Figure 14.13 illustrates the current view of our Galaxy's evolution. It explains, in broad terms at least, the Galactic structure we see today. Not unlike the star-formation scenario outlined in Chapter 11, it starts from a contracting cloud of pregalactic gas. ∞ *(Sec. 11.4)* When the first Galactic stars and globular clusters formed, the gas in our Galaxy had not yet accumulated into a thin disk. Instead, it was spread out over an irregular, and quite extended, region of space, spanning many tens of kiloparsecs in all directions. When the first stars formed, they were distributed throughout this volume (Figure 14.13b). Their distribution today (the Galactic halo) reflects that fact—it is an imprint of their birth. Many astronomers think that the very first stars formed even earlier, in smaller systems that later merged to create our Galaxy (Figure 14.13a). ∞ *(Sec. 16.4)* Probably many more stars were born during the mergers themselves, as interstellar gas clouds collided and began to collapse. The present-day halo would look much the same in either case.

Since those early times, rotation has flattened the gas in our Galaxy into a relatively thin disk (Figure 14.13c). Physically, this process is similar to the flattening of the solar nebula during the formation of the solar system, except on a vastly larger scale. ∞ *(Sec. 4.3)* Star formation in the halo ceased billions of

---

[3]Halo stars may in fact have some small net rotation about the Galactic center, but the rotational component of their motion is overwhelmed by the much larger random component. The motion of bulge stars also has a rotational component, larger than that of the halo but still smaller than the random component of stellar motion within the bulge.

▶ **Figure 14.13** **Milky Way Galaxy Formation**
(a) The Milky Way Galaxy possibly formed through the merger of several smaller systems. (b) Astronomers reason that early on, our Galaxy was irregularly shaped, with gas distributed throughout its volume. When stars formed during this stage, there was no preferred direction in which they moved and no preferred location in which they were found. Their orbits carried them throughout an extended three-dimensional volume surrounding the newborn Galaxy. (c) In time, the gas and dust fell to the Galactic plane and formed a spinning disk. The stars that had already formed were left behind, in the halo. (d) New stars forming in the disk inherit its overall rotation and so orbit the Galactic center on ordered, circular orbits.

years ago when the raw materials—interstellar gas and dust—fell to the Galactic plane. Ongoing star formation in the disk gives it its bluish tint, but the halo's short-lived blue stars have long since burned out, leaving only the long-lived red stars that give it its characteristic pinkish glow (Figure 14.13d). The Galactic halo is ancient, whereas the disk is full of youthful activity.

The chaotic orbits of the halo stars are also explained by this theory. When the halo developed, the irregularly shaped Galaxy was rotating only very slowly, so there was no strongly preferred direction in which matter tended to move. As a result, halo stars were free to travel along nearly any path once they formed (or when their parent systems merged). As the Galactic disk formed, however, conservation of angular momentum caused it to spin more rapidly. Stars forming from the gas and dust of the disk inherit its rotational motion and so move on well-defined circular orbits.

In principle, the structure of our Galaxy bears witness to the conditions that created it. In practice, however, the interpretation of the observations is made difficult by the sheer complexity of the system we inhabit and by the many competing physical processes that have modified its appearance since it formed. As a result, the early stages of the Milky Way are still quite poorly understood. We will return to the subject of galaxy formation in Chapters 15 and 16.

◀◀◀ **CONCEPT CHECK**

Why are there no young halo stars?

TABLE 14.1	Overall Properties of the Galactic Disk, Halo, and Bulge	
GALACTIC DISK	**GALACTIC HALO**	**GALACTIC BULGE**
Highly flattened	roughly spherical—mildly flattened	somewhat flattened—elongated in the plane of the disk (football-shaped)
Contains both young and old stars	contains old stars only	contains both young and old stars; more old stars at greater distances from the center
Contains gas and dust	contains no gas and dust	contains gas and dust, especially in the inner regions
Site of ongoing star formation	no star formation during the last 10 billion years	ongoing star formation in the inner regions
Gas and stars move in circular orbits in the Galactic plane	stars have random orbits in three dimensions	stars have largely random orbits, but with some net rotation about the Galactic center
Spiral arms (Sec. 14.5)	little discernible substructure; globular clusters, tidal streams (Sec. 14.3)	ring of gas and dust near center; central Galactic nucleus (Sec. 14.7)
Overall white coloration, with blue spiral arms	reddish in color	yellow-white

14.5 Galactic Spiral Arms

Radio studies provide perhaps the best direct evidence that we live in a spiral Galaxy. As illustrated in Figure 14.14, radio observations of interstellar gas, repeated in all directions in the Galactic plane, allow astronomers to map out the distribution of interstellar gas and paint a detailed picture of the Galactic disk.

The Galaxy's differential rotation means that gas clouds at different distances are moving at different speeds with respect to Earth, so the radiation received from them is Doppler shifted by different amounts. But how do we translate the Doppler measurements into estimates of the *distances* to the clouds emitting the radiation? Astronomers accomplish this by using all available data, coupled with our knowledge of Newtonian mechanics, to construct a *mathematical model* of the rotation of stars and gas throughout the Galactic disk. ∞ *(Sec. 2.7)* The model allows us to turn a measured velocity into a distance along the line of sight. As in so many areas of astronomy, theory and observations complement one another—the data refine the theoretical model, while the model in turn provides the framework needed to understand and interpret further observations. ∞ *(Sec. E.5)*

Figure 14.15 is an artist's conception (based on observational data) of the appearance of our Galaxy as seen from far above the disk. The figure clearly shows our Galaxy's **spiral arms**, pinwheel-like structures originating close to the Galactic bulge and extending outward throughout much of the Galactic disk. One of these arms, as far as we can tell, wraps around a large part of the disk and contains our Sun. Notice, incidentally, the scale markers on Figures 14.8, 14.9, and 14.15: the Galactic globular-cluster distribution (shown in Figure 14.8), the luminous stellar component of the disk (Figure 14.9), and the known spiral structure (Figure 14.15) all have roughly the same diameter—about 30 kpc. This diameter is fairly typical of spiral galaxies observed elsewhere in the universe.

The spiral arms in our Galaxy are made up of much more than just interstellar gas and dust. Studies of the Galactic disk within a kiloparsec or so of the Sun indicate that young stellar and prestellar objects—emission nebulae, O- and B-type stars, and recently formed open clusters—are also distributed in a spiral pattern that closely follows the distribution of interstellar clouds. The obvious conclusion is that the spiral arms are the part of the Galactic disk where star formation takes place. The brightness of the young stellar objects just listed is the main reason that the spiral arms of other galaxies are easily seen from afar (Figure 14.2b).

A central problem facing astronomers trying to understand spiral structure is how that structure persists over long periods of time. The basic issue is simple: differential rotation makes it impossible for any large-scale structure "tied" to the disk material to survive. Figure 14.16 shows how a spiral pattern consisting always of the same group of stars and gas clouds would necessarily "wind up" and disappear within a few hundred million years. How then do the Galaxy's spiral arms retain their structure over long periods of time despite differential rotation? A leading explanation for the existence of spiral arms holds that they are **spiral density waves**—coiled waves of gas compression that move through the Galactic disk, squeezing clouds of interstellar gas and triggering the process of star formation as they go. ∞ *(Sec. 11.4)* Like traffic

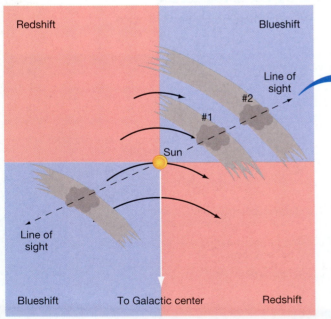

▼ **Figure 14.14 Gas in the Galactic Disk** Because the disk of our Galaxy is rotating differentially, signals from different clumps of hydrogen matter along any given line of sight are Doppler shifted by different amounts, allowing both densities and distances to be measured. The graph at right is a typical radio spectrum, showing how cloud #2, which is part of an outer spiral arm in the upper right quadrant, is blueshifted more than cloud #1, which is in an inner spiral arm. By repeating these observations in many different directions, astronomers map out the distribution of gas in our Galaxy.

▶ **Figure 14.15 Milky Way Spiral Structure** An artist's conception of our Milky Way Galaxy seen face-on. This illustration is based on data accumulated by legions of astronomers during the past few decades and includes radio and infrared maps of gas and dust in the Galactic disk. Painted from the perspective of an observer 100 kpc above the Galactic plane, the spiral arms are at their best-determined positions, apparently emanating from a bar whose length is twice its width. All the features are drawn to scale (except for the oversized yellow dot near the top, which represents our Sun). The two small blotches to the left are dwarf galaxies, called the Magellanic Clouds. We study them in Chapter 15. (L. Chaisson)

30 kpc

slowing as it passes an obstacle on the highway (see *Discovery 14-2*, p. 390), Galactic gas slows down and becomes denser as it passes through the wave. The spiral arms we observe are defined by the denser-than-normal clouds of gas thus created and by the new stars formed as a result of the spiral waves' passage.

This explanation of spiral structure avoids the problem of differential rotation because the wave pattern is not tied to any particular piece of the Galactic disk. The spirals we see are merely patterns moving through the disk, not great masses of matter being transported from place to place. The density wave moves through the collection of stars and gas making up the disk just as a sound wave moves through air or an ocean wave passes through water, compressing different parts of the disk at different times. Even though the rotation rate of the disk material varies with distance from the Galactic center, the wave itself remains intact, defining the Galaxy's spiral arms.

Over much of the visible portion of the Galactic disk (within about 15 kpc of the center), the spiral wave pattern is predicted to rotate *more slowly* than the stars and gas. Thus, as shown in Figure 14.17, Galactic material catches up with the wave, is temporarily slowed down and compressed by the wave's gravitational pull as it passes through, then continues on its way. (For a more down-to-earth example of an analogous process, see *Discovery 14-2*.)

As the density wave enters the arm from behind, the gas is compressed and forms stars. Dust lanes mark the regions of highest-density gas. The most prominent stars—the bright O- and B-type blue giants—live for only a short time, so emission nebulae and young star clusters are found only within the

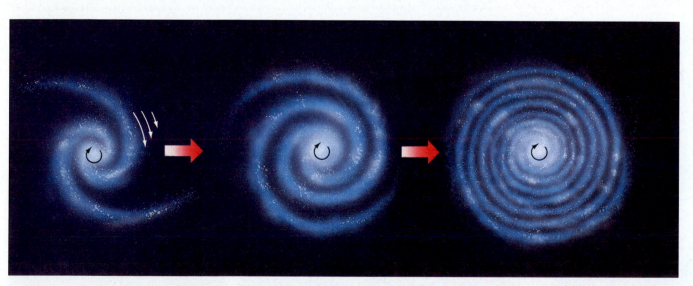

▲ **Figure 14.16 Differential Galactic Rotation** The disk of our Galaxy rotates differentially—stars close to the center take less time to orbit the Galactic center than those at greater distances. If spiral arms were somehow tied to the material of the Galactic disk, this differential rotation would cause the spiral pattern to wind up and disappear in a few hundred million years. Spiral arms would be too short-lived to be consistent with the numbers of spiral galaxies we observe today.

Young O, B stars

Dust Lane

High-density
gas and dust
behind arm

Emission
nebula

Older
stars

Galactic
disk

Spiral arm

→ Spiral arm
rotation

→ Disk/gas
motion

R I V U X G

◄ **Figure 14.17 Density-Wave Theory** This theory
holds that the spiral arms seen in our own and many other
galaxies are waves of gas compression and star formation
moving through the material of the galactic disk. In the
painting at the right, gas motion is indicated by red arrows
and arm motion by white arrows. Gas enters an arm from
behind, is compressed, and forms stars. The spiral pattern is
delineated by dust lanes, regions of high gas density, and
newly formed bright stars. The inset shows the spiral galaxy
NGC 1566, which displays many of the features just
described. (AURA)

arms, near their birthsites, just ahead of the dust
lanes. Their brightness emphasizes the spiral structure. Further downstream, ahead of the spiral arms,
we see mostly older stars and star clusters, which
have had enough time since their formation to out-
distance the wave and pull away from it. Over millions of years their random individual motions,
superimposed on the overall rotation around the
Galactic center, distort and eventually destroy their
original spiral configuration, and they become part
of the general disk population.

Note, incidentally, that although the spirals
shown in Figure 14.17 have two arms each, astronomers are not completely
certain how many arms make up the spiral structure in our own Galaxy (see
Figure 14.15). The theory makes no strong predictions on this point.

An alternative possibility is that the formation of stars drives the waves, instead of the other way around. Imagine a row of newly formed massive stars
somewhere in the disk. The emission nebula created when these stars form, and
the supernovae when they die, send shock waves through the surrounding gas,
possibly triggering new star formation. ∞ *(Sec. 12.7)* Thus, as illustrated in
Figure 14.18(a), the formation of one group of stars provides the mechanism for
the creation of more stars. Computer simulations suggest that it is possible for
the "wave" of star formation thus created to take on the form of a partial spiral

▼ **Figure 14.18 Self-Propagating Star Formation**
(a) In this theory of the formation of spiral arms, the shock
waves produced by the formation and later evolution of a
group of stars provide the trigger for new rounds of star
formation. We have used supernova explosions to illustrate the
point here, but the formation of emission nebulae and
planetary nebulae are also important. (b) This process may
well be responsible for the partial spiral arms seen in some
galaxies, such as this one catalogued as AM0644-741 and
shown here in true color. The distinct blue appearance derives
from the vast numbers of young stars that pepper its ill-defined
spiral arms, which resemble more of a ring. (STScI)

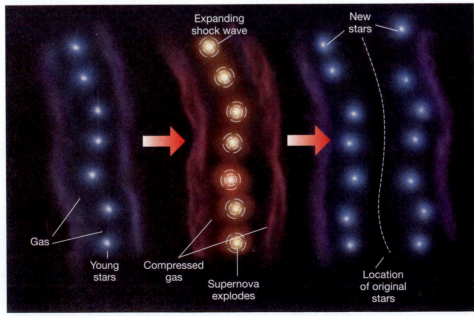

Expanding
shock wave

New
stars

Gas

Young
stars

Compressed
gas

Supernova
explodes

Location
of original
stars

(a)

(b)

R I V U X G

and for this pattern to persist for some time. However, this process, known as **self-propagating star formation,** can produce only pieces of spirals, as are seen in some galaxies (Figure 14.18b). It apparently cannot produce the galaxywide spiral arms seen in other galaxies and present in our own. It may well be that there is more than one process at work in the spectacular spirals we see.

An important question, but one that unfortunately is not answered by either of the two theories just described, is: Where do these spirals come from? What was responsible for generating the density wave in the first place or for creating the line of newborn stars whose evolution drives the advancing spiral arm? Scientists speculate that (1) instabilities in the gas near the galactic bulge, (2) the gravitational effects of nearby galaxies, or (3) the elongated shape of the bulge itself may have had a big enough influence on the disk to get the process going. The fact is that we still don't know for sure how galaxies—including our own—acquire such beautiful spiral arms.

◀◀◀ **CONCEPT CHECK**

Why can't spiral arms simply be clouds of gas and young stars orbiting the Galactic center?

14.6 The Mass of the Milky Way Galaxy

We can measure our Galaxy's mass by studying the motions of gas clouds and stars in the Galactic disk. Recall from Chapter 1 that Kepler's third law (as modified by Newton) connects the orbital period, orbit size, and masses of any two objects in orbit around one another: ∞ *(Sec. 1.4)*

$$\text{total mass(solar masses)} = \frac{\text{orbit size}(AU)^3}{\text{orbit period(years)}^2}.$$

The distance from the Sun to the Galactic center is about 8 kpc and the Sun's orbital period is 225 million years. Substituting these numbers into the equation, we obtain a mass of almost 10^{11} solar masses—100 *billion* times the mass of our Sun.

But what mass have we just measured? When we performed the analogous calculation in the case of a planet orbiting the Sun, there was no ambiguity: neglecting the planet's mass, the result of our calculation was the mass of the Sun. ∞ *(More Precisely 1-1)* However, the Galaxy's matter is not concentrated at the Galactic center (as the Sun's mass is concentrated at the center of the solar system). Instead, Galactic matter is distributed over a large volume of space. What portion of the Galaxy's mass controls the Sun's orbit? Isaac Newton answered this question three centuries ago: the Sun's orbital period is determined by the portion of the Galaxy that lies *within the orbit of the Sun* (Figure 14.19a). This is the mass computed from the above equation.

Dark Matter

Based largely on data obtained from radio observations, Figure 14.19(b) shows the Galactic **rotation curve,** which plots the rotation speed of the Galactic disk against distance from the Galactic center. Using this graph, we can repeat our earlier calculation to compute the total mass that lies within any given distance from the Galactic center. We find, for example, that the mass within about 15 kpc from the center—the volume defined by the globular clusters and the known spiral structure—is roughly 2×10^{11} solar masses, about twice the mass contained within the Sun's orbit. Does most of the matter in the Galaxy "cut off" at this point, where the luminosity drops off sharply? Surprisingly, the answer is no.

If all of the mass of the Galaxy were contained within the edge of the visible structure, Newton's laws of motion predict that the orbital speed of stars and gas

(a)

(b)

▲ **Figure 14.19 Weighing the Galaxy** (a) The orbital speed of a star or gas cloud moving around the Galactic center is determined only by the mass of the Galaxy lying inside the orbit (gray-shaded region). Thus, to measure the Galaxy's total mass, we must look for objects orbiting at large distances from the center. (b) The rotation curve for the Milky Way Galaxy plots rotation speed against distance from the Galactic center. We can use this curve to compute the mass of the Galaxy that lies within any given radius. The dashed curve is the rotation curve expected if the Galaxy "ended" abruptly at a radius of 15 kpc, the limit of most of the known spiral structure and the globular cluster distribution. The fact that the red curve does not follow this dashed line, but instead stays well above it, indicates that there must be additional unseen matter beyond that radius.

beyond 15 kpc would decrease with increasing distance from the Galactic center, just as the orbital speeds of the planets diminish as we move outward from the Sun. ∞ *(More Precisely 1-1)* The dashed line in Figure 14.19(b) indicates how the rotation curve should look in that case. However, the actual rotation curve is quite different. Far from declining at larger distances, it *rises* slightly out to the limits of our measurement capabilities. This implies that the amount of mass contained within successively larger radii continues to grow beyond the orbit of the Sun, apparently out to a distance of at least 50 kpc.

14-2 DISCOVERY

Density Waves

In the late 1960s, American astrophysicists C. C. Lin and Frank Shu proposed a way in which spiral arms in the Galaxy could persist for many Galactic rotations. They argued that the arms themselves contain no "permanent" matter. They should not be viewed as assemblages of stars, gas, and dust moving intact through the disk because such structures would quickly be destroyed by differential rotation. Instead, as described in the text, a spiral arm should be envisaged as a *density wave*—a wave of alternating compression and expansion sweeping through the Galaxy.

A wave in water builds up material temporarily in some places (crests) and lets it down in others (troughs). Similarly, as Galactic matter encounters a spiral density wave, the matter is compressed to form a region of higher than normal density. The matter enters the wave, is temporarily slowed down (by gravity) and compressed as it passes through, then continues on its way. This compression triggers the formation of new stars and nebulae. In this way, the spiral arms are formed and reformed repeatedly, without wrapping up.

The accompanying figure illustrates the formation of a density wave in a more familiar context—a traffic jam triggered by the presence of a repair crew moving slowly down the road. Cars slow down temporarily as they approach the crew, then speed up again as they pass the worksite and continue on their way. The result observed by a traffic helicopter flying overhead

is a region of high traffic density concentrated around the work crew and moving with it. An observer on the side of the road, however, sees that the jam never contains the same cars for very long. Cars constantly catch up to the bottleneck, move slowly through it, then speed up again, only to be replaced by more cars arriving from behind.

The traffic jam is analogous to the region of high stellar density in a Galactic spiral arm. Just as the traffic density wave is not tied to any particular group of cars, the spiral arms are not attached to any particular piece of disk material. Stars and gas enter a spiral arm, slow down for a while, then exit the arm and continue on their way around the Galactic center. The result is a moving region of high stellar and gas density, involving different parts of the disk at different times. Notice also that, just as in our Galaxy, the traffic jam wave moves more slowly than, and independently of, the overall traffic flow.

We can extend our traffic analogy a little further. Most drivers are well aware that the effects of a such a tie-up can persist long after the road crew has stopped work and gone home for the night. Similarly, spiral density waves can continue to move through the disk even after the disturbance that originally produced them has long since subsided. According to spiral density wave theory, this is precisely what has happened in the Milky Way Galaxy. Some disturbance in the past—an encounter with a satellite galaxy, perhaps, or the effect of the central bar—produced a wave that has been moving through the Galactic disk ever since.

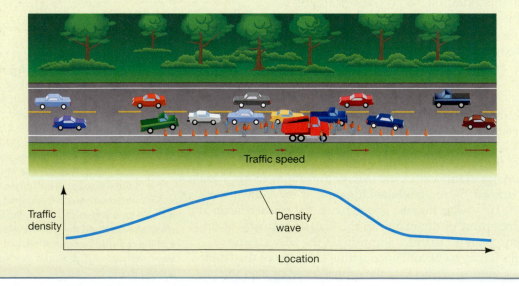

According to the equation above, the amount of mass within 50 kpc is approximately 6×10^{11} solar masses. Since 2×10^{11} solar masses lies within 15 kpc of the Galactic center, we have to conclude that at least twice as much mass lies *outside* the luminous part of our Galaxy—the part made up of stars, star clusters, and spiral arms—as lies inside!

Based on these observations, astronomers conclude that the luminous portion of the Milky Way Galaxy—the region outlined by the globular clusters and by the spiral arms—is merely the "tip of the Galactic iceberg." Our Galaxy is in reality very much larger. The luminous region is surrounded by an extensive, invisible **dark halo,** which dwarfs the inner halo of stars and globular clusters and extends well beyond the 15-kpc radius once thought to represent the limit of our Galaxy. But what is this dark halo made of? We do not detect enough stars or interstellar matter to account for the mass that our computations tell us must be there. We are inescapably drawn to the conclusion that most of the mass in our Galaxy (and, as we will see, of all galaxies) exists in the form of invisible **dark matter,** whose gravitational effects we can measure and quantify, but whose precise nature is not yet well understood.

The term *dark* here does not refer just to matter undetectable in visible light. The material has (so far) escaped detection at *all* electromagnetic wavelengths, from radio to gamma rays. Only by its gravitational pull do we know of its existence. Dark matter is not hydrogen gas (atomic or molecular), nor is it made up of ordinary stars. Given the amount of matter that must be accounted for, we would have been able to detect it with present-day equipment if it were in either of those forms. Its composition and its consequences for the evolution of galaxies and the universe are among the most important questions in astronomy today.

Many candidates have been suggested for this dark matter, although none is proven. Among the strongest "stellar" contenders are brown dwarfs and white dwarfs. ∞ *(Sec. 11.5, 12.3)* These objects could in principle exist in great numbers throughout the Galaxy, yet would be exceedingly hard to see. A radically different alternative is that the dark matter is made up of exotic *subatomic particles* that pervade the entire universe. In order to account for the properties of dark matter, these particles must have mass (to produce the observed gravitational effects) but otherwise interact hardly at all with "normal" matter (because otherwise we would be able to see them).[4] Many theoretical astrophysicists think that such particles could have been produced in abundance during the very earliest moments of our universe. If they survived to the present day, there might be enough of them to account for all the dark matter we think must be out there. These ideas are hard to test, however, because such particles would be very difficult to detect. Several detection experiments have been attempted, so far without success.

A few astronomers have proposed a very different explanation for the "dark matter problem," suggesting that its resolution may lie not in the nature of dark matter, but rather in a modification to Newton's law of gravity that increases the gravitational force on very large (Galactic) scales, doing away with the need for dark matter in the first place. We must emphasize that the vast majority of scientists do *not* accept this view in any way. However, the very fact that it has been proposed—and is being seriously discussed in scientific circles—is a testament to our current level of uncertainty. Dark matter is one of the great unsolved mysteries in astronomy today.

---

[4]One class of candidate particles satisfying these requirements has been dubbed *Weakly Interacting Massive Particles*, or WIMPs. Not to be outdone, astronomers searching for the more mundane stellar dark matter have labeled them *MAssive Compact Halo Objects*, or MACHOs. Who says astronomers don't have a sense of humor?

The Search for Stellar Dark Matter

Recently, researchers have obtained insight into the distribution of stellar dark matter by using a key element of Albert Einstein's theory of general relativity—the prediction that a beam of light can be deflected by a gravitational field, which has already been verified in the case of starlight passing close to the Sun. ∞ *(More Precisely 13-2)* Although this effect is small, it has the potential for making otherwise invisible stellar objects observable from Earth. Here's how.

Imagine looking at a distant star as a faint foreground object (such as a brown dwarf or a white dwarf) happens to cross your line of sight. As illustrated in Figure 14.20, the intervening object deflects a little more starlight than usual toward you, resulting in a temporary, but quite substantial, *brightening* of the distant star. In some ways, the effect is like the focusing of light by a lens, and the process is known as **gravitational lensing.** The foreground object is referred to as a *gravitational lens.* The amount of brightening and the duration of the effect depend on the mass, distance, and speed of the lensing object. Typically, the apparent brightness of the background star increases by a factor of two to five for a period of several weeks. Thus, even though the foreground object cannot be seen directly, its effect on the light of the background star makes it detectable.

Tutorial
Gravitational Lensing

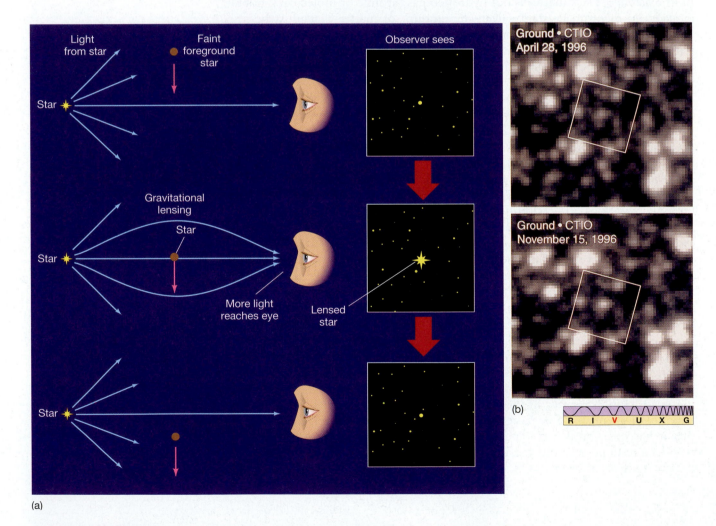

▲ **Figure 14.20 Gravitational Lensing** (a) Gravitational lensing by a faint foreground object (such as a brown dwarf) can temporarily cause a background star to brighten significantly, providing a means of detecting otherwise invisible stellar dark matter. (b) The brightening of a star during a lensing event, this one implying that a massive, but unseen, object passed in front of the unnamed star at the center of the two boxes imaged six months apart. (AURA)

Of course, the probability of one star passing almost directly in front of another, as seen from Earth, is extremely small. However, by observing millions of stars every few days over a period of years (using automated telescopes and high-speed computers to reduce the burden of coping with so much data), astronomers have so far seen 15 such events since these studies began in 1993. Subsequent observations are consistent with lensing by low-mass white dwarfs and suggest that such stars could account for at least half—but, it now seems, apparently not all—of the dark matter inferred from dynamical studies.

Bear in mind, though, that the identity of the dark matter is not necessarily an "all-or-nothing" proposition. It is perfectly conceivable that more than one type of dark matter exists. For example, it is quite possible that most of the dark matter in the inner (visible) parts of galaxies could be in the form of low-mass stars, while the dark matter farther out might be primarily in the form of exotic particles. We will return to this perplexing problem in later chapters.

◄◄◄ **CONCEPT CHECK** ▶

In what sense is dark matter "dark"?

14.7 The Galactic Center

Theory predicts that the Galactic bulge, and especially the region close to the Galactic center, should be densely populated with billions of stars. However, we are unable to see this region of our Galaxy—the interstellar medium in the Galactic disk shrouds what otherwise would be a stunning view. Figure 14.21 shows the (optical) view we do have of the region of the Milky Way toward the Galactic center, in the general direction of the constellation Sagittarius.

▲ **Figure 14.21 Galactic Center** Photograph of stellar and interstellar matter in the direction of the Galactic center. Because of heavy obscuration, even the largest optical telescopes can see no farther than one-tenth the distance to the center. To connect with previous figures, the M8 nebula can be seen at extreme top center. ∞ *(Sec. 11.2)* The field is roughly 20°, top to bottom, and is a continuation of the bottom part of Figure 11.5. The overlaid box indicates the location of the center of our Galaxy. The inset shows an adaptive optics near-infrared view of the dense stellar cluster surrounding the Galactic center, whose very core is indicated by the twin arrows. (AURA; ESO)

With the help of infrared and radio techniques we can peer more deeply into the central regions of our Galaxy than we can by optical means. Infrared observations (Figure 14.21 inset) show that the innermost parsec of our Galaxy harbors a dense cluster containing roughly 1 million stars. That's a stellar density about 100 *million* times greater than in our solar neighborhood, high enough that stars must experience frequent close encounters and even collisions. Infrared radiation has also been detected from what appear to be huge interstellar clouds rich in dust.

Radio observations of the region indicate a ring of molecular gas nearly 400 pc across, containing hundreds of thousands of solar masses of material and rotating around the Galactic center at about 100 km/s. The origin of this ring is unclear, although researchers think that the gravitational influence of our Galaxy's elongated, rotating bulge is involved, deflecting gas from farther out into the central regions. Higher-resolution radio studies show structure on all scales. Figure 14.22(b) shows a region called Sagittarius A. (The name simply means that it is the brightest radio source in Sagittarius.) It lies at the center of the boxed region in Figures 14.21 and 14.22(a)—and, we think, at the center

▲ **Figure 14.22 Galactic Center Close-up** (a) An infrared image of the region around the center of our Galaxy shows many bright stars packed into a relatively small volume. The average density of matter in this region is estimated to be about a million times that in the solar neighborhood. (b) The central portion of our Galaxy, as observed in the radio part of the spectrum. This image shows a region about 100 pc across surrounding the Galactic center (which lies within the bright blob at bottom right). The long-wavelength radio emission cuts through the Galaxy's dust, providing a view of matter in the immediate vicinity of the Galaxy's center. (c) A recent *Chandra* image showing the relation of a hot supernova remnant (red) and Sgr A*, the black hole at the very center. (d) The spiral pattern of radio emission arising from Sagittarius A, the center of the Galaxy. The data suggest a rotating ring of matter only a few parsecs across. All images are false-color, since they lie outside the visible spectrum. (UMass/Caltech; NRAO; NASA)

of our Galaxy. On a scale of about 30 pc, extended filaments can be seen. Their presence suggests to many astronomers that strong magnetic fields operate in the vicinity of the center, creating structures similar in appearance to (but much larger than) those observed on the active Sun. On even smaller scales (Figure 14.22c) *Chandra* observations indicate an extended region of hot X-ray-emitting gas, apparently associated with a supernova remnant, in addition to many other individual bright X-ray sources. And within that region lies a rotating ring or disk of molecular gas only a few parsecs across, with streams of matter spiraling inward toward the center (Figure 14.22d).

What could cause all this activity? An important clue comes from the Doppler broadening of infrared spectral lines emitted from the central swirling whirlpool of gas. The extent of the broadening indicates that the gas is moving very rapidly. In order to keep this gas in orbit, whatever is at the center must be extremely massive—several million solar masses. Given the twin requirements of large mass and small size, a leading contender is a supermassive black hole. The strong magnetic fields are thought to be generated within the accretion disk around the black hole as matter spirals inward, and may act as "particle accelerators," creating the extremely high-energy particles detected on Earth as *cosmic rays*. ∞ (*Sec. 13.3*) In the late 1990s, the *Compton Gamma Ray Observatory* found indirect evidence for a fountain of high-energy particles, possibly produced by violent processes close to the event horizon, gushing into the halo more than a thousand parsecs beyond the Galactic center. Astronomers have reason to suspect that similar events are occurring at the centers of many other galaxies.

At the very center of our Galaxy, at the heart of Sagittarius A, lies a remarkable object with the odd-sounding name Sgr A* (pronounced "saj ay star"). By the standards of the active galaxies to be studied in Chapter 15, this compact *Galactic nucleus* is not particularly energetic. Still, radio observations made during the past two decades, along with more recent X- and gamma-ray observations, suggest that it is nevertheless a pretty violent place. Its total energy output (at all wavelengths) is estimated to be 10^{33} W, which is more than a million times that of the Sun. VLBI observations using radio telescopes arrayed from Hawaii to Massachusetts imply that Sgr A* cannot be much larger than 10 AU, and it is probably a good deal smaller than that. ∞(*Sec. 3.4*) This size is consistent with the view that the energy source is a massive black hole. Figure 14.23 is perhaps the strongest evidence to date supporting the black hole picture. It shows a high-resolution infrared image of an 0.04-pc (8000 AU) field near the Galactic center, centered on Sgr A*. Using advanced adaptive-optics techniques on the Keck telescope and the VLT, U.S., and European researchers have created the first-ever diffraction-limited (0.05″ resolution) images of the region. ∞ (*Sec. 3.3*)

Remarkably, the image quality is good enough that the *proper motions* of several of the stars—their orbits around the Galactic center—can clearly be seen. The inset shows a series of observations of one of the brightest stars—called S2—over a 10-year period. The motion is consistent with an orbit around a massive object at the location of Sgr A*, in accordance with Newton's laws of motion. ∞ (*Sec. 1.4*) The solid curve on the figure shows the elliptical orbit that best fits the observations. It corresponds to a 15-year orbit with a semi-major axis of 950 AU, corresponding (from Kepler's third law, as modified by Newton) to a central mass of 3.7 million solar masses. Note that, even with this large mass, if Sgr A* is indeed a genuine

▼ **Figure 14.23 Orbits Near the Galactic Center** This extremely close-up map of the Galactic center was obtained by infrared adaptive optics, resulting in an ultra-high-resolution image of the innermost 0.1 pc of the Milky Way. The resolution is high enough that the orbits of individual stars can be tracked with confidence around the still-unseen Sgr A*. The inset shows the orbit of the innermost star in the frame, labeled S2, between 1992 and 2003. The solid line shows the best-fitting orbit for S2 around a black hole of 3.7 million solar masses, located at Sgr A*. (ESO)

(a)   50 kpc   (b)   5 kpc   (c)   500 pc

▲ **Figure 14.24 Galactic Center Zoom** Six artist's conceptions, each centered on the Galactic center and each increasing in resolution by a factor of 10. Frame (a) shows the same scene as Figure 14.15. Frame (f) is a rendition of a vast whirlpool within the innermost 0.5 parsec of our Galaxy. The data imaged in Figure 14.22 do not closely match these artistic renderings because the Figure 14.22 view is parallel to the Galactic disk—along the line of sight from the Sun to the Galactic center—whereas these six paintings portray a simplified view perpendicular to the disk, while progressively zooming down onto that disk.

black hole, the size of its event horizon is still only 0.08 AU. ∞ *(Sec. 13.5)* Such a small region, 8 kpc away, is currently unresolvable with any telescope now in existence.

Figure 14.24 places these findings into a simplified perspective—all of it drawn by an artist, yet based on real data. Each frame is centered on the Galaxy's core, and each increases in resolution by a factor of 10. Frame (a) renders the Galaxy's overall shape, as painted in Figure 14.15. This frame measures about 50 kpc across. Frame (b) spans a distance of 5 kpc and is nearly filled by the Galactic bar and the great circular sweep of the innermost spiral arm. Moving in to a 500-pc span, frame (c) depicts the 400-pc ring of nebulous matter mentioned earlier and some young dense star clusters, evidence of recent star formation near the Galactic center. The dark blobs represent giant molecular clouds, the pink patches emission nebulae associated with star formation within those clouds.

In frame (d), at 50 pc, a loose, pinkish region of ionized gas surrounds the thicker, reddish heart of the Galaxy. The colors rendered here, though invisible to our eyes, are inferred from measured temperatures of the gas. The sources of the energy responsible for this vast ionized region are frequent supernovae and other violent activity in the Galactic center. A central star cluster and another surrounding star-forming ring dominate the scene on this scale. Frame (e), spanning 5 pc, depicts the central cluster and ring in more detail, along with the tilted, spinning whirlpool of hot (10^4 K) gas surrounding the center of our Galaxy. The innermost part of this gigantic whirlpool is shown in frame (f), in which a swiftly spinning, white-hot disk of gas with temperatures in the millions of kelvins nearly engulfs the central black hole (artistically shown as a minute black dot). Small groups of stars within the central cluster are torn apart by the black hole's gravitational pull. The black hole itself, and even the stellar orbits shown in Figure 14.23, are too small to be pictured on this scale.

The last decade has seen an explosion in our knowledge of the innermost few parsecs of our Galaxy, and astronomers are working hard to decipher the clues hidden within its invisible radiation. Still, we are only now beginning to appreciate the full magnitude of this strange and complex new realm deep in the heart of the Milky Way.

CONCEPT CHECK ►►►

What is the most likely explanation of the energetic events observed at the Galactic center?

(d)   50 pc   (e)   5 pc   (f)   0.5 pc

CHAPTER REVIEW

Summary

1 A **galaxy** (p. 376) is a huge collection of stellar and interstellar matter isolated in space and bound together by its own gravity. Because we live within it, the **Galactic disk** (p. 376) of our own **Milky Way Galaxy** (p. 376) appears as a broad band of light across the sky—the Milky Way. Near the center, the Galactic disk thickens into the **Galactic bulge** (p. 376). The disk is surrounded by a roughly spherical **Galactic halo** (p. 377) of old stars and star clusters. Our Galaxy, like many others visible in the sky, is a **spiral galaxy** (p. 376).

2 The halo can be studied using **variable stars** (p. 378), whose luminosity changes with time. Two types of pulsating variable stars of particular importance to astronomers are **RR Lyrae variables** (p. 378) and **Cepheid variables** (p. 378). All RR Lyrae stars have roughly the same luminosity. For Cepheids, the luminosity can be determined using the **period–luminosity relationship** (p. 379). Knowing the luminosity, astronomers can apply the inverse-square law to determine the distance. The brightest Cepheids can be seen at distances of millions of parsecs, extending the cosmic distance ladder well beyond our own Galaxy. In the early twentieth century, Harlow Shapley used RR Lyrae stars to determine the distances to many of the Galaxy's globular clusters and found that they have a roughly spherical distribution in space, but the center of the sphere lies far from the Sun. The center of their distribution is close to the **Galactic center** (p. 380), about 8 kpc away.

3 Disk and halo stars differ in their spatial distributions, ages, colors, and orbital motion. The luminous portion of our Galaxy has

a diameter of about 30 kpc. In the vicinity of the Sun, the Galactic disk is about 300 pc thick. The halo lacks gas and dust, so no stars are forming there. All halo stars are old. The gas-rich disk is the site of current star formation and contains many young stars. Stars and gas within the Galactic disk move on roughly circular orbits around the Galactic center. Stars in the halo and bulge move on largely random three-dimensional orbits that pass repeatedly through the disk plane but have no preferred orientation. Halo stars appeared early on, before the Galactic disk took shape, when there was still no preferred orientation for their orbits. As the gas and dust formed a rotating disk, stars that formed in the disk inherited its overall spin and so moved on circular orbits in the Galactic plane, as they do today.

4 Radio observations clearly reveal the extent of our Galaxy's **spiral arms** (p. 386), regions of the densest interstellar gas where star formation is taking place. The spirals cannot be "tied" to the disk material, as the disk's differential rotation would have wound them up long ago. Instead, they may be **spiral density waves** (p. 386) that move through the disk, triggering star formation as they pass by. Alternatively, the spirals may arise from **self-propagating star formation** (p. 389), when shock waves produced by the formation and evolution of one generation of stars trigger the formation of the next.

5 The Galactic **rotation curve** (p. 389) plots the orbital speed of matter in the disk against distance from the Galactic center. By applying Newton's laws of motion, astronomers can determine the mass of the Galaxy. They find that the Galactic mass continues to increase beyond the radius defined by the globular clusters and the spiral structure we observe, indicating that our Galaxy has an invisible **dark halo** (p. 398). The **dark matter** (p. 398) making up this dark halo is of unknown composition. Leading candidates include low-mass stars and exotic subatomic particles. Recent attempts to detect stellar dark matter have used the fact that a faint foreground object can occasionally pass in front of a more distant star, deflecting the star's light and causing its apparent brightness to increase temporarily. This deflection is called **gravitational lensing** (p. 392).

6 Astronomers working at infrared and radio wavelengths have uncovered evidence for energetic activity within a few parsecs of the Galactic center. The leading explanation is that a black hole roughly 4 million times more massive than the Sun resides there. The hole lies at the center of a dense star cluster containing millions of stars, which is in turn surrounded by a star-forming disk of molecular gas. The observed activity is thought to be powered by accretion onto the black hole, as well as by supernova explosions in the cluster surrounding it.

Review and Discussion

1. What do globular clusters tell us about our Galaxy and our place within it?

2. How are Cepheid variables used in determining distances?

3. Roughly how far out into space can we use Cepheids to measure distance?

4. What important discoveries were made early in this century using RR Lyrae variables?

5. Why can't we study the central regions of the Galaxy using optical telescopes?

6. Of what use is radio astronomy in the study of Galactic structure? Why are radio studies often more useful than observations made in visible light?

7. Contrast the motions of disk and halo stars.

8. Explain why Galactic spiral arms are thought to be regions of recent and ongoing star formation.

9. Describe the behavior of interstellar gas as it passes through a spiral density wave.

10. What is self-propagating star formation?

11. What do the red stars in the Galactic halo tell us about the history of the Milky Way Galaxy?

12. What does the rotation curve of our Galaxy tell us about its total mass?

13. What evidence is there for dark matter in the Galaxy?

14. What are some possible explanations for dark matter?

15. Why do astronomers think that a supermassive black hole lies at the center of the Milky Way Galaxy?

Self-Test True or False?/Multiple Choice

1. Herschel's attempt to map the Milky Way by counting stars led to an inaccurate estimate of the Galaxy's size because he was unaware of absorption by interstellar dust. **(T/F)**

2. Cepheid variables can be used to determine the distances to the nearest galaxies. **(T/F)**

3. Globular clusters trace out the large-scale structure of the Galactic disk. **(T/F)**

4. The Galactic halo contains about as much gas and dust as the Galactic disk. **(T/F)**

5. The Galactic disk contains only old stars. **(T/F)**

6. Stars and gas in the Galactic disk move in roughly circular orbits around the Galactic center. **(T/F)**

7. Astronomers use 21-cm radiation to study Galactic molecular clouds. **(T/F)**

8. The most likely explanation of the high-speed motion of stars and gas near the Galactic center is that they are orbiting a supermassive black hole. **(T/F)**

9. In the Milky Way Galaxy, our Sun is located **(a)** near the Galactic center; **(b)** about halfway out from the center; **(c)** at the outer edge; **(d)** in the halo.

10. A telescope searching for young, bright stars would make the most discoveries if it were pointed **(a)** directly away from the Galactic center; **(b)** perpendicular to the Galactic disk; **(c)** within a spiral arm; **(d)** between spiral arms.

11. The first stars that formed in the Milky Way now **(a)** have chaotic orbits; **(b)** orbit in the Galactic plane; **(c)** orbit closest to the Galactic center; **(d)** orbit in the same direction as the Milky Way spins.

12. Stars in the outermost regions of the Milky Way Galaxy **(a)** are youngest; **(b)** orbit faster than astronomers would expect based on the Galactic mass we can see; **(c)** are more likely to explode as supernovae; **(d)** are more luminous than other stars.

13. Most of the mass of the Milky Way exists in the form of **(a)** stars; **(b)** gas; **(c)** dust; **(d)** dark matter.

14. According to Figure 14.6 (Period-Luminosity Plot), a Cepheid variable star with luminosity 1000 times that of the Sun has a pulsation period of roughly **(a)** 1 day; **(b)** 3 days; **(c)** 10 days; **(d)** 50 days.

15. Figure 14.19 (Galaxy Rotation Curve) tells us that **(a)** the Galaxy rotates like a solid body; **(b)** far from the center, the Galaxy rotates more slowly than we would expect based on the light we see; **(c)** far from the center, the Galaxy rotates more rapidly than we would expect based on the light we see; **(d)** there is no matter beyond about 15 kpc from the Galactic center.

Problems

Algorithmic versions of these questions are available in the Practice Problems module of the Companion Website.

The number of squares preceding each problem indicates its approximate level of difficulty.

1. ■ At one time some astronomers claimed that the Andromeda "nebula" (Figure 14.2a) was a star-forming region lying within our own Galaxy. Calculate the angular diameter of a prestellar nebula of radius 100 AU, lying 100 pc from Earth. ∞ *(More Precisely 1-1)* How close to Earth would the nebula have to lie for it to have the same angular diameter (roughly 6°) as Andromeda?

2. ■ What is the greatest distance at which an RR Lyrae star of absolute magnitude 0 could be seen by a telescope capable of detecting objects as faint as 20th magnitude? ∞ *(Sec. 10.2)*

3. ■ A typical Cepheid variable is 100 times brighter than a typical RR Lyrae star. On average, how much farther away than RR Lyrae stars can Cepheids be used as distance-measuring tools?

4. ■■ The brightest Cepheids have luminosities 30,000 times greater than that of the Sun. Taking the Sun's absolute magnitude to be 5, calculate their absolute magnitudes. The *Hubble Space Telescope* can just detect a star like the Sun at a distance of 100,000 pc. Neglecting interstellar absorption, how far away can *HST* see the brightest Cepheids?

5. ■■ Calculate the proper motion (transverse angular velocity, in arc seconds per year) of a globular cluster with a transverse velocity (relative to the Sun) of 200 km/s and a distance of 3 kpc. Do you think that this motion might be measurable? ∞ *(Sec. 10.1)*

6. ■■ Calculate the total mass of the Galaxy lying within 20 kpc of the Galactic center if the rotation speed at that radius is 240 km/s.

7. ■■■ Using the data presented in FIgure 14.19, calculate how long it takes the Sun to "lap" stars orbiting 15 kpc from the Galactic center. How long does it take matter at 5 kpc to lap us?

8. ■ A density wave made up of two spiral arms is moving through the Galactic disk. At the 8-kpc radius of the Sun's orbit around the Galactic center, the wave's speed is 120 km/s, and the Galactic rotation speed is 220 km/s. Calculate how many times the Sun has passed through a spiral arm since the Sun formed 4.6 billion years ago.

9. ■■ Given the data in the previous question and the fact that O-type stars live at most 10 million years before exploding as supernovae, calculate the maximum distance at which an O-type star (orbiting at the Sun's distance from the Galactic center) can be found from the density wave in which it formed.

10. ■■■ Material at an angular distance of 0.2″ from the Galactic center is observed to have an orbital speed of 1200 km/s. If the Sun's distance to the Galactic center is 8 kpc and the material's orbit is circular and is seen edge-on, calculate the radius of the orbit and the mass of the object around which the material is orbiting.

The Companion Website at www.prenhall.com/chaisson provides algorithmically generated versions of each chapter's Problems, in addition to an Animations and Videos gallery, a Tutorials gallery and Tutorials Quiz questions, a full eBook, an Images gallery, Practice Questions, Multiple Choice, and True or False quizzes, and Labeling exercises.

15

Normal and Active Galaxies

Building Blocks of the Universe

As our field of view expands to truly cosmic scales, the focus of our studies shifts dramatically. Planets become inconsequential, stars themselves mere points of hydrogen consumption. Now entire galaxies become the "atoms" from which the universe is built—distant realms completely unknown to scientists just a century ago. We know of literally millions of galaxies beyond our own. Most are smaller than the Milky Way, some comparable in size, a few much larger. Many are sites of explosive events far more energetic than anything ever witnessed in our own Galaxy. All are vast, gravitationally bound assemblages of stars, gas, dust, dark matter, and radiation separated from us by almost incomprehensibly large distances. The light we receive tonight from the most distant galaxies was emitted long before Earth existed. By studying the properties of galaxies and the violence that ensues when they collide, we gain insight into the history of our Galaxy and the universe in which we live.

Galaxies are among the grandest, most beautiful objects in the universe. They are colossal collections of typically a hundred billion stars, held together loosely by gravity. Despite their widespread existence in space, astronomers do not yet fully understand how galaxies originated. Here the graceful, winding arms of the majestic spiral galaxy, M51, also known as the Whirlpool Galaxy, resemble a curving staircase sweeping through space. This view, taken by the *Hubble Space Telescope*, shows the galaxy's yellowish central core, its reddish nebulae in the arms, and myriad stars throughout. (STScI)

LEARNING GOALS

Studying this chapter will enable you to:

1 Describe the basic properties of the main types of normal galaxies.

2 Discuss the distance-measurement techniques that enable astronomers to map the universe beyond the Milky Way.

3 Describe how galaxies are observed to clump into clusters.

4 State Hubble's law, and explain how it is used to determine distances to the most remote objects in the observable universe.

5 Specify the basic differences between active and normal galaxies.

6 Describe some important features of active galaxies.

7 Explain what drives the central engine thought to power all active galaxies.

Visit www.prenhall.com/chaisson for additional annotated images, animations, and links to related sites for this chapter.

15.1 Hubble's Galaxy Classification

Figure 15.1 shows a vast expanse of space lying about 100 million pc from Earth. Almost every patch or point of light in this figure is a separate galaxy— several hundred can be seen in just this one photograph. Even using a small telescope, images of galaxies look distinctly nonstellar. They have fuzzy edges, and many are quite elongated—not at all like the sharp, pointlike images normally associated with stars. Although it is difficult to tell from the photograph, some of the blobs of light in Figure 15.1 are spiral galaxies like the Milky Way Galaxy and Andromeda. Others, however, are definitely not spirals—no disks or spiral arms can be seen. Even when we take into account their different orientations in space, galaxies do *not* all look the same.

The American astronomer Edwin Hubble was the first to categorize galaxies in a comprehensive way. Working with the then recently completed 2.5-m optical telescope on Mount Wilson in California in 1924, he classified the galaxies he saw into four basic types—*spirals, barred spirals, ellipticals,* and *irregulars*—solely on the basis of their visual appearance. Many modifications and refinements have been incorporated over the years, but the basic **Hubble classification scheme** is still widely used today.

Spirals

We saw several examples of **spiral galaxies** in Chapter 14—for example, our own Milky Way Galaxy and our neighbor Andromeda. All galaxies of this type contain a flattened galactic disk in which spiral arms are found, a central galactic bulge with a dense nucleus, and an extended halo of faint, old stars. ∞ *(Sec. 14.3)* The stellar density (that is, the number of stars per unit volume) is greatest in the *galactic nucleus,* at the center of the bulge.

In Hubble's scheme, a spiral galaxy is denoted by the letter S and classified as type a, b, or c according to the size of its central bulge—Type Sa galaxies having the largest bulges, Type Sc the smallest (Figure 15.2). The tightness of the spiral pattern is quite well correlated with the size of the

▼ **Figure 15.1 Coma Cluster** (a) A collection of many galaxies, each consisting of hundreds of billions of stars. Called the Coma Cluster, this group of galaxies lies more than 100 million pc from Earth. (The blue spiked object at top right is a nearby star; virtually every other object in this image is a galaxy.) (b) A recent *Hubble Space Telescope* image of part of the cluster. (AURA; NASA)

(a)

R I V U X G

(b)

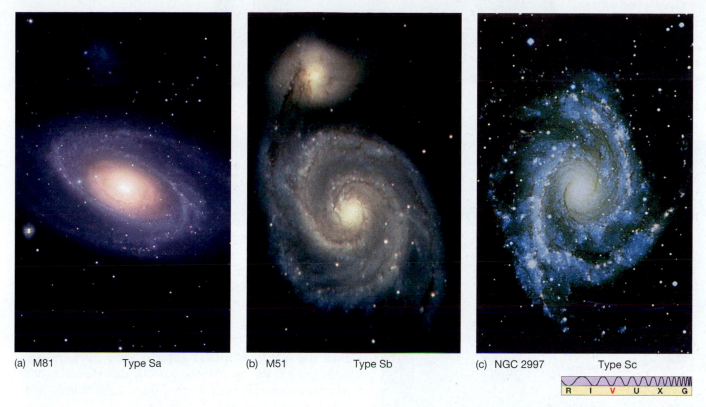

(a) M81   Type Sa   (b) M51   Type Sb   (c) NGC 2997   Type Sc

R   I   V   U   X   G

▲ **Figure 15.2 Spiral Galaxy Shapes**   Variation in shape among spiral galaxies. As we progress from type Sa to Sb to Sc, the bulges become smaller while the spiral arms tend to become less tightly wound. (R. Gendler; NOAO; D. Malin/AAT)

bulge (although the correspondence is not perfect). Type Sa spiral galaxies tend to have tightly wrapped, almost circular, spiral arms. Type Sb galaxies typically have more open spiral arms, while Type Sc spirals often have loose, poorly defined spiral structure. The arms also tend to become more "knotty," or clumped, in appearance as the spiral pattern becomes more open.

The bulges and halos of spiral galaxies contain large numbers of reddish old stars and globular clusters, similar to those observed in our own Galaxy and in Andromeda. ∞ *(Sec. 14.1)* Most of the light from spirals, however, comes from A- through G-type stars in the disk, giving these galaxies an overall whitish glow. Like the disk of the Milky Way, the flat galactic disks of typical spiral galaxies are rich in gas and dust. The 21-cm radio radiation emitted by spirals betrays the presence of interstellar gas, and obscuring dust lanes are clearly visible in many systems (see Figures 15.2b and c). Stars are forming within the spiral arms, which contain numerous emission nebulae and newly formed O- and B-type stars. ∞ *(Sec. 14.5)* The arms appear bluish because of the presence of bright blue O- and B-type stars there. Type Sc galaxies contain the most interstellar gas and dust, Sa galaxies the least.

Most spirals are not seen face-on as in Figure 15.2. Many are tilted with respect to our line of sight, making their spiral structure hard to discern. However, we do not need to see spiral arms to classify a galaxy as a spiral. The presence of the disk, with its gas, dust, and newborn stars, is sufficient. For example, the galaxy shown in Figure 15.3 is classified as a spiral because of the clear line of obscuring dust seen along its midplane. That dust is evident most clearly in the inset infrared image, which was acquired by the Earth-orbiting *Spitzer Space Telescope.*

A variation of the spiral category in Hubble's classification scheme is the **barred-spiral galaxy.** Barred spirals differ from ordinary spirals mainly by the presence of an elongated "bar" of stellar and interstellar matter passing through the center and extending beyond the bulge, into the disk. The spiral arms project from near the ends of the bar rather than from the bulge (as they do in normal spirals). Barred spirals are designated by the letters SB and are subdivided, like the ordinary spirals, into categories SBa, SBb, and SBc, depending on the size of the bulge. Again like ordinary spirals, the tightness of

The Sombrero Galaxy, a spiral system seen edge-on. Officially cataloged as M104, this galaxy has a dark band composed of interstellar gas and dust. The large size of this galaxy's central bulge marks it as type Sa, even though its spiral arms cannot be seen from our perspective. The inset shows this galaxy in the infrared part of the spectrum, highlighting its dust content in false-colored pink. (NASA)

the spiral pattern is correlated with bulge size. Figure 15.4 shows the variation among barred-spiral galaxies. In the case of the SBc category, it is often hard to tell where the bar ends and the spiral arms begin.

The discovery that the bulge of our own Galaxy is elongated suggests that the Milky Way may be a barred spiral, of type SBb or SBc. However, the full extent of the bar remains uncertain. Some astronomers place our Galaxy in an intermediate category lying between Hubble's spiral and barred spiral types. ∞ *(Sec. 14.3)*

Ellipticals

Unlike the spirals, **elliptical galaxies** have no spiral arms and, in most cases, no obvious galactic disk—in fact, other than a dense central nucleus, they often exhibit little internal structure of any kind. As with spirals, the stellar density increases sharply in the central nucleus. Denoted by the letter E, these systems are subdivided according to how elliptical they appear on the sky. The most circular are designated E0, slightly flattened systems are labeled E1, and so on, all the way to the most elongated ellipticals, of type E7 (Figure 15.5).

Notice, by the way, that the Hubble type of an elliptical galaxy depends both on the galaxy's intrinsic three-dimensional shape *and* on its orientation relative to the line of sight. Consider, for example, a spherical galaxy, a cigar-

▼ **Figure 15.4 Barred-Spiral Galaxy Shapes** Variation in shape among barred-spiral galaxies. The variation from SBa to SBc is similar to that for the spirals in Figure 15.2, except that now the spiral arms begin at either end of a bar through the galactic center. In frame (c), the bright star is a foreground object in our own Galaxy; the object at top center is another galaxy that is probably interacting with NGC 6872. (NASA; AAT; ESO)

(a) NGC 1300   Type SBa

(b) NGC 1365   Type SBb

(c) NGC 6872   Type SBc

▶ **Figure 15.5 Elliptical Galaxy Shapes** Variation in shape among elliptical galaxies. (a) The E1 galaxy M49 is nearly circular in appearance. (b) M84 is a slightly more elongated elliptical galaxy, classified as E3. Both galaxies lack spiral structure, and neither shows evidence of cool interstellar dust or gas, although each has an extensive X-ray halo of hot gas that extends far beyond the visible portion of the galaxy. (c) M110 is a dwarf elliptical companion to the much larger Andromeda Galaxy. (AURA; SAO; R. Gendler)

shaped galaxy seen end on, and a disk-shaped galaxy seen face on; all would appear circular on the sky. As a result, it is often difficult to decipher a galaxy's true shape from its visual appearance.

There is a large range in both the size and the number of stars contained in elliptical galaxies. The largest elliptical galaxies are much larger than our own Milky Way Galaxy. These *giant ellipticals* can range up to a few megaparsecs across and contain trillions of stars. At the other extreme, *dwarf ellipticals* may be as small as 1 kpc in diameter and contain fewer than a million stars. The significant observational differences between giant and dwarf ellipticals have led many astronomers to conclude that these galaxies are members of separate classes, with quite different formation histories and stellar content. The dwarfs are by far the most common type of ellipticals, outnumbering their brighter counterparts by about ten to one. However, most of the *mass* that exists in the form of elliptical galaxies is contained in the larger systems.

Lack of spiral arms is not the only difference between spirals and ellipticals. Most ellipticals also contain little or no cool gas and dust and display no evidence of young stars or ongoing star formation. Like the halo of our own Galaxy, ellipticals are made up mostly of old, reddish, low-mass stars. Again like the halo of our Galaxy, the orbits of stars in ellipticals are disordered, exhibiting little or no overall rotation; objects move in all directions, not in regular, circular paths as in our Galaxy's disk. Ellipticals differ from our Galaxy's halo in at least one important respect, however: X-ray observations reveal large amounts of very *hot* (several million kelvin) interstellar gas distributed throughout their interiors, often extending well beyond the visible portions of the galaxies.

Some giant ellipticals are exceptions to many of the foregoing general statements about elliptical galaxies, as they have been found to contain disks of gas and dust in which stars are forming. Astronomers believe that these systems may be the results of mergers between gas-rich galaxies (see *Discovery 15-1*). Indeed, galactic collisions may have played an important role in determining the appearance of many of the systems we observe today.

Intermediate between the E7 ellipticals and the Sa spirals in the Hubble classification is a class of galaxies that show evidence of a thin disk and a flattened bulge yet contain neither loose gas nor spiral arms. Two such objects are shown in Figure 15.6. These galaxies are called **S0 galaxies** ("S-zero") if no bar is evident and **SB0 galaxies** if a bar is present. They are also known as

(a) M49   Type E2

(b) M84   Type E3

(c) M110   Type E5

R I V U X G

(a) NGC 1201   Type S0   (b) NGC 2859   Type SB0

R I V U X G

◀ **Figure 15.6 S0 Galaxies** (a) S0 (or lenticular) galaxies contain a disk and a bulge, but no interstellar gas and no spiral arms. They are in many respects intermediate between E7 ellipticals and Sa spirals in their properties. (b) SB0 galaxies are similar to S0 galaxies, except for a bar of stellar material extending beyond the central bulge. (Palomar/Caltech) *lenticular* galaxies, because of their lens-shaped appearance. They look a little like spirals whose dust and gas have been stripped away, leaving behind just a stellar disk. Observations in recent years have shown that many normal elliptical galaxies have faint disks within them, like the S0 galaxies. As with the S0s, the origin of these disks is uncertain, but some researchers suspect that S0s and ellipticals may be closely related.

Irregulars

The final galaxy class identified by Hubble is a catch-all category—**irregular galaxies**—so named because their visual appearance does not allow us to place them into any of the other categories just discussed. Irregulars tend to be rich in interstellar matter and young, blue stars, but they lack any regular structure, such as well-defined spiral arms or central bulges. Irregular galaxies tend to be smaller than spirals but somewhat larger than dwarf ellipticals. They typically contain between 10^8 and 10^{10} stars. The smallest are called *dwarf irregulars.* As with elliptical galaxies, the dwarf type is the most common irregular. Dwarf ellipticals and dwarf irregulars occur in approximately equal numbers and together make up the vast majority of galaxies in the universe. They are often found close to a larger "parent" galaxy.

Irregular galaxies are divided into two subclasses—Irr I and Irr II. The Irr I galaxies often look like misshapen spirals. Figure 15.7 shows the **Magellanic Clouds**, a famous pair of Irr I galaxies that orbit the Milky Way Galaxy. They are shown to proper scale in Figure 14.15. Studies of Cepheid variables within the Clouds show them to be approximately 50 kpc from the center of our Galaxy. ∞ *(Sec. 14.2)* The Large Cloud contains about six billion solar masses of material and is a few kiloparsecs across. Both Magellanic Clouds contain lots of gas, dust, and blue stars (and the recent, well-documented supernova discussed in *Discovery 12-1*), indicating ongoing star formation. Both also contain many old stars and several old globular clusters, so we know that star formation has been going on there for a very long time.

The much rarer Irr II galaxies (Figure 15.8), in addition to their irregular shape, have other peculiarities, often exhibiting a distinctly explosive or filamentary appearance. Their appearance once led astronomers to suspect that violent events had occurred within them. However, it now seems more likely that in some (but probably not all) cases, we are seeing the result of a close encounter or collision between two previously "normal" systems.

The Hubble Sequence

Table 15.1 summarizes the basic characteristics of the various galaxy types. When he first developed his classification scheme, Hubble arranged the galaxy

▼ **Figure 15.7 Magellanic Clouds** The Magellanic Clouds are prominent features of the night sky in the Southern Hemisphere. Named for the sixteenth-century Portuguese explorer Ferdinand Magellan, whose around-the-world expedition first brought word of these fuzzy patches of light to Europe, they are dwarf irregular (Irr I) galaxies, gravitationally bound to our own Milky Way Galaxy. They orbit our Galaxy and accompany it on its trek through the cosmos. (a) The Clouds' relationship to one another in the southern sky. Both the Small (b) and the Large Magellanic Cloud (c) have distorted, irregular shapes, although some observers claim they can discern a single spiral arm in the Large Cloud. (F. Espenak; Harvard Observatory)

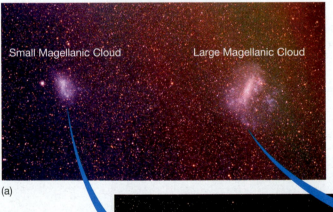

Small Magellanic Cloud   Large Magellanic Cloud

(a)

(b)

(c)

R I V U X G

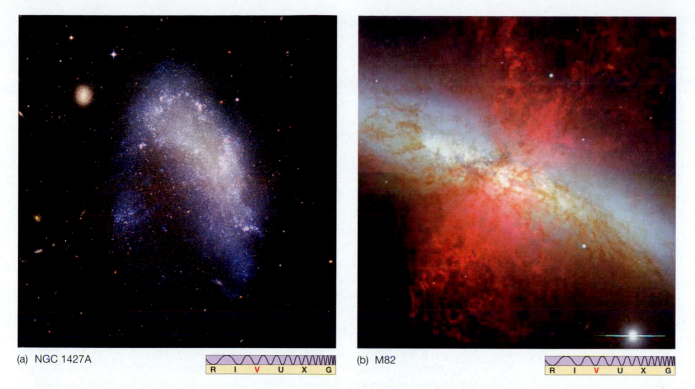

(a) NGC 1427A

R I V U X G

(b) M82

R I V U X G

▲ **Figure 15.8 Irregular Galaxy Shapes**   Some irregular (Irr II) galaxies. (a) The strangely shaped galaxy NGC 1427A is probably plunging headlong into a group of several other galaxies (not shown), causing huge rearrangements of its stars, gas, and dust. (b) The galaxy M82 seems to show an explosive appearance, probably the result of a recent galaxywide burst of star formation. (NASA; Subaru)

types into the "tuning fork" diagram shown in Figure 15.9. The variation in galaxy types across the diagram, from ellipticals to spirals to irregulars, is often referred to as the *Hubble sequence*.

Hubble's primary aim in creating this diagram was to indicate similarities in appearance among galaxies, but he also regarded the tuning fork as an evolutionary sequence from left to right, with E0 ellipticals evolving into flatter ellipticals and S0 systems, and ultimately forming disks and spiral arms. Indeed,

▼ **Figure 15.9 Galactic "Tuning Fork"**   Hubble's tuning fork diagram, showing his basic galaxy classification scheme. The placement of the four basic types of galaxies—ellipticals, spirals, barred spirals, and irregulars—in the diagram is suggestive, but the tuning fork has no physical meaning.

TABLE 15.1	Basic Galaxy Properties by Type		
	SPIRAL/BARRED SPIRAL (S/SB)	ELLIPTICAL (E)[1]	IRREGULAR (Irr)
Shape and structural properties	Highly flattened disk of stars and gas, containing spiral arms and thickening to central bulge. SB galaxies have an elongated central "bar" of stars and gas.	No disk. Stars smoothly distributed through an ellipsoidal volume. No obvious substructure other than a dense central nucleus.	No obvious structure. Irr II galaxies often have "explosive" appearance.
Stellar content	Disks contain both young and old stars; halos consist of old stars only.	Contain old stars only.	Contain both young and old stars.
Gas and dust	Disks contain substantial amounts of gas and dust; halos contain little of either.	Contain hot X-ray emitting gas, little or no cool gas and dust.	Very abundant in gas and dust.
Star formation	Ongoing star formation in spiral arms.	No significant star formation during the last 10 billion years.	Vigorous ongoing star formation.
Stellar motion	Gas and stars in disk move in circular orbits around the galactic center; halo stars have random orbits in three dimensions.	Stars have random orbits in three dimensions.	Stars and gas have very irregular orbits.

[1]As noted in the text, some giant ellipticals appear to be the result of mergers between gas-rich galaxies and are exceptions to many of the statements listed here.

CONCEPT CHECK ▶▶▶

In what ways are large spirals such as The Milky Way and Andromeda not representative of galaxies as a whole?

Hubble's terminology referring to ellipticals as "early-type" galaxies" and spirals as "late-type" is still widely used today. However, as far as modern astronomers can tell, there is *no* direct evolutionary connection of this sort along the Hubble sequence. Isolated normal galaxies do *not* evolve from one type to another. Spirals are not ellipticals with arms, nor are ellipticals simply spirals that have somehow shed their star-forming disks. In short, astronomers know of no simple parent–child relationship among Hubble types.

However, the key word here is *isolated*. As we will see, there is now strong observational evidence indicating that collisions and tidal interactions *between* galaxies are commonplace, and these may be the main physical processes driving galaxy evolution. We will return to this important subject in Chapter 16.

15.2 The Distribution of Galaxies in Space

Now that we have seen some of their basic properties, let us ask how galaxies are spread through the expanse of the universe beyond the Milky Way. Galaxies are not distributed uniformly in space. Rather, they tend to clump into still larger agglomerations of matter. As we will see, this uneven distribution is crucial in determining both their appearance and their evolution. As always in astronomy, our understanding hinges on our ability to tell how far away an object lies. We therefore begin by looking more closely at the means used by astronomers to measure distances to galaxies.

Extending the Distance Scale

Astronomers estimate that some 40 billion galaxies brighter than our own exist in the observable universe. Some reside close enough for the Cepheid variable technique to work—astronomers have detected and measured the periods of Cepheids in galaxies as far away as 25 Mpc. ∞ *(Sec. 14.2)* However, some galaxies contain no Cepheid stars (can you think of some reasons why this might be?), and in any case most known galaxies lie much farther away than 25 Mpc. Cepheid variables in very distant galaxies simply cannot be observed well enough, even through the world's most sensitive telescopes, to allow us to measure their apparent brightness and periods. To extend our distance-measurement ladder, therefore, we must find some new object to study. What individual objects are bright enough for us to observe at great distances?

▲ **Figure 15.10 Galaxy Rotation**   A galaxy's rotation causes some of the radiation it emits to be blueshifted and some to be redshifted (relative to what the emission would be from an unmoving source). From a distance, when the radiation from the galaxy is combined into a single beam and analyzed spectroscopically, the redshifted and blueshifted components combine to produce a broadening of the galaxy's spectral lines. The amount of broadening is a direct measure of the rotation speed of the galaxy, such as the one at the right, NGC 4603, about 100 million light-years away. (NASA)

One way in which researchers have tackled this problem is through observations of **standard candles**—easily recognizable astronomical objects whose luminosities are confidently known. The basic idea is very simple. Once an object is identified as a standard candle—by its appearance or by the shape of its light curve, say—its luminosity can be estimated. Comparison of the luminosity with the apparent brightness then gives the object's distance and hence the distance to the galaxy in which it resides. ∞ *(Sec. 10.2)* Note that, apart from the way in which the luminosity is determined, the Cepheid variable technique relies on identical reasoning. However, the term *standard candle* tends to be applied only to very bright objects.

To be most useful, a standard candle must (1) have a narrowly defined luminosity so that the uncertainty in estimating its brightness is small and (2) be bright enough to be seen at great distances. In recent years planetary nebulae and Type I supernovae have proved particularly reliable as standard candles. ∞ *(Sec. 12.3, 12.5)* The latter have remarkably consistent peak luminosities and are bright enough to be identified and measured out to distances of many hundreds of megaparsecs.

An important alternative to standard candles was discovered in the 1970s, when astronomers found a close correlation between the rotational speeds and the luminosities of spiral galaxies within a few tens of megaparsecs of the Milky Way Galaxy. Rotation speed is a measure of a spiral galaxy's total mass, so it is perhaps not surprising that this property should be related to luminosity. ∞ *(Sec. 14.5)* What *is* surprising, though, is how tight the correlation is. The **Tully–Fisher relation**, as it is now known (after its discoverers), allows us to obtain a remarkably accurate estimate of a spiral galaxy's luminosity simply by observing how fast it rotates (Figure 15.10). As usual, comparing the galaxy's (true) luminosity with its (observed) apparent brightness yields its distance.

The Tully–Fisher relation can be used to measure distances to spiral galaxies out to about 200 Mpc, beyond which the line broadening in Figure 15.10 becomes increasingly difficult to measure accurately. A somewhat similar connection, relating line broadening to a galaxy's *diameter*, exists for elliptical galaxies. Once the galaxy's diameter and angular size are known, its distance can be computed from elementary geometry. ∞ *(More Precisely E-2)* These methods bypass many of the standard candles often used by astronomers and so provide independent means of determining distances to faraway objects.

As indicated in Figure 15.11, standard candles and the Tully–Fisher relation form the fifth and sixth rungs of our cosmic distance ladder, introduced in Chapter 1 and expanded in Chapters 10 and 14. ∞ *(Sec. 1.4, 10.6, 14.2)* In

▲ **Figure 15.11 Extragalactic Distance Ladder**
An inverted pyramid summarizes the distance techniques used to study different realms of the universe. The techniques shown in the bottom four layers—radar ranging, stellar parallax, spectroscopic parallax, and variable stars—take us as far as the nearest galaxies. To go farther, we must use new techniques—the Tully–Fisher relation and the use of standard candles—based on distances determined by the four lowest techniques.

fact, they stand for perhaps a dozen or so related but separate techniques that astronomers have employed in their quest to map out the universe on large scales. Just as with the lower rungs, we calibrate the properties of these new techniques using distances measured by more local means. In this way, the distance-measurement process "bootstraps" itself to greater and greater distances. However, at the same time, the errors and uncertainties in each step accumulate, so the distances to the farthest objects are the least well known.

Clusters of Galaxies

Figure 15.12 sketches the locations of all the known major astronomical objects within about 1 Mpc of the Milky Way. Our Galaxy appears with its dozen or so satellite galaxies—including the two Magellanic Clouds discussed earlier and a recently discovered companion (labeled "Sagittarius" in the figure) lying almost within our own Galactic plane. The Andromeda Galaxy, lying 800 kpc from us, is also shown, surrounded by satellites of its own. Two of Andromeda's galactic neighbors are shown in the inset to the figure. M33 is a spiral, while M32 is a dwarf elliptical, easily seen in Figure 14.2(a) to the bottom right of Andromeda's central bulge.

All told, some 45 galaxies are known to populate our Galaxy's neighborhood. Three of them (the Milky Way, Andromeda, and M33) are spirals; the remainder are dwarf irregulars and dwarf ellipticals. Together, these galaxies form the **Local Group**—a new level of structure in the universe above the scale of our Galaxy. As indicated in Figure 15.12, the Local Group's diameter is a little over 1 Mpc. The Milky Way Galaxy and Andromeda are by far its largest members, and most of the smaller galaxies are gravitationally bound to

CONCEPT CHECK ▶▶▶

What are some of the problems in measuring accurately the distances to faraway galaxies?

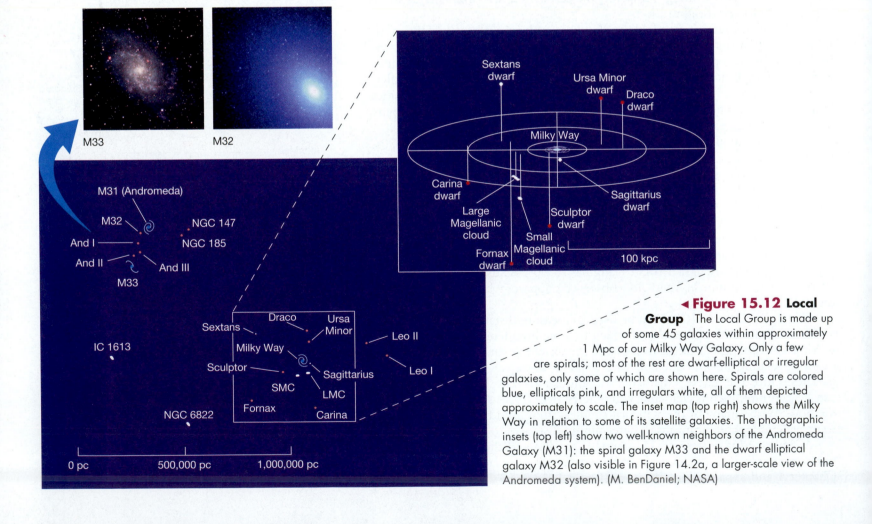

◀ **Figure 15.12 Local Group** The Local Group is made up of some 45 galaxies within approximately 1 Mpc of our Milky Way Galaxy. Only a few are spirals; most of the rest are dwarf-elliptical or irregular galaxies, only some of which are shown here. Spirals are colored blue, ellipticals pink, and irregulars white, all of them depicted approximately to scale. The inset map (top right) shows the Milky Way in relation to some of its satellite galaxies. The photographic insets (top left) show two well-known neighbors of the Andromeda Galaxy (M31): the spiral galaxy M33 and the dwarf elliptical galaxy M32 (also visible in Figure 14.2a, a larger-scale view of the Andromeda system). (M. BenDaniel; NASA)

▶ **Figure 15.13 Virgo Cluster**   The central region of the Virgo Cluster of galaxies, about 17 Mpc from Earth. Many large spiral and elliptical galaxies can be seen. The inset shows several galaxies surrounding the giant elliptical known as M86. An even bigger elliptical galaxy, M87 noted at bottom, will be discussed later in this chapter. (M. BenDaniel; AURA)

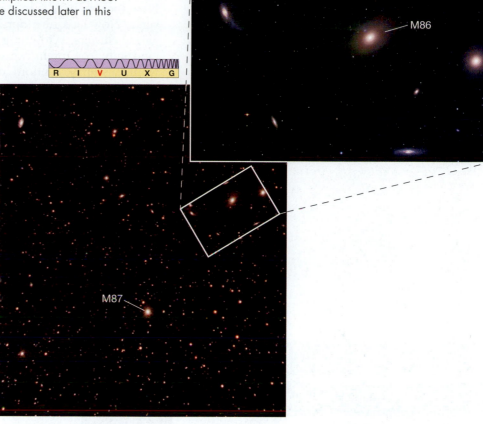

one or the other of them. The combined gravity of the galaxies in the Local Group binds them together, like stars in a star cluster, but on a millionfold larger scale. More generally, a group of galaxies held together by their mutual gravitational attraction is called a **galaxy cluster**.

Moving beyond the Local Group, the next large concentration of galaxies we come to is the *Virgo Cluster* (Figure 15.13), named for the constellation in which it is found. It lies some 18 Mpc from the Milky Way. The Virgo Cluster does not contain a mere 45 galaxies, however. Rather, it houses more than 2500 galaxies, bound by gravity into a tightly knit group about 3 Mpc across.

Wherever we look in the universe we find galaxies, and the majority of galaxies are members of galaxy clusters. Small clusters, such as the Local Group, contain only a few galaxies and are quite irregular in shape. Large, "rich" clusters like Virgo contain thousands of individual galaxies distributed fairly smoothly in space. The Coma cluster, shown in Figure 15.1, is another example of a rich cluster. It lies approximately 100 Mpc away. Figure 15.14 is a long-exposure photograph of a much more distant rich cluster, residing roughly 2 billion parsecs from Earth. A sizeable minority of galaxies (perhaps 20–30 percent) are not members of any cluster. They are apparently isolated systems, moving alone through intercluster space.

like fireflies in a moving jar...

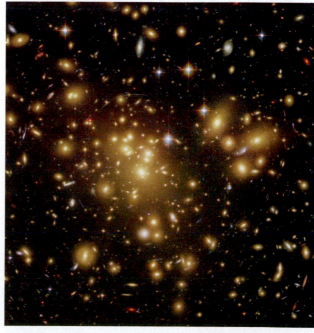

▶ **Figure 15.14 Distant Galaxy Cluster**   The galaxy cluster Abell 1689 contains huge numbers of galaxies and resides roughly 2 billion parsecs from Earth. Virtually every patch of light in this photograph is a separate galaxy. Thanks to the high resolution of the optics onboard the *Hubble Space Telescope*, we can now discern, even at this great distance, spiral structure in some of the galaxies. We also see many galaxies in collision—some tearing matter from one another, others merging into single systems. In addition to the frenzied activity of their constituent galaxies, the clusters themselves also move through space—like buzzing fireflies in a moving jar. (NASA)

15.3 Hubble's Law

Now that we have seen some basic properties of galaxies throughout the universe. Now let's turn our attention to the large-scale *motions* of galaxies and galaxy clusters. Within a galaxy cluster, individual galaxies move more or less randomly. You might expect that, on even larger scales, the clusters themselves would also have random, disordered motion—some clusters moving this way, some that. In fact, this is not the case. On the largest scales, galaxies and galaxy clusters alike move in a very *ordered* way.

Universal Recession

In 1912 the American astronomer Vesto M. Slipher, working under the direction of Percival Lowell, discovered that virtually every spiral galaxy he observed had a redshifted spectrum—it was *receding* from our Galaxy. ∞ *(Sec. 2.7)* In fact, with the exception of a few nearby systems, *every* known galaxy is part of a general motion away from us in all directions. Individual galaxies that are not part of galaxy clusters are steadily receding. Galaxy clusters, too, have an overall recessional motion, although their individual member galaxies move randomly with respect to one another. (Consider a jar full of fireflies that has been thrown into the air. The fireflies within the jar, like the galaxies within the cluster, have random motions due to their individual whims, but the jar as a whole, like the galaxy cluster, has some directed motion as well.)

Figure 15.15 shows the optical spectra of several galaxies, arranged in order of increasing distance from the Milky Way Galaxy. These spectra are redshifted, indicating that the galaxies are receding. Furthermore, the extent of the redshift increases progressively from top to bottom in the figure. There is a connection between Doppler shift and distance: the greater the distance, the greater the redshift. This trend holds for nearly all galaxies in the universe. (Two galaxies within our Local Group, including Andromeda, and a few galaxies in the Virgo Cluster display blueshifts and so are moving toward us, but this results from their local motions within their parent clusters. Recall the fireflies in the jar.)

Figure 15.16(a) shows recessional velocity plotted against distance for the galaxies of Figure 15.15. Figure 15.16(b) is a similar plot for several more galaxies within about one billion parsecs of Earth. Plots like these were first made by Edwin Hubble in the 1920s and now bear his name—*Hubble diagrams*. The data points generally fall close to a straight line, indicating that the rate at which a galaxy recedes is *directly proportional* to its distance from us. This rule is called **Hubble's law**. We could construct such a diagram for any group of galaxies, provided we could determine their distances

◀ Figure 15.15 Galaxy Spectra Optical spectra, shown at left, of several galaxies named on the right. Both the extent of the redshift (denoted by the horizontal red arrows) and the distance from the Milky Way Galaxy to each galaxy (numbers in center column) increase from top to bottom. The vertical yellow arrow in each spectrum highlights a particular spectral feature (a pair of dark absorption lines). The horizontal red arrows indicate how this feature shifts to longer wavelengths in the spectra of more distant galaxies. The many white vertical lines at the top and bottom of each spectrum are laboratory references. (Palomar/Caltech)

and velocities. The universal recession described by the Hubble diagram is sometimes called the *Hubble flow*.

The recessional motions of the galaxies prove that the cosmos is not steady and unchanging on the largest scales. The universe (actually, *space itself*—see Sec. 17.2) is expanding. However, let's be clear on just *what* is expanding and what is not. Hubble's law does not mean that humans, Earth, the solar system, or even individual galaxies and galaxy clusters are physically increasing in size. These groups of atoms, rocks, planets, stars, and galaxies are held together by their own internal forces and are not themselves getting bigger. Only the largest framework of the universe—the distances separating the galaxy clusters and superclusters—is expanding.

To distinguish recessional redshift from redshifts caused by motion *within* an object—for example, galaxy orbits within a cluster or explosive events in a galactic nucleus—the redshift resulting from the Hubble flow is called the **cosmological redshift**. Objects that lie so far away that they exhibit a large cosmological redshift are said to be at *cosmological distances*—distances comparable to the scale of the universe itself.

Hubble's law has some fairly dramatic implications. If nearly all galaxies show recessional velocities according to Hubble's law, then doesn't that mean that they all started their journey from a single point? If we could run time backward, wouldn't all the galaxies fly back to this one point, perhaps the site of some explosion in the remote past? The answer is yes—but not in the way you might expect! In Chapter 17 we will explore the ramifications of the Hubble flow for the past and future evolution of our universe. For now, however, we set aside its cosmic implications and use Hubble's law simply as a convenient distance-measuring tool.

Hubble's Constant

The constant of proportionality between recessional velocity and distance in Hubble's law is known as **Hubble's constant**, denoted by the symbol H_0. The data shown in Figure 15.16 then obey the equation:

$$\text{recessional velocity} = H_0 \times \text{distance}.$$

The value of Hubble's constant is the slope of the straight line—recessional velocity divided by distance—in Figure 15.16(b). Reading the numbers off the graph, this comes to roughly 70,000 km/s divided by 1000 Mpc, or 70 km/s/Mpc (kilometers per second per megaparsec, the most commonly used unit for H_0). Astronomers continually strive to refine the accuracy of the Hubble diagram and the resulting estimate of H_0 because Hubble's constant is one of the most fundamental quantities of nature—it specifies the rate of expansion of the entire cosmos.

The precise value of Hubble's constant is the subject of considerable debate. The most recent measurements, made by many different research groups, using different sets of galaxies and a wide variety of distance-measurement techniques, give results mainly between 50 and 80 km/s/Mpc. Most astronomers would be quite surprised if the true value of H_0 turned out to lie outside this range. However, the width of the quoted range is not the result of measurement uncertainties in any one method—there remain real, and as yet unresolved, inconsistencies between the different techniques currently in use. Infrared Tully–Fisher measurements and studies of Cepheid variables, the latter now extended to include the Virgo cluster by researchers using the *Hubble Space Telescope*, generally produce results at the high end of the range, 70–80 km/s/Mpc. ∞ *(Sec. 14.2)* However, visible-light Tully–Fisher studies and techniques using standard candles, including Type I supernovae, tend to return lower values, in the range 50–65 km/s/Mpc. Other methods give results scattered between 50 and 80 km/s/Mpc.

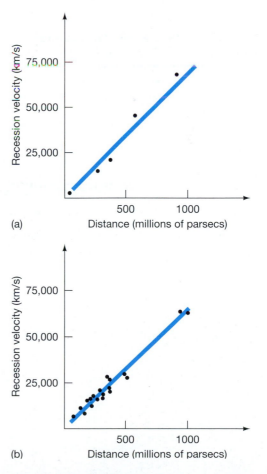

(a)

(b)

▲ **Figure 15.16 Hubble's Law** Plots of recessional velocity against distance (a) for the galaxies shown in Figure 15.15 and (b) for numerous other galaxies within about 1 billion pc of Earth.

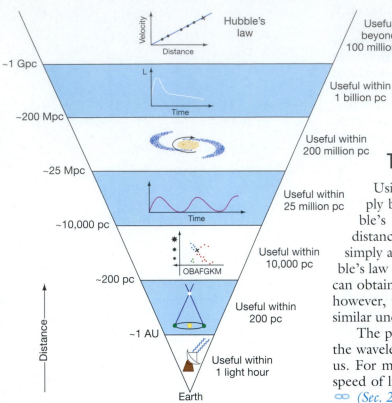

▲ **Figure 15.17 Cosmic Distance Ladder** Hubble's law tops the hierarchy of distance-measurement techniques. It is used to find the distances to astronomical objects all the way out to the limits of the observable universe.

CONCEPT CHECK ▶▶▶

How does the use of Hubble's Law differ from the other extragalactic distance-measurement techniques we have seen in this text?

For now, astronomers must simply live with this uncertainty. We will adopt $H_0 = 70$ km/s/Mpc (roughly in the middle of all recent results, and a value consistent with some precise measurements to be discussed in Chapter 17) as the best current estimate of Hubble's constant for the remainder of the text. Bear in mind, though, that there is considerable ambiguity—and dispute—among experts as to the true value of this very important number.[1]

The Top of the Distance Ladder

Using Hubble's law, we can derive the distance to a remote object simply by measuring the object's recessional velocity and dividing by Hubble's constant. Hubble's law thus tops our inverted pyramid of distance-measurement techniques (Figure 15.17). This seventh method simply assumes that Hubble's law holds. If this assumption is correct, Hubble's law enables us to measure great distances in the universe—so long as we can obtain an object's spectrum, we can determine how far away it is. Notice, however, that the uncertainty in Hubble's constant translates directly into a similar uncertainty in all distances determined by this method.

The precise astronomical definition of *redshift* is the fractional increase in the wavelength of a beam of radiation due to the source's motion away from us. For many nearby objects, recession velocities are small compared to the speed of light and the redshifts are correspondingly low—just a few percent. ∞ *(Sec. 2.7)* However, more distant objects can have recessional motions that are substantial fractions of the speed of light. The most distant objects thus far observed in the universe—some young galaxies and quasars—have redshifts of more than 6, meaning that their radiation has been stretched in wavelength not by a few percent, but *sevenfold*. Their ultraviolet spectral lines are shifted all the way into the infrared part of the spectrum! Such huge redshifts correspond to recessional velocities of more than 96 percent of the speed of light. Hubble's law implies that such objects lie more than 9000 Mpc away from us, as close to the limits of the observable universe as astronomers have yet been able to probe. (See *More Precisely 15-1* on p. 418, for a fuller discussion of redshifts and recession velocities comparable to the speed of light.)

The speed of light is finite. It takes time for light or any kind of radiation to travel from one point in space to another. The radiation that we now see from these most distant objects originated long ago. Incredibly, the radiation that astronomers now detect was emitted almost 14 billion years ago, well before our planet, our Sun, and perhaps even our Galaxy came into being.

15.4 Active Galactic Nuclei

The galaxies described in the previous sections—those falling into the various Hubble classes—are generally referred to as *normal* galaxies. More than 90 percent of all galaxies fall into this broad category. Their luminosities range from a million or so times that of the Sun for dwarf ellipticals and irregulars to more than a trillion solar luminosities for the largest giant ellipticals known. For comparison, in round numbers, the luminosity of the Milky Way Galaxy is

---

[1] It is worth noting that Hubble's original value for H_0 was about 500 km/s/Mpc, far higher than any currently accepted value, due almost entirely to errors in the cosmic distance scale at the time, particularly the calibrations of Cepheid variables and standard candles. The measured value dropped rapidly as the various observational errors were recognized and resolved and distance measurements became more reliable. Published estimates of H_0 entered the "modern" range in roughly the mid 1960s.

2×10^{10} solar luminosities, or roughly 10^{37} W. For our purposes, the term "bright" will be taken to mean anything more than about 10^{10} times the solar value. Our Galaxy is bright, but not abnormally so.

The substantial minority of bright galaxies—perhaps 20–25 percent, although some researchers would say more—which don't fit well into the Hubble scheme are of enormous interest to astronomers. Some of these galaxies are among the most energetic objects known, and all may represent an important, if intermittent, phase of galaxy evolution (see Sec. 16.3). Having luminosities sometimes thousands of times greater than that of the Milky Way, they are known collectively as **active galaxies**. At visible wavelengths, active galaxies generally *look* like galaxies—familiar components such as disks, bulges, stars, gas, and dust can often be identified. At other wavelengths, however, their unusual properties are much more apparent.

Galactic Radiation

Active galaxies differ fundamentally from normal galaxies in both their overall luminosities and in the *character* of the radiation they emit. Most of a normal galaxy's energy is emitted in or near the visible portion of the electromagnetic spectrum, much like the radiation from stars. Indeed, to a large extent, the light we see from a normal galaxy *is* just the accumulated light of its many component stars (each described approximately by a blackbody curve). ∞ *(Sec. 2.4)* By contrast, as illustrated schematically in Figure 15.18, the radiation from active galaxies does *not* peak in the visible. Most active galaxies do emit substantial amounts of visible radiation, but far more energy is emitted at invisible wavelengths, both longer and shorter than the visible range. Put another way, the radiation from active galaxies is *inconsistent* with what we would expect if it were the combined radiation of myriad stars. Their radiation is said to be *nonstellar*.

Many luminous galaxies with nonstellar emission are known to be *starburst galaxies*—previously normal systems currently experiencing widespread episodes of star formation, most likely as a result of interactions with a neighbor. The irregular galaxy M82 shown in Figure 15.8(b) is a prime example. We will study these important systems, and their role in galaxy evolution, in Chapter 16. For purposes of this text, however, we will use the term *active galaxy* to mean a system where the abnormal activity is related to violent events occurring in or near the galactic nucleus. Such systems are also known as **active galactic nuclei**.

Two important categories of active galaxies are *Seyfert galaxies* and *radio galaxies*. More extreme in their properties are the even more luminous *quasars*. Astronomers have conventionally distinguished between active galaxies and quasars based on their appearance, spectra, and distance from us. Quasars are generally so far away that little structure can be discerned, giving most quasars a "starlike" appearance. But improving observations now reveal "galactic" components in many quasars, so the distinction between quasars and active galaxies is not clear-cut. Most astronomers believe that quasars are simply an early stage of galaxy formation, and that the same basic processes power all of these active objects.

Seyfert Galaxies

In 1943 Carl Seyfert, an American optical astronomer studying spiral galaxies from Mount Wilson Observatory, discovered the type of active galaxy that now bears his name. **Seyfert galaxies** are a class of astronomical objects whose properties lie between those of normal galaxies and those of the most energetic active galaxies known.

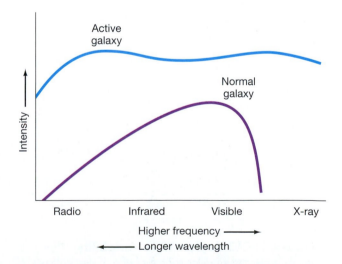

▲ **Figure 15.18 Galaxy Energy Distribution** The energy emitted by a normal galaxy differs significantly from that emitted by an active galaxy. This plot illustrates the general run of intensity for all galaxies of a particular type and does not represent any one individual galaxy.

R I V U X G

◄ **Figure 15.19** **Seyfert Galaxy** The Circinus galaxy, a Seyfert with a bright compact core, lies some 4 Mpc away—it is one of the closest active galaxies. (NASA)

Superficially, Seyferts resemble normal spiral galaxies (Figure 15.19). Indeed, the stars in a Seyfert's galactic disk and spiral arms produce about the same amount of visible radiation as the stars in a normal spiral galaxy. However, most of a Seyfert's energy is emitted from a small central region known as the **galactic nucleus**—the center of the overexposed white patch in Figure 15.19. The nucleus of a Seyfert galaxy is some 10,000 times brighter than the center of our own Galaxy. In fact, the brightest Seyfert nuclei are 10 times more energetic than the *entire* Milky Way.

Some Seyferts produce radiation spanning a broad range in wavelengths, from the infrared all the way through ultraviolet and even X-rays. However, the majority (about 75 percent) emit most of their energy in the infrared. Scientists believe that much of the high-energy radiation in these Seyferts is absorbed by dust in or near the nucleus, then reemitted as infrared radiation.

Seyfert spectral lines have many similarities to those observed toward the center of our own Galaxy. ∞ *(Sec. 14.7)* The lines are very broad, most likely indicating rapid (5000 km/s or more) internal motion within the nuclei. ∞ *(Sec. 2.8)* In addition, the energy emission often varies in time (Figure 15.20). A Seyfert's luminosity can double or halve within a fraction of a year. These rapid fluctuations in luminosity lead us to conclude that the source of energy emissions must be quite compact—simply put, an object cannot "flicker" in less time than radiation takes to cross it. ∞ *(Sec. 13.4)* The emitting region must therefore be less than one light-year across—an extraordinarily small region, considering the amount of energy emanating from it.

Radio Galaxies

As the name suggests, **radio galaxies** are active galaxies that emit large amounts of energy in the radio portion of the electromagnetic spectrum. They differ from Seyferts not only in the wavelengths at which they radiate, but also in both the appearance and the extent of their emitting regions.

Figure 15.21 shows the radio galaxy Centaurus A, which lies about 4 Mpc from Earth. Almost none of this galaxy's radio emission comes from a compact nucleus. Instead, the energy is released from two huge extended regions called **radio lobes**—roundish clouds of gas spanning about half a megaparsec, lying well beyond the visible galaxy.[2] Undetectable in visible light, the radio lobes of radio galaxies are truly enormous. From end to end, they typically span more than 10 times the size of the Milky Way Galaxy, comparable in scale to the entire Local Group.

Figure 15.22 shows the relationship between the galaxy's visible, radio, and X-ray emission. In visible light, Centaurus A is apparently a large E2 galaxy some 500 kpc in diameter, bisected by an irregular band of dust. Centaurus A is a member of a small cluster of galaxies, and numerical simulations suggest that this peculiar galaxy is probably the result of a collision between an elliptical galaxy and a smaller spiral galaxy about 500 million years ago (see *Discovery 15-1*). The radio lobes are roughly symmetrically placed, jutting out from the center of the visible galaxy, roughly perpendicular to the dust lane, suggesting that they consist of material ejected in opposite directions from the

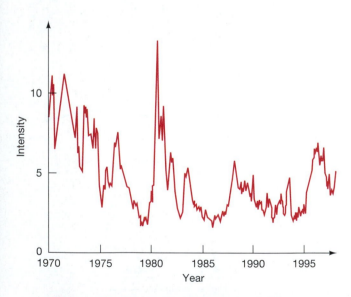

▲ **Figure 15.20** **Seyfert Time Variability** This graph illustrates the irregular variations of a Seyfert galaxy's luminosity over two decades. Because this Seyfert, called 3C 84, emits strongly in the radio part of the electromagnetic spectrum, these observations were made with large radio telescopes. The optical and X-ray luminosities vary as well. (NRAO)

[2]The term *visible galaxy* is commonly used to refer to those components of an active galaxy that emit visible "stellar" radiation, as opposed to the nonstellar and invisible "active" component of the galaxy's emission.

(a)

(b)

▲ **Figure 15.21 Centaurus A Radio Lobes** Radio galaxies, such as Centaurus A shown here optically in (a), often have giant radio-emitting lobes (b) extending a million parsecs or more beyond the central galaxy. The lobes cannot be imaged in visible light and are observable only with radio telescopes. The lobes are shown here in false color, with decreasing intensity from red to yellow to green to blue. (ESO; NRAO)

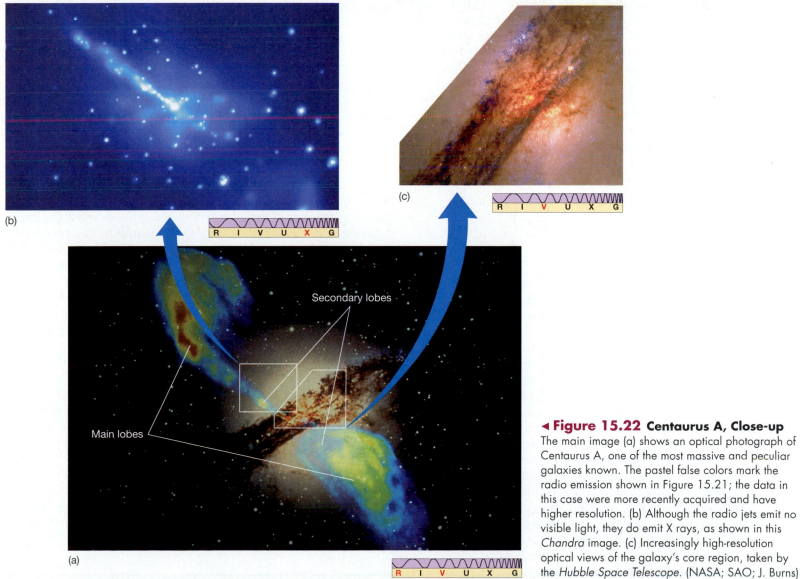

(b)

(c)

Secondary lobes

Main lobes

(a)

◀ **Figure 15.22 Centaurus A, Close-up**
The main image (a) shows an optical photograph of Centaurus A, one of the most massive and peculiar galaxies known. The pastel false colors mark the radio emission shown in Figure 15.21; the data in this case were more recently acquired and have higher resolution. (b) Although the radio jets emit no visible light, they do emit X rays, as shown in this *Chandra* image. (c) Increasingly high-resolution optical views of the galaxy's core region, taken by the *Hubble Space Telescope*. (NASA; SAO; J. Burns)

galactic nucleus. This conclusion is strengthened by the presence of a pair of smaller secondary lobes closer to the visible galaxy and by the presence of a roughly 1-kpc-long *jet* of matter in the galactic center, all aligned with the main lobes (and marked in Figure 15.22).

If the material was ejected from the nucleus at close to the speed of light, then Centaurus A's outer lobes were created a few hundred million years ago, quite possibly around the time of the elliptical/spiral merger thought to be responsible for the galaxy's odd optical appearance. The secondary lobes were expelled more recently. Apparently, some violent process at the center of Centaurus A—most probably triggered by the collision—started up around that

15-1 MORE PRECISELY

Relativistic Redshifts and Look-Back Time

When discussing very distant objects, astronomers usually talk about their redshifts rather than their distances. Indeed, it is very common for researchers to speak of an event occurring "at" a certain redshift—meaning that the light received today from that event is redshifted by the specified amount. Of course, because of Hubble's law, redshift and distance are equivalent to one another. However, redshift is the preferred quantity because it is a directly observable property of an object, whereas distance is derived from redshift using Hubble's constant, whose value is not accurately known.

The redshift of a beam of light is, by definition, the *fractional* increase in its wavelength resulting from the recessional motion of the source. ∞ *(Sec. 2.7)* A redshift of 1 corresponds to a *doubling* of the wavelength. Using the formula for the Doppler shift given previously, the redshift of radiation received from a source moving away from us with speed v is

$$\text{redshift} = \frac{\text{observed wavelength} - \text{true wavelength}}{\text{true wavelength}}$$

$$= \frac{\text{recessional velocity, } v}{\text{speed of light, } c}.$$

Let's illustrate this with two examples, rounding the speed of light, c, to 300,000 km/s. A galaxy at a distance of 100 Mpc has a recessional speed (by Hubble's law) of 70 km/s/Mpc × 100 Mpc = 7000 km/s. Its redshift therefore is 7000 km/s ÷ 300,000 km/s = 0.023. Conversely, an object that has a redshift of 0.05 has a recessional velocity of 0.05 × 300,000 km/s = 15,000 km/s and hence a distance of 15,000 km/s ÷ 70 km/s/Mpc = 210 Mpc.

Unfortunately, while the foregoing equation is correct for low speeds, it does not take into account the effects of relativity. As we saw in Chapter 13, the rules of everyday physics have to be modified when speeds begin to approach the speed of light, and the formula for the Doppler shift is no exception. ∞ *(More Precisely 13-1)* Our formula is valid for speeds much less than the speed of light, but when $v = c$ the redshift is not 1, as the equation suggests, but is in fact *infinite*. Radiation received from an object moving away from us at nearly the speed of light is redshifted to almost infinite wavelength.

Thus, do not be alarmed to find that many galaxies and quasars have redshifts greater than 1. This does not mean that they are receding faster than light! It simply means that their recessional speeds are relativistic—comparable to the speed of light—and the preceding simple formula is not applicable. Table 15.2 presents a conversion chart relating redshift, recession speed, and present distance. The column headed "v/c" gives recession velocities based on the Doppler effect, taking relativity properly into account (but see Sec. 17.2 for a more correct interpretation of the redshift). All values are based on reasonable assumptions and are usable even for $v \approx c$. We take Hubble's constant to be 70 km/s/Mpc and assume a flat universe (see Sec. 17.3 for details) in which matter (mostly dark) contributes just over one-quarter of the total density. The conversions in the table are used consistently throughout this text.

Because the universe is expanding, the "distance" to a galaxy is not very well defined—do we mean the distance when the galaxy emitted the light we see today, or the present distance (as listed in the table), even though we do not see the galaxy as it is today, or is some other measure more appropriate? Largely because of this ambiguity, astronomers prefer to work in terms of a quantity known as the *look-back time* (last column of Table 15.2), which is simply how long ago an object emitted the radiation we see today. While astronomers talk frequently about redshifts and sometimes about look-back times, they hardly ever talk of distances to high-redshift objects.

For nearby sources, the look-back time is numerically equal to the distance in light-years—the light we receive tonight from a galaxy at a distance of 100 million light-years was emitted 100 million years ago. However, for more distant objects, the look-back time and the present distance in light-years differ because of the expansion of the universe, and the divergence increases dramatically with increasing redshift (see Sec. 17.2). For example, a galaxy now located 15 billion light-years from Earth was much closer to us when it emitted the light we now see. Consequently, its light has taken considerably less than 15 billion years—in fact, a little less than 10 billion years—to reach us.

▶ **Figure 15.23 Cygnus A**   (a) Cygnus A also appears to be two galaxies in collision. (b) On a much larger scale, it displays radio-emitting lobes on either side of the optical image. The optical galaxy in (a) is about the size of the small dot at the center of (b). Note the thin line of radio-emitting material joining the right lobe to the central galaxy. The distance from one lobe to the other is approximately a million light-years. (NOAO; NRAO)

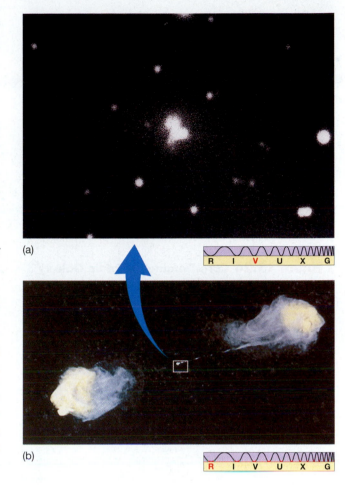

(a)

(b)

time and has been intermittently firing jets of matter out into intergalactic space ever since.

Centaurus A is a relatively low-luminosity source that happens to lie very close to us, astronomically speaking, making it particularly easy to study. Figure 15.23 shows a much more powerful emitter, called Cygnus A, lying roughly 250 Mpc from Earth. The high-resolution radio map in Figure 15.23(b) clearly shows two narrow, high-speed jets joining the radio lobes to the center of the visible galaxy (the dot at the center of the radio image). Notice also that, as with Centaurus A, Cygnus A is a member of a small group of galaxies, and the optical image (Figure 15.23a) appears to show a galaxy collision in progress.

The radio lobes of the brightest radio galaxies (such as Cygnus A) emit roughly 10 times more energy than the Milky Way Galaxy at all wavelengths, coincidentally about the same as the most luminous Seyfert nuclei. However, despite their names, radio galaxies actually radiate far more energy at shorter wavelengths. Their total energy output can be 100 times or more greater than their radio emission. Most of this energy comes from the nucleus of the visible galaxy. Having total luminosities up to 1000 times that of the Milky Way Galaxy, bright radio galaxies are among the most energetic objects known in the universe. Their radio emission is important to astronomers because it lets them study in detail the connection between the small-scale nucleus and the large-scale radio lobes.

TABLE 15.2	Redshift, Distance, and Look-Back Time			
REDSHIFT	v/c	PRESENT DISTANCE (Mpc)	(10^6 light-years)	LOOK-BACK TIME (millions of years)
0.000	0.000	0	0	0
0.010	0.010	42	137	137
0.025	0.025	105	343	338
0.050	0.049	209	682	665
0.100	0.095	413	1350	1290
0.250	0.220	999	3260	2920
0.500	0.385	1880	6140	5020
1.000	0.600	3320	10800	7730
2.000	0.800	5250	17100	10300
3.000	0.882	6460	21100	11500
4.000	0.923	7310	23800	12100
5.000	0.946	7940	25900	12500
6.000	0.960	8420	27500	12700
10.00	0.984	9660	31500	13200
50.00	0.999	12300	40100	13600
100.0	1.000	12900	42200	13700
∞	1.000	14600	47500	13700

▲ Figure 15.24 Core-Dominated Radio Galaxy
On this radio contour map of the radio galaxy M86, we can see that the radio emission comes from a bright central nucleus, which is surrounded by an extended, less-intense radio halo. The radio map is superimposed on an optical image of the galaxy and some of its neighbors, a wider-field version of which was shown previously in Figure 15.13. (Harvard-Smithsonian Center for Astrophysics)

Not all radio galaxies have obvious radio lobes. Figure 15.24 shows a *core-dominated* radio galaxy, most of whose energy is emitted from a small central nucleus (which radio astronomers refer to as the *core*) less than 1 pc across. Weaker radio emission comes from an extended region surrounding the nucleus. It is likely that all radio galaxies have jets and lobes, but what we observe depends on our perspective. As illustrated in Figure 15.25, viewed from the side we see the jets and lobes. However, if we view the jet almost head-on—in other words, looking *through* the lobe—we see a core-dominated system.

Common Features of Active Galaxies

Only about 10 percent of active galaxies are radio galaxies, and of the rest, most are not classified as Seyferts. Many other technical categories exist, but the basic properties just described for these two types of active galaxies are fairly representative of active galaxies in general. In all cases, a huge amount of energy, spanning much of the electromagnetic spectrum, is generated within—and usually also emitted from—a central compact nucleus.

Many active galaxies contain jets. Figure 15.26 presents several images of the giant elliptical galaxy M87, a prominent member of the Virgo Cluster (see Figure 15.13). A long time exposure (Figure 15.26a) shows a large, fuzzy ball of light—a fairly normal-looking E1 galaxy about 100 kpc across. A shorter exposure of M87 (Figure 15.26b), capturing only the galaxy's bright inner regions, reveals a long (2 kpc) thin jet of matter ejected from the galactic center at nearly the speed of light. Computer enhancement shows that it is made up of a series of distinct "blobs" more or less evenly spaced along its length, suggesting that the material was ejected during bursts of activity. The jet has also been imaged in the radio, infrared (Figure 15.26c), and X-ray regions of the spectrum.

Our location with respect to a jet also affects the *type* of radiation we see. The theory of relativity tells us that radiation emitted by particles moving close to the speed of light is strongly concentrated, or beamed, in the direction of motion. ∞ *(More Precisely 13-1)* As a result, if we happen to be directly in line with the beam, the radiation we receive is both very intense and Doppler shifted toward short wavelengths. ∞ *(Sec. 2.7)* The resulting object is called a **blazar** (see Figure 3.30). Much of the luminosity of the hundred or so known blazars is received in the form of X rays or gamma rays.

Finally, note that all the active galaxies described so far show signs of *interactions* with other galaxies. We have already discussed this in the context of Centaurus A and Cygnus A, and M87 lies near the center of the Virgo Cluster, having probably achieved its present size via mergers with other smaller galaxies. The Seyfert shown in Figure 15.19 also shows evidence for a burst of star formation within the past few million years—a clear signature of a recent galactic encounter (see Sec. 16.2).

Observer sees core-dominated radio galaxy

Radio lobe

Central energy source

Jets

Radio lobe

Observer sees lobe radio galaxy

◄ Figure 15.25 Radio Galaxy A central energy source produces high-speed jets of matter that interact with intergalactic gas to form radio lobes. The system may appear to us as either radio lobes or a core–dominated radio galaxy, depending on our location with respect to the jets and lobes.

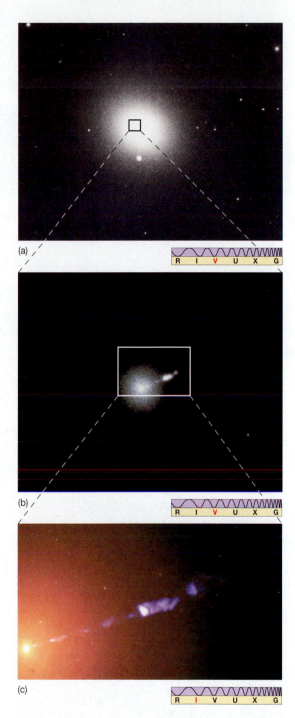

▶ **Figure 15.26** **M87 Jet** The giant elliptical galaxy M87 (also called Virgo A) is displayed here in a series of zooms. (a) A long optical exposure of the galaxy's halo and embedded central region. (b) A short optical exposure of its core and an intriguing jet of matter, on a somewhat smaller scale. (c) An infrared image of M87's jet, examined more closely compared with (b). The bright point at left in (c) marks the bright nucleus of the galaxy; the bright blob near the center of the image corresponds to the bright "knot" visible in the jet in (b). (NOAO; NASA)

Quasars

In the early days of radio astronomy, many radio sources were detected for which no corresponding visible object was known. By 1960 several hundred such sources were listed in the *Third Cambridge Catalog* and astronomers were scanning the skies in search of visible counterparts to these radio sources. Their job was made difficult both by the low resolution of the radio observations (which meant that the observers did not know exactly where to look) and by the faintness of these objects at visible wavelengths.

In 1960 astronomers detected what appeared to be a faint blue star at the location of the radio source 3C 48 (the 48th object on the third *Cambridge* list) and obtained its spectrum. Containing many unknown broad emission lines, the unusual spectrum defied interpretation. 3C 48 remained a unique curiosity until 1962, when another similar-looking, and similarly mysterious, faint blue object with "odd" spectral lines was discovered and identified with the radio source 3C 273 (Figure 15.27).

The following year saw a breakthrough when astronomers realized that the strongest unknown lines in 3C 273's spectrum were simply familiar spectral lines of hydrogen redshifted by a very unfamiliar amount—about 16 percent, corresponding to a recession velocity of 48,000 km/s. Figure 15.28 shows the spectrum of 3C 273. Some prominent emission lines and the extent of their redshift are marked on the diagram. Once the nature of the strange spectral lines was known, astronomers quickly found a similar explanation for the spectrum of 3C 48. Its 37 percent redshift implied that it is receding from

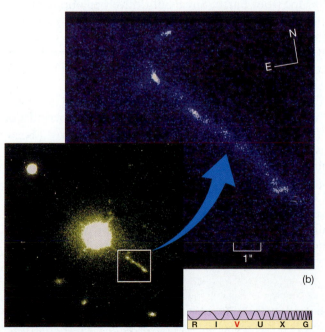

◀ **Figure 15.27** **Quasar 3C 273** (a) The bright quasar 3C 273 displays a luminous jet of matter, but the main body of the quasar is starlike in appearance. (b) The jet extends for about 30 kpc and can be seen better in this high-resolution image. (AURA)

Red — Hβ — Hγ — Hδ — Red shift — 3C 273 — Blue

603 nm — 502 nm — Hβ — Hγ — Hδ — 390 nm — Comparison spectrum

◄ Figure 15.28 Quasar Spectrum Optical spectrum of the distant quasar 3C 273. Notice both the redshift and the widths of the three hydrogen spectral lines marked as Hβ, Hγ, and Hδ. ∞ *(Sec. 2.6)* The redshift indicates the quasar's enormous distance. The width of the lines implies rapid internal motion within the quasar. (Adapted from Palomar/Caltech)

Earth at almost one-third the speed of light. These huge speeds mean that neither of these two objects can possibly be members of our Galaxy. In fact, their large redshifts mean that they are very far away indeed. Applying Hubble's law (with our adopted value of the Hubble constant $H_0 = 70$ km/s/Mpc), we obtain distances of 650 Mpc for 3C 273 and 1400 Mpc for 3C 48. (See *More Precisely 15-1* and Chapter 17 for an account of how these distances are determined and an explanation of what these large redshifts really mean.)

However, this explanation of the unusual spectra created an even deeper mystery. A simple calculation using the inverse-square law reveals that despite their unimpressive optical appearance, these faint "stars" are in fact the brightest known objects in the universe! ∞ *(Sec. 10.2)* 3C 273, for example, has a luminosity of about 10^{40} W, comparable to 20 trillion Suns or 1000 Milky Way Galaxies. More generally, quasars range in luminosity from around 10^{38} W—about the same as the brightest Seyferts—up to nearly 10^{42} W. A value of 10^{40} W (comparable to the luminosity of a bright radio galaxy) is fairly typical.

Clearly not stars (with such enormous luminosities), these objects became known as *quasi-stellar radio sources* (quasi-stellar means "starlike"), or **quasars**. Because we now know that not all such highly redshifted, starlike objects are strong radio sources, the term **quasi-stellar object** (or QSO) is more common today. However, the name quasar persists, and we will continue to use it here. More than 30,000 quasars are now known, and the numbers are increasing rapidly as large-scale surveys probe deeper and deeper into space. The distance to the *closest* quasar is 240 Mpc, the farthest lies some 9000 Mpc away. Most quasars lie well over 1000 Mpc from Earth. Since light travels at a finite speed, these faraway objects represent the universe as it was in the distant past. The implication is that most quasars date back to much earlier periods of galaxy formation and evolution rather than more recent times. The prevalence of these energetic objects at great distances tells us that the universe was once a much more violent place than it is today.

Quasars share many properties with Seyferts and radio galaxies. Their radiation is nonstellar and may vary irregularly in brightness over periods of months, weeks, days, or (in some cases) even hours. Some also show evidence of jets and extended emission features. Note the jet of luminous matter in 3C 273 (Figure 15.27), reminiscent of the jet in M87 and extending nearly 30 kpc from the center of the quasar. Figure 15.29 shows a quasar with radio lobes similar to those seen in Cygnus A (Figure 15.23b). Quasars have been observed in all parts of the electromagnetic spectrum, although many emit most of their energy in the infrared. As mentioned earlier, most astronomers think that quasars *are* in fact the intensely bright cores of distant active galaxies lying too far away for the galaxies themselves to be seen.

CONCEPT CHECK ▶▶▶

The energy emission from an active galactic nucleus does not resemble a blackbody curve. Why is this important?

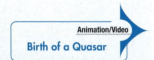

Animation/Video
Birth of a Quasar

Quasar

Lobes

R I V U X G

◄ Figure 15.29 Quasar Jets Radio image of the quasar 3C 175, showing radio jets feeding faint radio lobes. The bright (white) central object is the quasar, some 3000 Mpc away. The lobes themselves span approximately a million light-years—comparable in size to the radio galaxies discussed earlier. (NRAO)

15.5 The Central Engine of an Active Galaxy

The present consensus among astronomers is that, despite their differences in appearance and luminosity, Seyferts, radio galaxies, and quasars share a common energy-generation mechanism. As a class, active galactic nuclei have some or all of the following properties:

1. They have *high luminosities*, generally greater than the 10^{37} W characteristic of a bright normal galaxy.
2. Their energy emission is mostly *nonstellar*—it cannot be explained as the combined radiation of even trillions of stars.
3. Their energy output can be highly *variable*, implying that it is emitted from a small central nucleus much less than a parsec across.
4. They often exhibit *jets* and other signs of explosive activity.
5. Their optical spectra may show broad emission lines, indicating *rapid internal motion* within the energy-producing region.

The principal questions then are: How can such vast quantities of energy arise from these relatively small regions of space? Why is the radiation nonstellar? And what is the origin of the jets and extended radio-emitting lobes? We first consider how the energy is *produced*, then turn to the question of how it is actually *emitted* into intergalactic space.

Energy Production

To develop a feeling for the enormous emissions of active galaxies, consider for a moment an object having a luminosity of 10^{38} W. In and of itself, this energy output is not inconceivably large. The brightest giant ellipticals are comparably powerful. Thus, some 10^{12} stars—a few normal galaxies' worth of material—could *equivalently* power a typical active galaxy. However, in an active galaxy this energy production is packed into a region much less than a parsec in diameter!

The twin requirements of large energy generation and small physical size bring to mind our discussion of X-ray sources in Chapter 13. ∞ *(Sec. 13.3, 13.8)* The presence of the jets in M87 and 3C 273 and the radio lobes in Centaurus A and Cygnus A strengthen the connection, because similar phenomena have also been observed in some stellar X ray–emitting systems. Recall that the best current explanation for those "small-scale" phenomena involves the accretion of material onto a compact object—a neutron star or a black hole. Large amounts of energy are produced as matter spirals down onto the central object. In Chapter 14 we suggested that a similar mechanism, involving a *supermassive black hole*—having a mass of around 4 million suns—may also be responsible for the energetic radio and infrared emission observed at the center of our own Galaxy. ∞ *(Sec. 14.7)*

As illustrated in Figure 15.30, the leading model for the central engine of active galaxies is basically a scaled-up version of the same accretion process, only now the black holes involved are millions or even billions of times more massive than the

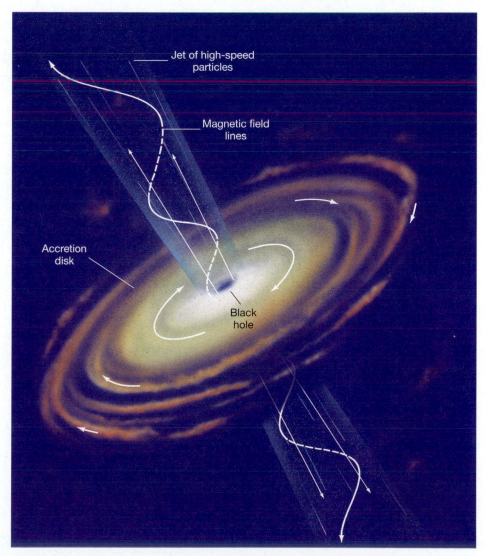

▼ **Figure 15.30 Active Galactic Nucleus** The leading theory for the energy source in active galactic nuclei holds that these objects are powered by material accreting onto a supermassive black hole. As matter spirals toward the hole, it heats up, producing large amounts of energy. At the same time, high-speed jets of gas may be ejected perpendicular to the accretion disk, forming the jets and lobes seen in many active objects. Magnetic fields generated in the disk are carried by the jets out to the radio lobes, where they play a crucial role in producing the observed radiation.

Jet of high-speed particles

Magnetic field lines

Accretion disk

Black hole

Sun. As with this model's smaller-scale counterparts, infalling gas forms an accretion disk and spirals down toward the black hole. It is heated to high temperatures by friction within the disk and emits large amounts of radiation as a result. In this case, however, the origin of the accreted gas is not a binary companion, as in stellar X-ray sources, but entire stars and clouds of interstellar gas, most likely diverted into the galactic center by an encounter with another galaxy, that come too close to the hole and are torn apart by its strong gravity.

Accretion is extremely efficient at converting infalling mass (in the form of gas) into energy (in the form of electromagnetic radiation). Detailed calculations indicate that as much as 10 or 20 percent of the total mass–energy of the infalling matter can be radiated away before it crosses the hole's event horizon and is lost forever. ∞ *(Sec. 13.5)* Since the total mass–energy of a star like the Sun—the mass times the speed of light squared—is about 2×10^{47} J, it follows that the 10^{38} W luminosity of a bright active galaxy can be accounted for by the consumption of "only" one solar mass of gas per decade by a billion-solar-mass black hole. More or less luminous active galaxies would require correspondingly more or less fuel. Thus, to power a 10^{40}-W quasar, which is 100 times brighter, the black hole simply consumes 100 times more fuel, or 10 stars per year. The central black hole of a 10^{36}-W Seyfert galaxy would devour only one Sun's worth of material every thousand years.

The small size of the emitting region is a direct consequence of the compact central black hole. Even a billion-solar-mass black hole has a radius of only 3×10^9 km, or 10^{-4} pc—about 20 AU—and theory suggests that the part of the accretion disk responsible for most of the emission would be much less than 1 pc across. ∞ *(Sec. 13.5)* Instabilities in the accretion disk can cause fluctuations in the energy released, leading to the variability observed in many objects. The broadening of the spectral lines observed in the nuclei of many active galaxies may result from the rapid orbital motion of the gas in the black hole's intense gravity.

The jets consist of material (mainly protons and electrons) blasted out into space—and completely out of the visible portion of the galaxy—from the inner regions of the disk. The details of how jets form remain uncertain, but there is a growing consensus among theorists that jets are a common feature of accretion flows, large and small. They are most likely formed by strong magnetic fields produced within the accretion disk itself, which accelerate charged particles to nearly the speed of light and eject them parallel to the disk's rotation axis. Figure 15.31 shows a *Hubble Space Telescope* image of a disk of gas and dust at the core of the radio galaxy NGC 4261 in the Virgo Cluster. Consistent with the theory just described, the disk is perpendicular to the huge jets emanating from the galaxy's center.

Figure 15.32 shows imaging and spectroscopic data from the center of M87, suggesting a rapidly rotating disk of matter orbiting the galaxy's center,

Animation/Video
Cosmic Jets

(a) R I V U X G   (b)   R I V U X G

◀ **Figure 15.31 Giant Elliptical Galaxy** (a) A combined optical/radio image of the giant elliptical galaxy NGC 4261, in the Virgo Cluster, shows a white visible galaxy at center, from which blue-orange (false color) radio lobes extend for about 60 kpc. (b) A close-up photograph of the galaxy's nucleus reveals a 100-pc-diameter disk surrounding a bright hub thought to harbor a black hole. (NRAO; NASA)

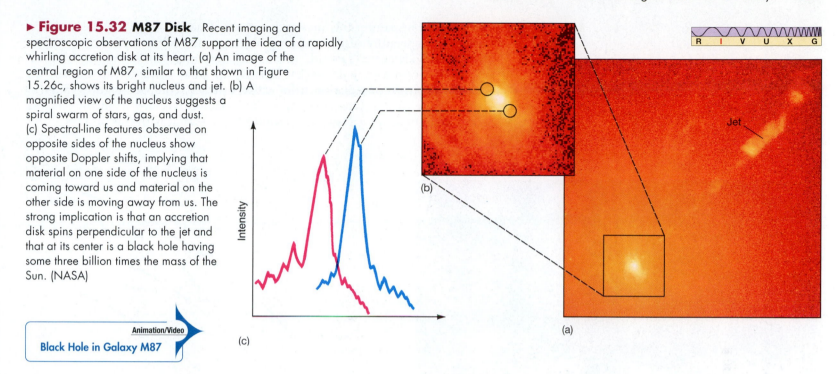

▶ **Figure 15.32 M87 Disk**   Recent imaging and spectroscopic observations of M87 support the idea of a rapidly whirling accretion disk at its heart. (a) An image of the central region of M87, similar to that shown in Figure 15.26c, shows its bright nucleus and jet. (b) A magnified view of the nucleus suggests a spiral swarm of stars, gas, and dust. (c) Spectral-line features observed on opposite sides of the nucleus show opposite Doppler shifts, implying that material on one side of the nucleus is coming toward us and material on the other side is moving away from us. The strong implication is that an accretion disk spins perpendicular to the jet and that at its center is a black hole having some three billion times the mass of the Sun. (NASA)

Animation/Video
Black Hole in Galaxy M87

again perpendicular to the jet. Measurements of the gas velocity on opposite sides of the disk indicate that the mass within a few parsecs of the center is approximately 3×10^9 solar masses—we assume that this is the mass of the central black hole. At M87's distance, *HST*'s resolution of 0.05 arc second corresponds to a scale of about 5 pc, so we are still far from seeing the (solar system–sized) central black hole itself, but the improved "circumstantial" evidence has convinced many astronomers of the basic correctness of the theory.

Energy Emission

Theory suggests that the radiation emitted by the hot accretion disk surrounding a supermassive black hole would span a broad range of wavelengths, from infrared through X rays, which accounts for the observed spectra of at least some active galactic nuclei. However, as mentioned earlier, it appears that in many cases the high-energy radiation emitted from the accretion disk itself is "reprocessed"—that is, absorbed and reemitted at longer, particularly infrared, wavelengths—by dust surrounding the nucleus before eventually reaching our detectors.

Researchers believe that the most likely site of this reprocessing is in a rather fat, donut-shaped ring of gas and dust surrounding the inner accretion disk where the energy is actually produced. As illustrated in Figure 15.33, if our line of sight to the black hole does not intersect the dusty donut, we see the "bare" energy source, emitting large amounts of high-energy radiation. If the donut intervenes, we see instead large amounts of infrared radiation reradiated from the dust.

A different reprocessing mechanism operates in many jets and radio lobes. It involves the *magnetic fields* possibly produced within the accretion disk and transported by the jets into intergalactic space (Figure 15.30). As sketched in Figure 15.34(a), whenever a charged particle (here an electron) encounters a magnetic field, the particle tends to spiral around the magnetic field lines. We have already encountered this idea in the discussions of Earth's magnetosphere and solar activity. ∞ *(Sec. 5.7, 9.4)*

As the particles whirl around, they emit electromagnetic radiation. ∞ *(Sec. 2.2)* The radiation produced in this way—called **synchrotron radiation**, after the type of particle accelerator in which it was first observed—is *nonthermal* in nature: this means that there is no link between the emission and the temperature of the radiating object, so the radiation is not described by a blackbody curve. Instead, its intensity decreases with increasing frequency, as shown in Figure 15.34(b). This is just what is needed to explain the overall spectrum of radiation from active galaxies (compare Figure 15.34(b) with Figure 15.18). Observations of the radiation received from the jets and radio lobes of active galaxies are completely consistent with this process.

(a)

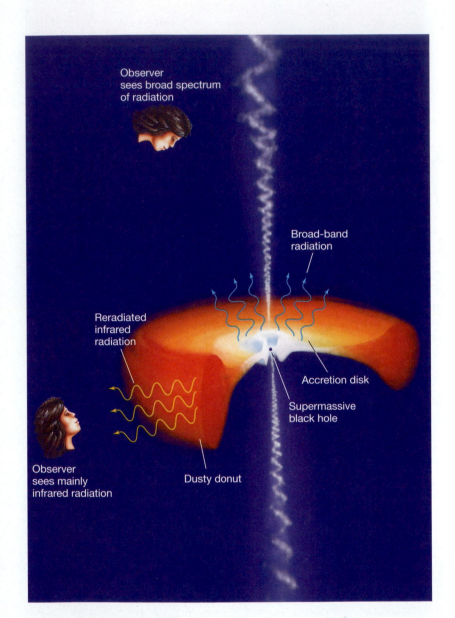

▲ **Figure 15.33 Dusty Donut** The accretion disk surrounding a massive black hole, drawn here with some artistic license, consists of hot gas at many different temperatures (hottest nearest the center). When viewed from above or below, the disk radiates a broad spectrum of electromagnetic energy extending into the X ray band. However, the dusty infalling gas that ultimately powers the system is thought to form a rather fat, donut-shaped region outside the accretion disk (shown here in dull red), which effectively absorbs much of the high-energy radiation reaching it, reemitting it mainly in the form of cooler, infrared radiation. Thus, when viewed from the side, strong infrared emission is observed. The appearance of the jets, radiating mostly in radio and X-rays, also depends on the viewing angle (see Figure 15.25). (Adapted from D. Berry)

Animation/Video
Supermassive Black Hole

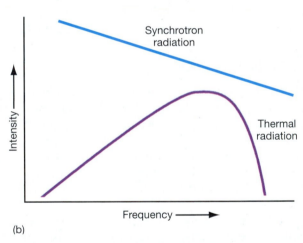

(b)

▲ **Figure 15.34 Nonthermal Radiation**
(a) Charged particles, especially fast-moving electrons (red), emit synchrotron radiation (blue) while spiraling in a magnetic field (black). This process is not confined to active galaxies. It occurs, on smaller scales, when charged particles interact with magnetism in Earth's Van Allen belts ∞ *(Sec. 5.6)*, when charged matter arches above sunspots on the Sun ∞ *(Sec. 9.4)*, in the vicinity of neutron stars ∞ *(Sec. 13.2)*, and at the center of our own Galaxy. ∞ *(Sec. 14.7)*
(b) Variation of the intensity of thermal and synchrotron (nonthermal) radiation with frequency. Thermal radiation, described by a blackbody curve, peaks at some frequency that depends on the temperature of the source. Nonthermal synchrotron radiation, by contrast, is more intense at low frequencies and is independent of the temperature of the emitting object. Compare this figure with Figure 15.18.

Eventually, the jet is slowed and stopped by the intergalactic medium, the flow becomes turbulent, and the magnetic field grows tangled. The result is a gigantic radio lobe emitting virtually all of its energy in the form of synchrotron radiation. Thus, even though the radio *emission* comes from an enormously extended volume of space that dwarfs the visible galaxy, the *source* of the energy is still the accretion disk—a billion billion times smaller in volume than the radio lobe—lying at the galactic center. The jets serve merely as a conduit to transport energy from the nucleus, where it is generated, into the lobes, where it is finally radiated into space.

What do active galaxies look like between active outbursts? What is their connection with the normal galaxies we see? To answer these important questions we must delve more deeply into the subject of *galaxy evolution*, to which we now turn in Chapter 16.

◄◄◄ CONCEPT CHECK

How does accretion onto a supermassive black hole power the energy emission from the extended radio lobes of a radio galaxy?

CHAPTER REVIEW

Summary

1 The **Hubble classification scheme** (p. 402) divides galaxies into several classes, depending on their appearance. **Spiral galaxies** (p. 402) have flattened disks, central bulges, and spiral arms. Their halos consist of old stars, whereas the gas-rich disks are the sites of ongoing star formation. **Barred-spiral galaxies** (p. 403) contain an extended "bar" of material projecting beyond the central bulge. **Elliptical galaxies** (p. 404) have no disk and contain little or no cool gas or dust, although very hot interstellar gas is observed. In most cases, they consist entirely of old stars. They range in size from dwarf ellipticals, which are much less massive than the Milky Way Galaxy, to giant ellipticals, which may contain trillions of stars. **S0** and **SB0 galaxies** (p. 405) are intermediate in their properties between ellipticals and spirals. **Irregular galaxies** (p. 406) are galaxies that do not fit into either of the other categories. Some may be the result of galaxy collisions or close encounters. Many irregulars are rich in gas and dust and are the sites of vigorous star formation.

2 Astronomers often use **standard candles** (p. 409) as distance-measuring tools. These are objects that are easily identifiable and whose luminosities lie in some reasonably well-defined range. Comparing luminosity and apparent brightness, astronomers determine the distance using the inverse-square law. An alternative approach is the **Tully–Fisher relation** (p. 409), an empirical correlation between rotational velocity and luminosity in spiral galaxies.

3 The Milky Way, Andromeda, and several other smaller galaxies form the **Local Group** (p. 410), a small **galaxy cluster** (p. 411). Galaxy clusters consist of a collection of galaxies orbiting one another, bound together by their own gravity. The nearest large galaxy cluster to the Local Group is the Virgo Cluster.

4 Distant galaxies are observed to be receding from the Milky Way at speeds proportional to their distances from us. This relationship is called **Hubble's law** (p. 412). The constant of proportionality in the law is **Hubble's constant** (p. 413). Its value is approximately 70 km/s/Mpc. Astronomers use Hubble's law to determine distances to the most remote objects in the universe simply by measuring their redshifts and converting the corresponding speed directly to a distance. The redshift associated with the Hubble expansion is called the **cosmological redshift** (p. 412).

5 **Active galaxies** (p. 415) are much more luminous than normal galaxies and have nonstellar spectra, emitting most of their energy outside the visible part of the electromagnetic spectrum. Often the nonstellar activity suggests rapid internal motion and is associated with a bright central **active galactic nucleus** (p. 415). Many active galaxies have high-speed, narrow jets of matter shooting out from their central nuclei. Often the the jets transport energy from the nucleus, where it is generated, to enormous **radio lobes** (p.416) lying far beyond the visible portion of the galaxy the lobes, where it is radiated into space. The jets often appear to be made up of distinct "blobs" of gas, suggesting that the process generating the energy is intermittent.

6 A **Seyfert galaxy** (p. 415) looks like a normal spiral, except that the Seyfert has an extremely bright central **galactic nucleus** (p. 416). Spectral lines from Seyfert nuclei are very broad, indicating rapid internal motion, and rapid variability implies that the source of the radiation is much less than one light-year across. **Radio galaxies** (p. 416) emit large amounts of energy in the radio part of the spectrum. The corresponding visible galaxy is usually elliptical. **Quasars** (p. 422), or **quasistellar objects** (p. 422) are the most luminous objects known. In visible light they appear starlike, and their spectra are usually substantially redshifted. All quasars are very distant, indicating that we see them as they were in the distant past.

7 The generally accepted explanation for the observed properties of all active galaxies is that the energy is generated by accretion of galactic gas onto a supermassive (million to billion-solar-mass) black hole lying in the galactic center. The small size of the accretion disk explains the compact extent of the emitting region, and the high-speed orbit of gas in the black hole's intense gravity accounts for the rapid motion observed. Much of the high-energy radiation emitted from the disk is absorbed by a fat, donut-shaped

region outside the disk, and reemitted in the form of infrared radiation. Typical active galaxy luminosities require the consumption of about one solar mass of material every few years. Some of the infalling matter is blasted out into space, producing magnetized jets that create and feed the radio lobes. Charged particles spiraling around the magnetic field lines produce **synchrotron radiation** (p. 425) whose spectrum is consistent with the nonstellar radiation observed in radio galaxies and jets.

Review and Discussion

1. What are the differences between the various types of spiral galaxy?

2. Compare and contrast an elliptical galaxy with the halo of the Milky Way.

3. Describe the four rungs in the distance-measurement ladder involved in determining the distance to a galaxy lying 5 Mpc away.

4. How does the Tully–Fisher relation allow astronomers to measure the distances to galaxies?

5. What is the Virgo Cluster?

6. What is Hubble's law, and how is it used by astronomers to measure distances to galaxies?

7. What is the most likely range of values for Hubble's constant? Why is the exact value uncertain?

8. Why do you think astronomers prefer to speak in terms of redshifts rather than distances to faraway objects?

9. Name two basic differences between normal and active galaxies.

10. What is the evidence that the radio lobes of some active galaxies consist of material ejected from the galaxy's nucleus?

11. How do astronomers know that the energy-producing region of an active galaxy must be very small?

12. Briefly describe the leading model for the central engine of an active galaxy.

13. How do astronomers account for the differences in the spectra observed from active galaxies?

14. What is synchrotron radiation, and what does it tell us about energy emission from active galaxies?

15. How do we know that quasars are extremely luminous?

Self-Test: True or False?/Multiple Choice

1. Most galaxies are spirals like the Milky Way. **(T/F)**

2. Most elliptical galaxies contain only old stars. **(T/F)**

3. Irregular galaxies, although small, often have lots of star formation taking place in them. **(T/F)**

4. Type I supernovae can be used to determine distances to galaxies. **(T/F)**

5. Most galaxies are receding from the Milky Way Galaxy. **(T/F)**

6. Hubble's law can be used to determine distances to the farthest objects in the universe. **(T/F)**

7. Active galaxies can emit thousands of times more energy than our own Galaxy. **(T/F)**

8. Using the method of standard candles, we could, in principle, estimate the distance to a campfire if we knew **(a)** the number of logs used; **(b)** the fire's temperature; **(c)** the length of time the fire has been burning; **(d)** the type of wood used in the fire.

9. Within 30 Mpc of the Sun, there are about **(a)** 3 galaxies; **(b)** 30 galaxies; **(c)** a few thousand galaxies; **(d)** a few million galaxies.

10. Hubble's law states that **(a)** more distant galaxies are younger; **(b)** the greater the distance to a galaxy, the greater is the

galaxy's redshift; **(c)** most galaxies are found in clusters; **(d)** the greater the distance to a galaxy, the fainter the galaxy appears.

11. If the light from a galaxy fluctuates in brightness very rapidly, the region producing the radiation must be **(a)** very large; **(b)** very small; **(c)** very hot; **(d)** rotating very rapidly.

12. Quasar spectra **(a)** are strongly redshifted; **(b)** show no spectral lines; **(c)** look like the spectra of stars; **(d)** contain emission lines from unknown elements.

13. Active galaxies are very luminous because they **(a)** are hot; **(b)** contain black holes in their cores; **(c)** are surrounded by hot gas; **(d)** emit jets.

14. If the galaxy in Figure 15.10 (Galaxy Rotation) were smaller and spinning more slowly, then, in order to represent it correctly, the figure should be redrawn to show **(a)** a greater blueshift; **(b)** a greater redshift; **(c)** a narrower combined line; **(d)** a larger combined amplitude.

15. According to Figure 15.18 (Galaxy Energy Distribution), active galaxies **(a)** emit most of their energy at long wavelengths; **(b)** emit very little energy at high frequencies; **(c)** emit large amounts of energy at all wavelengths; **(d)** emit most of their energy in the visible part of the spectrum.

Problems

Algorithmic versions of these questions are available in the Practice Problems module of the Companion Website.

The number of squares preceding each problem indicates its approximate level of difficulty.

1. ■ The Andromeda Galaxy is approaching our Galaxy with a radial velocity of 266 km/s. Given the galaxies' present separation of 800 kpc, and neglecting both the transverse component of the velocity and the effect of gravity in accelerating the motion, estimate when the two galaxies will collide.

2. ■■ A Cepheid variable star in the Virgo Cluster has an apparent magnitude of 26 and is observed to have a pulsation period of 20 days. Use these numbers and Figure 14.6 to estimate the distance to the Virgo Cluster. The absolute magnitude of the Sun is +4.85.

3. ■ A supernova of luminosity one billion times the luminosity of the Sun is used as a standard candle to measure the distance to a faraway galaxy. From Earth the supernova appears as bright as the Sun would appear from a distance of 10 kpc. What is the distance to the galaxy?

4. ■ According to Hubble's law, with $H_0 = 70$ km/s/Mpc, what is the recessional speed of a galaxy at a distance of 200 Mpc? How far away is a galaxy whose recessional speed is 4000 km/s? How would these answers change if (a) $H_0 = 50$ km/s/Mpc, (b) $H_0 = 80$ km/s/Mpc?

5. ■■ If $H_0 = 70$ km/s/Mpc, how long will it take for the distance from the Milky Way Galaxy to the Virgo Cluster to double? How long for the distance to the Coma Cluster (100 Mpc) to double?

6. ■■ A quasar is observed to have luminosity fluctuations and broadened emission lines indicating a speed of 5000 km/s within 0.1 light-years of its center. Assuming circular orbits, use Kepler's laws to estimate the mass within this radius. ∞ *(Sec. 14.6)*

7. ■ Assuming a jet speed of 75 percent of the speed of light, calculate the time taken for material in Cygnus A's jet to cover the 250 kpc between the galaxy's nucleus and its radio-emitting lobes.

8. ■■ Assuming the upper end of the efficiency range indicated in the text, calculate how much energy an active galaxy would generate if it consumed one Earth mass of material every day. Compare this value with the luminosity of the Sun.

9. ■ On the basis of the data presented in the text, calculate the orbital speed of material orbiting at a distance of 0.25 pc from the center of M87.

10. ■■■ Calculate the amount of energy received per unit area per unit time that would be observed at Earth from a 10^{37}-W Seyfert nucleus located at the Galactic center, 8 kpc away, neglecting the effects of absorption by interstellar dust. Using the data presented in Appendix 3, Table 3, compare with the energy received from Sirius A, the brightest star in the night sky. From what you know about active galaxy energy emission, is it reasonable to ignore interstellar absorption?

The Companion Website at www.prenhall.com/chaisson provides algorithmically generated versions of each chapter's Problems, in addition to an Animations and Videos gallery, a Tutorials gallery and Tutorials Quiz questions, a full eBook, an Images gallery, Practice Questions, Multiple Choice, and True or False quizzes, and Labeling exercises.

16

Galaxies and Dark Matter

The Large-Scale Structure of the Cosmos

On scales much larger than even the largest galaxy clusters, new levels of structure are revealed and a humbling new reality emerges. We may be star stuff, the product of countless cycles of stellar evolution, but we are not the stuff of the cosmos. The universe in the large is composed of matter fundamentally different from the familiar atoms and molecules that compose our bodies, our planet, our star and galaxy, and all the luminous matter we observe in the heavens. Only its gravity announces the presence of this strange kind of matter. By comparing and classifying the properties of galaxies near and far, astronomers have begun to understand their evolution. By mapping out their distribution in space, astronomers trace out the immense realms of the universe. Points of light in the uncharted darkness, they remind us that our position in the universe is no more special than that of a boat adrift at sea.

Active galaxies are much more energetic than the normal galaxy—the Milky Way—in which we live. The "central engines" powering them are thought to be supermassive black holes. This is a double image, superposing optical light (acquired by the Hubble Telescope in Earth orbit) on top of radio radiation (captured by the Very Large Array in New Mexico). At center (in white) is a giant, visible elliptical galaxy catalogued as NGC1316, which is probably devouring its small northern neighbor. The result is the complex radio emission (in orange), called Fornax A, that extends more than a million light-years across. (NRAO/STSCI)

LEARNING GOALS

Studying this chapter will enable you to:

1 Describe some of the methods used to determine the masses of distant galaxies.

2 Explain why astronomers think that most of the matter in the universe is invisible.

3 Discuss some theories of how galaxies form and evolve.

4 Explain the role of active galaxies in current theories of galactic evolution.

5 Summarize what is known about the large-scale distribution of galaxies in the universe.

6 Describe some techniques used by astronomers to probe the universe on large scales.

Visit www.prenhall.com/chaisson for additional annotated images, animations, and links to related sites for this chapter.

16.1 Dark Matter in the Universe

In Chapter 14 we saw how measurements of the orbital velocities of stars and gas in our own Galaxy reveal the presence of an extensive *dark-matter halo* surrounding the galaxy we see. ∞ *(Sec. 14.6)* Do other galaxies have similar dark halos? And what evidence do we have for dark matter on larger scales? To answer these questions, we need a way to determine the masses of galaxies and galaxy clusters.

How can we measure the masses of such large systems? Surely, we can neither count all their stars nor estimate their interstellar content very well. Galaxies are just too complex to take direct inventory of their material makeup. Instead, we must rely on indirect techniques. Despite their enormous sizes, galaxies and galaxy clusters obey the same physical laws as do the planets in our own solar system. To determine galaxy masses, we turn as usual to Newton's law of gravity.

Galaxy Masses

Astronomers can calculate the masses of some spiral galaxies by determining their *rotation curves*, which plot rotation speed versus distance from the galactic center (Figure 16.1a). ∞ *(Sec. 14.6)* The mass within any given radius then follows directly from Newton's laws. Rotation curves for a few nearby spirals are shown in Figure 16.1(b). They imply masses ranging from about 10^{11} to 5×10^{11} solar masses within about 25 kpc of the center—comparable to the results obtained for our own Galaxy using the same technique. For galaxies too distant for detailed rotation curves to be drawn, the overall rotation speed can still be inferred from the broadening of spectral lines, as described in the previous chapter. ∞ *(Sec. 15.2)*

This approach is useful for measuring the mass lying within about 50 kpc of a galaxy's center (the extent of the electromagnetic emission from stellar and interstellar material). To probe farther from the centers of galaxies, astronomers turn to *binary* galaxy systems (Figure 16.2a), whose components may lie hundreds of kiloparsecs apart. The orbital period of such a system is typically billions of years, far too long for the orbit to be accurately measured. However, by estimating the period and semi-major axis using the information available—the line-of-sight velocities and angular separation of the components—an approximate total mass can be derived. ∞ *(More Precisely 1-1)*

Galaxy masses obtained in this way are fairly uncertain. However, by combining many such measurements, astronomers can obtain quite reliable *statistical* information about galaxy masses. Most normal spirals (the Milky Way Galaxy included) and large ellipticals contain between 10^{11} and 10^{12} solar masses of material. Irregular galaxies often contain less mass, about 10^8 to 10^{10} times that of the Sun. Dwarf ellipticals and dwarf irregulars can contain as little as 10^6 or 10^7 solar masses of material.

We can use another statistical technique to derive the combined mass of all the galaxies within a galaxy cluster. As depicted in Figure 16.2(b), each galaxy within a cluster moves relative to all other cluster members, and we can estimate the cluster's mass simply by asking how massive it must be in order to bind its galaxies gravitationally. Typical cluster masses obtained in this way lie

(a)

(b)

◄ **Figure 16.1 Galaxy Rotation Curves** (a) By observing orbital velocities at many different distances from the center of a disk galaxy, astronomers can plot a rotation curve for the galaxy. This is M64, the "Evil Eye" galaxy, some 5 Mpc distant. (b) Rotation curves for some nearby spiral galaxies indicate masses of a few hundred billion times the mass of the Sun. The corresponding curve for our own Galaxy (from Figure 14.19b) is marked in red for comparison (NASA)

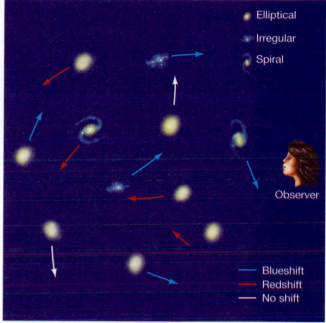

▶ Figure 16.2 Galaxy Masses (a) In a binary galaxy system, galaxy masses may be estimated by observing the orbit of one galaxy about the other. (b) The mass of a galaxy cluster may be estimated by observing the motion of many galaxies in the cluster and then estimating how much mass is needed to prevent the cluster from flying apart.

in the range of 10^{13}–10^{14} solar masses. Notice that this calculation gives us no information whatsoever about the masses of individual galaxies. It tells us only about the *total* mass of the entire cluster.

Visible Matter and Dark Halos

Radio studies indicate that the rotation curves of the spiral galaxies shown in Figure 16.1(b) remain flat (that is, do not decline and may even rise slightly) far beyond the visible images of the galaxies themselves, implying that these galaxies—and perhaps all spiral galaxies—have invisible dark halos similar to that surrounding the Milky Way. ∞ *(Sec. 14.5)* Spiral galaxies seem to contain from 3 to 10 times more mass than can be accounted for in the form of visible matter. Some studies of elliptical galaxies suggest similarly large dark halos surrounding them too, but these observations are still far from conclusive.

Curiously, some of the least massive galaxies seem to have the largest fractions of dark matter, raising the intriguing possibility of *dark galaxies*—systems comprised almost entirely of dark matter, emitting virtually no visible light. Direct observational evidence for such objects is naturally hard to obtain, although some astronomers have suggested that the peculiar appearance of the dwarf elliptical galaxy UGC 10214 (Figure 16.3), which has a stream of matter flowing out of it, apparently toward nothing, may be due to a dark companion.

Astronomers find even greater discrepancies between visible light and total mass when they study galaxy clusters. Calculated cluster masses range from 10 to nearly 100 times the mass suggested by the light emitted by individual cluster galaxies. Put another way, a lot more mass is needed to bind galaxy clusters than we can see. Thus the problem of dark matter exists not just in our own Galaxy but also in other galaxies and, to an even greater degree, in galaxy clusters too. In that case, we are compelled to accept the fact that *upward of 90 percent of the matter in the universe is dark.* As noted in Chapter 14, this matter is not just dark in the visible portion of the spectrum—it is undetected at *any* electromagnetic wavelength. ∞ *(Sec. 14.5)*

The dark matter in clusters apparently cannot simply be the accumulation of dark matter within individual galaxies. Even including the galaxies' dark halos, we still cannot account for all the dark matter in galaxy clusters. As we look on larger and larger scales, we find that a larger and larger fraction of the matter in the universe is dark.

Intracluster Gas

In addition to the luminous matter observed within the cluster galaxies themselves, astronomers also have evidence for large amounts of *intracluster gas*—superhot (more than 10 million K), diffuse intergalactic matter filling the space among the galaxies. Satellites orbiting above Earth's atmosphere have detected X rays from hot gas in many clusters. Figure 16.4 shows false-color X-ray images of two such systems. In each case, the X ray–emitting region is centered on, and comparable in size to, the visible cluster image.

▶ Figure 16.3 Dark Galaxy? Astronomers speculate that some galaxies may be composed almost entirely of dark matter, emitting virtually no visible light. Has the long stream of gas leaving this galaxy, called UGC 10214, been torn out by a close encounter with a dark companion at bottom right? (NASA/ESA)

R I V U X G

433

▶ **Figure 16.4** **Galaxy Cluster X-Ray Emission**
(a) X-ray image of Abell 85, an old, distant cluster of galaxies, taken by the *Einstein* X-ray Observatory. The cluster's X-ray emission is shown in orange. The green graphs display a smooth, peaked intensity profile centered on the cluster but not associated with individual galaxies. (b) A contour map of the X rays is superimposed on an optical photo, showing that its X rays are strongest around Abell 85's central supergiant galaxy. Images like these demonstrate that the space between the galaxies within galaxy clusters is filled with superheated gas. (c) Superposition of infrared and X-ray radiation from another distant galaxy cluster. The X rays are shown as a fuzzy, bluish cloud of hot gas filling the intracluster spaces among the galaxies. Part (d) is a longer infrared exposure of the central region, showing the richness of this cluster, which spans about a million parsecs. (NASA; AURA; ESA)

(a) R I V U X G

(b) R I V U X G

(c) R I V U X G

(d) R I V U X G

CONCEPT CHECK ▶▶▶

What assumptions are we making when we infer the mass of a galaxy cluster from observations of the spectra of its constituent galaxies?

Further evidence for intracluster gas can be found in the appearance of the radio lobes of some active galaxies. ∞ *(Sec. 15.4)* In some systems, known as *head–tail* radio galaxies, the lobes seem to form a "tail" behind the main part of the galaxy. For example, the lobes of radio galaxy NGC 1265, shown in Figure 16.5, appear to be "swept back" by some onrushing wind, and, indeed, this is the most likely explanation for the galaxy's appearance. If NGC 1265 were at rest, it would be just another double-lobe source, perhaps quite similar to Centaurus A (see Figure 15.21). However, the galaxy is traveling through the intergalactic medium of its parent galaxy cluster (known as the Perseus Cluster), and the outflowing matter forming the lobes tends to be left behind as NGC 1265 moves.

How much gas do these observations reveal? At least as much matter—and in a few cases substantially more—exists within clusters in the form of hot gas as is visible in the form of stars. This is a lot of material, but it still doesn't solve the dark-matter problem. To account for the total masses of galaxy clusters implied by dynamical studies, we would have to find from 10 to 100 times more mass in gas than in stars.

(a)

(b)

R I V U X G

R I V U X G

◄ **Figure 16.5** **Head–Tail Radio Galaxy**
(a) Radiograph, in false color, of the head–tail radio galaxy NGC 1265. (b) The same radio data, in contour form, superimposed on the optical image of the galaxy. Astronomers reason that this object is moving rapidly through space, trailing a "tail" behind as it goes. (NRAO; Palomar/Caltech)

16.2 Galaxy Collisions

Contemplating the congested confines of a rich galaxy cluster (such as Virgo or Coma, with thousands of member galaxies orbiting within a few megaparsecs), we might expect that collisions among galaxies would be common. ∞ *(Sec. 15.2)* Gas particles collide in our atmosphere, and hockey players collide in the rink. So, do galaxies in clusters collide, too? The answer is yes.

Figure 16.6 apparently shows the aftermath of a bull's-eye collision between a small galaxy (perhaps one of the two at the right, although that is by no means certain) and the larger galaxy at the left. The result is the "Cartwheel" galaxy, about 150 Mpc from Earth, its halo of young stars resembling a vast ripple in a pond. The ripple is most likely a density wave created by the passage of the smaller galaxy through the disk of the larger one. ∞ *(Sec. 14.5)* The disturbance is now spreading outward from the region of impact, creating new stars as it goes.

Figure 16.7 shows an example of a close encounter that hasn't (yet) led to an actual collision. Two spiral galaxies are apparently passing each other like majestic ships in the night. The larger and more massive galaxy on the left is called NGC 2207; the smaller one on the right is IC 2163. Analysis of this image suggests that IC 2163 is now swinging past NGC 2207 in a counterclockwise direction, having made a close approach some 40 million years ago. The two galaxies seem destined to undergo further close encounters, as IC 2163 apparently does not have enough energy to escape the gravitational pull of NGC 2207. In roughly a billion years, these two galaxies will probably merge into a single, massive galaxy.

No human will ever witness an entire galaxy collision, for it lasts many millions of years. However, modern computers can follow the event in a matter of hours. Large-scale simulations modeling in detail the gravitational interactions among stars and gas, and incorporating the best available models of gas dynamics, allow astronomers to better understand the effects of a collision on the galaxies involved and even estimate the eventual outcome of the interaction. The computer simulation shown in Figure 16.8(b) began with two colliding spiral galaxies, not so different from those shown in Figure 16.7, but the details of the original structure have been largely obliterated by the collision. Notice the similarity to the real image of NGC 4038/4039 (Figure 16.8a), the

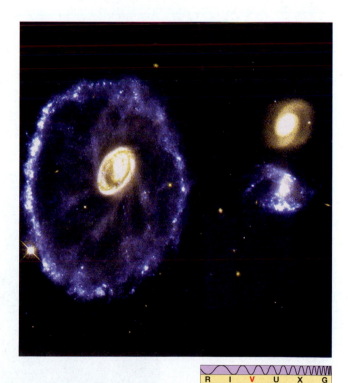

R I V U X G

▲ **Figure 16.6** **Cosmic Cartwheel**   The "Cartwheel" galaxy (left) appears to be the result of a collision (possibly with one of the smaller galaxies at right) that has led to an expanding ring of star formation moving outward through the galactic disk. (NASA)

▶ **Figure 16.7 Galaxy Encounter** This encounter between two spirals, NGC 2207 (left) and IC2163 (right), has already led to bursts of star formation in each. Eventually the two will merge, but not for a billion years or so. (NASA)

R I V U X G

so-called Antennae galaxies, which show extended tails, as well as double galactic centers only a few hundred parsecs across. Star formation induced by the collision is clearly traced by the blue light from thousands of young, hot stars. The simulations indicate that, as with the galaxies in Figure 16.7, ultimately the two galaxies will merge into one.

Galaxies in clusters apparently collide quite often. In the smaller groups, the galaxies' speeds are low enough that interacting galaxies tend to "stick together," and mergers, as shown in the computer simulation, are a very common outcome. In larger groups, galaxies move faster and tend to pass through one another without sticking. Either way, the encounters have substantial effects on the galaxies involved. Some researchers go so far as to suggest that most galaxies have been strongly influenced by collisions, in many cases in the relatively recent past. Incidentally, if we wait long enough, we will have an opportunity to see for ourselves what a galaxy collision is really like: Our nearest large neighbor, the Andromeda galaxy (see Figure 14.2), is currently approaching us at a velocity of 120 km/s. In a few billion years it will collide with the Milky Way, and we will be able to test astronomers' theories firsthand!

▶ **Figure 16.8 Galaxy Collision** (a) The "Antennae" galaxies collided a few tens of millions of years ago. The long tidal "tails" (black and white image at left) mark their final plunge. Strings of young, bright "super star clusters" (color image at right) are the result of violent shock waves produced in the gas disks of the two colliding galaxies. (b) A computer simulation of the encounter shows many of the same features as the real thing, strengthening the case that we really are seeing a collision in progress. (AURA; NASA; J. Barnes)

Super star clusters
Galaxy cores
Super star clusters

(a)

R I V U X G

(b)

Curiously, although a collision may wreak havoc on the large-scale structure of the galaxies involved, it has essentially no effect on the individual stars they contain. The stars within each galaxy just glide past one another. In contrast to galaxies in the cluster, the stars in a galaxy are so small compared with the distances between them, that when two galaxies collide, the star population merely doubles for a time, and the stars continue to have so much space that they do not run into each other. Collisions can rearrange the stellar and interstellar contents of each galaxy, often producing a spectacular burst of star formation that may be visible to enormous distances, but from the point of view of the stars, it's clear sailing.

Animation/Video
Collision of Two Spiral Galaxies

16.3 Galaxy Formation and Evolution

With Hubble's law as our indispensable guide to distances in the universe, and armed now with knowledge of the distribution of dark matter on galactic and larger scales, let's finally turn to the question of how galaxies came to be the way they are. Can we explain the different galaxy types we see? Astronomers know of no simple evolutionary connections among the various categories in the Hubble classification scheme. *(Sec. 15.1)* To answer the question, then, we must understand how galaxies formed.

Unfortunately, compared with the theories of star formation and stellar evolution, the theory of galaxy formation is still very much in its infancy. Galaxies are much more complex than stars, they are harder to observe, and the observations are harder to interpret. In addition, we have only a partial understanding of conditions in the universe immediately preceding galaxy formation, quite unlike the corresponding situation for stars. *(Sec. 11.2)* Finally, whereas stars almost never collide with one another, with the result that most single stars and binaries evolve in isolation, galaxies may suffer numerous collisions and mergers during their lives, making it much harder to decipher their pasts. *(Discovery 15-1)* Nevertheless, some general ideas have begun to gain widespread acceptance, and we can offer some insights into the processes responsible for the galaxies we see.

Mergers and Acquisitions

The seeds of galaxy formation were sown in the very early universe, when small density fluctuations in the primordial matter began to grow (see Section 17.7). Our discussion here begins with these "pregalactic" blobs of gas already formed. The masses of these fragments were quite small—only a few million solar masses, comparable to the masses of the smallest present-day dwarf galaxies, which may in fact be remnants of this early time. Most astronomers think that galaxies grew by repeated *merging* of smaller objects, as illustrated in Figure 16.9(a). Contrast this with the process of star formation, in which a large cloud fragments into smaller pieces that eventually become stars. *(Sec. 11.3)*

(c)

(b)

R I V U X G

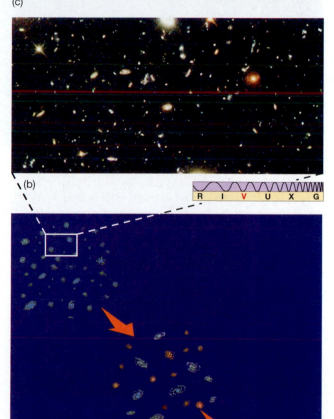

▶ **Figure 16.9 Galaxy Formation**   (a) The present view of galaxy formation holds that large systems were built up from smaller ones through collisions and mergers, as shown schematically in this drawing. (b) This photograph, one of the deepest ever taken of the universe, provides "fossil evidence" for hundreds of galaxy shards and fragments, up to 5000 Mpc distant. (c) Enlargements of selected portions of (b) reveal rich (billion-star) "star clusters," all lying within a relatively small volume of space (about 1 Mpc across). Their proximity to one another suggests that we may be seeing a group of pregalactic fragments about to merge to form a galaxy. The events pictured took place about 10 billion years ago. (NASA)

(a)

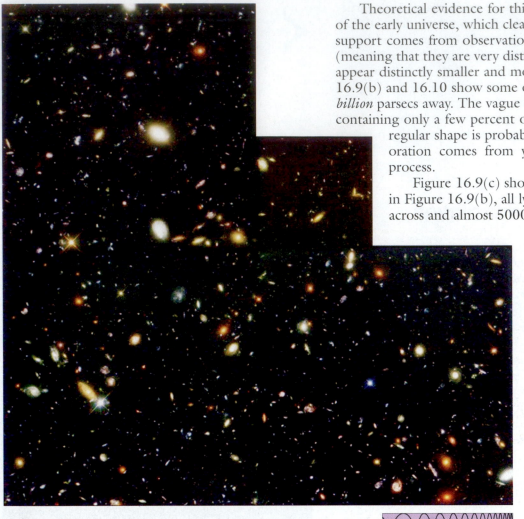

Theoretical evidence for this picture is provided by computer simulations of the early universe, which clearly show merging taking place. Further strong support comes from observations that indicate that galaxies at large redshifts (meaning that they are very distant and the light we see was emitted long ago) appear distinctly smaller and more irregular than those found nearby. Figures 16.9(b) and 16.10 show some of these images, which include objects up to 5 *billion* parsecs away. The vague bluish patches are separate small galaxies, each containing only a few percent of the mass of the Milky Way Galaxy. Their irregular shape is probably the result of galaxy mergers; the bluish coloration comes from young stars that formed during the merger process.

Figure 16.9(c) shows more detailed views of some of the objects in Figure 16.9(b), all lying in the same region of space, about 1 Mpc across and almost 5000 Mpc from Earth. Each blob seems to contain several billion stars spread throughout a distorted spheroid about a kiloparsec across. Their decidedly bluish tint suggests that active star formation is already under way. We see them as they were nearly 10 billion years ago, a group of young galaxies possibly poised to merge into one or more larger objects.

Galaxy Interactions

Left alone, a galaxy will evolve slowly and fairly steadily as interstellar clouds of gas and dust are turned into new generations of stars, and main sequence stars evolve into giants and ultimately into compact remnants—white dwarfs, neutron stars, and black holes. ∞ *(Sec. 11.3, Sec. 13.5)* The galaxy's overall color, composition, and appearance change in a more or less predictable way as the cycle of stellar evolution recycles and enriches the galaxy's interstellar matter. ∞ *(Sec. 12.7)*

▲ **Figure 16.10 Hubble Deep Field** Numerous small, irregularly shaped young galaxies can be seen in this very deep optical image. Known as the Hubble Deep Field-North, this image, made with an exposure of approximately 100 hours, captured objects as faint as 30th magnitude. Redshift measurements (see *Sec. 15.3*) indicate that some of these galaxies lie well over 1000 Mpc from Earth. Their size, color, and irregular appearance support the theory that galaxies grew by mergers and were smaller and less regular in the past. The field of view is about two arc minutes across. (NASA)

R I V U X G

But many—perhaps most—galaxies are not alone. They reside in small groups and clusters, and, as we have just seen, their orderly "internal" evolution may be significantly complicated by external events—close encounters, mergers, and the accumulation of smaller satellite galaxies over extended periods of time. As described in Section 16.2, these interactions can rearrange a galaxy's internal structure and trigger a sudden, intense burst of star formation. The result is a **starburst galaxy**; some examples are shown in Figure 16.11. Encounters may also divert fuel to a central black hole, powering violent activity in some galactic nuclei. ∞ *(Sec. 15.5)* Thus, starbursts and nuclear activity are key indicators of interactions and mergers between galaxies.

Careful studies of starburst galaxies and active galactic nuclei indicate that most such encounters probably took place long ago—in galaxies having redshifts greater than about 1, meaning that we see them as they were roughly 10 billion years ago (see Table 15.2, p. 419). Nevertheless, the galaxy interactions observed in the local universe are the extension of this same basic process into the present day. We have ample evidence that galaxies have evolved in response to external factors long after the first pregalactic fragments formed and merged.

Computer simulations performed over the past decade have shown that the extensive dark-matter halos surrounding most, if not all, galaxies are crucial to galaxy interactions. Consider first two galaxies orbiting one another—a binary galaxy system. As they orbit, the galaxies interact with each other's dark

Animation/Video
Hubble Deep Field Zoom Sequence

Animation/Video
Starburst Galxaxy

CONCEPT CHECK ▶▶▶

Why would we expect the evolution of galaxies in voids to differ from that of galaxies in clusters?

▶ **Figure 16.11 Starburst Galaxies** (a) This interacting galaxy pair (IC 694, at left, and NGC 3690) shows starbursts now under way in both galaxies—hence the bluish tint. Such intense, short-lived bursts probably last for no more than a few tens of millions of years— a small fraction of a typical galaxy's lifetime. (b) This infrared image of a pair of starburst galaxies (called Arp 107) shows numerous young star clusters arrayed as though along a pearl necklace. The colors are false, representing different infrared wavelengths observed by the *Spitzer Space Telescope*. (c) The peculiar (Irr II) galaxy NGC 1275 contains a system of long filaments (in its outskirts) that seem to be exploding outward into space. Its blue blobs, as revealed by the *Hubble Space Telescope*, are probably young globular clusters formed by the collision of two galaxies. (W. Keel; NASA)

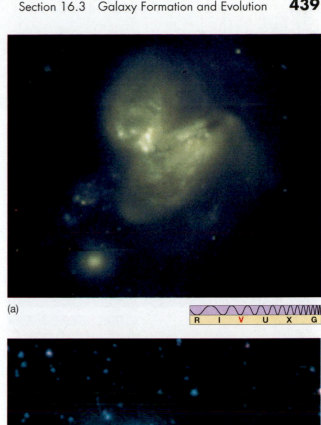

(a)

halos, one galaxy stripping halo material from the other by tidal forces. The freed matter is redistributed between the galaxies or is entirely lost from the binary system. In either case, the interaction changes the orbits of the galaxies, causing them to spiral toward one another and eventually to merge.

If one galaxy of the pair happens to have a much lower mass than the other, the process is colloquially termed *galactic cannibalism*. Such cannibalism might explain why supermassive galaxies are often found at the cores of rich galaxy clusters. Having dined on their companions, they now lie at the center of the cluster, waiting for more food to arrive. Figure 16.12 is a remarkable combination of images that has apparently captured this process at work. Closer to home, the Magellanic Clouds (see Figure 15.7) will eventually suffer a similar fate at the center of the Milky Way.

Now consider two interacting disk galaxies, one a little smaller than the other but each having a mass comparable to the Milky Way Galaxy. As shown in the computer-generated frames of Figure 16.13, the smaller galaxy can substantially distort the larger one, causing spiral arms to appear where none existed before. The entire event requires several hundred million years—a span of evolution that a supercomputer can model in minutes. The final frame of Figure 16.13 looks remarkably similar to the double galaxy shown in Figure 15.2(b), demonstrating how the two galaxies might have interacted millions of years ago, and how spiral arms might have been created or enhanced as a result.

Making the Hubble Sequence

If galaxies form and evolve by repeated mergers, can we account for the Hubble sequence and, specifically, differences between spirals and ellipticals? ∞ *(Sec. 15.1)* The details are still far from certain but, remarkably, the answer now seems to be a qualified yes. Collisions and close encounters are random events and do not represent a genuine evolutionary sequence linking all galaxies. However, computer simulations suggest a plausible way in which the observed Hubble types might have arisen, starting from a universe populated only by irregular, gas-rich galaxy fragments.

The simulations reveal that a *major merger*—a collision between galaxies of comparable size—can destroy a spiral galaxy's disk, creating a galaxy-wide starburst episode. The violence of the merger and the effects of subsequent supernovae eject much of the remaining gas into intergalactic space, creating the hot intracluster gas noted in Section 16.1. Once the burst of star formation has subsided, the resulting object looks very much like an elliptical galaxy. The elliptical's hot X-ray halo is the last vestige of the original spiral's disk. The Irr II galaxy shown in Figure 16.11(c) may be an example of this phenomenon. The blue blobs are thought to be young star clusters formed during the starburst, and the explosive appearance suggests that we are witnessing the gas and dust being ejected.

The simulations also indicate that *minor mergers*, in which a small galaxy interacts with and ultimately is absorbed by a larger one, generally leave the

(b)

(c)

▲ Figure 16.12 Galactic Cannibalism This computer-enhanced, false-color composite optical photograph of the galaxy cluster Abell 2199 is thought to show an example of galactic cannibalism. The large central galaxy of the cluster (itself 120 kpc along its long axis) is displayed with a superimposed "window." (This results from a shorter time exposure, which shows only the brightest objects that fall within the frame.) Within the core of the large galaxy are several smaller galaxies (the three bright yellow images at center) apparently already "eaten" and now being "digested" (that is, being torn apart and becoming part of the larger system). Other small galaxies swarm on the outskirts of the swelling galaxy, almost certainly to be eaten, too. (SAO)

larger galaxy intact, with more or less the same Hubble type as it had before the merger. This is the most likely way for large spirals to grow—in particular, our own Galaxy probably formed in this way. The "streams" of halo stars in our own Galaxy, all with similar orbits and composition, may well be the stellar remnants of such mergers in the past. ∞ *(Sec. 14.3)* The small Sagittarius dwarf galaxy (Figure 15.12) and the Magellanic Clouds (Figure 15.7) will probably be the Milky Way's next victims.

Supporting evidence for this general picture comes from observations that spiral galaxies are relatively rare in regions of high galaxy density, such as the central regions of rich galaxy clusters. This is consistent with the view that their fragile disks are easily destroyed by collisions, which are more common in dense galactic environments. Spirals also seem to be more common at larger redshifts (that is, in the past), implying that their numbers are decreasing with time, presumably also as the result of collisions. However, nothing in this area of astronomy is clear-cut. Astronomers know of numerous isolated elliptical galaxies in low-density regions of the universe that are hard to explain as the result of mergers.

In principle, the starbursts associated with galaxy mergers leave their imprint on the star-formation history of the universe in a way that can be correlated with the properties of galaxies. As a result, studies of star formation in distant galaxies have become a very important way of testing and quantifying the details of the entire merger scenario.

16.4 Black Holes and Active Galaxies

Now let's ask how quasars and active galaxies fit into the framework of galaxy evolution just described. The fact that quasars are more common at great distances from us demonstrates that they were much more prevalent in the past than they are today. ∞ *(Sec. 15.4)* Quasars have been observed with redshifts of up to 6.4 (the current record, as of early 2006), so the process must have started at least 13 billion years ago (see Table 15.2). However, most quasars have redshifts between 2 and 3, corresponding to an epoch some 2 billion years later. Most astronomers agree that quasars represent an early stage of galaxy evolution—an "adolescent" phase of development, prone to frequent flare-ups and "rebellion" before settling into more steady "adulthood." This view is reinforced by the fact that the same black-hole energy-generation mechanism can account for the luminosities of quasars, active galaxies, and the central regions of normal galaxies like our own.

▼ Figure 16.13 Galaxy Interaction Galaxies can change their shapes long after their formation. In this computer-generated sequence, two galaxies closely interact over several hundred million years. The smaller galaxy, in red, has gravitationally disrupted the larger galaxy, in blue, changing it into a spiral galaxy. Compare the result of this supercomputer simulation with Figure 15.2(b), a photograph of M51 and its small companion. (J. Barnes & L. Hernquist)

Animation/Video
Interacting Galaxies

Time

The Quasar Epoch

In Chapter 15 we presented the standard model of active galactic nuclei accepted by most astronomers—accretion of gas onto a supermassive black hole. ∞ *(Sec. 15.5)* But where did these supermassive black holes come from? To be honest, the answer is not yet known. The processes whereby the first billion-solar-mass black holes formed early in the history of the universe are not fully understood. However, the accretion responsible for the energy emission also naturally accounts for the mass of the black hole, and simple estimates suggest that the accretion rates needed to power the quasars are generally consistent with black-hole masses inferred by other means.

Since the brightest known quasars devour about a thousand solar masses of material every year, it is unlikely that they could maintain their luminosity for very long—even a million years would require a billion solar masses, enough to account for the most massive black holes known. ∞ *(Sec. 15.4)* This suggests that a typical quasar spends only a fairly short amount of time in its highly luminous phase—perhaps only a few million years in some cases—before running out of fuel. Thus, most quasars were relatively brief events that occurred long ago.

To make a quasar, we need a black hole and enough fuel to power it. While fuel was abundant at early times in the universe's history in the form of gas and newly formed stars, black holes were not. They had yet to form, most probably by the same basic stellar evolutionary processes we saw in Chapter 12, although the details are still not known. ∞ *(Sec. 12.4)* The building blocks of the supermassive black holes that would ultimately power the quasars may well have been relatively small black holes having masses tens or perhaps a few hundreds of times the mass of the Sun. These small black holes sank to the center of their still-forming parent galaxy and merged to form a single, more massive black hole.

As galaxies merged, so, too, did their central black holes, and eventually supermassive (1-million- to 1-billion-solar mass) black holes existed in the centers of many young galaxies. Some supermassive black holes may have formed directly by the gravitational collapse of the dense central regions of a protogalactic fragment or perhaps by accretion or a rapid series of mergers in a particularly dense region of the universe. These events resulted in the earliest (redshift-6) quasars known, shining brightly 13 billion years ago. However, in most cases, all those mergers took time—roughly another 2 billion years. By then (at redshifts between 2 and 3, roughly 11 billion years ago), many supermassive black holes had formed, and there was still plenty of merger-driven fuel available to power them. This was the height of the "quasar epoch" in the universe.

Until recently, astronomers were confident that black holes would merge when their parent galaxies collided, but they had no direct evidence of the process—no image of two black holes "caught in the act." In 2002, the *Chandra* X-ray observatory discovered a binary black hole—two supermassive objects, each having a mass a few tens of millions of times that of the Sun—in the center of the ultraluminous starburst galaxy NGC 6240, itself the product of a galaxy merger some 30 million years ago. Figure 16.14 shows optical and X-ray views of the system. The black holes are the two blue-white objects near the center of the (false-color) X-ray image. Orbiting just 1000 pc apart, they are losing energy through interactions with stars and gas and are predicted to merge in about 400 million years. NGC 6240 lies just 120 Mpc from Earth, so we are far from seeing a quasar merger in the early universe. Nevertheless, astronomers think that events similar to this must have occurred countless times billions of years ago, as galaxies collided and quasars blazed.

Distant galaxies are generally much fainter than their bright quasar cores. As a result, until quite recently, astronomers were hard pressed to discern any

(a)

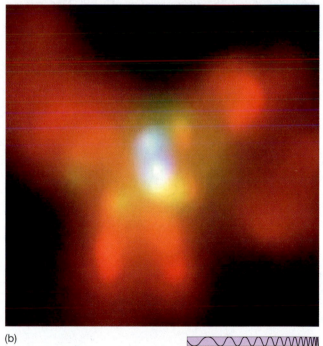

(b)

▲ **Figure 16.14 Binary Black Hole**   These (a) optical (*Hubble*) and (b) X-ray (*Chandra*) images of the starburst galaxy NGC 6420 show two supermassive black holes (the blue-white objects near the center of the X-ray image) orbiting about 1 kpc apart. According to theoretical estimates, they will merge in about 400 million years, releasing an intense burst of gravitational radiation in the process. The colors of the optical image are real; the false colors in the X-ray image indicate a range of energies in the X-ray band. ∞ *(Discovery 13-1)* (NASA)

▲ **Figure 16.15 Quasar Host Galaxies** These long-exposure *Hubble Space Telescope* images of distant quasars clearly show the young host galaxies in which the quasars reside, lending strong support to the view that quasars represent an early, very luminous phase of galactic evolution. The quasar at the left is the best example, having the catalog name PG0052 + 251 and residing roughly 690 Mpc from Earth. Note that several of the quasars appear to be associated with interacting galaxies, consistent with current theories of galaxy evolution. (NASA)

galactic structure in quasar images. Since the mid-1990s, several groups of astronomers have used the *Hubble Space Telescope* to search for the "host" galaxies of moderately distant quasars. After removing the bright quasar core from the *HST* images and carefully analyzing the remnant light, the researchers have reported that, in every case studied—several dozen quasars so far—a host galaxy can be seen enveloping the quasar. Figure 16.15 shows some of the longest quasar exposures ever taken. Even without sophisticated computer processing, the hosts are clearly visible.

As we saw in Chapter 15, the connection between active galaxies and galaxy clusters is well established, and many relatively nearby quasars are also known to be members of clusters. ∞ *(Sec. 15.4)* The link is less clear-cut for the most distant quasars, however, simply because they are so far away that other cluster members are very faint and extremely hard to see. However, as the number of known quasars continues to increase, evidence for quasar clustering (and presumably, therefore, for quasar membership in young galaxy clusters) mounts. Thus, as best we can tell, quasar activity—and, in fact, galaxy activity of *all* sorts—is intimately related to interactions and collisions in galaxy clusters.

Active and Normal Galaxies

Early on, frequent mergers may have replenished the quasar's fuel supply, extending its luminous lifetime. However, as the merger rate declined, these systems spent less and less of their time in the "bright" phase. The rapid decline in the number of bright quasars roughly 10 billion years ago marks the end of the quasar epoch. Today, the number of quasars has dropped virtually to zero (recall that the nearest lies hundreds of megaparsecs away). ∞ *(Sec. 15.4)*

Large black holes do not simply vanish. If a galaxy contained a bright quasar 10 billion years ago, the black hole responsible for all that youthful activity must still be present in the center of the galaxy today. We see some of these black holes as active galaxies, but since the difference between a normal galaxy and an active one seems to be mainly a matter of fuel supply, the implication is that many—perhaps most—normal galaxies should also contain black holes. Do we have any evidence that this in fact the case? The answer appears to be yes. In recent years, astronomers have amassed evidence that many bright normal galaxies contain black holes at their centers. Our own Galaxy is a case in point. ∞ *(Sec. 14.7)* The 3- to 4-million-solar mass black hole at the center of the Milky Way is not currently active, but if fresh fuel were supplied (say, by a star coming too close to the hole's intense gravitational field), it might well become a (relatively weak) active galactic nucleus.

Perhaps the most compelling evidence for a supermassive black hole at the center of a normal galaxy comes from a radio study of NGC 4258, a spiral galaxy about 6 Mpc away. Using the Very Long Baseline Array, a continentwide interferometer comprising 10 radio telescopes, a U.S.–Japanese team has achieved an angular resolution hundreds of times better than that attain-

▶ **Figure 16.16 Galactic Black Hole**   A network of radio telescopes has probed the core of the spiral galaxy NGC 4258, shown here in the light of mostly hydrogen emission. Within the innermost 0.2 pc (inset), observations of Doppler-shifted molecular clouds (designated by red, green, and blue dots) show that they obey Kepler's third law perfectly, and have revealed a slightly warped disk of rotating gas (shown here in an artist's conception). At the center of the disk presumably lurks a huge black hole. (J. Moran)

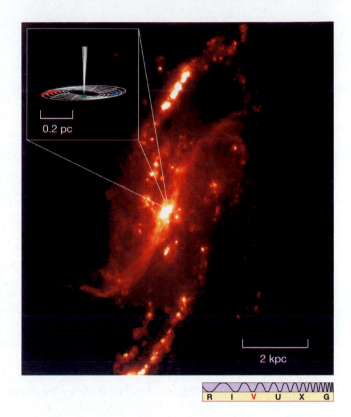

able with the *Hubble* telescope. ∞ *(Sec. 3.4)* The observations reveal a group of molecular clouds swirling in an organized fashion about the galaxy's center. Doppler measurements indicate a slightly warped, spinning disk centered precisely on the galaxy's heart (Figure 16.16). The rotation speeds imply the presence of more than 40 million solar masses packed into a region less than 0.2 pc across.

Discoveries such as these blur the distinction between normal and active galaxies. When the fuel runs out and a quasar shuts down, its central black hole remains behind, its energy output reduced to a relative trickle. But the black holes at the hearts of normal galaxies are simply quiescent, awaiting another interaction to trigger a new active outburst. Occasionally, two nearby galaxies may interact with each other, causing a flood of new fuel to be directed toward the central black hole of one or both. The engine starts up for a while, giving rise to the nearby active galaxies—radio galaxies, Seyferts, and others—we observe.

Should this general picture be correct, it follows that many relatively nearby galaxies (but probably *not* our own Milky Way, whose central black hole is even now only a paltry 3–4 million solar masses) must once have been brilliant quasars. ∞ *(Sec. 14.7)* Perhaps some alien astronomer, thousands of megaparsecs away, is at this very moment observing the progenitor of M87 in the Virgo Cluster—seeing it as it was billions of years ago—and is commenting on its enormous luminosity, nonstellar spectrum, and high-speed jets and wondering what exotic physical process can account for its violent activity! ∞ *(Sec. 14.4)*

Active Galaxy Evolution

The galaxy merger scenario, combined with local observations of supermassive black holes, may also provide some insight into the connection between activity and galaxy type. Astronomers have found that the largest black holes tend to be found in the most massive galaxies (Figure 16.17). Furthermore, it is likely that the most massive black holes power the brightest active galactic nuclei. In that case, we would expect that the most luminous nuclei should reside in the largest galaxies, which probably came into being via "major" mergers of other large galaxies. Since the products of such mergers are elliptical galaxies, we have a plausible explanation of why the brightest active galaxies—the radio galaxies—should be associated with large ellipticals. ∞ *(Sec. 15.4)* Furthermore, the path to spiral galaxies would necessarily have entailed a series of mergers involving smaller galaxies, resulting in the less violent Seyferts. Some possible (but unproven) evolutionary connections among quasars, active, and normal galaxies are illustrated in Figure 16.18.

▶ **Figure 16.17 Black Hole Masses**   Careful observations of nearby normal and active galaxies reveal that the mass of the central black hole is quite well correlated with the mass of the galactic bulge. In this diagram, each point represents a different galaxy. The straight line is the best fit to the data, implying a black-hole mass of 1/200 the mass of the bulge. (Data courtesy L. Ferrarese)

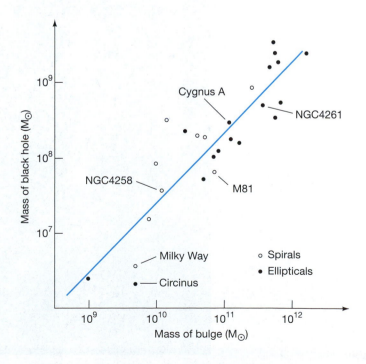

When active galactic nuclei—and especially quasars—were first discovered, their extreme properties defied conventional explanation. Initially, the idea of supermassive (million- to billion-solar mass) black holes in galaxies was just one of several competing, and very different, hypotheses advanced to account for the enormous luminosities and small sizes of those baffling objects. Some astronomers suggested instead that the energy source might be multiple supernova explosions, others that perhaps some exotic form of matter-antimatter annihilation was taking place. A few went so far as to propose that these inexplicable objects demanded an even more radical explanation: that Hubble's law was incorrect and that the large redshifts of quasars did not mean that they were very distant, but that their redshifts had some other, unknown cause.

However, as observational evidence mounted, the other hypotheses were abandoned one by one, and massive black holes in galactic nuclei became first the leading, and eventually the standard, theory of active galaxies. As often happens in science, a theory once itself considered extreme is now the accepted explanation for these phenomena. Far from threatening the laws of physics, as some astronomers once feared, active galaxies are now an integral part of our understanding of how galaxies form and evolve. The synthesis of studies of normal and active galaxies, galaxy formation, and large-scale structure is one of the great triumphs of extragalactic astronomy.

CONCEPT CHECK ▶▶▶

Does *every* galaxy have the potential for activity?

▼ **Figure 16.18 Galaxy Evolution** Some possible evolutionary sequences for galaxies, beginning with galaxy mergers leading to the highly luminous quasars, decreasing in violence through the radio and Seyfert galaxies, and ending with normal ellipticals and spirals. The central black holes that powered the early activity are still there at later times; they simply run out of fuel as time goes on.

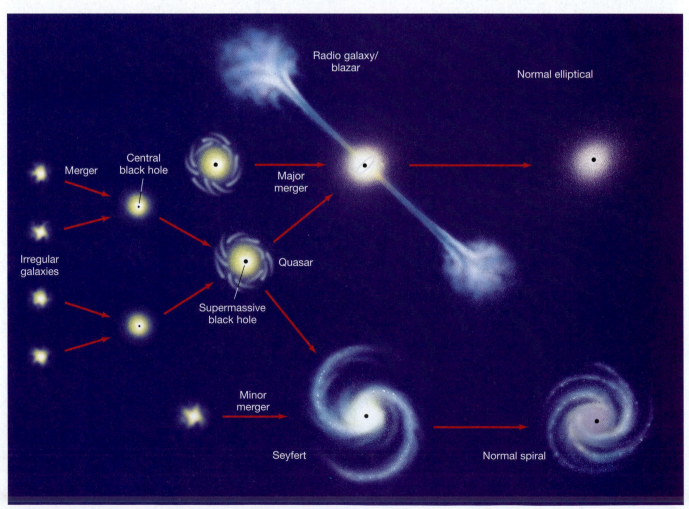

16.5 The Universe on Very Large Scales

Many galaxies, including our own, are members of galaxy clusters—mega-parsec-sized structures held together by their own gravity. ∞ *(Sec. 15.2)* Our own small cluster is called the Local Group. Figure 16.19 shows the locations of the Virgo Cluster, the closest "large" cluster, and of several other well-defined clusters in our cosmic neighborhood. The region displayed is about 70 Mpc across. Each point in the figure represents an entire galaxy whose distance has been determined by one of the methods described in Chapter 15.

Clusters of Clusters

Does the universe have even greater groupings of matter, or do galaxy clusters top the cosmic hierarchy? Most astronomers have concluded that the galaxy clusters are themselves clustered, forming titanic agglomerations of matter known as **superclusters**.

Together, the galaxies and clusters shown in Figure 16.19 form the *Local Supercluster*, also known as the Virgo Supercluster. Aside from the Virgo Cluster itself, it contains the Local Group and numerous other clusters lying within about 20–30 Mpc of Virgo. Most of the galaxies depicted in the figure are fairly large spirals and ellipticals; the fainter irregulars and dwarfs are not included in the diagram. Figure 16.20 shows a three-dimensional rendering of an even wider view, illustrating the Virgo Supercluster (near the center) relative to other "nearby" galaxy superclusters within a vast imaginary rectangle roughly 100 Mpc on its short side.

All told, the Local Supercluster is about 40–50 Mpc across, contains some 10^{15} solar masses of material (several tens of thousands of galaxies), and is very irregular in shape. The Local Supercluster is significantly elongated perpendicular to the line joining the Milky Way to Virgo, with its center lying near the Virgo Cluster. By now it should perhaps come as no surprise that the Local Group is *not* found at the heart of the Local Supercluster—we live far off in the periphery, about 18 Mpc from the center.

Redshift Surveys

The farther we peer into deep space, the more galaxies, clusters of galaxies, and superclusters we see. Is there structure on even larger scales? To answer this question, astronomers use Hubble's law to map out the distribution of galaxies in the universe and have developed indirect techniques to probe the dark recesses of intergalactic space.

Figure 16.21 shows part of an early survey of the universe performed by astronomers at Harvard University in the 1980s. Using Hubble's law as a distance indicator, the team systematically mapped out the locations of galaxies within about 200 Mpc of the Milky Way in a series of wedge-shaped "slices," each 6° thick, starting in the northern sky. The first slice (Figure 16.21) covered a region of the sky containing the Coma Cluster (see Figure 15.1), which happens to lie in a direction almost perpendicular to our Galaxy's plane. Because redshift is used as the primary distance indicator, these studies are known as *redshift surveys*.

The most striking feature of maps such as this is that the distribution of galaxies on very large scales is decidedly nonrandom. The galaxies appear to be

▲ **Figure 16.19 Local Supercluster**   The locations of numerous galaxies and galaxy clusters in the vicinity of Virgo. More than 4500 galaxies are represented here, and several prominent galaxy clusters are labeled. The diagram shows the Virgo Supercluster roughly as it appears from the direction of our own Galaxy, which is located approximately 20 Mpc (two grid squares) above the page. Galaxies are false-colored according to the local galaxy density—white and yellow indicate the most congested regions, green less dense regions, and blue the least dense. Notice the supercluster's irregular, elongated shape. (Data courtesy B. Tully, U. Hawaii; visualization by S. Levy, NCSA)

Animation/Video
Cluster Merger

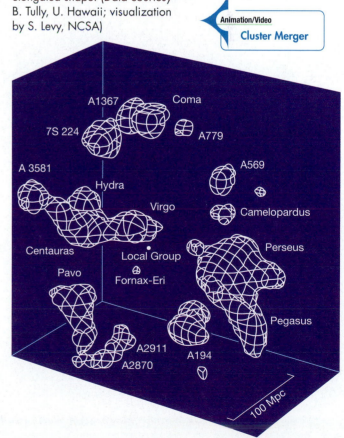

▶ **Figure 16.20 Virgo Supercluster in 3-D**   The elongated structure of the Virgo supercluster (left center) is mapped relative to other superclusters within about 100 Mpc of the Milky Way. Individual galaxies are not shown; rather, smoothed contour plots outline galaxy clusters, each named or numbered for its most prominent member.

Great Wall

Coma cluster

Voids

Milky Way Galaxy

Velocity (km/s)

Distance (Mpc)

15,000

10,000

5,000

0

200

100

0

...like beads on a string

◄ **Figure 16.21 Galaxy Survey** The first slice of a survey of the universe, covering 1732 galaxies out to an approximate distance of 200 Mpc, clearly shows that galaxies and clusters are not randomly distributed on large scales. Instead, they appear to have a filamentary structure, surrounding vast, nearly empty voids—often resembling beads on a string. We assume $H_0 = 70$ km/s/Mpc. The slice covers 6° of the sky in the direction out of the plane of the page. (Harvard-Smithsonian Center for Astrophysics)

arranged in a network of strings, or filaments, surrounding large, relatively empty regions of space known as **voids**. The biggest voids measure some 100 Mpc across. The most likely explanation for the voids and filamentary structure in Figure 16.21 is that the galaxies and galaxy clusters are spread across the surfaces of vast "bubbles" in space. The voids are the interiors of these gigantic bubbles. The galaxies seem to be distributed like beads on strings only because of the way our slice of the universe cuts through the bubbles. Like suds on soapy water, these bubbles fill the entire universe. The densest clusters and superclusters lie in regions where several bubbles meet. The elongated shape of the Virgo Supercluster (see Figure 16.19) is a local example of this same filamentary structure.

Most theorists think that this "frothy" distribution of galaxies, and in fact all structure on scales larger than a few megaparsecs, traces its origin directly to conditions in the very earliest stages of the universe (Chapter 17). As such, studies of large-scale structure are vital to our efforts to understand the origin and nature of the cosmos itself. The idea that the filaments are the intersection of the survey slice with much larger structures (the bubble surfaces) was confirmed when the next three slices of the survey, lying above and below the first, were completed. The region of Figure 16.21 indicated by the red outline was found to continue through both the other slices. This extended sheet of galaxies, which has come to be known as the *Great Wall*, measures at least 70 Mpc (out of the plane of the page) by 200 Mpc (across the page). It is one of the largest known structures in the universe.

Figure 16.22 shows a more recent redshift survey, considerably larger than the one presented in Figure 16.21. This survey includes nearly 24,000 galaxies within about 750 Mpc of the Milky Way. Numerous voids and "Great Wall-like" filaments can be seen (some are marked), but, apart from the general falloff in numbers of galaxies at large distances—basically because the more distant galaxies are harder to see due to the inverse-square law—there is no obvious evidence for any structures on scales *larger* than about 200 Mpc. Careful statistical analysis and even larger surveys (see *Discovery 16-1*) confirm this impression. Apparently, voids and walls represent the largest structures in the universe. We will return to the far-reaching implications of this fact in Chapter 17.

Quasar Absorption Lines

How can we probe the structure of the universe on very large scales? As we have seen, much of the matter is dark, and even the "luminous" component is so faint that it is hard to detect at large distances. One way to study large-scale structure is to take advantage of the great distances, pointlike appearance, and large luminosities of quasars, much the same way as astronomers use bright stars to probe the interstellar medium near the Sun. ∞ *(Sec. 11.1, 15.3)* Since quasars are so far away, light traveling from a quasar to Earth has a pretty good chance of passing through or near something "interesting" en route. By analyzing quasar images and spectra, it is possible to piece together a partial picture of the intervening space.

In addition to their own strongly redshifted spectra, many quasars also show additional absorption features that are redshifted by substantially *less* than the lines from the quasar itself. For example, the quasar PHL 938 has an emission-line redshift of 1.954, placing it at a distance of some 5200 Mpc, but it also shows absorption lines having redshifts of just 0.613. These are interpreted as arising from intervening gas that is much closer to us (only about 2200 Mpc away) than the quasar itself. Most probably this gas is part of an otherwise invisible galaxy lying along the line of sight. Quasar spectra, then, afford astronomers a means of probing previously undetected parts of the universe.

The absorption lines of atomic hydrogen are of particular interest, since hydrogen makes up so much of all matter in the cosmos. Specifically, hydrogen's ultraviolet (122 nm) "Lyman alpha" line, associated with transitions between the ground and first excited states, is often used in this context. ∞ *(Sec. 2.6)* As illustrated in Figure 16.23, when astronomers observe the spectrum of a high-redshift quasar, they typically see a "forest" of absorption lines starting at the (redshifted) wavelength of the quasar's own Lyman-alpha emission line and extending to shorter wavelengths. These lines are interpreted as Lyman-alpha absorption features produced by gas clouds in foreground structures—galaxies, clusters, and so on—giving astronomers crucial information about the distribution of matter along the line of sight.

Quasar light thus explores an otherwise invisible component of cosmic gas. In principle, every intervening cloud of atomic hydrogen leaves its own characteristic imprint on the quasar's spectrum, in a form that lets us explore the distribution of matter in the universe. By comparing these *Lyman-alpha forests* with the results of simulations, astronomers hope to refine many key elements of the theories of galaxy formation and the evolution of large-scale structure.

Quasar "Mirages"

In 1979 astronomers were surprised to discover what appeared to be a binary quasar—two quasars with exactly the same redshift and very similar spectra, separated by only a few arc seconds on the sky. Remarkable as the discovery of such a binary would have been, the truth about this pair of quasars turned out to be even more amazing. Closer study of the quasars' radio emission revealed that they were *not* two distinct objects. Instead, they were two separate images of the *same* quasar! Optical views of such a *twin quasar* are shown in Figure 16.24.

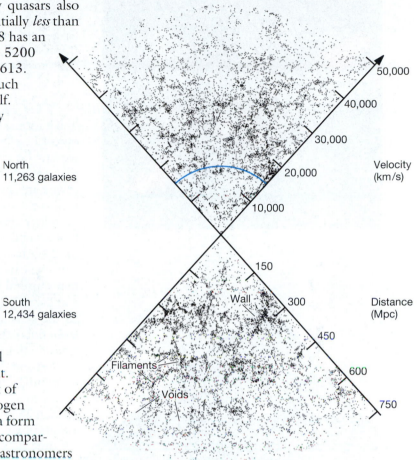

▲ Figure 16.22 The Universe on Large Scales This large-scale galaxy survey, carried out at the Las Campanas Observatory in Chile, consists of 23,697 galaxies within about 1000 Mpc, in two 80° × 4.5° wedges of the sky. Many voids and "walls" on scales of up to 100–200 Mpc can be seen, but no larger structures are evident. For scale, the extent of the survey shown in Figure 16.21 is marked as a thin blue arc in the northern sky.

◄ Figure 16.23 Absorption Line "Forest" When light from a distant quasar passes through a foreground cloud of atomic hydrogen, the absorption lines produced bear the cloud's redshift. The huge number of absorption lines in the spectrum of this quasar (called QSO 1422 + 2309) are the ultraviolet Lyman-alpha lines from hundreds of clouds of foreground hydrogen gas, each redshifted by a slightly different amount (but less than the quasar itself). The peak at left marks the Lyman-alpha emission line from the quasar, emitted at 122 nm but redshifted here to a wavelength of 564 nm, in the visible range.

◄ **Figure 16.24 Twin Quasar** This twin quasar (designated AC114 and located about 2 billion parsecs away) is not two separate objects at all. Instead, the two large "blobs" (at upper left and lower right) are images of the same object, created by a gravitational lens. The lensing galaxy itself is probably not visible in this image—the two objects near the center of the frame are thought to be unrelated galaxies in a foreground cluster. (NASA)

What could produce such a "doubling" of a quasar image? The answer is gravitational lensing—the deflection and focusing of light from a background object by the gravity of some foreground body (Figure 16.25). In Chapter 14 we saw how lensing by compact objects in the halo of the Milky Way Galaxy may amplify the light from a distant star, allowing astronomers to detect otherwise invisible stellar dark matter. ∞ *(Sec. 14.5)* In the case of quasars, the idea is the same, except that the foreground lensing object is an entire galaxy or galaxy cluster and the deflection of the light is so great (a few arc seconds) that several separate images of the quasar may be formed, as shown in Figure 16.26. About two dozen such gravitational lenses are known.

The existence of these multiple images provides astronomers with a number of useful observational tools. First, lensing by a foreground galaxy tends to amplify the light of the quasar, as just mentioned, making it easier to observe. At the same time, *microlensing* by individual stars within the galaxy may cause large fluctuations in the quasar's brightness, allowing astronomers to study the galaxy's stellar content.

Second, because the light rays forming the images usually follow paths of different lengths, there is often a *time delay*, ranging from days to years, between them. This delay provides advance notice of interesting events, such as sudden changes in the quasar's brightness—if one image flares up, in time the other(s) will too, giving astronomers a second chance to study the event. The time delay also allows astronomers to determine the *distance* to the lensing galaxy. This method provides an alternative means of measuring Hubble's constant that is independent of any of the techniques discussed previously. The average value of H_0 reported by workers using this approach is roughly 65 km/s/Mpc, a little less than the value used throughout this text.

Mapping Dark Matter

Astronomers have extended the ideas first learned from studies of quasars to use lensing of any distant object to probe the universe. Distant faint irregular galaxies—the raw material of the

▶ **Figure 16.25 Gravitational Lens** When light from a distant object passes close to a galaxy or cluster of galaxies along the line of sight, the image of the background object (here, the quasar) can sometimes be split into two or more separate images (A and B). The foreground object is a gravitational lens.

Image A

Quasar

Image B

Lensing galaxy

Observer on earth

16-1 DISCOVERY

The Sloan Digital Sky Survey

Many of the photographs used in this book—not to mention most of the headline-grabbing imagery found in the popular media—come from large, high-profile, and usually very expensive, instruments such as NASA's *Hubble Space Telescope* and the European Southern Observatory's *Very Large Telescope* in Chile. ∞ *(Sec. 3.3, 3.4)* Their spectacular views of deep space have revolutionized our view of the universe, yet a less well-known, considerably cheaper, but no less ambitious, project currently underway may, in the long run, have every bit as great an impact on astronomy and our understanding of the cosmos.

The Sloan Digital Sky Survey (SDSS) is a five year project, now extended for three more years, designed to systematically map out a quarter of the entire sky on a scale and at a level of precision never before attempted. By the end of the first phase of the project in 2005, it had cataloged more than 100 million celestial objects, recording their apparent brightnesses at five different colors (wavelength ranges) spread across the optical part of the spectrum. In addition, spectroscopic follow-up observations have determined redshifts and hence distances to 1 million galaxies and 100,000 quasars. These data are being used to construct even more detailed redshift surveys than those described in the text and to probe the structure of the universe on very large scales. The sensitivity of the survey is such that it can detect bright galaxies like our own out to distances of more than 1 billion parsecs. Very bright objects, such as quasars and young starburst galaxies, are detectable almost throughout the entire observable universe.

The first figure shows the Sloan Survey telescope, a special-purpose 3.5-m instrument sited in Apache Point Observatory, near Sunspot, New Mexico. The telescope is not space-based,

does not employ active or adaptive optics, and cannot probe as deep (i.e., far) into space as larger instruments. How can it possibly compete with these other systems? The answer is that, unlike most other large telescopes in current use, where hundreds or even thousands of observers share the instrument and compete for its time, the SDSS telescope was designed specifically for the purpose of the survey. It has a wide field of view and is dedicated to the task, carrying out observations of the sky on *every* clear night during the duration of the project.

The use of a single instrument night after night, combined with tight quality controls on which nights' data are actually incorporated into the survey (nights with poor seeing or other problematic conditions are discarded and the observations repeated) mean that the end-product is a database of exceptionally high quality and uniformity spanning an enormous volume of space—a monumental achievement and an indispensable tool for cosmology. The survey field of view covers much of the sky away from the Galactic plane in the north, together with three broad "wedges" in the south.

Archiving images and spectra on millions of galaxies produces a lot of data. The current survey consists of roughly 15 *trillion* bytes of information—comparable to the entire Library of Congress! As of late 2005, data for roughly two-thirds of the survey area, comprising 180 million objects (670,000 with measured spectra), have been released to the public. The second figure shows the distant galaxy NGC 5792 and a bright red star much closer to us, in fact in our own Milky Way Galaxy, just one of hundreds of thousands of images that make up the full dataset. Among recent highlights, SDSS has detected the largest known structure in the universe, observed the most distant known galaxies and quasars, and has been instrumental in pinning down the key observational parameters describing our universe (see Chapter 17).

SDSS will impact astronomy in areas as diverse as the large-scale structure of the universe, the origin and evolution of galaxies, the nature of dark matter, the structure of the Milky Way, and the properties and distribution of interstellar matter. Its uniform, accurate, and detailed database is likely to be used by generations of scientists for decades to come.

▶ **Figure 16.26 Einstein Cross**
(a) The Einstein Cross, a multiply-imaged quasar. In this *Hubble* view, spanning only a couple of arc seconds, four separate images of the same quasar have been produced by the galaxy at the center. (b) A simplified artist's conception of what might be occurring here, with Earth at right and the distant quasar at left. (NASA; D. Berry)

(a)   R I V U X G   (b)

universe, if current theories are correct (see Section 16.4)—are of particular interest here, as they are far more common than quasars, so they provide much better coverage of the sky. By studying the lensing of background quasars and galaxies by foreground galaxy clusters, astronomers can obtain a better understanding of the distribution of dark matter on large scales.

Figures 16.27(a) and (b) show how the images of faint, background galaxies are bent into arcs by the gravity of a foreground galaxy cluster. The degree of bending allows the total mass of the cluster (*including* the dark matter) to be measured. The (mostly blue) loop- and arc-shaped features visible in Figure 16.27(b) are thought to be multiple images of a single, distant (unseen) spiral or ring-shaped galaxy, lensed by the foreground galaxy cluster (the yellow-red blobs in the image). Astronomers are now trying to delicately "reassemble" an undistorted image of the distant object by retracing the light rays back to a common point.

▼ **Figure 16.27 Galaxy Cluster Lensing** (a) This spectacular example of gravitational lensing shows more than a hundred faint arcs from very distant galaxies. The wispy pattern spread across the foreground galaxy cluster (A 2218, about a billion parsecs distant) resembles a spider's web, but it is really an illusion caused by A2218's gravitational field, which deflects the light from background galaxies and distorts their appearance. By measuring the extent of the distortion, scientists can estimate the mass of the intervening cluster. (b) An approximately true-color image of the galaxy cluster known only by its catalog name, 0024 + 1654, residing some 1.5 billion pc away. The reddish-yellow blobs are mostly normal elliptical galaxy members of the cluster. The bluish looplike features are images of a single background galaxy. (NASA)

(a)   R I V U X G   (b)   R I V U X G

(a)

R I V U X G

(b)

▲ **Figure 16.28 Dark Matter Map** By measuring distortions in the images of background objects astronomers can produce maps of dark matter in the universe. (a) An optical view of a region of the sky containing a small galaxy cluster—the clump of yellowish galaxies near the center of the frame (indicated by the arrow). The fuzzy blue specks scattered across the frame are the much more distant background galaxies. (b) The distribution of dark matter in and near the visible cluster, on the same scale as (a), derived by analyzing the images of the background galaxies. (Data courtesy J. A. Tyson, Bell Labs; NSF/NOAO)

It is even possible to reconstruct the foreground dark mass distribution by carefully analyzing the distortions of the background objects, providing a means of tracing out the distribution of matter on scales far larger than have previously been possible. Figure 16.28(a) is an optical image of a (hard to see) foreground galaxy cluster, set against a background of much fainter distant galaxies. Figure 16.28(b) is the reconstructed dark-matter image, revealing the presence of dark mass many megaparsecs from the cluster center, extending far beyond the visible galaxies. Evidently the dark matter fraction is even greater on supercluster scales than it is on the scales of individual galaxies or galaxy clusters. Notice also the elongated structure of the dark-matter distribution, reminiscent of the Virgo Supercluster and filamentary structure seen in large-scale galaxy surveys. ∞ *(Sec. 14.5)*

◄◄◄ **CONCEPT CHECK**

How do observations of distant quasars tell us about the structure of the universe closer to home?

CHAPTER REVIEW

Summary

① The masses of nearby spiral galaxies can be determined by studying their rotation curves. Astronomers also use studies of binary galaxies and galaxy clusters to obtain statistical mass estimates of the galaxies involved.

② Measurements of galaxy and cluster masses reveal the presence of large amounts of dark matter. The fraction of dark matter grows as the scale under consideration increases. More than 90 percent of the mass in the universe is dark. Substantial quantities of hot X ray–emitting gas have been detected among the galaxies in many clusters, but not enough to account for the dark matter inferred from dynamical studies.

③ Researchers know of no simple evolutionary sequence that links spiral, elliptical, and irregular galaxies. Most astronomers think that large galaxies formed by the merger of smaller ones, and collisions and mergers of galaxies play very important roles in galactic evolution. A **starburst galaxy** (p. 438) may result when a galaxy has a close encounter or a collision with a neighbor. The strong tidal distortions caused by the encounter compress galactic gas, resulting in a widespread burst of star formation. Mergers between spirals most likely result in elliptical galaxies.

④ Quasars, active galaxies, and normal galaxies may represent an evolutionary sequence. When galaxies began to form and merge, conditions may have been suitable for the formation of large black holes at their centers, and a highly luminous quasar could have been the result. The brightest quasars consume so much fuel that their energy-emitting lifetimes must be quite short. As the fuel supply diminished, the quasar dimmed, and the galaxy in which it was embedded became intermittently visible as an active galaxy. At even later times, the nucleus became virtually inactive, and a normal galaxy was all that remained. Many normal galaxies have been found to contain central massive black holes, suggesting that most galaxies in clusters have the capacity for activity if they should interact with a neighbor.

⑤ Galaxy clusters themselves tend to clump together into **superclusters** (p. 445). The Virgo Cluster, the Local Group, and several other nearby clusters form the Local Supercluster. On even larger scales, galaxies and galaxy clusters are arranged on the surfaces of enormous "bubbles" of matter surrounding vast low-density regions called **voids** (p. 446). The origin of this structure is thought to be closely related to conditions in the very earliest epochs of the universe.

⑥ Quasars can be used as probes of the universe along the line of sight. Some quasars have been observed to have double or multiple images. These result from   gravitational lensing, in which the gravitational field of a foreground galaxy or galaxy cluster bends and focuses the light from the more distant quasar. Analysis of the images of distant galaxies, distorted by the gravitational effect of a foreground cluster, provides a means of determining the masses of galaxy clusters—including the dark matter—far beyond the information that optical images of the galaxies themselves afford.

Review and Discussion

1. Describe two techniques for measuring the mass of a galaxy.
2. Why do astronomers think that galaxy clusters contain more mass than we can see?
3. Why do some clusters of galaxies emit X rays?
4. What evidence do we have that galaxies collide with one another?
5. Describe the role of collisions in the formation and evolution of galaxies.
6. Do you think that collisions between galaxies constitute "evolution" in the same sense as the evolution of stars?
7. What are starburst galaxies, and what do they have to do with galaxy evolution?
8. Why do astronomers think that quasars represent an early stage of galactic evolution?
9. Why does the theory of galaxy evolution suggest that there should be supermassive black holes at the centers of many normal galaxies?
10. What evidence do we have for supermassive black holes in galaxies?
11. How might a normal galaxy become active?
12. What is a redshift survey?
13. What are voids? What is the distribution of galactic matter on very large (more than 100 Mpc) scales?
14. How can observations of distant quasars be used to probe the space between us and them?
15. How do astronomers "see" dark matter?

Self-Test: True or False?/Multiple Choice

1. Intergalactic gas in galaxy clusters emits large amounts of energy in the form of radio waves. **(T/F)**
2. Distant galaxies appear to be much larger than those nearby. **(T/F)**
3. Galaxy collisions are rare, and have little or no effect on the stars and interstellar gas in the galaxies involved. **(T/F)**
4. The quasar stage of a galaxy ends because the central black hole swallows up all the matter around it. **(T/F)**
5. Elliptical galaxies may be formed by mergers between spiral systems. **(T/F)**
6. The fact that a typical quasar would consume an entire galaxy's worth of mass in 10 billion years suggests that quasar lifetimes are relatively long. **(T/F)**
7. On the largest scales, galaxies in the universe appear to be arranged on huge sheets surrounding nearly empty voids. **(T/F)**

8. The image of a distant quasar can be split into several images by the gravitational field of a foreground cluster along the line of sight. **(T/F)**

9. A spiral galaxy containing substantial amounts of dark matter will **(a)** appear darker; **(b)** spin faster; **(c)** repel other galaxies; **(d)** have more tightly wound arms.

10. According to X-ray observations, the space between galaxies in a galaxy cluster is **(a)** completely devoid of matter; **(b)** very cold; **(c)** very hot; **(d)** filled with faint stars.

11. The fraction of mass of the universe made up of dark matter is **(a)** zero; **(b)** less than 10 percent; **(b)** roughly 50 percent; **(d)** more than 90 percent.

12. In current theories of galaxy evolution, quasars occur **(a)** early in the evolutionary sequence; **(b)** near the Milky Way; **(c)** when elliptical galaxies merge; **(d)** late in the evolutionary sequence.

13. Many nearby galaxies **(a)** may be much more active in the future; **(b)** contain quasars; **(c)** have radio lobes; **(d)** may have been much more active in the past.

14. If light from a distant quasar did not pass through any intervening atomic hydrogen clouds, then Figure 16.23 (Absorption Line "Forest") would have to be redrawn to show **(a)** more absorption features; **(b)** fewer absorption features; **(c)** a single, large absorption feature; **(d)** more features at short wavelengths, but fewer at long wavelengths.

15. If Figure 16.25 (Gravitational Lens) were to be redrawn using a more massive lensing galaxy, the quasar images would be **(a)** farther apart; **(b)** closer together; **(c)** fainter; **(d)** redder.

Problems

Algorithmic versions of these questions are available in the Practice Problems module of the Companion Website.

The number of squares preceding each problem indicates its approximate level of difficulty.

1. ■ Two galaxies orbit one another at a separation of 500 kpc. Their orbital period is estimated to be 30 billion years. Use Kepler's third law to find the total mass of the pair. ∞ *(Sec. 14.6)*

2. ■ Based on the data in Figure 16.1, estimate the mass of the galaxy NGC 4984 inside 20 kpc.

3. ■■■ A small satellite galaxy is moving in a circular orbit around a much more massive parent and just happens to be moving exactly parallel to the line of sight as seen from Earth. The recession velocities of the satellite and the parent galaxy are measured to be 6450 km/s and 6500 km/s, respectively, and the two galaxies are separated by an angle of 0.1° on the sky. Assuming $H_0 = 70$ km/s/Mpc, calculate the mass of the parent galaxy.

4. ■ Calculate the average speed of hydrogen nuclei (protons) in a gas of temperature 20 million K. ∞ *(More Precisely 5-1)* Compare this with the speed of a galaxy moving in a circular orbit of radius 1 Mpc around the center of a cluster of mass 10^{14} solar masses.

5. ■ In a galaxy collision, two similar-sized galaxies pass through each other with a combined relative velocity of 1500 km/s. If each galaxy is 100 kpc across, how long does the event last?

6. ■■ A certain quasar has a redshift of 0.25 and an apparent magnitude of 13. Using the data from Table 15-2, calculate the

quasar's absolute magnitude and hence its luminosity. ∞ *(More Precisely 10-1)* Compare the apparent brightness of the quasar, viewed from a distance of 10 pc, with that of the Sun as seen from Earth.

7. ■ What are the absolute magnitude and luminosity of a quasar with a redshift of 5 and an apparent magnitude of 22?

8. ■■ Assuming an energy-generation efficiency (that is, the ratio of energy released to total mass-energy consumed) of 20 percent, calculate how long a 10^{40}-W quasar can shine if a total of 10^8 solar masses of fuel is available.

9. ■■ The spectrum of a quasar with a redshift of 0.20 contains two sets of absorption lines, redshifted by 0.15 and 0.155, respectively. If $H_0 = 70$ km/s/Mpc, estimate the distance between the intervening galaxies responsible for the two sets of lines.

10. ■■■ Light from a distant quasar is deflected through an angle of 3" by an intervening lensing galaxy and is subsequently detected on Earth (see Figure 16.25). If Earth, the galaxy, and the quasar are all aligned, the quasar's redshift is 3.0 (see Table 15.2), and the galaxy lies midway between Earth and the quasar, calculate the minimum distance between the light ray and the center of the galaxy.

The Companion Website at www.prenhall.com/chaisson provides algorithmically generated versions of each chapter's Problems, in addition to an Animations and Videos gallery, a Tutorials gallery and Tutorials Quiz questions, a full eBook, an Images gallery, Practice Questions, Multiple Choice, and True or False quizzes, and Labeling exercises.

17

Cosmology

The Big Bang and the Fate of the Universe

The universe began in a fiery expansion some 14 billion years ago, and out of this maelstrom emerged all the energy that would later form galaxies, stars, and planets. The story of the origin and fate of all these systems comprises the subject of cosmology. This image—called the Ultra Deep Field—was taken with the Advanced Camera for Surveys aboard the *Hubble Space Telescope*. More than a thousand galaxies are crowded into this one image, displaying many different types and shapes. In all, astronomers estimate that the observable universe contains about 40 billion such galaxies. (NASA/ESA)

O ur field of view now extends for billions of parsecs into space and billions of years back in time. We have asked and answered many questions about the structure and evolution of planets, stars, and galaxies. At last we are in a position to address the central issues of the biggest puzzle of all: How big is the universe? How long has it been around, and how long will it last? What was its origin, and what will be its fate? What is it made of? Many cultures have asked these questions, in one form or another, and have developed their own cosmologies—theories about the nature, origin, and destiny of the universe—to answer them. In this chapter, we see how modern scientific cosmology addresses these important issues and what it has to tell us about the universe we inhabit. After more than 10,000 years of civilization, science may be ready to provide some insight into the ultimate origin of all things.

LEARNING GOALS

Studying this chapter will enable you to:

1 State the cosmological principle, and explain its significance and observational underpinnings.

2 Describe the Big Bang theory of the expanding universe.

3 Explain the relationship between the future of the universe and the overall geometry of space.

4 Discuss the factors that determine whether the universe will expand forever.

5 Describe the cosmic microwave background and its importance to the science of cosmology.

6 Explain how nuclei and atoms emerged from the primeval fireball.

7 Summarize the horizon and flatness problems, and describe how the theory of cosmic inflation solves them.

8 Discuss the formation of large-scale structure in the cosmos, and present observational evidence for the leading theory of structure formation.

Visit www.prenhall.com/chaisson for additional annotated images, animations, and links to related sites for this chapter.

17.1 The Universe on the Largest Scales

Galaxy surveys have revealed the existence of structures in the universe up to 200 Mpc across. ∞ *(Sec. 16.5)* Are there even larger structures? Does the hierarchy of clustering of stars into galaxies and galaxies into galaxy clusters, superclusters, voids, and filaments continue forever, on larger and larger scales? Perhaps surprisingly, given the trend we have just described, most astronomers think the answer is *no*.

We saw in Chapter 16 how astronomers use *redshift survey* to explore the large-scale distribution of galaxies in space, combining position on the sky with distances inferred from Hubble's law to construct three-dimensional maps of the universe on truly "cosmic" scales. ∞ *(Sec. 16.5)* Figure 17.1 is a map similar in concept to those shown previously, but based on data from the most extensive redshift survey to date—the Sloan Digital Sky Survey. ∞ *(Discovery 16-1)* The figure plots the positions of some 67,000 galaxies lying in a 120° wide "slice" of the sky within a few degrees of the celestial equator. The distances shown on the right axis assume a Hubble constant of $H_0 = 70$ km/s/Mpc. ∞ *(Sec. 15.3)*

Figure 17.1 extends out to a distance of almost 1000 Mpc. This is comparable to the extent of Figure 16.22, but because the Sloan survey includes much fainter galaxies, this map contains nearly three times more points than the earlier figure, making structure somewhat easier to discern, particularly at large distances. The extended "filament" of galaxies near the center of the wedge, some 300 Mpc from Earth, is called the *Sloan Great Wall*. Discovered in 2003 and measuring some 250 Mpc long by 50 Mpc thick, it is currently the largest known structure in the universe. However, on larger scales, apart from the general falloff in numbers of galaxies at large distances—basically because more distant galaxies are harder to see due to the inverse-square law—there is no obvious evidence for any structures on scales larger than about 200 Mpc. ∞ *(Sec. 10.2)* Careful statistical analysis of the galaxy distribution confirms this impression—even the Sloan Great Wall can be explained statistically as a chance superposition of smaller structures.

▼ **Figure 17.1** **Galaxy Survey** This map of the universe is drawn using data from the Sloan Digital Sky Survey. ∞ *(Discovery 16-1)* It shows the locations of 66,976 galaxies lying within 12 degrees of the celestial equator and extends out to a distance of almost 1000 Mpc (or a redshift of about 0.25). The largest known structure in the universe, the "Sloan Great Wall," is marked, stretching nearly 300 Mpc across the center of the frame. There is no evidence for any structure on larger scales. (Astrophysical Research Consortium/SDSS Collaboration)

The results of this and other large-scale surveys strongly suggest that the universe is **homogeneous** (the same everywhere) on scales greater than a few hundred megaparsecs. In other words, if we took a huge cube—300 Mpc on a side, say—and placed it anywhere in the universe, its overall contents would look much the same no matter where it was centered. The universe also appears to be **isotropic** (the same in all directions) on these scales. Excluding directions that are obscured by our Galaxy, we count roughly the same number of galaxies per square degree in any patch of the sky we choose to observe, provided we look deep (far) enough that local inhomogeneities don't distort our sample.

We will turn to the *origin* of large-scale cosmic structure in Section 17.8. For most of this chapter, however, we focus on the *absence* of structure on the very largest scales to frame our discussion. In the science of **cosmology**—the study of the structure and evolution of the entire universe—researchers generally assume that the universe is homogeneous and isotropic on sufficiently large scales. No one knows if these assumptions are precisely correct, but we can at least say that they are consistent with current observations. In this chapter we simply assume that they hold. The twin assumptions of homogeneity and isotropy are known as the **cosmological principle**.

The cosmological principle has some very far-reaching implications. For example, it implies that there can be no *edge* to the universe because that would violate the assumption of homogeneity. And there can be no *center* because then the universe would not look the same in all directions from any noncentral point, violating the assumption of isotropy. This is the familiar Copernican principle expanded to truly cosmic proportions—not only are we not central to the universe, but *no one* can be central, because *the universe has no center*!

17.2 The Expanding Universe

Every time you go outside at night and notice that the sky is dark, you are making a profound cosmological observation. Here's why.

Olbers's Paradox

Let's assume for a moment that, in addition to being homogeneous and isotropic, the universe is also infinite in spatial extent and unchanging in time. On average, then, the universe is uniformly populated with galaxies filled with stars. In that case, as illustrated in Figure 17.2, when you look up at the night sky, your line of sight must *eventually* encounter a star. The star may lie at an enormous distance, in some remote galaxy, but the laws of probability dictate that, sooner or later, any line drawn outward from Earth will run into a bright stellar surface.

Of course, faraway stars appear fainter than those nearby because of the inverse-square law. ∞ *(Sec. 10.2)* However, they are also much more numerous—the number of stars we see at any given distance in fact increases as the *square* of the distance (just consider the area of a sphere of increasing radius). Thus the diminishing brightnesses of distant stars are exactly balanced by their increasing numbers, and stars at all distances contribute equally to the total amount of light received on Earth.

This fact has a dramatic implication: No matter where you look, the sky should be as bright as the surface of a star—the entire night sky should be as brilliant as the surface of the Sun! The obvious difference between this prediction and the actual appearance of the night sky is known as **Olbers's paradox**, after the nineteenth-century German astronomer Heinrich Olbers, who popularized the idea.

◀◀◀ **CONCEPT CHECK**

In what sense is the universe homogeneous and isotropic?

...every sight line intersects a tree

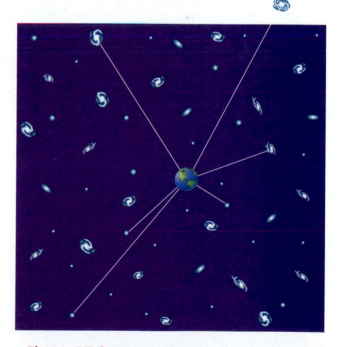

▲ **Figure 17.2 Olbers's Paradox**   If the universe were homogeneous, isotropic, infinite in extent, and unchanging, then any line of sight from Earth should eventually run into a star and the entire night sky should be bright. This obvious contradiction of the facts is known as Olbers's paradox. An analogy is a huge forest, where every sight line eventually intersects a tree.

So why is the sky dark at night? Given that the universe is homogeneous and isotropic, then one (or both) of the other two assumptions must be false. Either the universe is finite in extent, or it evolves in time. In fact, the answer involves a little of each and is intimately tied to the behavior of the universe on the largest scales.

The Birth of the Universe

We have seen that all the galaxies in the universe are rushing away from us as described by Hubble's law:

$$\text{recession velocity} = H_0 \times \text{distance},$$

where we take Hubble's constant H_0 to be 70 km/s/Mpc. ∞ *(Sec. 15.3)* We have used this relationship as a convenient means of determining the distances to galaxies and quasars, but it is much more than that.

Assuming that all velocities have remained constant in time, we can ask how long it has taken for any given galaxy to reach its present distance from us. The answer follows from Hubble's law. The time taken is simply the distance traveled divided by the velocity:

$$\text{time} = \frac{\text{distance}}{\text{velocity}}$$

$$= \frac{\text{distance}}{H_0 \times \text{distance}} \quad \text{(using Hubble's law for the velocity)}$$

$$= \frac{1}{H_0}.$$

For $H_0 = 70$ km/s/Mpc, this time is about 14 billion years.[1] Notice that it is *independent* of the distance—galaxies twice as far away are moving twice as fast, so the time they took to cross the intervening distance is the same.

Hubble's law therefore implies that, at some time in the past—14 billion years ago, according to the foregoing simple calculation (see also Section 17.4)—*all* the galaxies in the universe lay right on top of one another. In fact, the almost unanimous view among astronomers is that *everything* in the universe—matter and radiation alike—was confined to a single point at that instant. Then the point exploded, flying apart at high speeds. The present locations and velocities of the galaxies are a direct consequence of that primordial blast. This gargantuan explosion, involving everything in the cosmos, is known as the **Big Bang**. It marked the beginning of the universe.

The Big Bang provides the resolution of Olbers's paradox. Whether the universe is actually finite or infinite in extent is irrelevant, at least as far as the appearance of the night sky is concerned. The key point is that we see only a *finite* part of it—the region lying within roughly 14 billion light-years of us. What lies beyond is unknown—its light has not had time to reach us.

Where Was the Big Bang?

Now we know *when* the Big Bang occurred. Is there any way of telling *where*? The observed recession of the galaxies described by Hubble's law implies that all the galaxies exploded from a point some time in the past. Wasn't that point, then, different from the rest of the universe, violating the assumption of homogeneity expressed in the cosmological principle? The answer is a definite *no*!

To understand why there is no center to the expansion, we must make a great leap in our perception of the universe. If we were to imagine the Big

---

[1]Since 1 Mpc = 3.1×10^{19} km, it follows that $1/H_0 = 3.1 \times 10^{19}$ km ÷ 70 km/s = 4.4×10^{17} s = 14 billion yr.

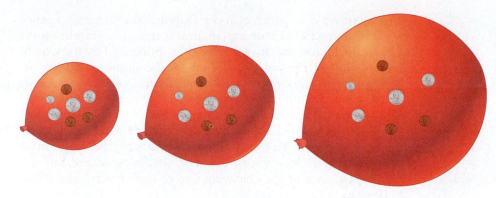

◄ **Figure 17.3** **Receding Galaxies**   Coins taped to the surface of a spherical balloon recede from one another as the balloon inflates (left to right). Similarly, galaxies recede from one another as the universe expands. As the coins recede, the distance between any two of them increases, and the rate of increase of this distance is proportional to the distance between them. Thus, the balloon expands according to Hubble's law.

Bang as simply an enormous explosion that spewed matter out into space, ultimately to form the galaxies we see, then the foregoing reasoning would be quite correct. The universe would have a center and an edge, and the cosmological principle would not apply. But the Big Bang was *not* an explosion in an otherwise featureless, empty universe. The only way that we can have Hubble's law *and* retain the cosmological principle is to realize that the Big Bang involved the entire universe—not just the matter and radiation within it, but the universe *itself*. In other words, the galaxies are not flying apart into the rest of the universe. The universe itself is expanding. Like raisins in a loaf of raisin bread that move apart as the bread expands in an oven, the galaxies are just along for the ride.

Let's reconsider some of our earlier statements in this new light. We now recognize that Hubble's law describes the expansion of the universe itself. Apart from small-scale individual random motions, galaxies are not moving with respect to the fabric of space. The component of the galaxies' motion that makes up the Hubble flow is really an expansion of space itself. The expanding universe remains homogeneous at all times. There is no "empty space" beyond the galaxies into which they rush. At the time of the Big Bang, the galaxies did not reside at a point located at some well-defined place within the universe. The *entire universe* was a point. The Big Bang happened *everywhere* at once.

To illustrate these ideas, imagine an ordinary balloon with coins taped to its surface, as shown in Figure 17.3. (Better yet, do the experiment yourself.) The coins represent galaxies and the two-dimensional surface of the balloon represents the "fabric" of our three-dimensional universe. The cosmological principle applies here because every point on the balloon looks pretty much the same as every other. Imagine yourself as a resident of one of the coin "galaxies" on the leftmost balloon, and note your position relative to your neighbors. As the balloon inflates (that is, as the universe expands), the other galaxies recede from you and more distant galaxies recede more rapidly.

Regardless of which galaxy you chose to consider, you would see all the other galaxies receding from you, but nothing is special or peculiar about this fact. There is no center to the expansion and no position that can be identified as the location from which the universal expansion began. Everyone sees an overall expansion described by Hubble's law, with the same value of Hubble's constant in all cases. Now imagine letting the balloon deflate, running the "universe" backward from the present time to the moment of the Big Bang. *All* the galaxies (coins) would arrive at the same place at the same time at the instant the balloon reached zero size, but no single point on the balloon could be said to be *the* place where that occurred.

The Cosmological Redshift

Up to now, we have explained the cosmological redshift of galaxies as a Doppler shift, a consequence of their motion relative to us. ∞ *(Sec. 15.3)* However, we have just argued that the galaxies are not moving with respect to

▲ **Figure 17.4 Cosmological Redshift** As the universe expands, photons of radiation are stretched in wavelength, giving rise to the cosmological redshift.

CONCEPT CHECK ▶▶▶

Why does Hubble's law imply a Big Bang?

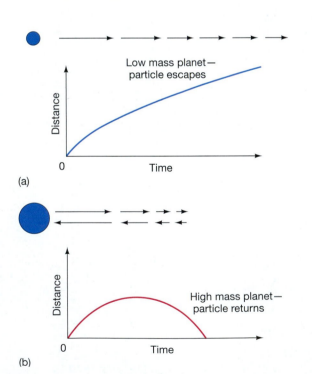

▲ **Figure 17.5 Escape Speed** (a) A spacecraft (arrow) leaving a planet (blue ball) with a speed greater than the escape speed follows an unbound trajectory and escapes. The graph shows the distance between the ship and the planet as a function of time. (b) If the launch speed is less than the escape speed, the ship eventually drops back to the planet. Its distance from the planet first rises, then falls.

the universe, in which case the Doppler interpretation cannot be correct. The true explanation is that, as a photon moves through space, its wavelength is influenced by the expansion of the universe. Think of the photon as being attached to the expanding fabric of space, so its wavelength expands along with the universe, as illustrated in Figure 17.4. Although it is common in astronomy to refer to the cosmological redshift in terms of recessional velocity, bear in mind that, strictly speaking, this is not the right thing to do. The cosmological redshift is a consequence of the changing size of the universe and is *not* related to velocity at all.

The redshift of a photon measures the amount by which the universe has expanded since that photon was emitted. For example, when we measure the light from a quasar to have a redshift of 5, it means that the observed wavelength is six times (one plus the redshift) greater than the wavelength at the time of emission, which means that the light was emitted at a time when the universe was just one-sixth its present size (and we are observing the quasar as it was at that time). ∞ *(More Precisely 15-1)* In general, the larger a photon's redshift, the smaller the universe was at the time the photon was emitted, and so the longer ago that emission occurred. Because the universe expands with time and redshift is related to that expansion, cosmologists routinely use redshift as a convenient means of expressing time.

17.3 Cosmic Dynamics and the Geometry of Space

Will the universe expand forever? This fundamental question about the fate of the universe has been at the heart of cosmology since Hubble's law was first discovered. Until the late 1990s the prevailing view among cosmologists was that the answer would most likely be found by determining the extent to which gravity would slow, and perhaps ultimately reverse, the current expansion. However, it now appears that the answer is rather more subtle—and perhaps a lot more profound in its implications—than was hitherto thought.

Two Futures

Let's begin with another analogy. Assume for the moment that gravity is the only force affecting large-scale motion in the universe, and consider a rocket ship launched from the surface of a planet. Until relatively recently, this scenario was much more than just an analogy—this basic picture and its implications represented the conventional wisdom among cosmologists. However, as we will see in Section 17.4, new observations have forced fundamental changes in astronomers' view of the universe. Nevertheless, the simplified view we now present is a convenient starting point, as it allows us to define some basic ideas and terminology.

What are the likely outcomes of the rocket ship's motion? According to Newtonian mechanics, there are just two basic possibilities, depending on the launch speed of the ship relative to the escape speed of the planet. ∞ *(More Precisely 5-1)* If the launch speed is high enough, it will exceed the planet's escape speed, and the ship will never return to the surface. The speed will diminish because of the planet's gravitational pull, but it will never reach zero. The spacecraft leaves the planet on an unbound trajectory, as illustrated in Figure 17.5(a). Alternatively, if the launch speed is lower than the escape speed, the ship will reach a maximum distance from the planet and then fall back to the surface. Its bound trajectory is shown in Figure 17.5(b).

Similar reasoning applies to the expansion of the universe. Imagine two galaxies at some known distance from one another, their present relative velocity given by Hubble's law. The same two basic possibilities exist for these galaxies as for our rocket ship. The distance between them can increase forever, or it can increase for a while and then start to decrease. What's more, the cosmological principle says that whatever the outcome, it must be the same for *any* two galaxies—in other words, the same statement applies to the universe *as a whole*. Thus, as illustrated in Figure 17.6, the universe has only two options: It can continue to expand forever, or the present expansion will someday stop and turn around into a contraction. The two curves in the figure are drawn so that they pass through the same point at the present time. Both are possible descriptions of the universe, given its current size and expansion rate.

What determines which of these possibilities will actually occur? In the case of a rocket ship of fixed launch speed, the *mass* of the planet determines whether or not escape will occur—a more massive planet has a higher escape speed, making it less likely that the rocket can escape. For the universe, the corresponding factor is the *density* of the cosmos. A high-density universe contains enough mass to stop the expansion and eventually cause a recollapse. Such a universe is destined ultimately to shrink toward a superdense, superhot singularity much like the one from which it originated. Some astronomers call this the "Big Crunch." A low-density universe, conversely, will expand forever. In that case, in time, even with the most powerful telescopes, we will see no galaxies in the sky beyond the Local Group (which is not itself expanding). The rest of the observable universe will appear dark, the distant galaxies too faint to be seen.

The dividing line between these outcomes—the density corresponding to a universe in which *gravity acting alone* would be just sufficient to halt the present expansion—is called the universe's **critical density**. For $H_0 = 70$ km/s/Mpc, the critical density is about 9×10^{-27} kg/m^3. That's an extraordinarily low density—just five hydrogen atoms per cubic meter, a volume the size of a typical household closet. In more "cosmological" terms, it corresponds to about 0.1 Milky Way Galaxy (including the dark matter) per cubic megaparsec.

We will see in a moment that the separation between never-ending expansion and cosmic collapse is not quite as straightforward as the foregoing simple reasoning would suggest. Several independent lines of evidence now indicate that gravity is *not* the only influence on the dynamics of the universe on large scales (Section 17.4). As a result, while the "futures" just described are still the only two possibilities for the long-term evolution of the universe, the distinction between them turns out to be more than just a matter of density alone. Nevertheless, the density of the universe—or, more precisely, the *ratio of the total density to the critical value*—is a vitally important quantity in cosmology.

The Shape of the Universe

We have reverted to the familiar notions of Newtonian mechanics and gravity and moved away from general relativity and the more correct concept of warped spacetime, because speaking in Newtonian terms makes our discussion of the evolution of the universe easier to understand. ∞ *(Sec. 1.4, 13.5)* However, general relativity makes some predictions about the universe that do not have a simple description within Newton's theory.

Foremost among these non-Newtonian predictions is the fact that space is *curved*, and the degree of curvature is determined by the *total density* of the cosmos. ∞ *(Sec. 13.5)* But there is an important clarification to be made here. In our earlier discussion, we did not specify exactly what we meant by "density." Our discussion implicitly focused on the density of matter, but according to the theory of relativity, mass (m) and energy (E) are equivalent, connected by Einstein's famous relation $E = mc^2$. ∞ *(More Precisely 13-1)* What, then, actually contributes to the warping of space? General relativity's

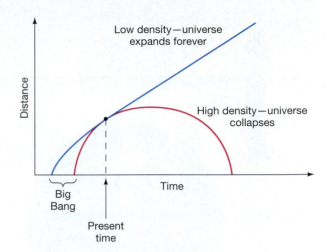

▲ **Figure 17.6 Model Universes** Distance between two galaxies as a function of time in each of the two basic universes discussed in the text: a low-density universe that expands forever and a high-density cosmos that collapses. The point where the two curves touch represents the present time.

◀◀◀ **CONCEPT CHECK**

What are the two basic possibilities for the future expansion of the universe?

▲ Figure 17.7 Einstein's Curve Ball In a closed universe a beam of light launched in one direction might return someday from the opposite direction after circling the universe, just as motion in a "straight line" on Earth's surface will eventually encircle the globe.

answer is clear: Both matter *and* energy must be taken into account, with energy properly "converted" into matter units via division by the square of the speed of light. The "density" of the universe includes not just the atoms and molecules that make up the familiar normal matter around us, but also the invisible dark matter that dominates the masses of galaxies and galaxy clusters, as well as *everything* that carries energy—photons, relativistic neutrinos, gravity waves, and anything else we can think of.

In a homogeneous universe, the curvature (on sufficiently large scales) must be the same everywhere, so there are really only three distinct possibilities for the large-scale geometry of space. (For more information on the different types of geometry involved, see *More Precisely 17-1*.) General relativity tells us that the geometry of the universe depends only on the ratio of the density of the universe to the critical density (as defined in the previous section). Cosmologists conventionally call the ratio of the universe's actual density to the critical value the *cosmic density parameter* and denote it by the symbol Ω_0 ("omega nought"). In terms of this quantity, then, a universe with density equal to the critical value has $\Omega_0 = 1$, a "low-density" cosmos has Ω_0 less than 1, and a "high-density" universe has Ω_0 greater than 1.

In a high-density universe (Ω_0 greater than 1), space is curved so much that it bends back on itself and closes off, making the universe *finite* in size. Such a universe is known as a **closed universe**. It is difficult to visualize a three-dimensional volume uniformly arching back on itself in this way, but the two-dimensional version is well known: It is the surface of a sphere, like the balloon we discussed earlier. Figure 17.3, then, is the two-dimensional likeness of a three-dimensional closed universe. Like the surface of a sphere, a closed universe has no boundary, yet it is finite in extent.[2] One remarkable property of a closed universe is illustrated in Figure 17.7: A flashlight beam shining in some direction in space might eventually traverse the entire universe and return from the opposite direction.

The surface of a sphere curves, loosely speaking, "in the same direction" no matter which way we move from a given point. A sphere is said to have *positive curvature*. However, if the density of the universe is below the critical value, then the curvature of space is qualitatively quite different from that of a sphere. The two-dimensional surface corresponding to such a space curves like a saddle and is said to have *negative curvature*. Most people have seen a saddle—it curves "up" in one direction and "down" in another, but no one has ever seen a uniformly negatively curved surface, for the good reason that it cannot be constructed in three-dimensional space! It is just "too big" to fit. A low-density (Ω_0 less than 1), saddle-curved universe is infinite in extent and is usually called an **open universe**.

The intermediate case, when the density is precisely equal to the critical density (i.e., $\Omega_0 = 1$), is the easiest to visualize. This **critical universe** has no curvature. It is said to be "flat," and it is infinite in extent. In this case, and only in this case, the geometry of space on large scales is precisely the familiar Euclidean geometry taught in high school. Apart from its overall expansion, this is basically the universe that Newton knew.

Euclidean geometry—the geometry of flat space—is familiar to most of us because it is a good description of space in the vicinity of Earth. It is the geometry of everyday experience. Does this imply that the universe is flat, which would in turn mean that it has exactly the critical density? The answer is no. Just as a flat street map is a good representation of a city, even though we know Earth is really a sphere, Euclidean geometry is a good description of space within the solar system, or even the Galaxy, because the curvature of the uni-

CONCEPT CHECK ▶▶▶

How is the curvature of space related to the size and density of the universe?

[2] Notice that for the sphere analogy to work, we must imagine ourselves as two-dimensional "flatlanders" who cannot visualize or experience in any way the third dimension perpendicular to the sphere's surface. Flatlanders are confined to the sphere's surface, just as we are confined to the three-dimensional volume of our universe.

verse is negligible on scales smaller than about 1000 Mpc. Only on the very largest scales would the geometrical effects we have just described become evident. However, if the results of the previous section are indeed correct, then the density is in fact critical and the universe really is flat (Euclidean) on the largest scales.

17-1 MORE PRECISELY

Curved Space

Euclidean geometry is the geometry of flat space—the geometry taught in high schools everywhere. Set forth by Euclid, one of the most famous of the ancient Greek mathematicians, it is the geometry of everyday experience. Houses are built with flat floors. Writing tablets and blackboards are also flat. We work easily with flat, straight objects because a straight line is the shortest distance between any two points.

In constructing houses or any other straight-walled buildings on the surface of Earth, some other basic axioms of Euclid's geometry also apply: parallel lines never meet even when extended to infinity; the angles of any triangle always sum to 180°; the circumference of a circle equals π times its diameter. If these rules were not obeyed, walls and roof would never meet to form a house!

In reality, though, the geometry of Earth's surface is not flat. It is curved. We live on a sphere, and on its surface Euclidean geometry breaks down. Instead, the rules are those of *Riemannian geometry*, named after the nineteenth-century German mathematician Georg Friedrich Riemann. There are no parallel lines (or curves) on a sphere's surface—any "straight" lines drawn on the surface and around the full circumference will eventually intersect. The sum of a triangle's angles always *exceeds* 180°. (For example, in the 90°-90°-90° triangle shown in the figure, the sum is 270°.) And the circumference of a circle is *less* than π times its diameter.

Riemannian geometry on the curved surface of a sphere differs greatly from the flat-space geometry of Euclid. The two are approximately the same only if we confine ourselves to a small patch on the surface. If the patch is much smaller than the sphere's radius, the surface surrounding any point looks flat, and Euclidean geometry is approximately valid. This is why we can draw a usable map of our home, our city, even our state, on a flat sheet of paper, but an accurate map of the entire Earth must be drawn on a globe.

When we work with larger portions of Earth, we must abandon Euclidean geometry. World navigators are well aware of this. Aircraft do not fly along what you might regard as a straight-line path from one point to another. Instead they follow a *great circle*, which, on the curved surface of a sphere, is the shortest distance between two points. For example, a flight from Los Angeles to London does not proceed directly across the United States and the Atlantic Ocean as you might expect from looking at a flat map. Instead, it goes far to the north, over Canada and Greenland, above the Arctic Circle, finally

coming in over Scotland for a landing at London. This is the shortest path between the two cities. Look at a globe of Earth (or see the accompanying figure) to verify that this is indeed the case.

The *positively curved* space of Riemann is not the only possible departure from flat space. Another is the *negatively curved* space first studied by Nikolai Ivanovich Lobachevsky, a nineteenth-century Russian mathematician. In this geometry, there are an *infinite* number of lines through any given point parallel to another line, the sum of a triangle's angles is *less* than 180° (see the figure), and the circumference of a circle is *greater* than π times its diameter. Instead of the flat surface of a plane or the curved surface of a sphere, this type of space is described by the surface of a curved saddle. It is a difficult geometry to visualize!

Even though the universe on the largest scales may be curved as we have just described, most of the local realm of the *three*-dimensional universe (including the solar system, the neighboring stars, and even our Milky Way Galaxy) is correctly described by Euclidean geometry.

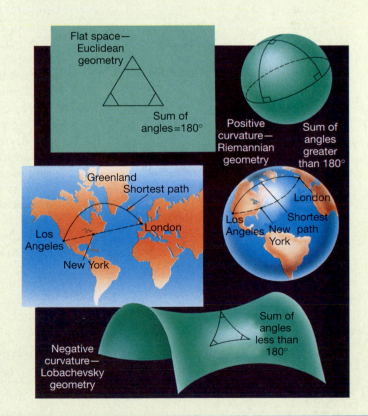

Flat space—
Euclidean
geometry

Sum of
angles=180°

Positive
curvature—
Riemannian
geometry

Sum of
angles
greater
than 180°

Greenland
Shortest path

London

Los
Angeles

New York

London

Los
Angeles

Shortest
path

New
York

Negative
curvature—
Lobachevsky
geometry

Sum of
angles
less than
180°

17.4 The Fate of the Cosmos

Is there any way for us to determine which of the futures we have described actually applies to our universe (apart from just waiting to find out)? Will the universe end as a small, dense point much like that from which it began? Or will it expand forever? And can we hope to measure the geometry of the vast cosmos we inhabit? Finding answers to these questions has been the dream of astronomers for decades. We are fortunate to live at a time when astronomers can subject these questions to intensive observational tests and come up with definite answers—even though they aren't what most cosmologists expected! Let's begin by looking at the density of the universe (or, equivalently, the cosmic density parameter Ω_0).

The Density of the Universe

How might we determine the average density of the universe? On the face of it, it would seem simple—just measure the average mass of the galaxies residing within a large parcel of space, calculate the volume of that space, and compute the total mass density. When astronomers do this, they usually find a little less than 10^{-28} kg/m^3 in the form of luminous matter. Largely independent of whether the chosen region contains only a few galaxies or a rich galaxy cluster, the resulting density is about the same, within a factor of two or three. Galaxy counts thus yield a value of Ω_0 of only a few percent. If that measure were correct, and galaxies were all that existed, then we would live in a low-density open universe destined to expand forever.

But there is a catch. Most of the matter in the universe is *dark*—it exists in the form of invisible material that has been detected only through its gravitational effect in galaxies and galaxy clusters. ∞ *(Sec. 14.6, 16.1)* We currently do not know *what* the dark matter is, but we *do* know that it is there. Galaxies may contain as much as 10 times more dark matter than luminous material, and the figure for galaxy clusters is even higher—perhaps as much as 95 percent of the total mass in clusters is invisible. Even though we cannot see it, dark matter contributes to the average density of the universe and plays its part in opposing the cosmic expansion. Including all the dark matter that is known to exist in galaxies and galaxy clusters increases the value of Ω_0 to about 0.25.

Unfortunately, the distribution of dark matter on larger scales is not very well known. We can infer its presence in galaxies and galaxy clusters, but we are largely ignorant of its extent in larger structures. One of the few mass estimates for an object much larger than a supercluster comes from optical and infrared observations of the overall motion of galaxies within the Local Supercluster. The measured velocities suggest the presence of a nearby huge accumulation of mass known as the *Great Attractor*, with a total mass of about 10^{17} solar masses and a size of 100–150 Mpc. Its average density may be quite close to the critical value.

However, it seems that, when averaged over the entire universe, even objects such as the Great Attractor don't raise the overall cosmic density by much. In short, there doesn't seem to be much additional dark matter "tucked away" on very large scales. Most cosmologists agree that the density of matter (luminous plus dark) in the universe is not much greater than about 30 percent of the critical value—*not* enough to halt the universe's current expansion.

Cosmic Acceleration

Determining the density of the universe is an example of a *local* measurement that provides an estimate of Ω_0. But the result we obtain depends on just how local our measurement is—as we have just seen, there are many uncertainties

▶ **Figure 17.8 Accelerating Universe** (a) Observations of distant supernovae allow astronomers to measure changes in the rate of expansion of the universe. In a decelerating universe (purple and red curves), redshifts of distant objects are greater than would be predicted from Hubble's law (black curve). The reverse is true for an accelerating universe. The points showing observations of some 50 supernovae strongly suggest that the cosmic expansion is accelerating. The vertical scale shows redshift; for small velocities, redshift is just velocity divided by the speed of light. ∞ *(More Precisely 15-1)* (b) The bottom frames show three of the above-graphed supernovae (marked by arrows) that exploded in distant galaxies when the universe was nearly half its current age. The top frames show the same areas prior to the detonation of those supernovae, which were originally discovered in 1997 during a ground-based survey with the Canada–France–Hawaii Telescope on Mauna Kea. (P. Garnavich/Harvard-Smithsonian Center for Astrophysics/NASA)

in the result, especially on large scales. In an attempt to get around this problem, astronomers have devised alternative methods that rely instead on *global* measurements, covering much larger regions of the observable universe. In principle, such global tests should indicate the universe's overall density, not just its value in our cosmic neighborhood.

One such global method is based on observations of Type I (carbon detonation) supernovae. ∞ *(Sec. 12.5)* Recall that these objects are both very bright and have a remarkably narrow spread in luminosities, making them particularly useful as standard candles. ∞ *(Sec. 15.2)* They can be used as probes of the universe because, by measuring their distances (*without* using Hubble's law) and their redshifts, we can determine the rate of cosmic expansion in the distant past. Here's how the method works.

Suppose the universe is decelerating, as we would expect if gravity were slowing its expansion. Then, because the expansion rate is decreasing, objects at great distances—that is, objects that emitted their radiation long ago—should appear to be receding *faster* than Hubble's law predicts. Figure 17.8 illustrates this concept. If the universal expansion were constant in time, recessional velocity and distance would be related by the black line in the figure. (The line is not quite straight because it takes the expansion of the universe properly into account in computing the distance.) ∞ *(More Precisely 15-1)* In a decelerating universe, the velocities of distant objects should lie *above* the black curve, and the deviation from that curve is greater for a denser universe, in which gravity has been more effective at slowing the expansion.

How does theory compare with reality? In the late 1990s two groups of astronomers announced the results of independent, systematic surveys of distant supernovae. Some of these supernovae are shown in Figure 17.8(b); the data are marked on Figure 17.8(a). Far from clarifying the picture of cosmic deceleration, however, these findings seem to indicate that the expansion of the universe is not slowing, but actually accelerating! According to the supernova data, galaxies at large distances are receding *less* rapidly than Hubble's law would predict. The deviations from the decelerating curves appear small in the figure, but they are statistically very significant, and both groups report similar findings.

These observations are *inconsistent* with the standard "gravity only" Big Bang theory just described and have sparked a major revision of our view of the cosmos. The measurements are difficult, and the results depend quite sensitively on just how "standard" the supernovae luminosities really are. Not surprisingly, since so much hangs on this measurement, the reliability of the supernova measurement technique has been the subject of intense scrutiny by cosmologists. However, no convincing argument against the method has yet been put forward, so there is no reason to believe that we are somehow being "fooled" by nature. As far as we can tell, the measurements are good, and the acceleration is real.

What could cause an overall acceleration of the universe? Frankly, cosmologists do not know, although several possibilities have been suggested.

Whatever it is, the mysterious cosmic force causing the universe to accelerate is neither matter nor radiation. Although it carries energy, it exerts an overall *repulsive* effect on the universe, causing empty space to expand. It has come to be known simply as **dark energy**, and it is perhaps *the* leading puzzle in astronomy today.

One leading candidate is an additional "vacuum pressure" force associated with empty space and effective only on very large scales. Known simply as the **cosmological constant**, it has a long and checkered history (see *Discovery 17-1*). Its influence increases as the universe expands. Thus it was negligible at early times, but today it may be the major factor controlling the cosmic expansion. Note, however, that while models including this force can fit the observational data but, astronomers as yet have *no* clear physical interpretation of what it actually means—it is neither required nor explained by any known law of physics.

17-1 DISCOVERY

Einstein and the Cosmological Constant

Even the greatest minds are fallible. The first scientist to apply general relativity to the universe was (not surprisingly), the theory's inventor, Albert Einstein. When he derived and solved the equations describing the behavior of the universe, Einstein discovered that they predicted a universe that evolved in time. But in 1917 neither he nor anyone else knew about the expansion of the universe, as described by Hubble's law, which would not be discovered for another 10 years. ∞ *(Sec. 15.3)* At the time Einstein, like most scientists, believed that the universe was static—that is, unchanging and everlasting. The discovery that there was no static solution to his equations seemed to Einstein a near-fatal flaw in his new theory.

To bring his theory into line with his beliefs, Einstein tinkered with his equations, introducing a "fudge factor" now known as the cosmological constant. The accompanying figure shows the effect of introducing a cosmological constant into the equations describing the expansion of a critical-density universe. One possible solution is a "coasting" universe, whose radius does indeed remain constant for an indefinite period of time. Einstein took this to be the static universe he expected. Instead of predicting an evolving cosmos, which would have been one of general relativity's greatest triumphs, Einstein yielded to a preconceived notion of the way the universe "should be," unsupported by observational evidence. Later, when the expansion of the universe was discovered and Einstein's equations—without the cosmological constant—were found to describe it perfectly, he declared that the cosmological constant was the biggest mistake of his scientific career.

For many researchers—Einstein included—the main problem with the cosmological constant was (and still is) the fact that it had no clear physical interpretation. Einstein introduced it to fix what he thought was a problem with his equations, but he discarded it immediately once he realized that no problem actually existed. Scientists are very reluctant to introduce unknown quantities into their equations purely to make the results "come out right." As a result, the cosmological constant fell out of favor among astronomers for many years. However, as described in the text, it has apparently been completely rehabilitated and is now identified as a leading candidate for the "dark energy" whose existence is inferred from studies of the universe on very large scales.

Attempts are underway to construct theories of dark energy that preserve the character of the cosmological constant, yet also account for its value in some natural way, perhaps somehow coupling it to the density of matter. Before we make too many sweeping statements about the role of the cosmological constant in cosmology, we should probably remember the experience of its inventor and bear in mind that—at least for now—its physical meaning remains completely unknown.

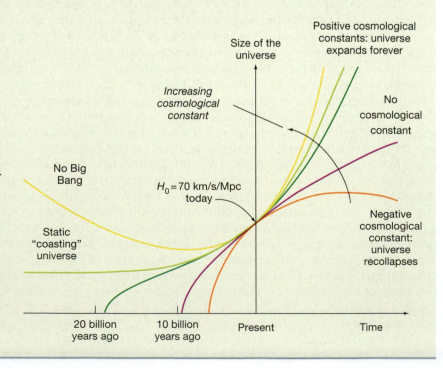

If confirmed, the magnitude of the cosmic acceleration implies that the amount of dark energy in the universe may exceed the total mass-energy of matter (luminous and dark) by a substantial margin. Furthermore, since the effect of gravity *weakens* as the expansion proceeds, it follows that, once the cosmological constant begins to dominate, gravity can never catch up, and the universe will continue to accelerate at an ever-increasing pace. Thus, despite the considerable uncertainty as to the nature of dark energy, we can at least make one definite statement: By opposing the attractive force of gravity, the repulsive effect of the dark energy strengthens our earlier conclusion that the universe will expand forever.

Cosmic Composition

In addition to measurements of density and acceleration, astronomers have several other means of estimating the "cosmological parameters" that describe the large-scale properties of our universe. Theoretical studies of the very early universe (Section 17.7) strongly suggest that the geometry of the universe should be precisely flat—and hence the density should be exactly critical. These ideas first became widespread in the 1980s and for many years seemed to be at odds with observations indicating a cosmic matter density of just 20–30 percent of the critical value, even taking the dark matter into account. The cosmic acceleration just discussed actually resolves that conflict, albeit at the price of introducing yet another unknown component into the cosmic mix.

Recent detailed measurements of the radiation field known to fill the entire cosmos (Sections 17.5 and 17.8) support the theoretical prediction of critical density and are also consistent with the dark energy inferred from the supernova studies. Further support comes from careful analysis of galaxy surveys such as those discussed in Section 17.1, which allow astronomers to measure the growth of large-scale structure in the universe. Simply put, the more mass there is in the universe, the easier it is for clusters, superclusters, walls, and voids to grow as gravity gathers matter into larger and larger clumps, so structure measurements constrain the value of Ω_0.

Remarkably, all the approaches just described yield consistent results! As of early 2006, the consensus among cosmologists is that the universe is of precisely critical density, $\Omega_0 = 1$, but this density is made up of both matter (mostly dark) and dark energy (suitably converted into mass units using the relation $E = mc^2$). ∞ *(Sec. 9.5)* Note that, while dark energy *opposes* the force of gravity in its effect on cosmic expansion, it still *adds* to the total density (and Ω_0) in its effect on the curvature of space. Based on all available data, the best estimate is that matter accounts for 27 percent of the total, dark energy for the remaining 73 percent. This is the assumption underlying Table 15.1 and used consistently throughout this book. ∞ *(More Precisely 15-1)*

Note that such a universe will expand forever and, the heavy machinery of general relativity and curved spacetime notwithstanding, is perfectly flat (Figure 17.9)—an irony that would no doubt have amused Newton!

▲ **Figure 17.9 Geometry of the Universe** As best we can tell, the universe on the largest scales is geometrically flat—governed by the same familiar Euclidean geometry taught in high schools.

The Age of the Universe

We have at least one other independent (that is, noncosmological) way of testing this important conclusion. In Section 17.2, when we estimated the age of the universe from the accepted value of Hubble's constant, we made the assumption that the expansion speed of the cosmos was constant in the past. However, as we have just seen, this is a considerable oversimplification. Gravity tends to slow the universe's expansion, while dark energy acts to accelerate it. In the absence of a cosmological constant, the universe expanded faster in the past than it does today, so the assumption of constant expansion rate leads to an overestimate of the universe's age—such a universe is younger than the

▶ **Figure 17.10 Cosmic Age** The age of a universe without a cosmological constant is always less than $1/H_0$ and decreases for larger values of the present-day density. The existence of a repulsive cosmological constant increases the age of the cosmos.

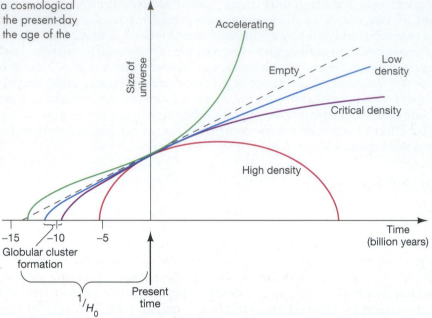

14 billion years calculated earlier. Conversely, the repulsive effect of dark energy tends to increase the age of the cosmos.

Figure 17.10 illustrates these points. It is similar to Figure 17.6, except that here we have added three curves, including one (labeled "Empty") corresponding to constant expansion rate at the present value, and another to an accelerating universe with the "best estimate" parameters given above. The age of a critical-density universe with no cosmological constant (whose curve is also added here) is about 9 billion years. A low-density, unbound universe is older than 9 billion years but still less than 14 billion years old. The age corresponding to the accelerating universe is about 14 billion years, coincidentally very close to the value calculated for constant expansion.

How does this compare with an age estimated by other means? On the basis of the theory of stellar evolution, the oldest globular clusters formed about 12 billion years ago, and most are estimated to be between 10 and 12 billion years old. ∞ *(Sec. 12.6)* This range is indicated on Figure 17.10. These ancient star clusters are thought to have formed at around the same time as our Galaxy, so they date the epoch of galaxy formation. More important, they can't be older than the universe! Figure 17.10 shows that globular cluster ages are consistent with a 14 billion-year-old cosmos, and even allow a couple of billion years for galaxies to grow, as discussed in Chapter 16. ∞ *(Sec. 16.3)* Note also that the cluster ages are *not* consistent with a critical-density universe without dark energy. This independent check of a key prediction is an important piece of evidence supporting the modern version of the Big Bang theory.

Thus, for $H_0 = 70$ km/s/Mpc, our current best guess of the history of the universe places the Big Bang at 14 billion years ago. The first quasars appeared about 13 billion years ago (at a redshift of 6), the peak quasar epoch (redshifts 2–3) occurred during the next 1 billion years, and the oldest known stars in our Galaxy formed during the 2 billion years after that. ∞ *(Sec. 16.3)* Even though astronomers do not presently understand the nature of dark energy, the good agreement between so many separate lines of reasoning has convinced many that the dark-matter/dark-energy Big Bang theory just described is the correct description of the universe. But astronomers aren't ready to relax just yet—the history of this subject suggests that there may be a few more unexpected twists and turns in the road before the details are finally resolved.

Bear in mind as we proceed through the remainder of this chapter that the Big Bang *is* a scientific theory and as such is continually being challenged and scrutinized. ∞ *(Sec. E.5)* It makes testable predictions about the state and history of the cosmos, and must change—or fail—if these predictions are found to be at odds with observations. The supernova observations just described are a case in point. The rest of chis chapter highlights several key points that provide strong support in favor of the Big Bang theory of the universe.

◄◄◄ **CONCEPT CHECK** ▶

Why do astronomers think the universe will expand forever?

17.5 The Early Universe

We now turn from studies of the far future of our universe to the quest to understand its distant past. Just how far back in time can we probe? Is there any way to study the universe beyond the most remote quasar? How close can we come to perceiving directly the edge of time, the very origin of the universe?

The Cosmic Microwave Background

A partial answer to these questions was discovered by accident in 1964, during an experiment designed to improve America's telephone system. As part of a project to identify and eliminate unwanted interference in satellite communications, Arno Penzias and Robert Wilson, scientists at Bell Telephone Laboratories in New Jersey, were studying the Milky Way's emission at microwave (radio) wavelengths, using the horn-shaped antenna shown in Figure 17.11. In their data they noticed a bothersome background "hiss"—a little like the background static on an AM radio station. Regardless of where and when they pointed their antenna, the hiss persisted. Never diminishing or intensifying, the weak signal was detectable at any time of the day and on any day of the year, apparently filling all of space.

Eventually, after discussions with colleagues at Bell Labs and theorists at nearby Princeton University, the two experimentalists realized that the origin of the mysterious static was nothing less than the fiery creation of the universe itself. The radio hiss that Penzias and Wilson detected is now known as the **cosmic microwave background**. Their discovery won them the 1978 Nobel Prize in physics.

In fact, theorists had predicted the existence and general properties of the microwave background well before its discovery. In addition to being extremely dense, they reasoned, the early universe must also have been very *hot*, and so, shortly after the Big Bang, the universe must have been filled with extremely high-energy thermal radiation—gamma rays of very short wavelength. As illustrated in Figure 17.12, this intensely hot primordial radiation has been redshifted by the expansion of the universe from gamma rays, through optical and infrared wavelengths, all the way into the radio (microwave) range of the electromagnetic spectrum. ∞ *(Sec. 2.4)* The radiation observed today is the "fossil remnant" of the primeval fireball that existed at those very early times. The discovery of the microwave background was thus a powerful piece of evidence in support of the Big Bang theory of the universe and was instrumental in convincing many astronomers of the basic correctness of the idea.

Researchers at Princeton confirmed the existence of the microwave background and estimated its temperature at about 3 K. However, this part of the electromagnetic spectrum happens to be very difficult to observe from the ground, and it was 25 years before astronomers could demonstrate conclusively that the radiation was indeed described by a blackbody curve. ∞ *(Sec. 2.3)*

▲ **Figure 17.11 Microwave Background Discoverers** This "sugarscoop" antenna was built to communicate with Earth-orbiting satellites. Pictured are Robert Wilson (left) and Arno Penzias, who used it to discover the 2.7 K cosmic background radiation. (Bell Photographers, Inc.)

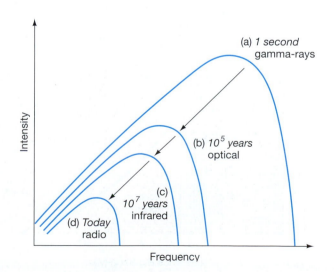

▶ **Figure 17.12 Cosmic Blackbody Curves** Theoretically derived blackbody curves for the entire universe (a) 1 second after the Big Bang, (b) 100,000 years after the Big Bang, (c) 10 million years after the Big Bang, and (d) today, approximately 14 billion years after the Big Bang.

▲ **Figure 17.13 Microwave Background Spectrum** The intensity of the cosmic background radiation, as measured by the *COBE* satellite, agrees very well with theory. The curve is the best fit to the data, corresponding to a temperature of 2.725 K. The experimental errors in this remarkably accurate observation are smaller than the dots representing the data points.

CONCEPT CHECK ▶▶▶

When was the cosmic microwave background formed? Why is it still a blackbody curve?

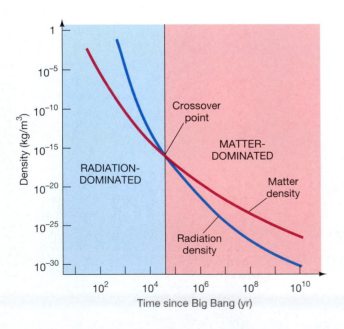

◀ **Figure 17.14 Radiation–Matter Dominance** As the universe expanded, the number of both matter particles and photons per unit volume decreased. However, the photons were also reduced in energy by the cosmological redshift, in turn reducing their equivalent mass, and hence their density, still further. As a result, the density of radiation fell faster than the density of matter as the universe grew. Tracing the curves back from the densities observed today, we see that radiation dominated matter at early times, before the crossover point.

In 1989 the *Cosmic Background Explorer* satellite—*COBE* for short—measured the intensity of the microwave background at wavelengths straddling the peak, from half a millimeter to about 1 cm. The results are shown in Figure 17.13. The solid line is the blackbody curve that best fits the *COBE* data. The near-perfect fit corresponds to a universal temperature of just over 2.7 K.

A striking aspect of the cosmic microwave background is its high degree of *isotropy*. When we correct for Earth's motion through space (which causes the microwave background to appear a little hotter than average in front of us and slightly cooler behind, by the Doppler effect), the intensity of the radiation is virtually constant (to about one part in 10^5) from one direction in the sky to another.

Remarkably, the cosmic microwave background contains more energy than has been emitted by all the stars and galaxies that have ever existed in the history of the universe. Stars and galaxies, though very intense sources of radiation, occupy only a small fraction of space. When their energy is averaged out over the volume of the entire cosmos, it falls short of the energy in the microwave background by at least a factor of 10. For our present purposes, then, the cosmic microwave background is the only significant form of radiation in the universe.

Matter and Radiation

Is matter the dominant constituent of the present universe, or does radiation also play an important role on the largest scales? We can answer this question by comparing the energy contained in the microwave background with the total energy in the form of matter.

To make the comparison we must first convert the energy in the microwave background into an equivalent mass density by calculating the number of microwave background photons in any cubic meter of space, then converting the total energy of these photons to a mass using $E = mc^2$. When we do this, we find that the density of the microwave background is about 5×10^{-31} kg/m^3. The overall density of matter in the universe is not known with certainty, but as we have seen, it is thought to be a few tenths of the critical density of 9×10^{-27} kg/m^3. Thus, *at the present moment* the density of matter in the universe greatly exceeds the density of radiation. In cosmological terminology (which predates the possibility of dark energy in the cosmos), we say that we live in a **matter-dominated** universe.

Was the universe always matter-dominated? To address this question, we must ask how the densities of both matter and radiation change as the universe expands. Both decrease, as the expansion dilutes the numbers of atoms and photons alike, but the radiation is also diminished in energy by the cosmological redshift, so its equivalent density falls faster than the density of matter as the universe grows (Figure 17.14). Thus, even though today the radiation density is much less than the density of matter, there must have been a time in the past when they were equal. Before that time, radiation was the main constituent of the cosmos. The universe is said to have been **radiation dominated** then.

According to current theory, the density associated with the dark energy remains *constant* as the universe expands. Thus, even though it dominates today, dark energy was *unimportant* in the early universe.

17.6 The Formation of Nuclei and Atoms

According to the Big Bang theory, at the very earliest times the cosmos consisted entirely of radiation. During the first minute or so of the universe's existence, temperatures were high enough that individual photons of radiation had sufficient energy to transform themselves into matter in the form of elementary particles. This period saw the creation of all the basic building blocks of matter we know today—both "normal" matter, such as protons, neutrons, and electrons, and the "exotic" particles that may compose the dark matter in galaxies and galaxy clusters. ∞ *(Sec. 14.5)* Since then, matter has evolved, clumping together into more and more complex structures forming the nuclei, atoms, planets, stars, galaxies, and large-scale structure we know today, but no *new* matter has been created. *Everything we see around us formed out of radiation as the early universe expanded and cooled.*

Helium Formation

Astronomers find that no matter where they look, and no matter how low a star's abundance of heavy elements may be, there is a minimum amount of helium—a little under 25 percent by mass—in all stars. This is the case even in stars containing very few heavy elements, indicating that the helium already existed when those ancient stars formed. The accepted explanation is that this base level of helium is *primordial*—that is, it was created not by fusion in stars but during the early, hot epochs of the universe, before any stars had formed. The production of elements heavier than hydrogen by nuclear fusion shortly after the Big Bang is called **primordial nucleosynthesis.**

As the temperature of the universe fell below about 900 million K, roughly two minutes after the Big Bang, conditions became favorable for protons and neutrons to fuse to form deuterium.[3] Prior to that time, deuterium nuclei were broken apart by high-energy gamma rays as quickly as they were created. Once the deuterium was at last able to survive the radiation background, other fusion reactions quickly converted it into heavier elements, especially helium-4. In just a few minutes most of the available neutrons were consumed, leaving a universe whose "normal" matter content was primarily hydrogen and helium, with a trace of deuterium. Figure 17.15 illustrates some of the reactions responsible for helium formation. Contrast it with Figure 9.26, which depicts how helium is formed today in the cores of main-sequence stars such as the Sun. ∞ *(Sec. 9.5)*

This early burst of fusion did not last long. Conditions in the cooling, thinning universe became less and less conducive to further fusion reactions as time went by. For all practical purposes, primordial nucleosynthesis stopped at helium-4. Detailed calculations show that, by about 15 minutes after the Big Bang, the cosmic elemental abundance was set, and helium accounted for approximately one-quarter of the total mass of normal matter in the universe. The remaining 75 percent was hydrogen. It would be almost a billion years before nuclear reactions in stars would change these figures. The agreement between this theoretical calculation and actual observations of helium and deuterium in stars is another key point in favor of the Big Bang theory.

Nucleosynthesis and the Composition of the Universe

Although most deuterium was quickly burned into helium as soon as it formed, a small amount was left over when the primordial nuclear reactions

---

[3]Recall from Chapter 9 that deuterium is simply a heavy form of hydrogen. Its nucleus contains one proton and one neutron. ∞ *(Sec. 9.5)*

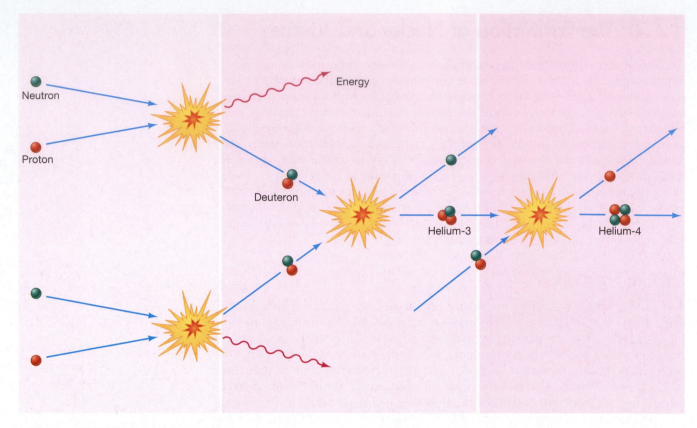

▲ **Figure 17.15 Helium Formation** Some of the reaction sequences that led to the formation of helium in the early universe. (A deuteron is a nucleus of deuterium, the heavy form of hydrogen.) Compare this figure with Figure 9.26, which depicts the proton–proton chain in the Sun.

ceased. Nucleosynthesis calculations indicate that the amount of deuterium remaining is a very sensitive indicator of the present-day density of matter in the universe: The *denser* the universe is today, the more matter (protons and neutrons) there was at those early times to react with the deuterium and the *less* deuterium was left when nucleosynthesis ended. Comparison between theoretical calculations of the production of deuterium in the early universe and the observed abundance of deuterium in stars and the interstellar medium implies a present-day density of *at most* 3 or 4 percent of the critical value.

But there is an important qualification. Primordial nucleosynthesis depends only on the presence of *protons* and *neutrons* in the early universe. Thus, measurements of the abundance of helium and deuterium tell us only about the density of "normal" matter—matter made up of protons and neutrons—in the cosmos. This finding has a momentous implication for the overall composition of the universe. Since the total matter density is about one-third the critical value, if the density of normal matter is only a few percent of this density, then we are forced to conclude that not only is most of the matter in the universe dark, but most of the dark matter is *not* composed of protons and neutrons. The bulk of the matter in the universe apparently exists in the form of elusive subatomic particles whose nature we do not fully understand and whose very existence has yet to be conclusively demonstrated in laboratory experiments. ∞ *(Sec. 14.5)*

For the sake of brevity, we will adopt from here on the convention that the term *dark matter* refers only to these unknown particles and not to "stellar" dark matter, such as brown or white dwarfs, which are made of normal matter.

CONCEPT CHECK ▶▶▶

How do we know that most of the dark matter in the universe is not of "normal" composition?

The Formation of Atoms

When the universe was tens of thousands of years old, matter (consisting of electrons, protons, helium nuclei, and dark matter) began to dominate over radiation. The temperature at that time was several tens of thousands of kelvins—too hot for atoms of hydrogen to exist, although some helium ions may already have formed. During the next few hundred thousand years, the universe expanded by another factor of 10, the temperature dropped to a few thousand kelvins, and electrons and nuclei combined to form neutral atoms. By the time the temperature had fallen to 3000 K, the universe consisted of atoms, photons, and dark matter.

The period during which nuclei and electrons combined to form atoms is often called the epoch of **decoupling**, for it was during this period that the radiation background parted company with normal matter. At early times, when matter was ionized, the universe was filled with large numbers of free electrons, which interacted frequently with electromagnetic radiation of all wavelengths. A photon could not travel far before encountering an electron and scattering off it. The universe was *opaque* to radiation. However, when the electrons combined with nuclei to form atoms of hydrogen and helium, only certain wavelengths of radiation—those corresponding to the spectral lines of those atoms—could interact with matter. ∞ *(Sec. 2.6)* Light of all other wavelengths could travel virtually forever without being absorbed. The universe became nearly *transparent*. After that time, as the universe expanded, the radiation simply cooled, eventually to become the microwave background we observe today.

The microwave photons now detected on Earth have been traveling through the universe ever since they decoupled. According to the models that best fit the observational data, their last interaction with matter (at the epoch of decoupling) occurred when the universe was about 400,000 years old and roughly 1100 times smaller (and hotter) than it is today—that is, at a redshift of 1100. As illustrated in Figure 17.16, the epoch of atom formation created a kind of "photosphere" in the universe, often referred to as the *surface of last scattering*, completely surrounding Earth at a distance of approximately 14,000 Mpc, corresponding to a redshift of 1100. ∞ *(More Precisely 15-1)* On our side of the photosphere—that is, since decoupling—the universe is transparent. On the far side—before decoupling—it was opaque. Thus, by observing the microwave background, we are probing conditions in the universe almost all the way back in time to the Big Bang itself, in much the same way as studying sunlight tells us about the surface layers of the Sun.

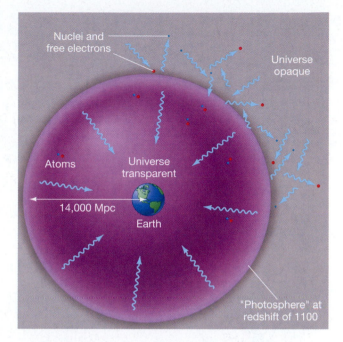

▲ **Figure 17.16 Radiation–Matter Decoupling** When atoms formed, the universe became virtually transparent to radiation. Thus, observations of the cosmic background radiation allow us to study conditions in the universe around a time at a redshift of 1100, when the temperature dropped below about 3000 K. For an explanation of how we can see a region of space 14,000 Mpc (46 billion light-years) away when the universe is just 14 billion years old, see *More Precisely 15-1*.

17.7 Cosmic Inflation

The Horizon and Flatness Problems

In the late 1970s, cosmologists were confronted with two nagging problems that had no easy explanation within the standard Big Bang model. The first is known as the **horizon problem**. Imagine observing the microwave background in two opposite directions on the sky, as illustrated in Figure 17.17. As we have just seen, in doing so we are actually observing two distant regions of the universe, marked A and B on the figure, where the radiation background

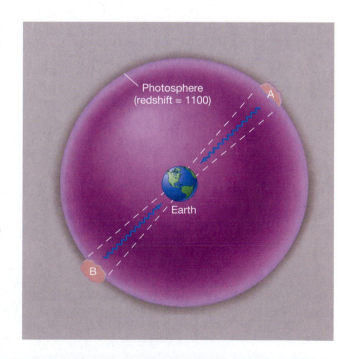

▶ **Figure 17.17 Horizon Problem** The isotropy of the microwave background indicates that regions A and B in the universe were very similar to each another when the radiation we observe left them, but there has not been enough time since the Big Bang for them ever to have interacted with one another. Why then should they look the same?

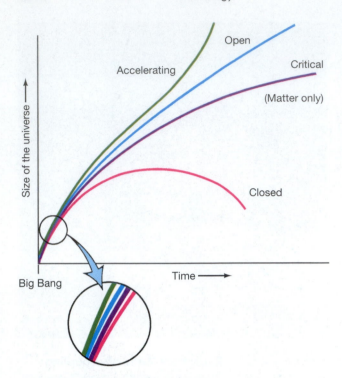

▲ Figure 17.18 Flatness Problem If the universe deviates even slightly from critical density, that deviation grows rapidly in time. For the universe to be as close to critical as it is today, it must have differed from the critical density in the past by only a tiny amount.

last interacted with matter. The fact that the background is known to be *isotropic* to high accuracy means that regions A and B must have had very similar densities and temperatures at the time the radiation we see left them.

The problem is, within the Big Bang theory as just described, there is no particular reason *why* these regions should have been so similar. At the time in question, a few hundred thousand years after the Big Bang, regions A and B were separated by many megaparsecs, and no signal—sound waves, light rays, or anything else—had had time to travel from one to the other. In cosmological terminology, each region lay outside the other's *horizon*. But in that case, with no possibility of information or energy exchange between these regions, there was no way for them to "know" that they were supposed to look the same. The only alternative is that they simply started off looking alike—an assumption that cosmologists are very unwilling to make.

The second problem with the standard Big Bang model is called the **flatness problem**. Whatever the exact value of the cosmic density, it appears to be fairly close to the critical value. In terms of spacetime curvature, the universe is remarkably close to being *flat*. But again, there is no good reason *why* the universe should have formed with a density very close to critical. Why not a millionth or a million times that value? Furthermore, as can be seen in Figure 17.18, a universe that starts off close to, but not exactly on, the critical curve soon deviates greatly from it, so if the universe is close to critical now, it must have *extremely* close to critical in the past. For example, even if Ω_0 were as low as 0.1 today (considerably less than the density of "known" dark matter), the departure from the critical density at the time of nucleosynthesis would have been only one part in 10^{15} (a thousand trillion).

These observations constitute "problems" because cosmologists want to be able to *explain* the present condition of the universe, not just accept it "as is." They would prefer to resolve the horizon and flatness problems in terms of physical processes that could have taken a universe with no special properties and caused it to evolve into the cosmos we now see. The resolution of both problems takes us back in time even earlier than nucleosynthesis or the formation of any of the elementary particles we know today—back, in fact, almost to the instant of the Big Bang itself.

The Epoch of Inflation

In the 1970s and 1980s, theoretical physicists succeeded in combining, or *unifying*, the three nongravitational forces in the universe—electromagnetism and the strong and weak forces—into a single all-encompassing "superforce." ∞ *(More Precisely 9-1)* A general prediction of the **Grand Unified Theories** (GUTs) that describe this superforce is that the three forces are unified and indistinguishable from one another *only* at enormously high energies—corresponding to temperatures in excess of 10^{28} K. At lower temperatures, the superforce splits into three, revealing its separate electromagnetic, strong, and weak characters.

In the early 1980s, cosmologists discovered that Grand Unified Theories had some remarkable implications for the very early universe. About 10^{-34} s after the Big Bang, as temperatures fell below 10^{28} K and the basic forces of nature became reorganized—a little like a gas liquefying or water freezing as the temperature drops—the universe briefly entered a very odd and unstable high-energy state that physicists call the "false vacuum." In essence, the universe remained in the "unified" condition a little too long, like water that has been cooled below freezing but has not yet turned to ice.

This had dramatic consequences. For a short while, empty space acquired an enormous *pressure* (similar in effect but much greater in magnitude than the pressure associated with the cosmological constant), which temporarily overcame the pull of gravity and accelerated the expansion of the universe at an enormous rate. The pressure remained constant as the cosmos expanded, and the acceleration grew more and more rapid with time—in fact, the size of the universe *doubled* every 10^{-34} s or so. This period of unchecked cosmic expansion, illustrated in Figure 17.19, is known as the epoch of **inflation**.

Eventually, the universe returned to the lower-energy "true vacuum" state. Regions of normal space began to appear within the false vacuum and rapidly spread to include the entire cosmos. With the return of the true vacuum, inflation stopped, and the universe resumed its normal, relatively leisurely expansion. The whole episode lasted a mere 10^{-32} s, but during that time the universe swelled in size by the incredible factor of about 10^{50}.

Implications for the Universe

Inflation provides a natural solution for the horizon and flatness problems. The horizon problem is solved because inflation took regions of the universe that had already had time to communicate with one another—and so had established similar physical properties—and then dragged them far apart, well out of communication range of one another. In effect, the universe expanded much faster than the speed of light during the inflationary epoch, so what was once well within the horizon now lies far beyond it. (Relativity restricts matter and energy to speeds less than the speed of light, but imposes no such limit on the universe as a whole.) Regions A and B in Figure 17.17 have been out of contact with one another since 10^{-32} s after the Big Bang, but they *were* in contact before then. Their properties are the same today because they were the same long ago, before inflation separated them.

Figure 17.20 illustrates how inflation solves the flatness problem. Imagine that you are a 1-mm-long ant sitting on the surface of an expanding balloon. When the balloon is just a few centimeters across, you can easily see that its surface is curved because its circumference is comparable to your own size. However, as the balloon expands, the curvature becomes less pronounced. When the balloon is a few kilometers across, your "ant-sized" patch of the surface looks quite flat, just as Earth's surface looks flat to us. Now imagine that the balloon expands 100 trillion trillion trillion trillion times. Your local patch of the surface would be completely indistinguishable from a perfectly flat plane. Thus, any curvature the universe may have had before inflation has been expanded so much that space is now perfectly flat on all scales we can ever hope to observe.

Notice that this resolution to the flatness problem—the universe appears close to being flat because the universe *is* in fact precisely flat, to very high accuracy—has a very important consequence. Because the universe is geometrically flat, relativity tells us the total density must be exactly equal to the critical value: $\Omega_0 = 1$ (Section 17.3). This is the key result that led us to conclude in Section 17.4 that dark energy—whatever it is—must dominate the density of the universe. Thus, not only is most matter dark (Section 17.6), but *most of the cosmic density isn't made up of matter at all*. In a sense, this is the ultimate statement of the Copernican principle: Not only is Earth in no way central to the cosmos, but the stuff of which we are made is unrepresentative of matter in general, and matter itself is the minority constituent of the universe! ∞ *(Sec. 1.1)*

▶ **Figure 17.20 Inflation and the Flatness Problem**
Inflation solves the flatness problem by taking a curved surface, here represented by the surface of the expanding balloon, and expanding it enormously in size. To an ant on the surface, the balloon looks perfectly flat when the expansion is complete.

▲ **Figure 17.19 Cosmic Inflation** During the period of inflation the universe expanded enormously in a very short time. Afterward, it resumed its earlier "normal" expansion rate, except that now the size of the cosmos was about 10^{50} times bigger than it was before inflation.

Even though inflation solves the horizon and flatness problems in a quite convincing way, for nearly two decades after it was first proposed the theory was resisted by many astronomers. The main reason was that its prediction of $\Omega_0 = 1$ was clearly at odds with the growing evidence that the density of matter in the universe was no more than 30 percent or so of the critical value. Many cosmologists had in fact considered the possibility that a cosmological constant might provide a way to account for the remaining 70 percent of the cosmic density, but without independent corroboration, a conclusive case could not be made. That is why the supernova observations were so important—by demonstrating acceleration in the cosmic expansion rate, they established independent evidence for the effects of dark energy and in doing so reconciled inflation with the discrepant observations.

Physicists will probably never create in terrestrial laboratories conditions even remotely similar to those that existed in the universe during the inflationary epoch. Creation of a false vacuum is (safely) beyond our reach. Nevertheless, cosmic inflation seems to be a natural consequence of many Grand Unified Theories, and it explains two otherwise intractable problems within the Big Bang theory. For these reasons, despite the lack of direct evidence for the process, inflation theory has become an integral part of modern cosmology. It makes definite, testable predictions about the large-scale geometry and structure of the universe that are critically important to current theories of galaxy formation. As we will see in the next section, astronomers are now subjecting these predictions to rigorous scrutiny.

CONCEPT CHECK ▶▶▶

Does the theory of inflation itself say anything about the composition of the universe?

17.8 The Formation of Large-Scale Structure in the Universe

Animation/Video
Cosmic Structure

As far as we can tell, all of the present large-scale structure in the universe grew from small *inhomogeneities*—slight deviations from perfectly uniform density—that existed in the early universe. Denser than average clumps of matter contracted under the influence of gravity and merged with other clumps, eventually reaching the point where stars began to form and luminous galaxies appeared. ∞ *(Sec. 16.3)*

Most cosmologists think that the intense background radiation in the early universe effectively prevented clumps of normal matter from forming or growing. Just as in a star, the pressure of the radiation stabilized the clumps against the inward pull of their own gravity. ∞ *(Sec. 12.1)* Instead, the first structures to appear did so in the *dark* matter that makes up the bulk of matter in the universe. While the true nature of dark matter remains uncertain, its defining property is the fact that it interacts only very weakly with both normal matter and radiation. Consequently, clumps of dark matter were unaffected by the radiation background and began to grow as soon as matter first began to dominate the cosmic density.

Thus, as illustrated in Figure 17.21, dark matter determined the overall distribution of mass in the universe and clumped to form the observed large-scale structure. Then, at later times, normal matter was drawn by gravity into the regions of highest density, eventually forming galaxies and galaxy clusters. This picture explains why so much dark matter is found outside the visible galaxies. The luminous material is strongly concentrated near the density peaks and dominates the dark matter there, but the rest of the universe is essentially devoid of normal matter. Like foam on the crest of an ocean wave, the universe we see is only a tiny fraction of the total.

Figure 17.22 shows the results of a recent supercomputer simulation of a "best-bet" universe consisting of 4 percent normal matter, 23 percent dark matter, and 73 percent dark energy (in the form of a cosmological constant).

▶ Figure 17.21 Structure Formation   (a) The universe started out as a mixture of (mostly) dark and normal matter. (b) A few thousand years after the Big Bang, the dark matter began to clump. (c) Eventually, the dark matter formed large structures (represented here by the two high-density peaks) into which normal matter flowed, ultimately to form the galaxies we see today. The three frames at right represent the densities of dark matter (red) and normal matter (yellow) in the graph at left.

∞ *(Sec. 17.3)* The similarities with real observations of cosmic structure, shown in Figures 16.21, 16.22, and 17.1, are striking, and more detailed statistical analysis shows that these models in fact agree extremely well with reality. Although calculations like this cannot prove that cold dark-matter models are the correct description of the universe, the agreement in detail between models and reality strongly favor the cosmological constant/cold dark-matter model of the cosmos.

Although dark matter does not interact directly with photons, the background radiation is influenced slightly by the *gravity* of the growing dark clumps. This causes a slight gravitational redshift in the background radiation that varies from place to place depending on the dark-matter density. As a result, a key prediction of dark-matter models is that there should be tiny "ripples" in the microwave background—temperature variations of only a few parts per million from place to place on the sky. In 1992, after almost two years of careful observation, the *COBE* team announced that the expected ripples had been detected. The temperature variations are tiny—only 30–40 millionths of a kelvin—but they are there. The *COBE* results are displayed as a temperature map of the microwave sky in Figure 17.23.

The ripples seen by *COBE*, combined with computer simulations such as that shown in Figure 17.22, predict present-day structure that is consistent with the superclusters, voids, filaments and Great Walls we see around us. While the *COBE* data were limited by relatively low (roughly 7°) resolution, detailed analysis of the ripples also supports the key prediction of inflation theory—that the universe is of exactly critical

Dark matter

Normal matter

Space ⟶
(a) Time = 1 second

Dark matter

Normal matter

Space ⟶
(b) Time = 1000 years

Dark matter

Galaxies/clusters (normal matter)

Space ⟶
(c) Time = 10^8 years

◀ Figure 17.22 Structure Simulated   Three views of a (present-day) 100 × 100 × 100-Mpc cube in a simulated cold dark-matter universe with $\Omega_0 = 1$. The cube expands with the universe, so that it encloses the same material at all times. The three frames show the growth of structure resulting from the growth of small-density fluctuations in the very early universe, displayed at three different times after the Big Bang. (V. Springel/Max-Planck Institut fur Astrophysik)

1 billion years

4 billion years

14 billion years

▲ **Figure 17.23 Cosmic Microwave Background Map** *COBE* map of temperature fluctuations in the cosmic microwave background over the entire sky. Hotter-than-average regions are shown in yellow, cooler-than-average regions in blue. The total temperature range shown is ±200 millionths of a kelvin. The temperature variation due to Earth's motion has been subtracted, as has the radio emission from the Milky Way Galaxy, and temperature deviations from the average are displayed. (NASA)

density and hence spatially flat. For these reasons, the *COBE* observations rank alongside the discovery of the microwave background itself in terms of their importance to the field of cosmology.

The decade following the end of the *COBE* mission in late 1993 has seen dramatic improvements in both the resolution and sensitivity of microwave background measurements. Figure 17.24 shows results returned by two recent experiments. Both have radically improved our view of the microwave background, confirming and refining the *COBE* results.

Figure 17.24 shows an all-sky map made by the *Wilkinson Microwave Anisotropy Probe* (*WMAP*), launched by NASA in 2001. Permanently stationed some 1.5 million km outside Earth's orbit along the Sun–Earth line and always pointing away from the Sun to keep its delicate heat-sensitive detectors in shadow, *WMAP* completes a scan of the entire sky every six months. The instrument's angular resolution is roughly 20′–30′, some 20 times finer than that of *COBE* (compare Figures 17.23 and 17.24), allowing extraordinarily detailed measurements of many cosmological parameters to be made. The inset in Figure 17.24 shows a smaller-scale (just 2° wide), but even higher resolution (7′) image returned by *Cosmic Background Imager*, a ground-based microwave telescope located high in the Chilean Andes. (Recall from Chapter 2 that the microwave part of the spectrum is only partly transparent to electromagnetic radiation, so microwave detectors must be placed above as much of Earth's absorbing atmosphere as possible.) ∞ *(Sec. 2.3)*

Both maps show temperature fluctuations of a few hundred microkelvins, with a characteristic angular scale of about 1°. This temperature range is larger than the fluctuations seen by *COBE* because *COBE*'s low resolution effectively averaged the data over a large area of the sky.

▼ **Figure 17.24 Early Structure** The entire microwave sky, as seen by the *WMAP* spacecraft at frequencies between 22 and 90 GHz (wavelengths ranging from 14 to 3 mm). This new map may be compared directly with the *COBE* map shown in Figure 17.23. The effective resolution of this WMAP multiwavelength survey is about 20–30′ (compared with 7° for *COBE*). The inset at right shows a 7′ resolution map of a roughly 2° patch of sky, obtained by the ground-based *Cosmic Background Imager* instrument at 30 GHz (1-cm wavelength). The bright blobs are slightly denser-than-average regions of the universe at an age of roughly 400,000 years; they will eventually collapse to form clusters of galaxies.

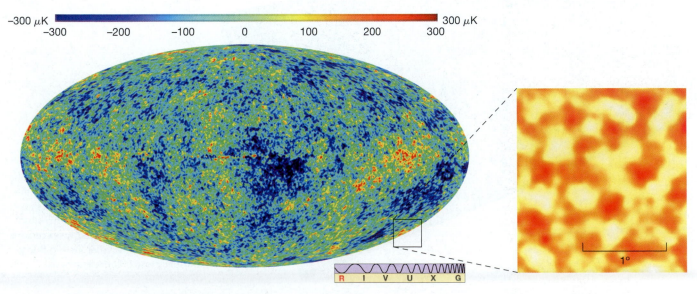

The angular scale of the fluctuations is important because detailed calculations reveal that this is a *direct* measure of Ω_0, largely independent of other cosmological "details" (such as the fraction of dark matter or the nature of the dark energy). The slightly less than 1° "blob-to-blob" angular scale, evident to the eye in the inset of Figure 17.24 and confirmed by more quantitative means, provides the strongest evidence yet that Ω_0 is very close to 1, with an extremely small margin of error, in full agreement with the inflation prediction. Each of the irregular high-temperature regions in these images represents a slightly denser than average clump of dark matter that will one day collapse to form a supercluster-sized clump of galaxies. According to the *WMAP* data, we are seeing the universe here when it was just 380,000 years old, yet the future development of the cosmos is already imprinted in these tiny ripples.

Ongoing and upcoming space-based observations will further refine these measurements, improving their errors and expanding the coverage to the entire sky. The first decade of the twenty-first century may well see the basic parameters of the universe measured (even if not fully understood) to an accuracy only dreamed of just a few years ago.

◄◄◄ **CONCEPT CHECK**

Why is dark matter important to galaxy formation? How do observations of the microwave background support this view?

CHAPTER REVIEW

Summary

1 On scales larger than a few hundred megaparsecs, the universe appears roughly **homogeneous** (p. 457) (the same everywhere) and **isotropic** (p. 457) (the same in all directions). In **cosmology** (p. 457)—the study of the universe as a whole—researchers usually assume that the universe is homogeneous and isotropic on large scales. This assumption is known as the **cosmological principle** (p. 457).

2 If the universe were homogeneous, isotropic, infinite, and unchanging, then the night sky would be bright because any line of sight would eventually intercept a star. The fact that the night sky is instead dark is called **Olbers's paradox.** (p. 457) Its resolution is that we see only a finite part of the universe—the region within which light has had time to reach us since the universe formed. Tracing the observed motions of galaxies back in time implies that, about 14 billion years ago, the universe consisted of a single point that expanded rapidly in the **Big Bang** (p. 458). Space itself was compressed to a point at that instant—the Big Bang happened everywhere at once. The cosmological redshift occurs as a photon's wavelength is "stretched" by cosmic expansion. The extent of the observed redshift is a direct measure of the expansion of the universe since the photon was emitted.

3 There are only two possible outcomes to the current expansion: Either the universe will expand forever, or it will recollapse to a point. The **critical density** (p. 461) is the density of matter needed for gravity to overcome the present expansion and cause the universe to collapse. The total density of the universe (including matter, radiation, and dark energy) determines the geometry of the universe on the largest scales, as described by general relativity. In a high-

(greater-than-critical) density universe, the curvature of space is sufficiently large that the universe "bends back" on itself and is finite in extent, somewhat like the surface of a sphere. Such a universe is said to be a **closed universe** (p. 462). A low-density **open universe** (p. 462) is infinite in extent and has a "saddle-shaped" geometry. The **critical universe** (p. 462) has a density precisely equal to the critical value and is spatially flat.

4 Luminous matter by itself contributes only about 1 percent of the critical density. When dark matter in galaxies and clusters is taken into account, the figure rises to 20 or 30 percent. Observations of distant supernovae suggest that the expansion of the universe may be accelerating, possibly driven by the effects of a force commonly called **dark energy** (p. 466). One candidate for this dark energy is the **cosmological constant** (p. 466), a repulsive force that may exist throughout all space. Its physical nature is unknown. Other, independent observations are consistent with the idea that the universe is flat—that is, of exactly critical density—with (mostly dark) matter making up 27 percent of the total and dark energy making up the rest. Such a universe will expand forever.

5 The **cosmic microwave background** (p. 469) is isotropic blackbody radiation that fills the entire universe. Its present temperature is about 3 K. Its existence is evidence that the universe expanded from a hot, dense state. As the universe has expanded, the initially high-energy radiation has been redshifted to lower and lower temperatures. At the present time, the density of matter in the universe greatly exceeds the equivalent mass density of radiation. We live in a **matter-dominated** (p. 470) universe. The matter density was much greater in the past, when the universe was smaller. However, because radiation is redshifted as the universe expands, the density of radiation was greater still. The early universe was **radiation dominated** (p. 470).

6 All of the hydrogen in the universe is primordial, formed from radiation as the hot early universe expanded and cooled. Most of the helium observed in the universe today is also primordial, created by fusion between protons and neutrons a few minutes after the Big Bang. This is known as **primordial nucleosynthesis** (p. 471). Other, heavier elements were formed much later, in the cores of stars. Detailed studies of this process indicate that "normal" matter can account for at most 3 percent of the critical density. By the time the universe was about 1500 times smaller than it is today, the temperature had become low enough for the first atoms to form. At that time, the (then visible) radiation background **decoupled** (p. 473) from the matter. The photons that now make up the microwave background have been traveling freely through space ever since.

7 According to modern **Grand Unified Theories** (p. 474), the three nongravitational forces of nature first began to display their separate characters about 10^{-34} s after the Big Bang. A brief period of rapid cosmic expansion called the epoch of **inflation**

(p. 474) ensued, during which the size of the universe increased by a factor of about 10^{50}. The **horizon problem** (p. 473) is the fact that, according to the standard (that is, non-inflationary) Big Bang model, there is no good reason for widely separated parts of the universe to be as similar as they are. Inflation solves the horizon problem by taking a small homogeneous patch of the early universe and expanding it enormously. Inflation also solves the **flatness problem** (p. 474), which is the fact that there is no obvious reason why the density of the universe is so close to critical. Inflation implies that the cosmic density is almost exactly critical.

8 The large-scale structure observed in the universe today formed when density inhomogeneities in the dark matter clumped and grew to create the "skeleton" of the structure now observed. Normal matter then flowed into the densest regions of space, eventually forming the galaxies we now see. "Ripples" in the microwave background are the imprint of these early inhomogeneities on the radiation field and lend strong support to the inflationary prediction that we live in a flat, critical-density universe.

Review and Discussion

1. What evidence do we have that there is no structure in the universe on very large scales? How large is "very large"?

2. What is the cosmological principle?

3. What is Olbers's paradox? How is it resolved?

4. Explain how an accurate measure of Hubble's constant can lead to an estimate of the age of the universe.

5. Why isn't it correct to say that the expansion of the universe involves galaxies flying outward into empty space?

6. Why are recent observations of distant supernovae so important to cosmology?

7. Where and when did the Big Bang occur?

8. How does the cosmological redshift relate to the expansion of the universe?

9. What is the cosmic microwave background, and why is it so significant?

10. Why do all stars, regardless of their abundance of heavy elements, contain at least 25 percent helium by mass?

11. What is Ω_0? How do measurements of the cosmic deuterium abundance provide a lower limit on Ω_0?

12. When did the universe become transparent to radiation?

13. What is cosmic inflation? How does inflation solve the horizon problem? The flatness problem?

14. What is the connection between dark matter and the formation of large-scale structure in the universe?

15. What prediction of inflation theory has since been verified by observations of the microwave background?

Conceptual Self-Test: True or False/Multiple Choice

1. Deep surveys of the universe indicate that the largest structures in space are about 50 Mpc in size. **(T/F)**

2. If the universe had an edge, that fact would violate the assumption of isotropy in the cosmological principle. **(T/F)**

3. Hubble's law implies that the universe will expand forever. **(T/F)**

4. The cosmological redshift is a direct measure of cosmic expansion. **(T/F)**

5. The cosmic microwave background is the highly redshifted radiation of the early Big Bang. **(T/F)**

6. Observations suggest that the density of the universe is made up mostly of dark matter. **(T/F)**

7. The microwave background radiation last interacted with matter around the time of decoupling. **(T/F)**

8. The theory of inflation predicts that the density of the universe is exactly equal to the critical density. **(T/F)**

9. Olbers's paradox is resolved by (**a**) the finite size of the universe; (**b**) the finite age of the universe; (**c**) light from distant galaxies being redshifted so we can't see them; (**d**) the fact that there is an edge to the universe.

10. The galactic distances used to measure the acceleration of the universe are determined by observations of (**a**) trigonometric parallax; (**b**) Hubble's law; (**c**) Cepheid variable stars; (**d**) exploding white dwarfs.

11. On the basis of our current best estimate of the present mass density of the universe, astronomers think that **(a)** the universe is finite in extent and will expand forever; **(b)** the universe is finite in extent and will eventually collapse; **(c)** the universe is infinite in extent and will expand forever; **(d)** the universe is infinite in extent and will eventually collapse.

12. The age of the universe is estimated to be **(a)** less than Earth's age; **(b)** the same as the age of the Sun; **(c)** the same as the age of the Milky Way Galaxy; **(d)** greater than the age of the Milky Way Galaxy.

13. The horizon problem in the standard Big Bang model is solved by having the universe **(a)** accelerate; **(b)** inflate rapidly early in its existence, **(c)** have tiny, but significant fluctuations in temperature; **(d)** be geometrically flat.

14. The data points in Figure 17.8 **(a)** prove that the universe is not expanding; **(b)** imply that the expansion is decelerating faster than expected; **(c)** allow a measurement of Hubble's constant; **(d)** indicate that the redshifts of distant galaxies are greater than would be expected if gravity alone were acting.

15. According to Figure 17.14, when the universe was 10 years old, the density of radiation was **(a)** much greater than that of matter; **(b)** much less than that of matter; **(c)** comparable to that of matter; **(d)** unknown.

Problems

Algorithmic versions of these questions are available in the Practice Problems module of the Companion Website.

The number of squares preceding each problem indicates its approximate level of difficulty.

1. ■ What is the greatest distance at which a galaxy survey sensitive to objects as faint as 20th magnitude could detect a galaxy as bright as the Milky Way (absolute magnitude −20)? ∞ *(More Precisely 10-1)*

2. ■■ If the entire universe were filled with Milky Way-like galaxies, with an average density of 0.1 per cubic megaparsec, calculate the total number of galaxies observable by the survey in the previous question if it covered the entire sky.

3. ■■ Eight galaxies are located at the corners of a cube. The present distance from each galaxy to its nearest neighbor is 10 Mpc, and the entire cube is expanding according to Hubble's law, with $H_0 = 70$ km/s/Mpc. Calculate the recession velocity (magnitude *and* direction) of one corner of the cube relative to the opposite corner.

4. ■ According to the standard Big Bang theory (neglecting any effects of cosmic acceleration), what is the *maximum* possible age of the universe if $H_0 = 50$ km/s/Mpc? 70 km/s/Mpc? 80 km/s/Mpc?

5. ■ For $H_0 = 70$ km/s/Mpc, the critical density is 9×10^{-27} kg/m³. **(a)** How much mass does that correspond to in a volume of 1 AU³? **(b)** How large a cube would be required to enclose one Earth mass of material?

6. ■■ The Virgo Cluster is observed to have a recessional velocity of 1200 km/s. Assuming $H_0 = 70$ km/s/Mpc and a critical-density universe, calculate the total mass contained within a sphere centered on Virgo and just enclosing the Milky Way Galaxy. What is the escape speed from the surface of this sphere? ∞ *(More Precisely 5-1)*

7. ■■ **(a)** What is the present peak wavelength of the cosmic microwave background? Calculate the size of the universe relative to the present size when the radiation background peaked in **(b)** the infrared, at $10 \mu m$, **(c)** in the ultraviolet, at 100 nm, and **(d)** in the gamma-ray, at 1 nm.

8. ■■ What were the matter density and the equivalent mass density of the cosmic radiation field when the universe was one-thousandth its present size? Assume critical density, and don't forget the cosmological redshift.

9. ■ What was the distance between the points that would some-day become the center of the Milky Way and the center of the Coma Cluster at the time of decoupling? (The present distance to Coma is 100 Mpc.)

10. ■■■ The "blobs" evident in the inset to Figure 17.24 are about 20′ (arcminutes). If those blobs represent clumps of matter around the time of decoupling (redshift = 1100), and assuming Euclidean geometry, estimate the size of these clumps at the time of decoupling. ∞ *(More Precisely 15-1)*

The Companion Website at www.prenhall.com/chaisson provides algorithmically generated versions of each chapter's Problems, in addition to an Animations and Videos gallery, a Tutorials gallery and Tutorials Quiz questions, a full eBook, an Images gallery, Practice Questions, Multiple Choice, and True or False quizzes, and Labeling exercises.

18

Life in the Universe

Are We Alone?

Are we unique? Is life on our planet the only example of life in the universe? If so, then what would be the implications of that momentous discovery? If not, then how and where should we look for other intelligent beings? These are difficult questions, for the subject of extraterrestrial life is one on which we have no data, but they are important questions, with profound implications for the human species. Earth is the only place in the universe where we know for certain that life exists. In this final chapter we take a look at how humans evolved on Earth and then consider whether those evolutionary steps might have happened elsewhere. Having done that, we assess the likelihood of our having Galactic neighbors and consider how we might learn about them if they exist.

Is there intelligent life elsewhere in the universe? This fanciful painting by noted space artist Dana Berry suggests a plurality of civilizations—some perhaps extinct—on alien worlds well beyond our own solar system. Despite blockbuster movies, science-fiction novels, and a host of claims for extraterrestrial contact, astronomers have found no unambiguous evidence for life, intelligent or otherwise, anywhere else in the universe. (*Galaxyrise Over Alien Planet*, courtesy D. Berry)

LEARNING GOALS

Studying this chapter will enable you to:

1 Summarize the process of cosmic evolution as it is currently understood.

2 Evaluate the chances of finding life in the solar system.

3 Summarize the various probabilities used to estimate the number of advanced civilizations that might exist in our Galaxy.

4 Discuss some of the techniques we might use to search for extraterrestrials and to communicate with them.

Visit www.prenhall.com/chaisson for additional annotated images, animations, and links to related sites for this chapter.

482

18.1 Cosmic Evolution

Figure 18.1 identifies seven major phases in the history of the universe: *particulate, galactic, stellar, planetary, chemical, biological,* and *cultural* evolution. Taken together, these evolutionary stages make up the grand sweep of **cosmic evolution**—the continuous transformation of matter and energy that has led to the appearance of life and civilization on Earth. The first four phases represent, in reverse order, the contents of this book up to this point. We now expand our field of view beyond astronomy to include the other three.

From the Big Bang to the evolution of intelligence and culture, the universe has evolved from simplicity to complexity. We are the result of an incredibly complex chain of events that spanned billions of years. Were those events random, making us unique, or are they in some sense *natural* and technological civilization inevitable? Put another way, are we alone in the universe, or are we just one among countless other intelligent life forms?

Life in the Universe

The distinction between the living and the nonliving is perhaps not as obvious as we might at first think. Generally speaking, scientists regard the following as characteristics of living organisms: (1) they can *react* to their environment and can often heal themselves when damaged; (2) they can *grow* by taking in nourishment from their surroundings; (3) they can *reproduce*, passing along some of their own characteristics to their offspring; and (4) they have the capacity for genetic change and can therefore *evolve* from generation to generation to adapt to a changing environment.

These rules are not hard and fast, and a good working definition of life is elusive. Stars, for example, react to the gravity of their neighbors, grow by accretion, generate energy, and "reproduce" by triggering the formation of new stars, but no one would suggest that they are alive. A virus is inert, often crystalline in form, when isolated from living organisms, but, once inside a living system, it exhibits all the properties of life, seizing control of the cell and using the cell's own genetic machinery to *grow* and reproduce. Most researchers think that the distinction between living and nonliving is more one of structure and complexity than a simple checklist of rules.

▶ **Figure 18.1 Arrow of Time** Some highlights of cosmic history are noted along this arrow of time, from the beginning of the universe to the present. Along the bottom of the arrow are seven "windows" outlining the major phases of cosmic evolution: evolution of primal energy into elementary particles; of atoms into galaxies and stars; of stars into heavy elements; of elements into solid, rocky planets; of those same elements into the molecular building blocks of life; of those molecules into life itself; and of advanced life forms into intelligence, culture, and technological civilization. (D. Berry)

▶ **Figure 18.2** **DNA Molecule**   DNA (deoxyribonucleic acid) is the molecule containing all the genetic information needed for a living organism to reproduce and survive. Often consisting of literally tens of billions of individual atoms, its double-helix structure allows it to "unzip," exposing its internal structure to control the creation of proteins needed for a cell to function. The ordering of its constituent parts is unique to each individual organism. The letters represent the 4 types of nucleotide bases that participate in DNA: Adenine, Cytosine, Guanine and Thymine.

The general case in favor of extraterrestrial life is summed up in what are sometimes called the *assumptions of mediocrity*: (1) because life on Earth depends on just a few basic molecules, (2) because the elements that make up these molecules are (to a greater or lesser extent) common to all stars, and (3) if the laws of science we know apply to the entire universe (as we have supposed throughout this book), then—given sufficient time—life must have originated elsewhere in the cosmos. The opposing view maintains that intelligent life on Earth is the product of a series of extremely fortunate accidents—astronomical, geological, chemical, and biological events unlikely to have occurred anywhere else in the universe. The purpose of this chapter is to examine some of the scientific arguments for and against these viewpoints.

Life on Earth

What information do we have about the earliest stages of planet Earth? Unfortunately, not very much. Geological hints about the first billion years or so were largely erased by violent surface activity as volcanoes erupted and meteorites bombarded our planet; subsequent erosion by wind and water has seen to it that little evidence has survived to the present day. However, the early Earth was probably barren, with shallow lifeless seas washing upon grassless, treeless continents. Outgassing from our planet's interior through volcanoes, fissures, and geysers produced an atmosphere rich in hydrogen, nitrogen, and carbon compounds and poor in free oxygen. As Earth cooled, ammonia, methane, carbon dioxide, and water formed. The stage was set for the appearance of life.

The surface of the young Earth was a very violent place. Natural radioactivity, lightning, volcanism, solar ultraviolet radiation, and meteoritic impacts all provided large amounts of energy that eventually shaped the ammonia, methane, carbon dioxide, and water into more complex molecules known as **amino acids** and **nucleotide bases**—organic (carbon-based) molecules that are the building blocks of life as we know it. Amino acids build *proteins*, and proteins control *metabolism*, the daily utilization of food and energy by means of which organisms stay alive and carry out their vital activities. Sequences of nucleotide bases form *genes*—parts of the DNA molecule—that direct the synthesis of proteins and hence determine the characteristics of organisms (Figure 18.2). These same genes also carry an organism's hereditary characteristics from one generation to the next. In all living creatures on Earth—from bacteria to amoebas to humans—genes mastermind life, while proteins maintain it.

The first experimental demonstration that complex molecules could have evolved naturally from simpler ingredients found on the primitive Earth came in 1953. Scientists Harold Urey and Stanley Miller, using laboratory equipment somewhat similar to that shown in Figure 18.3, took a mixture of the materials thought to be present on Earth long ago—a "primordial soup" of water, methane, carbon dioxide, and ammonia—and energized it by passing an electrical discharge ("lightning") through it. After a few days, they analyzed their mixture and found that it contained many of the same amino acids found today in all living things on Earth. About a decade later scientists succeeded in constructing nucleotide bases in a similar manner. These experiments have been repeated in many different forms, with more realistic mixtures of gases and a variety of energy sources, but always with the same basic outcomes.

▲ **Figure 18.3** **Miller-Urey Experiment**   This chemical apparatus is designed to synthesize complex biochemical molecules by energizing a mixture of simple chemicals. A mixture of gases (ammonia, methane, carbon dioxide, water vapor) is placed in the upper bulb to simulate Earth's primordial atmosphere and then energized by spark-discharge electrodes. After about a week, amino acids and other complex molecules are found in the trap at the bottom, which simulates the primordial oceans into which heavy molecules produced in the overlying atmosphere would have fallen.

(a)

2.5 μ

(b)   (c)

10μ

◄ **Figure 18.4 Chemical Evolution** (a) These carbon-rich, proteinlike droplets display the clustering of as many as a billion amino acid molecules in a liquid. Droplets can "grow," and parts of droplets can separate from the "parent" to become new individual droplets (such as at A, B, C). The scale of 2.5 microns noted here is 1/4000 of a centimeter. (b) This photograph, taken through a microscope, shows a fossilized organism found in sediments radioactively dated as 2 billion years old. This primitive system possesses concentric spheres or walls connected by smaller spheroids. The roundish fossils here measure about a thousandth of a centimeter in diameter. (c) For comparison, modern blue-green algae, on approximately the same scale. (S. Fox; E. Barghoorn)

None of these experiments has ever produced a living organism, and the molecules created are far less complex than the strand of DNA illustrated in Figure 18.2. Nevertheless, they demonstrate conclusively that "biological" molecules can be synthesized by strictly nonbiological means, using raw materials available on the early Earth. More advanced experiments have fashioned proteinlike blobs (Figure 18.4a) that behave to some extent like true biological cells. They resist dissolution in water and tend to cluster into small droplets called *microspheres*—a little like oil globules floating on the surface of water. The walls of these laboratory-made droplets permit the inward passage of small molecules, which then combine within the droplet to produce more complex molecules too large to pass back out through the walls. As the droplets "grow," they tend to "reproduce," forming smaller droplets.

Can we consider these proteinlike microspheres to be alive? Almost certainly not. The microspheres contain many of the basic ingredients needed to form life but are not life itself. They lack the hereditary molecule DNA. However, they do have similarities to ancient cells found in the fossil record (Figure 18.4b). Thus, while no actual living cells have yet been created "from scratch," many biochemists feel that many key steps in *chemical evolution*—the chain of events leading from simple nonbiological molecules to the building blocks of life itself—have now been demonstrated.

An Interstellar Origin?

Recently, a dissenting view has emerged. Some scientists argue that there may not have been sufficient energy available to power the chemical reactions in Earth's primitive atmosphere, and there may not have been enough raw material for the reactions to occur at a significant rate in any case. Instead, they suggest, most of the organic material that combined to form the first living cells was produced in *interstellar space* and subsequently arrived on Earth in the form of interplanetary dust and meteors that did not burn up during their descent through the atmosphere.

To test this hypothesis, NASA researchers carried out their own version of the Urey-Miller experiment, exposing an icy mixture of water, methanol, ammonia, and carbon monoxide—representative of many interstellar grains—to

Earth's origin   Life's origin

4   3

ultraviolet radiation to simulate the energy from a nearby newborn star. As shown in Figure 18.5, when they later placed the irradiated ice in water and examined the results, they found that it had formed droplets, surrounded by membranes, containing complex organic molecules. No amino acids, proteins, or DNA were found in the mix, but the results, repeated numerous times, clearly show that even the harsh, cold vacuum of interstellar space can be a suitable medium for the formation of complex molecules and primitive cellular structures.

Interstellar molecular clouds are known to contain many complex molecules, and large amounts of organic material were detected on comets Halley and Hale-Bopp when they last visited the inner solar system, so the idea that organic matter is constantly raining down on Earth from space in the form of interplanetary debris is quite plausible. ∞ *(Sec. 4.2)* Furthermore, the presence of water on Earth is generally taken to indicate an intense cometary bombardment early in our planet's history. ∞ *(Sec. 4.3)* Perhaps those comets carried complex molecules, too. However, whether this was the *primary* means by which those molecules appeared on Earth remains unclear.

Diversity and Culture

However the basic materials appeared on Earth, we know that life *did* appear. The fossil record chronicles how life on Earth became widespread and diversified over time. The study of fossil remains shows the initial appearance about 3.5 billion years ago of simple one-celled organisms, such as blue-green algae. These were followed about 2 billion years ago by more complex one-celled creatures like the amoeba. Multicellular organisms did not appear until about 1 billion years ago, after which a wide variety of increasingly complex organisms flourished—among them insects, reptiles, mammals, and humans. Figure 18.6 illustrates some of the key developments in the evolution of life on our planet.

The fossil record leaves no doubt that biological organisms have changed over time—all scientists accept the reality of *biological evolution*. As conditions on Earth have shifted and Earth's surface has evolved, those organisms that could best take advantage of their new surroundings succeeded and thrived—often at the expense of those organisms that could not make the necessary adjustments and consequently became extinct.

In the opinion of many anthropologists, intelligence, like any other highly advantageous trait, is strongly favored by natural selection. The social cooperation that went with coordinated hunting efforts was an important competitive advantage that developed as brain size increased. Perhaps most important of all was the development of language. Experience and ideas, stored in the

▲ **Figure 18.5 Interstellar Globules** These oily, hollow droplets rich in organic molecules were made by exposing a freezing mixture of primordial matter to harsh ultraviolet radiation. The larger ones span about 10 microns across and, when immersed in water, show cell-like membrane structure. Although they are not alive, they bolster the idea that life on Earth could have come from space. (NASA)

▼ **Figure 18.6 Life On Earth** A simplified timeline of the origin and evolution of life on our planet. The scale begins at the left with the origin of Earth about 4.6 billion years ago, extending linearly to the present at right. Notice how life forms most familiar to us emerged relatively recently in the history of Earth.

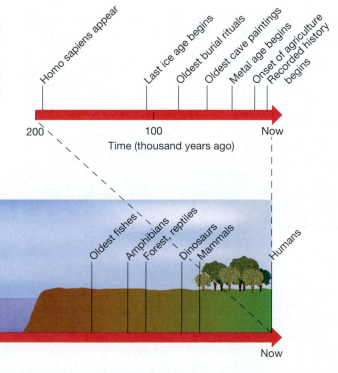

brain as memories, could be passed down from one generation to the next. A new kind of evolution had begun, namely, *cultural evolution*—the changes in the ideas and behavior of society. Our more recent ancestors created, within about 10,000 years, the entirety of human civilization.

To put all this into historical perspective, let's imagine the entire lifetime of Earth to be 46 years rather than 4.6 billion years. On this scale, we have no reliable record of the first decade of our planet's existence. Life originated at least 35 years ago, when Earth was about 10 years old. Our planet's middle age is still largely a mystery. Not until about 6 years ago did abundant life flourish throughout Earth's oceans. Life came ashore about 4 years ago, and plants and animals mastered the land only about 2 years ago. Dinosaurs reached their peak about 1 year ago, only to die suddenly about four months later. ∞ *(Discovery 4-1)* Humanlike apes changed into apelike humans only last week, and the latest ice ages occurred only a few days ago. *Homo sapiens*—our species—did not emerge until about four hours ago. Agriculture was invented within the last hour, and the Renaissance—along with all of modern science—is just three minutes old!

18.2 Life in the Solar System

Simple one-celled life forms reigned supreme on Earth for most of our planet's history. It took time—a great deal of time—for life to emerge from the oceans, to evolve into simple plants, to continue to evolve into complex animals, and to develop intelligence, culture, and technology. Have those (or similar) events occurred elsewhere in the solar system? Let's try to assess what little evidence we have on the subject.

Life as We Know It

"Life as we know it" is generally taken to mean carbon-based life that originated in a liquid-water environment—in other words, life on Earth. We must always be careful in our studies not to be misled by this "Earth-centric" viewpoint—it is all too easy to attribute human or Earth-based characteristics to other life forms—but since we have no extraterrestrial examples to guide us, at the very least this is a reasonable starting point. Thus we begin by asking, "Might Earth-like life exist (or have existed) on some other body in our planetary system?"

The Moon and Mercury lack liquid water, protective atmospheres, and magnetic fields, and so are subjected to fierce bombardment by solar ultraviolet radiation, the solar wind, meteoroids, and cosmic rays. Simple molecules could not survive in such hostile environments. On the other hand, Venus has far too much protective atmosphere. Its dense, dry, scorchingly hot atmospheric blanket effectively rules it out as a possible abode for life.

The jovian planets have no solid surfaces, and Pluto and most of the moons of the outer planets are too cold. However, the possibility of liquid water below Europa's icy surface has refueled speculation about the development of life there, making this moon of Jupiter a prime candidate for future exploration. ∞ *(Sec. 8.1)* For now, at least, Europa is high on the priority list for missions by both NASA and the European Space Agency. Saturn's moon Titan, with its atmosphere of methane, ammonia, and nitrogen gases, apparent geological activity, and possibly liquid hydrocarbon lakes or oceans on its surface, is conceivably another site where life might have arisen. The most recent results from the *Cassini-Huygens* mission suggest that Titan's frigid surface conditions are inhospitable for anything familiar to us, but (as we discuss below) scientists are finding more and more examples of terrestrial organisms able to survive and thrive in environments once regarded as uninhabitable. ∞ *(Sec. 8.2)*

CONCEPT CHECK ▶▶▶

Has chemical evolution been verified in the laboratory?

10μ

R I V U X G

▲ **Figure 18.7 Murchison Meteorite** The Murchison meteorite contains relatively large amounts of amino acids and other organic material, indicating that chemical evolution of some sort has occurred beyond our own planet. In this magnified view of a fragment from the meteorite, the arrow points to a microscopic sphere of organic matter. (Harvard-Smithsonian Center for Astrophysics)

What about the cometary and meteoritic debris that orbits within our solar system? As we have just seen, comets contain many of the basic ingredients for life, but they spend almost all of their time frozen, far from the Sun. A small fraction of the meteorites that survive the plunge to Earth's surface also contain organic compounds. A meteorite fell near Murchison, Australia, in 1969 that was located soon after crashing to the ground. This well-studied meteorite contains many of the amino acids normally found in living cells (Figure 18.7). The moderately large molecules found in meteorites and in interstellar clouds are our only evidence that chemical evolution has occurred elsewhere in the universe. However, most researchers regard this organic matter as prebiotic—matter that could eventually lead to life, but that has not yet done so.

The planet most likely to harbor life (or to have harbored it in the past) seems to be Mars. The red planet is harsh by Earth standards—liquid water is scarce, the atmosphere is thin, and the lack of magnetism and an ozone layer allows high-energy solar particles and radiation to reach the surface unabated. But the Martian atmosphere was once thicker, and the surface warmer and much wetter. ∞ *(Sec. 6.8)* There is strong photographic evidence from orbiters such as *Viking* and *Mars Global Surveyor* for flowing and standing water on Mars in the distant (and perhaps even relatively recent) past and, in 2004, the European *Mars Express* orbiter confirmed the long-hypothesized presence of water ice at the Martian poles, while NASA's *Opportunity* rover reported strong geological evidence that the region around its landing site was once "drenched" with water for an extended period. The *Viking* landers found no evidence of life, but they landed on the safest Martian terrain, not in the most interesting regions, such as near the moist polar caps. ∞ *(Discovery 6-2) Mars Pathfinder* also surveyed part of the Martian surface, again without finding any evidence of present or past life.

Some scientists think that a different type of biology might be operating, or might have operated, on the Martian surface. They suggest that Martian microbes capable of digesting oxygen-rich compounds in the Martian soil could explain the *Viking* results. This speculation will be greatly strengthened if recent announcements of fossilized bacteria in meteorites originating on Mars are confirmed. ∞ *(Discovery 6-2)* The consensus among biologists and chemists today is that Mars does not house any life similar to that on Earth, but a solid verdict regarding present or past life on Mars will likely not be reached until we have thoroughly explored our intriguing neighbor.

In considering the emergence of life under adversity, we should perhaps not be too quick to rule out an environment based solely on its extreme properties. Figure 18.8 shows a very hostile environment on a deep-ocean floor, where hydrothermal vents spew forth boiling hot water from vertical tubes a few meters tall. The conditions are quite unlike anything on our planet's surface, yet life thrives in an environment rich in sulfur, poor in oxygen, and completely dark. Such underground hot springs might conceivably exist on alien worlds, raising the possibility of life forms with much greater diversity over a much wider range of conditions than those known to us on Earth.

In recent years scientists have discovered many instances of so-called "extremophiles"—life forms that have adapted to live in extreme environments. The superheated hydrothermal vents in Figure 18.8 are one example, but extremophiles have also been found in frigid lakes buried deep under the Antarctic glaciers, in the dark, oxygen-poor and salt-rich floor of the Mediterranean Sea, in the mineral-rich superalkaline environment of California's Mono Lake, and even in the hydrogen-rich volcanic darkness far below Earth's

▼ **Figure 18.8 Hydrothermal Vents**   A small, two-person submarine (the *Alvin*, partly seen at bottom) took this picture of a hot spring, or "black smoker"—one of many along the mid-ocean ridge in the eastern Pacific Ocean. As hot water rich in sulfur pours out of the top of the vent's tube (near center), black clouds billow forth, providing a strange environment for many life forms thriving near the vent. The inset shows a close-up of the vent base, where extremophilic life thrives, including as seen here giant red tube worms and huge crabs. (WHOI)

CONCEPT CHECK ▶▶▶

Which solar system bodies (other than Earth) are the leading candidates in the search for extraterrestrial life, and why?

crust. In many cases these organisms have evolved to create the energy they need by purely chemical means, rather than relying on sunlight. These environments may present conditions not so different from those found on Mars, Europa, or Titan, suggesting that "life as we know it" might well be able to thrive in these hostile, alien worlds.

18.3 Intelligent Life in the Galaxy

With humans apparently the only intelligent life in the solar system, we must broaden our search for extraterrestrial intelligence to other stars, perhaps even other galaxies. At such distances, though, we have little hope of detecting life with current equipment. Instead, we must ask, How likely is it that life in any form—carbon-based, silicon-based, water-based, ammonia-based, or something we cannot even dream of—exists? Let's look at some numbers to develop statistical estimates of the probability of life elsewhere in the universe.

The Drake Equation

An early approach to this statistical problem is known as the **Drake equation**, after Frank Drake, the U.S. astronomer who pioneered this analysis:

| number of technological, intelligent civilizations now present in the Galaxy | = | rate of star formation, averaged over the lifetime of the Galaxy | × | fraction of stars having planetary systems | × | average number of habitable planets within those planetary systems | × | fraction of those habitable planets on which life arises | × | fraction of those life-bearing planets on which intelligence evolves | × | fraction of those intelligent-life planets that develop technological society | × | average lifetime of a technologically competent civilization. |

Of course, several of the factors in this equation are largely a matter of opinion. We do not have nearly enough information to determine—even approximately—all of them, so the Drake equation cannot give us a hard-and-fast answer. Its real value is that it subdivides a large and very difficult question into smaller pieces that we can attempt to answer separately. Figure 18.9 illustrates how, as our requirements become more and more stringent, only a small fraction of star systems in the Milky Way Galaxy have the advanced qualities specified by the combination of factors on the right-hand side of the equation.

Let's examine the factors in the equation one by one and make some educated guesses about their values. Bear in mind, though, that if you ask two scientists for their best estimates of any given one, you will likely get two very different answers!

Rate of Star Formation

We can estimate the average number of stars forming each year in the Galaxy simply by noting that at least 100 billion stars now shine in it. Dividing this number by the roughly 10-billion-year lifetime of the Galaxy, we obtain a formation rate of 10 stars per year. This may be an overestimate because astronomers think that fewer stars are forming now than formed during earlier epochs, when more interstellar gas was available. ∞ *(Sec. 16.3)* However, we do know that stars are forming today, and our estimate does not include stars that formed in the past and have since died, so our value of 10 stars per year is probably reasonable when averaged over the lifetime of the Milky Way Galaxy.

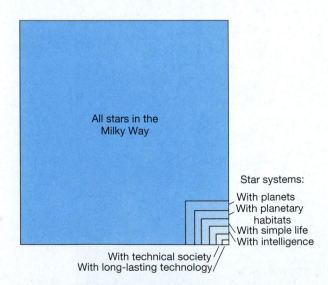

All stars in the Milky Way

Star systems:
With planets
With planetary habitats
With simple life
With intelligence
With technical society
With long-lasting technology

▲ **Figure 18.9 Drake Equation** Of all the star systems in our Milky Way Galaxy (represented by the largest box), progressively fewer and fewer have each of the qualities typical of a long-lasting technological society (represented by the smallest box at the lower right corner).

Fraction of Stars Having Planetary Systems

Many astronomers regard planet formation as a natural result of the star-formation process. If the condensation theory is correct, and if there is nothing special about our Sun, as we have argued throughout this book, we would expect many stars to have a planetary system of some sort. ∞ *(Sec. 4.3)*

No planets like our own have yet been *seen* orbiting any other star. The light reflected by an Earth-like planet circling even the closest star would be too faint to detect with current equipment, although planned orbiting telescopes may soon be able to image Jupiter-sized and perhaps smaller planets orbiting the nearest stars. However, there is now overwhelming evidence for planets orbiting other stars. ∞ *(Sec. 4.4)* The planets found so far are Jupiter-sized rather than Earth-sized, and generally have rather eccentric orbits, but then again these are the only planets we could have detected with current instruments. Many astronomers are confident that an Earth-like planet (on an Earth-like orbit) will one day be detected, and plans to build the necessary equipment are well under way.

Accepting the condensation theory and its consequences, and without being either too conservative or naively optimistic, we assign a value near 1 to this factor—that is, we think that essentially all stars have planetary systems.

Number of Habitable Planets per Planetary System

What determines the feasibility of life on a given planet? Temperature is perhaps the single most important factor. However, the possibility of catastrophic external events, such as cometary impacts or even distant supernovae, must also be considered. ∞ *(Discovery 4-1, Sec. 12.4)*

The surface temperature of a planet depends on two things: the planet's distance from its parent star and the thickness of its atmosphere. Planets with a nearby parent star (but not too close) and some atmosphere (though not too thick) should be reasonably warm, like Earth or Mars. Planets far from the star and with no atmosphere, like Pluto, will surely be cold by our standards. And planets too close to the star and with a thick atmosphere, like Venus, will be very hot indeed.

Figure 18.10 illustrates how a three-dimensional zone of "comfortable" temperatures—often called a *habitable zone*—surrounds every star. (The zones are indicated as rings in this two-dimensional figure.) It represents the range of distances within which a planet of mass and composition similar to those of Earth would have a surface temperature between the freezing and boiling points of water. (Our Earth-based bias is plainly evident here!) The hotter the star, the larger this zone. For example, an A- or an F-type star has a rather large habitable zone. G-, K-, and M-type stars have successively smaller zones. O- and B-type stars are not considered here because they are not expected to last long enough for life to develop, even if they do have planets.

Three planets—Venus, Earth, and Mars—reside in or near the habitable zone surrounding our Sun. Venus is too hot because of its thick atmosphere and proximity to the Sun. Mars is a little too cold because its atmosphere is too thin and it is too far from the Sun. But if the orbits of Venus and Mars were swapped—not inconceivable, since chance played such a large role in the formation of the terrestrial planets—then both of these nearby planets might conceivably have evolved surface conditions resembling those on Earth. ∞ *(Sec. 4.3, 6.8)* In that case our solar system would have had three habitable planets instead of one. Proximity to a giant planet may also render a moon (such as Europa) habitable, the planet's tidal heating making up for the lack of sunlight. ∞ *(Sec. 8.1)*

A planet moving on a "habitable" orbit may still be rendered uninhabitable by external events. Many scientists think that the outer planets in our own solar system are critical to the habitability of the inner worlds, both by

▲ **Figure 18.10 Stellar Habitable Zones**   The extent of the habitable zone is much larger around a hot star than around a cool one. For a star like the Sun (a G-type star), the zone extends from about 0.85 AU to 2.0 AU. For a hotter F-type star, the range is 1.2 to 2.8 AU. For a cool M-type star, only Earth-like planets orbiting between about 0.02 and 0.06 AU would be habitable.

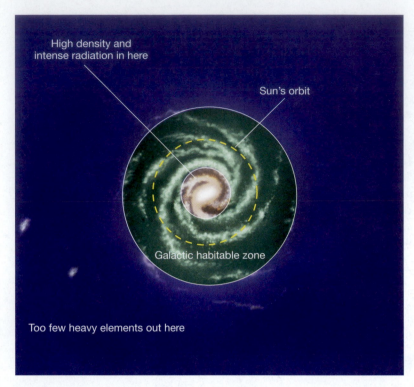

High density and
intense radiation in here

Sun's orbit

Galactic habitable zone

Too few heavy elements out here

◄ Figure 18.11 Galactic Habitable Zone Some regions of the Galaxy may be more conducive to life than others. Too far from the Galactic center, there may not be enough heavy elements for terrestrial planets to form or technological society to evolve. Too close, the radiative or gravitational effects of nearby stars may render life impossible. The result is a ring-shaped habitable zone, colored here in green. This zone presumably encloses the orbit of the Sun, but its radial extent in both directions is uncertain.

stabilizing their orbits and by protecting them from cometary impacts, deflecting would-be impactors away from the inner part of the solar system. The theory presented in Chapter 4 suggests that a star with inner terrestrial planets on stable orbits would probably also need the jovian worlds to safeguard their survival. ∞ *(Sec. 4.3)* However, observations of extrasolar planets are not yet sufficiently refined to determine the fraction of stars having "outer planet" systems like our own. ∞ *(Sec. 4.4)*

Other external forces may also influence a planet's survival. Some researchers have suggested that there is a *Galactic habitable zone* for stars in general, outside of which conditions are unfavorable for life (see Figure 18.11). Far from the Galactic center, the star formation rate is low and few cycles of star formation have occurred, so there are insufficient heavy elements to form terrestrial planets or populate them with technological civilizations. ∞ *(Sec. 12.7)* Too close, and the radiation from bright stars and supernovae in the crowded inner part of the Galaxy might be detrimental to life. More important, the gravitational effects of nearby stars may send frequent showers of comets from the Oort cloud into the inner regions of a planetary system, impacting the terrestrial planets and terminating any chain of evolution that might lead to intelligent life.

Thus, to estimate the number of habitable planets per planetary system, we must first take inventory of how many stars of each type shine in the Galactic habitable zone, then calculate the sizes of their stellar habitable zones and estimate the number of planets likely to be found there. We must eliminate almost all of the 10 percent of surveyed stars around which planets have so far been observed, and presumably a similar fraction of stars in general. The large jovian planets seen in most cases have eccentric orbits that would destabilize the motion of any inner terrestrial world, either ejecting it completely from the system or making conditions so extreme that the chances for the development of life are severely reduced. ∞ *(Sec. 4.4)*

Finally, we exclude the majority of binary-star systems because a planet's orbit within the habitable zone of a binary would be unstable in many cases, as illustrated in Figure 18.12. Given the properties of binaries in our Galaxy, habitable planetary orbits would probably be unstable in most cases, so there would not be time for life to develop.

The inner and outer radii of the Galactic habitable zone are not known with any certainty, and we still have insufficient data about most stars to make any definitive statement about their planetary systems. Taking these uncertainties into account as best we can, we assign a value of 1/10 to this factor in our equation. In other words, we believe that, on average, there is one potentially habitable planet for every 10 planetary systems that might exist in our Galaxy. Single F-, G-, and K-type stars are the best candidates.

(a)

(b)

(c)

◄ Figure 18.12 Binary-Star Planets In binary-star systems, planets are restricted to only a few kinds of orbits that are gravitationally stable. (a) This orbit is stable only if the planet lies very close to its parent star, so the gravity of the other star is negligible. (b) A planet circulating at a great distance about both stars in an elliptical orbit. This orbit is stable only if it lies far from both stars. (c) Another possible but unstable path interweaves between the two stars in a figure-eight pattern.

Fraction of Habitable Planets on Which Life Arises

The number of possible combinations of atoms is incredibly large. If the chemical reactions that led to the complex molecules that make up living organisms occurred completely at random, then it is extremely unlikely that those molecules could have formed at all. In that case, life is extraordinarily rare, this factor is close to zero, and we are probably alone in the Galaxy, perhaps even in the entire universe.

However, laboratory experiments (like the Urey-Miller experiment described earlier) seem to suggest that certain chemical combinations are strongly favored over others—that is, the reactions are not random. Of the billions upon billions of basic organic groupings that could possibly occur on Earth from the random combination of all sorts of simple atoms and molecules, only about 1500 actually do occur. Furthermore, these 1500 organic groups of terrestrial biology are made from only about 50 simple "building blocks" (including the amino acids and nucleotide bases mentioned earlier). This suggests that molecules critical to life may not be assembled by pure chance. Apparently, additional factors are at work at the microscopic level. If a relatively small number of chemical "evolutionary tracks" is likely to exist, then the formation of complex molecules—and hence, we assume, life—becomes much more likely, given sufficient time.

To assign a very low value to this factor in the equation is to believe that life arises randomly and rarely. To assign a value close to 1 is to believe that life is inevitable, given the proper ingredients, a suitable environment, and a long enough period of time. No easy experiment can distinguish between these extreme alternatives, and there is little or no middle ground. To many researchers, the discovery of life—past or present—on Mars, Europa, Titan, or any other object in our solar system would convert the appearance of life from an unlikely miracle to a virtual certainty throughout the Galaxy. In the absence of objective evidence, we will simply take the optimistic view and adopt a value of 1.

Fraction of Life-Bearing Planets on Which Intelligence Arises

As with the evolution of life, the appearance of a well-developed brain is a very unlikely event if only random chance is involved. However, biological evolution through natural selection is a mechanism that generates apparently highly improbable results by singling out and refining useful characteristics. Organisms that profitably use adaptations can develop more complex behavior, and complex behavior provides organisms with the *variety* of choices needed for more advanced development.

One school of thought maintains that, given enough time, intelligence is inevitable. In this view, assuming that natural selection is a universal phenomenon, at least one organism on a planet will always rise to the level of "intelligent life." If this is correct, then the fifth factor in the Drake equation equals or nearly equals 1.

Others argue that there is only one known case of intelligence, and that case is life on Earth. For 2.5 billion years—from the start of life about 3.5 billion years ago to the first appearance of multicellular organisms about 1 billion years ago—life did not advance beyond the one-celled stage. Life remained simple and dumb, but it survived. If this latter view is correct, then the fifth factor in our equation is very small, and we are faced with the depressing prospect that humans may be the smartest form of life anywhere in the Galaxy. As with the previous factor, we will be optimistic and adopt a value of 1 here.

Fraction of Planets on Which Intelligent Life Develops and Uses Technology

To evaluate the sixth factor of our equation, we need to estimate the probability that intelligent life eventually develops technological competence. Should the rise of technology be inevitable, this factor is close to 1, given long enough periods of time. If it is not inevitable—if intelligent life can somehow "avoid" developing technology (as dolphins on Earth seem to have done)—then this factor could be much less than 1. The latter possibility envisions a universe possibly teeming with civilizations but very few among them ever becoming technologically competent. Perhaps only one managed it—ours.

Again, it is difficult to decide conclusively between these two views. We don't know how many (if any) prehistoric Earth cultures either failed to develop technology or rejected its use. We do know that tool-using societies arose independently at several places on Earth, including Mesopotamia, India, China, Egypt, Mexico, and Peru. Because so many of these ancient "technological" cultures originated *independently* at about the same time, we conclude that the chances are good that some sort of technological society will inevitably develop, given some basic intelligence and enough time. Accordingly, we take this factor to be close to 1.

Average Lifetime of a Technological Civilization

The reliability of these estimates declines markedly from the first to the last factor of the Drake equation. Our knowledge of astronomy allows us to make a reasonably good stab at the first two, but even the third is hard to evaluate, and the values adopted for the last few are more wishful thinking than science. The final factor is totally unknown. There is only one example of such a civilization that we are aware of—humans on planet Earth. Our own civilization has presently survived in its technological state for only about 100 years, and how long we will be around before a human-made catastrophe or a planetwide natural disaster ends it all is impossible to tell. ∞ *(Discovery 4-1)*

One thing is certain: If the correct value for *any one factor* in the equation is very small, then few technological civilizations now exist in the Galaxy. If the pessimistic view of the development of life or of intelligence is correct, then we are unique, and that is the end of our story. However, if both life and intelligence are inevitable consequences of chemical and biological evolution, as many scientists think, and if intelligent life always becomes technological, then we can plug the higher, more optimistic values into the Drake equation. In that case, combining our estimates for the other six factors (and noting that, conveniently, $10 \times 1 \times 1/10 \times 1 \times 1 \times 1 = 1$), we can say

$$\begin{matrix} \text{number of technological} \\ \text{intelligent civilizations} \\ \text{now present in the Milky} \\ \text{Way Galaxy} \end{matrix} = \begin{matrix} \text{average lifetime of a} \\ \text{technologically} \\ \text{competent civilization,} \\ \text{in years.} \end{matrix}$$

Thus, if advanced civilizations typically survive for 1000 years, there should be 1000 of them currently in existence scattered throughout the Galaxy. If they survive for a million years, on average, we would expect there to be a million of them in the Galaxy, and so on.

Note that even setting aside language and cultural issues, the sheer size of the Galaxy presents a significant hurdle to communication between technological civilizations. The minimum requirement for a two-way conversation is that we can send a signal and receive a reply in a time shorter than our own lifetime.

If the lifetime is short, then civilizations are literally few and far between—small in number, according to the Drake equation, and scattered over the vastness of the Milky Way—and the distances between them (in light-years) are much greater than their lifetimes (in years). In that case, two-way communication, even at the speed of light, will be impossible. However, as the lifetime increases, the distances get smaller as the Galaxy becomes more crowded, and the prospects improve.

Taking into account the size, shape, and distribution of stars in the Galactic disk (why do we exclude the halo?) and under the assumptions made above, we find that unless the life expectancy of a civilization is *at least* a few thousand years, it is unlikely to have time to communicate with even its nearest neighbor.

If most of the factors are largely a matter of opinion, how does the Drake equation assist astronomers in refining their search for extraterrestrial life?

18.4 The Search for Extraterrestrial Intelligence

Let's continue our optimistic assessment of the prospects for life and assume that civilizations enjoy a long stay on their parent planet once their initial technological "teething problems" are past. In that case, they may be plentiful in the Galaxy. How might we become aware of their existence?

Meeting Our Neighbors

For definiteness, let's assume that the average lifetime of a technological civilization is 1 million years—only 1 percent of the reign of the dinosaurs, but 100 times longer than our civilization has survived thus far, and 10,000 times longer than human society has so far been in the "technological" state. The Drake equation then tells us that there are 1 million such civilizations in the Galaxy, and we can then estimate the average distance between these civilizations to be some 30 pc, or about 100 light-years. Thus, any two-way communication with our neighbors will take at least 200 years (100 years for the message to reach the planet and another 100 years for the reply to travel back to us)—a long time by human standards, but comfortably less than the lifetime of the civilization.

Might we hope to visit our neighbors by developing the capability of traveling far outside our solar system? This may never be a practical possibility. At a speed of 50 km/s, the speed of the current fastest space probes, the round-trip to even the nearest Sun-like star, Alpha Centauri, would take about 50,000 years. The journey to the nearest technological neighbor (assuming a distance of 100 pc) and back would take 1 million years—the entire lifetime of our civilization! Interstellar travel at these speeds is clearly not feasible. Speeding up our ships to near the speed of light would reduce the travel time, but this is far beyond our present technology.

Actually, our civilization has already launched some interstellar probes, although they have no specific stellar destination. Figure 18.13 is a reproduction of a plaque mounted on board the *Pioneer 10* spacecraft launched in the mid-1970s and now well beyond the orbit of Pluto, on its way out of the solar system. Similar information was also included aboard the *Voyager* probes launched in 1978. While these spacecraft would be incapable of reporting back to Earth the news that they had encountered an alien culture, scientists hope that the civilization on the other end would be able to unravel most of its contents using the universal language of mathematics. The caption to Figure 18.13 notes how the aliens might discover from where and when the *Pioneer* and *Voyager* probes were launched.

▲ **Figure 18.13** *Pioneer 10* **Plaque** A replica of a plaque mounted on board the *Pioneer 10* spacecraft. Included are a scale drawing of the spacecraft, a man, and a woman; a diagram of the hydrogen atom undergoing a change in energy (top left); a starburst pattern representing various pulsars and the frequencies of their radio waves that can be used to estimate when the craft was launched (middle left); and a depiction of the solar system, showing that the spacecraft departed the third planet from the Sun and passed the fifth planet on its way into outer space (bottom). All the drawings have computer-coded (binary) markings from which actual sizes, distances, and times can be derived. (NASA)

▶ **Figure 18.14 Earth's Radio Leakage** Radio radiation now leaks from Earth into space because of the daily activities of our technological civilization. (a) FM radio and television transmitters broadcast their energy parallel to Earth's surface, sending a great "sheet" of electromagnetic radiation into interstellar space and producing the strongest signal in any given direction when they happen to lie on Earth's horizon, as seen from that direction. (The more common AM broadcasts are trapped below our ionosphere, so those signals never leave Earth.) (b) Because the great majority of transmitters are clustered in the eastern United States and western Europe, a distant observer would detect blasts of radiation from Earth as our planet rotates each day.

▲ **Figure 18.15 Water Hole** The "water hole" is bounded by the natural emission frequencies of the hydrogen (H) atom (21-cm wavelength) and the hydroxyl (OH) molecule (18-cm wavelength). The topmost solid (blue) curve sums the natural emissions of our Galaxy (dashed line on left side of diagram, labeled "Galactic background") and Earth's atmosphere (dotted line on right side of diagram, denoted by various chemical symbols). This sum is minimized near the water hole frequencies. Perhaps all intelligent civilizations conduct their interstellar communications within this quiet "electromagnetic oasis."

CONCEPT CHECK ▶▶▶

What assumptions go into the identification of the water hole as a likely place to search for extraterrestrial signals?

Radio Searches

A cheaper and much more practical alternative is to try to make contact with extraterrestrials using electromagnetic radiation, the fastest known means of transferring information from one place to another. Because light and other short-wavelength radiation are heavily scattered while moving through dusty interstellar space, long-wavelength radio radiation seems to be the best choice. ∞ *(Sec. 2.3)* We do not attempt to broadcast to all nearby candidate stars, however—that would be far too expensive and extremely inefficient. Instead, radio telescopes on Earth listen *passively* for radio signals emitted by other civilizations.

In what direction should we aim our radio telescopes? The answer to this question at least is fairly easy. On the basis of our earlier reasoning, we should target all F-, G-, and K-type stars in our vicinity. But are extraterrestrials broadcasting radio signals? If they are not, this search technique will obviously fail. If they are, how do we distinguish their artificially generated radio signals from signals emitted naturally by interstellar gas clouds? This depends on whether the signals are produced deliberately or are simply "waste radiation" escaping from a planet.

Consider first how "waste" radio wavelengths originating on Earth would look to extraterrestrials. Figure 18.14 shows the pattern of radio signals we broadcast into space. From the viewpoint of a distant observer, the spinning Earth emits a bright flash of radio radiation every few hours as the most technological regions of our planet rise or set. In fact, Earth is now a more intense radio emitter than the Sun. Our radio and television transmissions race out into space and have been doing so since the invention of these technologies nearly seven decades ago. Another civilization as advanced as ours might have constructed devices capable of detecting this radiation. If any sufficiently advanced (and sufficiently interested) civilization resides within about 65 light-years (20 pc) of Earth, then we have already announced our presence to them.

Of course, it may very well be that, having discovered cable and fiber-optics technology, most civilizations' indiscriminate transmissions cease after a few decades. In that case, radio silence becomes the hallmark of intelligence, and we must find an alternative means of locating our neighbors.

The Water Hole

Now let us suppose that a civilization has decided to assist searchers by actively broadcasting its presence to the rest of the Galaxy. At what frequency should we listen for such an extraterrestrial beacon? The electromagnetic spectrum is enormous; the radio domain alone is vast. To hope to detect a signal at some unknown radio frequency is like searching for a needle in a haystack. Are some frequencies more likely than others to carry alien transmissions?

Some basic arguments suggest that civilizations might well communicate at a wavelength near 20 cm. The basic building blocks of the universe, namely, hydrogen atoms, radiate naturally at a wavelength of 21 cm. ∞ *(Sec. 11.2)* Also, one of the simplest molecules, hydroxyl (OH), radiates near 18 cm. Together, these two substances form water (H_2O). Arguing that water is likely to be the evolution medium for life anywhere, and that radio radiation travels through the disk of our Galaxy with the least absorption by interstellar gas and dust, some researchers have proposed that the interval between 18 cm and 21 cm is the best wavelength range for civilizations to transmit or listen. Called the **water hole,** this radio interval might serve as an "oasis" where all advanced Galactic civilizations would gather to conduct their electromagnetic business.

This water-hole frequency interval is only a guess, of course, but its use is supported by other arguments. Figure 18.15 shows the water hole's location in the electromagnetic spectrum and plots the amount of natural emission from our Galaxy and from Earth's atmosphere. The 18- to 21-cm range lies within the quietest part of the spectrum, where the Galactic "static" from stars and interstellar clouds happens to be minimized. Furthermore, the atmospheres of typical planets are also expected to interfere least at these wavelengths. Thus, the water hole seems like a good choice for the frequency of an interstellar beacon, although we cannot be sure of this reasoning until contact is actually achieved.

A few radio searches are now in progress at frequencies in and around the water hole. One of the most sensitive and comprehensive projects in the ongoing search for extraterrestrial intelligence (known to many by its acronym SETI) was Project Phoenix, carried out during the late 1990s. Large radio antennas, such as that shown in Figure 18.16(a), were used to search millions of channels simultaneously in the 1–3 GHz range. Figure 18.16(b) shows what a typical narrow-band, 1-Hz signal—a potential "signature" of an intelligent transmission—would look like on a computer monitor. However, this observation was merely a test to detect the weak, redshifted radio signal emitted by the *Pioneer 10* robot, now receding into the outer realm of our solar system—a sign of intelligence, but one that we put there. Nothing resembling an extraterrestrial signal has yet been detected.

The space surrounding all of us could be, right now, flooded with radio signals from extraterrestrial civilizations. If only we knew the proper direction and frequency, we might be able to make one of the most startling discoveries of all time. The result would provide whole new opportunities to study the cosmic evolution of energy, matter, and life throughout the universe.

(a)

(b)

▲ **Figure 18.16** **Project Phoenix** (a) This large radio telescope at the National Radio Astronomy Observatory in Green Bank, West Virginia, was used by Project Phoenix during the 1990s to search for extraterrestrial intelligent signals. (b) This is a typical recording of an alien signal—here, as a test, the Doppler-shifted broadcast from the *Pioneer 10* spacecraft, now well beyond the orbit of Pluto. The diagonal line across the computer monitor, in contrast to the random noise in the background, betrays the presence of an intelligent signal in the data stream. (NRAO; SETI Institute)

CHAPTER REVIEW

Summary

1 **Cosmic evolution** (p. 484) is the continuous process that has led to the appearance of galaxies, stars, planets, and life on Earth. Living organisms may be characterized by their ability to react to their environment, to grow by taking in nutrition from their surroundings, to reproduce, passing along some of their own characteristics to their offspring, and to evolve in response to a changing environment. Organisms that can best take advantage of their new surroundings succeed at the expense of those organisms that cannot make the necessary adjustments. Intelligence is strongly favored by natural selection.

2 Powered by natural energy sources, reactions between simple molecules in the oceans of the primitive Earth may have led to the formation of **amino acids** (p. 485) and **nucleotide bases** (p. 485), the basic molecules of life. Alternatively, some complex molecules may have been formed in interstellar space and then delivered to Earth by meteors or comets. The best hope for life beyond Earth in the solar system is the planet Mars, although no evidence for living organisms has been found. Jupiter's Europa and Saturn's Titan may also be possibilities, but conditions on both those bodies are harsh by terrestrial standards.

3 The **Drake equation** (p. 490) provides a means of estimating the probability of other intelligent life in the Galaxy. The astronomical terms in the equation are the Galactic star-formation rate, the likelihood of planets, and the number of habitable planets. Chemical and biological terms are the probability of life appearing

and the probability that it subsequently develops intelligence. Cultural and political terms are the probability that intelligence leads to technology and the lifetime of a technological civilization. Taking an optimistic view of the development of life and intelligence leads to the conclusion that the total number of technologically competent civilizations in the Galaxy is approximately equal to the lifetime of a typical civilization, expressed in years.

4 A technological civilization would probably "announce" itself to the universe by the radio and television signals it emits into space. Observed from afar, our planet would appear as a radio source with a 24-hour period, as different regions of the planet rise and set. The **water hole** (p. 497) is a region in the radio range of the electromagnetic spectrum, near the 21-cm line of hydrogen and the 18-cm line of hydroxyl, where natural emissions from the Galaxy happen to be minimal. Many researchers regard this as the best part of the spectrum for communications purposes.

Review and Discussion

1. Why is life difficult to define?

2. What is chemical evolution?

3. What is the Urey-Miller experiment? What important organic molecules were produced in this experiment?

4. What are the basic ingredients from which biological molecules formed on Earth?

5. How do we know anything at all about the early episodes of life on Earth?

6. What is the role of language in cultural evolution?

7. Where else, besides Earth, have organic molecules been found?

8. Where—besides the planet Mars—might we find signs of life in our solar system?

9. Do we know whether Mars ever had life at any time during its past? What argues in favor of the position that it may once have harbored life?

10. What is generally meant by "life as we know it"? What other forms of life might be possible?

11. How many of the terms in the Drake equation are known with any degree of certainty? Which factor is least well known?

12. What is the relationship between the average lifetime of galactic civilizations and the possibility of our someday communicating with them?

13. How would Earth appear, at radio wavelengths, to extraterrestrial astronomers?

14. What are the advantages in using radio waves for communication over interstellar distances?

15. What is the water hole? What advantage does it have over other parts of the radio spectrum?

Conceptual Self-Test: True or False?/Multiple Choice

1. The definition of life requires only that, to be considered "alive," you must be able to reproduce. **(T/F)**

2. Laboratory experiments have created living cells from nonbiological molecules. **(T/F)**

3. Dinosaurs lived on Earth for more than a thousand times longer than human civilization has existed to date. **(T/F)**

4. The *Viking* landers on Mars discovered microscopic evidence of life but found no fossil evidence. **(T/F)**

5. The development of life and intelligence on Earth are extremely unlikely if chance is the only evolutionary factor involved. **(T/F)**

6. We have no direct evidence for Earth-like planets orbiting other stars. **(T/F)**

7. Our civilization has already launched probes into interstellar space and broadcast our presence to our neighbors. **(T/F)**

8. The chemical elements that form the basic molecules needed for life are found **(a)** in the cores of Sun-like stars; **(b)** commonly throughout the cosmos; **(c)** only on planets that have liquid water; **(d)** only on Earth.

9. Fossil records of early life forms on Earth suggest that life began about **(a)** 6000 years ago; **(b)** 65 million years ago; **(c)** 3.5 billion years ago; **(d)** 14 billion years ago.

10. The least-well-known factor in the Drake equation is **(a)** the rate of star formation; **(b)** the average number of habitable planets within planetary systems; **(c)** the average lifetime of a technologically competent civilization; **(d)** the diameter of the Milky Way Galaxy.

11. We don't expect life on planets orbiting B-type stars because the star **(a)** has too much gravity; **(b)** is too short-lived for life to

evolve; **(c)** is at too low a temperature to sustain life; **(d)** would have only gas giant planets.

12. NASA's Space Shuttle orbits Earth at about 17,500 mph. If it traveled to the next Sun-like star at that speed, the trip would take at least **(a)** one week; **(b)** one decade; **(c)** one century; **(d)** 100 millennia.

13. The strongest radio-wavelength emission in the solar system is **(a)** human-made signals from Earth; **(b)** the Sun; **(c)** the Moon; **(d)** Jupiter.

14. From the data shown in Figure 18.10 (Stellar Habitable Zones), the habitable zone surrounding a main sequence K-type star **(a)** cannot be determined; **(b)** extends from roughly 1 to 2 AU from the star; **(c)** is larger than that of an F-type star; **(d)** is larger than that of an M-type star.

15. If Figure 18.14 (Earth's Radio Leakage) were to be redrawn for a planet spinning twice as fast, the new jagged line would be **(a)** unchanged; **(b)** taller; **(c)** stretched out horizontally; **(d)** compressed horizontally.

Problems

 Algorithmic versions of these questions are available in the Practice Problems module of the Companion Website.

The number of squares preceding each problem indicates its approximate level of difficulty.

1. ■ As described in the text, imagine that Earth's 4.6-billion-year age is compressed to just 46 years. What would be your age, in seconds? How long ago (in seconds) was the end of World War II, the Declaration of Independence, Columbus's discovery of the New World? How long ago did the extinction of the dinosaurs occur (in days)?

2. ■■■ According to the inverse-square law, a planet receives energy from its parent star at a rate proportional to the star's luminosity and inversely proportional to the square of the planet's distance from the star. ∞ *(Sec. 10.2)* According to Stefan's law, the rate at which the planet radiates energy into space is proportional to the fourth power of its surface temperature. ∞ *(Sec. 2.4)* In equilibrium, the two rates are equal. Based on this information and given the fact that (taking into account the greenhouse effect) the Sun's habitable zone extends from 0.6 AU to 1.5 AU, estimate the extent of the habitable zone surrounding a K-type main-sequence star of luminosity one-tenth the luminosity of the Sun.

3. ■ Based on the numbers presented in the text, and assuming an average lifetime of 5 billion years for suitable stars, estimate the total number of habitable planets in the Galaxy.

4. ■■ A planet orbits one component of a binary-star system at a distance of 1 AU (see Figure 18.12a). If both stars have the same mass and their orbit is circular, estimate the minimum distance between the stars for the tidal force due to the companion not to exceed a "safe" 0.01 percent of the gravitational force between the planet and its parent star.

5. ■ Suppose that each of the "fraction" terms in the Drake equation turns out to have a value of 1/10, that stars form at an average rate of 20 per year, and that each star with a planetary system has exactly one habitable planet orbiting it. Estimate the present number of technological civilizations in the Milky Way Galaxy if the average lifetime of a civilization is **(a)** 100 years, **(b)** 10,000 years, **(c)** 1 million years.

6. ■■■ Adopting the estimate from the text that the number of technological civilizations in the Milky Way Galaxy is equal to the average lifetime of a civilization, in years, it follows that the distance to our nearest neighbor decreases as the average lifetime increases. Assuming that civilizations are uniformly spread over a two-dimensional Galactic disk of radius 15 kpc, and all have the same lifetime, calculate the *minimum* lifetime for which a two-way radio communication with our nearest neighbor would be possible before our civilization ends. Repeat the calculation for a round-trip personal visit, using current-technology spacecraft that travel at 50 km/s.

7. ■ How fast would a spacecraft have to travel in order to complete the trip from Earth to Alpha Centauri (a distance of 1.3 pc) and back in less than an average human lifetime (80 years, say)?

8. ■ Assuming that there are 10,000 FM radio stations on Earth, each transmitting at a power level of 50 kW, calculate the total radio luminosity of Earth in the FM band. Compare this value with the roughly 10^6 W radiated by the Sun in the same frequency range.

9. ■ Convert the water hole's wavelengths to frequencies. For practical reasons, any search of the water hole must be broken up into channels, much as you find on a television, except these channels are very narrow in radio frequency, about 100 Hz wide. How many channels must astronomers search in the water hole?

10. ■ There are 20,000 stars within 100 light-years that are to be searched for radio communications. How long will the search take if one hour is spent looking at each star? What if one day is spent per star?

 The Companion Website at www.prenhall.com/chaisson provides algorithmically generated versions of each chapter's Problems, in addition to an Animations and Videos gallery, a Tutorials gallery and Tutorials Quiz questions, a full eBook, an Images gallery, Practice Questions, Multiple Choice, and True or False quizzes, and Labeling exercises.

Appendix 1
Scientific Notation

The objects studied by astronomers range in size from the smallest particles to the largest expanse of matter we know—the entire universe. Subatomic particles have sizes of about 0.000000000000001 meter, while galaxies typically measure some 1,000,000,000,000,000,000,000 meters across. The most distant known objects in the universe lie on the order of 100,000,000,000,000,000,000,000,000 meters from Earth.

Obviously, writing all those zeros is both cumbersome and inconvenient. More important, it is also very easy to make an error—write down one zero too many or too few and your calculations become hopelessly wrong! To avoid this, scientists always write large numbers using a shorthand notation in which the number of zeros following or preceding the decimal point is denoted by a superscript power, or *exponent*, of 10. The exponent is simply the number of places between the first significant (nonzero) digit in the number (reading from left to right) and the decimal point. Thus 1 is 10^0, 10 is 10^1, 100 is 10^2, 1000 is 10^3, and so on. For numbers less than 1, with zeros between the decimal point and the first significant digit, the exponent is negative: 0.1 is 10^{-1}, 0.01 is 10^{-2}, 0.001 is 10^{-3}, and so on. Using this notation we can shorten the number describing subatomic particles to 10^{-15} meter and write the number describing the size of a galaxy as 10^{21} meters.

More complicated numbers are expressed as a combination of a power of 10 and a multiplying factor. This factor is conventionally chosen to be a number between 1 and 10, starting with the first significant digit in the original number. For example, 150,000,000,000 meters (the distance from Earth to the Sun, in round numbers) can be more concisely written at 1.5×10^{11} meters, 0.000000025 meters as 2.5×10^{-8} meter, and so on. The exponent is simply the number of places the decimal point must be moved *to the left* to obtain the multiplying factor.

Some other examples of scientific notation are:

- the approximate distance to the Andromeda Galaxy = 2,500,000 light-years = 2.5×10^6 light-years

- the size of a hydrogen atom = 0.00000000005 meter = 5×10^{-11} meter

- the diameter of the Sun = 1,392,000 kilometers = 1.392×10^6 kilometers

- the U.S. national debt (as of midnight, January 1, 2006) = \$8,131,059,000,000.00 = \$8.131059 trillion = 8.131059×10^{12} dollars.

In addition to providing a simpler way of expressing very large or very small numbers, this notation also makes it easier to do basic arithmetic. The rule for multiplication of numbers expressed in this way is simple: Just multiply the factors and add the exponents. Similarly for division: Divide the factors and subtract the exponents. Thus, 3.5×10^{-2} multiplied by 2.0×10^3 is simply $(3.5 \times 2.0) \times 10^{-2+3} = 7.0 \times 10^1$—that is, 70. Again, 5×10^6 divided by 2×10^4 is just $(5/2) \times 10^{6-4}$, or 2.5×10^2 (= 250). Applying these rules to unit conversions, 200,000 nanometers is $200,000 \times 10^{-9}$ meter (since 1 nanometer = 10^{-9} meter; see Appendix 2), or $(2 \times 10^5) \times 10^{-9}$ meter, or $2 \times 10^{5-9} = 2 \times 10^{-4}$ meter = 0.2 mm. Verify these rules yourself with a few examples of your own. The advantages of this notation when considering astronomical objects will soon become obvious.

Scientists often use "rounded-off" versions of numbers, both for simplicity and for ease of calculation. For example, we will usually write the diameter of the Sun as 1.4×10^6 kilometers, instead of the more precise number given earlier. Similarly, the diameter of Earth is 12,756 kilometers, or 1.2756×10^4 kilometers, but for "ballpark" estimates, we really don't need so many digits, and the more approximate number 1.3×10^4 kilometers will suffice. Very often, we perform rough calculations using only the first one or two significant digits in a number, and that may be all that is necessary to make a particular point. For example, to support the statement, "The Sun is much larger than Earth," we need only say that the ratio of the two diameters is roughly 1.4×10^6 divided by 1.3×10^4. Since 1.4/1.3 is close to 1, the ratio is approximately $10^6/10^4 = 10^2$, or 100. The essential fact here is that the ratio is much larger than 1; calculating it to greater accuracy (to get 109.13) would give us no additional *useful* information. This technique of stripping away the arithmetic details to get to the essence of a calculation is very common in astronomy, and we use it frequently throughout this text.

Appendix 2
Astronomical Measurement

Astronomers use many different kinds of units in their work, simply because no single system of units will do. Rather than the *Système Internationale* (SI), or meter-kilogram-second (MKS), metric system used in most high school and college science classes, many professional astronomers still prefer the older centimeter-gram-second (CGS) system. However, astronomers also commonly introduce new units when convenient. For example, when discussing stars, the mass and radius of the Sun are often used as reference points. The solar mass, written as $M_\odot$, is equal to 2.0×10^{33} g, or 2.0×10^{30} kg (since 1 kg = 1000 g). The solar radius, $R_\odot$, is equal to 700,000 km, or 7.0×10^8 m (1 km = 1000 m). The subscript $\odot$ always stands for Sun. Similarly, the subscript $\oplus$ always stands for Earth. In this book, we try to use the units that astronomers commonly use in any given context, but we also give the "standard" SI equivalents where appropriate.

Of particular importance are the units of length astronomers use. On small scales, the *angstrom* (1 Å = 10^{-10} m = 10^{-8} cm), the *nanometer* (1 nm = 10^{-9} m = 10^{-7} cm), and the *micron* (1 μm = 10^{-6} m = 10^{-4} cm) are used. Distances within the solar system are usually expressed in terms of the *astronomical unit* (AU),

the mean distance between Earth and the Sun. One AU is approximately equal to 150,000,000 km, or 1.5×10^{11} m. On larger scales, the *light-year* (1 ly = 9.5×10^{15} m = $9.5 \times 10^{12} km$) and the *parsec* (1 pc = 3.1×10^{16} m = 3.1×10^{13} km = 3.3 ly) are commonly used. Still larger distances use the regular prefixes of the metric system: *kilo* for one thousand and *mega* for one million. Thus 1 kiloparsec (kpc) = 10^3 pc = 3.1×10^{19} m, 10 megaparsecs (*Mpc*) = $10^7 pc$ = 3.1×10^{23} m, and so on.

Astronomers use units that make sense within a context, and as contexts change, so do the units. For example, we might measure densities in grams per cubic centimeter (g/cm^3), in atoms per cubic meter (atoms/m^3), or even in solar masses per cubic megaparsec ($M_\odot$/Mpc^3), depending on the circumstances. The important thing to know is that once you understand the units, you can convert freely from one set to another. For example, the radius of the Sun could equally well be written as $R_\odot$ = 6.96×10^8 m, or 6.96×10^{10} cm, or 109 $R_\oplus$, or 4.65×10^{-3} AU, or even 7.36×10^{-8} ly—whichever happens to be most useful. Some of the more common units used in astronomy, and the contexts in which they are most likely to be encountered, are listed below.

Length:

1 angstrom (Å)	= 10^{-10} m	atomic physics,
1 nanometer (nm)	= 10^{-9} m	spectroscopy
1 micron (mm)	= 10^{-6} m	interstellar dust and gas
1 centimeter (cm)	= 0.01 m	in widespread use
1 meter (m)	= 100 cm	throughout all
1 kilometer (km)	= 1000 m = 10^5 cm	astronomy
Earth radius ($R_\oplus$)	= 6378 km	planetary astronomy
Solar radius ($R_\odot$)	= 6.96×10^8 m	solar system,
1 astronomical unit (AU)	= 1.496×10^{11} m	stellar evolution
1 light-year (ly)	= 9.46×10^{15} m = 63,200 AU	galactic astronomy,
1 parsec (pc)	= 3.09×10^{16} m = 206,000 AU = 3.26 ly	stars and star clusters
1 kiloparsec (kpc)	= 1000 pc	galaxies, galaxy
1 megaparsec (Mpc)	= 1000 kpc	clusters, cosmology

Mass:

1 gram (g)		in widespread use in
1 kilogram (kg)	= 1000 g	many different areas
Earth mass ($M_\oplus$)	= 5.98×10^{24} kg	planetary astronomy
Solar mass ($M_\odot$)	= 1.99×10^{30} kg	"standard" unit for all mass scales larger than Earth

Time:

1 second (s)		in widespread use throughout astronomy
1 hour (h)	= 3600 s	planetary and stellar
1 day (d)	= 86,400 s	scales
1 year (yr)	= 3.16×10^7 s	virtually all processes occurring on scales larger than a star

Appendix 3
Tables

TABLE 1 Some Useful Constants and Physical Measurements[1]

astronomical unit	$1 \text{ AU} = 1.496 \times 10^8 \text{ km} (1.5 \times 10^8 \text{ km})$
light-year	$1 \text{ ly} = 9.46 \times 10^{12} \text{ km} (10^{13} \text{ km}; \text{ about 6 trillion miles})$
parsec	$1 \text{ pc} = 3.09 \times 10^{13} \text{ km} = 3.3 \text{ ly}$
speed of light	$c = 299{,}792.458 \text{ km/s} (3 \times 10^5 \text{ km/s})$
Stefan–Boltzmann Constant	$\sigma = 5.67 \times 10^{-8} \text{ W/m}^2 \cdot \text{K}^4$
gravitational constant	$G = 6.67 \times 10^{-11} \text{ N m}^2/\text{kg}^2$
mass of Earth	$M_\oplus = 5.97 \times 10^{24} \text{ kg} (6 \times 10^{24} \text{ kg}; 6000 \text{ billion billion tons})$
radius of Earth	$R_\oplus = 6378 \text{ km} (6500 \text{ km})$
mass of the Sun	$M_\odot = 1.99 \times 10^{30} \text{ kg} (2 \times 10^{30} \text{ kg})$
radius of the Sun	$R_\odot = 6.96 \times 10^5 \text{ km} (7 \times 10^5 \text{ km})$
luminosity of the Sun	$L_\odot = 3.90 \times 10^{26} \text{ W} (4 \times 10^{26} \text{ W})$
effective temperature of the Sun	$T_\odot = 5778 \text{ K} (5800 \text{ K})$
Hubble constant	$H_0 \approx 70 \text{ km/s/Mpc}$
mass of an electron	$m_e = 9.11 \times 10^{-31} \text{ kg}$
mass of a proton	$m_p = 1.67 \times 10^{-27} \text{ kg}$

[1] The rounded-off values used in the text are shown above in parentheses.

CONVERSIONS BETWEEN COMMON ENGLISH AND METRIC UNITS

ENGLISH	METRIC
1 inch	= 2.54 centimeters (cm)
1 foot (ft)	= 0.3048 meters (m)
1 mile	= 1.609 kilometers (km)
1 pound (lb)	= 453.6 grams (g) or 0.4536 kilograms (kg) (on Earth)

TABLE 2A   Planetary Data: Orbital Properties

PLANET	SEMI-MAJOR AXIS (AU)	SEMI-MAJOR AXIS (10^6 km)	SIDEREAL PERIOD (tropical years)	MEAN ORBITAL SPEED (km/s)	ORBITAL ECCENTRICITY	INCLINATION TO THE ECLIPTIC (degrees)
Mercury	0.39	57.9	0.241	47.9	0.206	7.00
Venus	0.72	108.2	0.612	35.0	0.007	3.39
Earth	1.00	149.6	1.00	29.8	0.017	0.01
Mars	1.52	227.9	1.88	24.1	0.093	1.85
Jupiter	5.20	778.4	11.86	13.1	0.048	1.31
Saturn	9.54	1427	29.42	9.65	0.054	2.49
Uranus	19.19	2871	83.75	6.80	0.047	0.77
Neptune	30.07	4498	163.7	5.43	0.009	1.77

TABLE 2B   Planetary Data: Physical Properties

PLANET	EQUATORIAL RADIUS (km)	EQUATORIAL RADIUS (Earth = 1)	MASS (kg)	MASS (Earth = 1)	MEAN DENSITY (kg/m^3)	SIDEREAL ROTATION PERIOD (solar days)[1]
Mercury	2440	0.38	3.30×10^{23}	0.055	5430	58.6
Venus	6052	0.95	4.87×10^{24}	0.82	5240	−243.0
Earth	6378	1.00	5.97×10^{24}	1.00	5520	0.9973
Mars	3394	0.53	6.42×10^{23}	0.11	3930	1.026
Jupiter	71,492	11.21	1.90×10^{27}	317.8	1330	0.41
Saturn	60,268	9.45	5.68×10^{26}	95.2	690	0.44
Uranus	25,559	4.01	8.68×10^{25}	14.5	1270	−0.72
Neptune	24,766	3.88	1.02×10^{26}	17.1	1640	0.67

PLANET	AXIAL TILT (degrees)	SURFACE GRAVITY (Earth = 1)	ESCAPE SPEED (km/s)	SURFACE TEMPERATURE[2] (K)	NUMBER OF MOONS[3]
Mercury	0.0	0.38	4.2	100 to 700	0
Venus	177.4	0.91	10.4	730	0
Earth	23.5	1.00	11.2	290	1
Mars	24.0	0.38	5.0	180 to 270	2
Jupiter	3.1	2.53	60	124	16
Saturn	26.7	1.07	36	97	18
Uranus	97.9	0.91	21	58	27
Neptune	29.6	1.14	24	59	13

[1]A negative sign indicates retrograde rotation.
[2]Temperature is effective temperature for jovian planets.
[3]Moons more than 10 km in diameter

TABLE 3   The Twenty Brightest Stars in Earth's Night Sky

NAME	STAR	SPECTRAL TYPE* A	B	PARALLAX (arc seconds)	DISTANCE (pc)	APPARENT VISUAL MAGNITUDE* A	B
Sirius	αCMa	A1V	wd†	0.379	2.6	−1.44	+8.4
Canopus	αCar	F0Ib–II		0.010	96	−0.62	
Arcturus	αBoo	K2III		0.089	11	−0.05	
Rigel Kentaurus (Alpha Centauri)	αCen	G2V	K0V	0.742	1.3	−0.01	+1.4
Vega	αLyr	A0V		0.129	7.8	+0.03	
Capella	αAur	GIII	M1V	0.077	13	+0.08	+10.2
Rigel	βOri	B8Ia	B9	0.0042	240	+0.18	+6.6
Procyon	αCMi	F5IV–V	Wd†	0.286	3.5	+0.40	+10.7
Betelgeuse	αOri	M2Iab		0.0076	130	+0.45	
Achernar	αEri	B5V		0.023	44	+0.45	
Hadar	βCen	B1III	?	0.0062	160	+0.61	+4
Altair	αAql	A7IV–V		0.194	5.1	+0.76	
Acrux	αCru	B1IV	B3	0.010	98	+0.77	+1.9
Aldebaran	αTau	K5III	M2V	0.050	20	+0.87	+13
Spica	αVir	B1V	B2V	0.012	80	+0.98	2.1
Antares	αSco	M1Ib	B4V	0.005	190	+1.06	+5.1
Pollux	βGem	K0III		0.097	10	+1.16	
Formalhaut	αPsA	A3V	?	0.130	7.7	+1.17	+6.5
Deneb	αCyg	A2Ia		0.0010	990	+1.25	
Mimosa	βCru	B1IV		0.0093	110	+1.25	

NAME	VISUAL LUMINOSITY* (Sun = 1) A	B	ABSOLUTE VISUAL MAGNITUDE[1] A	B	PROPER MOTION (arc seconds/yr)	TRANSVERSE VELOCITY (km/s)	RADIAL VELOCITY (km/s)
Sirius	22	0.0025	+1.5	+11.3	1.33	16.7	−7.6‡
Canopus	1.4×10^4		−5.5		0.02	9.1	20.5
Arcturus	110		−0.3		2.28	119	−5.2
Rigel Kentaurus	1.6	0.45	+4.3	+5.7	3.68	22.7	−24.6
Vega	50		+0.6		0.34	12.6	−13.9
Capella	130	0.01	−0.5	+9.6	0.44	27.1	30.2‡
Rigel	4.1×10^4	110	−6.7	−0.3	0.00	1.2	20.7
Procyon	7.2	0.0006	+2.7	+13.0	1.25	20.7	−3.2
Betelgeuse	9700		−5.1		0.03	18.5	21.0
Achernar	1100		−2.8		0.10	20.9	19
Hadar	1.3×10^4	560	−5.4	−2.0	0.04	30.3	−12
Altair	11		+2.2		0.66	16.3	−26.3
Acrux	4100	2200	−4.2	−3.5	0.04	22.8	−11.2
Aldebaran	150	0.002	−0.6	+11.5	0.20	19.0	54.1
Spica	2200	780	−3.5	−2.4	0.05	19.0	1.0
Antares	1.1×10^4	290	−5.3	−1.3	0.03	27.0	−3.2
Pollux	31		+1.1		0.62	29.4	3.3
Formalhaut	17	0.13	+1.7	+7.1	0.37	13.5	6.5
Deneb	2.6×10^5		−8.7		0.003	14.1	−4.6
Mimosa	3200		−3.9		0.05	26.1	—

* Energy output in the visible part of the spectrum; A and B columns identify individual components of binary-star systems.

† "wd" stands for "white dwarf."

‡ Average value of variable velocity

TABLE 4   The Twenty Nearest Stars

NAME	SPECTRAL TYPE		PARALLAX (arc seconds)	DISTANCE (pc)	APPARENT VISUAL MAGNITUDE*	
	A	B			A	B
Sun	G2V				−26.74	
Proxima Centauri	M5		0.772	1.30	+11.01	
Alpha Centauri	G2V	K1V	0.742	1.35	−0.01	+1.35
Barnard's Star	M5V		0.549	1.82	+9.54	
Wolf 359	M8V		0.421	2.38	+13.53	
Lalande 21185	M2V		0.397	2.52	+7.50	
UV Ceti	M6V	M6V	0.387	2.58	+12.52	+13.02
Sirius	A1V	wd†	0.379	2.64	−1.44	+8.4
Ross 154	M5V		0.345	2.90	+10.45	
Ross 248	M6V		0.314	3.18	+12.29	
ε Eridani	K2V		0.311	3.22	+3.72	
Ross 128	M5V		0.298	3.36	+11.10	
61 Cygni	K5V	K7V	0.294	3.40	+5.22	+6.03
ε Indi	K5V		0.291	3.44	+4.68	
Grm 34	M1V	M6V	0.290	3.45	+8.08	+11.06
Luyten 789-6	M6V		0.290	3.45	+12.18	
Procyon	F5IV–V	wd†	0.286	3.50	+0.40	+10.7
Σ 2398	M4V	M5V	0.285	3.55	+8.90	+9.69
Lacaille 9352	M2V		0.279	3.58	+7.35	
G51-15	MV		0.278	3.60	+14.81	

NAME	VISUAL LUMINOSITY* (Sun = 1)		ABSOLUTE VISUAL MAGNITUDE*		PROPER MOTION (arc seconds/yr)	TRANSVERSE VELOCITY (km/s)	RADIAL VELOCITY (km/s)
	A	B	A	B			
Sun	1.0		+4.83				
Proxima Centauri	5.6×10^{-5}		+15.4		3.86	23.8	−16
Alpha Centauri	1.6	0.45	+4.3	+5.7	3.68	23.2	−22
Barnard's Star	4.3×10^{-4}		+13.2		10.34	89.7	−108
Wolf 359	1.8×10^{-5}		+16.7		4.70	53.0	+13
Lalande 21185	0.0055		+10.5		4.78	57.1	−84
UV Ceti	5.4×10^{-5}	0.00004	+15.5	+16.0	3.36	41.1	+30
Sirius	22	0.0025	+1.5	+11.3	1.33	16.7	−8
Ross 154	4.8×10^{-4}		+13.3		0.72	9.9	−4
Ross 248	1.1×10^{-4}		+14.8		1.58	23.8	−81
ε Eridani	0.29		+6.2		0.98	15.3	+16
Ross 128	3.6×10^{-4}		+13.5		1.37	21.8	−13
61 Cygni	0.082	0.039	+7.6	+8.4	5.22	84.1	−64
ε Indi	0.14		+7.0		4.69	76.5	−40
Grm 34	0.0061	0.00039	+10.4	+13.4	2.89	47.3	+17
Luyten 789-6	1.4×10^{-4}		+14.6		3.26	53.3	−60
Procyon	7.2	0.00055	+2.7	+13.0	1.25	2.8	−3
Σ 2398	0.0030	0.0015	+11.2	+11.9	2.28	38.4	+5
Lacaille 9352	0.013		+9.6		6.90	117	+10
G51-15	1.4×10^{-5}		+17.0		1.26	21.5	—

* A and B columns identify individual components of binary-star systems.

† "wd" stands for "white dwarf."

Glossary

Key terms that are boldface in the text are followed by a page reference in the Glossary.

A

A ring One of three Saturnian rings visible from Earth. The A ring is farthest from the planet and is separated from the B ring by the Cassini Division. (p. 226)

aberration of starlight Small shift in the observed direction to a star, caused by Earth's motion perpendicular to the line of sight.

absolute brightness The apparent brightness a star would have if it were placed at a standard distance of 10 parsecs from Earth.

absolute magnitude The apparent magnitude a star would have if it were placed at a standard distance of 10 parsecs from Earth. (p. 271)

Absolute Zero The lowest possible temperature that can be obtained; all thermal motion ceases at this temperature.

absorption line Dark line in an otherwise continuous bright spectrum, where light within one narrow frequency range has been removed. (p. 56)

abundance Relative amounts of different elements in a gas.

acceleration The rate of change of velocity of a moving object. (p. 35)

accretion Gradual growth of bodies, such as planets, by the accumulation of other, smaller bodies. (p. 121)

accretion disk Flat disk of matter spiraling down onto the surface of a neutron star or black hole. Often, the matter originated on the surface of a companion star in a binary-star system. (p. 327)

active galactic nucleus Region of intense emission at the center of an active galaxy, responsible for virtually all of the galaxy's nonstellar luminosity. (p. 415)

active galaxies The most energetic galaxies, which can emit hundreds or thousands of times more energy per second than the Milky Way, mostly in the form of long-wavelength nonthermal radiation. (p. 415)

active optics Collection of techniques used to increase the resolution of ground-based telescopes. Minute modifications are made to the overall configuration of an instrument as its temperature and orientation change; used to maintain the best possible focus at all times. (p. 81)

active region Region of the photosphere of the Sun surrounding a sunspot group, which can erupt violently and unpredictably. During sunspot maximum, the number of active regions is also a maximum. (p. 254)

active Sun The unpredictable aspects of the Sun's behavior, such as sudden explosive outbursts of radiation in the form of prominences and flares.

adaptive optics Technique used to increase the resolution of a telescope by deforming the shape of the mirror's surface under computer control while a measurement is being taken; used to undo the effects of atmospheric turbulence. (p. 82)

aerosol Suspension of liquid or solid particles in air.

alpha particle A helium-4 nucleus.

alpha process Process occurring at high temperatures, in which high-energy photons split heavy nuclei to form helium nuclei.

ALSEP Acronym for Apollo Lunar Surface Experiments Package.

altimeter Instrument used to determine altitude.

amino acids Organic molecules that form the basis for building the proteins that direct metabolism in living creatures. (p. 485)

amor asteroid Asteroid that crosses only the orbit of Mars.

amplitude The maximum deviation of a wave above or below zero point. (p. 45)

angstrom Distance unit equal to 0.1 nanometers, or one tenbillionth of a meter.

angular diameter Angle made between the top (or one edge) of an object, the observer and the bottom (or opposite edge) of the object.

angular distance Angular separation between two objects as seen by some observer.

angular momentum Tendency of an object to keep rotating; proportional to the mass, radius, and rotation speed of the body.

angular resolution The ability of a telescope to distinguish between adjacent objects in the sky. (p. 79)

annular eclipse Solar eclipse occurring at a time when the Moon is far enough away from Earth that it fails to cover the disk of the Sun completely, leaving a ring of sunlight visible around its edge. (p. 16)

antiparallel Configuration of the electron and proton in a hydrogen (or other) atom when their spin axes are parallel but the two rotate in opposite directions.

antiparticle A particle of the same mass but opposite in all other respects (e.g., charge) to a given particle; when a particle and its antiparticle come into contact, they annihilate and release energy in the form of gamma rays.

aphelion The point on the elliptical path of an object in orbit about the Sun that is most distant from the Sun. (p. 32)

Apollo asteroid *See* Earth-crossing asteroid.

apparent brightness The brightness that a star appears to have, as measured by an observer on Earth. (p. 268)

apparent magnitude The apparent brightness of a star, expressed using the magnitude scale. (p. 270)

association Small grouping of (typically 100 or less) bright stars, spanning up to a few tens of parsecs across, usually rich in very young stars. (p. 310)

Assumption of Mediocrity Statements suggesting that the development of life on Earth did not require any unusual circumstances, suggesting that extraterrestrial life may be common.

asteroid One of thousands of very small members of the solar system orbiting the Sun between the orbits of Mars and Jupiter. Often referred to as "minor planets." (p. 103)

asteroid belt Region of the solar system, between the orbits of Mars and Jupiter, in which most asteroids are found. (p. 104)

asthenosphere Layer of Earth's interior, just below the lithosphere, over which the surface plates slide.

astrology Pseudoscience that purports to use the positions of the planets, sun, and moon to predict daily events and human destiny.

astronomical unit (AU) The average distance of Earth from the Sun. Precise radar measurements yield a value for the AU of 149,603,500 km. (p. 33)

astronomy Branch of science dedicated to the study of everything in the universe that lies above Earth's atmosphere. (p. 4)

asymptotic-giant branch Path on the Hertzsprung–Russell diagram corresponding to the changes that a star undergoes after helium burning ceases in the core. At this stage, the carbon core shrinks and drives the expansion of the envelope, and the star becomes a swollen red giant for a second time.

aten asteroid Earth-crossing asteroid with semimajor axis less than 1 AU.

atmosphere Layer of gas confined close to a planet's surface by the force of gravity. (p. 134)

atom Building block of matter, composed of positively charged protons and neutral neutrons in the nucleus surrounded by negatively charged electrons. (p. 58)

atomic epoch Period after decoupling when the first simple atoms and molecules formed.

aurora Event that occurs when atmospheric molecules are excited by incoming charged particles from the solar wind, then emit energy as they fall back to their ground states. Aurorae generally occur at high latitudes, near the north and south magnetic poles. (p. 154)

autumnal equinox Date on which the Sun crosses the celestial equator moving southward, occurring on or near September 22. (p. 11)

B

B ring One of three Saturnian rings visible from Earth. The B ring is the brightest of the three, and lies just past the Cassini Division, closer to the planet than the A ring. (p. 226)

background noise Unwanted light in an image, from unresolved sources in the telescope's field of view, scattered light from the atmosphere, or instrumental "hiss" in the detector itself.

barred-spiral galaxy Spiral galaxy in which a bar of material passes through the center of the galaxy, with the spiral arms beginning near the ends of the bar. (p. 403)

basalt Solidified lava; an iron-magnesium-silicate mixture.

baseline The distance between two observing locations used for the purposes of triangulation measurements. The larger the baseline, the better the resolution attainable. (p. 17)

belt Dark, low-pressure region in the atmosphere of a jovian planet, where gas flows downward.

Big Bang Event that cosmologists consider the beginning of the universe, in which all matter and radiation in the entire universe came into being. (p. 458)

Big Crunch Point of final collapse of a bound universe.

binary asteroid Asteroids that have a partner in orbit around it.

binary pulsar Binary system in which both components are pulsars.

binary-star system A system that consists of two stars in orbit about their common center of mass, held together by their mutual gravitational attraction. Most stars are found in binary-star systems. (p. 281)

biological evolution Change in a population of biological organisms over time.

bipolar flow Jets of material expelled from a protostar perpendicular to the surrounding protostellar disk.

blackbody curve The characteristic way in which the intensity of radiation emitted by a hot object depends on frequency. The frequency at which the emitted intensity is highest is an indication of the temperature of the radiating object. Also referred to as the Planck curve. (p. 52)

black dwarf The endpoint of the evolution of an isolated, low-mass star. After the white-dwarf stage, the star cools to the point where it is a dark "clinker" in interstellar space. (p. 326)

black hole A region of space where the pull of gravity is so great that nothing—not even light—can escape. A possible outcome of the evolution of a very massive star. (p. 357)

blazar Particularly intense active galactic nucleus in which the observer's line of sight happens to lie directly along the axis of a high-speed jet of particles emitted from the active region. (p. 420)

blue giant Large, hot, bright star at the upper-left end of the main sequence on the Hertzsprung–Russell diagram. Its name comes from its color and size. (p. 277)

blueshift Motion-induced changes in the observed wavelength from a source that is moving toward us. Relative approaching motion between the object and the observer causes the wavelength to appear shorter (and hence bluer) than if there were no motion at all.

blue straggler Star found on the main sequence of the Hertzsprung–Russell diagram, but which should already have evolved off the main sequence, given its location on the diagram; thought to have formed from mergers of lower mass stars.

blue supergiant The very largest of the large, hot, bright stars at the uppermost-left end of the main sequence on the Hertzsprung–Russell diagram. (p. 277)

Bohr model First theory of the hydrogen atom to explain the observed spectral lines. This model rests on three ideas: that there is a state of lowest energy for the electron, that there is a maximum energy beyond which the electron is no longer bound to the nucleus, and that within these two energies the electron can only exist in certain energy levels. (p. 58)

bok globule Dense, compact cloud of interstellar dust and gas on its way to forming one or more stars.

boson Particle that exerts or mediates forces between elementary particles in quantum physics.

bound trajectory Path of an object with launch speed low enough that it cannot escape the gravitational pull of a planet.

brown dwarf Remnants of fragments of collapsing gas and dust that did not contain enough mass to initiate core nuclear fusion. Such objects are then frozen somewhere along their pre-main-sequence contraction phase, continually cooling into compact dark objects. Because of their small size and low temperature they are extremely difficult to detect observationally. (p. 309)

brown oval Feature of Jupiter's atmosphere that appears only at latitudes near 20 degrees N, this structure is a long-lived hole in the clouds that allows us to look down into Jupiter's lower atmosphere. (p. 200)

C

C ring One of three Saturnian rings visible from Earth. The C ring lies closest to the planet and is relatively thin compared to the A and B rings. (p. 226)

caldera Crater that forms at the summit of a volcano.

capture theory (Moon) Theory suggesting that the Moon formed far from Earth but was later captured by it.

carbonaceous asteroid The darkest, or least reflective, type of asteroid, containing large amounts of carbon.

carbon-based molecule Molecule containing atoms of carbon.

carbon-detonation supernova *See* Type I supernova.

cascade Process of deexcitation in which an excited electron moves down through energy states one at a time.

Cassegrain telescope A type of reflecting telescope in which incoming light hits the primary mirror and is then reflected upward toward the prime focus, where a secondary mirror reflects the light back down through a small hole in the main mirror into a detector or eyepiece. (p. 72)

Cassini Division A relatively empty gap in Saturn's ring system, discovered in 1675 by Giovanni Cassini. It is now known to contain a number of thin ringlets. (p. 226)

cataclysmic variable Collective name for novae and supernovae.

catalyst Something that causes or helps a reaction to occur, but is not itself consumed as part of the reaction.

catastrophic theory A theory that invokes statistically unlikely accidental events to account for observations.

celestial coordinates Pair of quantities—right ascension and declination—similar to longitude and latitude on Earth, used to pinpoint locations of objects on the celestial sphere. (p. 7)

celestial equator The projection of Earth's equator onto the celestial sphere. (p. 6)

celestial mechanics Study of the motions of bodies, such as planets and stars, that interact via gravity.

celestial pole Projection of Earth's North or South pole onto the celestial sphere.

celestial sphere Imaginary sphere surrounding Earth to which all objects in the sky were once considered to be attached. (p. 5)

Celsius Temperature scale in which the freezing point of water is 0 degrees and the boiling point of water is 100 degrees.

center of mass The "average" position in space of a collection of massive bodies, weighted by their masses. For an isolated system this point moves with constant velocity, according to Newtonian mechanics. (p. 37)

centigrade *See* Celsius.

centripetal force (literally "center seeking") Force directed toward the center of a body's orbit.

centroid Average position of the material in an object; in spectroscopy, the center of a spectral line.

Cepheid variable Star whose luminosity varies in a characteristic way, with a rapid rise in brightness followed by a slower decline. The period of a Cepheid variable star is related to its luminosity, so a determination of this period can be used to obtain an estimate of the star's distance. (p. 378)

chandrasekhar mass Maximum possible mass of a white dwarf.

chaotic rotation Unpredictable tumbling motion that nonspherical bodies in eccentric orbits, such as Saturn's satellite Hyperion, can exhibit. No amount of observation of an object rotating chaotically will ever show a well-defined period.

charge-coupled device (CCD) An electronic device used for data acquisition; composed of many tiny pixels, each of which records a buildup of charge to measure the amount of light striking it. (p. 76)

chemical bond Force holding atoms together to form a molecule.

chemosynthesis Analog of photosynthesis that operates in total darkness.

chromatic aberration The tendency for a lens to focus red and blue light differently, causing images to become blurred.

chromosphere The Sun's lower atmosphere, lying just above the visible atmosphere. (p. 242)

circumnavigation Traveling all the way around an object.

cirrus High-level clouds composed of ice or methane crystals.

closed universe Geometry that the universe as a whole would have if the density of matter is above the critical value. A closed universe is finite in extent and has no edge, like the surface of a sphere. It has enough mass to stop the present expansion and will eventually collapse. (p. 462)

CNO cycle Chain of reactions that converts hydrogen into helium using carbon, nitrogen, and oxygen as catalysts.

cocoon nebula Bright infrared source in which a surrounding cloud of gas and dust absorb ultraviolet radiation from a hot star and reemits it in the infrared.

cold dark matter Class of dark-matter candidates made up of very heavy particles, possibly formed in the very early universe.

collecting area The total area of a telescope capable of capturing incoming radiation. The larger the telescope, the greater its collecting area, and the fainter the objects it can detect. (p. 77)

collisional broadening Broadening of spectral lines due to collisions between atoms, most often seen in dense gases.

color index A convenient method of quantifying a star's color by comparing its apparent brightness as measured through different filters. If the star's radiation is well described by a blackbody spectrum, the ratio of its blue intensity (B) to its visual intensity (V) is a measure of the object's surface temperature.

color–magnitude diagram A way of plotting stellar properties, in which absolute magnitude is plotted against color index.

coma An effect occurring during the formation of an off-axis image in a telescope. Stars whose light enters the telescope at a large angle acquire cometlike tails on their images. The brightest part of a comet, often referred to as the "head." (p. 108)

comet A small body, composed mainly of ice and dust, in an elliptical orbit about the Sun. As it comes close to the Sun, some of its material is vaporized to form a gaseous head and extended tail. (p. 103)

common envelope Outer layer of gas in a contact binary.

comparative planetology The systematic study of the similarities and differences among the planets, with the goal of obtaining deeper insight into how the solar system formed and has evolved in time.

composition The mixture of atoms and molecules that make up an object.

condensation nuclei Dust grains in the interstellar medium which act as seeds around which other material can cluster. The presence of dust was very important in causing matter to clump during the formation of the solar system. (p. 120)

condensation theory Currently favored model of solar system formation which combines features of the old nebular theory with new information about interstellar dust grains, which acted as condensation nuclei. (p. 120)

conjunction Orbital configuration in which a planet lies in the same direction as the Sun, as seen from Earth.

conservation of mass and energy *See* law of conservation of mass and energy.

constellation A human grouping of stars in the night sky into a recognizable pattern. (p. 4)

constituents *See* composition.

contact binary A binary-star system in which both stars have expanded to fill their Roche lobes and the surfaces of the two stars merge. The binary system now consists of two nuclear burning stellar cores surrounded by a continuous common envelope.

continental drift The movement of the continents around Earth's surface.

continuous spectra Spectra in which the radiation is distributed over all frequencies, not just a few specific frequency ranges. A prime example is the blackbody radiation emitted by a hot, dense body. (p. 55)

convection Churning motion resulting from the constant upwelling of warm fluid and the concurrent downward flow of cooler material to take its place. (p. 139)

convection zone Region of the Sun's interior, lying just below the surface, where the material of the Sun is in constant convection motion. This region extends into the solar interior to a depth of about 20,000 km. (p. 242)

co-orbital satellites Satellites sharing the same orbit around a planet.

Copernican Principle The removal of Earth from any position of cosmic significance.

Copernican revolution The realization, toward the end of the sixteenth century, that Earth is not at the center of the universe. (p. 28)

core The central region of Earth, surrounded by the mantle. (p. 134); The central region of any planet or star. (p. 242)

core-accretion theory Theory that the jovian planets formed when icy protoplanetary cores became massive enough to capture gas directly from the solar nebula. *See* gravitational instability theory.

core-collapse supernova *See* Type II supernova.

core hydrogen burning The energy burning stage for mainsequence stars, in which the helium is produced by hydrogen fusion in the central region of the star. A typical star spends up to 90 percent of its lifetime in hydrostatic equilibrium brought about by the balance between gravity and the energy generated by core hydrogen burning. (p. 318)

cornea (eye) The curved transparent layer covering the front part of the eye.

corona One of numerous large, roughly circular regions on the surface of Venus, thought to have been caused by upwelling mantle material causing the planet's crust to bulge outward (plural, *coronae*). (p. 170)

corona The tenuous outer atmosphere of the Sun, which lies just above the chromosphere and, at great distances, turns into the solar wind. (p. 242)

coronal hole Vast regions of the Sun's atmosphere where the density of matter is about 10 times lower than average. The gas there streams freely into space at high speeds, escaping the Sun completely. (p. 255)

coronal mass ejection Giant magnetic "bubble" of ionized gas that separates from the rest of the solar atmosphere and escapes into interplanetary space. (p. 255)

corpuscular theory Early particle theory of light.

cosmic density parameter Ratio of the universe's actual density to the critical value corresponding to zero curvature.

cosmic distance scale Collection of indirect distance-measurement techniques that astronomers use to measure distances in the universe. (p. 17)

cosmic evolution The collection of the seven major phases of the history of the universe; namely particulate, galactic, stellar, planetary, chemical, biological, and cultural evolution. (p. 484)

cosmic microwave background The almost perfectly isotropic radio signal that is the electromagnetic remnant of the Big Bang. (p. 469)

cosmic ray Very energetic subatomic particle arriving at Earth from elsewhere in the Galaxy.

cosmological constant Quantity originally introduced by Einstein into general relativity to make his equations describe a static universe. Now one of several candidates for the repulsive "dark energy" force responsible for the observed cosmic acceleration. (p. 466)

cosmological distance Distance comparable to the scale of the universe.

cosmological principle Two assumptions that make up the basis of cosmology, namely that the universe is homogeneous and isotropic on sufficiently large scales. (p. 457)

cosmological redshift The component of the redshift of an object that is due only to the Hubble flow of the universe. (p. 413)

cosmology The study of the structure and evolution of the entire universe. (p. 467)

cosmos The universe.

coudé focus Focus produced far from the telescope using a series of mirrors. Allows the use of heavy and/or finely tuned equipment to analyze the image.

crater Bowl-shaped depression on the surface of a planet or moon, resulting from a collision with interplanetary debris. (p. 150)

crescent Appearance of the Moon (or a planet) when less than half of the body's hemisphere is visible from Earth.

crest Maximum departure of a wave above its undisturbed state.

critical density The cosmic density corresponding to the dividing line between a universe that recollapses and one that expands forever. (p. 461)

critical universe Universe in which the density of matter is exactly equal to the critical density. The universe is infinite in extent and has zero curvature. The expansion will continue forever, but will approach an expansion speed of zero. (p. 462)

crust Layer of Earth which contains the solid continents and the seafloor. (p. 134)

C-type asteroid *See* carbonaceous asteroid.

cultural evolution Change in the ideas and behavior of a society over time.

current sheet Flat sheet on Jupiter's magnetic equator where most of the charged particles in the magnetosphere lie due to the planet's rapid rotation.

D

D ring Collection of very faint, thin rings, extending from the inner edge of the C ring down nearly to the cloud tops of Saturn. This region contains so few particles that it is completely invisible from Earth. (p. 229)

dark dust cloud A large cloud, often many parsecs across, which contains gas and dust in a ratio of about 1012 gas atoms for every dust particle. Typical densities are a few tens or hundreds of millions of particles per cubic meter. (p. 297)

dark energy Generic name given to the unknown cosmic force field thought to be responsible for the observed acceleration of the Hubble expansion. (p. 466)

dark halo Region of a galaxy beyond the visible halo where dark matter is believed to reside. (p. 391)

dark matter Term used to describe the mass in galaxies and clusters whose existence we infer from rotation curves and other techniques, but that has not been confirmed by observations at any electromagnetic wavelength. (p. 391)

dark matter particle Particle undetectable at any electromagnetic wavelength, but can be inferred from its gravitational influence.

Daughter/Fission theory Theory suggesting that the Moon originated out of Earth.

declination Celestial coordinate used to measure latitude above or below the celestial equator on the celestial sphere. (p. 7)

decoupling Event in the early universe when atoms first formed, after which photons could propagate freely through space. (p. 437)

deferent A construct of the geocentric model of the solar system which was needed to explain observed planetary motions. A deferent is a large circle encircling Earth, on which an epicycle moves. (p. 27)

degree Unit of angular measure. There are 360 degrees in one complete circle.

density A measure of the compactness of the matter within an object, computed by dividing the mass of the object by its volume. Units are kilograms per cubic meter (kg/m^3), or grams per cubic centimeter (g/cm^3). (p. 101)

detached binary Binary system where each star lies within its respective Roche lobe.

detector noise Readings produced by an instrument even when it is not observing anything; produced by the electronic components within the detector itself.

deuterium A form of hydrogen with an extra neutron in its nucleus.

deuterium bottleneck Period in the early universe between the start of deuterium production and the time when the universe was cool enough for deuterium to survive.

deuteron An isotope of hydrogen in which a neutron is bound to the proton in the nucleus. Often called "heavy hydrogen" because of the extra mass of the neutron. (p. 258)

differential rotation The tendency for a gaseous sphere, such as a jovian planet or the Sun, to rotate at a different rate at the equator than at the poles. More generally, a condition where the angular speed varies with location within an object. (p. 196)

differentiation Variation in the density and composition of a body, such as Earth, with low density material on the surface and higher density material in the core. (p. 145)

diffraction The ability of waves to bend around corners. The diffraction of light establishes its wave nature. (p. 46)

diffraction grating Sheet of transparent material with many closely spaced parallel lines ruled on it, designed to separate white light into a spectrum.

diffraction-limited resolution Theoretical resolution that a telescope can have due to diffraction of light at the telescope's aperture. Depends on the wavelength of radiation and the diameter of the telescope's mirror.

direct motion *See* prograde motion.

distance modulus Difference between the apparent and absolute magnitude of an object; equivalent to distance, by the inverse-square law.

diurnal motion Apparent daily motion of the stars, caused by Earth's rotation.

DNA Deoxyribonucleic acid, the molecule that carries genetic information and determine the characteristics of a living organism.

Doppler effect Any motion-induced change in the observed wavelength (or frequency) of a wave. (p. 63)

double-line spectroscopic binary Binary system in which spectral lines of both stars can be distinguished and seen to shift back and forth as the stars orbit one another.

double-star system System containing two stars in orbit around one another.

Drake equation Expression that gives an estimate of the probability that intelligence exists elsewhere in the galaxy, based on a number of supposedly necessary conditions for intelligent life to develop. (p. 490)

dust grain An interstellar dust particle, roughly 10^{-7} m in size, comparable to the wavelength of visible light. (p. 29)

dust lane A lane of dark, obscuring interstellar dust in an emission nebula or galaxy. (p. 295)

dust tail The component of a comet's tail that is composed of dust particles. (p. 109)

dwarf Any star with radius comparable to, or smaller than, that of the Sun (including the Sun itself). (p. 276)

dwarf elliptical Elliptical galaxy as small as 1 kiloparsec across, containing only a few million stars.

dwarf galaxy Small galaxy containing a few million stars.

dwarf irregular Small irregular galaxy containing only a few million stars.

dynamo theory Theory that explains planetary and stellar magnetic fields in terms of rotating, conducting material flowing in an object's interior.

E

E ring A faint ring, well outside the main ring system of Saturn, which was discovered by *Voyager* and is believed to be associated with volcanism on the moon Enceladus. (p. 229)

Earth-asteroid An asteroid whose orbit crosses that of Earth. Earth-crossing asteroids are also called Apollo asteroids, after the first asteroid of this type discovered. (p. 104)

earthquake A sudden dislocation of rocky material near Earth's surface. (p. 143)

eccentricity A measure of the flatness of an ellipse, equal to the distance between the two foci divided by the length of the major axis. (p. 32)

eclipse Event during which one body passes in front of another, so that the light from the occulted body is blocked. (p. 14)

eclipse season Times of the year when the Moon lies in the same plane as Earth and Sun, so that eclipses are possible.

eclipse year Time interval between successive orbital configurations in which the line of nodes of the Moon's orbit points toward the Sun.

eclipsing binary Rare binary-star system that is aligned in such a way that from Earth we observe one star pass in front of the other, eclipsing the other star. (p. 282)

ecliptic The apparent path of the Sun, relative to the stars on the celestial sphere, over the course of a year. (p. 9)

effective temperature Temperature of a blackbody of the same radius and luminosity as a given star or planet.

ejecta (planetary) Material thrown outward by a meteoroid impact.

ejecta (stellar) Material thrown into space by a nova or supernova.

electric field A field extending outward in all directions from a charged particle, such as a proton or an electron. The electric field determines the electric force exerted by the particle on all other charged particles in the universe; the strength of the electric field decreases with increasing distance from the charge according to an inverse-square law. (p. 46)

electromagnetic energy Energy carried in the form of rapidly fluctuating electric and magnetic fields.

electromagnetic radiation Another term for light, electromagnetic radiation transfers energy and information from one place to another. (p. 44)

electromagnetic spectrum The complete range of electromagnetic radiation, from radio waves to gamma rays, including the visible spectrum. All types of electromagnetic radiation are basically the same phenomenon, differing only by wavelength, and all move at the speed of light. (p. 50)

electromagnetism The union of electricity and magnetism, which do not exist as independent quantities but are in reality two aspects of a single physical phenomenon. (p. 47)

electron An elementary particle with a negative electric charge; one of the components of the atom. (p. 46)

electron degeneracy pressure The pressure produced by the resistance of electrons to further compression once they are squeezed to the point of contact.

electrostatic force Force between electrically charged objects.

electroweak force Unification of the weak electromagnetic forces.

element Matter made up of one particular atom. The number of protons in the nucleus of the atom determines which element it represents. (p. 62)

elementary particle Technically, a particle that cannot be subdivided into component parts; however, the term is also often used to refer to particles such as protons and neutrons, which are themselves made up of quarks.

ellipse Geometric figure resembling an elongated circle. An ellipse is characterized by its degree of flatness, or eccentricity, and the length of its long axis. In general, bound orbits of objects moving under gravity are elliptical. (p. 32)

elliptical galaxy Category of galaxy in which the stars are distributed in an elliptical shape on the sky, ranging from highly elongated to nearly circular in appearance. (p. 404)

elongation Angular distance between a planet and the Sun.

emission line Bright line in a specific location of the spectrum of radiating material, corresponding to emission of light at a certain frequency. A heated gas in a glass container produces emission lines in its spectrum. (p. 55)

emission nebula A glowing cloud of hot interstellar gas. The gas glows as a result of one or more nearby young stars which ionize the gas. Since the gas is mostly hydrogen, the emitted radiation falls predominantly in the red region of the spectrum, because of the hydrogen-alpha emission line. (p. 292)

emission spectrum The pattern of spectral emission lines produced by an element. Each element has its own unique emission spectrum. (p. 55)

empirical Discovery based on observational evidence (rather than from theory).

Encke gap A small gap in Saturn's A ring. (p. 226)

energy flux Energy per unit area per unit time radiated by a star (or recorded by a detector).

epicycle A construct of the geocentric model of the solar system which was necessary to explain observed planetary motions. Each planet rides on a small epicycle whose center in turn rides on a larger circle (the deferent). (p. 27)

epoch of inflation Short period of unchecked cosmic expansion early in the history of the universe. During inflation, the universe swelled in size by a factor of about 10^{50}.

equinox *See* vernal equinox, autumnal equinox.

escape speed The speed necessary for one object to escape the gravitational pull of another. Anything that moves away from a gravitating body with more than the escape speed will never return. (p. 134)

euclidean geometry Geometry of flat space.

event horizon Imaginary spherical surface surrounding a collapsing star, with radius equal to the Schwarzschild radius, within which no event can be seen, heard, or known about by an outside observer. (p. 358)

evolutionary theory A theory which explains observations in a series of gradual steps, explainable in terms of well-established physical principles.

evolutionary track A graphical representation of a star's life as a path on the Hertzsprung–Russell diagram. (p. 305)

excited state State of an atom when one of its electrons is in a higher energy orbital than the ground state. Atoms can become excited by absorbing a photon of a specific energy, or by colliding with a nearby atom. (p. 59)

extinction The dimming of starlight as it passes through the interstellar medium.

extrasolar planet Planet orbiting a star other than the Sun. (p. 116)

extremophilic Adjective describing organisms that can survive in very harsh environments.

eyepiece Secondary lens through which an observer views an image. This lens is often chosen to magnify the image.

F

F ring Faint narrow outer ring of Saturn, discovered by *Pioneer 11* in 1979. The F ring lies just inside the Roche limit of Saturn and was found by *Voyager 1* to be made up of several ring strands apparently braided together. (p. 229)

Fahrenheit Temperature scale in which the freezing point of water is 32 degrees and the boiling point of water is 212 degrees.

false vacuum Region of the universe that remained in the "unified" state after the strong and electroweak forces separated; one possible cause of cosmic inflation at very early times.

fault line Dislocation on a planet's surface, often indicating the boundary between two plates.

field line Imaginary line indicating the direction of an electric or magnetic field.

fireball Large meteor that burns up brightly and sometimes explosively in Earth's atmosphere.

firmament Old-fashioned term for the heavens (i.e., the sky).

flare Explosive event occurring in or near an active region on the Sun. (p. 254)

flatness problem One of two conceptual problems with the standard Big Bang model, which is that there is no natural way to explain why the density of the universe is so close to the critical density. (p. 474)

fluidized ejecta The ejecta blankets around some Martian craters, which apparently indicate that the ejected material was liquid at the time the crater formed.

fluorescence Phenomenon where an atom absorbs energy, then radiates photons of lower energy as it cascades back to the ground state; in astronomy, often produced as ultraviolet photons from a hot young star are absorbed by a neutral gas, causing some of the gas atoms to become excited and give off an optical (red) glow.

flyby Unbound trajectory of a spacecraft around a planet or other body.

focal length Distance from a mirror or the center of a lens to the focus.

focus One of two special points within an ellipse, whose separation from each other indicates the eccentricity. In a bound orbit, planets orbit in ellipses with the Sun at one focus. (p. 32)

forbidden line A spectral line seen in emission nebulae, but not seen in laboratory experiments because, under laboratory conditions, collisions kick the electron in question into some other state before emission can occur.

force Action on an object that causes its momentum to change. The rate at which the momentum changes is numerically equal to the force. (p. 35)

fragmentation The breaking up of a large object into many smaller pieces (for example, as the result of high-speed collisions between planetesimals and protoplanets in the early solar system). (p. 121)

Fraunhofer lines The collection of over 600 absorption lines in the spectrum of the Sun, first categorized by Joseph Fraunhofer in 1812.

frequency The number of wave crests passing any given point per unit time. (p. 45)

full When the full hemisphere of the Moon or a planet can be seen from Earth.

full Moon Phase of the Moon in which it appears as a complete circular disk in the sky. (p. 12)

fusion *See* nuclear fusion.

G

G ring Faint, narrow ring of Saturn, discovered by *Pioneer 11* and lying just outside the F ring.

galactic bulge Thick distribution of warm gas and stars around the galactic center. (p. 376)

galactic cannibalism A galaxy merger in which a larger galaxy consumes a smaller one.

galactic center The center of the Milky Way, or any other galaxy. The point about which the disk of a spiral galaxy rotates. (p. 380)

galactic disk Flattened region of gas and dust that bisects the galactic halo in a spiral galaxy. This is the region of active star formation. (p. 376)

galactic epoch Period from 100 million to 3 billion years after the Big Bang when large agglomerations of matter (galaxies and galaxy clusters) formed and grew.

galactic habitable zone Region of a galaxy in which conditions are conducive to the development of life.

galactic halo Region of a galaxy extending far above and below the galactic disk, where globular clusters and other old stars reside. (p. 377)

galactic nucleus Small, central, high-density region of a galaxy. Almost all the radiation from active galaxies is generated within the nucleus. (p. 416)

galactic rotation curve Plot of rotation speed versus distance from the center of a galaxy.

galactic year Time taken for objects at the distance of the Sun (about 8 kpc) to orbit the center of the Galaxy, roughly 225 million years.

galaxy Gravitationally bound collection of a large number of stars. The Sun is a star in the Milky Way Galaxy. (p. 4)

galaxy cluster A collection of galaxies held together by their mutual gravitational attraction. (p. 411)

Galilean Moon The four large moons of Jupiter discovered by Galileo Galilei.

Galilean satellites The four brightest and largest moons of Jupiter (Io, Europa, Ganymede, Callisto), named after Galileo Galilei, the seventeenth-century astronomer who first observed them.

gamma ray Region of the electromagnetic spectrum, far beyond the visible spectrum, corresponding to radiation of very high frequency and very short wavelength. (p. 44)

gamma-ray burst Object that radiates tremendous amounts of energy in the form of gamma rays, possibly due to the collision and merger of two neutron stars initially in orbit around one another. (p. 352)

gamma-ray spectrograph Spectrograph designed to work at gamma-ray wavelengths. Used to map the abundances of certain elements on the Moon and Mars.

gaseous Composed of gas.

gas-exchange experiment Experiment to look for life on Mars. A nutrient broth was offered to Martian soil specimens. If there were life in the soil, gases would be created as the broth was digested.

gene Sequence of nucleotide bases in the DNA molecule that determines the characteristics of a living organism.

general theory of relativity Theory proposed by Einstein to incorporate gravity into the framework of special relativity (p. 357).

geocentric model A model of the solar system that holds that Earth is at the center of the universe and all other bodies are in orbit around it. The earliest theories of the solar system were geocentric. (p. 27)

giant A star with a radius between 10 and 100 times that of the Sun. (p. 276)

giant elliptical Elliptical galaxy up to a few megaparsecs across, containing trillions of stars.

gibbous Appearance of the Moon (or a planet) when more than half (but not all) of the body's hemisphere is visible from Earth.

globular cluster Tightly bound, roughly spherical collection of hundreds of thousands, and sometimes millions, of stars spanning about 50 parsecs. Globular clusters are distributed in the halos around the Milky Way and other galaxies. (p. 310)

gluon Particle that exerts or mediates the strong force in quantum physics.

gradient Rate of change of some quantity (e.g., temperature or composition) with respect to location in space.

Grand Unified Theories Class of theories describing the behavior of the single force that results from unification of the strong, weak, and electromagnetic forces in the early universe. (p. 474)

granite Igneous rock, containing silicon and aluminum, that makes up most of Earth's crust.

granulation Mottled appearance of the solar surface, caused by rising (hot) and falling (cool) material in convective cells just below the photosphere. (p. 247)

gravitational force Force exerted on one body by another due to the effect of gravity. The force is directly proportional to the masses of both bodies involved and inversely proportional to the square of the distance between them. (p. 36)

gravitational instability theory Theory that the jovian planets formed directly from the solar nebula via instabilities in the gas leading to gravitational contraction. *See* core-accretion theory.

gravitational lensing The effect induced on the image of a distant object by a massive foreground object. Light from the distant object is bent into two or more separate images. (p. 392)

gravitational radiation Radiation resulting from rapid changes in a body's gravitational field.

gravitational redshift A prediction of Einstein's general theory of relativity. Photons lose energy as they escape the gravitational field of a massive object. Because a photon's energy is proportional to its frequency, a photon that loses energy suffers a decrease in frequency, which corresponds to an increase, or redshift, in wavelength. (p. 364)

graviton Particle carrying the gravitational field in theories attempting to unify gravity and quantum mechanics.

gravity The attractive effect that any massive object has on all other massive objects. The greater the mass of the object, the stronger its gravitational pull. (p. 36)

gravity assist Using gravity to change the flight path of a satellite or spacecraft.

gravity wave Gravitational counterpart of an electromagnetic wave.

great attractor A huge accumulation of mass in the relatively nearby universe (within about 200 Mpc of the Milky Way).

Great Dark Spot Prominent storm system in the atmosphere of Neptune observed by *Voyager 2*, near the equator of the planet. The system was comparable in size to Earth. (p. 204)

Great Red Spot A large, high-pressure, long-lived storm system visible in the atmosphere of Jupiter. The Red Spot is roughly twice the size of Earth. (p. 198)

great wall Extended sheet of galaxies measuring at least 200 megaparsecs across; one of the largest known structures in the universe.

greenhouse effect The partial trapping of solar radiation by a planetary atmosphere, similar to the trapping of heat in a greenhouse. (p. 140)

greenhouse gas Gas (such as carbon dioxide or water vapor) that efficiently absorbs infrared radiation.

ground state The lowest energy state that an electron can have within an atom. (p. 58)

GUT epoch Period when gravity separated from the other three forces of nature.

gyroscope System of rotating wheels that allows a spacecraft to maintain a fixed orientation in space.

H

habitable zone Three-dimensional zone of comfortable temperature (corresponding to liquid water) that surrounds every star.

half-life The amount of time it takes for half of the initial amount of a radioactive substance to decay into something else.

hayashi track Evolutionary track followed by a protostar during the final pre-main-sequence phase before nuclear fusion begins.

heat Thermal energy, the energy of an object due to the random motion of its component atoms or molecules.

heat death End point of a bound universe, in which all matter and life are destined to be incinerated.

heavy element In astronomical terms, any element heavier than hydrogen and helium.

heliocentric model A model of the solar system that is centered on the Sun, with Earth in motion about the Sun. (p. 28)

helioseismology The study of conditions far below the Sun's surface through the analysis of internal "sound" waves that repeatedly cross the solar interior. (p. 244)

helium-burning shell Shell of burning helium gas surrounding a nonburning stellar core of carbon ash.

helium capture The formation of heavy elements by the capture of a helium nucleus. For example, carbon can form heavier elements by fusion with other carbon nuclei, but it is much more likely to occur by helium capture, which requires less energy.

helium flash An explosive event in the post-main-sequence evolution of a low-mass star. When helium fusion begins in a dense stellar core, the burning is explosive in nature. It continues until the energy released is enough to expand the core, at which point the star achieves stable equilibrium again. (p. 32)

helium precipitation Mechanism responsible for the low abundance of helium of Saturn's atmosphere. Helium condenses in the upper layers to form a mist, which rains down toward Saturn's interior, just as water vapor forms into rain in the atmosphere of Earth.

helium shell flash Condition in which the helium-burning shell in the core of a star cannot respond to rapidly changing conditions within it, leading to a sudden temperature rise and a dramatic increase in nuclear reaction rates.

Hertzsprung–Russell (H-R) diagram A plot of luminosity against temperature (or spectral class) for a group of stars. (p. 276)

high-energy astronomy Astronomy using X- or gamma-ray radiation rather than optical radiation.

high-energy telescope Telescope designed to detect X- and gamma radiation. (p. 90)

highlands Relatively light-colored regions on the surface of the Moon that are elevated several kilometers above the maria. Also called terrae. (p. 457)

high-mass star Star with a mass more than 8 times that of the Sun; progenitor of a neutron star or black hole.

HI region Region of space containing primarily neutral hydrogen.

HII region Region of space containing primarily ionized hydrogen.

homogeneity Assumed property of the universe such that the number of galaxies in an imaginary large cube of the universe is the same no matter where in the universe the cube is placed. (p. 457)

horizon problem One of two conceptual problems with the standard Big Bang model, which is that some regions of the universe that have very similar properties are too far apart to have exchanged information within the age of the universe. (p. 473)

horizontal branch Region of the Hertzsprung–Russell diagram where post-main-sequence stars again reach hydrostatic equilibrium. At this point, the star is burning helium in its core and fusing hydrogen in a shell surrounding the core. (p. 321)

hot dark matter A class of candidates for the dark matter in the universe, composed of lightweight particles such as neutrinos, much less massive than the electron.

hot Jupiter A massive, gaseous planet orbiting very close to its parent star.

hot longitudes (Mercury) Two opposite points on Mercury's equator where the Sun is directly overhead at perihelion.

Hubble classification scheme Method of classifying galaxies according to their appearance, developed by Edwin Hubble. (p. 402)

Hubble diagram Plot of galactic recession velocity versus distance; evidence for an expanding universe.

Hubble flow Universal recession described by the Hubble diagram and quantified by the Hubble Law.

Hubble's constant The constant of proportionality that gives the relation between recessional velocity and distance in Hubble's law. (p. 413)

Hubble's law Law that relates the observed velocity of recession of a galaxy to its distance from us. The velocity of recession of a galaxy is directly proportional to its distance away. (p. 412)

hydrocarbon Molecules consisting solely of hydrogen and carbon.

hydrogen envelope An invisible sheath of gas engulfing the coma of a comet, usually distorted by the solar wind and extending across millions of kilometers of space. (p. 108)

hydrogen shell burning Fusion of hydrogen in a shell that is driven by contraction and heating of the helium core. Once hydrogen is depleted in the core of a star, hydrogen burning stops and the core contracts due to gravity, causing the temperature to rise, heating the surrounding layers of hydrogen in the star, and increasing the burning rate there. (p. 320)

hydrosphere Layer of Earth that contains the liquid oceans and accounts for roughly 70 percent of Earth's total surface area. (p. 134)

hydrostatic equilibrium Condition in a star or other fluid body in which gravity's inward pull is exactly balanced by internal forces due to pressure. (p. 244)

hyperbola Curve formed when a plane intersects a cone at a small angle to the axis of the cone.

hypernova Explosion where a massive star undergoes core collapse and forms a black hole and a gamma-ray burst. *See* supernova.

I

igneous Rocks formed from molten material.

image The optical representation of an object produced when from the object is reflected or refracted by a mirror or lens. (p. 70)

Impact Theory (Moon) Combination of the Capture and Daughter theories, suggesting that the Moon formed after an impact which dislodged some of Earth's mantle and placed it in orbit.

inertia The tendency of an object to continue moving at the same speed and in the same direction, unless acted upon by a force. (p. 35)

inferior conjunction Orbital configuration in which an inferior planet (Mercury or Venus) lies closest to Earth.

inflation *See* epoch of inflation. (p. 475)

infrared Region of the electromagnetic spectrum just outside the visible range, corresponding to light of a slightly longer wavelength than red light. (p. 44)

infrared waves Electromagnetic radiation with wavelength in the infrared part of the spectrum. (p. 44)

infrared telescope Telescope designed to detect infrared radiation. Many such telescopes are designed to be lightweight so that they can be carried above (most of) Earth's atmosphere by balloons, airplanes, or satellites. (p. 88)

inhomogeneity Deviation from perfectly uniform density; in cosmology, inhomogeneities in the universe are ultimately due to quantum fluctuations before inflation.

inner core The central part of Earth's core, believed to be solid, and composed mainly of nickel and iron. (p. 144)

instability strip Part of the Hertzsprung–Russell diagram where pulsating post-main-sequence stars are found.

intensity A basic property of electromagnetic radiation that specifies the amount or strength of the radiation. (p. 51)

intercloud medium Superheated bubbles of hot gas extending far into interstellar space.

intercrater plains Regions on the surface of Mercury that do not show extensive cratering but are relatively smooth. (p. 167)

interference The ability of two or more waves to interact in such a way that they either reinforce or cancel each other. (p. 46)

interferometer Collection of two or more telescopes working together as a team, observing the same object at the same time and at the same wavelength. The effective diameter of an interferometer is equal to the distance between its outermost telescopes. (p. 86)

interferometry Technique in widespread use to dramatically improve the resolution of radio and infrared maps. Several telescopes observe the object simultaneously, and a computer analyzes how the signals interfere with each other. (p. 85)

intermediate-mass black hole Black hole having a mass 100–1000 times greater than the mass of the Sun.

interplanetary matter Matter in the solar system that is not part of a planet or moon—cosmic "debris."

interstellar dust Microscopic dust grains that populate space between the stars, having their origins in the ejected matter of long-dead stars.

interstellar gas cloud A large cloud of gas found in the space among the stars.

interstellar medium The matter between stars, composed of two components, gas and dust, intermixed throughout all of space. (p. 290)

intrinsic variable Star that varies in appearance due to internal processes (rather than, say, interaction with another star).

inverse-square law The law that a field follows if its strength decreases with the square of the distance. Fields that follow the inverse-square law decrease rapidly in strength as the distance increases, but never quite reach zero. (p. 36)

Io plasma torus Doughnut-shaped region of energetic ionized particles, emitted by the volcanoes on Jupiter's moon Io and swept up by Jupiter's magnetic field.

ion An atom that has lost one or more of its electrons. (p. 59)

ionization state Term describing the number of electrons missing from an atom: I refers to a neutral atom, II refers to an atom missing one electron, and so on.

ion tail Thin stream of ionized gas that is pushed away from the head of a comet by the solar wind. It extends directly away from the Sun. Often referred to as a plasma tail. (p. 109)

ionosphere Layer in Earth's atmosphere above about 100 km where the atmosphere is significantly ionized and conducts electricity. (p. 139)

irregular galaxy A galaxy that does not fit into any of the other major categories in the Hubble classification scheme. (p. 406)

isotopes Nuclei containing the same number of protons but different numbers of neutrons. Most elements can exist in several isotopic forms. A common example of an isotope is deuterium, which differs from normal hydrogen by the presence of an extra neutron in the nucleus.

sotropic Assumed property of the universe such that the universe looks the same in every direction. (p. 457)

J

jet stream Relatively strong winds in the upper atmosphere channeled into a narrow stream by a planet's rotation. Normally refers to a horizontal, high-altitude wind.

joule The SI unit of energy.

jovian planet One of the four giant outer planets of the solar system, resembling Jupiter in physical and chemical composition. (p. 102)

K

Kelvin Scale Temperature scale in which absolute zero is at 0 K; a change of 1 kelvin is the same as a change of 1 degree Celsius.

Kelvin-Helmholtz contraction phase Evolutionary track followed by a star during the protostar phase.

Kepler's laws of planetary motion Three laws, based on precise observations of the motions of the planets by Tycho Brahe, that summarize the motions of the planets about the Sun.

kinetic energy Energy of an object due to its motion.

Kirchhoff's laws Three rules governing the formation of different types of spectra. (p. 57)

Kirkwood gaps Gaps in the spacings of orbital semimajor axes of asteroids in the asteroid belt, produced by dynamical resonances with nearby planets, especially Jupiter.

Kuiper belt A region in the plane of the solar system outside the orbit of Neptune where most short-period comets are thought to originate. (p. 111)

Kuiper-belt object Small icy body orbiting in the Kuiper belt.

L

labeled-release experiment Experiment to look for life on Mars. Radioactive carbon compounds were added to Martian soil specimens. Scientists looked for indications that the carbon had been eaten or inhaled.

Lagrangian point One of five special points in the plane of two massive bodies orbiting one another, where a third body of negligible mass can remain in equilibrium.

lander Spacecraft that lands on the object it is studying.

laser ranging Method of determining the distance to an object by firing a laser beam at it and measuring the time taken for the light to return.

lava dome Volcanic formation formed when lava oozes out of fissures in a planet's surface, creating the dome, and then withdraws, causing the crust to crack and subside.

law of conservation of mass and energy A fundamental law of modern physics which states that the sum of mass and energy must always remain constant in any physical process. In fusion reactions, the lost mass is converted into energy, primarily in the form of electromagnetic radiation. (p. 257)

laws of planetary motion Three laws derived by Kepler describing the motion of the planets around the Sun. (p. 32)

leap year Year in which an additional day is inserted into the calendar in order to keep the calendar year synchronized with Earth's orbit around the Sun.

lens Optical instrument made of glass or some other transparent material, shaped so that, as a parallel beam of light passes through it, the rays are refracted so as to pass through a single focal point.

lens (eye) The part of the eye that refracts light onto the retina.

lepton Greek word for light, referring in particle physics to low-mass particles such as electrons, muons, and neutrinos that interact via the weak force.

lepton epoch Period when the light elementary particles (leptons) were in thermal equilibrium with the cosmic radiation field.

lidar Light Detection and Ranging—a device that uses laser-ranging to measure distance.

light *See* electromagnetic radiation.

light curve The variation in brightness of a star with time. (p. 282)

light element In astronomical terms, hydrogen and helium.

light-gathering power Amount of light a telescope can view and focus, proportional to the area of the primary mirror.

lighthouse model The leading explanation for pulsars. A small region of the neutron star, near one of the magnetic poles, emits a steady stream of radiation that sweeps past Earth each time the star rotates. The period of the pulses is the star's rotation period. (p. 347)

light-year The distance that light, moving at a constant speed of 300,000 km/s, travels in one year. One light-year is about 10 trillion kilometers.

limb Edge of a lunar, planetary, or solar disk.

linear momentum Tendency of an object to keep moving in a straight line with constant velocity; the product of the object's mass and velocity.

line of nodes Intersection of the plane of the Moon's orbit with Earth's orbital plane.

line of sight technique Method of probing interstellar clouds by observing their effects on the spectra of background stars.

lithosphere Earth's crust and a small portion of the upper mantle that make up Earth's plates. This layer of Earth undergoes tectonic activity.

Local Bubble The particular low-density intercloud region surrounding the Sun.

Local Group The small galaxy cluster that includes the Milky Way Galaxy. (p. 410)

local supercluster Collection of galaxies and clusters centered on the Virgo cluster. *See* also supercluster.

logarithm The power to which 10 must be raised to produce a given number.

logarithmic scale Scale using the logarithm of a number rather than the number itself; commonly used to compress a large range of data into more manageable form.

look-back time Time in the past when an object emitted the radiation we see today.

low-mass star Star with a mass less than 8 times that of the Sun; progenitor of a white dwarf.

luminosity One of the basic properties used to characterize stars, luminosity is defined as the total energy radiated by star each second, at all wavelengths. (p. 243)

luminosity class A classification scheme that groups stars according to the width of their spectral lines. For a group of stars with the same temperature, luminosity class differentiates between supergiants, giants, main-sequence stars, and subdwarfs. (p. 281)

lunar dust *See* regolith.

lunar eclipse Celestial event during which the moon passes through the shadow of Earth, temporarily darkening its surface. (p. 14)

lunar phase The appearance of the Moon at different points along its orbit.

Lyman-alpha forest Collection of lines in an object's spectra starting at the redshifted wavelength of the object's own Lyman-alpha emission line and extending to shorter wavelengths; produced by gas in galaxies along the line of sight.

M

macroscopic Large enough to be visible by the unaided eye.

Magellanic Clouds Two small irregular galaxies that are gravitationally bound to the Milky Way Galaxy. (p. 406)

magnetic field Field that accompanies any changing electric field and governs the influence of magnetized objects on one another. (p. 47)

magnetic poles Points on a planet where the planetary magnetic field lines intersect the planet's surface vertically.

magnetism The presence of a magnetic field.

magnetometer Instrument that measures magnetic field strength.

magnetopause Boundary between a planet's magnetosphere and the solar wind.

magnetosphere A zone of charged particles trapped by a planet's magnetic field, lying above the atmosphere. (p. 134)

magnitude scale A system of ranking stars by apparent brightness, developed by the Greek astronomer Hipparchus. Originally, the brightest stars in the sky were categorized as being of first magnitude, while the faintest stars visible to the naked eye were classified as sixth magnitude. The scheme has since been extended to cover stars and galaxies too faint to be seen by the unaided eye. Increasing magnitude means fainter stars, and a difference of five magnitudes corresponds to a factor of 100 in apparent brightness. (p. 270)

main sequence Well-defined band on the Hertzsprung–Russell diagram on which most stars are found, running from the top left of the diagram to the bottom right. (p. 276)

main-sequence turnoff Special point on the Hertzsprung–Russell diagram for a cluster, indicative of the cluster's age. If all the stars in the cluster are plotted, the lower mass stars will trace out the main sequence up to the point where stars begin to evolve off the main sequence toward the red giant branch. The point where stars are just beginning to evolve off is the main-sequence turnoff. (p. 336)

major axis The long axis of an ellipse.

mantle Layer of Earth just interior to the crust. (p. 134)

mare Relatively dark-colored and smooth region on the surface of the Moon (plural: *maria*). (p. 149)

marginally bound universe Universe that will expand forever but at an increasingly slow rate.

mass A measure of the total amount of matter contained within an object. (p. 35)

mass function Relation between the component masses of a single-line spectroscopic binary.

massive compact halo object (MACHO) Collective name for "stellar" candidates for dark matter, including brown dwarfs, white dwarfs, and low-mass red dwarfs.

mass–luminosity relation The dependence of the luminosity of a main-sequence star on its mass. The luminosity increases roughly as the mass raised to the third power.

mass–radius relation The dependence of the radius of a main-sequence star on its mass. The radius rises roughly in proportion to the mass.

mass transfer Process by which one star in a binary system transfers matter onto the other.

mass-transfer binary *See* semi-detached binary.

matter Anything having mass.

matter-antimatter annihilation Reaction in which matter and antimatter annihilate to produce high energy gamma rays. *See also* antiparticle.

matter-dominated universe A universe in which the density of matter exceeds the density of radiation. The present-day universe is matter dominated. (p. 470)

matter era (Current) era following the Radiation era when the universe is larger and cooler, and matter is the dominant constituent of the universe.

Maunder minimum Lengthy period of solar inactivity that extended from 1645 to 1715.

mean solar day Average length of time from one noon to the next, taken over the course of a year—24 hours.

medium Material through which a wave, such as sound, travels.

meridian An imaginary line on the celestial sphere through the north and south celestial poles, passing directly overhead at a given location.

mesosphere Region of Earth's atmosphere lying between the stratosphere and the ionosphere, 50–80 km above Earth's surface. (p. 139)

Messier object Member of a list of "fuzzy" objects compiled by astronomer Charles Messier in the eighteenth century.

metabolism The daily utilization of food and energy by which organisms stay alive.

metallic Composed of metal or metal compounds.

metamorphic Rocks created from existing rocks exposed to extremes of temperature or pressure.

meteor Bright streak in the sky, often referred to as a "shooting star," resulting from a small piece of interplanetary debris entering Earth's atmosphere and heating air molecules, which emit light as they return to their ground states. (p. 112)

meteor shower Event during which many meteors can be seen each hour, caused by the yearly passage of Earth through the debris spread along the orbit of a comet.

meteorite Any part of a meteoroid that survives passage through the atmosphere and lands on the surface of Earth. (p. 113)

meteoroid Chunk of interplanetary debris prior to encountering Earth's atmosphere. (p. 103)

meteoroid swarm Pebble-sized cometary fragments dislodged from the main body, moving in nearly the same orbit as the parent comet. (p. 113)

microlensing Gravitational lensing by individual stars in a galaxy.

micrometeoroid Relatively small chunks of interplanetary debris ranging from dust-particle size to pebble-sized fragments. (p. 113)

microsphere Small droplet of protein-like material that resists dissolution in water.

midocean ridge Place where two plates are moving apart, allowing fresh magma to well up.

Milky Way Galaxy The spiral galaxy in which the Sun resides. The disk of our Galaxy is visible in the night sky as the faint band of light known as the Milky Way. (p. 376)

millisecond pulsar A pulsar whose period indicates that the neutron star is rotating nearly 1000 times each second. The most likely explanation for these rapid rotators is that the neutron star has been spun up by drawing in matter from a companion star. (p. 350)

molecular cloud A cold, dense interstellar cloud which contains a high fraction of molecules. It is widely believed that the relatively high density of dust particles in these clouds plays an important role in the formation and preservation of the molecules. (p. 300)

molecular cloud complex Collection of molecular clouds that spans as much as 50 parsecs and may contain enough material to make millions of Sun-size stars. (p. 300)

molecule A tightly bound collection of atoms held together by the atoms' electromagnetic fields. Molecules, like atoms, emit and absorb photons at specific wavelengths. (p. 62)

molten In liquid form due to high temperatures.

moon A small body in orbit around a planet.

mosaic (photograph) Composite photograph made up of many smaller images.

M-type asteroid Asteroid containing large fractions of nickel and iron.

multiple star system Group of two or more stars in orbit around one another.

muon A type of lepton (along with the electron and tau).

N

naked singularity A singularity that is not hidden behind an event horizon.

nanobacterium Very small bacterium with diameter in the nanometer range.

nanometer One billionth of a meter.

neap tide The smallest tides, occurring when the Earth-Moon line is perpendicular to the Earth-Sun line at the first and third quarters.

nebula General term used for any "fuzzy" patch on the sky, either light or dark. (pp. 118, 294)

nebular theory One of the earliest models of solar system formation, dating back to Descartes, in which a large cloud of gas began to collapse under its own gravity to form the Sun and planets. (p. 118)

nebulosity "Fuzziness," usually in the context of an extended or gaseous astronomical object.

neon-oxygen white dwarf White dwarf formed from a low-mass star with a mass close to the "high-mass" limit, in which neon and oxygen form in the core.

neutrino Virtually massless and chargeless particle that is one of the products of fusion reactions in the Sun. Neutrinos move at close to the speed of light, and interact with matter hardly at all. (p. 258)

neutrino oscillations Possible solution to the solar neutrino problem, in which the neutrino has a very tiny mass. In this case, the correct number of neutrinos can be produced in the solar core, but on their way to Earth some can "oscillate," or become transformed into other particles, and thus go undetected. (p. 260)

neutron An elementary particle with roughly the same mass as a proton, but which is electrically neutral. Along with protons, neutrons form the nuclei of atoms. (p. 62)

neutron capture The primary mechanism by which very massive nuclei are formed in the violent aftermath of a supernova. Instead of fusion of like nuclei, heavy elements are created by the addition of more and more neutrons to existing nuclei.

neutron degeneracy pressure Pressure due to the Pauli exclusion principle, arising when neutrons are forced to come into close contact.

neutronization Process occurring at high densities, in which protons and electrons are crushed together to form neutrons and neutrinos.

neutron spectrometer Instrument designed to search for water ice by looking for hydrogen.

neutron star A dense ball of neutrons that remains at the core of a star after a supernova explosion has destroyed the rest of the star. Typical neutron stars are about 20 km across, and contain more mass than the Sun. (p. 346)

new Moon Phase of the moon during which none of the lunar disk is visible. (p. 12)

Newtonian mechanics The basic laws of motion, postulated by Newton, which are sufficient to explain and quantify virtually all of the complex dynamical behavior found on Earth and elsewhere in the universe. (p. 35)

Newtonian telescope A reflecting telescope in which incoming light is intercepted before it reaches the prime focus and is deflected into an eyepiece at the side of the instrument. (p. 72)

nodes Two points on the Moon's orbit when it crosses the ecliptic.

nonrelativistic Speeds that are much less than the speed of light.

nonthermal spectrum Continuous spectrum not well described by a blackbody.

north celestial pole Point on the celestial sphere directly above Earth's North Pole. (p. 6)

Northern and Southern Lights Colorful displays produced when atmospheric molecules, excited by charged particles from the Van Allen belts, fall back to their ground state.

nova A star that suddenly increases in brightness, often by a factor of as much as 10,000, then slowly fades back to its original luminosity. A nova is the result of an explosion on the surface of a white-dwarf star, caused by matter falling onto its surface from the atmosphere of a binary companion. (p. 327)

nuclear binding energy Energy that must be supplied to split an atomic nucleus into neutrons and protons.

nuclear epoch Period when protons and neutrons fused to form heavier nuclei.

nuclear fusion Mechanism of energy generation in the core of the Sun, in which light nuclei are combined, or fused, into heavier ones, releasing energy in the process. (p. 257)

nuclear reaction Reaction in which two nuclei combine to form other nuclei, often releasing energy in the process. *See* also fusion.

nucleotide base An organic molecule, the building block of genes that pass on hereditary characteristics from one generation of living creatures to the next. (p. 485)

nucleus Dense, central region of an atom, containing both protons and neutrons, and orbited by one or more electrons. (p. 58); The solid region of ice and dust that composes the central region of the head of a comet. (p. 108)

nucleus The dense central core of a galaxy. (p. 395)

O

obscuration Blockage of light by pockets of interstellar dust and gas.

Olbers's paradox A thought experiment suggesting that if the universe were homogeneous, infinite, and unchanging, the entire night sky would be as bright as the surface of the Sun. (p. 457)

Oort cloud Spherical halo of material surrounding the solar system out to a distance of about 50,000 AU; where most comets reside. (p. 111)

opacity A quantity that measures a material's ability to block electromagnetic radiation. Opacity is the opposite of transparency. (p. 50)

open cluster Loosely bound collection of tens to hundreds of stars, a few parsecs across, generally found in the plane of the Milky Way. (p. 310)

open universe Geometry that the universe would have if the density of matter were less than the critical value. In an open universe there is not enough matter to halt the expansion of the universe. An open universe is infinite in extent. (p. 462)

opposition Orbital configuration in which a planet lies in the opposite direction from the Sun, as seen from Earth.

optical double Chance superposition in which two stars appear to lie close together but are actually widely separated.

optical telescope Telescope designed to observe electromagnetic radiation at optical wavelengths.

orbital One of several energy states in which an electron can exist in an atom.

orbital period Time taken for a body to complete one full orbit around another.

orbiter Spacecraft that orbits an object to make observations.

organic compound Chemical compound (molecule) containing a significant fraction of carbon atoms; the basis of living organisms.

outer core The outermost part of Earth's core, believed to be liquid and composed mainly of nickel and iron. (p. 144)

outflow channel Surface feature on Mars, evidence that liquid water once existed there in great quantity; believed to be the relics of catastrophic flooding about 3 billion years ago. Found only in the equatorial regions of the planet. (p. 175)

outgassing Production of atmospheric gases (carbon dioxide, water vapor, methane, and sulphur dioxide) by volcanic activity.

ozone layer Layer of Earth's atmosphere at an altitude of 20–50 km where incoming ultraviolet solar radiation is absorbed by oxygen, ozone, and nitrogen in the atmosphere. (p. 139)

P

parallax The apparent motion of a relatively close object with respect to a more distant background as the location of the observer changes. (p. 18)

parsec The distance at which a star must lie in order for its measured parallax to be exactly 1 arc second; 1 parsec equals 206,000 AU. (p. 266)

partial eclipse Celestial event during which only a part of the occulted body is blocked from view. (p. 14)

particle A body with mass but of negligible dimension.

particle accelerator Device used to accelerate subatomic particles to relativistic speeds.

particle-antiparticle pair Pair of particles (e.g., an electron and a positron) produced by two photons of sufficiently high energy.

particle detector Experimental equipment that allows particles and antiparticles to be detected and identified.

Pauli exclusion principle A rule of quantum mechanics that prohibits electrons in dense gas from being squeezed too close together.

penumbra Portion of the shadow cast by an eclipsing object in which the eclipse is seen as partial. (p. 15)

penumbra The outer region of a sunspot, surrounding the umbra, which is not as dark and not as cool as the central region. (p. 250)

perihelion The closest approach to the Sun of any object in orbit about it. (p. 32)

period The time needed for an orbiting body to complete one revolution about another body. (p. 33)

period–luminosity relation A relation between the pulsation period of a Cepheid variable and its absolute brightness. Measurement of the pulsation period allows the distance of the star to be determined. (p. 379)

permafrost Layer of permanently frozen water ice believed to lie just under the surface of Mars. (p. 176)

phase Appearance of the sunlit face of the Moon at different points along its orbit, as seen from Earth. (p. 12)

photodisintegration Process occurring at high temperatures, in which individual photons have enough energy to split a heavy nucleus (e.g., iron) into lighter nuclei.

photoelectric effect Emission of an electron from a surface when a photon of electromagnetic radiation is absorbed.

photoevaporation Process in which a cloud in the vicinity of a newborn hot star is dispersed by the star's radiation.

photometer A device that measures the total amount of light received in all or part of the image.

photometry Branch of observational astronomy in which the brightness of a source is measured through each of a set of standard filters.

photomicrograph Photograph taken through a microscope.

photon Individual packet of electromagnetic energy that makes up electromagnetic radiation. (p. 59)

photosphere The visible surface of the Sun, lying just above the uppermost layer of the Sun's interior, and just below the chromosphere. (p. 242)

photosynthesis Process by which plants manufacture carbohydrates and oxygen from carbon dioxide and water using chlorophyll and sunlight as the energy source.

pixel One of many tiny picture elements, organized into an array, making up a digital image.

Planck curve *See* blackbody curve.

Planck epoch Period from the beginning of the Universe to roughly 10^{-43} second, when the laws of physics are not understood.

Planck's constant Fundamental physical constant relating the energy of a photon to its radiation frequency (color).

planet One of eight major bodies that orbit the Sun, visible to us by reflected sunlight. (p. 4)

planetary nebula The ejected envelope of a red-giant star, spread over a volume roughly the size of our solar system. (p. 323)

planetary ring system Material organized into thin, flat rings encircling a giant planet, such as Saturn.

planetesimal Term given to objects in the early solar system that had reached the size of small moons, at which point their gravitational fields were strong enough to begin influencing their neighbors. (p. 121)

plasma A gas in which the constituent atoms are completely ionized.

plate tectonics The motions of regions of Earth's lithosphere, which drift with respect to one another. Also known as continental drift. (p. 147)

plutino Kuiper-belt object whose orbital period (like that of Pluto) is in a 3:2 resonance with the orbit of Neptune.

polarity A measure of the direction of the solar magnetic field in a sunspot. Conventionally, lines coming out of the surface are labeled "S" while those going into the surface are labeled "N." (p. 251)

polarization The alignment of the electric fields of emitted photons, which are generally emitted with random orientations.

Population I and II stars Classification scheme for stars based on the abundance of heavy elements. Within the Milky Way, Population I refers to young disk stars and Population II refers to old halo stars.

positron Atomic particle with properties identical to those a negatively charged electron, except for its positive charge. This positron is the antiparticle of the electron. Positrons and electrons annihilate one another when they meet, producing pure energy in the form of gamma rays. (p. 258)

prebiotic compound Molecule that can combine with others to form the building blocks of life.

precession The slow change in the direction of the rotation axis of a spinning object, caused by some external gravitational influence. (p. 11)

primary atmosphere The chemical components that would have formed Earth's atmosphere. (p. 184)

primary mirror Mirror placed at the prime focus of a telescope (*see* prime focus).

prime focus The point in a reflecting telescope where the mirror focuses incoming light to a point. (p. 70)

prime-focus image Image formed at the prime focus of a telescope.

primordial matter Matter created during the early, hot epochs of the universe.

primordial nucleosynthesis The production of elements heavier than hydrogen by nuclear fusion in the high temperatures and densities that existed in the early universe. (p. 471)

principle of cosmic censorship A proposition to separate the unexplained physics near a singularity from the rest of the well-behaved universe. The principle states that nature always hides any singularity, such as a black hole, inside an event horizon, which insulates the rest of the universe from seeing it.

progenitor "Ancestor" star of a given object; e.g., the star that existed before a supernova explosion is the supernova's progenitor.

prograde motion Motion across the sky in the eastward direction.

prominence Loop or sheet of glowing gas ejected from an active region on the solar surface, which then moves through the inner parts of the corona under the influence of the Sun's magnetic field. (p. 254)

proper motion The angular movement of a star across the sky, as seen from Earth, measured in seconds of arc per year. This movement is a result of the star's actual motion through space. (p. 268)

protein Molecule made up of amino acids that controls metabolism.

proton An elementary particle carrying a positive electric charge, a component of all atomic nuclei. The number of protons in the nucleus of an atom dictates what type of atom it is. (p. 46)

proton–proton chain The chain of fusion reactions, leading from hydrogen to helium, that powers main-sequence stars. (p. 258)

protoplanet Clump of material, formed in the early stages of solar system formation, that was the forerunner of the planets we see today. (p. 121)

protostar Stage in star formation when the interior of a collapsing fragment of gas is sufficiently hot and dense that it becomes opaque to its own radiation. The protostar is the dense region at the center of the fragment. (p. 303)

protosun The central accumulation of material in the early stages of solar system formations, the forerunner of the present-day Sun. (p. 121)

Ptolemaic model Geocentric solar system model, developed by the second century astronomer Claudius Ptolemy. It predicted with great accuracy the positions of the then known planets. (p. 27)

pulsar Object that emits radiation in the form of rapid pulses with a characteristic pulse period and duration. Charged particles, accelerated by the magnetic field of a rapidly rotating neutron star, flow along the magnetic field lines, producing radiation that beams outward as the star spins on its axis. (p. 346)

pulsating variable star A star whose luminosity varies in a predictable, periodic way. (p. 378)

P-waves Pressure waves from an earthquake that travel rapidly through liquids and solids.

pyrolific-release experiment Experiment to look for life on Mars. Radioactively tagged carbon dioxide was added to Martian soil specimens. Scientists looked for indications that the radioactive material had been absorbed.

Q

quantization The fact that light and matter on small scales behave in a discontinuous manner, and manifest themselves in the form of tiny "packets" of energy, called quanta.

quantum fluctuation Temporary random change in the amount of energy at a point in space.

quantum gravity Theory combining general relativity with quantum mechanics.

quantum mechanics The laws of physics as they apply on atomic scales.

quark A fundamental matter particle that interacts via the strong force; basic constituent of protons and neutrons.

quark epoch Period when all heavy elementary particles (composed of quarks) were in thermal equilibrium with the cosmic radiation field.

quarter Moon Lunar phase in which the Moon appears as a half disk. (p. 12)

quasar Starlike radio source with an observed redshift that indicates an extremely large distance from Earth. The brightest nucleus of a distant active galaxy. (p. 422)

quasi-stellar object (QSO) *See* quasar.

quiescent prominence Prominence that persists for days or weeks, hovering high above the solar photosphere.

quiet Sun The underlying predictable elements of the Sun's behavior, such as its average photospheric temperature, which do not change in time.

R

radar Acronym for RAdio Detection And Ranging. Radio waves are bounced off an object, and the time taken for the echo to return indicates its distance. (p. 34)

radial motion Motion along a particular line of sight, which induces apparent changes in the wavelength (or frequency) of radiation received.

radial velocity Component of a star's velocity along the line of sight.

radian Angular measure equivalent to $180/\pi = 57.3$ degrees.

radiant Constellation from which a meteor shower appears to come.

radiation A way in which energy is transferred from place to place in the form of a wave. Light is a form of electromagnetic radiation.

radiation darkening The effect of chemical reactions that result when high-energy particles strike the icy surfaces of objects in the outer solar system. The reactions lead to a buildup of a dark layer of material.

radiation-dominated universe Early epoch in the universe, when the equivalent density of radiation in the cosmos exceeded the density of matter. (p. 470)

radiation era The first few thousand years after the Big Bang when the universe was small, dense, and dominated by radiation.

radiation zone Region of the Sun's interior where extremely high temperatures guarantee that the gas is completely ionized. Photons only occasionally interact with electrons, and travel through this region with relative ease. (p. 242)

radio Region of the electromagnetic spectrum corresponding to radiation of the longest wavelengths. (p. 44)

radio galaxy Type of active galaxy that emits most of its energy in the form of long-wavelength radiation. (p. 416)

radiograph Image made from observations at radio wavelengths.

radio lobe Roundish extended region of radio-emitting gas, lying well beyond the center of a radio galaxy. (p. 416)

radio telescope Large instrument designed to detect radiation from space at radio wavelengths. (p. 83)

radio wave Electromagnetic radiation with wavelength in the radio part of the spectrum (p. 44)

radioactivity The release of energy by rare, heavy elements when their nuclei decay into lighter nuclei. (p. 145)

radius–luminosity–temperature relationship A mathematical proportionality, arising from Stefan's Law, which allows astronomers to indirectly determine the radius of a star once its luminosity and temperature are known. (p. 275)

rapid mass transfer Mass transfer in a binary system that proceeds at a rapid and unstable rate, transferring most of the mass of one star onto the other.

ray The path taken by a beam of radiation.

Rayleigh scattering Scattering of light by particles in the atmosphere.

recession velocity Rate at which two objects are separating from one another.

recurrent nova Star that "goes nova" several times over the course of a few decades.

red dwarf Small, cool faint star at the lower-right end of the main sequence on the Hertzsprung–Russell diagram. (p. 277)

red giant A giant star whose surface temperature is relatively low so that it glows red. (p. 276)

red-giant branch The section of the evolutionary track of a star corresponding to intense hydrogen shell burning, which drives a steady expansion and cooling of the outer envelope of the star. As the star gets larger in radius and its surface temperature cools, it becomes a red giant. (p. 320)

red-giant region The upper-right corner of the Hertzsprung–Russell diagram, where red-giant stars are found. (p. 279)

redshift Motion-induced change in the wavelength of light emitted from a source moving away from us. The relative recessional motion causes the wave to have an observed wavelength longer (and hence redder) than it would if it were not moving.

redshift survey Three-dimensional survey of galaxies, using redshift to determine distance.

red supergiant An extremely luminous red star. Often found on the asymptotic-giant branch of the Hertzsprung–Russell diagram. (pp. 276, 328)

reddening Dimming of starlight by interstellar matter, which tends to scatter higher-frequency (blue) components of the radiation more efficiently than the lower-frequency (red) components. (p. 290)

reflecting telescope A telescope which uses a mirror to gather and focus light from a distant object. (p. 70)

reflection nebula Bluish nebula caused by starlight scattering from dust particles in an interstellar cloud located just off the line of sight between Earth and a bright star.

refracting telescope A telescope that uses a lens to gather and focus light from a distant object. (p. 70)

refraction The tendency of a wave to bend as it passes from one transparent medium to another. (p. 70)

regolith Surface dust on the moon, several tens of meters thick in places, caused by billions of years of meteoritic bombardment.

relativistic Speeds comparable to the speed of light.

relativistic fireball Leading explanation of a gamma-ray burst in which an expanding region of superhot gas radiates in the gamma-ray part of the spectrum.

residual cap Portion of Martian polar ice caps that remains permanently frozen, undergoing no seasonal variations.

resonance Circumstance in which two characteristic times are related in some simple way, e.g., an asteroid with an orbital period exactly half that of Jupiter.

retina (eye) The back part of the eye onto which light is focused by the lens.

retrograde motion Backward, westward loop traced out by a planet with respect to the fixed stars. (p. 26)

revolution Orbital motion of one body about another, such as Earth about the Sun. (p. 8)

revolving *See* revolution.

Riemannian geometry Geometry of positively curved space (such as the surface of a sphere).

right ascension Celestial coordinate used to measure longitude on the celestial sphere. The zero point is the position of the Sun at the vernal equinox. (p. 7)

rille A ditch on the surface of the Moon where molten lava flowed in the past.

ring *See* planetary ring system.

ringlet Narrow region in Saturn's planetary ring system where the density of ring particles is high. *Voyager* discovered that the rings visible from Earth are actually composed of tens of thousands of ringlets. (p. 228)

Roche limit Often called the tidal stability limit, the Roche limit gives the distance from a planet at which the tidal force (due to the planet) between adjacent objects exceeds their mutual attraction. Objects within this limit are unlikely to accumulate into larger objects. The rings of Saturn occupy the region within Saturn's Roche limit. (p. 228)

Roche lobe An imaginary surface around a star. Each star in a binary system can be pictured as being surrounded by a teardrop-shaped zone of gravitational influence, the Roche lobe. Any material within the Roche lobe of a star can be considered to be part of that star. During evolution, one member of the binary system can expand so that it overflows its own Roche lobe and begins to transfer matter onto the other star.

rock Material made predominantly from compounds of silicon and oxygen.

rock cycle Process by which surface rock on Earth is continuously redistributed and transformed from one type into another.

rotation Spinning motion of a body about an axis. (p. 6)

rotation curve Plot of the orbital speed of disk material in a galaxy against its distance from the galactic center. Analysis of rotation curves of spiral galaxies indicates the existence of dark matter. (p. 389)

R-process "Rapid" process in which many neutrons are captured by a nucleus during a supernova explosion.

RR Lyrae variable Variable star whose luminosity changes in a characteristic way. All RR Lyrae stars have more or less the same average luminosity. (p. 378)

runaway greenhouse effect A process in which the heating of a planet leads to an increase in its atmosphere's ability to retain heat and thus to further heating, causing extreme changes in the temperature of the surface and the composition of the atmosphere. (p. 185)

runoff channel Riverlike surface feature on Mars, evidence that liquid water once existed there in great quantities. They are found in the southern highlands and are thought to have been formed by water that flowed nearly 4 billion years ago. (p. 175)

S

S0 galaxy Galaxy that shows evidence of a thin disk and a bulge, but which has no spiral arms and contains little or no gas. (p. 405)

Sagittarius A/Sgr A Strong radio source corresponding to the supermassive black hole at the center of the Milky Way.

Saros cycle Time interval between successive occurrences of the "same" solar eclipse, equal to 18 years, 11.3 days.

satellite A small body orbiting another larger body.

SB0 galaxy An S0-type galaxy whose disk shows evidence of a bar. (p. 405)

scarp Surface feature on Mercury believed to be the result of cooling and shrinking of the crust forming a wrinkle on the face of the planet. (p. 167)

Schmidt telescope Type of telescope having a very wide field of view, allowing large areas of the sky to be observed at once.

Schwarzschild radius The distance from the center of an object such that, if all the mass were compressed within that region, the escape speed would equal the speed of light. Once a stellar remnant collapses within this radius, light cannot escape and the object is no longer visible. (p. 358)

scientific method The set of rules used to guide science, based on the idea that scientific "laws" be continually tested, and modified or replaced if found inadequate. (p. 19)

scientific notation Expressing large and small numbers using power-of-10 notation.

seasonal cap Portion of Martian polar ice caps that is subject to seasonal variations, growing and shrinking once each Martian year.

seasons Changes in average temperature and length of day that result from the tilt of Earth's (or any planet's) axis with respect to the plane of its orbit. (p. 10)

secondary atmosphere The chemicals that composed Earth's atmosphere after the planet's formation, once volcanic activity outgassed chemicals from the interior. (p. 184)

sedimentary Rocks formed from the buildup of sediment.

seeing A term used to describe the ease with which good telescopic observations can be made from Earth's surface, given the blurring effects of atmospheric turbulence. (p. 81)

seeing disk Roughly circular region on a detector over which a star's pointlike images is spread, due to atmospheric turbulence. (p. 81)

seismic wave A wave that travels outward from the site of an earthquake through Earth. (p. 143)

seismology The study of earthquakes and the waves they produce in Earth's interior.

seismometer Equipment designed to detect and measure the strength of earthquakes (or quakes on any other planet).

selection effect Observational bias in which a measured property of a collection of objects is due to the way in which the measurement was made, rather than being intrinsic to the objects themselves. (p. 126)

self-propagating star formation Mode of star formation in which shock waves produced by the formation and evolution of one generation of stars triggers the formation of the next. (p. 389)

semi-detached Binary system where one star lies within its Roche lobe but the other fills its Roche lobe and is transferring matter onto the first star.

semi-major axis One-half of the major axis of an ellipse. The semimajor axis is the way in which the size of an ellipse is usually quantified. (p. 32)

SETI Acronym for Search for Extraterrestrial Intelligence.

Seyfert galaxy Type of active galaxy whose emission comes from a very small region within the nucleus of an otherwise normal-looking spiral system. (p. 415)

shepherd satellite Satellite whose gravitational effect on a ring helps preserve the ring's shape. Examples are two satellites of Saturn, Prometheus and Pandora, whose orbits lie on either side of the F ring. (p. 229)

shield volcano A volcano produced by repeated nonexplosive eruptions of lava, creating a gradually sloping, shield-shaped low dome. Often contains a caldera at its summit. (p. 170)

shock wave Wave of matter, which may be generated by a newborn star or supernova, that pushes material outward into the surrounding molecular cloud. The material tends to pile up, forming a rapidly moving shell of dense gas.

short-period comet Comet with orbital period less than 200 years.

SI Système International, the international system of metric units used to define mass, length, time, etc.

sidereal day The time needed between successive risings of a given star. (p. 8)

sidereal month Time required for the Moon to complete one trip around the celestial sphere. (p. 14)

sidereal year The time required for the constellations to complete one cycle around the sky and return to their starting points, as seen from a given point on Earth. Earth's orbital period around the Sun is one sidereal year. (p. 11)

single-line spectroscopic binary Binary system in which one star is too faint for its spectrum to be distinguished, so only the spectrum of the brighter star can be seen to shift back and forth as the stars orbit one another.

singularity A point in the universe where the density of matter and the gravitational field are infinite, such as at the center of a black hole. (p. 365)

sister/coformation theory (Moon) Theory suggesting that the Moon formed as a separate object close to Earth.

soft landing Use of rockets, parachutes, or packaging to break the fall of a space probe as it lands on a planet.

solar constant The amount of solar energy reaching Earth per unit area per unit time, approximately 1400 W/m^2. (p. 243)

solar core The region at the center of the Sun, with a radius of nearly 200,000 km, where powerful nuclear reactions generate the Sun's energy output.

solar cycle The 22-year period that is needed for both the average number of spots and the Sun's magnetic polarity to repeat themselves. The Sun's polarity reverses on each new 11-year sunspot cycle. (p. 253)

solar day The period of time between the instant when the Sun is directly overhead (i.e., noon) to the next time it is directly overhead. (p. 7)

solar eclipse Celestial event during which the new Moon passes directly between Earth and the Sun, temporarily blocking the Sun's light. (p. 14)

solar interior The region of the Sun between the solar core and the photosphere.

solar maximum Point of the sunspot cycle during which many spots are seen. They are generally confined to regions in each hemisphere, between about 15 and 20 degrees latitude.

solar minimum Point of the sunspot cycle during which only a few spots are seen. They are generally confined to narrow regions in each hemisphere at about 25–30 degrees latitude.

solar nebula The swirling gas surrounding the early Sun during the epoch of solar system formation, also referred to as the primitive solar system. (p. 118)

solar neutrino problem The discrepancy between the theoretically predicted flux of neutrinos streaming from the Sun as a result of fusion reactions in the core and the flux that is actually observed. The observed number of neutrinos is only about half the predicted number. (p. 260)

solar system The Sun and all the bodies that orbit it—Mercury, Venus, Earth, Mars, Jupiter, Saturn, Uranus, Neptune, Pluto, their moons, the asteroids, and the comets. (p. 100)

solar wind An outward flow of fast-moving charged particles from the Sun. (pp. 109, 242)

solstice *See* summer solstice, winter solstice.

south celestial pole Point on the celestial sphere directly above Earth's South Pole. (p. 6)

spacetime Single entity combining space and time in special and general relativity.

spatial resolution The dimension of the smallest detail that can be seen in an image.

special relativity Theory proposed by Einstein to deal with the preferred status of the speed of light.

speckle interferometry Technique whereby many short-exposure images of a star are combined to make a high-resolution map of the star's surface.

spectral class Classification scheme, based on the strength of stellar spectral lines, which is an indication of the temperature of a star. (p. 274)

spectral window Wavelength range in which Earth's atmosphere is transparent.

spectrograph Instrument used to produce detailed spectra of stars. Usually, a spectrograph records a spectrum on a CCD detector, for computer analysis.

spectrometer Instrument used to produce detailed spectra of stars. Usually, a spectrograph records a spectrum on a photographic plate, or more recently, in electronic form on a computer.

spectroscope Instrument used to view a light source so that it is split into its component colors. (p. 55)

spectroscopic binary A binary-star system which appears as a single star from Earth, but whose spectral lines show back-and-forth Doppler shifts as two stars orbit one another. (p. 282)

spectroscopic parallax Method of determining the distance to a star by measuring its temperature and then determining its absolute brightness by comparing with a standard Hertzsprung–Russell diagram. The absolute and apparent brightness of the star give the star's distance from Earth. (p. 280)

spectroscopy The study of the way in which atoms absorb and emit electromagnetic radiation. Spectroscopy allows astronomers to determine the chemical composition of stars. (p. 57)

spectrum The separation of light into its component colors.

speed Distance moved per unit time, independent of direction. *See* also velocity.

speed of light The fastest possible speed, according to the currently known laws of physics. Electromagnetic radiation exists in the form of waves or photons moving at the speed of light. (p. 47)

spicule Small solar storm that expels jets of hot matter into the Sun's lower atmosphere.

spin–orbit resonance State that a body is said to be in if its rotation period and its orbital period are related in some simple way.

spiral arm Distribution of material in a galaxy forming a pinwheel-shaped design, beginning near the galactic center. (p. 386)

spiral density wave A wave of matter formed in the plane of planetary rings, similar to ripples on the surface of a pond, which wrap around the rings forming spiral patterns similar to grooves in a record disk. Spiral density waves can lead to the appearance of ringlets. (p. 386)

spiral galaxy Galaxy composed of a flattened, star-forming disk component which may have spiral arms and a large central galactic bulge. (pp. 376, 402)

spiral density wave A proposed explanation for the existence of galactic spiral arms, in which coiled waves of gas compression move through the galactic disk, triggering star formation.

spiral nebula Historical name for spiral galaxies, describing their appearance.

spring tide The largest tides, occurring when the Sun, Moon, and Earth are aligned at new and full Moon.

S-process "Slow" process in which neutrons are captured by a nucleus; the rate is typically one neutron capture per year.

standard candle Any object with an easily recognizable appearance and known luminosity, which can be used in estimating distances. Supernovae, which all have the same peak luminosity (depending on type) are good examples of standard candles and are used to determine distances to other galaxies. (p. 409)

standard solar model A self-consistent picture of the Sun, developed by incorporating the important physical processes that are believed to be important in determining the Sun's internal structure into a computer program. The results of the program are then compared with observations of the Sun and modifications are made to the model. The standard solar model, which enjoys widespread acceptance, is the result of this process. (p. 243)

standard time System of dividing Earth's surface into 24 time zones, with all clocks in each zone keeping the same time.

star A glowing ball of gas held together by its own gravity and powered by nuclear fusion in its core. (p. 242)

star cluster A grouping of anywhere from a dozen to a million stars which formed at the same time from the same cloud of interstellar gas. Stars in clusters are useful to aid our understanding of stellar evolution because, within a given cluster, stars are all roughly the same age and chemical composition and lie at roughly the same distance from Earth. (p. 309)

starburst galaxy Galaxy in which a violent event, such as near-collision, has caused a sudden, intense burst of star formation in the recent past. (p. 438)

Stefan's law Relation that gives the total energy emitted per square centimeter of its surface per second by an object of a given temperature. Stefan's law shows that the energy emitted increases rapidly with an increase in temperature, proportional to the temperature raised to the fourth power. (p. 52)

stellar epoch Most recent period, when stars, planets, and life have appeared in the universe.

stellar nucleosynthesis The formation of heavy elements by the fusion of lighter nuclei in the hearts of stars. Except for hydrogen and helium, all other elements in our universe resulted from stellar nucleosynthesis.

stellar occultation The dimming of starlight produced when a solar system object such as a planet, moon, or ring passes directly in front of a star. (p. 230)

stratosphere The portion of Earth's atmosphere lying above the troposphere, extending up to an altitude of 40–50 km. (p. 139)

string theory Theory that interprets all particles and forces in terms of particular modes of vibration of submicroscopic strings.

strong nuclear force Short-range force responsible for binding atomic nuclei together. The strongest of the four fundamental forces of nature. (p. 257)

S-type asteroid Asteroid made up mainly of silicate or rocky material.

subatomic particle Particle smaller than the size of an atomic nucleus.

subduction zone Place where two plates meet and one slides under the other.

subgiant branch The section of the evolutionary track of a star that corresponds to changes that occur just after hydrogen is depleted in the core, and core hydrogen burning ceases. Shell hydrogen burning heats the outer layers of the star, which causes a general expansion of the stellar envelope. (p. 320)

sublimation Process by which element changes from the solid to the gaseous state, without becoming liquid.

summer solstice Point on the ecliptic where the Sun is at its northernmost point above the celestial equator, occurring on or near June 21. (p. 10)

sunspot An Earth-sized dark blemish found on the surface of the Sun. The dark color of the sunspot indicates that it is a region of lower temperature than its surroundings. (p. 250)

sunspot cycle The fairly regular pattern that the number and distribution of sunspots follows, in which the average number of spots reaches a maximum every 11 or so years, then fall off to almost zero. (p. 252)

supercluster Grouping of several clusters of galaxies into a larger, but not necessarily gravitationally bound, unit. (p. 445)

superforce An attempt to combine the strong and electroweak forces into one single force.

supergiant A star with a radius between 100 and 1000 times that of the Sun. (p. 276)

supergranulation Large-scale flow pattern on the surface of the Sun, consisting of cells measuring up to 30,000 km across, believed to be the imprint of large convective cells deep in the solar interior. (p. 247)

superior conjunction Orbital configuration in which an inferior planet (Mercury or Venus) lies farthest from Earth (on the opposite side of the Sun).

supermassive black hole Black hole having a mass a million to a billion times greater than the mass of the Sun; usually found in the central nucleus of a galaxy.

supernova Explosive death of a star, caused by the sudden onset of nuclear burning (Type I), or an enormously energetic shock wave (Type II). One of the most energetic events of the universe, a supernova may temporarily outshine the rest of the galaxy in which it resides. (p. 331)

supernova remnant The scattered glowing remains from a supernova that occurred in the past. The Crab Nebula is one of the best-studied supernova remnants. (p. 333)

supersymmetric relic Massive particles that should have been created in the Big Bang if supersymmetry is correct.

surface gravity The acceleration due to gravity at the surface of a star or planet. (p. 134)

S-waves Shear waves from an earthquake, which can travel only through solid material and move more slowly than p-waves.

synchrotron radiation Type of nonthermal radiation produced by high-speed charged particles, such as electrons, as they are accelerated in a strong magnetic field. (p. 425)

synchronous orbit State of an object when its period of rotations is exactly equal to its average orbital period. The Moon is in a synchronous orbit, and so presents the same face toward Earth at all times. (p. 136)

synodic month Time required for the Moon to complete a full cycle of phases. (p. 14)

synodic period Time required for a body to return to the same apparent position relative to the Sun, taking Earth's own motion into account; (for a planet) the time between one closest approach to Earth and the next.

T

T Tauri star Protostar in the late stages of formation, often exhibiting violent surface activity. T Tauri stars have been observed to brighten noticeably in a short period of time, consistent with the idea of rapid evolution during this final phase of stellar formation. (p. 306)

tail Component of a comet that consists of material streaming away from the main body, sometimes spanning hundreds of millions of kilometers. May be composed of dust or ionized gases. (p. 108)

tau A type of lepton (along with the electron and muon).

tectonic fracture Cracks on a planet's surface, in particular on the surface of Mars, caused by internal geological activity.

telescope Instrument used to capture as many photons as possible from a given region of the sky and concentrate them into a focused beam for analysis. (p. 70)

temperature A measure of the amount of heat in an object, and an indication of the speed of the particles that comprise it. (p. 51)

tenuous Thin, having low density.

terminator The line separating night from day on the surface of the Moon or a planet.

terrae *See* highlands.

terrestrial planet One of the four innermost planets of the solar system, resembling Earth in general physical and chemical properties. (p. 102)

theories of relativity Einstein's theories, on which much of modern physics rests. Two essential facts of the theory are that nothing can travel faster than the speed of light and that everything, including light, is affected by gravity.

theory A framework of ideas and assumptions used to explain some set of observations and make predictions about the real world. (p. 19)

thermal equilibrium Condition in which new particle-antiparticle pairs are created from photons at the same rate as pairs annihilate one another to produce new photons.

thick disk Region of a spiral galaxy where an intermediate population of stars resides, younger than the halo stars but older than stars in the disk.

threshold temperature Critical temperature above which pair production is possible and below which pair production cannot occur.

tidal bulge Elongation of Earth caused by the difference between the gravitational force on the side nearest the Moon and the force on the side farthest from the Moon. The long axis of the tidal bulge points toward the Moon. More generally, the deformation of any body produced by the tidal effect of a nearby gravitating object. (p. 135)

tidal force The variation in one body's gravitational force from place to place across another body—for example, the variation of the Moon's gravity across Earth. (p. 135)

tidal locking Circumstance in which tidal forces have caused a moon to rotate at exactly the same rate at which it revolves around its parent planet, so that the moon always keeps the same face turned toward the planet.

tidal stability limit The minimum distance within which a moon can approach a planet before being torn apart by the planet's tidal force.

tides Rising and falling motion of terrestrial bodies of water, exhibiting daily, monthly, and yearly cycles. Ocean tides on Earth are caused by the competing gravitational pull of the Moon and Sun on different parts of Earth. (p. 135)

time dilation A prediction of the theory of relativity, closely related to the gravitational reshift. To an outside observer, a clock lowered into a strong gravitational field will appear to run slow. (p. 365)

time zone Region on Earth in which all clocks keep the same time, regardless of the precise position of the Sun in the sky, for consistency in travel and communications.

total eclipse Celestial event during which one body is completely blocked from view by another. (p. 14)

transit Orbital configuration where a planet (i.e., Mercury or Venus) passes between Earth and the Sun.

transition zone The region of rapid temperature increases that separates the Sun's chromosphere from the corona. (p. 242)

transverse motion Motion perpendicular to a particular line of sight, which does not result in Doppler shift in radiation received.

transverse velocity Component of star's velocity perpendicular to the line of sight.

triangulation Method of determining distance based on the principles of geometry. A distant object is sighted from two well-separated locations. The distance between the two locations and the angle between the line joining them and the line to the distant object are all that are necessary to ascertain the object's distance. (p. 17)

triple-alpha process The creation of carbon-12 by the fusion of three helium-4 nuclei (alpha particles). Helium-burning stars occupy a region of the Hertzsprung–Russell diagram known as the horizontal branch.

triple star system Three stars that orbit one another, bound together by gravity.

Trojan asteroid One of two groups of asteroids which orbit at the same distance from the Sun as Jupiter, 60 degrees ahead of and behind the planet. (p. 104)

tropical year The time interval between one vernal equinox and the next. (p. 11)

troposphere The portion of Earth's atmosphere from the surface to about 15 km. (p. 139)

trough Maximum departure of a wave below its undisturbed state.

true space motion True motion of a star, taking into account both its transverse and radial motion according to the Pythagorean theorem.

Tully-Fisher relation A relation used to determine the absolute luminosity of a spiral galaxy. The rotational velocity, measured from the broadening of spectral lines, is related to the total mass, and hence the total luminosity. (p. 409)

turnoff mass The mass of a star that is just now evolving off the main sequence in a star cluster.

21-centimeter line Spectral line in the radio region of the electromagnetic spectrum, associated with a change of spin of the electron in a hydrogen atom. (p. 299)

21-centimeter radiation Radio radiation emitted when an electron in the ground state of a hydrogen atom flips its spin to become parallel to the spin of the proton in the nucleus.

twin quasar Quasar that is seen twice at different locations in the sky due to gravitational lensing.

Type I supernova One possible explosive death of a star. A white dwarf in a binary-star system can accrete enough mass that it cannot support its own weight. The star collapses and temperatures become high enough for carbon fusion to occur. Fusion begins throughout the white dwarf almost simultaneously and an explosion results. (p. 331)

Type II supernova One possible explosive death of a star, in which the highly evolved stellar core rapidly implodes and then explodes, destroying the surrounding star. (p. 331)

U

ultraviolet Region of the electromagnetic spectrum, just beyond the visible range, corresponding to wavelengths slightly shorter than blue light. (p. 44)

ultraviolet telescope A telescope that is designed to collect radiation in the ultraviolet part of the spectrum. Earth's atmosphere is partially opaque to these wavelengths, so ultraviolet telescopes are put on rockets, balloons, and satellites to get high above most or all of the atmosphere. (p. 90)

umbra Central region of the shadow cast by an eclipsing body.

umbra The central region of a sunspot, which is its darkest and coolest part. (p. 15)

unbound An orbit which does not stay in a specific region of space, but where an object escapes the gravitational field of another. Typical unbound orbits are hyperbolic in shape.

unbound trajectory Path of an object with launch speed high enough that it can escape the gravitational pull of a planet.

uncompressed density The density a body would have in the absence of any compression due to its own gravity.

universal time Mean solar time at the Greenwich meridian.

universe The totality of all space, time, matter, and energy. (p. 4)

unstable nucleus Nucleus that cannot exist indefinitely, but rather must eventually decay into other particles or nuclei.

upwelling Upward motion of material having temperature higher than the surrounding medium.

V

Van Allen belts At least two doughnut-shaped regions of magnetically trapped, charged particles high above Earth's atmosphere. (p. 153)

variable star A star whose luminosity changes with time. (p. 378)

velocity Displacement (distance plus direction) per unit time. *See* also speed.

vernal equinox Date on which the Sun crosses the celestial equator moving northward, occurring on or near March 21. (p. 11)

visible light The small range of the electromagnetic spectrum that human eyes perceive as light. The visible spectrum ranges from about 400–700 nm, corresponding to blue through red light. (p. 44)

visible spectrum The small range of the electromagnetic spectrum that human eyes perceive as light. The visible spectrum ranges from about 4000–7000 angstroms, corresponding to blue through red light.

visual binary A binary-star system in which both members are resolvable from Earth. (p. 282)

void Large, relatively empty region of the universe around which superclusters and "walls" of galaxies are organized. (p. 446)

volcano Upwelling of hot lava from below Earth's crust to the planet's surface.

W

wane (referring to the Moon or a planet) To shrink. The Moon appears to wane, or shrink in size, for two weeks after full Moon.

warm longitudes (Mercury) Two opposite points on Mercury's equator where the Sun is directly overhead at aphelion. Cooler than the hot longitudes by 150 degrees.

water hole The radio interval between 18 cm and 21 cm, the respective wavelengths at which hydroxyl (OH) and hydrogen (H) radiate, in which intelligent civilizations might conceivably send their communication signals. (p. 497)

water volcano Volcano that ejects water (molten ice) rather than lava (molten rock) under cold conditions.

watt/kilowatt Unit of power: 1 watt (W) is the emission of 1 joule (J) per second; 1 kilowatt (kW) is 1000 watts.

wave A pattern that repeats itself cyclically in both time and space. Waves are characterized by the speed at which they move, their frequency, and their wavelength. (p. 44)

wavelength The distance from one wave crest to the next, at a given instant in time. (p. 45)

wave period The amount of time required for a wave to repeat itself at a specific point in space. (p. 45)

wave theory of radiation Description of light as a continuous wave phenomenon, rather than as a stream of individual particles.

wax (referring to the Moon or a planet) To grow. The Moon appears to wax, or grow in size, for two weeks after new Moon.

weakly interacting massive particle (WIMP) Class of subatomic particles that might have been produced early in the history of the universe; dark-matter candidates.

weak nuclear force Short-range force, weaker than both electromagnetism and the strong force, but much stronger than gravity; responsible for certain nuclear reactions and radioactive decays. (p. 258)

weight The gravitational force exerted on you by Earth (or the planet on which you happen to be standing).

weird terrain A region on the surface of Mercury with oddly rippled features. This feature is thought to be the result of a strong impact which occurred on the other side of the planet, and sent seismic waves traveling around the planet, converging in the weird region.

white dwarf A dwarf star with sufficiently high surface temperature that it glows white. (p. 276)

white-dwarf region The bottom-left corner of the Hertzsprung–Russell diagram, where white-dwarf stars are found. (p. 278)

white oval Light-colored region near the Great Red Spot in Jupiter's atmosphere. Like the red spot, such regions are apparently rotating storm systems. (p. 200)

Wien's law Relation between the wavelength at which a blackbody curve peaks and the temperature of the emitter. The peak wavelength is inversely proportional to the temperature, so the hotter the object, the bluer its radiation. (p. 52)

winter solstice Point on the ecliptic where the Sun is at its southernmost point below the celestial equator, occurring on or near December 21. (p. 10)

wispy terrain Prominent light-colored streaks on Rhea, one of Saturn's moons.

X

X-ray Region of the electromagnetic spectrum corresponding to radiation of high frequency and short wavelength, far beyond the visible spectrum. (p. 44)

X-ray burster X-ray source that radiates thousands of times more energy than our Sun in short bursts lasting only a few seconds. A neutron star in a binary system accretes matter onto its surface until temperatures reach the level needed for hydrogen fusion to occur. The result is a sudden period of rapid nuclear burning and release of energy. (p. 350)

X-ray nova Nova explosion detected at X-ray wavelengths.

Z

Zeeman effect Broadening or splitting of spectral lines due to the presence of a magnetic field.

zero-age main sequence The region on the Hertzsprung–Russell diagram, as predicted by theoretical models, where stars are located at the onset of nuclear burning in their cores. (p. 308)

zodiac The twelve constellations on the celestial sphere through which the Sun appears to pass during the course of a year. (p. 9)

zonal flow Alternating regions of westward and eastward flow, roughly symmetrical about the equator of Jupiter, associated with the belts and zones in the planet's atmosphere. (p. 199)

zone Bright, high-pressure region in the atmosphere of a jovian planet, where gas flows upward.

Answers to Concept Check Questions

Exploring/Introduction

1. (p. 7) (1) Because the celestial sphere provides a natural means of specifying the locations of stars on the sky. Celestial coordinates are directly related to Earth's orientation in space but are independent of Earth's rotation. (2) Distance information is lost.   2. (p. 11) Because of the tilt of Earth's rotation axis, the Sun is highest in the northern sky during northern summer, and the days are longest.   3. (p. 15) (1) The angular size of the Moon would remain the same, but that of the Sun would be halved, making it easier for the Moon to eclipse the Sun. We would expect to see total or partial eclipses, but no annular ones. (2) If the distance were halved, the angular size of the Sun would double, and total eclipses would never be seen, only partial or annular ones.   4. (p. 18) Because astronomical objects are too distant for direct ("measuring tape") measurements, so we must rely on indirect means and mathematical reasoning.   5 (p. 21) A theory can never become proven "fact," because it can always be invalidated, or forced to change, by a single contradictory observation. However, once a theory's predictions have been repeatedly verified by experiments over many years, it is often widely regarded as "true."

Chapter 1

1. (p. 28) In the geocentric view, retrograde motion is a real backward motion of a planet as it moves on its epicycle. In the heliocentric view, the backward motion is only apparent, caused by Earth "overtaking" the planet in its orbit.   2. (p. 31) The discovery of the phases of Venus could not be reconciled with the geocentric model. Observations of Jupiter's moons proved that some objects in the universe did not orbit Earth.   3. (p. 32) Galileo was an experimentalist; Kepler was a theorist. Galileo found observational evidence supporting the Copernican theory; Kepler found an empirical description of the motions of the planets within the Copernican picture that greatly simplified the theoretical view of the solar system.   4. (p. 33) Because those planets were unknown to Kepler. The fact that the laws apply to the outer planets then constitutes a prediction subsequently verified by observation.   5. (p. 34) Because Kepler determined the overall geometry of the solar system by triangulation using Earth's orbit as a baseline, so all distances were known only relative to the scale of Earth's orbit—the astronomical unit.   6. (p. 39) In the absence of any force, a planet would move in a straight line with constant velocity (Newton's first law) and therefore tends to move along the tangent to its orbital path. The Sun's gravity causes the planet to accelerate toward the Sun (Newton's second law), bending its trajectory into the orbit we observe.

Chapter 2

1. (p. 46) A wave is a way in which energy or information is transferred from place to place via a repetitive, regularly varying disturbance of some sort. A wave is characterized by its wavelength, frequency, velocity, and amplitude: the product of the first two always equals the third.   2. (p. 48) Light is an electromagnetic wave produced by accelerating charged particles. All waves on the list are characterized by their wave periods, frequencies, and wavelengths, and all carry energy and information from one location to another. Unlike waves in water or air, however, light waves require no physical medium through which to propagate.   3. (p. 50) All are electromagnetic radiation and travel at the speed of light. In physical terms, they differ only in frequency (or wavelength), although their effects on our bodies (or our detectors) differ greatly.   4. (p. 54) As the switch is turned and the temperature of the filament rises, the bulb's brightness increases rapidly, by Stefan's law, and its color shifts, by Wien's law, from infrared to red to yellow to white.   5. (p. 58) They are characteristic frequencies (wavelengths) at which matter absorbs or emits photons of electromagnetic radiation. They are unique to each atom or molecule and thus provide a means of identifying the gas producing them.   6. (p. 61) Electron orbits can occur only at certain specific energies, and there is a ground state that has the lowest possible energy. Planetary orbits can have any energy. Planets can stay in any orbit indefinitely; in an atom the electron must eventually fall to the ground state, emitting electromagnetic radiation in the process. Planets may reasonably be thought of as having specific trajectories around the Sun—there is no ambiguity as to "where" a planet is. Electrons in an atom are smeared out into an electron cloud, and we can only talk of the electron's location in probabilistic terms.   7. (p. 62) Spectral lines correspond to transitions between specific orbitals within an atom. The structure of an atom determines the energies of these orbitals, and hence the possible transitions, and thus the energies (colors) of the photons involved.

8. (p. 64) According to *More Precisely 1-2,* measuring masses in astronomy usually entails measuring the orbital speed of one object—a companion star or a planet, perhaps—around another. In most cases, the Doppler effect is an astronomer's only way of making such measurements.   9. (p. 65) Because, with few exceptions, spectral analysis is the only way we have of determining the physical conditions—composition, temperature, density, velocity, etc.,—in a distant object. Without spectral analysis, astronomers would know next to nothing about the properties of stars and galaxies.

Chapter 3

1. (p. 77) Because large reflecting telescopes are easier to design, build, and maintain than refracting instruments.   2. (p. 80) The need to gather as much light as possible, and the need to achieve the highest possible angular resolution.   3. (p. 82) The atmosphere absorbs some radiation arriving from space, and turbulent motion blurs the incoming rays. To reduce or overcome the effects of atmospheric absorption, instruments are placed on high mountains or in space. To compensate for atmospheric turbulence, adaptive optics techniques probe the air above the observing site and adjust the mirror surface accordingly to try to recover the undistorted image.   4. (p. 86) Radio astronomy opens up a new window on the universe, allowing astronomers to study known objects in new ways and to discover new objects that would otherwise be completely unobservable. The faintness of the sources is addressed by combining the largest available telescopes with sensitive detectors. Resolution can be greatly improved by the use of interferometry, which combines the signals from two or more separate telescopes to create the effect of a single instrument of much larger diameter.   5. (p. 92) Benefits: They are above the atmosphere, so they are unaffected by seeing or absorption; they can also make round-the-clock observations of the entire sky. Drawbacks: Cost/smaller size, inaccessibility, vulnerability to damage by radiation and cosmic rays.

Chapter 4

1. (p. 103) Because the two classes of planets differ in almost every physical property—orbit, mass, radius, composition, existence of rings, number of moons.   2. (p. 107) Similarities: All orbit in the inner solar system, roughly in the ecliptic plane, and are solid bodies of generally "terrestrial" composition. Differences: Asteroids are much smaller than the terrestrial planets, are generally less evolved than the planets, and move on much less regular orbits.   3. (p. 110) Most comets never come close enough to the Sun for us to see them.   4. (p. 115) Because it is thought to be much less evolved than material now found in planets, and hence it is a better indicator of conditions in the early solar system.   5. (p. 117) Because it must explain certain general features of solar-system architecture, while accommodating the fact that there are exceptions to many of them.   6. (p. 125) Yes. If a star formed with a disk of matter around it then, even if it doesn't have planets like Earth, the basic processes of condensation and accretion would probably have occurred there, too.   7. (p. 128) Because the search techniques are most sensitive to massive planets orbiting close to their parent stars, which are exactly what have been observed.

Chapter 5

1. (p. 138) A tidal force is the *variation* in one body's gravitational force from place to place across another. Tidal forces tend to deform a body rather than causing an overall acceleration, and they decrease rapidly with distance.   2. (p. 143) It raises Earth's average surface temperature above the freezing point of water, which was critical for the development of life on our planet. Should the greenhouse effect continue to strengthen, however, it may conceivably cause catastrophic climate changes on Earth.   3. (p. 146) We would have far less detailed direct (from volcanoes) or indirect (from seismic studies following earthquakes) information on our planet's interior.   4. (p. 149) Convection currents in the upper mantle cause portions of Earth's crust—plates—to slide around on the surface. As the plates move and interact, they are responsible for volcanism, earthquakes, the formation of mountain ranges and ocean trenches, and the creation and destruction of oceans and continents.   5. (p. 153) The maria are younger, denser, and are much less heavily cratered than the highlands.   6. (p. 153) It tells us that the planet has a conducting, liquid core in which the magnetic field is continuously generated.   7. (p. 157) The Moon's small size, which is the main reason that the Moon (1) has lost its initial heat, making it geologically dead today, and (2) has insufficient gravity to retain an atmosphere.

Chapter 6

1. (p. 164) The Sun's tidal force has caused Mercury's rotation period to become exactly 2/3 of the orbital period, and the rotation axis to be oriented exactly perpendicular to the planet's orbit plane. A synchronous orbit is not possible, because of Mercury's eccentric orbit around the Sun.   2. (p. 166) Venus's atmosphere is much more massive, hotter, and denser than that of Earth, and is composed almost entirely of carbon dioxide.   3. (p. 167) Scarps are thought to have been formed when the planet's iron core cooled and contracted, causing the crust to crack. Faults on Earth are the result of tectonic activity.   4. (p. 170) No. They are mainly shield volcanoes formed by lava upwelling through "hot spots" in the crust and are not associated with plate tectonics (which is not observed on Venus).   5. (p. 179) Observations reveal evidence for dried-up rivers, outflow channels, and what may be ancient oceans, strongly suggesting that the surface was once warm enough and the atmosphere thick enough for large bodies of liquid water to exist on the planet's surface. Today most of the remaining water on Mars is frozen, underground or at the poles.   6. (p. 184) Venus probably has a large liquid metal core, but it rotates too slowly for a planetary dynamo to operate. Mars rotates rapidly but lacks a liquid metal core.   7. (p. 187) Both might have climates quite similar to that on Earth, as the greenhouse effect would probably not have run away on Venus, and the water in the Martian atmosphere might not have frozen and become permafrost.

Chapter 7

1. (p. 195) Uranus's orbit was observed to deviate from a perfect ellipse, leading astronomers to try to compute the mass and location of the body responsible for those discrepancies. That body was Neptune.   2. (p. 198) The magnetic field is generated by the motion of electrically conducting liquid in the deep interior and therefore presumably shares the rotation of that region of the planet.   3. (p. 203) Like weather systems on Earth, the belts and zones are regions of high and low pressure and are associated with convective motion. However, unlike storms on Earth, they wrap all the way around the planet because of Jupiter's rapid rotation. In addition, the clouds are arranged in three distinct layers, and the bright colors are the result of cloud chemistry unlike anything operating in Earth's atmosphere. Jupiter's spots are somewhat similar to hurricanes on Earth, but they are far larger and longer-lived.   4. (p. 204) Uranus's low temperature means that the planet's clouds lie deeper in the atmosphere, making them harder to distinguish from Earth.   5. (p. 209) They all combine the two essential ingredients needed to operate a planetary dynamo: (1) rapid rotation, in all cases faster than Earth's rotation, and (2) conducting liquid interiors: metallic hydrogen in the cases of Jupiter and Saturn, and "slushy" ice in the cases of Uranus and Neptune.

Chapter 8

1. (p. 219) Jupiter's gravitational field, via its tidal effect on the moons.   2. (p. 223) The haze layers high in the moon's dense atmosphere are opaque to visible light. The *Huygens* probe landed on the surface, and *Cassini*'s radar and infrared sensors can penetrate the haze.   3. (p. 224) They are all very heavily cratered—a result of the heavy bombardment that marked the clearing of the comets from the outer solar system.   4. (p. 231) The Roche limit is the radius inside of which a moon will be torn apart by tidal forces. Planetary rings are found inside the Roche limit, (most) moons outside.   5. (p. 232) They have similar masses, radii, composition, and perhaps also similar origins in the Kuiper Belt.   6. (p. 235) Because of its similarities to the known Kuiper Belt objects, as well as to the icy moons of the outer planets, it is now more commonly regarded as a large member of the Kuiper Belt rather than as a small planet.

Chapter 9

1. (p. 243) When we multiply the solar constant by the total area to obtain the solar luminosity, we are implicitly assuming that the same amount of energy reaches every square meter of the large sphere in Figure 9.3.   2. (p. 246) The energy may be carried in the form of (1) radiation, where energy travels in the form of light, and (2) convection, where energy is carried by physical motion of upwelling solar gas.   3. (p. 249) The coronal spectrum shows (1) emission lines of (2) highly ionized elements, implying high temperature.   4. (p. 256) There is a strong field, with a well defined east-west organization, just below the surface. The field direction in the southern hemisphere is opposite than in the north. However, the details of the fields are very complex.   5. (p. 261) Because the Sun shines by nuclear fusion, which converts mass into energy as hydrogen turns into helium.

Chapter 10

1. (p. 267) Because stars are so far away that their parallaxes relative to any baseline on Earth are too small to measure accurately.   2. (p. 271) Nothing—We need to know their distances before the luminosities (or absolute magnitudes) can be determined.   3. (p. 275) Because temperature controls which excited states the star's atoms and ions are in, and hence which atomic transitions are possible.   4. (p. 276) Yes, by using the radius–luminosity–temperature relationship, but only if we can find a method of determining the luminosity that doesn't depend on the

inverse-square law (Sec. 10.2).   5. (p. 279) Because giants are intrinsically very luminous and can be seen to much greater distances than the more common main-sequence or white dwarfs.   6. (p. 280) All stars would be further away, but their measured spectral types and apparent brightnesses would presumably be unchanged, so their luminosities would be greater than previously thought. The main sequence would therefore move vertically upward in the H–R diagram. We would then use larger luminosities in the method of spectroscopic parallax, so distances inferred by that method would also increase.   7. (p. 284) We don't. We assume that their masses are the same as similar stars found in binaries.

Chapter 11

1. (p. 292) Because the scale of interstellar space is so large that even very low densities can accumulate to a large amount of obscuring matter along the line of sight to a distant star.   2. (p. 296) Because the UV light is absorbed by hydrogen gas in the surrounding cloud, ionizing it to form the emission nebulae. The red light is Hα radiation, part of the visible hydrogen spectrum emitted as electrons, and protons recombine to form hydrogen atoms.   3. (p. 301) Because the cloud's main constituent, hydrogen, is very hard to observe, so astronomers must use other molecules as tracers of the cloud's properties.   4. (p. 306) (1) The existence of a photosphere, meaning that the inner part of the cloud becomes opaque to its own radiation, signaling the slowing of the collapse phase. (2) Nuclear fusion in the core, and equilibrium between pressure and gravity.   5. (p. 308) No. Different parts of the main sequence correspond to stars of different masses. A typical star stays at roughly the same location on the main sequence of most of its lifetime.   6. (p. 312) Because stars form at different rates—high-mass stars reach the main sequence and start disrupting the parent cloud long before lower-mass stars have finished forming.

Chapter 12

1. (p. 319) Hydrostatic equilibrium means that, if some property of the Sun changes a little, the star's structure adjusts to compensate. Small changes in the Sun's internal temperature or pressure will not lead to large changes in its radius or luminosity.   2. (p. 321) Because the nonburning inner core, unsupported by fusion, begins to shrink, releasing gravitational energy, heating the overlying layers, causing them to burn more vigorously, and increasing the luminosity.   3. (p. 327) Because iron cannot fuse to produce energy. As a result, no further nuclear reactions are possible, and the core's equilibrium cannot be restored.   4. (p. 330) Because the two types of supernova differ in their spectra and their light curves, making it impossible to explain them in terms of a single phenomenon.   5. (p. 334) Because a star cluster gives us a "snapshot" of stars of many different masses—but of the same age and initial composition—allowing us to directly test the predictions of the theory.   6. (p. 339) Because it is responsible for creating and dispersing all the heavy elements out of which we are made. In addition, it may also have played an important role in triggering the collapse of the interstellar cloud from which our solar system formed.

Chapter 13

1. (p. 346) No. Only Type-II supernovae. According to theory, the rebounding central core of the original star becomes a neutron star.   2. (p. 348) Because (1) not all supernovae form neutron stars, (2) the emission is beamed, and may not be observable from Earth, and (3) pulsars spin down and become too faint to observe after a few tens of millions of years.   3. (p. 352) Some X-ray sources are binaries containing accreting neutron stars, which may be in the process of being spun up to form millisecond pulsars.   4. (p. 355) They are energetic bursts of gamma rays, roughly uniformly distributed on the sky, occurring roughly once per day. They pose a challenge because they are very distant, and hence extremely luminous, but their energy comes from a region less than a few hundred kilometers across.   5. (p. 358) Its gravity becomes so strong that even light cannot escape, and it becomes a black hole.   6. (p. 363) Newton's theory describes gravity as a force produced by a massive object that influences all other massive objects. Einstein's relativity describes gravity as a curvature of spacetime produced by a massive object; that curvature then determines the trajectories of all particles—matter or radiation—in the universe.   7. (p. 367) Because the object would appear to take infinitely long to reach the event horizon, and its light would be infinitely redshifted by the time it got there.   8. (p. 369) By observing their gravitational effects on other objects, and from the X-rays emitted when matter falls into them.

Chapter 14

1. (p. 377) The Milky Way is the thin plane of the Galactic disk, seen from within. When our line of sight lies in the plane of the Galaxy, we see many stars blurring into a continuous band. In other directions we see darkness.   2. (p. 380) No, because even the brightest Cepheids are unobservable at distances of more than a kiloparsec or so through the obscuration of interstellar dust.   3. (p. 384) They differ greatly in their spatial distributions of stars, the types and ages of stars found in each, the amount of interstellar gas they contain, and in the orbital motions of

the stars comprising them.   4. (p. 385) Because all the gas and dust in the halo fell to the Galactic disk billions of years ago, so halo star formation has long since ceased.   5. (p. 389) Because differential rotation would destroy the spiral structure within a few hundred million years.   6. (p. 393) Its emission has not been observed at any electromagnetic wavelength. Its presence is inferred from its gravitational effect on stars and gas orbiting the Galactic center.   7. (p. 396) Observations of rapidly moving stars, gas, and the variability of the radiation emitted suggest the presence of a roughly 3 million solar-mass black hole.

Chapter 15

1. (p. 408) Most galaxies are not large spirals—the most common galaxy types are dwarf ellipticals and dwarf irregulars.   2. (p. 410) Distance-measurement techniques ultimately rely upon the existence of very bright objects whose luminosities can be inferred by other means. Such objects become increasingly hard to find and calibrate the farther we look out into intergalactic space. 3. (p. 414) It doesn't use the inverse-square law. The other methods all provide a way of determining the luminosity of a distant object, which then is converted to a distance using the inverse-square law. Hubble's law gives a direct connection between redshift and distance.   4. (p. 422) Because it means that the energy source cannot simply be the summed energy of a huge number of stars—some other mechanism must be at work.   5. (p. 427) The energy is generated in an accretion disk in the central nucleus of the visible galaxy, then transported by jets out of the galaxy and into the lobes, where it is eventually emitted by the synchrotron process in the form of radio waves.

Chapter 16

1. (p. 434) First, that the galaxies are gravitationally bound to the cluster. Second, and more fundamentally, that the laws of physics as we know them in the solar system—gravity, atomic structure, the Doppler effect—all apply on very large scales, and to systems possibly containing a lot of dark matter.   2. (p. 438) Galaxies in voids are much less likely to encounter other galaxies, so their evolution will be dictated more by "internal" processes—star formation and stellar evolution—than galaxies in clusters, which will be more affected by interactions with other galaxies.   3. (p. 444) Probably not. There may well be galaxies which do not harbor central black holes, and in any case, only galaxies in clusters are likely to experience the encounters that trigger activity.   4. (p. 451) Light traveling from these objects to Earth is influenced by cosmic structure all along the line of sight. Light rays are deflected by the gravitational field of intervening concentrations of mass, and gas along the line of sight produces absorption lines whose redshifts tell us the distance at which each feature formed. The light received on Earth thus gives astronomers a "core sample" through the universe, from which detailed information can be extracted.

Chapter 17

1. (p. 460) When viewed on very large scales—more than 300 Mpc—the distribution of galaxies seems to be roughly the same everywhere and in all directions. (In addition, the microwave background radiation field is observed to be isotropic to very high precision.)   2. (p. 461) Because, tracing the motion

backward in time, it implies that all galaxies, and in fact everything in the entire universe, were located at a single point at the same instant in the past. The cosmological principle does not allow that point to be special, leaving a Big Bang as the only possibility.   3. (p. 462) A low-density universe has negative curvature and is infinite in extent; a critical-density universe is spatially flat (Euclidean) and is also infinite in extent; a high-density universe has positive curvature and is finite in extent.   4. (p. 469) There doesn't seem to be enough matter to halt the collapse, and in addition, the observed cosmic acceleration suggests the existence of a large-scale repulsive force in the cosmos that also opposes recollapse.   5. (p. 470) It was formed at the time of the Big Bang. It is the electromagnetic remnant of the primeval fireball. It started off as a blackbody curve because it was the thermal emission of the hot, young cosmos. It has retained this form because the cosmological redshift has stretched all photons by the same factor, preserving the shape of the blackbody curve.   6. (p. 472) The amount of deuterium observed in the universe today implies that the present density of normal matter is at most a few percent of the critical value—much less than the density of dark matter inferred from dynamical studies.   7. (p. 476) Not directly. Inflation implies that the universe is flat, and hence that the total cosmic density equals the critical value. However, the matter density seems to be only about one third of the critical value, and the density of electromagnetic radiation (the microwave background) is a tiny fraction of the critical density. Together these statements suggest that the remaining density is in the form of "dark energy," as described in Section 17.3. 8. (p. 479) Because dark matter is largely unaffected by the background radiation field, it was able to clump earlier than normal matter, forming the structures within which luminous matter subsequently formed the galaxies we see today. The gravity of the early dark-matter clumps caused tiny fluctuations in the temperature of the cosmic microwave background. The ripples we see today in the microwave background are the beginning of galaxy formation long ago.

Chapter 18

1. (p. 488) Not yet. The formation of complex molecules from simple ingredients by nonbiological processes has been repeatedly demonstrated, but no living cell or self-replicating molecule has ever been created.   2. (p. 490) Mars remains the most likely site because of its generally Earthlike makeup and because several lines of evidence suggest that the planet was warmer in the past and may have had liquid water on its surface. Europa and Titan also have properties that might have been conducive to the emergence of living organisms—liquid water under Europa's icy crust and an atmosphere and many complex molecules on Titan. 3. (p. 495) It breaks a complex problem up into simpler "astronomical," "biochemical," "anthropological," and "cultural" pieces, which can be analyzed separately. The analysis itself helps identify the types of stars where a search might be most fruitful.   4. (p. 496) We assume that a technological civilization will opt to work in the radio part of the spectrum, where Galactic absorption is least, and will choose a region where natural Galactic background "static" is minimized. We also assume that the 21-cm line of hydrogen will have significance to them, since hydrogen is the most common element in the universe, and that hydroxyl will also have special meaning if they evolved in a water environment as we did.

Answers to Self-Test Questions

Exploring/Introduction

True or False?/Multiple Choice   1) F   2) F   3) T   4) F   5) T   6) T   7) F
8) b   9) b   10) c   11) a   12) c   13) d   14) b   15) a
Problems   1) Scorpius   2) 29.9 km, 1.08×10^5 km, 2.58×10^6 km   3) It
would decrease by 8 minutes   4) (a) 33 arc minutes, (b) 33 arc seconds, (c) 0.55
arc seconds; 56.4 minutes   5) 1.0 km/s   6) (a) 57,300 km, (b) 3.44×10^6 km,
(c) 2.06×10^8 km   7) 0.9 degrees, for a 1.5 cm thumb at a distance of 1m
8) 12,000 km   9) 391   10) 0

Chapter 1

True or False?/Multiple Choice   1) F   2) F   3) F   4) F   5) F   6) F   7) F
8) T   9) d   10) c   11) c   12) a   13) c   14) b   15) b
Problems   1) (a) 110 km , (b) 44,000 km, (c) 370,000 km   2) 700 s   3) 0.4
degrees retrograde   4) 3 AU, 0.333, 5.2 yr   5) 35 AU   6) Pluto
perihelion = 29.65 AU, Neptune perihelion = 29.80 AU   7) 9.42×10^{-4}
solar = 1.88×10^{27} kg   8) 9.50, 7.33, 1.49 m/s²   9) For a 70 kg person,
force = 684 N = 154 pounds; we call it the person's weight   10) 1.7 km/s

Chapter 2

True or False?/Multiple Choice   1) F   2) T   3) T   4) F   5) T   6) T   7) T
8) a   9) d   10) a   11) b   12) c   13) d   14) a   15) d
Problems   1) 1480 m/s   2) 3 m   3) 23 Hz; radio   4) A; 3.25; 112   5) 310;
9.4 microns; infrared   6) 6.4×10^7 W/m²; 3.9×10^{26} W   7) 3×10^{10}
8) Second excited state: 3 possibilities: $2 \to 1$ (Hα, 656 nm), $2 \to$ ground
(Lyβ, 103 nm), $1 \to$ ground (Lyα, 122 nm); Third excited state: 6 possibilities:
those already listed, plus $3 \to 2$ (called Paschen α, 1876 nm), $3 \to 1$ (Hβ,
486 nm), $3 \to$ ground (Lyγ, 97.3 nm)   9) 137 km/s approaching
10) 1.94×10^{27} kg

Chapter 3

True or False?/Multiple Choice   1) F   2) F   3) T   4) T   5) F   6) F   7) F
8) d   9) d   10) c   11) a   12) d   13) b   14) b   15) c
Problems   1) 0.3 arc seconds; 6.8 pixels   2) 580 μm; longer than the
$3 - 200$ μm operating range   3) 6.7 minutes; 1.7 minutes   4) (a) 0.25″,
(b) 0.01″   5) 15 m   6) 160 light years, using the formula in the text to compute
the resolution   7) 3.3 minutes   8) (a) 36, (b) 0.61, (c) 0.012 light years
9) 14.1 m, 16 m   10) (a) 0.003 arc seconds, (b) 0.005 arc seconds

Chapter 4

True or False?/Multiple Choice   1) T   2) F   3) F   4) F   5) T   6) T   7) F
8) d   9) b   10) d   11) a   12) c   13) b   14) d   15) c
Problems   1) 2×10^{21} kg, 0.03 percent of Earth's mass; 40 km   2) 1.1 AU, 2.0
AU   3) 3 kg   4) 1, 0.1, 0.01, 0.001 times Ceres's mass, or 1.6×10^{21} kg,
1.6×10^{20} kg, 1.6×10^{19} kg, 1.6×10^{18} kg.   5) (a) 11 million years, (b) 50
AU   6) (a) 1.9×10^{43}, 7.8×10^{42}, and 2.7×10^{40} kg·m²/s, (b) 1.0×10^{31}
kg·m²/s   7) It doubles.   8) 3.8 million   9) 2×10^8; once every 2.5 years
10) 0.56

Chapter 5

True or False?/Multiple Choice   1) T   2) T   3) T   4) F   5) T   6) F   7) F
8) b   9) d   10) b   11) b   12) d   13) d   14) a   15) c
Problems   1) 5.3 m/s² = 0.55 times the actual value, 8.3 km/s   2) 30 kg
3) 21 m = 70 feet   4) 1.6×10^{-7} 5) (a) 8.4×10^{-3}, (b) 0.16, (c) 0.84,
(d) 7.0×10^{-3} 6) 5.0×10^{18} kg = 8.4×10^{-7} times Earth's mass   7) 45
percent   8) 43 minutes   9) Need at least 4.8×10^5 such craters, requiring at
least 4.8 trillion years; rate must have been greater by a factor of about 1000
10) 93 m

Chapter 6

True or False?/Multiple Choice   1) T   2) F   3) T   4) T   5) T   6) F   7) F
8) b   9) b   10) d   11) c   12) b   13) b   14) a   15) c
Problems   1) 8.8 minutes   2) 4.0×10^{23} kg, (for a temperature of 700 K) or
roughly 22 percent greater than it is now   3) 4.9×10^{20} kg, 98 times more
massive than Earth's atmosphere, 10^{-4} times the mass of Venus   4) 400 km/h,
250 mph   5) (a) 12,100 km, (b) yes—the smallest detectable feature would be
about 20 km across, much smaller than the largest impact features   6) 197
minutes, 94 minutes   7) 28 kg, assuming I weigh 70 kg on Earth

8) 10 m/s = 22 mph   9) 2)7×10^{17} kg; 4.5×10^{-7} times the mass of Mars;
5.5×10^{-4} times the mass of Venus's atmosphere   10) 16′, 2.7′; no—the
minimum angular diameter of the Sun, as seen from Mars, is 19′

Chapter 7

True or False?/Multiple Choice   1) F   2) T   3) F   4) T   5) T   6) F   7) T
8) T   9) a   10) d   11) d   12) b   13) c   14) a   15) b
Problems   1) 150 km, or 0.002 planetary radii; 1100 km, or 0.04 planetary radii
2) 2.2×10^{17} newton; 1.4×10^{21} newton, or 6400 times larger   3) 4200 s =
1.17 hours; 84,000 km   4) 6×10^{-5} Jupiter masses, or 0.019 Earth masses
5) —   6) 1.5 percent of the actual mass   7) 10.5 days   8) 74 K
9) 1.7×10^{25} kg (2.8 Earth masses); 81 percent   10) 49 μm; infrared

Chapter 8

True or False?/Multiple Choice   1) T   2) T   3) T   4) T   5) F   6) T   7) T
8) F   9) a   10) c   11) c   12) b   13) d   14) c   15) b
Problems   1) 20.3 km/s, 7.6 km/s; Saturn is much more massive than Earth!
2) 2.3 Jupiter radii   3) For Io: acceleration difference = 6.3×10^{-3} m/s²,
surface gravity = 1.8 m/s² , ratio = 0.0035 = 0.35 percent; For Moon:
acceleration difference = 2.2×10^{-5} m/s², surface gravity = 1.6 m/s²,
ratio = 1.4×10^{-5} 4) 1.35 m/s² (Earth: 9.80 m/s²); 2.6 km/s   5) Io: 36′,
Europa: 18′, Ganymede: 18′, Callisto: 9′, Sun: 6′. Yes   6) 38 km radius
7) 1.1×10^{18} 8) 6.8 times farther out   9) For a weight of 70 kg, 4.7 kg on
Pluto, 2.1 kg on Charon   10) 11.1 hours; 28.700 km; 2.8 orbits

Chapter 9

True or False?/Multiple Choice   1) T   2) F   3) F   4) F   5) T   6) F   7) F
8) c   9) c   10) a   11) b   12) c   13) a   14) b   15) c
Problems   1) (a) 14,600 W/m²; (b) 52 W/m²   2) (a) 3000 km, (b) 1500,
(c) 1/33 of the 167-minute orbital period   3) (a) 0.29 nm (hard X-ray), (b) 29
nm (far UV), (c) 290 nm (near UV)   4) 1000 s = 17 min, comparable to the
granule lifetime   5) 36 percent of the average solar value (64 percent less)
6) (a) radiation mass loss = 4.3 million tons/s ≈ twice the mass loss due to the
solar wind; (b) 30 trillion years   7) 83 days   8) 300,000 years   9) 74 billion
years   10) 4×10^{28}

Chapter 10

True or False?/Multiple Choice   1) F   2) F   3) F   4) F   5) T   6) F   7) F
8) d   9) b   10) b   11) c   12) d   13) c   14) b   15) b
Problems   1) 77 pc; 0.39″   2) 47 km/s; 56 km/s   3) (a) 80 times the
luminosity of the Sun; (b) 2 solar radii   4) B is 3 times farther away   5) A is 10
times farther away   6) 3.3×10^{-10} W/m²; 2.3×10^{-13} times the solar constant
7) factor of 16 trillion   8) 320 pc   9) 3.5, 2.3 solar masses   10) (a) 200 billion
years, (b) 1 billion years, (c) 100 million years

Chapter 11

True or False?/Multiple Choice   1) F   2) F   3) F   4) T   5) F   6) T   7) T
8) d   9) b   10) c   11) d   12) b   13) b   14) c   15) b
Problems   1) 1.9 grams   2) 1428.6 MHz for a wavelength of 21 cm (but note
that the precise wavelength is 21.11 cm, corresponding to frequency of 1420.4
MHz); frequency: 1428.8—1428.2 MHz, wavelength: 20.9965—21.0053 cm
3) 0.017 pc = 3500 AU   4) 4100, 9.0   5) escape speeds (km/s): 1.8, 1.1, 1.1,
0.80; average molecular speeds: 13.6, 14.0, 14.6, 14.2; No   6) yes, barely: the
escape speed is 0.93 km/s, molecular speed is 0.35 km/s   7) (a) 200, (b) 2.3
solar radii   8) 8   9) 10^{-6} solar luminosities   10) 37 pc

Chapter 12

True or False?/Multiple Choice   1) T   2) T   3) F   4) T   5) F   6) T   7) T
8) F   9) a   10) b   11) a   12) b   13) b   14) c   15) d
Problems   1) 3.5×10^7 kg/m³; 5.7×10^{-4} kg/m³, smaller by a factor of
1.6×10^{-11}; solar central density is 150,000 kg/m³, smaller than the white dwarf
density by a factor of 4.3×10^{-3} 2) 220 pc   3) (a) 2.9 solar masses, (b) 1.7
solar masses   4) 7.1 years, 63,000 years   5) 7300 km/s, 500,000 Earth gravities
6) 20, 3.2 Mpc; 1000 Mpc   7) 0.4 pc; no—there are no O or B stars (in fact no
stars at all) within that distance of the Sun   8) apparent magnitude is −14.1, 1.6

magnitudes (4 times) brighter than the full Moon, 9.7 magnitudes (7600 times) brighter than Venus; yes—according to Appendix 3, several O and B stars, as well as Betelgeuse and some other red supergiants, lie within that distance   9) roughly 1000 km/s; not too bad an assumption—the Nebula is moving too fast to be affected much by gravity, although it is probably slowing down as it runs into the interstellar medium   10) 1.2×10^{44} J; 220 Earth masses

Chapter 13

True or False?/Multiple Choice   1) T   2) T   3) T   4) F   5) T   6) F   7) F   8) b   9) a   10) b   11) c   12) b   13) c   14) c   15) b
Problems   1) 1 million revolutions per day, or 11.6 revolutions per second   2) 3×10^{14} times higher—2.1×10^{16} kg for a 70-kg human; Moon's mass is greater by a factor of 3.5 million; asteroid mass is roughly 1.6×10^{12} kg (assuming spherical shape and density 3000 kg/m^3), smaller by a factor of 13,000   3) 1.9×10^{12} m/s^2, or 190 billion Earth gravities; 190,000 km/s, or 64 percent of the speed of light; 300,000 km/s   4) 1.6×10^{-5}, 1.6×10^{3}, 1.6×10^{11} solar luminosities; ultraviolet, X rays, gamma rays; they might be very bright immediately after formation, but would become very faint as they cool; the coolest could be plotted at the bottom left of the diagram, as a very faint O-type "star"   5) 2.4×10^{44} J, roughly twice the Sun's lifetime energy output; 2.4×10^{34} J; 1.4×10^{25} J   6) 140 μs   7) 3 million km = 4.3 solar radii; 3 billion km = 20 AU   8) 2×10^{10} m/s^2 = 2×10^{9} g; 2×10^{-3} g; 2×10^{-9} g   9) 1760 km   10) 3×10^{7} km = 0.2 AU

Chapter 14

True or False?/Multiple Choice   1) T   2) T   3) F   4) F   5) F   6) T   7) F   8) T   9) b   10) c   11) a   12) b   13) d   14) b   15) c
Problems   1) 2.0″, much less than the angular diameter of Andromeda   2) 100 kpc   3) 10 times   4) −6.2; 17 Mpc   5) 0.014″/yr; proper motion has in fact been measured for several globular clusters   6) 2.7×10^{11} solar masses   7) 570 million years; 460 million years   8) 18   9) 1 kpc   10) 1600 AU, 2.6×10^{6} solar masses

Chapter 15

True or False?/Multiple Choice   1) F   2) T   3) T   4) T   5) T   6) T   7) T   8) a   9) c   10) b   11) b   12) a   13) b   14) c   15) c
Problems   1) 3 billion years   2) 17 Mpc, taking the luminosity of the Cepheid to be 10^4 times that of the Sun   3) 320 Mpc   4) 14,000 km/s, 57 Mpc; 10,000 km/s, 80 Mpc; 16,000 km/s, 50 Mpc   5) 14 billion years; 14 billion years   6) 1.8×10^{8} solar masses   7) 1.1 million years   8) 1.3×10^{36} W = 3.2 billion solar luminosities   9) 7200 km/s   10) Seyfert: 1.3×10^{-5} W/m^2; Sirius A: 1.0×10^{-7} W/m^2; smaller by a factor of 120; yes, if much of the emission from the nucleus is in the infrared

Chapter 16

True or False?/Multiple Choice   1) F   2) F   3) F   4) T   5) T   6) F   7) T   8) T   9) b   10) c   11) d   12) a   13) d   14) b   15) a
Problems   1) 1.2×10^{12} solar masses   2) 5.6×10^{11} solar masses, assuming a velocity of 350 km/s   3) 9.3×10^{10} solar masses   4) 700 km/s, comparable to the 660 km/s speed of the galaxy in the cluster   5) 130 million years   6) −27.2; 6.4 trillion solar luminosities; apparent magnitude of the Sun = −26.8, so the quasar at 10 pc is 0.4 magnitudes, or a factor of 1.4 times, brighter than the Sun at 1 AU   7) −22.5; 8.3×10^{10} solar luminosities   8) 11 million years   9) 21 Mpc   10) 48 kpc

Chapter 17

True or False?/Multiple Choice   1) F   2) F   3) F   4) T   5) T   6) F   7) T   8) T   9) a   10) d   11) c   12) d   13) b   14) c   15) a
Problems   1) 1000 Mpc   2) 4×10^{8} 3) 1200 km/s, directly away from the opposite corner   4) 20, 14, 12 billion years   5) (a) 30,000 tons, (b) 2.8 pc   6) 3×10^{15} solar masses; 1200 km/s   7) (a) 1.1 mm, (b) 0.0093, (c) 9.3×10^{-5}, (d) 9.3×10^{-7} 8) 9×10^{-18} kg/m^3; 5×10^{-19} kg/m^3 9) 91 kpc 10) 74 kpc

Chapter 18

True or False?/Multiple Choice   1) F   2) F   3) T   4) F   5) T   6) T   7) T   8) b   9) c   10) c   11) b   12) d   13) a   14) d   15) d
Problems   1) 6.3 seconds (for a 20 year old reader); 18 seconds (in 2003); 72 seconds; 161 seconds; 237 days   2) 0.19–0.47 AU   3) 5 billion   4) 27 AU   5) (a) 0.2, (b) 20, (c) 2000   6) 3100 years, 3.1 million years   7) 32,000 km/s   8) 5×10^{8} W, 500 times the emission of the Sun   9) 1.43×10^{9} to 1.67×10^{9} Hz; 2.4×10^{6} channels   10) 2.3 years; 55 years

INDEX

Star Charts

Have you ever become lost in an unfamiliar city or state? Chances are you used two things to get around: a map and some signposts. In much the same way, these two items can help you find your way around the night sky in any season. Fortunately, in addition to the seasonal Star Charts on the following pages, the sky provides us with two major signposts. Each seasonal description will talk about the Big Dipper—a group of seven bright stars that dominates the constellation Ursa Major the Great Bear. Meanwhile, the constellation Orion the Hunter plays a key role in finding your way around the sky from late autumn until early spring.

Each chart portrays the sky as seen from near 35° north latitude at the times shown at the top of the page. Located just outside the chart are the four directions: north, south, east, and west. To find stars above your horizon, hold the map overhead and orient it so a direction label matches the direction you're facing. The stars above the map's horizon now match what's in the sky.

WINTER SPRING

N

E W

SUMMER AUTUMN

S

Exploring the Winter Sky

Winter finds the Big Dipper climbing the northeastern sky, with the three stars of its handle pointing toward the horizon and the four stars of its bowl standing highest. The entire sky rotates around a point near Polaris, a 2nd-magnitude star found by extending a line from the uppermost pair of stars in the bowl across the sky to the left of the Dipper. Polaris also performs two other valuable functions: The altitude of the star above the horizon equals your latitude north of the equator, and a straight line dropped from the star to the horizon.

Turn around with your back to the Dipper and you'll be facing the diamond-studded winter sky. The second great signpost in the sky, Orion the Hunter, is central to the brilliant scene. Three closely spaced, 2nd-magnitude stars form a straight line that represents the unmistakable belt of Orion. Extending the imaginary line joining these stars to the upper right leads to Taurus the Bull and its orangish 1st-magnitude star, Aldebaran. Reverse the direction of your gaze to the belt's lower left and you cannot miss Sirius the Dog Star—brightest in all the heavens at magnitude −1.5.

Now move perpendicular to the belt from its westernmost star, Mintaka, and find at the upper left of Orion the red supergiant star Betelgeuse. Nearly a thousand times the Sun's diameter, Betelgeuse marks one shoulder of Orion. Continuing this line brings you to a pair of bright stars, Castor and Pollux. Two lines of fainter stars extend from this pair back toward Orion—these represent Gemini the Twins. At the northeastern corner of this constellation lies the beautiful open star cluster M35. Head south of the belt instead and your gaze will fall on the blue supergiant star Rigel, Orion's other luminary.

Above Orion and nearly overhead on winter evenings is brilliant Capella, in Auriga the Charioteer. Extending a line through the shoulders of Orion to the east leads you to Procyon in Canis Minor the Little Dog. Once you have these principal stars mastered, using the chart to discover the fainter constellations will be a whole lot easier. Take your time, and enjoy the journey. Before leaving Orion, however, aim your binoculars at the line of stars below the belt. The fuzzy "star" in the middle is actually the glorious Orion Nebula (M42), a stellar nursery illuminated by bright, newly formed stars.

WINTER

2 a.m. on December 1; midnight on January 1; 10 p.m. on February 1

Exploring the Spring Sky

The Big Dipper, our signpost in the sky, swings high overhead during the spring and lies just north of the center of the chart. This season of rejuvenation encourages us to move outdoors with the milder temperatures, and with the new season a new set of stars beckons us.

Follow the arc of stars outlining the handle of the Dipper away from the bowl and you will land on brilliant Arcturus. This orangish star dominates the spring sky in the kite-shaped constellation Boötes the Herdsman. Well to the west of Boötes lies Leo the Lion. You can find its brightest star, Regulus, by using the pointers of the Dipper in reverse. Regulus lies at the base of a group of stars shaped like a sickle or backward question mark, which represents the head of the lion.

Midway between Regulus and Pollux in Gemini, which is now sinking in the west, is the diminutive group Cancer the Crab. Centered in this group is a hazy patch of light that binoculars reveal as the Beehive star cluster (M44).

To the southeast of Leo lies the realm of the galaxies and the constellation Virgo the Maiden. Virgo's brightest star, Spica, shines at magnitude 1.0.

During springtime, the Milky Way lies level with the horizon, and it's easy to visualize that we are looking out of the plane of our galaxy. In the direction of Virgo, Leo, Coma Berenices, and Ursa Major lie thousands of galaxies whose light is unhindered by intervening dust in our own galaxy. However, all these galaxies are elusive to the untrained eye and require binoculars or a telescope to be seen.

Boötes lies on the eastern border of this galaxy haven. Midway between Arcturus and Vega, the bright "summer" star rising in the northeast, is a region where no star shines brighter than 2nd magnitude. A semicircle of stars represents Corona Borealis the Northern Crown, and adjacent to it is a large region that houses Hercules the Strongman, the fifth-biggest constellation in the sky. It is here we can find the northern sky's brightest globular star cluster, M13. A naked-eye object from a dark site, it looks spectacular when viewed through a telescope.

Returning to Ursa Major, check the second-to-last star in the Dipper's handle. Most people will see it as double, while binoculars show this easily. The pair is called Mizar and Alcor, and they lie just 0.2° apart. A telescope reveals Mizar itself to be a double. Its companion star shines at magnitude 4.0 and lies 14 arcseconds away.

SPRING

1 a.m. on March 1; 11 p.m. on April 1; 9 p.m. on May 1. Add one hour for daylight-saving time

Exploring the Summer Sky

The richness of the summer sky is exemplified by the splendor of the Milky Way. Stretching from the northern horizon in Perseus, through the cross-shaped constellation Cygnus overhead, and down to Sagittarius in the south, the Milky Way is packed with riches. These riches include star clusters, nebulae, double stars, and variable stars.

Let's start with the Big Dipper, our perennial signpost, which now lies in the northwest with its handle still pointing toward Arcturus. High overhead, and the first star to appear after sunset, is Vega in Lyra the Harp. Vega forms one corner of the summer triangle, a conspicuous asterism of three stars. Near Vega lies the famous double-double, Epsilon (ϵ) Lyrae. Two 5th-magnitude stars lie just over 3 arcminutes apart and can be split when viewed through binoculars. Each of these two stars is also double, but you need a telescope to split them.

To the east of Vega lies the triangle's second star: Deneb in Cygnus the Swan (some see a cross in this pattern). Deneb marks the tail of this graceful bird, the cross represents its outstretched wings, and the base of the cross denotes its head, which is marked by the incomparable double star Albireo. Albireo matches a 3rd-magnitude yellow star and a 5th-magnitude blue star and offers the finest color contrast anywhere in the sky. Deneb is a supergiant star that pumps out enough light to equal 60,000 Suns. Also notice that the Milky Way splits into two parts in Cygnus, a giant rift caused by interstellar dust blocking starlight from beyond.

Altair, the third star of the summer triangle and the one farthest south, is the second brightest of the three. Lying 17 light-years away, it's the brightest star in the constellation Aquila the Eagle.

Frequently overlooked to the north of Deneb lies the constellation Cepheus the King. Shaped rather like a bishop's hat, the southern corner of Cepheus is marked by a compact triangle of stars that includes Delta (Δ) Cephei. This famous star is the prototype of the Cepheid variable stars used to determine the distances to some of the nearer galaxies. It varies regularly from magnitude 3.6 to 4.3 and back again with a 5.37-day period.

Hugging the southern horizon, the constellations Sagittarius the Archer and Scorpius the Scorpion lie in the thickest part of the Milky Way. Scorpius's brightest star, Antares, is a red supergiant star whose name means "rival of Mars" and derives from its similarity to the planet in both color and brightness.

SUMMER

1 a.m. on June 1; 11 p.m. on July 1; 9 p.m. on August 1. Add one hour for daylight-saving time

Exploring the Autumn Sky

The cool nights of autumn are here to remind us the chill of winter is not far off. Along with the cool air, the brilliant stars of the summer triangle descend in the west to be replaced with a rather bland-looking region of sky. But don't let initial appearances deceive you. Hidden in the fall sky are gems equal to summertime.

The Big Dipper swings low this season, and for parts of the Southern United States it actually sets. Cassiopeia the Queen, a group of five bright stars in the shape of a "W" or "M," reaches its highest point overhead, the same spot the Big Dipper reached six months ago. To the east of Cassiopeia, Perseus the Hero rises high. Nestled between these two groups is the wondrous Double Cluster—NGC 869 and NGC 884—a fantastic sight in binoculars or a low-power telescope.

Our view to the south of the Milky Way is a window out of the plane of our galaxy in the opposite direction to that visible in spring. This allows us to look at the Local Group of galaxies. Due south of Cassiopeia is the Andromeda Galaxy (M31), a 4th-magnitude smudge of light that passes directly overhead around 9 p.m. in mid-November. Farther south, between Andromeda and Triangulum, lies M33, a sprawling face-on spiral galaxy best seen in binoculars or a rich-field telescope.

The Great Square of Pegasus passes just south of the zenith. Four 2nd- and 3rd-magnitude stars form the square, but few stars can be seen inside of it. If you draw a line between the two stars on the west side of the square and extend it southward, you'll find 1st-magniude Fomalhaut in Piscis Austrinus the Southern Fish. Fomalhaut is the solitary bright star low in the south. Using the eastern side of the square as a pointer to the south brings you to Diphda in the large, faint constellation of Cetus the Whale.

To the east of the Square lies the Pleiades star cluster (M45) in Taurus, which reminds us of the forthcoming winter. By late evening in October and early evening in December, Taurus and Orion have both cleared the horizon and Gemini is rising in the northeast. In concert with the reappearance of winter constellations, the view to the northwest finds summertime's Cygnus and Lyra about to set. The autumn season is a great transition period, both on Earth and in the sky, and a fine time to experience the subtleties of these constellations.

AUTUMN

1 a.m. on September 1; 11 p.m. on October 1; 9 p.m. on November 1. Add one hour for daylight-saving time